Feature Paper Collection in the Section 'Energy Science and Technology'

Feature Paper Collection in the Section 'Energy Science and Technology'

Guest Editors

Demis Pandelidis
Katrzyna Wartalska
Martyna Grzegorzek

MDPI

Basel • Beijing • Wuhan • Barcelona • Belgrade • Novi Sad • Cluj • Manchester

Guest Editors
Demis Pandelidis
Department of Mechanical and Power Engineering
Wroclaw University of Technology
Wroclaw
Poland

Katrzyna Wartalska
Department of Environmental Engineering
Wroclaw University of Technology
Wroclaw
Poland

Martyna Grzegorzek
Department of Environmental Engineering
Wroclaw University of Technology
Wroclaw
Poland

Editorial Office
MDPI AG
Grosspeteranlage 5
4052 Basel, Switzerland

This is a reprint of the Special Issue, published open access by the journal *Applied Sciences* (ISSN 2076-3417), freely accessible at: https://www.mdpi.com/journal/applsci/special_issues/LCK6VK99CH.

For citation purposes, cite each article independently as indicated on the article page online and as indicated below:

Lastname, A.A.; Lastname, B.B. Article Title. *Journal Name* **Year**, *Volume Number*, Page Range.

ISBN 978-3-7258-6244-3 (Hbk)
ISBN 978-3-7258-6245-0 (PDF)
https://doi.org/10.3390/books978-3-7258-6245-0

Cover image courtesy of Demis Pandelidis

Contents

About the Editors

Demis Pandelidis

Demis Pandelidis received his MS in 2011 and PhD in 2016 from Wrocław University of Science and Technology. His work focuses on applying water phase-change phenomena to energy-intensive applications—including air and water cooling, distillation, and desalination. With over 12 years' experience developing energy-efficient technologies, he has led eight scientific and twelve industrial R&D projects and contributed to ten more as a consultant. He has commercialized multiple solutions, such as the BHR energy-recovery unit with Brookvent Group, the Stellar air-conditioning unit, and a district-heating air-handling unit with Moltres Energy. He is the co-founder of Baryon Inc., an advanced technology firm that received the U.S. Department of Energy's BENEFIT 2020 award. Dr. Pandelidis holds five patents in renewable-energy technologies and has authored nearly 200 publications in the field. He also developed the world's first evaporative cooler adapted for operation in humid climates.

Katrzyna Wartalska

Katrzyna Wartalska has worked at the Department of Water Supply and Sewerage, Faculty of Environmental Engineering, Wrocław University of Science and Technology, in a research-and-teaching position (assistant professor) since October 2018. She has participated in the following projects: • "Evaporative Water Purification" (project funded by the National Centre for Research and Development). Role: project performer (04.2022–12.2023). • "Protection of the Natural Water Resources of Polanica-Zdrój through Intelligent Water and Wastewater Management" (project funded by the National Centre for Research and Development). Role: organizational manager (05.2021–09.2023). • "OPTIMA: Open Practices, Transparency and Integrity for Modern Academia" (Programme: Erasmus+). Role: project performer (09.2020–10.2023).

Martyna Grzegorzek

Martyna Grzegorzek, M.Sc. Eng., is a Graduate of the Faculty of Environmental Engineering at Wrocław University of Science and Technology, specializing in water supply, wastewater disposal, and waste management. She obtained a doctorate in technical sciences at Wrocław University of Science and Technology, Faculty of Environmental Engineering, in the field of environmental engineering, mining, and power engineering, submitting her doctoral dissertation on the "Removal of Fluoride from Aqueous Solutions in Integrated Electromembrane Systems." From April 2022 to December 2024, she worked as a project performer in "Evaporative Water Purification (WOW)", funded by the National Centre for Research and Development (NCBR). Since 1 October 2018, she has been employed at Wrocław University of Science and Technology (Faculty of Environmental Engineering) in a research-and-teaching position (assistant professor). Her main research interests include water and wastewater quality (including stormwater), water and wastewater treatment, alternative water sources, and membrane processes.

applied sciences

MDPI

Editorial

Editorial for the Special Issue "Feature Paper Collection in the Section 'Energy Science and Technology'"

Demis Pandelidis

Department of Mechanical and Energy Engineering, Wroclaw University of Science and Technology, 50-370 Wroclaw, Poland; demis.pandelidis@pwr.edu.pl; Tel.: +48-500353863

1. Introduction

Energy science and technology sit at the heart of today's decarbonization challenge [1–4]. As global demand grows, society needs order-of-magnitude gains in efficiency, rapid integration of renewables, and deployable control and forecasting tools that safeguard reliability and health [5–7]. This Special Issue collects feature papers spanning materials, systems modeling, controls, and field-ready solutions—covering solar and wind, thermal storage and cooling, hydrogen production and use, power electronics, transport, buildings, and power-quality/public health interfaces.

2. Overview of Contributions

Nomandela, Mnguni, and Raji validate grid-connected operation of a large-scale wind power plant via real-time simulation aligned with renewable energy grid codes, showing improved busbar voltages and offering a transferable methodology for future compliance studies. Zhou, Zeng, Zheng, Wang, and Zhou introduce a de-stationarized Crossformer for remaining useful life prediction in PEM fuel cells—combining adaptive normalization, non-stationary attention, crafted features, and a similarity-focused loss—to outperform Transformer/Crossformer baselines on real-vehicle data. Teixeira, Cerveira, Pires, and Baptista integrate LSTM with genetic algorithms for short- to medium-term weather forecasting, and Emexidis and Gkonis benchmark IoT-driven models for wind-energy prediction, highlighting Random Forest's robustness. Gabaldón, Ruiz-Abellón, Martínez, and Guillamón propose aggregated supercapacitor–battery models for hybrid locomotives to enable pre-feasibility sizing and improved energy recovery on non-electrified lines. Santana-Méndez, García-Afonso, and González-Díaz model Tenerife's pathway to 100% renewable supply for power and road transport, quantifying the scale of renewables, storage, dispatchables, and electrolysis required. Pukalskas and co-authors assess dual-fuel gasoline–hydrogen operation across WLTC, finding efficiency gains and lower incomplete-combustion products alongside reduced CO_2. Mendiara, Filsouf, Adánez-Rubio, Izquierdo, and Abad evaluate Ti-doped manganese oxide oxygen carriers for chemical looping, demonstrating durable oxygen release/regeneration and stable fluidization with an estimated multi-thousand-hour lifetime. Fernando and Naser develop a Eulerian–Eulerian CFD approach to predict iron-particle agglomeration in DRI conditions, aligning with CFD-DEM and experiments. Kočí, Maděra, and Černý present an on-site sensor self-adjustment technique for maintenance-free heat-flow monitoring in building envelopes. Marmol, Zamiri, Purraji, Murillo, Tuñón Díaz, Vazquez, and de Castro mitigate non-minimum-phase behavior in DC–DC boost converters via a dual-control scheme—short-horizon FCS-MPC for fast dynamics with a PI controller that tunes the current reference for steady-state accuracy (validated in simulation and HIL). Gonzalez-Rodriguez, Roldan-Fernandez, Serrano-Gonzalez, and Muñoz-Díez accelerate offshore wind-farm layout optimization by reframing wake interactions as power-deficit aggregation with a conservative-field analogy, achieving substantial runtime reductions in irregular concession areas. Khalil provides a comparative viability analysis across current and emerging hydrogen production pathways using quantitative and qualitative criteria, identifying electrolysis and biomass-waste valorization as the most promising near-term routes to DOE cost targets. Criollo and co-authors show how reduced-order electrolyzer models preserve techno-economic fidelity while cutting compute cost. Skibko and co-authors report a biogas-plant case where undersized feed-in cables caused large voltage excursions; practical fixes restored dual-generator operation and improved stability. Litzbarski and colleagues measure low-frequency EMF near power infrastructure, contrasting ICNIRP guidelines with stricter national limits and advocating localized exposure assessments. Li et al. proposed a source–storage–load collaborative scheduling strategy for aluminum electrolysis, using refined load models (including temperature–output–power coupling and crane–production dynamics) and a multi-objective formulation to cut costs and emissions while maximizing renewable use. An improved heuristic algorithm solves the model. Simulations show, versus single-objective optimization, up to 1677.35 t CO_2 reduction and +12.11% renewable use at 67.8% RE penetration, and up to 8260.97 t CO_2 reduction and +18.86% renewable use at 76.2%, without harming production stability. Bucur et al. used 2D time-resolved simulations (ANSYS Fluent, Dynamic Mesh 6DOF) to test passive flow control and blade material choices for H-Darrieus VAWTs. Blades with inclined cavities on the extrados accelerate faster and achieve higher tip-speed ratios and power coefficients than baseline NACA0012, especially at higher wind speeds. Lightweight 3D-printed polycarbonate outperforms resin, indicating that combining cavity-modified profiles with low-mass materials can markedly improve VAWT start-up and overall efficiency, guiding future experiments and prototypes.

3. Outlook

Future research should pair scalable optimization with explicit optimality bounds and uncertainty quantification; advance validated digital twins via data reconciliation and hybrid modeling for plant-level monitoring and dispatch; couple thermal storage with low-exergy cooling in buildings and industry; and link comparative TEA/LCA for hydrogen with operational control strategies and infrastructure planning, particularly in isolated or weakly interconnected systems. Finally, engineering-policy co-design will remain essential to harmonize grid codes, exposure standards, and public acceptance.

Funding: Research conducted by Demis Pandelidis is carried out under the " Destylacja Punktu Rosy/Dew Point Distillation (DPD)" project within the FENG.02.02-IP.05-0247/23 program, First Team of the Foundation for Polish Science, co-financed by the European Union under the European Funds for Smart Economy 2021–2027 (FENG).

Conflicts of Interest: The authors declare no conflict of interest.

List of Contributions:

1. Nomandela S.; Mnguni M.E.S.; Raji A.K. A Case Study on the Validation of Renewable Energy Grid Code Compliance for a Large-Scale Wind Power Plant Grid-Connected Mode of Operation in Real-Time Simulation. *Appl. Sci.* **2025**, *15*, 5521. https://doi.org/10.3390/app15105521.
2. Zhou N.; Zeng H.; Zheng Z.; Wang K.; Zhou J. PEMFC RUL Prediction for Non-Stationary Time Series Based on Crossformer Model. *Appl. Sci.* **2025**, *15*, 2515. https://doi.org/10.3390/app15052515.

3. Teixeira R.; Cerveira A.; Pires E.J.S.; Baptista J. Enhancing Weather Forecasting Integrating LSTM and GA. *Appl. Sci.* **2024**, *14*, 5769. https://doi.org/10.3390/app14135769.
4. Gabaldón A.; Ruiz-Abellón M.C.; Martínez F.; Guillamón A. Demand Management in Hybrid Locomotives Through Aggregated Models of Supercapacitors and Railway Units. *Appl. Sci.* **2025**, *15*, 2412. https://doi.org/10.3390/app15052412.
5. Pukalskas, S.; Vipartas, T.; Rimkus, A.; Kriaučiūnas, D.; Žaglinskis, J.; Stravinskas, S.; Ušinskas, A.; Juknelevičius, R.; Mejeras, G.; Žuraulis, V.; et al. Numerical Modelling Assessment of the Impact of Hydrogen on the Energy and Environmental Performance of a Car Using Dual Fuel (Gasoline–Hydrogen). *Appl. Sci.* **2025**, *15*, 1939. https://doi.org/10.3390/app15041939.
6. Mendiara T.; Filsouf A.; Adánez-Rubio I.; Izquierdo M.T.; Abad A. Performance of Granulated Ti-Dopped Manganese Oxide Oxygen Carriers in Chemical Looping Processes. *Appl. Sci.* **2025**, *15*, 750. https://doi.org/10.3390/app15020750.
7. Khalil Y.F. A Comparative Science-Based Viability Assessment Among Current and Emerging Hydrogen Production Technologies. *Appl. Sci.* **2025**, *15*, 497. https://doi.org/10.3390/app150 20497.
8. Xu L.; You J.; Wu J.; Liu Y.; Zhang D. Parameter Correction on Waste Heat Recovery System of a Gas Turbine Using Supercritical CO2 Based on Data Reconciliation. *Appl. Sci.* **2025**, *15*, 248. https://doi.org/10.3390/app15010248.
9. Gonzalez-Rodriguez A.G.; Roldan-Fernandez J.M.; Serrano-Gonzalez J.; Muñoz-Díez J.V. Gradient Descent Algorithm with Greedy Repositioning Using Power Deficit Aggregation of Wakes to Accelerate the Offshore Wind Farm Layout Optimization Problem in Irregular Concession Areas. *Appl. Sci.* **2024**, *14*, 11331. https://doi.org/10.3390/app142311331.
10. Criollo, A.; Minchala-Avila, L.I.; Benavides, D.; Ochoa-Correa, D.; Tostado-Véliz, M.; Meteab, W.K.; Jurado, F. Green Hydrogen Production—Fidelity in Simulation Models for Technical–Economic Analysis. *Appl. Sci.* **2024**, *14*, 10720. https://doi.org/10.3390/app142210720.
11. Marmol, A.; Zamiri, E.; Purraji, M.; Murillo, D.; Díaz, J.T.; Vazquez, A.; de Castro, A. Dual Control Strategy for Non-Minimum Phase Behavior Mitigation in DC-DC Boost Converters Using Finite Control Set Model Predictive Control and Proportional–Integral Controllers. *Appl. Sci.* **2024**, *14*, 10318. https://doi.org/10.3390/app142210318.
12. Emexidis C.; Gkonis P. The Integration of Internet of Things and Machine Learning for Energy Prediction of Wind Turbines. *Appl. Sci.* **2024**, *14*, 10276. https://doi.org/10.3390/app142210276.
13. Moravič M.; Marasová D.; Kaššay P.; Ozdoba, M.; Lopot, F.; Bortnowski, P. Experimental Verification of a Compressor Drive Simulation Model to Minimize Dangerous Vibrations. *Appl. Sci.* **2024**, *14*, 10164. https://doi.org/10.3390/app142210164.
14. Santana-Méndez I.; García-Afonso Ó.; González-Díaz B. Full Road Transport Sector Transition Towards 100% Autonomous Renewable Energy Supply in Isolated Systems: Tenerife Island Test Case. *Appl. Sci.* **2024**, *14*, 9734. https://doi.org/10.3390/app14219734.
15. Litzbarski L.S.; Olesz M.; Redlarski G.; Tojza, P.M.; Żak, A.; Gifuni, E.; Cieślikowska, Z.; Czapliński, M. The Assessment of the Influence of Low-Frequency Electromagnetic Fields Originated from the Power Infrastructure on Humans' Health. *Appl. Sci.* **2024**, *14*, 9668. https://doi.org/10.3390/app14219668.
16. Fernando W.D.S.; Naser J. Eularian–Eularian Model for Agglomeration Behavior of Combusted Iron Particles. *Appl. Sci.* **2024**, *14*, 7829. https://doi.org/10.3390/app14177829.
17. Kočí J.; Maděra J.; Černý R. On-Site Sensor Sensitivity Adjustment Technique for a Maintenance-Free Heat Flow Monitoring in Building Systems. *Appl. Sci.* **2024**, *14*, 7323. https://doi.org/10.3 390/app14167323.
18. Skibko Z.; Borusiewicz A.; Romaniuk W.; Pietruszynska, M.; Milewska, A.; Marczuk, A. Voltage Problems on Farms with Agricultural Biogas Plants—A Case Study. *Appl. Sci.* **2024**, *14*, 7003. https://doi.org/10.3390/app14167003.
19. Li J.; Chen Y.; Liu G.; Han R. Low-Carbon Economic Collaborative Scheduling Strategy for Aluminum Electrolysis Loads with a High Proportion of Renewable Energy Integration. *Appl. Sci.* **2025**, *15*, 10919. https://doi.org/10.3390/app152010919.

20. Bucur I.O.; Crunțeanu D.E.; Dombrovschi M.C. Performance Optimization of Vertical Axis Wind Turbines Through Passive Flow Control and Material Selection: A Dynamic Mesh Study. *Appl. Sci.* **2025**, *15*, 11251. https://doi.org/10.3390/app152011251.

References

1. Kim, J.; Sovacool, B.K.; Bazilian, M.; Griffiths, S.; Yang, M. Energy, material, and resource efficiency for industrial decarbonization: A review. *Energy Res. Soc. Sci.* **2024**, *112*, 103521. [CrossRef]
2. Ringel, M. Is the updated Energy Efficiency Directive fit for purpose? A review. *Energy Res. Soc. Sci.* **2025**, *127*, 104244. [CrossRef]
3. Awuzie, B.; Ngowi, A.; Aghimien, D. Towards built-environment decarbonisation: A review of recent advances and pathways. *Energy Build.* **2024**, *319*, 114491. [CrossRef]
4. Myint, N.N.; Shafique, M.; Zhou, X.; Zheng, Z. Net zero carbon buildings: A review on recent advances and techniques. *Case Stud. Constr. Mater.* **2025**, *22*, e04200.
5. Jafari, M.; Botterud, A.; Sakti, A. Decarbonizing power systems: A critical review of the role of energy storage systems. *Renew. Sustain. Energy Rev.* **2022**, *158*, 112077. [CrossRef]
6. Liang, Y.; Li, C.; Liu, Z.; Wang, X.; Zeng, F.; Yuan, X.; Pan, Y. Decarbonization potentials of embodied energy use and operational processes in buildings: A review from the life-cycle perspective. *Heliyon* **2023**, *9*, e20190. [CrossRef]
7. Yang, S.; Yi, W.; Zong, H. A systematic review of retrofitting buildings for sustainability and decarbonization. *Sustain. Cities Soc.* **2025**, *131*, 106702. [CrossRef]

applied sciences

MDPI

Article

Enhancing Weather Forecasting Integrating LSTM and GA

Rita Teixeira [1], Adelaide Cerveira [2,3], Eduardo J. Solteiro Pires [1,3] and José Baptista [1,3,*]

1 Department of Engineering, University of Trás-os-Montes and Alto Douro, 5000-801 Vila Real, Portugal; rmteixeira@utad.pt (R.T.); epires@utad.pt (E.J.S.P.)
2 Department of Mathematics, University of Trás-os-Montes and Alto Douro, 5000-801 Vila Real, Portugal; cerveira@utad.pt
3 INEC-TEC UTAD Pole, University of Trás-os-Montes and Alto Douro, 5000-801 Vila Real, Portugal
* Correspondence: baptista@utad.pt

Abstract: Several sectors, such as agriculture and renewable energy systems, rely heavily on weather variables that are characterized by intermittent patterns. Many studies use regression and deep learning methods for weather forecasting to deal with this variability. This research employs regression models to estimate missing historical data and three different time horizons, incorporating long short-term memory (LSTM) to forecast short- to medium-term weather conditions at Quinta de Santa Bárbara in the Douro region. Additionally, a genetic algorithm (GA) is used to optimize the LSTM hyperparameters. The results obtained show that the proposed optimized LSTM effectively reduced the evaluation metrics across different time horizons. The obtained results underscore the importance of accurate weather forecasting in making important decisions in various sectors.

Keywords: forecast horizon; weather variables; regression; long short-term memory; genetic algorithm

1. Introduction

Weather variables play a vital role in many sectors and activities, affecting various aspects of our daily lives. From aviation to energy production, weather conditions have a significant impact on the performance and results of various sectors. Agriculture and renewable energy production are two sectors that are highly dependent on weather variables [1].

Weather variability presents significant challenges related with the variability and unpredictability of the data, making it difficult to rely on consistent patterns. Weather forecasting plays an important role in improving the performance and efficiency of different sectors, including agriculture, energy, and transportation [2,3]. However, there are some problems associated with weather forecasts that can affect their accuracy such as traditional forecasting methods may struggle to accurately predict sudden changes, leading to less reliable weather forecasts; historical data can be incomplete or inaccurate, impacting the ability to model future conditions accurately; weather variables can have complex iterations between them and capturing these interactions in forecasting models is challenging, often requiring sophisticated algorithms and computational resources.

In fact, the accuracy and reliability of weather forecasts have a huge impact on decision making, enabling effective planning, risk mitigation, and optimization of operations. For example, understanding and accurately predicting weather variables is critical to optimizing renewable energy production, reducing costs and minimizing environmental impact. By accurately forecasting weather conditions, grid operators can anticipate fluctuations in renewable energy output and plan for backup power sources or storage solutions to maintain a reliable supply of electricity.

In the agricultural sector, weather variables play a crucial role in shaping farming practices, yield forecasting, and crop management. Farmers rely on a combination of weather variables such as temperature, rainfall, humidity, and wind patterns to make informed decisions about planting, irrigation, and harvesting. In fact, understanding weather patterns allows farmers to optimize their operations, mitigate risk, and ensure the overall health and productivity of their crops. On the other hand, solar and wind energy production is highly dependent on weather conditions, which determine the availability and efficiency of these sustainable energy systems. Karevan and Suykens [4] emphasized the importance of weather forecasting in these areas because of its impact on productivity, resource allocation, and risk management. They proposed a transductive long short-term memory (T-LSTM) method to extract local information in weather time series.

Regarding the historical data used in forecasts, missing values affect the accuracy of forecasts. For this reason, it is important use resourceful and flexible techniques. Van Buuren [5] emphasized the importance of addressing missing data in statistical analysis, and highlighted the advantages of the multiple imputation (MI) approach and Doreswamy et al. [6] discussed the significance of handling missing data in weather datasets, which is crucial for accurate climate forecasting impacting various sectors. The paper proposes a new technique for addressing missing values in weather data using machine learning algorithms, such as kernel ridge regression, linear regression, random forest, or support vector machine (SVM) imputation. In this paper, regarding the characteristics of the places under studies, we chose to use regression methods to estimate missing values due to their ability to leverage relationships between variables.

Long short-term memory (LSTM) networks are a type of recurrent neural network (RNN) designed to model sequential data, making them highly suitable for weather forecasting. Since, weather data involve complex, non-linear interactions between various variables, such as, wind speed or direction, LSTMs are capable of modeling these non-linear relationships more effectively than traditional linear models, but also can effectively capture long-term dependencies in sequential data, like the temperature [7,8].

The accuracy of LSTM forecasts can be influenced by various factors. Viedma et al. [9] highlight the impact of classical time series pre-processing methods, with seasonal-component removal being particularly effective, the importance of data quality, model architecture, and hyperparameter tuning. The use of genetic algorithms (GA) to optimize the parameters of LSTM forecasting methods has been shown to significantly improve their accuracy. The authors in [10,11] both demonstrate the effectiveness of this approach in short-term load forecasting, with the first achieving a small mean absolute percentage error and the second improving prediction accuracy by 63%. Bouktif et al. [12] extend this to medium- and long-term forecasting, highlighting the ability of the LSTM model, optimized with GAs, to capture the characteristics of complex time series and reduce forecasting errors. These studies collectively underscore the importance of using GAs to enhance the accuracy of LSTM forecasting methods.

With this in mind, the main objective of this work was to analyze short- to medium-term weather forecasting. To this end, combined methods of regression and LSTM techniques were proposed and later evaluated to assess their accuracy and efficiency. Therefore, the main contributions of this work are (1) estimation of historical data considering regression techniques; (2) short- and medium-term weather prediction using LSTM networks; (3) evaluation of the model's forecasting task considering specific metrics.

The subsequent parts of the paper are organized as follows: Section 2 presents some work already developed regarding the proposed theme; Section 3 describes the sites under study, the methodology, presenting the methods considered, their mathematical formulation, the evaluation metrics used, and the case studies considered; Section 4 presents the main results obtained, discussing the relevant points according to the results of accuracy and efficiency of the evaluation metrics; finally, Section 5 presents the main conclusions of the paper and future work.

2. Related Work

Several works have studied the importance and accuracy of weather forecasting in the literature. Regression and machine learning models are two of the main techniques used, and some of the work is described below. Regression models are a type of statistical model commonly used in forecasting techniques. In effect, statistical models use historical data to predict new values by adjusting the parameters of the model by analyzing the differences between the observed and predicted values [13]. Linear regression models analyze the linear relationship between dependent and independent variables, improving the accuracy and reducing the computational time of the model [14]. According to Jassena and Kovoor [1], there are two main types of linear regression models: simple and multiple. Simple linear regression describes the linear relationship between one dependent variable and one independent variable. On the other hand, multiple linear regression (MLR) analyses the linear relationship between a dependent variable and more than one independent variable. In Ansarifar et al. [15], different regression models such as MLR or least absolute shrinkage and selection operator (LASSO) regression were proposed to continuously update weather and management data with future scenarios to obtain weekly crop yield forecasts. The results showed that the proposed model achieved accurate and timely crop yield predictions compared to machine learning algorithms. In Barriguinha et al. [16], a systematic review was presented that analyzed multiple linear regression (MLR), as well as partial least squares regression, and random forest regression, to estimate grape and wine production by predicting climatic conditions such as temperature or precipitation. Overall, the paper highlighted that the accuracy and reliability of regression models depend on the quality and quantity of data, the selection of appropriate predictors, and the validation of the model, but can be useful for estimating vineyard yield.

Similarly, in the renewable energy context, regression methods are often used to predict climate variables. Amir and Zaheeruddin [17] analyzed the short-term prediction of solar radiation and wind speed using regression models to assess the linear relationship between the variables. The proposed model can effectively capture the dependencies between the variables with an accuracy of 98.2%. Ridge regression models have been proposed in [18,19] to predict solar radiation and wind speed. In the former, the main objective was to reduce the problems associated with the intermittency of these variables, thus reducing the associated forecast errors. The results showed that the proposed model had a correlation coefficient of 0.903 and was able to improve the accuracy and generalization of the forecasts.

As regression methods are not as efficient at dealing with non-linear data and high variability, machine learning methods, in particular deep learning (DL), are emerging as a solution [20]. In fact, DL can automatically learn features from data, instead of considering traditional selection techniques [21].

Extreme gradient boosting (XGBoost) is a powerful and versatile machine learning algorithm that has demonstrated exceptional performance in various applications. It is particularly effective in modeling complex systems, achieving high prediction accuracy, and handling large-scale data. It has been successfully applied in various fields, including weather forecasting for agriculture and renewable energy production. Phan et al. and Wadhwa and Tiwary [22,23] demonstrated the effectiveness of XGBoost in improving the accuracy of short-term solar power and wind power forecasts. Another useful method is transformer networks. Transformer networks have been shown to significantly improve the accuracy of weather variable forecasts. In [24], a hybrid system was introduced, H-Transformer, which outperformed other models in predicting renewable energy production. Nascimento et al. [25] presented a transformer-based model with wavelet transform for wind speed and wind energy forecasting, showing superior performance compared to a baseline LSTM model. However, Walczewski et al. [26] found that the performance of transformer-based models varied depending on the specific forecasting task and the data provided, with the LSTM model outperforming all others in predicting onshore wind and photovoltaic energy.

Recurrent neural networks (RNNs) are widely used in weather forecasting due to their computational power, sophisticated algorithms, and storage capacity [27]. However, RNNs have a drawback when it comes to handling long-term correlations, as they suffer from gradient loss and explosion [28]. One solution to this problem is to use LSTMs, which are a type of RNN-based model. They excel at solving both short- and long-term problems by effectively predicting and storing temporal data [29,30].

Thirunavukkaras et al. [31] proposed an LSTM method to efficiently predict solar radiation 15 min ahead in certain areas with high intermittency. The proposed model outperformed other models such as auto-regressive or persistence by about 40%, improving the accuracy and efficiency of the model. Similarly, short-term wind energy forecasts are prone to accuracy problems due to the intermittent nature of wind [32]. Thus, an LSTM was proposed to deal with linear and non-linear components of time series of wind power generated by offshore turbines. The proposed model increased the prediction accuracy by 13.2% compared to a baseline model.

Finally, in the agricultural sector, LSTM networks are also being considered for weather forecasting. For example, in Salehin et al. [33] an LSTM was used to accurately determine the amount of rainfall needed to increase crop yields and reduce agricultural costs. The model takes into account weather variables such as temperature, humidity, and wind speed to make predictions and reports an accuracy of 76% in predicting rainfall.

3. Data and Methods

The methodology developed in this paper involves a two-step approach: first, regression models are employed to estimate missing values in the input dataset; second, LSTM networks are utilized to forecast weather variables for a specified future horizon. The dataset and methods employed are detailed below.

3.1. Dataset

This work considers weather data collected from two locations: Pinhão town and Quinta de Santa Bárbara. Quinta de Santa Bárbara is located at coordinates 41.172753° N, −7.549335° W. The data were collected from a sensor located in Quinta de Santa Bárbara between September 8 and 30 December 2022, totaling a sampling length of 114 days. However, for a more robust forecast, a larger input dataset is advantageous to improve the validation, stability, generalization, and accuracy of the forecasts. To this end, climate data from the Portuguese Institute for Sea and Atmosphere (IPMA) station, located at coordinates 41.172775° N, −7.548972° W, were also used, as it is in the same geographical area as the study site, as shown in Figure 1. This additional dataset comprises a sample of 4748 days, covering the period from 1 January 2010, to 31 December 2022.

When analyzing the data from Pinhão, reading errors of the sensor were eliminated for each of the variables studied. Since the data were collected daily, any outlier value for a particular day was replaced with the average value for that day in the other years.

In turn, the range of values measured at Quinta de Santa Bárbara was presented at 15-minute intervals. Thus, for each of the variables studied, the hourly values were averaged, and then, the average of these values was used to obtain the daily values.

3.2. Forecasting with Regression Models

Given the data from the two locations, it was necessary to find a relation between the variables measured at Pinhão and Quinta de Santa Bárbara, in order to predict the missing values from the measurements at Quinta de Santa Bárbara. The regression methods selected for this study included MLR [19], LASSO, and ridge regressions [34,35].

Figure 1. Location of the studied site.

The MLR is given by

$$Y = \beta_0 + \beta_1 \cdot x_1 + \beta_2 \cdot x_2 + \cdots + \beta_n \cdot x_n \quad (1)$$

where Y is the dependent variable, which corresponds to the missing values between 2010 and 2022 at Quinta de Santa Bárbara; $x_i, i = 1, 2, \ldots, n$ are the independent variables, corresponding to the measurements in Pinhão; β_0 is the constant term, β_i with $i = 1, 2, \ldots$; and n are the regression coefficients. For the LASSO and ridge models, the L1 and L2 regularization terms are added to Equation (1). Therefore, in order to take into account the advantages of these three methods, a combined model was considered to predict the missing data.

3.3. *Forecasting with Long Short-Term Memory*

Once the data from Quinta de Santa Bárbara between 2010 and 2022 had been estimated, LSTM networks were used to predict the variables under study for the short- to medium-term future.

According to [36], LSTM neural networks consist of three key components: input gate i_t, forgetting gate f_t, and output gate o_t. These components regulate the storage and removal of information within the network. This makes it possible to control when a memory unit retains previous information and maintains a constant error during the back-propagation process [27]. LSTMs have a cell state and a hidden state. The cell state propagates through the input and output sequences, while the temporal information is processed by the three gates responsible for incorporating, filtering, and selecting relevant information for output. The mathematical formulation outlining the LSTM structure can be represented by Equations (2)–(7) [32]:

$$f_t = \sigma(W_f[h_{t-1}, x_t] + b_f) \quad (2)$$

$$i_t = \sigma(W_i[h_{t-1}, x_t] + b_i) \quad (3)$$

$$o(t) = \sigma(W_o[h_{t-1}, x_t] + b_o) \quad (4)$$

$$c_t' = \tanh(W_s[h_{t-1}, x_t] + b_s) \quad (5)$$

$$c_t = c_{t-1} \odot f_t + g_t \odot i_t \quad (6)$$

$$h_t = \tanh(c_t) \odot i_t \quad (7)$$

where $[h_{t-1}, x_t]$ represents the input signal, which consists of the input neuron x_t at time t and the cell state c_{t-1} at time $t-1$. W_f, W_i, W_o, and W_s denote the weights that connect the forgetting gate, input gate, output gate, and cell state, respectively, to the input signal. b_f, b_i, b_o, and b_s represent the biases, and σ denotes the activation function. Additionally, h_t corresponds to the output value at time t. The operator $\odot$ represents the element-wise multiplication.

In this paper, instead of using tanh as the activation function, the rectified linear unit (ReLU) was considered. In brief, it returns the input value if it is positive, and zero if it is negative, introducing non-linearity to the model.

Since the hyperparameters that characterize an LSTM affect the accuracy of forecasts, a GA was considered to improve the performance. The fitness function uses the mean square error (MSE) between the predicted and observed values to guide the GA in finding the best LSTM parameter combination.

3.4. Genetic Algorithm

Genetic algorithm (GA) is a search and optimization technique inspired by natural selection. The GA simulates the evolution of a population of chromosomes, each one representing a potential solution to a problem. These chromosomes or individuals are scored according to their fitness, which determines their likelihood of being selected for reproduction. During the reproduction phase, genes from high-performing chromosomes are usually selected, combined, and mutated to produce offspring "solutions" potentially better than their progenitors. This iterative selection, reproduction, and mutation process allows the population to evolve towards increasingly optimal solutions over time. Genetic algorithms are powerful methods for solving problems, particularly where no traditional algorithm exists or where traditional optimization methods struggle due to large solution spaces and intricate variable relationships.

3.5. Evaluation Metrics

The coefficient of determination R^2 was used to evaluate the regression models, and the MSE, root mean square error (RMSE), and mean absolute error (MAE) were considered to evaluate the LSTM models.

The statistical measure R^2, where $0 \leq R^2 \leq 1$, assess the goodness of fit of the model and measure the predictive power of the independent variables, Equation (8). The higher the R^2 value, the better the model fits the data.

$$R^2 = 1 - \frac{SS_{res}}{SS_{tot}} \tag{8}$$

where n is the sample size, SS_{res} is the sum of squared residuals, $SS_{res} = \sum_{i=1}^{n}(y_i - y_i')^2$, SS_{tot} is the sum of squared differences between the observed values and the mean of observed values, $SS_{tot} = \sum_{i=1}^{n}(y_i - \bar{y})^2$, y_i is the observed value, and y_i' is the predicted value.

The MSE measures the average of the squared differences between the observed and predicted values:

$$\text{MSE} = \frac{1}{n}\sum_{i=1}^{n}(y_i - y_i')^2 \tag{9}$$

The MAE measures the average of the absolute differences between observed and predicted values:

$$\text{MAE} = \frac{1}{n}\sum_{i=1}^{n}|y_i - y_i'| \tag{10}$$

Finally, the RMSE takes into account the square root of the differences between the observed and predicted values:

$$\text{RMSE} = \sqrt{\frac{1}{n}\sum_{i=1}^{n}(y_i - y_i')^2} \quad (11)$$

These metrics therefore play an important role in the case studies that have been developed, as they make it possible to compare the different results obtained. They were calculated considering the forecast and observed data from the testing dataset for the time range of predictions.

Figure 2 shows the methodology applied in this paper. Considering the two locations, the datasets were pre-processed. After that, regression models were used to estimate the missing values for Quinta de Santa Bárbara between 2010 and 2022, considering the data obtained from Pinhão. After obtaining the new dataset for Quinta de Santa Bárbara, an LSTM network was considered to forecast short- to medium-term weather variables. A GA was also considered to optimize the hyperparameters of the network. Finally, the proposed evaluation metrics allow us to evaluate and compare the forecast results of each proposed network.

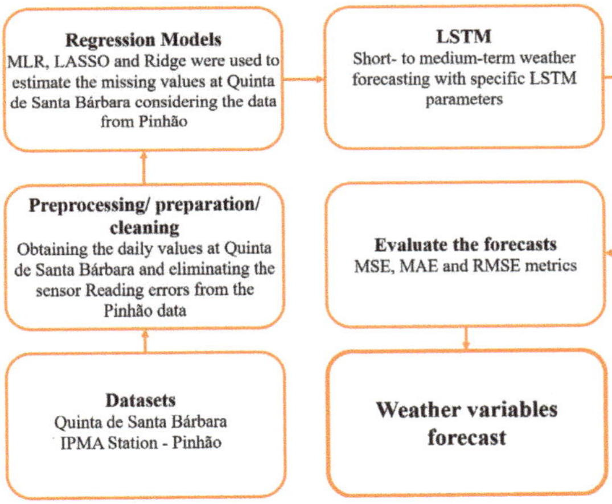

Figure 2. Applied methodology, considering the datasets and methods used for short- to medium-term weather forecasting.

The following cases have been proposed to evaluate the accuracy of weather forecasting using LSTM networks for short- and medium-term horizons:

- Case I: Short- and medium-term forecasting using LSTM.
- Case II: Short- and medium-term forecasting using an optimized LSTM with a GA.

4. Results and Discussion

This section presents the results obtained in relation to the estimated values and the forecast weather variables. Regression models were used to estimate and fill the missing data for Quinta de Santa Bárbara. Subsequently, LSTM networks were employed to forecast weather variables for 3 to 15 days.

The models used in this study were built using the Python 3.11 programming language [37]. The LSTM networks were implemented using the TensorFlow tool [38], a widely used framework for deep learning applications in Python, and the regression models were developed using the Scikit-learn library [39].

4.1. Forecasting with Regression Models

The linear relation between the two sites was analyzed for each of the variables in the time interval common to both sets of data, i.e., between 8 September and 30 December 2022.

To take advantage of the regression techniques described in Section 3, a combined model was developed to predict missing data. The measurements recorded at Quinta de Santa Bárbara were used as dependent variables, while those from Pinhão were considered as independent variables.

Figure 3 shows the scatter plots of the observed values (blue dots) and the trend line that fits the data.

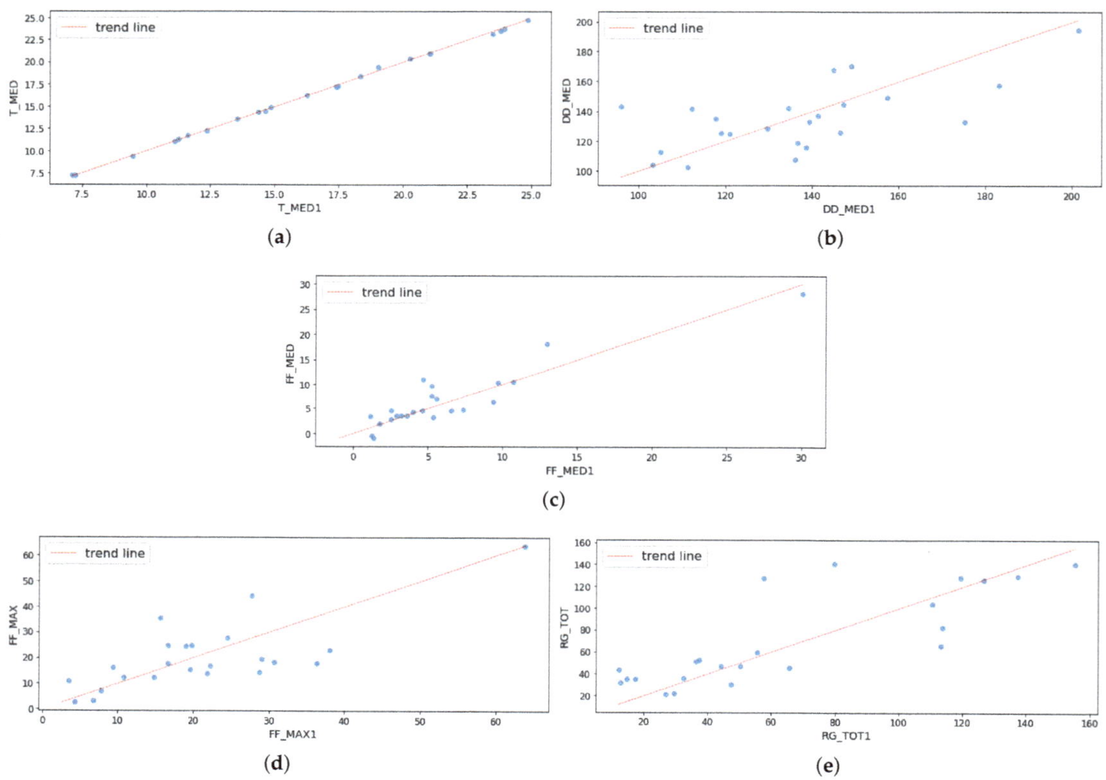

Figure 3. Linear relationship between the values observed in Pinhão and Quinta de Santa Bárbara: (**a**) daily average temperature, (**b**) daily wind direction, (**c**) daily average wind speed, (**d**) daily instantaneous wind speed, and (**e**) daily solar radiation.

For the daily average temperature (FF_{MED}), the values observed in the two locations have a strong linear relation. In turn, for the other variables, the observed values are dispersed around the trend line, which also shows that the proposed model is capable of making accurate predictions for these variables. However, in order to verify whether the MLR model fits the dataset, we evaluated the R^2 values and its residuals.

Table 1 shows the R^2 values for each of the variables under study. Analyzing the R^2 values, it can be seen that they are relatively high for all the variables except the daily wind direction (DD_{MED}), which had a value less than 0.5.

Table 1. R^2 values for each dependent variable studied.

Variable	Description	R^2
T_MED	Daily average temperature	0.999
DD_MED	Daily wind direction	0.477
FF_MED	Daily average wind speed	0.821
FF_MAX	Daily instantaneous wind speed	0.686
RG_TOT	Daily solar radiation	0.754

The residuals of an MLR are analyzed by constructing scatter plots, which are a useful tool for evaluating the fit of these models. Its representation consists of a graph, where the y-axis shows the normalized residuals and the x-axis shows the normalized residuals; Figure 4. In general, the results are randomly dispersed, with a mean of zero and a constant variance, as intended.

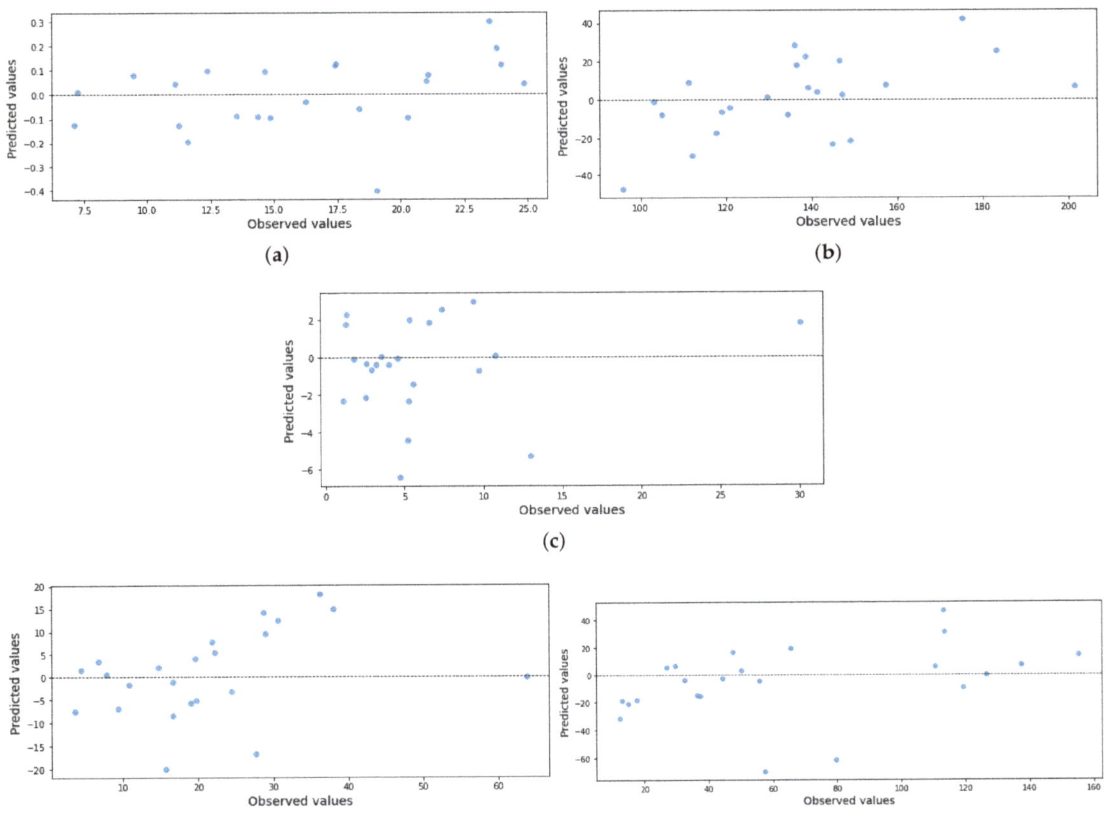

Figure 4. Graphical representation of MLR residuals: (**a**) daily average temperature, (**b**) daily wind direction, (**c**) daily average wind speed, (**d**) daily instantaneous wind speed, and (**e**) daily solar radiation.

Having these results, the p-values of the coefficients were analyzed to determine whether a given independent variable had a significant effect on the dependent variables. Therefore, the p-values were analyzed for each of the dependent variables and the coefficients that did not meet the condition $(P > |t|) < 0.05$ were excluded. Once the coefficients had been excluded, the best regression method for predicting each variable was chosen from among the MLR, ridge, and LASSO methods. Table 1 also shows the values of the coefficients of determination (R^2) for the new regression methods chosen.

Table 2 shows the method used for each variable, as well as the coefficients used to calculate the values corresponding to each variable between 2010 and 2022 in Quinta de Santa Bárbara. It can also be seen that the dependent variables do not depend on all the independent variables. This means that for each variable to be calculated, the methods select the variables that are highly correlated with the variable to be determined.

Table 2. Coefficients for the calculation of dependent variables.

Variable	Method	Coefficients					Intercept
		T_{MED1}	DD_{MED1}	FF_{MED1}	FF_{MAX1}	RG_{TOT1}	
T_{MED}	Linear	1.0190	0	0	0	−0.0032	−0.1887
DD_{MED}	LASSO	0	0	24.9277	0	0	96.7312
FF_{MED}	Ridge	0	0.0124	7.7472	0	0	−7.2055
FF_{MAX}	LASSO	0	0	7.2770	1.6848	0	−5.3412
RG_{TOT}	Linear	1.1333	0.0957	0	0	0.4822	−22.3257

Finally, using the coefficients provided in Table 2, the mathematical models to calculate the predicted daily values for each variable are given by Equations (12)–(16).

$$T_{MED} = -0.1887 + 1.0190 \cdot T_{MED1} - 0.0032 \cdot RG_{TOT1} \quad (12)$$

$$DD_{MED} = 96.7312 + 24.9277 \cdot FF_{MED1} \quad (13)$$

$$FF_{MED} = -7.2055 + 0.0124 \cdot DD_{MED1} - 7.7472 \cdot FF_{MED1} \quad (14)$$

$$FF_{MAX} = -5.3412 + 7.277 \cdot FF_{MED1} + 1.6848 \cdot FF_{MAX1} \quad (15)$$

$$RG_{TOT} = -22.3257 + 1.1333 \cdot T_{MED1} + 0.0957 \cdot DD_{MED1} + 0.4822 \cdot RG_{TOT1} \quad (16)$$

where T_{MED} is the predicted average temperature, DD_{MED} is the predicted average wind direction, FF_{MED} is the predicted average wind speed, FF_{MAX} is the predicted maximum instant wind speed, and RG_{TOT} is the predicted solar radiation for Quinta de Santa Bárbara, and T_{MED1}, DD_{MED1},FF_{MED1}, FF_{MAX1}, and RG_{TOT1} are the correspondent weather variables observed in Pinhão.

Finally, the input dataset was divided into training, validation, and testing sets. The main purpose of selecting these datasets is to assess the accuracy of the models' performance and also to ensure that they can be generalized to new data. For the variables under study, the values were divided into training, validation, and testing sets, with a split of 70%, 15%, and 15%, respectively, as shown in Figure 5.

4.2. Case I

In this case, each variable was forecast for 3 to 15 days, using an LSTM with specific hyperparameters characterized in Table 3. Each LSTM network had a ReLU activation function, a dense layer with 1 neuron, to obtain the output value, a dropout layer equal to 0.2, with 10 neurons, and they were optimized using Adam optimization with a learning rate of 0.001 in order to minimize the MSE.

Table 3. LSTM's parameters.

Parameter	Value
Number of LSTM layers	1
Number of LSTM units	50
Batch size	32
Number of epochs	40

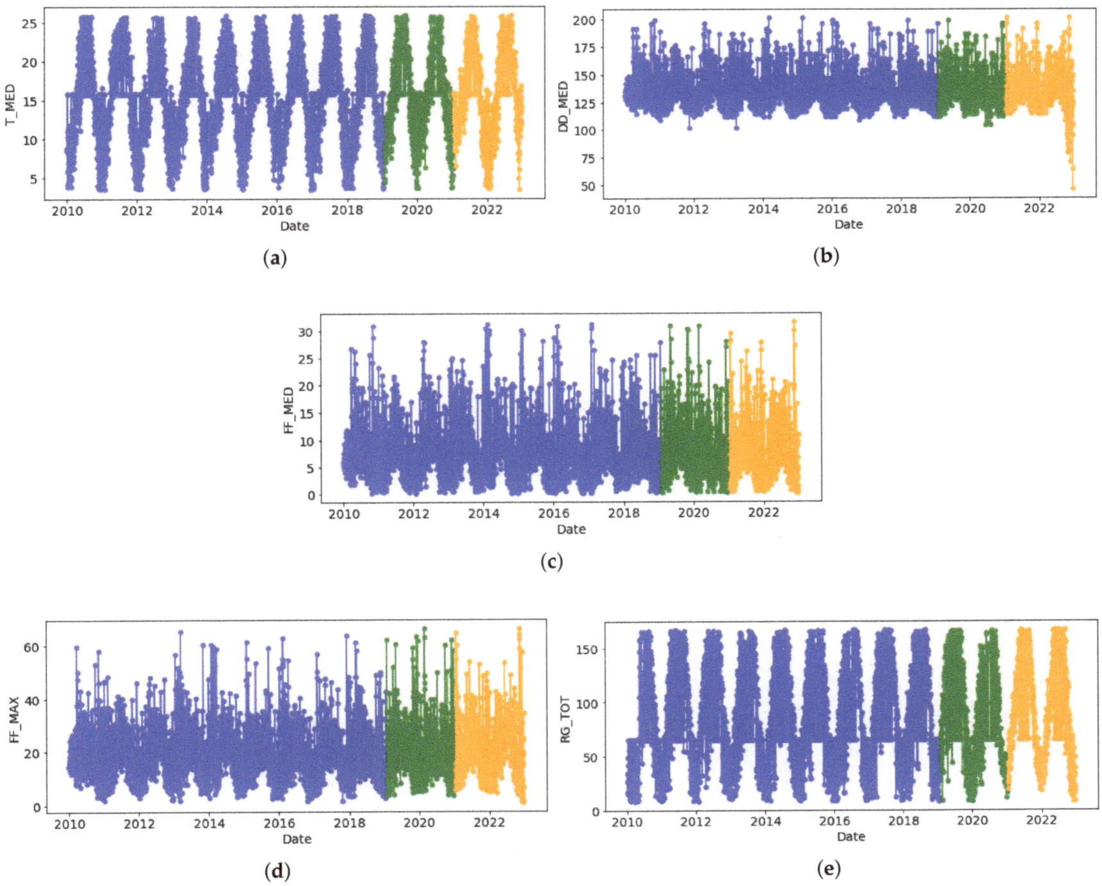

Figure 5. Train (blue), validation (green), and testing (yellow) sets: (**a**) daily average temperature, (**b**) daily wind direction, (**c**) daily average wind speed, (**d**) daily instantaneous wind speed, and (**e**) daily solar radiation.

Figures 6–8 show the results for the 3-, 7-, and 15-day forecasts, respectively. For each variable, the last 15 days of the test set and the predicted values for that period are also shown. In general, the predicted variables showed a similar behavior for all time periods, without significant fluctuations compared to the testing data.

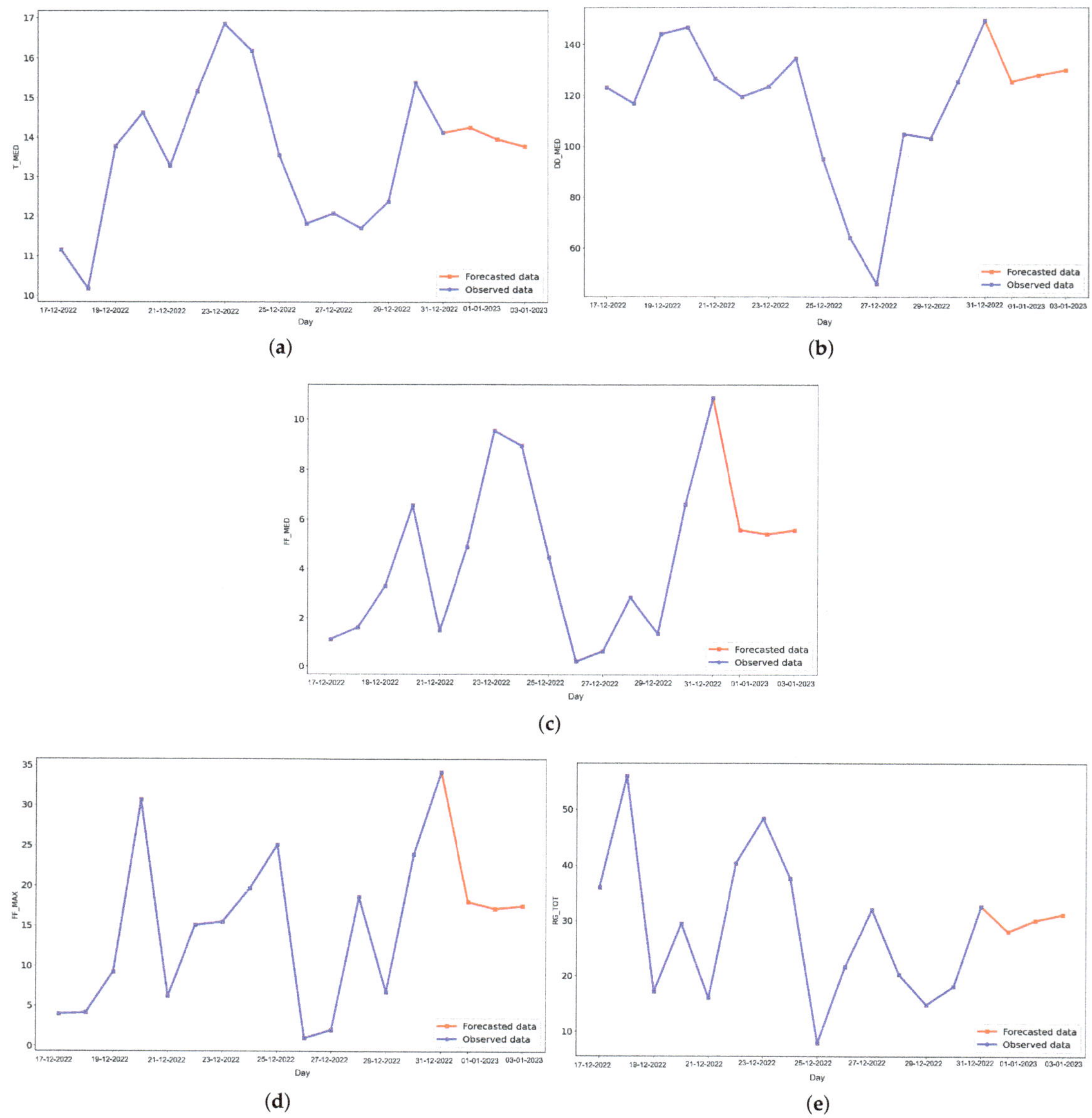

Figure 6. Last 15 days of the testing set (blue) and the predicted data (red) for 3 days ahead in the future for (**a**) daily average temperature, (**b**) daily average wind direction, (**c**) daily average wind speed, (**d**) daily instant wind speed, and (**e**) daily solar radiation predicted values (red), using an LSTM.

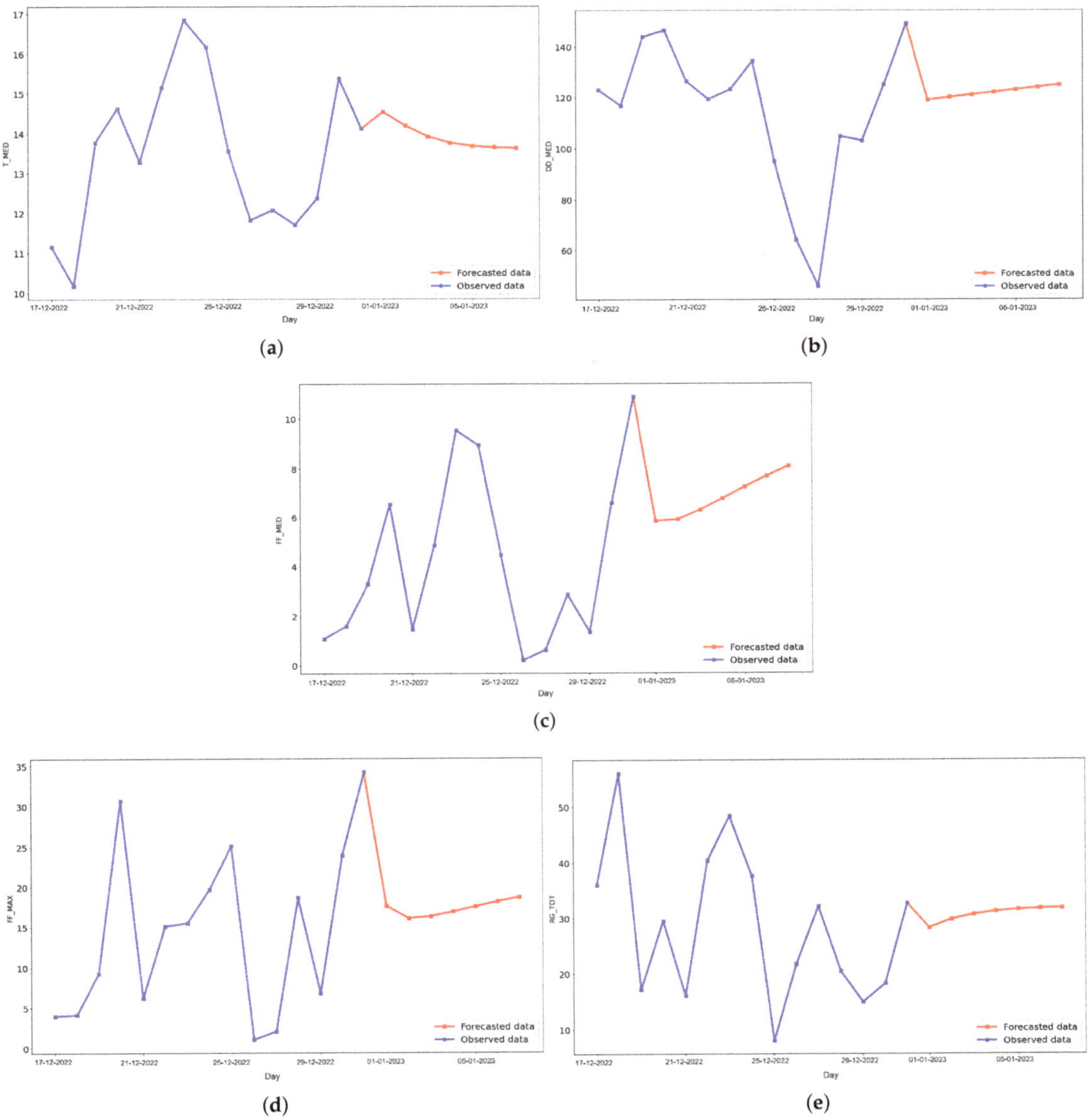

Figure 7. Last 15 days of the testing set (blue) and the predicted data (red) for 7 days ahead in the future for (**a**) daily average temperature, (**b**) daily average wind direction, (**c**) daily average wind speed, (**d**) daily instant wind speed, and (**e**) daily solar radiation predicted values (red), using an LSTM.

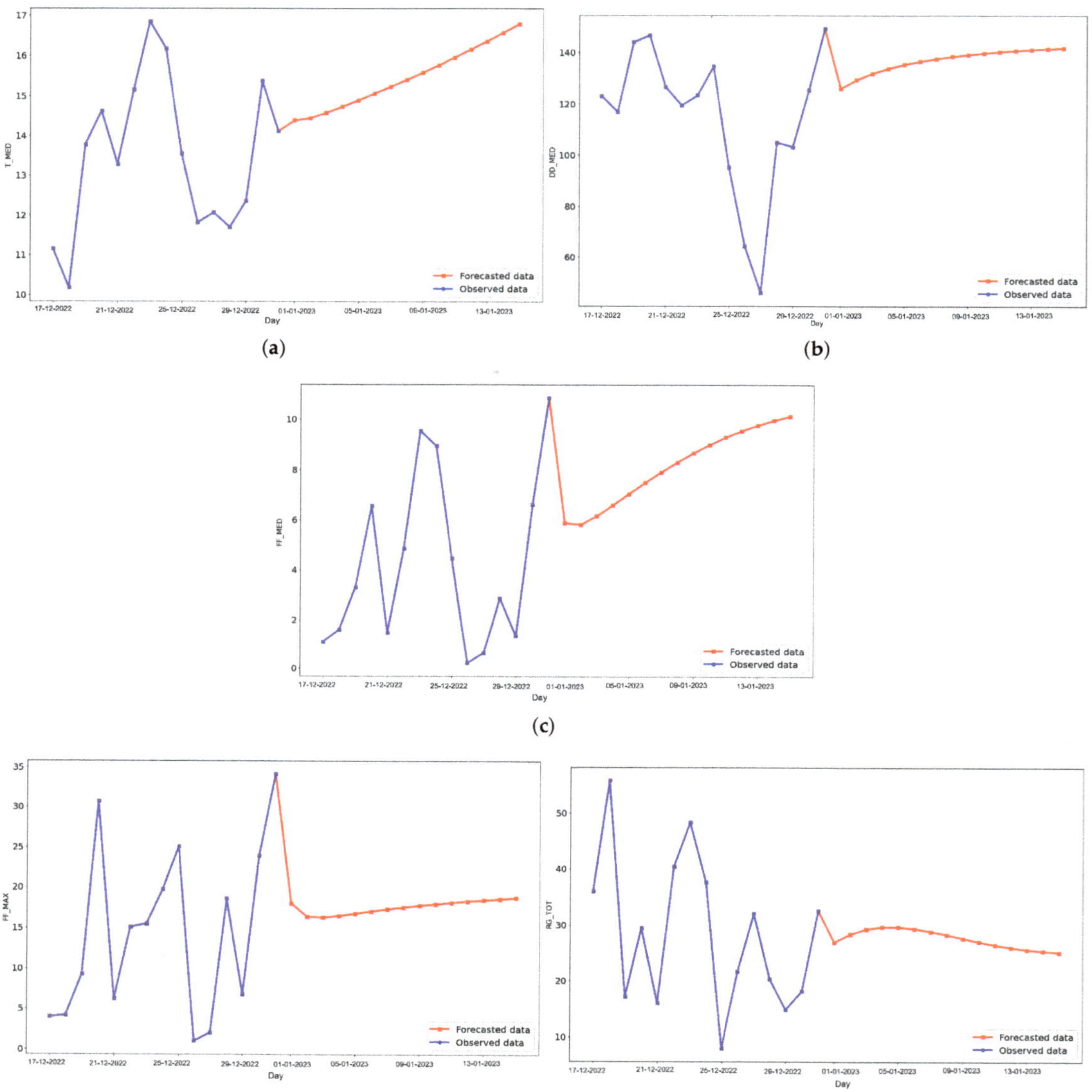

Figure 8. Last 15 days of the testing set (blue) and the predicted data (red) for 15 days ahead in the future for (**a**) daily average temperature, (**b**) daily average wind direction, (**c**) daily average wind speed, (**d**) daily instant wind speed, and (**e**) daily solar radiation predicted values (red), using an LSTM.

4.3. Case II

In this case, the LSTM hyperparameters were optimized with a GA method. The purpose of using this algorithm was to find the best LSTM hyperparameters that minimize the MSE of the network (Table 4). Table 5 shows the GA parameters used in this paper.

Table 4. Range of LSTM hyperparameters.

Parameter	Value
Number of LSTM layers	[1, 3]
Number of LSTM units	[50, 150]
Batch size	[16, 64]
Number of epochs	[20, 40]

Table 5. Genetic algorithm parameters.

Parameter	Value
Maximum number of iterations	5
Population size	10
Mutation rate	0.1
Elit rate	0.1
Crossover rate	0.5
Parents portion	0.3
Crossover type	Two point

Table 6 shows the LSTM hyperparameter optimization results for each variable. These results correspond to the parameters that minimize the MSE metric for each variable, according to the time period of the forecasts.

Table 6. Optimization results, case II.

	Variable	LSTM Layers	LSTM Units	Batch Size	Epochs
3 days	T_MED	3	87	49	34
	DD_MED	2	70	38	20
	FF_MED	3	98	50	27
	FF_MAX	3	50	17	21
	RG_TOT	1	76	41	31
7 days	T_MED	1	86	34	39
	DD_MED	2	91	21	21
	FF_MED	3	57	31	21
	FF_MAX	2	72	16	30
	RG_TOT	1	116	16	30
15 days	T_MED	1	118	31	30
	DD_MED	2	120	20	39
	FF_MED	2	124	54	29
	FF_MAX	3	134	25	20
	RG_TOT	2	109	42	32

Figures 9–11 show the prediction results for case II. In general, the behavior of each variable is different for each time period chosen. In this case, the predicted values behave similarly to the observed data, with a more fluctuating shape than in case I.

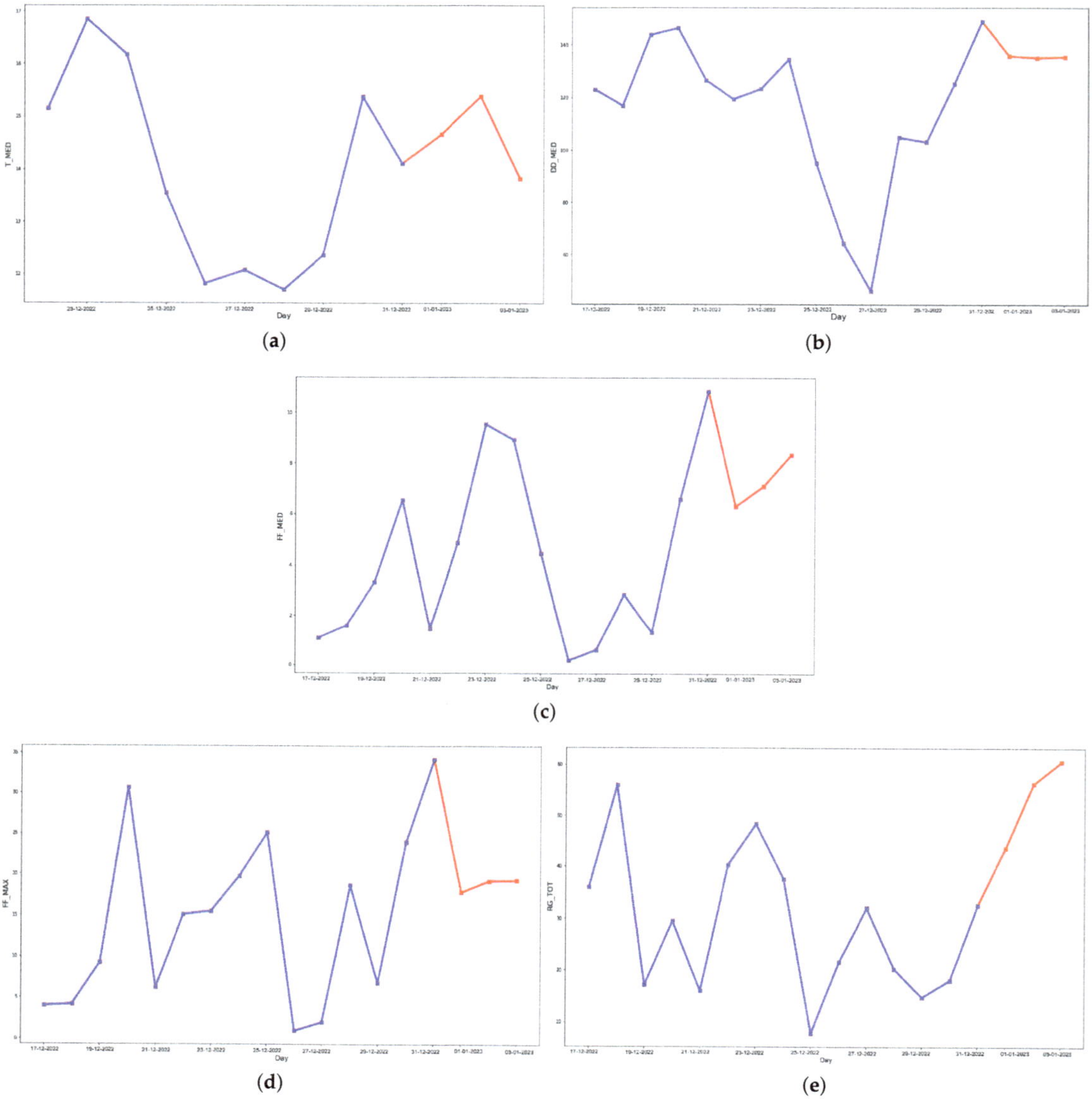

Figure 9. Last 15 days of the testing set (blue) and the predicted data (red) for 3 days ahead in the future for (**a**) daily average temperature, (**b**) daily average wind direction, (**c**) daily average wind speed, (**d**) daily instant wind speed, and (**e**) daily solar radiation predicted values (red), using an optimized LSTM.

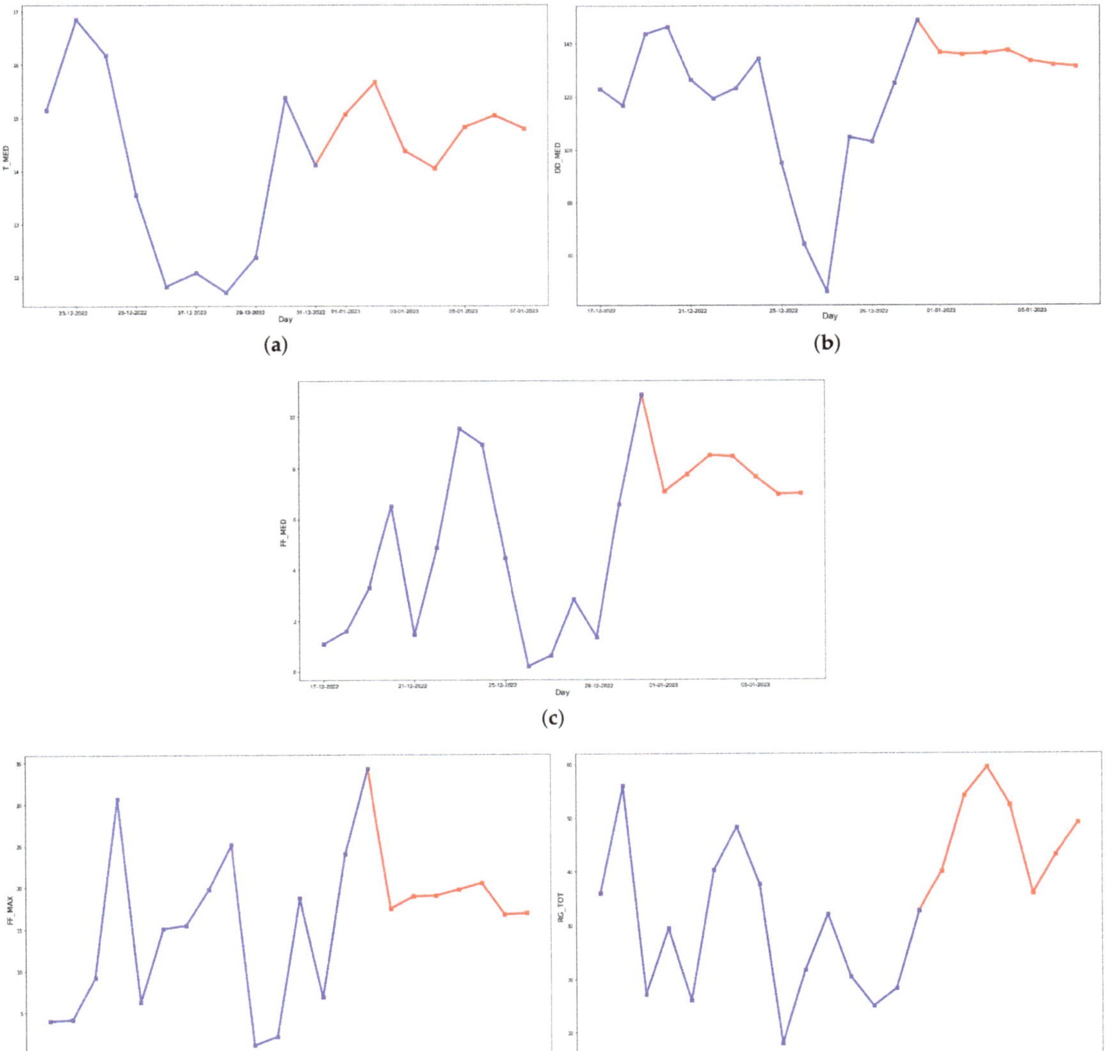

Figure 10. Last 15 days of the testing set (blue) and the predicted data (red) for 7 days ahead in the future for (**a**) daily average temperature, (**b**) daily average wind direction, (**c**) daily average wind speed, (**d**) daily instant wind speed, and (**e**) daily solar radiation predicted values (red), using an optimized LSTM.

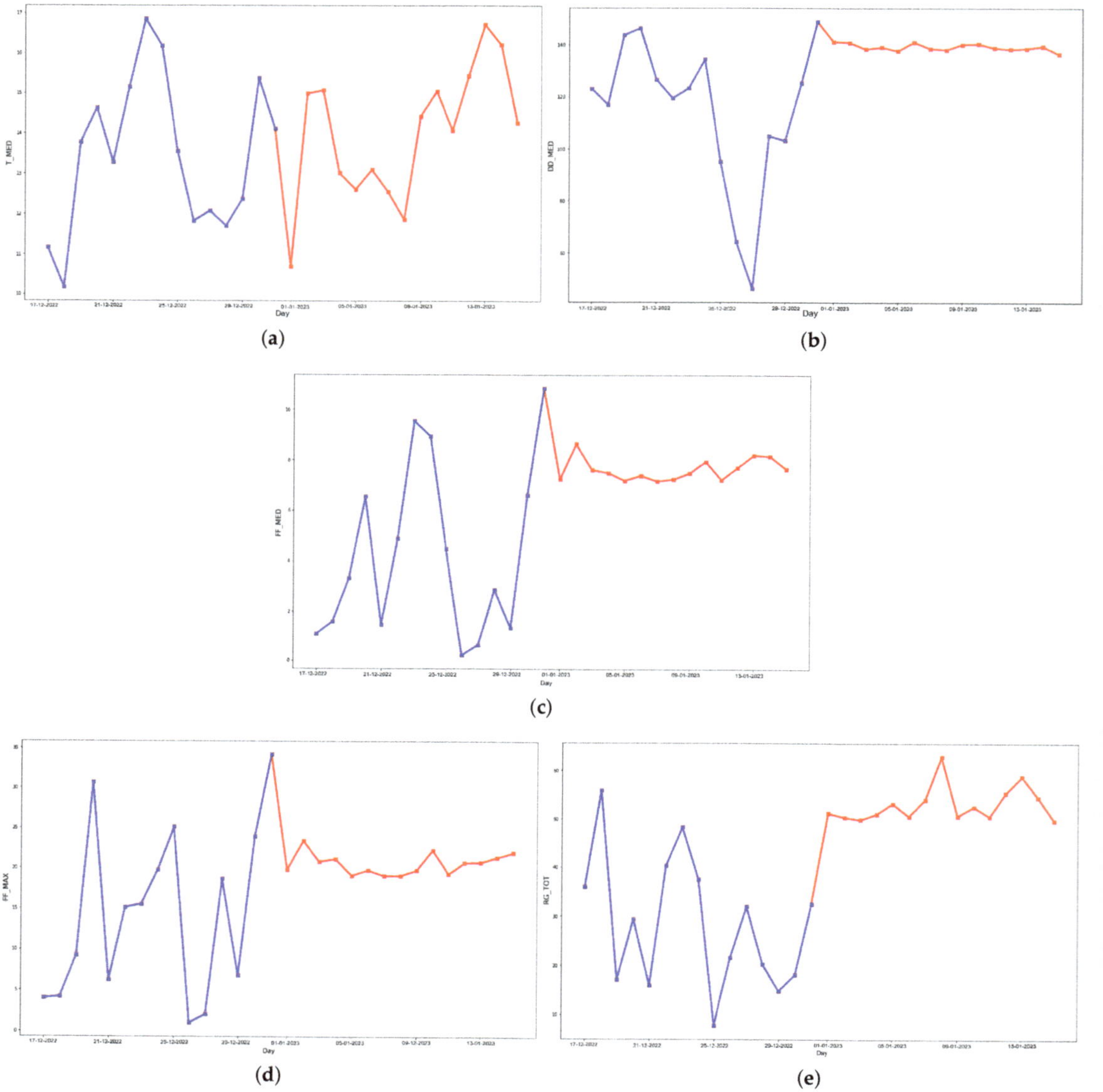

Figure 11. Last 15 days of the testing set (blue) and the predicted data (red) for 15 days ahead in the future for (**a**) daily average temperature, (**b**) daily average wind direction, (**c**) daily average wind speed, (**d**) daily instant wind speed, and (**e**) daily solar radiation predicted values (red), using an optimized LSTM.

4.4. Results Discussion

The evaluation metrics obtained for each proposed case were compared with two benchmark methods: XGBoost and transformer. Similarly to the proposed cases, a GA was also applied to these methods. The average numerical results obtained for each case and for the benchmark method, are presented in Table 7. The evaluation metrics represent each metric's average daily value according to the time range studied.

Table 7. Average evaluation metrics for climatic variables by each scenario and benchmark method.

		T_MED	DD_MED	3 Days FF_MED	FF_MAX	RG_TOT
Case I	MSE	0.0182	0.1512	0.1512	0.3555	0.0126
	MAE	0.1266	0.3756	0.575	0.5811	0.1014
	RMSE	0.1348	0.3888	0.6166	0.5962	0.1123
Case II	MSE	0.0453	0.0082	0.0085	0.0648	0.0071
	MAE	0.1926	0.0846	0.0863	0.2306	0.0733
	RMSE	0.2129	0.0903	0.0924	0.2545	0.0841
XGBoost	MSE	45.9822	307.0847	33.0072	102.5079	3432.2502
	MAE	5.7409	13.2791	4.3613	7.8988	51.9629
	RMSE	6.781	17.5238	5.7452	10.1246	58.5854
XGBoost-GA	MSE	46.8364	352.1824	33.6135	119.3655	3208.583
	MAE	5.7277	14.1451	4.5121	8.4562	50.3719
	RMSE	6.8437	18.7665	5.7977	10.9255	56.6444
Transformer	MSE	0.0026	0.0061	0.0053	0.0305	0.001
	MAE	0.0415	0.0696	0.0612	0.1622	0.0318
	RMSE	0.0506	0.07832	0.0726	0.1747	0.0321
Transformer-GA	MSE	0.0024	0.0057	0.0048	0.0253	0.0011
	MAE	0.0412	0.0655	0.0575	0.1393	0.0312
	RMSE	0.0492	0.0757	0.0695	0.1589	0.0335
		T_MED	DD_MED	7 days FF_MED	FF_MAX	RG_TOT
Case I	MSE	0.0078	0.0867	0.2034	0.1985	0.0205
	MAE	0.0778	0.2505	0.3774	0.385	0.1205
	RMSE	0.0882	0.2944	0.451	0.4456	0.1433
Case II	MSE	0.003	0.0328	0.0681	0.1633	0.0069
	MAE	0.0778	0.1596	0.2211	0.3682	0.0708
	RMSE	0.0882	0.1812	0.261	0.4042	0.08291
XGBoost	MSE	45.9822	307.0847	33.0072	102.5079	3432.2502
	MAE	5.7409	13.2791	4.3613	7.8988	51.9629
	RMSE	6.781	17.5238	5.7452	10.1246	58.5854
XGBoost-GA	MSE	46.9001	352.1685	33.6112	115.7482	3333.5516
	MAE	5.7328	14.1547	4.5106	8.4012	51.0867
	RMSE	6.8484	18.7662	5.7975	10.7586	57.7369
Transformer	MSE	0.003	0.034	0.0052	0.0291	0.0035
	MAE	0.0518	0.1569	0.0643	0.1513	0.0502
	RMSE	0.0548	0.1896	0.072	0.1706	0.0591
Transformer-GA	MSE	0.0037	0.0395	0.0059	0.0285	0.0024
	MAE	0.0546	0.1633	0.0675	0.1439	0.0392
	RMSE	0.0611	0.1988	0.0766	0.1689	0.0493
		T_MED	DD_MED	15 days FF_MED	FF_MAX	RG_TOT
Case I	MSE	0.0081	0.047	0.107	0.1116	0.0239
	MAE	0.073	0.1528	0.2305	0.2623	0.1368
	RMSE	0.0898	0.2166	0.3271	0.3341	0.1547
Case II	MSE	0.005	0.0337	0.0818	0.1254	0.0042
	MAE	0.0602	0.1408	0.2179	0.2919	0.054
	RMSE	0.0706	0.1835	0.286	0.3542	0.0647
XGBoost	MSE	45.9822	307.0847	33.0072	102.5079	3432.2502
	MAE	5.7409	13.2791	4.3613	7.8988	51.9629
	RMSE	6.781	17.5238	5.7452	10.1246	58.5854
XGBoost-GA	MSE	46.7768	352.1688	33.6143	118.2286	3316.611
	MAE	6.8394	18.7662	5.7978	10.8733	57.59
	RMSE	5.7214	14.1526	4.5125	8.4312	50.9951
Transformer	MSE	0.0057	0.0169	0.0082	0.0324	0.0044
	MAE	0.0575	0.1065	0.0783	0.1521	0.0563
	RMSE	0.0754	0.1301	0.0905	0.1799	0.0665
Transformer-GA	MSE	0.0059	0.0242	0.0053	0.0245	0.0046
	MAE	0.0634	0.1554	0.0726	0.1566	0.0677
	RMSE	0.077	0.1554	0.0726	0.1566	0.0676

Regarding the predicted variables, it can be seen that they generally have good results for the metrics, with values less than 1, which indicates that the proposed models are capable of predicting climate variables in the short and medium term. In general, when comparing the case studies considered, it can be seen that optimizing the characteristic parameters of the LSTM improved the metric values for all prediction horizons. The use of the GA minimized the MSE value for all variables, regardless of the horizon considered. For example, for a 3-day forecast, the MSE, MAE, and RMSE values for temperature (T_MED) decreased by 97%, 81%, and 81%, respectively, for mean wind speed (FF_MED) by 94%, 85%, and 85%, and for solar radiation (RG_TOT) by 44%, 28%, and 25%, respectively. On the other hand, for a 7-day forecast horizon, the values for T_MED decreased by 63%, 34%, and 52%, for FF_MED by 67%, 41%, and 42% and for RG_TOT by 66%, 41%, and 42%, respectively. Finally, for 15 days, the MSE, MAE, and RMSE values decreased by 38%, 18%, and 22% for T_MED, by 25%, 5%, and 13% for FF_MED, and by 82%, 61%, and 22% for RG_TOT.

Comparing our results with the benchmark methods it can be observed that both cases perform better than the XGBoost methods, and perform as well as the transformer methods. In relation to the results obtained with GA, the optimization of XGBoost does not significantly improve its performance, and both models struggle with higher errors.

Overall, the results show that transformers and LSTM methods are advanced neural network architectures capable of handling sequential data and capturing long-term dependencies. In fact, LSTM models, while computationally intensive, can be generally efficient, particularly for smaller datasets. Therefore, LSTMs are a good solution for weather forecasting. They are likely to provide accurate predictions, especially when optimized with techniques like GA.

5. Conclusions and Future Work

Accurate weather forecasts are crucial for decision making and planning in various sectors. Understanding weather patterns is essential for making informed decisions, optimizing operations, and mitigating risks in industries such as agriculture and renewable energy.

Solar radiation and temperature are more stable compared to wind speed, which can be unpredictable. Our numerical results show that regression models are better at estimating average temperature than wind-related variables, highlighting the reliability of these weather elements.

When it comes to LSTM predictions, the results are generally smoother and less variable compared to the input data. This is due to LSTM's ability to capture and learn long-term dependencies within a sequence of data using memory cells and gates to store past information and selectively forget or update it as new data are introduced. Finally, using a GA method to optimize the LSTM hyperparameters minimized the evaluation metrics, especially the MSE value, improving the accuracy of the LSTM in forecasting weather variables.

In LSTM predictions, the results are usually smoother and have less variation compared to the input data. This is because LSTM can capture and learn long-term dependencies in a sequence of data by using memory cells and gates to store past information while selectively forgetting or updating it as new data are introduced. Additionally, using a GA approach to fine-tune LSTM hyperparameters led to a reduction in evaluation metrics, especially the MSE value, thereby improving the accuracy of the LSTM in forecasting weather variables.

The evaluation of the metrics revealed promising results. The LSTM-based models consistently provided accurate and precise weather forecasts, regardless of the time horizon considered. This underscores the effectiveness and reliability of LSTM technology in weather forecasting.

The quality and quantity of available data posed a significant challenge in this work. Techniques for data augmentation and improvement may be explored in future studies.

On the other hand, the computational cost of training the models emerged as a critical consideration. Investigating more efficient training algorithms, transfer learning, or utilizing distributed computing resources could be valuable work for future research.

Author Contributions: Conceptualization, R.T., A.C., E.J.S.P. and J.B.; methodology, R.T., A.C., E.J.S.P. and J.B.; software, R.T., A.C., E.J.S.P. and J.B.; validation, R.T., A.C., E.J.S.P. and J.B.; formal analysis, R.T., A.C., E.J.S.P. and J.B.; investigation, R.T., A.C., E.J.S.P. and J.B.; resources, R.T., A.C., E.J.S.P. and J.B.; data curation, R.T., A.C., E.J.S.P. and J.B.; writing—original draft preparation, R.T., A.C., E.J.S.P. and J.B.; writing—review and editing, R.T., A.C., E.J.S.P. and J.B.; visualization, R.T., A.C., E.J.S.P. and J.B.; supervision, R.T., A.C., E.J.S.P. and J.B.; project administration, R.T., A.C., E.J.S.P. and J.B.; funding acquisition, R.T., A.C., E.J.S.P. and J.B. All authors have read and agreed to the published version of the manuscript.

Funding: This research received no external funding.

Data Availability Statement: Data are contained within the article.

Acknowledgments: This research was developed in line with the Vine and Wine Portugal—Driving Sustainable Growth Through Smart Innovation Mobilizing Agenda (ID: C644866286-011).

Conflicts of Interest: The authors declare no conflicts of interest.

Abbreviations

The following abbreviations are used in this manuscript:

DL	Deep learning
GA	Genetic algorithm
IPMA	Portuguese Institute for Sea and Atmosphere
LASSO	Least absolute shrinkage and selection operator
LSTM	Long short-term memory
ML	Machine learning
MAE	Mean absolute error
MSE	Mean square error
MLR	Multiple linear regression
RMSE	Root mean square error
ReLU	Rectified linear unit
XGBoost	Extreme gradient boosting

References

1. Jaseena, K.; Kovoor, B.C. Deterministic weather forecasting models based on intelligent predictors: A survey. *J. King Saud Univ. Comput. Inf. Sci.* **2022**, *34*, 3393–3412. [CrossRef]
2. Shekana, S.; Mulugeta, A.; Sharma, D.P. Weather Variability Forecasting Model through Data Mining Techniques. *Int. J. Adv. Comput. Sci. Appl.* **2020**, *11*, 31–41. [CrossRef]
3. Jain, H.; Jain, R. Big data in weather forecasting: Applications and challenges. In Proceedings of the 2017 International Conference on Big Data Analytics and Computational Intelligence (ICBDAC), Chirala, India, 23–25 March 2017; pp. 138–142. [CrossRef]
4. Karevan, Z.; Suykens, J.A. Transductive LSTM for time-series prediction: An application to weather forecasting. *Neural Netw.* **2020**, *125*, 1–9. [CrossRef] [PubMed]
5. van Buuren, S. *Flexible Imputation of Missing Data*, 2nd ed.; CRC Press: Boca Raton, FL, USA, 2018.
6. Doreswamy, I.G.; Manjunatha, B. Performance evaluation of predictive models for missing data imputation in weather data. In Proceedings of the 2017 International Conference on Advances in Computing, Communications and Informatics (ICACCI), Udupi, India, 13–16 September 2017; pp. 1327–1334. [CrossRef]
7. Mittal, S.; Sangwan, O.P. Big Data Analytics using Deep LSTM Networks: A Case Study for Weather Prediction. *Adv. Sci. Technol. Eng. Syst. J.* **2020**, *5*, 133–137. [CrossRef]
8. Zaytar, A.; Amrani, C.E. Sequence to Sequence Weather Forecasting with Long Short-Term Memory Recurrent Neural Networks. *Int. J. Comput. Appl.* **2016**, *143*, 7–11.
9. Trujillo Viedma, D.; Rivera Rivas, A.J.; Charte Ojeda, F.; del Jesus Díaz, M.J. A First Approximation to the Effects of Classical Time Series Preprocessing Methods on LSTM Accuracy. In Proceedings of the Advances in Computational Intelligence, San Luis Potosi, Mexico, 27 October 2019; Rojas, I., Joya, G., Catala, A., Eds.; Springer: Cham, Switzerland, 2019; pp. 270–280.
10. Santra, A.S.; Lin, J.L. Integrating Long Short-Term Memory and Genetic Algorithm for Short-Term Load Forecasting. *Energies* **2019**, *12*, 2040. [CrossRef]

11. Li, W.; Zang, C.; Liu, D.; Zeng, P. Short-term Load Forecasting of Long-short Term Memory Neural Network Based on Genetic Algorithm. In Proceedings of the 2020 IEEE 4th Conference on Energy Internet and Energy System Integration (EI2), Wuhan, China, 30 October–1 November 2020; pp. 2518–2522. [CrossRef]
12. Bouktif, S.; Fiaz, A.; Ouni, A.; Serhani, M.A. Optimal Deep Learning LSTM Model for Electric Load Forecasting using Feature Selection and Genetic Algorithm: Comparison with Machine Learning Approaches. *Energies* **2018**, *11*, 1636. [CrossRef]
13. Jiang, P.; Liu, Z.; Niu, X.; Zhang, L. A combined forecasting system based on statistical method, artificial neural networks, and deep learning methods for short-term wind speed forecasting. *Energy* **2021**, *217*, 119361. [CrossRef]
14. Maulud, D.; Abdulazeez, A.M. A Review on Linear Regression Comprehensive in Machine Learning. *J. Appl. Sci. Technol. Trends* **2020**, *1*, 140–147. [CrossRef]
15. Ansarifar, J.; Wang, L.; Archontoulis, S. An interaction regression model for crop yield prediction. *Sci. Rep.* **2021**, *11*, 17754. [CrossRef]
16. Barriguinha, A.; de Castro Neto, M.; Gil, A. Vineyard Yield Estimation, Prediction, and Forecasting: A Systematic Literature Review. *Agronomy* **2021**, *11*, 1789. [CrossRef]
17. Amir, M.; Zaheeruddin; Haque, A. Intelligent based hybrid renewable energy resources forecasting and real time power demand management system for resilient energy systems. *Sci. Prog.* **2022**, *105*, 00368504221132144. [CrossRef] [PubMed]
18. Carneiro, T.C.; Rocha, P.A.; Carvalho, P.C.; Fernández-Ramírez, L.M. Ridge regression ensemble of machine learning models applied to solar and wind forecasting in Brazil and Spain. *Appl. Energy* **2022**, *314*, 118936. [CrossRef]
19. Zheng, Y.; Ge, Y.; Muhsen, S.; Wang, S.; Elkamchouchi, D.H.; Ali, E.; Ali, H.E. New ridge regression, artificial neural networks and support vector machine for wind speed prediction. *Adv. Eng. Softw.* **2023**, *179*, 103426. [CrossRef]
20. Nam, K.; Hwangbo, S.; Yoo, C. A deep learning-based forecasting model for renewable energy scenarios to guide sustainable energy policy: A case study of Korea. *Renew. Sustain. Energy Rev.* **2020**, *122*, 109725. [CrossRef]
21. Zhou, H.; Zhang, Y.; Yang, L.; Liu, Q.; Yan, K.; Du, Y. Short-Term Photovoltaic Power Forecasting Based on Long Short Term Memory Neural Network and Attention Mechanism. *IEEE Access* **2019**, *7*, 78063–78074. [CrossRef]
22. Phan, Q.T.; Wu, Y.K.; Phan, Q.D. Short-term Solar Power Forecasting Using XGBoost with Numerical Weather Prediction. In Proceedings of the 2021 IEEE International Future Energy Electronics Conference (IFEEC), Taipei, Taiwan, 16–19 November 2021; pp. 1–6. [CrossRef]
23. Wadhwa, S.; Tiwari, R.G. Machine Learning-based Weather Prediction: A Comparative Study of Regression and Classification Algorithms. In Proceedings of the 2023 International Conference in Advances in Power, Signal, and Information Technology (APSIT), Bhopal, India, 28–29 June 2023; pp. 487–492. [CrossRef]
24. Galindo Padilha, G.A.; Ko, J.; Jung, J.J.; de Mattos Neto, P.S.G. Transformer-Based Hybrid Forecasting Model for Multivariate Renewable Energy. *Appl. Sci.* **2022**, *12*, 10985. [CrossRef]
25. Nascimento, E.G.S.; de Melo, T.A.; Moreira, D.M. A transformer-based deep neural network with wavelet transform for forecasting wind speed and wind energy. *Energy* **2023**, *278*, 127678. [CrossRef]
26. Walczewski, M.J.; Wöhrle, H. Prediction of Electricity Generation Using Onshore Wind and Solar Energy in Germany. *Energies* **2024**, *17*, 844. [CrossRef]
27. Quiñones, J.J.; Pineda, L.R.; Ostanek, J.; Castillo, L. Towards smart energy management for community microgrids: Leveraging deep learning in probabilistic forecasting of renewable energy sources. *Energy Convers. Manag.* **2023**, *293*, 117440. [CrossRef]
28. Cui, Y.; Chen, Z.; He, Y.; Xiong, X.; Li, F. An algorithm for forecasting day-ahead wind power via novel long short-term memory and wind power ramp events. *Energy* **2023**, *263*, 125888. [CrossRef]
29. Altan, A.; Karasu, S.; Zio, E. A new hybrid model for wind speed forecasting combining long short-term memory neural network, decomposition methods and grey wolf optimizer. *Appl. Soft Comput.* **2021**, *100*, 106996. [CrossRef]
30. Van Houdt, G.; Mosquera, C.; Nápoles, G. A Review on the Long Short-Term Memory Model. *Artif. Intell. Rev.* **2020**, *53*, 5929–5955. [CrossRef]
31. Thirunavukkarasu, G.S.; Kalair, A.R.; Seyedmahmoudian, M.; Jamei, E.; Horan, B.; Mekhilef, S.; Stojcevski, A. Very Short-Term Solar Irradiance Forecasting using Multilayered Long-Short Term Memory. In Proceedings of the 2022 7th International Conference on Smart and Sustainable Technologies (SpliTech), Split/Bol, Croatia, 5–8 July 2022; pp. 1–8. [CrossRef]
32. Zhang, W.; Lin, Z.; Liu, X. Short-term offshore wind power forecasting—A hybrid model based on Discrete Wavelet Transform (DWT), Seasonal Autoregressive Integrated Moving Average (SARIMA), and deep-learning-based Long Short-Term Memory (LSTM). *Renew. Energy* **2022**, *185*, 611–628. [CrossRef]
33. Salehin, I.; Talha, I.M.; Mehedi Hasan, M.; Dip, S.T.; Saifuzzaman, M.; Moon, N.N. An Artificial Intelligence Based Rainfall Prediction Using LSTM and Neural Network. In Proceedings of the 2020 IEEE International Women in Engineering (WIE) Conference on Electrical and Computer Engineering (WIECON-ECE), Bhubaneswar, India, 26–27 December 2020; pp. 5–8. [CrossRef]
34. Markovics, D.; Mayer, M.J. Comparison of machine learning methods for photovoltaic power forecasting based on numerical weather prediction. *Renew. Sustain. Energy Rev.* **2022**, *161*, 112364. [CrossRef]
35. Yang, D.; Yang, G.; Liu, B. Combining quantiles of calibrated solar forecasts from ensemble numerical weather prediction. *Renew. Energy* **2023**, *215*, 118993. [CrossRef]
36. Venkatachalam, K.; Trojovský, P.; Pamucar, D.; Bacanin, N.; Simic, V. DWFH: An improved data-driven deep weather forecasting hybrid model using Transductive Long Short Term Memory (T-LSTM). *Expert Syst. Appl.* **2023**, *213*, 119270. [CrossRef]

37. Van Rossum, G.; Drake, F.L. *Python 3 Reference Manual;* CreateSpace: Scotts Valley, CA, USA, 2009.
38. Abadi, M.; Agarwal, A.; Barham, P.; Brevdo, E.; Chen, Z.; Citro, C.; Corrado, G.S.; Davis, A.; Dean, J.; Devin, M.; et al. TensorFlow: A system for large-scale machine learning In Proceedings of the 12th USENIX Conference on Operating Systems Design and Implementation (OSDI'16), Savannah, GA, USA, 2–4 November 2016; pp. 265–283.
39. Pedregosa, F.; Varoquaux, G.; Gramfort, A.; Michel, V.; Thirion, B.; Grisel, O.; Blondel, M.; Prettenhofer, P.; Weiss, R.; Dubourg, V.; et al. Scikit-learn: Machine learning in Python. *J. Mach. Learn. Res.* **2011**, *12*, 2825–2830.

applied sciences

MDPI

Article

Voltage Problems on Farms with Agricultural Biogas Plants—A Case Study

Zbigniew Skibko [1], Andrzej Borusiewicz [2,3,*], Wacław Romaniuk [4], Marta Pietruszynska [5], Anna Milewska [6] and Andrzej Marczuk [7]

1 Faculty of Electrical Engineering, Bialystok University of Technology, Wiejska 45, 15-351 Bialystok, Poland; z.skibko@pb.edu.pl
2 Department of Agronomy, Modern Technology and Informatics, International Academy of Applied Sciences in Lomza, 18-402 Lomza, Poland
3 University of Social and Economics in Gdansk, ul. Rajska 6, 80-850 Gdansk, Poland
4 Institute of Technology and Life Sciences–National Research Institute, Hrabska 3, 05-090 Falenty, Poland; w.romaniuk@itp.edu.pl
5 Department of Ophthalmology, Medical University of Bialystok Clinical Hospital, Skłodowskiej-Curie 24A, 15-276 Bialystok, Poland; marta@pietruszynska.pl
6 Institute of Economics and Finance, Department of Finance, Division of Public Finance, Banking and Law, Warsaw University of Life Sciences, Nowoursynowska 166, 02-787 Warsaw, Poland; anna_milewska1@sggw.edu.pl
7 Department of Agricultural, Forestry and Transport Machines, University of Life Sciences in Lublin, Głęboka 28, 20-612 Lublin, Poland; andrzej.marczuk@up.lublin.pl
* Correspondence: andrzej.borusiewicz@mans.edu.pl

Abstract: Constructing agricultural microbial gasification plants near livestock farms is essential for technical, economic, and environmental reasons. Utilising substrates from these farms allows for producing electricity, heat, and environmentally friendly manure. However, biogas plants often face technical challenges. This study evaluates the power quality of an agricultural biogas plant on a dairy farm. It was found that the plant was connected via a cable with an insufficient conductor cross-section, leading to significant voltage overshoots exceeding 14.6%, which prevented the activation of the second generator. Both generators could operate after replacing the feed-in cable, but considerable fluctuations in the feed-in voltage persisted. Further measurements indicated the need for changes in the digester design. Specifically, replacing the current two mixers with more lower-powered mixers operating alternately was proposed. Sharing these solutions more broadly can help prevent similar issues in future microbial gas plant constructions and optimise electricity production.

Keywords: power quality; agricultural biogas plant; farm; voltage fluctuation; power supply reliability; costs; profitability

1. Introduction

In recent years, renewable energy generation has become crucial for the sustainable development of countries, regions, and individuals. The use of feedstocks to produce biogas and biomethane holds significant potential, potentially meeting about 20% of the current global gas demand. Agricultural and industrial waste negatively impacts health and the environment, emitting significant greenhouse gases such as methane and nitrous oxide [1,2]. Methane fermentation is one of the most effective technologies for converting organic waste into renewable energy. Agricultural biogas plants help reduce agriculture's environmental impact, and the digestate obtained from the processed substrates can be effectively used as an organic fertiliser [3,4].

Additionally, biogas can mitigate the harmful effects of fossil fuel use [5], significantly increasing atmospheric CO_2 concentrations [6]. Biogas is an alternative energy source that generates electricity, heat, and cooling, and it can replace natural gas derived from fossil fuels [7]. Biogas produced in biogas plants mainly contains methane (CH_4) and carbon dioxide (CO_2) [8]. It also contains smaller proportions of water, nitrogen, hydrogen sulfide, ammonia, oxygen, siloxanes, and other particles [9]. The composition of biogas depends primarily on the substrates used and the technology employed [10].

When building a biogas plant, the location plays a key role, which mainly depends on the availability of input materials (type of farm), the availability of the electricity grid, and the distance from urbanised areas (due to the potential impact of the biogas plant on noise and odours) [11–13]. From the point of view of sustainable agricultural development, the most important are micro biogas plants, built directly on farms and fed with waste (e.g., slurry) generated on the farm. These measures allow for efficient and rational use of resources at an optimal cost level. A farm with a micro biogas plant is seen as modern and environmentally friendly. The profitability of agricultural biogas plants located at cattle farms depends mainly on the price of electricity and on having the capital to purchase and maintain an electrical generator [14,15]. In addition, suitable technical conditions for connecting such an energy source at the location are necessary [16]. As herd size increases, there is a noticeable decrease in the cost of building a biogas plant per cow [17]. When calculating the profitability of an investment in an agricultural biogas plant, the equipment's maintenance and servicing costs should also be considered [18,19]. Various numerical methods, such as AHP [20], Vikor [21], and the numerical taxonomy method [22], can be used to assess the profitability of the investment. The payback period for the construction of an agricultural microbial gas plant is 6–7 years [23] (investments with a payback of up to 8 years are considered financially beneficial [24,25]). In comparison, the payback for a P.V. plant is within 5–6 years [26].

Agricultural biogas plants enable the generation of electricity heat and cooling and have several other advantages, among which are:

- the possibility of using biogas as a fuel for internal combustion engines [27];
- an increase in regional/national energy security through the development of distributed energy [28];
- a reduction in the cost of disposing of organic waste [29];
- mitigation of adverse climate change [30];
- use of local energy resources [31];
- production of wholesome manure [32].

Despite the apparent increase in biogas production, there is often a lack of relevant knowledge and technology in agriculture [33]. Therefore, an essential action is to broker information from agriculture to the technology industry, covering all legal, environmental, administrative, organisational, and logistical aspects, to implement as many biogas plant projects as possible successfully [34]. Many adverse effects, including financial losses, result from investors' ignorance, who often violate the basic principles of biotechnology when selecting substrates for their installations [35]. Existing installations are constantly being improved in terms of their production efficiency and environmental impact [6]. The local use of the generated energy avoids the losses associated with the transmission of energy over longer distances [36].

Biogas power plants, being renewable energy sources, are characterised by the stability of the generated power over time (concerning wind or P.V. power plants). Previous authors' studies on agricultural biogas plants connected to the Polish medium-voltage grid confirm this characteristic [8]. However, there is a lack of analysis in the literature of the results of studies that would present an in-depth analysis of the impact of generation in biogas plants on the voltage of the electrical system supplying consumers installed on farms. Power plants connected to the electricity grid can cause an increase in voltage at their point of connection [37]. This increase in network voltage stresses the insulation of electrical equipment, reducing its service life and, in extreme cases, can damage it [38]. Voltage variability depends on the structure of the grid and the power of the connected power plants [39]. In the case of P.V. plants, as the generation level increases, the number of voltage oscillations increases, which can cause instability—short-term voltage stability is strongly influenced by the disconnection of P.V. generation and the resulting voltage drop [36]. The main factors affecting static voltage stability are the power and location of the sources and how they are controlled [40]. Better voltage stability occurs near the transformer station. The variation in the voltage value depends not only on the actual power generated but also on the quality of the current connections, the length and cross-section of the wires connecting the micro installations to the main switchgear and the resistive parameters of the low-voltage network [41]. When the line is not loaded, and the distributed energy sources operate at total capacity, the voltage at each point in the system is higher than the voltage at the transformer terminals. This situation worsens as one moves away from the transformer. The effect of generation on mains voltage values is described in detail for P.V. systems. It is not easy to find publications that describe the impact of a biogas plant (as a stable energy source) on the voltage parameters of the power grid.

The impact of microbial biogas plants on the parameters of the supply voltage is all the more relevant because, in Poland, farmers with small agricultural biogas plants have reported problems with the operation of specific consumers, including milking clusters. The authors have filled this research gap by conducting a detailed study of one Polish microbial gas plant. The main objective of the research was to analyse the causes of voltage problems on the farm and to propose and apply a possible solution to this issue. Particular analysis was carried out on the parameters determining the quality of electricity generated in the agricultural micro biogas plant. Presenting the proposed solutions to a broader audience may make it possible in the future to avoid mistakes made in the construction of micro gas plants and even influence the companies building their components to optimise the amount and quality of energy fed into the grid.

2. Materials and Methods

The study was conducted at a Polish agricultural biogas plant directly connected to a farm installation. The primary substrate used in this biogas plant was manure from approximately 120 dairy cows. The biogas plant comprises three main facilities: a digester, an engine house, and a digestate tank. The digester had a diameter of 15.5 m and a wall height of 3.7 m, with a total height of 8.7 m, including the biogas tank. During regular operation, the digester held approximately 590 cubic meters of substrate, mixed by two agitators with a rated power of 9 kW. The engine room was constructed as a container housing an internal combustion engine connected to a generator and other technical and control equipment. Biogas purification was achieved using a gas filter and an activated carbon filter. The biogas produced was combusted in two WG1605 combustion engines, each with a power output of 20 kW and an efficiency of 97%. A type 4P/IE2 asynchronous generator, operating at 1500 rpm, was coupled to the engines, which were also equipped with a water-cooling circuit.

The digestate, the post-production biogas material, was stored in an open concrete tank. A schematic of the biogas plant is shown in Figure 1. The biogas plant was connected to the farm's main meter switchgear via a 198-meter-long YKY 4 × 25 mm^2 cable. The farm's connection power was 40 kW, limited by overcurrent protection C63. The meter switchgear was supplied from the 63 kVA 15/0.4 kV transformer station by a 413 m long overhead line with 50 mm^2 conductors. The overhead transformer station was supplied from the main power point of the electricity system by a 6000 m long overhead line made of steel and aluminium conductors with a cross-section of 50 mm^2. According to information obtained from the Distribution System Operator, the power system had a short-circuit capacity of 102 MVA.

The owner of the biogas plant reported the following problems with its operation:

Due to voltage exceedances at the agricultural biogas plant's connection point, it is impossible to run the second generator.

Increasingly frequent cluster malfunction—voltage error displayed.

The slurry stayed in the fermenter for an average of two to three weeks but no less than 12 days. Daily, a portion of the digested slurry (regulated by the biogas pressure sensor) was pumped out of the digester into the digestion tank. When the lower biogas pressure threshold value was exceeded, a new slurry was pumped into the digester (from the collection pit) to replenish the digestate.

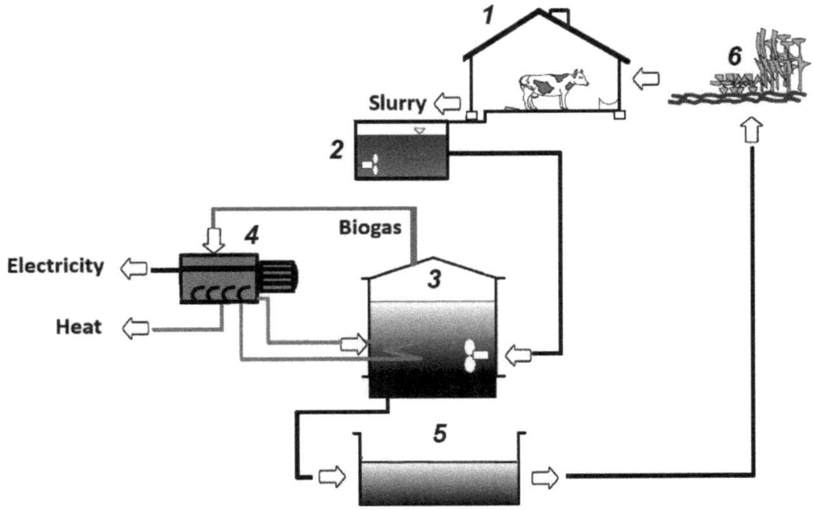

Figure 1. Diagram of the tested agricultural biogas plant. 1—livestock facility, 2—slurry tank, 3—digester with biogas tank, 4—cogeneration unit, 5—digestate tank, 6—agricultural field.

Power quality is defined as a set of parameters describing the characteristics of the power supply process to the user under normal operating conditions, characterising the supply voltage, and determining the continuity of the power supply to the consumer. The following parameters describing power quality were analysed in the study [42,43]:

- **Mains frequency** is the number of repetitions in the time waveform of the fundamental component of the supply voltage measured over a specified time interval. The frequency deviation is the difference between a given value and the rated frequency value exhibited during regular power system operation over at least a few seconds. The frequency deviation should, in most cases, not exceed +/− 1% of the rated grid frequency.
- **Voltage deviation (slow voltage variation)** is the difference between the actual and rated mains voltage values. In most cases, the free voltage variation should not exceed ± 10% of the rated mains voltage value.
- **Voltage fluctuations (rapid changes in voltage)**. Indicators that characterise voltage fluctuations include:
    - The amplitude of voltage fluctuations is expressed as the ratio of the voltage variation value to the rated voltage. In most cases, this value should not exceed 3%,
    - frequency of voltage fluctuation amplitudes or, in the case of periodic fluctuations, the frequency of voltage fluctuations;
    Short-term flicker index P_{st} (index indicating the annoyance of flickering light over a few minutes). $P_{st} = 1$ is the conventional threshold for the annoyance of light flicker;
    - the long-term flicker index P_{lt} (an index indicating the annoyance of flickering light over a long period, of the order of a few hours). In most cases, the index P_{lt}

should not exceed 1. The value of the index is determined from successive values of P_{st}, according to the relation:

$$P_{lt} = \sqrt[3]{\frac{\sum_{i=1}^{12} P_{st_i}^3}{12}} \quad (1)$$

in which P_{sti} (i = 1, 2, 3, ... 12) are successive values of the short-term flicker indices P_{st}.

- **Voltage asymmetry**—unequal voltage values and/or unequal angles between successive phase voltages. The asymmetry of the system of supply voltages results, among other things, in the appearance of symmetrical components of the opposite order. The parameter describing this condition is the voltage asymmetry factor $\alpha_{U\%}$ (opposite voltage asymmetry) [36,44]:

$$\alpha_{U\%} = \left|\frac{\underline{U}_2}{\underline{U}_1}\right| \cdot 100\% \quad (2)$$

  where: U_1, U_2—the composite values of the symmetrical components of the consensual and opposite order of the voltage.
  In most cases, the voltage asymmetry factor should not exceed 2%.
- **The distortion of the voltage waveform**, defined by the total harmonic distortion factor (THD_U) [45]:

$$THD_U = \frac{\sqrt{\sum_{h=2}^{\infty} U_h^2}}{U_1} \cdot 100\% \quad (3)$$

  where: U_1—rms value of voltage for the first harmonic, U_h—rms value of voltage for the h-th harmonic, h—order of harmonic.

In most cases, the voltage distortion coefficient THD_U in low-voltage networks must not exceed 8%.

At the agricultural biogas plant under study, electricity was the primary end product, and its production was maintained at the highest possible level. The tests were carried out using a SONEL PQM-701 portable power quality parameter analyser, performing measurements in accordance with Class A of the EN 61000-4-30 standard [46]. The recorder has a legalisation certificate issued by the Research and Calibration Laboratory in Świdnica. The recording of electricity parameters was carried out with 1-minute averaging and recording of the measurement results. The measurement of voltage/current was carried out with an accuracy of ±0.1% of the rated voltage/current, and the measurement of voltage unbalance with an accuracy of ±0.15% (absolute error). The maximum error of the harmonic measurement was ±0.05% U_n for voltage and ±0.15% I_n for current.

3. Results and Discussion

3.1. Analysis of the Supply Network Parameters of the Agricultural Biogas Plant under Study

The recorded grid frequency values are shown in Figure 2. Although frequent changes in frequency were noted, the maximum deviation from the rated value was slight, not exceeding 0.18%. However, a particular recurring pattern is worth noting: the lowest recorded frequency values occurred in the afternoon during peak electricity network load and increased as the load decreased.

The waveforms of the variation in voltage levels and the percentage deviation of the voltage from the rated voltage recorded during the tests are shown in Figures 3 and 4.

The analysis of the recorded voltage waveforms reveals a cyclic pattern of changes similar to those observed in frequency. Voltage fluctuates with changes in the energy system's load, decreasing as the load increases. Increases in voltage deviation values accompany these voltage decreases. The voltage varied from approximately 200 V to nearly 260 V. Such significant fluctuations can cause interference with electronic equipment.

The recorded values of the long-term flicker annoyance factor (P_{lt}) are included in Figure 5.

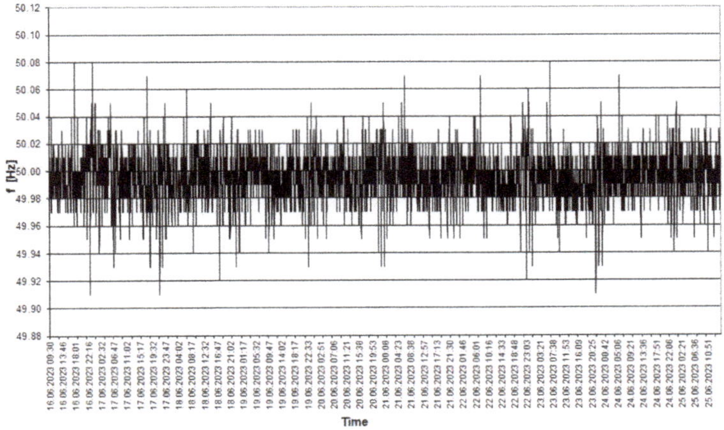

Figure 2. Frequency value variation recorded at the connection point of an agricultural biogas plant.

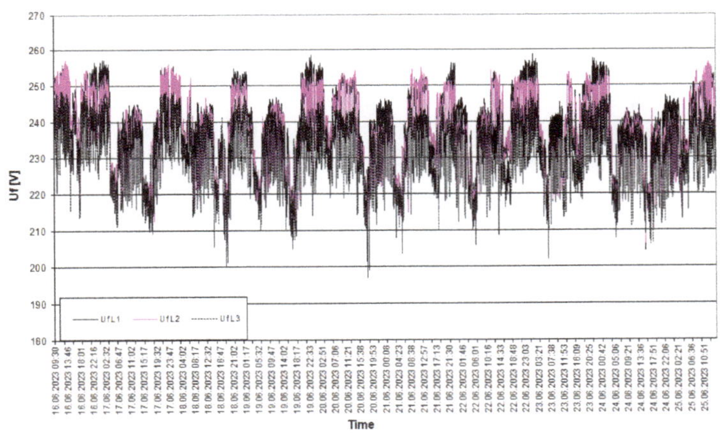

Figure 3. Voltage variation recorded at the connection point of an agricultural biogas plant.

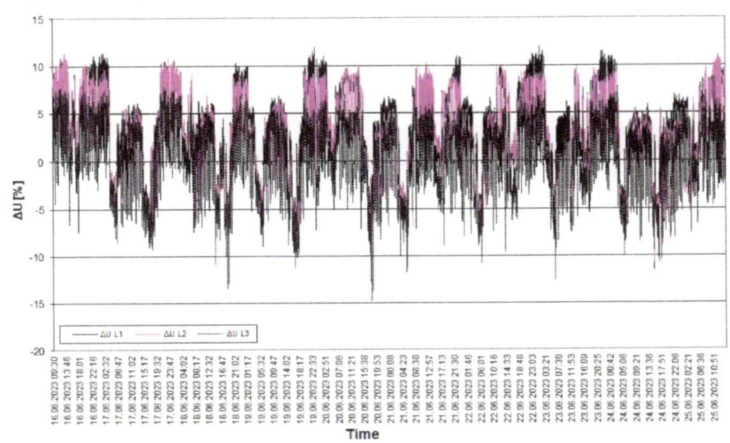

Figure 4. Variation in voltage deviation values recorded at the connection point of an agricultural biogas plant.

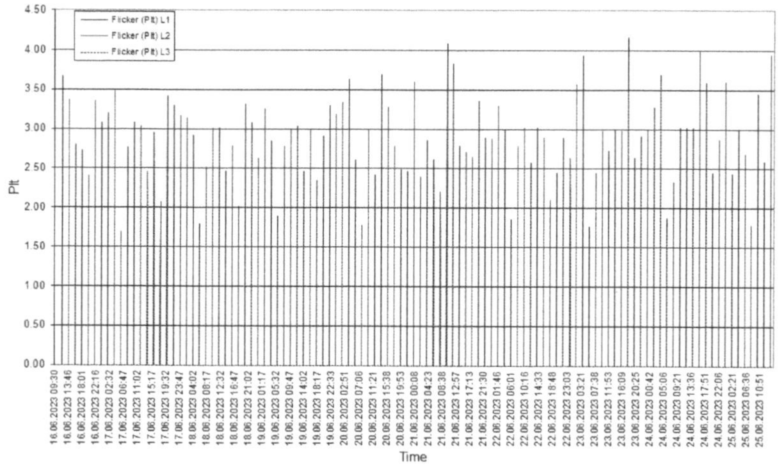

Figure 5. Variation in the long-term flicker nuisance factor (P_{lt}) recorded at the connection point of an agricultural biogas plant.

The values of the long-term flicker annoyance index also vary cyclically. The lowest P_{lt} index occurs at night, and the highest in the afternoon. It is usually related to the presence of so-called unstable loads (loads that consume electricity that are significant in value and variable in time) in the power system.

The voltage asymmetry factor also varies cyclically (Figure 6). The recorded waveform usually reaches a maximum in the afternoon, while the minimum occurs at night.

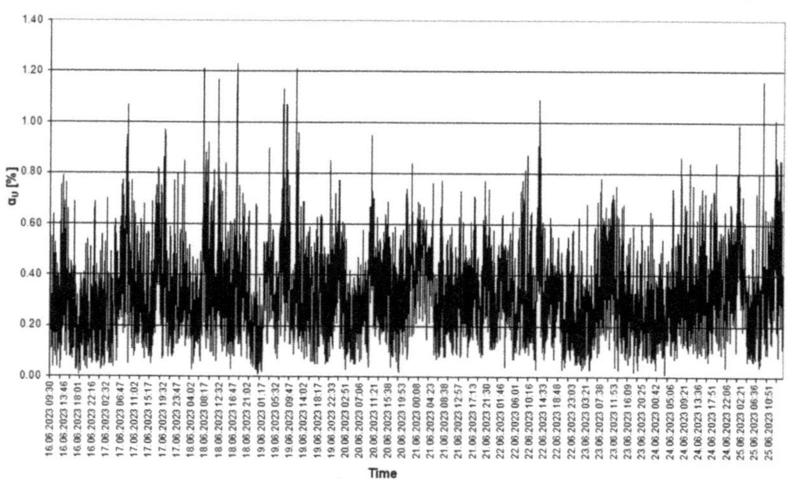

Figure 6. Variation in the voltage asymmetry factor (α_U) recorded at the connection point of the agricultural biogas plant.

The recorded THD_U values for each phase are shown in Figure 7.

The values of the total voltage distortion coefficient recorded at the biogas plant also vary in the diurnal system, mainly due to the cyclic operation of equipment drawing current distorted from the sinusoidal waveform from the grid. The highest values were recorded mainly in the afternoon.

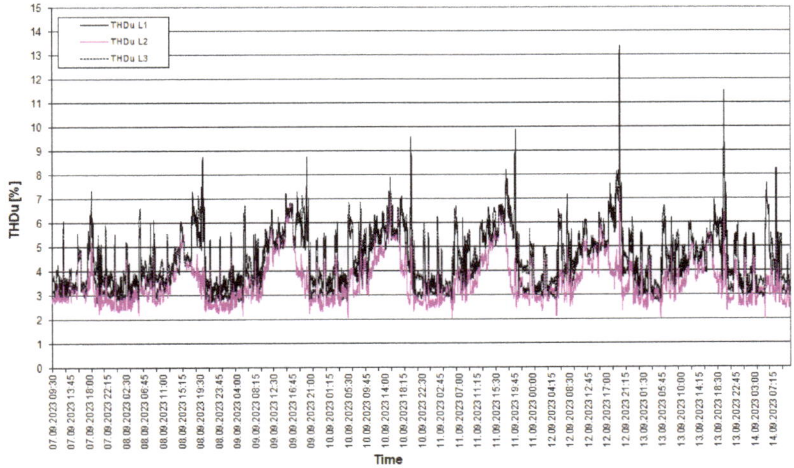

Figure 7. The course of variation in the total voltage distortion coefficient (THD_U) was recorded at the connection point of the agricultural biogas plant.

The results of the analysis of the active power levels generated by the agricultural biogas plant are summarised in Figure 8.

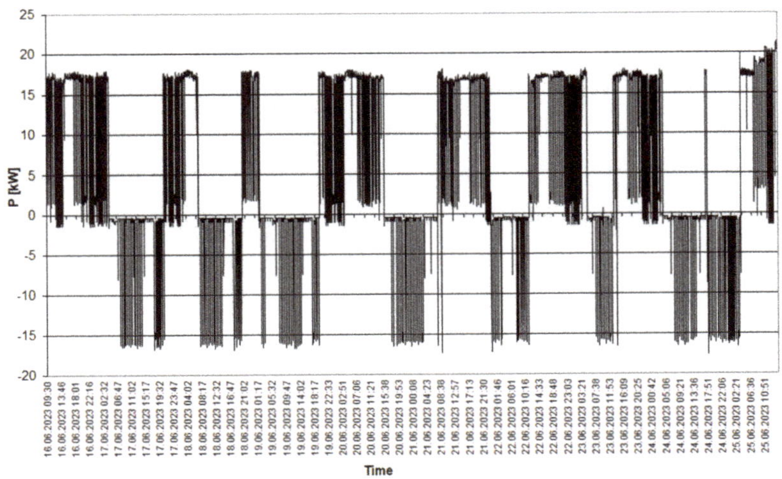

Figure 8. The recorded course of variation in the three-phase active power value generated at an agricultural biogas plant.

Figure 8 and information from the owner indicate that only one turbogenerator is operational at the agricultural biogas plant. The data also show long periods when the biogas plant is not generating electricity (the power is negative). The times when the agitator is running are visible on the waveform, with the recorded power dropping close to zero when the generator is running or around −17 kW when no energy production occurs. Therefore, it can be inferred that one 20 kW generator is necessary to meet the peak needs of the agricultural biogas plant. Without the agitator running, the biogas plant consumes about 1 kW (for control systems, monitoring, etc.).

The inability to switch on the second generator is due to the mains voltage exceeding the permissible value of 253 V. Even with one generator running, voltage exceedances were recorded (Figure 3), causing the generator unit to disconnect from the grid to protect other electrical equipment sensitive to voltage spikes. These voltage overshoots are also evident in the voltage deviation waveform (Figure 4), which repeatedly exceeds the regulatory limit of 10%. To enable the operation of the second generator, an analysis of the power supply system should be conducted to prevent voltage overshoots.

3.2. Analysis of the Feeding System of the Biogas Plant under Study

To carry out the analyses, a computer model of the power supply system for the agricultural biogas plant was created using the Neplan program, a global standard in power system analysis. The system data described in Section 2 were used for the calculations.

The original power supply system for the biogas plant consisted of a 198-meter-long cable line using YKY 4 $\times$ 25 mm^2. As shown in Figure 9, this cross-section does not ensure that the voltage remains within legal limits. The cable cross-section should be no smaller than 61 mm^2 to meet the voltage requirements, with the closest standardised cross-section being 70 mm^2. Changing the cross-section of the cable reduces the power loss from 1.82 kW to 0.65 kW. Assuming the biogas plant operates 8400 h annually, this change reduces annual energy losses by 8 MWh. An electricity price of €95/MWh results in an annual cost saving of €760. A cable with a more considerable cross-section costs €23 per meter. Considering only the reduction in power losses on the power cable, the upgrade from a 25 mm^2 to a 70 mm^2 conductor will pay for itself in 6 years.

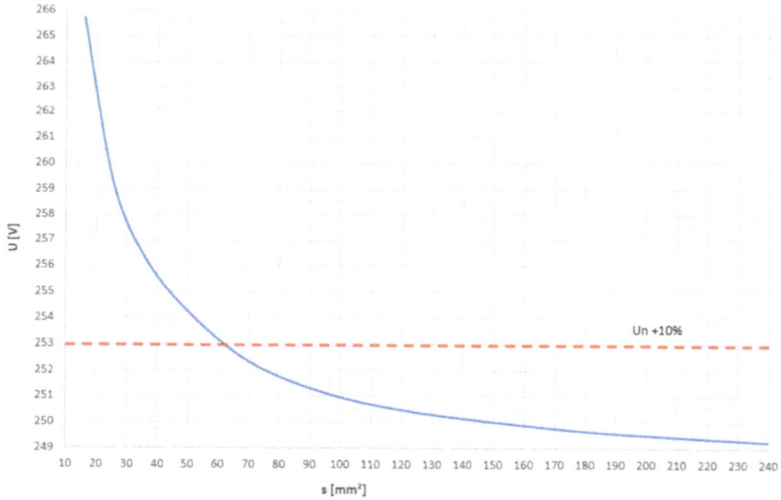

Figure 9. Dependence of the voltage value (U) occurring at the connection point of an agricultural biogas plant as a function of the cross-sectional area (s) of the working conductor of the supply cable.

3.3. Analysis of the Main Supply Parameters of the Studied Agricultural Biogas Plant after Changing the Supply Cable

The recorded values of the mains frequency after the power system conversion are shown in Figure 10.

Analysing the course of the variation in the frequency value recorded after the power cable change, repeatability analogous to that before the conversion can be observed.

Analysing the recorded voltage waveforms (Figure 11) and voltage deviation (Figure 12), as in the previous case, it is possible to notice a cyclic variation. However, the variations are noticeably more minor—it is much less frequent for the voltage to take values above 250 V, and only one measurement was recorded in which the voltage dropped below 210 V. This is reflected in the voltage deviation values (Figure 12), which were within $\pm$10% of the rated voltage throughout the recorded period.

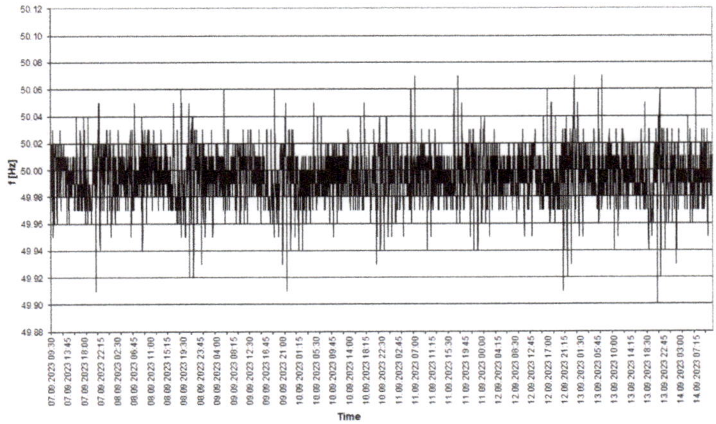

Figure 10. Frequency variation waveform recorded after changing the power cable.

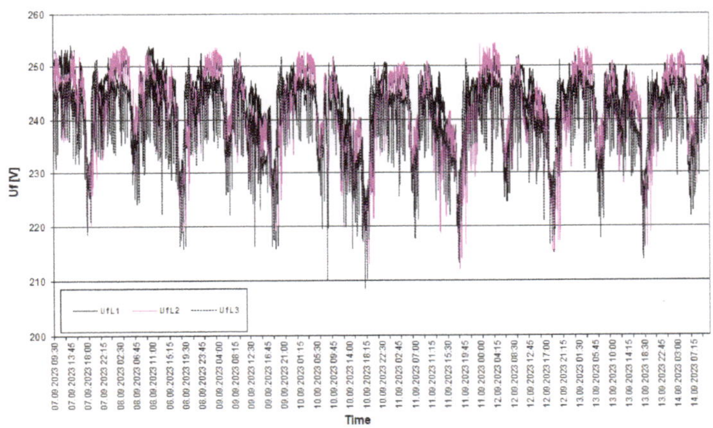

Figure 11. Voltage variation waveform recorded after changing the power cable.

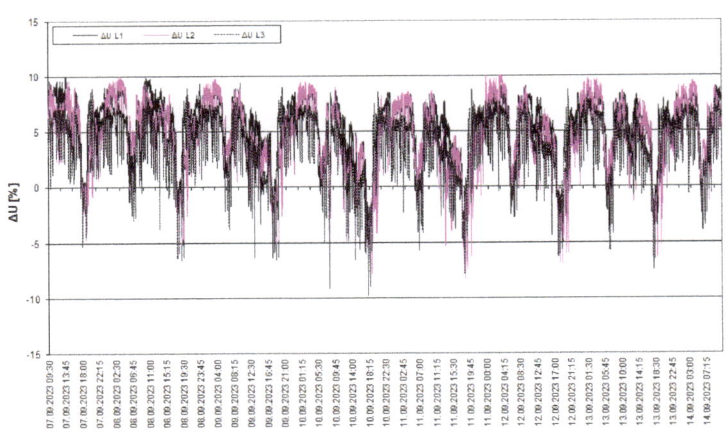

Figure 12. Voltage deviation variation waveform recorded after changing the power cable.

The unstable operation of the agricultural biogas plant is reflected in the recorded values of the flicker nuisance index (Figure 13). These are nearly three times lower than the values found in the system before the power cable change (Figure 5).

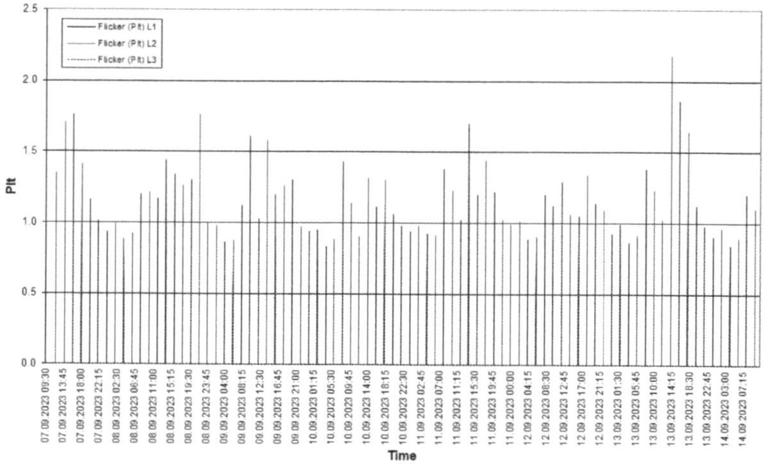

Figure 13. Variation waveform of the long-term flicker nuisance factor (P_{lt}) recorded after changing the power cable.

As with the original system, a cyclic variation in the voltage asymmetry factor was also recorded after changing the power cable (Figure 14). This does not show any significant differences in the values achieved, but comparing the two waveforms (Figures 6 and 14), it is clear that the waveform recorded after the power cable change is less 'jagged'.

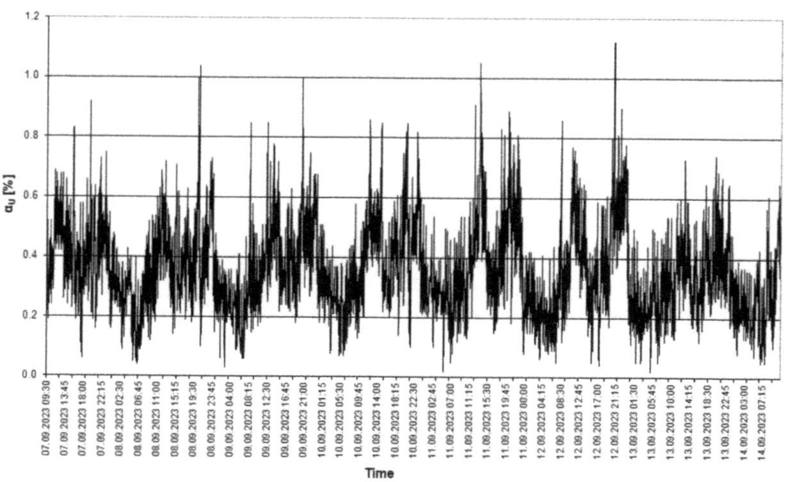

Figure 14. Variation waveform of the voltage asymmetry factor (α_U) recorded after changing the power cable.

As in the original system, the maximum values of the voltage distortion factor occur in the afternoon and evening (Figure 15). However, these values are almost twice as low as the first biogas plant studied.

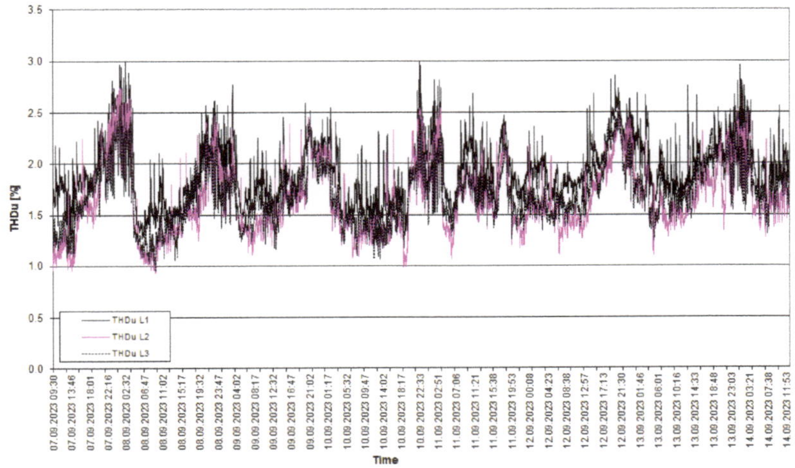

Figure 15. The waveform of the total voltage distortion coefficient (THD_U) variation was recorded after changing the power cable.

The maximum changes in the recorded active power values of the agricultural biogas plant after changing the feed cable (Figure 16) were due to the operation of the two digester mixers. Notably, at no point during the measurement did the active power reach the generator's rated power. This discrepancy was caused by the energy consumption of the equipment installed in the biogas plant.

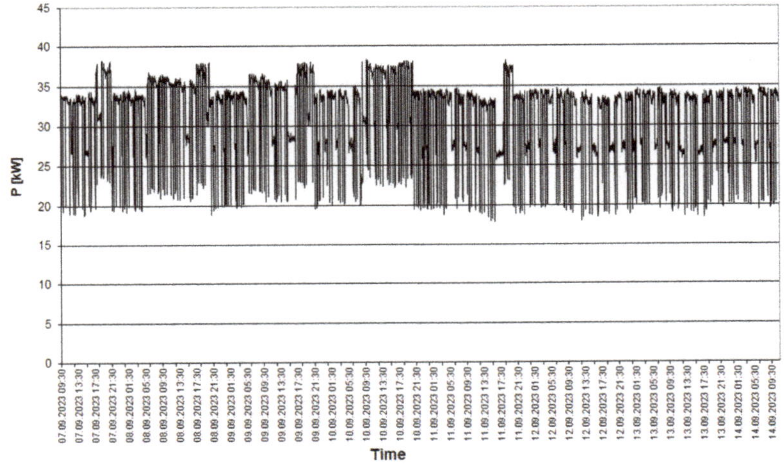

Figure 16. Recording of the course of the variation in the three-phase active power value generated in an agricultural biogas plant.

3.4. Analysis of the Main Supply Parameters of the Studied Agricultural Biogas Plant after Changing the Supply Cable

The value standardised in Polish regulations [42,43] is the 95% quantile. It is defined in the standard [42] as the highest value obtained from 95% of the recording time during a week. A summary of the recorded supply voltage frequency values is shown in Table 1.

Table 1. Results of statistical frequency analysis.

Parameter	Before the Changes		After Changes	
	Frequency [Hz]	**Deviation from Rated Value [%]**	**Frequency [Hz]**	**Deviation from Rated Value [%]**
Average value	49.995	−0.010	49.996	−0.008
Minimum value	49.910	−0.180	49.900	−0.200
Maximum value	50.080	0.160	50.070	0.140
Quantile 95	49.970	0.040	49.970	0.040

A summary of the measured voltage levels in each phase of the system supplying the agricultural biogas plant is shown in Table 2. Several relationships emerge from analysing these values. Noticeable voltage asymmetry exists in the individual phases. The lowest voltage occurred in the third phase, likely due to the asymmetrical connection of single-phase loads to the grid. However, in each phase, the voltage recorded after changing the power cable was lower than that of the primary circuit, translating into voltage deviation values. After the power cable change, although momentary exceedances of the permissible voltage deviation values were registered, the regulatory normal quantile of 95% slightly exceeded 9%. The operation of the biogas plant causes an increase in voltage at its point of connection, which, as seen from the measurements, is kept close to the upper limit allowed by regulations. Switching on the digester mixer reduces generation by almost half, resulting in a drop in voltage values. Turning it off causes the voltage to rise again; this cycle repeats every 30 minutes. This cyclic, step-by-step change in voltage is not beneficial for both the generated electricity's quality and consumer equipment's durability. It is mainly noticeable as a change in the intensity of the lighting installed on the farm and can lead to malfunctioning of voltage-sensitive equipment, potentially causing deactivation or even damage.

Table 2. Results of statistical analysis of tension levels.

Parameter	L1 Phase		L2 Phase		L3 Phase	
	Phase Voltage [V]	**Deviation from the Rated Voltage [%]**	**Phase Voltage [V]**	**Deviation from the Rated Voltage [%]**	**Phase Voltage [V]**	**Deviation from the Rated Voltage [%]**
Before the changes						
Average value	238.913	3.452	238.120	3.109	233.096	0.934
Minimum value	203.260	−11.986	200.130	−13.341	197.070	−14.666
Maximum value	258.650	11.999	257.120	11.336	248.810	7.738
Quantile 95	253.480	9.760	252.110	10.167	244.450	10.850
After changes						
Average value	242.655	5.073	241.368	4.515	240.101	3.967
Minimum value	214.320	−7.197	212.130	−8.145	208.580	−9.682
Maximum value	254.120	10.037	254.280	10.106	251.760	9.015
Quantile 95	250.980	8.678	251.737	9.005	247.690	7.253

Table 2 provides the answer to the malfunction of the milking apparatus—its shutdown. Before the power cable was replaced, the lowest recorded voltage value was only 197 V, below the usually accepted voltage limit of 200 V for milking clusters. Consequently, the clusters reported a voltage error and went into emergency mode. After replacing the power cable, the lowest recorded voltage value was 208 V, higher than the minimum value allowed by regulations. As a result, there were no more emergency shutdowns of the consumers during the test period.

A summary of the results of the statistical analysis of the values of the voltage asymmetry coefficient in the power supply system of the biogas plant under study is presented in Table 3. As was shown during the analysis of the recorded waveforms of variation in the asymmetry coefficients, the analysis of the values presented in Table 3 did not reveal any significant differences in the values of this parameter before and after the change of the power supply cable.

Table 3. Results of statistical analysis of voltage asymmetry coefficients.

Parameter	Before the Changes α_U [%]	After Changes α_U [%]
Average value	0.315	0.350
Minimum value	0.010	0.020
Maximum value	1.230	1.120
Quantile 95	0.610	0.620

A summary of the results of the statistical analysis of the long-term flicker nuisance value (P_{lt}) at the connection point of the agricultural biogas plant during the recording period is presented in Table 4. The values presented there show that the P_{lt} factor in each analysed system exceeds the permissible value (1). However, replacing the power cable reduced the value of the 95% quantile by almost three times.

Table 4. Results of statistical analysis of the long-term flicker annoyance index values (P_{lt}).

	Before the Changes			After Changes		
Parameter	P_{ltL1}	P_{ltL2}	P_{ltL3}	P_{ltL1}	P_{ltL2}	P_{ltL3}
Average value	2.688	2.762	2.887	0.993	1.141	1.074
Minimum value	1.540	1.620	1.690	0.720	0.830	0.790
Maximum value	3.840	3.960	4.160	2.180	1.830	1.860
Quantile 95	3.500	3.580	3.767	1.384	1.691	1.440

In the primary system, non-compliance with the voltage distortion requirements from the sinusoidal waveform was observed (Table 5). The maximum THD_U value is almost 1.7 times higher than the required (8%) [42,43]. The higher harmonics cause additional voltage and power losses (increased cable heating) in the biogas plant's power supply system. These overshoots are correlated with voltage drops occurring in the electricity network. Changing the power cable reduced the value of the voltage distortion factor by almost half.

Table 5. Results of statistical analysis of the coefficients of higher harmonic content of voltage—THD_U.

	Before the Changes			After Changes		
Parameter	THD_{UL1} [%]	THD_{UL2} [%]	THD_{UL3} [%]	THD_{UL1} [%]	THD_{UL2} [%]	THD_{UL3} [%]
Average value	4.366	3.641	4.693	2.114	1.821	1.925
Minimum value	2.360	2.020	2.820	1.220	1.010	1.010
Maximum value	13.380	7.190	10.750	7.330	6.050	6.290
Quantile 95	8.557	5.470	7.550	2.760	2.530	2.700

The issue of supply voltage is particularly evident in photovoltaic (P.V.) plants, where increased generation often causes the voltage at the plant's connection point to exceed current regulatory limits. This challenge arises due to the high prevalence of P.V. systems. On the other hand, agricultural biogas plants, including micro gas plants, are generally considered stable energy sources with minimal voltage issues. However, the authors' research indicates that these plants can also face difficulties maintaining voltage within regulatory limits. The analysis presented in this article applies to other facilities connected to the low-voltage network.

To determine the maximum connectable power of a source, it is essential to understand the impedance of the supply system and the voltage level maintained at the transformer station and to apply well-known power and voltage loss relationships. The technical solutions proposed in this paper aim to mitigate voltage level issues in micro gas plants. For biogas plants fed by longer lines, more significant, cross-sectional area cables are recommended to reduce voltage drops and power losses.

4. Conclusions

The authors' investigations revealed that the power supply voltage issues the owner reported were technical. Specifically, the problem with starting the second generator stemmed from an incorrect cable cross-sectional area used to supply the biogas plant. The inadequate cable size led to significant voltage drops, resulting in voltage levels at the connection point that significantly exceeded permissible limits. Replacing the power cable allowed both generators to operate continuously but did not fully resolve the voltage issues.

The operation of the digester mixers caused the most substantial voltage fluctuations. When both mixers ran simultaneously, the power input to the electrical system dropped by an average of approximately 14.2 kW. Increasing the number of mixers while reducing their power ratings is recommended to address this. Additionally, operating the mixers in cycles—where only one mixer runs at a time—would help stabilise power consumption and reduce voltage fluctuations at the biogas plant's connection point. Lower-powered motors would also minimise inrush currents and reduce voltage variations during startup. The exact number and power rating of the mixers will depend on the size and design of each mixer.

A well-sized agitator should be large enough to mix the fermentation mass thoroughly but small enough to be driven by a less powerful motor. Alternating between lower-powered mixers should mitigate voltage fluctuations caused by the activation and deactivation of individual devices. With appropriately selected mixing components, no significant changes in biogas production efficiency are expected. The authors are researching the optimal parameters for devices (agitator and motor) to maximise efficiency while minimising electricity consumption.

Oversizing the generator capacity (relative to the connection capacity) would cover auto consumption and enable power close to the connection capacity to be fed into the grid. The costs of oversizing the generator, already at the design stage of the biogas plant, are not high compared to the costs incurred when replacing the already installed device. Instead of two generators with a capacity of 20 kW each, it is recommended to install units with a capacity of 22 kW, which follow the manufacturer's range of biogas generators. At the same time, it is necessary to simulate voltage changes caused by the operation of an agricultural biogas plant already at the design stage. It is also possible to post-measure the network parameters at the planned biogas plant connection point. These measurements allow well-known formulas to determine the maximum connectable power to meet the system's voltage requirements. It will enable the proper operation of not only the biogas plant itself but also other equipment installed on the farm. Their smooth functioning will be essential regarding sound management's organisational, financial, and environmental aspects.

Author Contributions: Conceptualisation, A.B. and Z.S.; methodology, Z.S. and M.P.; software, A.M. (Anna Milewska); validation, A.B., Z.S. and W.R.; formal analysis, W.R.; investigation, A.M. (Anna Milewska); resources, A.M. (Andrzej Marczuk); data curation, Z.S. and A.M. (Andrzej Marczuk); writing—original draft preparation, Z.S. and M.P.; writing—review and editing, A.B. and W.R.; visualisation, A.M. (Anna Milewska); supervision, Z.S. and A.M. (Andrzej Marczuk); project administration, W.R. and A.M. (Andrzej Marczuk); funding acquisition, M.P. and A.B. All authors have read and agreed to the published version of the manuscript.

Funding: This research received no external funding.

Informed Consent Statement: Not applicable.

Data Availability Statement: The original contributions presented in the study are included in the article, further inquiries can be directed to the corresponding author.

Conflicts of Interest: The authors declare no conflicts of interest.

References

1. Monteny, G.-J.; Bannink, A.; Chadwick, D. Greenhouse gas abatement strategies for animal husbandry. *Agric. Ecosyst. Environ.* **2006**, *112*, 163–170. [CrossRef]
2. Scheftelowitz, M.; Thrän, D. Unlocking the energy potential of manure-An assessment of the biogas production potential at the farm level in Germany. *Agriculture* **2016**, *6*, 20. [CrossRef]
3. Akyürek, Z. Potential of biogas energy from animal waste in the Mediterranean region of Turkey. *J. Energy Syst.* **2018**, *2*, 160–167. [CrossRef]
4. Saracevic, E.; Frühauf, S.; Miltner, A.; Karnpakdee, K.; Munk, B.; Lebuhn, M.; Wlcek, B.; Leber, J.; Lizasoain, J.; Friedl, A.; et al. Utilisation of food and agricultural residues for a flexible biogas production: Process stability and effects on needed biogas storage capacities. *Energies* **2019**, *12*, 2678. [CrossRef]
5. Friedlingstein, P.; O'sullivan, M.; Jones, M.W.; Andrew, R.M.; Hauck, J.; Olsen, A.; Zaehle, S. Global carbon budget. *Earth Sys. Sci. Data* **2020**, *12*, 3269–3340. [CrossRef]
6. Korberg, A.D.; Skov, I.R.; Mathiesen, B.V. The role of biogas and biogas-derived fuels in Denmark's 100% renewable energy system. *Energy* **2020**, *199*, 117426. [CrossRef]
7. Angelidaki, I.; Treu, L.; Tsapekos, P.; Luo, G.; Campanaro, S.; Wenzel, H.; Kougias, P.G. Biogas upgrading and utilisation: Current status and perspectives. *Biotechnol. Adv.* **2018**, *36*, 452–466. [CrossRef] [PubMed]
8. Tymińska, M.; Skibko, Z.; Borusiewicz, A. The Effect of Agricultural Biogas Plants on the Quality of Farm Energy Supply. *Energies* **2023**, *16*, 4600. [CrossRef]
9. Meneses-Quelal, O.; Velázquez-Martí, B. Pretreatment of animal manure biomass to improve biogas production: A review. *Energies* **2020**, *13*, 3573. [CrossRef]
10. Kaltschmitt, M.; Scholwin, F.; Gattermann, H.; Schattauer, A.; Weiland, P. *Biogas-Production Utilisation*; Institut Für Energetik und Umwelt Gmbh: Leipzig, Germany, 2005.
11. Lansing, S.; Botero, R.; Martin, J.F. Waste treatment and biogas quality in small-scale agricultural digesters. *Bioresour. Technol.* **2008**, *99*, 5881–5890. [CrossRef]
12. Czekała, W.; Nowak, M.; Bojarski, W. Anaerobic Digestion and Composting as Methods of Bio-Waste Management. *Agric. Eng.* **2023**, *27*, 173–186. [CrossRef]
13. Derehajło, S.; Tymińska, M.; Skibko, Z.; Borusiewicz, A.; Romaniuk, W.; Kuboń, M.; Olech, E.; Koszel, M. Heavy Metal Content in Substrates in Agricultural Biogas Plants. *Agric. Eng.* **2023**, *27*, 315–329. [CrossRef]
14. Giesy, R.; Wilkie, A.; De Vries, A.; Nordstedt, R. Economic Feasibility of Anaerobic Digestion To Produce Electricity on Florida Dairy Farms1. *EDIS* **2006**. [CrossRef]
15. Gloy, B.A.; Dressler, J.B. Financial barriers to the adoption of anaerobic digestion on U.S. livestock operations. *Agric. Financ. Rev.* **2010**, *70*, 157–168. [CrossRef]
16. Soljan, Z.; Holdyński, G.; Zajkowski, M. Decomposition of the load's current supplied from a sinusoidal and asymmetrical voltage source in accordance with the Currents' Physical Components (CPC) Theory. In Proceedings of the 2019 20th International Scientific Conference on Electric Power Engineering (EPE), Kouty nad Desnou, Czech Republic, 15–17 May 2019; Rusek, S., Goňo, R., Eds.; IEEE: Piscataway, NJ, USA, 2019; pp. 389–394. [CrossRef]
17. Klavon, K.H.; Lansing, S.A.; Mulbry, W.; Moss, A.R.; Felton, G. Economic analysis of small-scale agricultural digesters in the United States. *Biomass Bioenergy* **2013**, *54*, 36–45. [CrossRef]
18. Larina, Y.; Galchynska, J.; Kucheruk, P.; Zghurska, O.; Ortina, G.; Al-Nadzhar, F.; Marusei, T.; Kuboń, M.; Dzieniszewski, G. Estimation of the Domestic Agricultural Sector Potential for the Growth of Energy Cultures for Bioenergy Fuel Production. *Agric. Eng.* **2021**, *25*, 73–82. [CrossRef]
19. Dach, J.; Boniecki, P.; Przybył, J.; Janczak, D.; Lewicki, A.; Czekała, W.; Witaszek, K.; Rodríguez Carmona, P.C.; Cieślik, M. Energetic efficiency analysis of the agricultural biogas plant in 250 kWe experimental installation. *Energy* **2014**, *69*, 34–38. [CrossRef]

20. Saaty, T.L.; Vargas, L.G.; Dellmann, K.L. The allocation of intangible resources: The analytic hierarchy process and linear programming. *Socioecon. Plann. Sci.* **2003**, *37*, 169–184. [CrossRef]
21. San Cristobal, J.R. Multi-criteria decision-making in the selection of a renewable energy project in Spain: The Vikor method. Renew. *Energy* **2011**, *36*, 498–502. [CrossRef]
22. Stoltmann, A.; Bućko, P. Analysis of the Biogas Plant Location Using Analytic Hierarchy Process (ahp) and Numerical Taxonomy Methods-Methods Comparison. Zeszyty Naukowe Wydziału Elektrotechniki i Automatyki Politechniki Gdańskiej Nr 53. 2017. Available online: https://www.google.com/url?sa=t&source=web&rct=j&opi=89978449&url=https://yadda.icm.edu.pl/baztech/element/bwmeta1.element.baztech-291c516d-5758-4f23-815e-62df0afc8864/c/ZN_WEIAPG_53-29.pdf&ved=2ahUKEwjyp4rTxuWFAxU6PxAIHeHoDagQFnoECBsQAQ&usg=AOvVaw1nDzQXR50HzVDrOTZ0pMa1 (accessed on 7 July 2024). (In Polish).
23. Mazurkiewicz, J. The Impact of Manure Use for Energy Purposes on the Economic Balance of a Dairy Farm. *Energies* **2023**, *16*, 6735. [CrossRef]
24. Vilas Bôas, T.F.; Barros, R.M.; Pinto, J.A.; dos Santos, I.F.S.; Lora, E.E.S.; Andrade, R.V.; Tiago Filho, G.L.; Almeida, K.A.; de Oliveira Machado, G. Energy Potential from the Generation of Biogas from Anaerobic Digestion of Olive Oil Extraction Wastes in Brazil. *Clean. Waste Syst.* **2023**, *4*, 100083. [CrossRef]
25. Furtado, L.A.; Guerreiro Ribeiro, S.; Pradelle, F.; Parise, J.A.R. Modeling and Techno-Economic Analysis of a Hybrid Sugarcane Plant Fed by Vinasse Biogas and Bagasse Surplus for Electricity Generation. *J. Clean. Prod.* **2023**, *413*, 137511. [CrossRef]
26. Wind, Solar Payback Times Under a Year in Some Parts of the World, Says Rystad. Available online: https://www.pv-magazine.com/2022/10/14/wind-solar-payback-times-under-a-year-in-some-parts-of-world-says-rystad/ (accessed on 23 April 2024).
27. Hagman, L.; Blumenthal, A.; Eklund, M.; Svensson, N. The role of biogas solutions in sustainable biorefineries. *J. Clean. Prod.* **2018**, *172*, 3982–3989. [CrossRef]
28. Kuboń, M.; Skibko, Z.; Tabor, S.; Malaga-Toboła, U.; Borusiewicz, A.; Romaniuk, W.; Zarajczyk, J.; Neuberger, P. Analysis of Voltage Distortions in the Power Grid Arising from Agricultural Biogas Plant Operation. *Energies* **2023**, *16*, 6189. [CrossRef]
29. Burg, V.; Bowman, G.; Haubensak, M.; Baier, U.; Thees, O. Valorization of an untapped resource: Energy and greenhouse gas emissions benefits of converting manure to biogas through anaerobic digestion. *Resour. Conserv. Recycl.* **2018**, *136*, 53–62. [CrossRef]
30. Tamburini, E.; Gaglio, M.; Castaldelli, G.; Fano, E.A. Biogas from Agri-Food and Agricultural Waste Can Appreciate Agro-Ecosystem Services: The Case Study of Emilia Romagna Region. *Sustainability* **2020**, *12*, 8392. [CrossRef]
31. Jeung, J.H.; Chung, W.J.; Chang, S.W. Evaluation of Anaerobic Co-Digestion to Enhance the Efficiency of Livestock Manure Anaerobic Digestion. *Sustainability* **2019**, *11*, 7170. [CrossRef]
32. Mao, C.; Feng, Y.; Wang, X.; Ren, G. Review on research achievements of biogas from anaerobic digestion. *Renew. Sustain. Energy Rev.* **2015**, *45*, 540–555. [CrossRef]
33. Rao, P.V.; Baral, S.S.; Dey, R.; Mutnuri, S. Biogas generation potential by anaerobic digestion for sustainable energy development in India. *Renew. Sustain. Energy Rev.* **2010**, *14*, 2086–2094. [CrossRef]
34. Garrison, M.; Richard, T. Methane and Manure: Feasibility Analysis of Price and Policy Alternatives. *Trans. ASAE* **2005**, *48*, 1287–1294. [CrossRef]
35. Chen, Y.; Yang, G.; Sweeney, S.; Feng, Y. Household biogas use in rural China: A study of opportunities and constraints. *Renew. Sustain. Energy Rev.* **2010**, *14*, 545–549. [CrossRef]
36. Hołdyński, G.; Skibko, Z.; Borusiewicz, A. Analysis of the Influence of Load on the Value of Zero-Voltage Asymmetry in Medium-Voltage Networks Operating with Renewable Energy Sources. *Energies* **2023**, *16*, 580. [CrossRef]
37. Alshamrani, A.; Majumder, P.; Das, A.; Hezam, I.M.; Božanić, D. An Integrated BWM-TOPSIS-I Approach to Determine the Ranking of Alternatives and Application of Sustainability Analysis of Renewable Energy. *Axioms* **2023**, *12*, 159. [CrossRef]
38. Gonzalez, P.; Romero-Cadaval, E.; Gonzalez, E.; Guerrero, M.A. Impact of grid-connected photovoltaic system in the power quality of a distribution network. In *Technological Innovation for Sustainability: Second IFIP WG 5.5/SOCOLNET Doctoral Conference on Computing, Electrical and Industrial Systems, DoCEIS 2011, Costa de Caparica, Portugal, 21-23 February 2011*; Proceedings 2; Springer: Berlin/Heidelberg, Germany, 2011; pp. 466–473. [CrossRef]
39. Ma, C.; Xiong, W.; Tang, Z.; Li, Z.; Xiong, Y.; Wang, Q. Distributed MPC-Based Voltage Control for Active Distribution Networks Considering Uncertainty of Distributed Energy Resources. *Electronics* **2024**, *13*, 2748. [CrossRef]
40. Uddin, M.; Rahman, M.; Hossain, T.R.; Rahman, H. Assessment by Simulation of Different Topological Integration of Solar Photovoltaic Plants in Medium Voltage Distribution Networks. *Pertanika J. Sci. Technol.* **2021**, *29*, 1159–1169. [CrossRef]
41. Alam, S.; Al-Ismail, F.S.; Salem, A.; Abido, M.A. High-Level Penetration of Renewable Energy Sources Into Grid Utility: Challenges and Solutions. *IEEE Access* **2020**, *8*, 190277–190299. [CrossRef]
42. *EN 50160:2010*; Supply Voltage Parameters for Public Distribution Networks. E.U.: Maastricht, The Netherlands, 2010.
43. Regulation of the Minister of Economy of May 4, 2007, on detailed conditions for the operation of the electric power system (in Poland). *Dz.U.* 2007 nr 93 poz. 623. Available online: https://isap.sejm.gov.pl/isap.nsf/DocDetails.xsp?id=WDU20070930623 (accessed on 7 July 2024).
44. Robak, S.; Pawlicki, A.; Pawlicki, B. The analysis of the voltage and current asymmetry in the power transmission lines. *Przegląd Elektrotechniczny* **2014**, *90*, 26. [CrossRef]

45. Kosicki, Ł.; Typańska, D. The research of current and voltage distortions generated by luminaries with the light emitting diodes. *Electr. Eng.* **2017**, *92*, 215–226. [CrossRef]
46. *IEC 61000-4-30:2015+AMD1*; 2021 CSV Consolidated Version, Electromagnetic compatibility (EMC)-Part 4-30: Testing and Measurement Techniques-Power Quality Measurement Methods. Available online: https://webstore.iec.ch/en/publication/68642 (accessed on 7 July 2024).

applied sciences

MDPI

Article

On-Site Sensor Sensitivity Adjustment Technique for a Maintenance-Free Heat Flow Monitoring in Building Systems

Jan Kočí *, Jiří Maděra and Robert Černý

Department of Materials Engineering and Chemistry, Faculty of Civil Engineering, Czech Technical University in Prague, Thákurova 7, Prague 6, 166 29 Prague, Czech Republic
* Correspondence: jan.koci@fsv.cvut.cz; Tel.: +420-2-2435-7130

Abstract: In this paper, an advanced solution for measuring heat flow through opaque building elements is presented. The solution is based on the implementation of a computer-aided technique for continuous monitoring of heat flow sensor performance and silent checking of their accuracy. In principle, the technique provides an ex-post compensation of potential deviations and inaccuracies detected during the measurement, which can be done without interfering with the ongoing experiment. As a consequence, traditional 'non-smart' sensors can be turned into advanced sensors with self-sensing or self-adjustment features at nearly zero additional costs. The high efficiency of the proposed approach was validated against experimental data obtained from an independent set of advanced high-sensitive sensors. Considering the validation results, the proposed technique brings an entirely new potential for maintenance-free applications for thermal performance monitoring in the building sector, typically for long-term experiments or measurements under dynamic environments.

Keywords: heat flux; automatic sensitivity adjustment; maintenance-free measurement; critical experiment; building engineering; decision-making process

1. Introduction

In the thermal assessment of buildings, various types of heat gains and losses constitute the overall balance of the object. Its character, whether positive or negative, depends mostly on the thermal performance of the building enclosure [1,2]. Therefore, the heat transfer analysis of this building element should receive special attention to provide an essential part of the data that forms a considerable contribution to the overall heat balance of buildings.

Heat flux probes and sensors considered for applications in the building sector need to meet specific requirements, such as relatively low values of heat flux, alternating directions of heat flux, surface roughness of investigated walls, or exposure to radiative heat flux. Various types of sensors are available on the market, differing from each other by measuring techniques and/or construction principles. Gardon [3–5], Schmidt-Boelter gauges [6–8], thin plate sensors [9,10], or radiometers [11,12] represent the most common types of non-optical sensors. An extensive review of heat flux measuring techniques was provided by Childs et al. [13]. Another report [14] provided a review of non-optical detectors for heat flux measurement in fires. Assuming that heat flux can be induced by different origins, the essential criterion of sensor selection should be made with respect to the environment, in which the sensor is intended to be placed in. Basically, the heat flux mode or possible combination of multiple modes (convective, radiative, or conductive), including the range of anticipated quantities, should be known in advance to provide a background for making a qualified decision on sensor selection. For example, Gardon and Schmidt-Boelter gauges were designed for the measurement of both convective and radiative heat fluxes. As such, they are suitable for fire- or flame-induced heat transfer measurement. Pure radiative transfer, such as solar radiation, can be measured by pyranometers [15]. The conductive heat transfer is typical for building physics applications in the case that opaque building elements are the subject of investigation. As the sensors are mostly mounted on the wall surface, flat plate sensors or foil sensors suit best for these kind of applications. Those sensors are represented by thermopiles working on the principle of the Seebeck effect; that is, they generate an electrical output signal that is proportional to the heat flux passing through the sensor. Before application, each transducer needs to be calibrated to determine its sensitivity, that is, the electrical output to the passing heat flux ratio. In most cases, the initial calibration is performed by the manufacturer and the user is provided with the calibration protocol. Also, the default sensitivity is very often printed somewhere on the sensor body to be easily located and adopted by the user.

The common practice for the calibration of heat flux transducers is based on inducing steady-state conductive heat flux in the controlled environment by means of the guarded hot plate [16] or the heat flow meter apparatus [17,18]. Procedures for calibrating heat flux probes that are relatively thin in comparison with their planar dimensions are well described in ASTM C1130 [19]. This practice in conjunction with other test methods serves as a valuable handbook for the determination of the sensitivity of heat flux transducers under both laboratory and in-situ conditions.

For the laboratory calibration, the conjunction with Test Methods C177 [16], C518 [17], C1363 [20], C1114 [21], or ISO 8302 [22] is well appreciated, while three different techniques are recognized: (i) ideally-guarded, (ii) embedded, and (iii) surface mounted. As described in Section 4.1 of ASTM C1130 [19], the ideally protected technique (i) is used for determining the baseline calibration assuming that only homogeneous and thermally characterized materials are used in the test stack (series of layers) promoting one-dimensional heat flow. The (ii) embedded technique is based on the same principle as (i); however, material layers identical or comparable to the building construction to be studied under application are used, instead. The surface-mounted calibration technique (iii) represents the most complex approach as it needs to address and incorporate all environmental effects that may cause the lateral heat flow in the vicinity of the applied sensor.

The conjunction of ASTM C1130 with C1046 [23] allows the calibration results to be used for in situ measurement of heat flux through opaque building elements with subsequent analysis of measurement data according to ASTM C1155 [24]. The in-situ measurement practice provides practical information, recommendations, and tips to ensure potential simplicity and ease of application of heat flux sensors. A comprehensive study to support the continuous development of precision and bias statements for ASTM Practice C 1130 was published by Zarr et al. [25], in which the authors provided a complex analysis of common issues that are typical for building applications. In that study, sensor equivalency, linearity and temperature effects, uncertainty analysis, and suitability test method were analyzed, which can partially address some of the typical issues associated with heat flux measurements in the building sector. According to the practice [19], the heat flux transducer that is to be mounted on the building envelope construction should be calibrated using a hot box that is oriented similarly to the measuring site. Additionally, this approach requires a specific configuration and sample preparation involving a homogeneous specimen of known thermal resistance, temperature sensor, and guarded mask to obtain reliable results. Also, the calibration should be performed at levels of temperature and relative humidity consistent with the end-use conditions.

Varying conditions during in-situ measurement may bring significant uncertainties to the measurement, therefore periodic performance checks should be carried out to reflect those fluctuations and to minimize measurement errors. However, such checks, according to the above-described practices, require unmounting the sensor thus interfering with the ongoing experiment. Some initiatives to address the varying sensitivity has come directly from the sensor producers, which documents the awareness of this issue. This initiative resulted in the introduction of new commercial gauges that were capable of self-calibration without the need for sensor removal in certain applications [26,27]. The advanced solution is based on a novel design of the sensors with integrated heaters in sensor bodies to generate a controlled heat flux allowing a subsequent auto-calibration.

Moreover, it is worth mentioning that some of the heat flux plates for measurement in buildings were originally developed for meteorological applications to measure heat fluxes in soils, which predetermines the self-calibration practice. Such practice presumes that the sensor is surrounded by soil during measurement. Typically, when the built-in heater is activated to generate the prescribed heat flux, the surrounding soil with equal properties ensures that half of the flux will be transferred upwards, while the other half will go downwards. This is an essential assumption for sensor calibration [28]. If the advanced sensors are mounted on the wall, surface-specific conditions must be ensured, similar to those in laboratory calibration using ASTM practices. This can bring some serious limitations or drawbacks or even exclude the in-situ self-calibrations of advanced sensors in the building sector.

An alternative to the standard calibration procedures might be represented by the approach proposed in this paper, which turns the traditional sensors applied in the building sector into advanced ones. The approach allows for the continuous measure of the heat flow without a need to interfere with the ongoing experiment. Moreover, the heat flux sensor may even stay mounted on the wall surface providing the edge over advanced sensors with integrated heaters and turning the non-smart sensors into smart ones. The term "smart" refers to the capability of the proposed method to effectively substitute the heat source element that is present in the standard calibration procedure at any time without extending or upgrading the experimental apparatus. The conjunction of experimental and computational methods used in the proposed approach makes it possible to accommodate specific requirements of the calibration procedure and facilitate various applications of the sensors in heat flux analyses in the building sector. As such it can contribute to the state-of-the-art and extend the numerical methods for the in-situ data analysis as mentioned in [19]. To demonstrate the capability of the proposed method, a "free-running" residential building with varying indoor environments was selected for long-term monitoring of environmental variables and wall performance. Then, the experimental data were analyzed and complemented with the outputs of a computational analysis to reveal potential deviations in the sensor performance using the ex-post scheme. If some deviation was detected then the complex experimental-computational datasets allowed for a compensation of errors resulting from the varying sensitivity of the probes, thus, consistently ensuring a high level of accuracy of heat flux measurement with a standard "non-smart" experimental equipment. This could be achieved by recording and analyzing both raw sensor data and end-user outputs in the form of heat flux variations. To validate the findings presented in this paper, the experimental setup was extended in the final stage of the field monitoring by the acquisition of high-sensitive flat plate sensors with the film heater integrated into the sensor body. Such a configuration allowed the proposed technique to be directly compared against the independent data, which is the essential precondition for performing the independent validation.

2. Materials and Methods

2.1. Theoretical Background

A one-dimensional problem is considered for the analysis of transient heat flux on the interior surface of a building wall (see Figure 1), where q_{IN} represents the rate of heat transfer per unit area, i.e., the heat flux, from the interior to the surface element, q_{OUT} represents the heat flux from the surface element to the wall, and x is the thickness of wall element.

Assuming that thickness of the element is infinitesimally small, i.e., $x \to 0$, then thermal storage in the element can be practically neglected implying that

$$q_{\text{IN}} \approx q_{\text{OUT}}, \tag{1}$$

at any time. In Equation (1), q_{IN} ($\text{W}{\cdot}\text{m}^{-2}$) is the heat flux from the interior to the wall surface, q_{OUT} ($\text{W}{\cdot}\text{m}^{-2}$) is the heat flux from the wall surface to the wall. For this case, q_{IN} can be expressed using the Newton's law of cooling as

$$q_{IN} = \alpha\,(T_s - T_i), \tag{2}$$

where α ($W{\cdot}m^{-2}{\cdot}K^{-1}$) is the heat transfer coefficient, T_i (°C) is the ambient wall (indoor) temperature, and T_s (°C) is the interior surface temperature of the wall. The heat flux from the surface element to the wall, q_{OUT}, can be expressed using Fourier's equation as

$$q_{OUT} = \lambda\,\Delta T/\Delta x, \tag{3}$$

where λ ($W\,m^{-1}\,K^{-1}$) is the thermal conductivity, ΔT (°C) is the temperature difference of the element surfaces in the direction of the heat flow, Δx (m) is the thickness of the element in the main direction of the heat flux. When a validated computational model is integrated then the heat transport in the surface element can be included to define a new computational-experimental domain allowing for a complex analysis, which might help to extend the current methods or design new principles for heat flux measurement and building performance simulations.

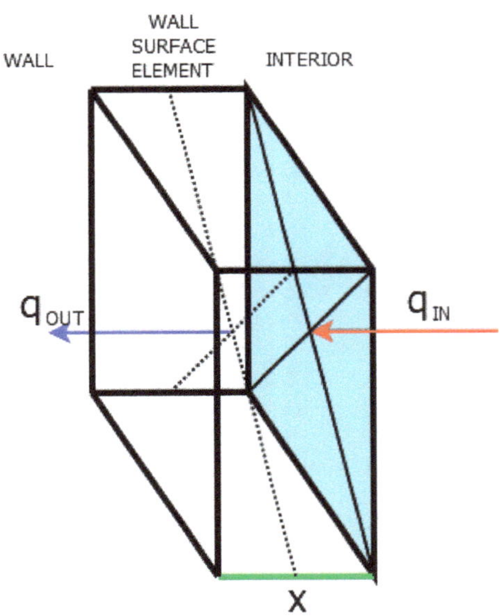

Figure 1. Scheme of heat transfer in the surface element of building wall.

For the computational modeling of heat transfer, which can provide thermal performance data of the wall to form a complex dataset for the performance check procedure, the modified Künzel's model of heat and moisture transport was used [29,30]. This model was developed for applications in building engineering and validated several times in the past [31]. A detailed information on the modifications of the original Künzel's model [32] was given in [30]. The following equations were used to govern the heat and moisture transport:

$$\frac{dH}{dT}\frac{\partial T}{\partial t} = \mathrm{div}(\lambda \mathrm{grad}T) + L_v \mathrm{div}(\delta_p \mathrm{grad}p_v), \tag{4}$$

$$\left[\rho_w \frac{\mathrm{d}w}{\mathrm{d}p_v} + (\psi - w)\frac{M}{RT}\right]\frac{\partial p_v}{\partial t} = \mathrm{div}\left[D_g \mathrm{grad}p_v\right], \tag{5}$$

where H ($J{\cdot}m^{-3}$) is the enthalpy density, T (K) is the absolute temperature, t (s) is the time, D_g (s) is the global moisture transport function, L_v ($J{\cdot}kg^{-1}$) is the latent heat of evaporation of water, δ_p (s) is the water vapor diffusion permeability, M (kg·mol^{-1}) is the molar mass of water vapor, p_v (Pa) is the partial pressure of water vapor in the porous space, w (m^3·m^{-3}) is the moisture content by volume, ρ_w (kg·m^{-3}) is the density of water, ψ (–) is the total open porosity, and R (J·K^{-1}·mol^{-1}) is the universal gas constant. More details and insights into model formulation and parameter description can be found in [30]. Involving moisture transport in the computational modeling was motivated by the effort to increase the calculation accuracy, particularly to obtain thermal conductivity λ as a moisture-dependent variable $\lambda = \lambda(w)$. The indoor boundary conditions for temperature and relative humidity were obtained from the experimental data, as well as outdoor boundary conditions that were defined using outdoor climatic data. The material properties were adopted from the previous research [33,34] and from the material database maintained by the Department of Materials Engineering and Chemistry, Faculty of Civil Engineering, Czech Technical University in Prague.

2.2. Design of the Adaptive Self-Adjustment Procedure

The sensitivity adjustment procedure was based on a complex analysis of time-resolved heat fluxes $q_{IN}(t)$ and $q_{OUT}(t)$ obtained by both experimental measurements and computational simulation of heat transfer in a studied wall. In the computational simulation, the material composition and orientation of the wall, material properties, and outdoor and indoor environmental data were used as input parameters. At first, the optimization of the computational model parameters was done to validate the capability of the model to simulate the real environment in this particular case. The computational model was optimized by a simple modification of the heat transfer coefficient α on the interior surface of the wall, in order to obtain verified $q_{IN}(t)$ and $q_{OUT}(t)$ data through the surface temperature T_s. For this purpose, a surface element with very low thickness (x = 1~2 mm) was used in spatial discretization of the computational model to meet preconditions set in Equation (1). The complex sensitivity adjustment experiment was designed using the following steps:

1. Heat transfer coefficient α of the computational model is optimized by fitting the simulated surface temperature, $T_{s,sim}(t)$, to that determined experimentally, $T_{s,exp}$ (t);
2. Heat flux from the interior to the wall surface, $q_{IN}(t)$, is calculated from the experimental data using Equation (2) and the optimized heat transfer coefficient α from the previous step;
3. Heat flux from the wall surface element to the wall, $q_{OUT}(t)$, is calculated from the computational data using Equation (3) and compared with $q_{IN}(t)$ from the previous step to validate the setup and accuracy of the computational model;
4. $q_{IN}(t)$ and $q_{OUT}(t)$ are averaged to form a new variable, $q_{REF}(t)$, having a sufficient robustness as it combines both experimental and computational outputs;
5. Sensitivity of heat flux sensors (thermopiles) is adjusted using raw U(t) (voltage) data and $q(t)$ from the previous step and custom sensitivity function S is identified for each thermopile;
6. Raw data provided by the heat flux sensors in the period are adopted and transformed into heat flux using S function from the previous step;
7. The heat flux obtained in the previous step is compared with experimentally and computationally derived data to verify the adjusted sensitivity;
8. The verified heat flux obtained in the previous step is validated against the reference heat flux obtained from an independent sensor.

The scheme of the designed approach is shown in Figure 2. In general, the experimental measurements can be divided into three phases that correspond to the purposes of data acquisition, processing, and application. The entire procedure can be applied repeatedly during ongoing experiments, providing a tool for an adaptive adjustment which is most effective in the scenarios with changing environmental conditions. The execution of Phase 1 is required to be carried out at least once upon the beginning of the adjusting procedure. Repeated executions of this phase are optional. However, they should be done always when environmental conditions significantly deviate from those acting during the last execution. After initialization of the adjustment procedure, the Phase 2 is carried out to obtain updated sensitivity of the heat flux sensor that will correspond with the actual environmental conditions. With the sensitivity adjusted in this way, the experiment may continue running while Phase 3 is being continuously executed in the background to verify the performance of the heat flux sensor. If some deviation is detected, i.e., substantial difference between measured heat flux and $q(t)$ calculated from the computational model and observed quantities, a repeated execution of Phase 2 and 3 is desired. Phase 4, which represents the independent validation of the proposed technique, is executed for the scientific purposes in this paper only, in order to confirm the applicability of the proposed method. Once the validation is successfully completed, it can stay outside the self-adjustment procedure in practical applications. More details of the validation procedure will be given in the further sections of this paper.

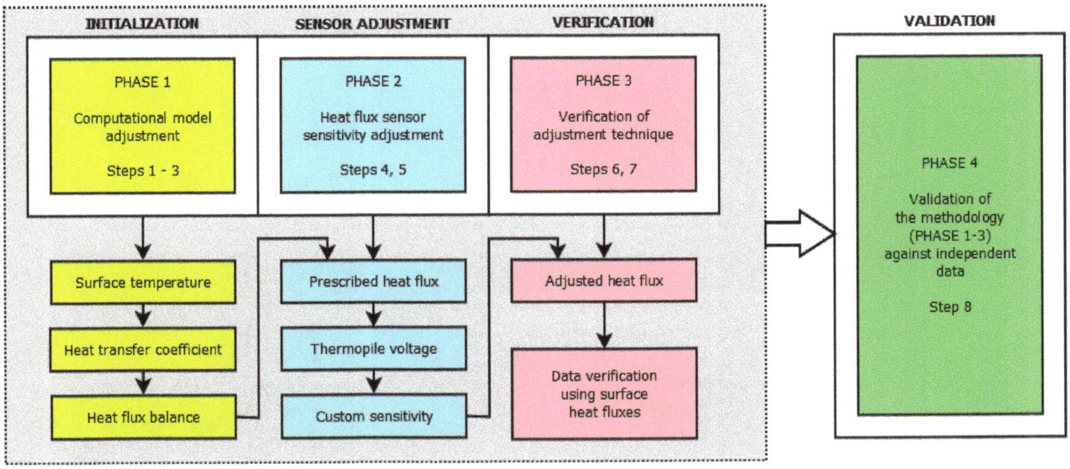

Figure 2. Scheme of the adjustment approach with an independent validation.

The ability of the smart adjustment approach to respond quickly to changing conditions does not place strict demands on the robustness of training data. Therefore, smaller datasets can be selected for the computational model initial setup and sensor sensitivity re-adjustment. Moreover, the scheme of the proposed method allows the initialization phase (Phase 1) to be executed on the same datasets as used for sensor adjustment (Phase 2). This would ensure that high accuracy of input data needed for the sensor sensitivity adjustment itself will be kept during the ongoing experiment without a significant increase of time required for calibration. The timing of individual adjustment periods may be driven either by the rate of changes of the ambient environment, monitoring of selected experimental conditions, or the accuracy of the heat flux sensor; they can also be scheduled regularly in user-defined periods. The smart approach thus brings an entirely new potential for maintenance-free applications, which can be fully automated upon the implementation of decision-making rules.

2.3. Self-Adjustment Procedure in Real Conditions

For the demonstration of the proposed technique in real conditions, a residential building in Pulovice (distr. Karlovy Vary), Czech Republic was selected. All necessary data and parameters including climatic variables, structural aspects, material properties, and indoor environmental variables were acquired during that experiment. The summary of data collected within the field survey and experimental measurements is shown in Figure 3.

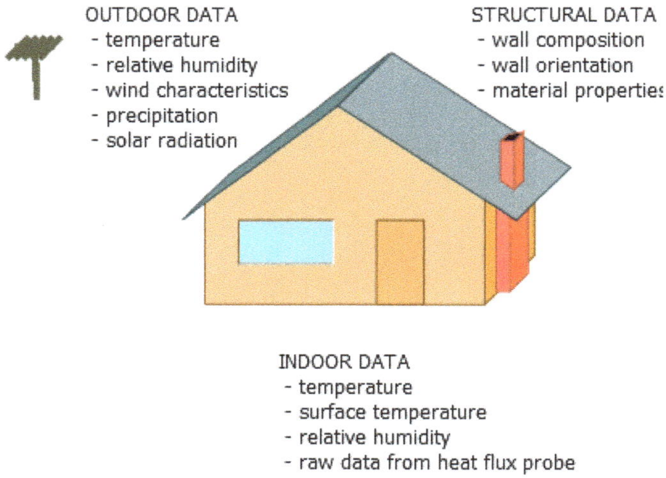

Figure 3. Experimental measurement and input data overview.

The realization of the in-situ experiment including a description of the experimental equipment is documented in Figure 4.

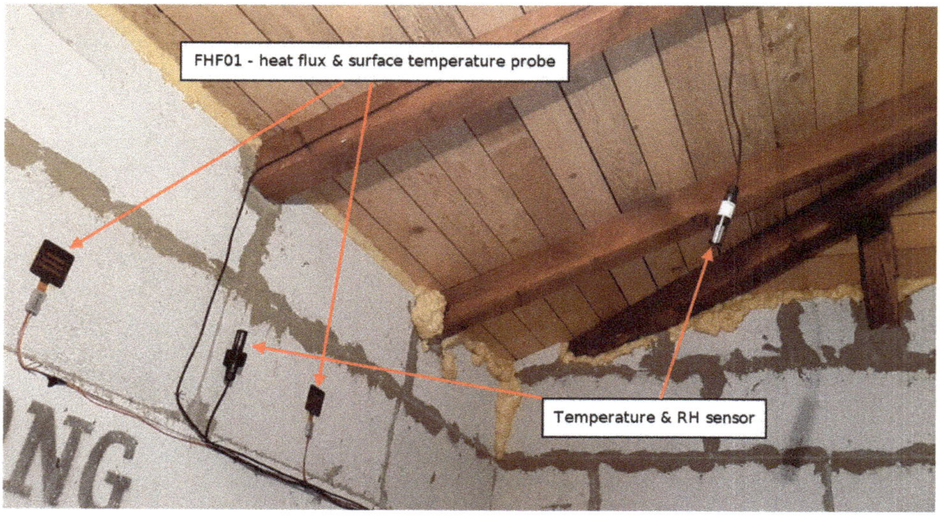

Figure 4. The realization of the in-situ experiment.

The wall was built from autoclaved aerated concrete (AAC) blocks with a thickness of 500 mm which were rendered with a plaster on the exterior surface only. The missing interior render helped to better analyze the background of the heat transport processes in the surface layers as the material characteristics of AAC are known very well and the potential sources of uncertainties could be significantly reduced by a simple visual check. In this way, sources of inhomogeneities or hidden materials in the wall composition could be excluded, which was very appreciating for the adjustment of the computational model. Otherwise, a pre-experimental site inspection using, for example, infrared thermography would be needed as required by C1046 [23].

The outdoor data were collected using the Davis Vanatge Pro2 weather station in 30-min intervals. At the same time resolution, the indoor data were collected using a MS6D Datalogger (by Comet System Ltd., Bratislava, Czech Republic) equipped with two T/RH sensors and two thin-foil heat flux probes Hukseflux FHF01.

Hukseflux sensors are considered thermopiles [35,36], i.e., they consist of a series of thermocouples that measure the temperature difference across the sensor body. The FHF01 is a very thin and flexible foil sensor for general-purpose heat flux measurements equipped with the type T thermocouple to monitor the surface temperature along with the heat flux. The sensor is designed for the measurements of heat flux in the range from −200 to 200 W·m^{-2} with a nominal sensitivity of 4.5×10^{-6} V·m^2·W^{-1}

Taking into account specific building-related conditions, such as surface roughness, various types of mountings, and low thermal conductivity of materials, or the conjunction of those factors, the default factory calibration should not always be suitable for heat flux measurement under such conditions. Therefore, a custom sensitivity adjustment approach is presented with the experimental setup described above, which does not require any additional equipment (such as foil heaters) or actions (sensor removal or replacement). Since all probes are classified as thermopiles, i.e., they produce raw data in the form of time-resolved voltage; a user-defined adjustment procedure seems to be optimal to get sufficiently accurate results for each specific application. To guarantee that the overall performance of the building element will be measured, and representative site conditions will be recorded, multiple heat flux sensors were involved in the experimental setup. The summary of input data and parameters that are necessary for both experimental measurement, computational simulation, and in-situ data analysis are shown in Table 1. Here, the following symbols are used: ρ_v (kg·m^{-3}) for the bulk density, ρ (kg·m^{-3}) for the matrix density (kg·m^{-3}), and c (J·kg^{-1}·K^{-1}) is the specific heat capacity.

Table 1. Summary of input data used in the simulation and calibration experiment.

	Parameter	AAC	Ext. Plaster
Simulation	ρ_v (kg·m^{-3})	500	1244
	ρ (kg·m^{-3})	2206	2415
	ψ (−)	0.31	0.34
	λ (W·m^{-1}·K^{-1})	0.12	0.28
	c (J·kg^{-1}·K^{-1})	1050	1054
	Boundary conditions	Determined experimentally	
Experiment	Wall orientation	North	
	Wall thickness	500 mm	
	Exterior plaster thickness	10 mm	

2.4. Validation of the Proposed Self-Adjustment Approach

For the independent validation, a series of additional high-sensitive heat flux sensors was acquired and installed on site. In particular, two flat-plate sensors Hukseflux HFP01 and one sensor HFP01SC were mounted on the wall together with the surface temperature sensor Pt1000TG7 (see Figure 5).

HFP01/HFP01SC is a ceramic-plastic composite sensor with a built-in thermopile. The application areas of those sensors include building physics, thermal comfort, or energy budget measurements, as well as geotechnics, meteorology, or other scientific applications based on heat transfer measurement. The sensor is designed for measurements of heat flux in the range from −2000 to 2000 W·m^{-2} with a nominal sensitivity of 60×10^{-6} V·m^2·W^{-1}. HFP01SC has identical parameters to HFP01 regarding the heat flux measurement. The only difference between those probes is that HFP01SC contains the film heater to generate prescribed heat flux allowing the sensor to perform self-test and/or self-calibrate.

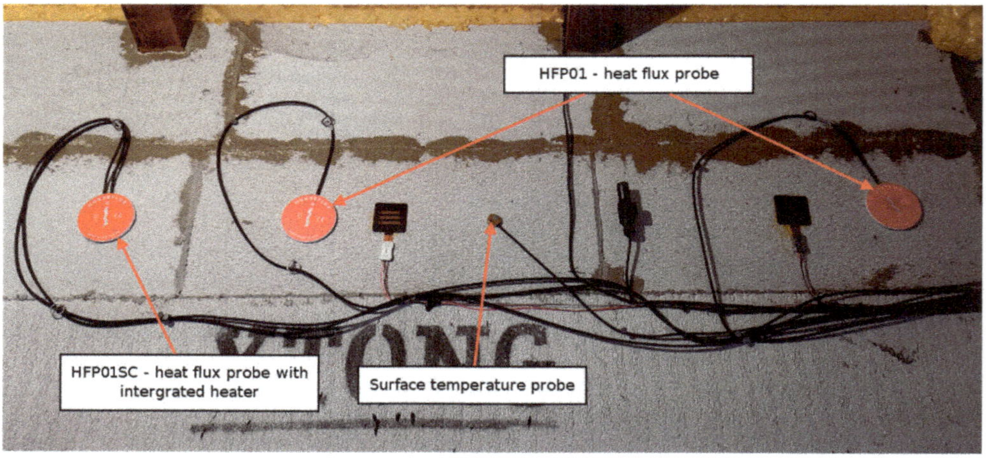

Figure 5. Extension of the experimental setup for the validation of the proposed technique.

To prove the efficiency and applicability of the proposed approach, a completely new dataset was obtained using high-sensitive HFP01 and HFP01SC probes in the final stage of the field measurement. Although the sensitivity of HFP sensors is sufficiently high, the HFP01SC was implemented in the measurement to confirm the data obtained by HFP01 during the validation procedure. Having the independent data from those sensitive sensors that stay apart of the sensitivity adjustment procedure described in the previous subsection, the independent validation could be performed by a simple confrontation of the proposed technique with sensor performances. Since the proposed technique potentially aims at the creation of a new alternative to the advanced type of sensors with self-calibration feature, it comes directly from the idea that independent data should be obtained from that kind of sensor. Therefore, additional smart sensors with integrated heaters were acquired, installed on the site, and self-calibrated to provide reference data for the validation.

In principle, the validation is based on a direct comparison of $q(t)$ (obtained using steps 1–4 of the proposed sensor adjustment technique as described in Section 2.3) with the time-resolved reference heat flux (recorded on HFP01 and HFP01SC probes). Although the HFP01 ones are high-sensitive sensors, which is essential for obtaining reliable reference data, the HFP01SC was implemented as well to confirm that reliability through on-site self-calibration. Such self-calibration applied to sensors with integrated heaters is based on measuring the difference in voltage output of the sensor before and after the heating period, and the heat flux generated by the heater to determine the new sensitivity. If the sensors need to stay mounted on the wall, the equal distribution of heat flux must be ensured allowing to use same approximation as dedicated for sensors buried in soils. To achieve this, the sensor was temporarily sandwiched by an AAC block, the same as those which the perimeter wall of the residential object was made of. Therefore, a temporary shelf was mounted on the wall and the sandwiching block was placed on top of it and gently pushed against the sensor. The good thermal contact between upper side of the sensor and temporary sandwiching block was achieved by using pressure-sensitive tape and heat sink grease. Also, the thermal conductivities of both wall and AAC block were checked using an ISOMET device to confirm that the thermal properties correspond to each other, so that Equation (6) can be applied directly without any further compensations. Assuming that the electrically generated heat flux is distributed equally to both sides of the sensor (i.e., half of the heat flux passes the sensor), the new sensitivity can be calculated [37] as

$$S_{SC} = \frac{2U_{SC}}{q_{SC}} = \frac{2U_{SC}R^2{}_{current}A_{heater}}{U^2{}_{current}R_{heater}}, \quad (6)$$

where S_{SC} sensor sensitivity obtained using the self-calibration technique ($V{\cdot}W^{-1}{\cdot}m^2$), U_{SC} (V) is the difference in the voltage output of the sensor before and after the heating period during self-calibration, q_{SC} ($W{\cdot}m^{-2}$) is the heat flux generated by the integrated heater, R_{heater} (Ω) is the nominal resistance of the film heater integrated in the heat flux probe, $R_{current}$ (Ω) is the resistance of the current sensing resistor used for self-calibration, A_{heater} (m^2) is the heating area of the heater integrated in the heat flux sensor body, $U_{current}$ (V) is the voltage measured to estimate the current passing through the heater. The details and the scheme of the electrical connection of the sensor is given in the product documentation [37]. The values of R_{heater}, A_{heater} are provided in the calibration certificate delivered along with the heat flux probe. Typically, the measurement of voltage generated by the sensor is taken at the times of t = 0, 180 and 360 s and U_{SC} is calculated as

$$U_{SC} = \left| U_{t=180} - \frac{(U_{t=0} + U_{t=360})}{2} \right|, \tag{7}$$

where U (V) is the voltage signal recorded at the thermopile.

3. Results

The results of experimental measurements and their subsequent processing are shown in the following subsections. The presented outputs of the adjustment procedure should help to analyze its performance and draw conclusions and recommendations for the application of the proposed approach. The experimental measurement was carried out between 10 August 2020 (Day 1) and 16 May 2021 (Day 279) and several sequences from this period were selected for the demonstration of applicability of the proposed sensor sensitivity adjustment method. For the validation, new data were obtained between 1 March 2022 (Day 568) and 17 March 2022 (Day 584). The in-situ experimental data and its processing in the sensor adjustment procedure is shown in Figure 6 and summarized in Table 2. The validation data are presented separately in Section 3.4.

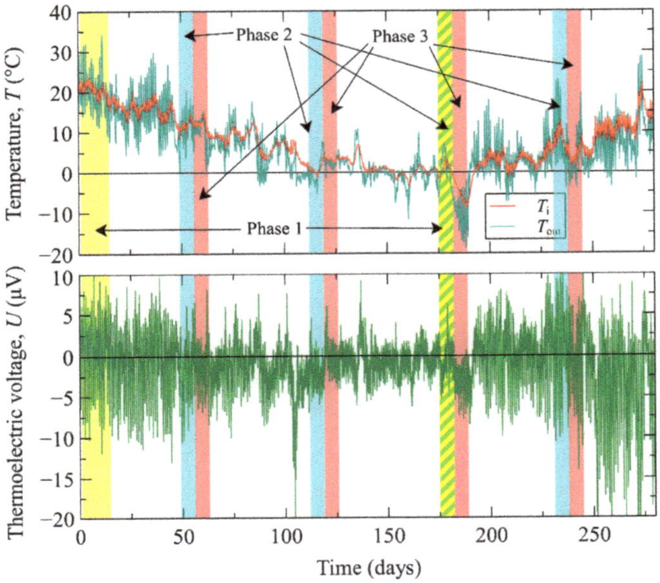

Figure 6. Experimental data for the demonstration of sensor calibration technique.

Table 2. Summary of experimental data used in the calibration.

	Description	Period
Phase 1	Initialization 1 Initialization 2	11 August 2020–24 August 2020 1 February 2021–7 February 2021
Phase 2	Sensor adjustment 1 Sensor adjustment 2 Sensor adjustment 3 Sensor adjustment 4	28 September 2020–4 October 2020 31 November 2020–6 December 2020 1 February 2021–7 February 2021 29 March 2021–4 April 2021
Phase 3	Performance verification 1 Performance verification 2 Performance verification 3 Performance verification 4	5 October 2020–12 October 2020 7 December 2020–14 December 2020 8 February 2021–15 February 2021 5 April 2021–12 April 2021
Phase 4	Validation	1 March 2022–17 March 2022

3.1. Phase 1—Initialization

In the initialization phase, the computational model was adjusted to optimize its performance under local conditions of the in-situ experiment. In this phase, the interior surface temperature T_s was the subject of experimental-computational analysis. The procedure of fitting simulated data $T_{s,sim}$ to those determined experimentally; $T_{s,exp}$, was used to adjust the heat transfer coefficient α, the free parameter of the model describing the rate of heat exchange on the solid/gas interface. The entire procedure was driven by searching for the minimum value of the Root-Mean-Square-Error (*RMSE*) between experimentally determined and simulated surface temperature variations. For the model adjustment, two periods were selected for demonstration. As mentioned in Section 2.2, the first adjustment (compulsory) was done on the dataset collected at the very beginning of the experiment in the period between 11 August 2020 and 24 August 2020 (Day 1–Day 14 of the experiment), in which 308 hourly outputs constituting time-resolved surface temperature variation were analyzed. The second adjustment of the computational model (optional) was done in the period between 1 February 2021 and 7 February 2021 (Day 175–Day 181 of the experiment, 169 hourly outputs) to compensate for the effects of a long-term measurement by adapting the model to the potentially new conditions that might be differing from those at the beginning of the experiment.

The *RMSE* of temperature profiles vs. heat transfer coefficient function is shown in Figure 7. The heat transfer coefficients in these particular periods were determined as α_1 = 10.75 W·m^{-2}·K^{-1} and α_2 = 13.75 W·m^{-2}·K^{-1}, respectively, indicating that environmental conditions slightly changed during the measurement. Therefore, the reinitialization should bring an increased accuracy to the subsequent re-adjustment procedures and experimental measurement, eventually.

The computational model performance during the ongoing experiment is presented in Figure 8. Based on the visual comparison of simulated temperature, $T_{s,sim}$, and experimentally determined temperature, $T_{s,exp}$, the computational model setup can be considered appropriate for further use in this research, and the optimized heat transfer coefficients could be used for the subsequent heat flux analyses using experimental data. The need for performing the re-initialization is documented by the difference in average indoor temperature in Initialization 1 and 2 which is more than 20 °C. Also, the daily oscillation of indoor temperature during Initialization 1 is apparent, which underlines the necessity of re-adjustment of the computational model to address different experimental environments.

3.2. Phase 2—Sensor Sensitivity Adjustment

With α calibrated from the previous phase, the verification of heat fluxes from both experimental and computational outputs could be done. Theoretically, if the computational model is perfectly accurate and the experimental measurement comes without noise and errors, then there should be an almost perfect match between the experimentally (q_{IN})

and computationally (q_{OUT}) determined heat fluxes (see Figure 1). Also, if the factory calibration of the heat flux probe is reasonably accurate under local experimental conditions, the measured heat fluxes (q_{EXP}) should match with those derived from computational-experimental data indicating that a specific calibration experiment is not necessary at all. On the other hand, if the derived heat fluxes, q_{IN} and q_{OUT}, correspond to each other while the measured heat flux q_{EXP} significantly differs from them, it will clearly justify the necessity for additional calibration. As mentioned previously, the sensor adjustment was performed for four different periods (see Table 1).

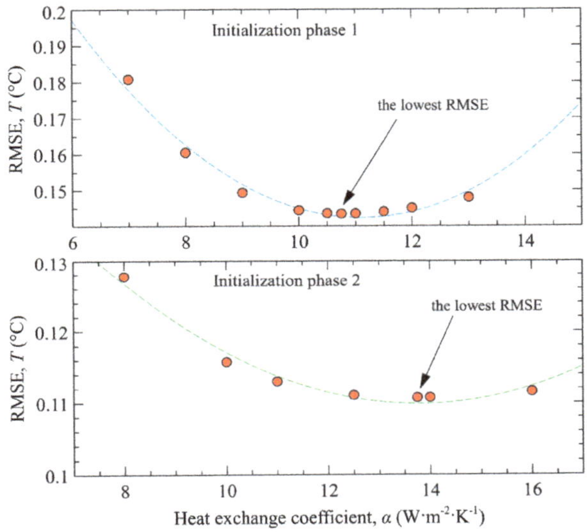

Figure 7. Adjustment of the computational model.

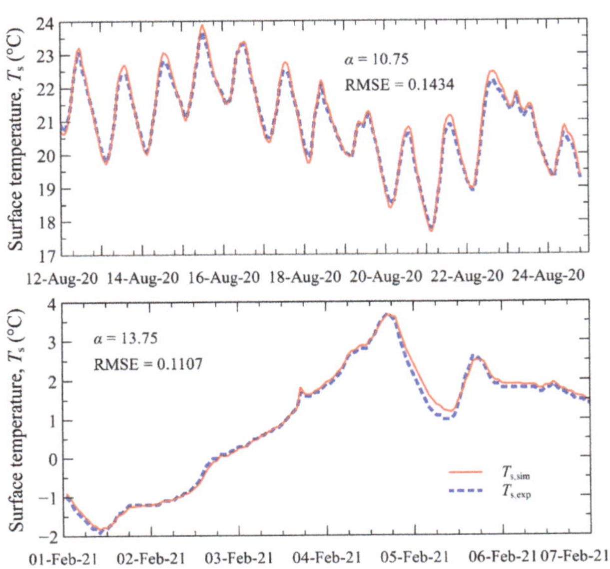

Figure 8. Adjusted performance of the computational model and its comparison with experimental data.

In Figure 9, there is a comparison of heat fluxes derived from experimental (q_{IN}, see Equation (2)) and simulation data (q_{OUT}, see Equation (3)), together with the heat flux measured using the heat flux sensor FHF01 with default sensitivity declared by the producer. The negative values of heat flux represent heat losses, while the positive ones stand for heat gains. The data can be effectively used as a pre-calibration check as well as a computational model adjustment back-check. If the computational model is well calibrated, a significant agreement between q_{IN} and q_{OUT} should be apparent. Any discrepancies detected would indicate insufficient performance of the computational model, which should be prevented by executing of the reinitialization phase. Figure 9 shows that the heat fluxes calculated from experimental and computational data using Equations (2) and (3) corresponded to each other (blue and red lines) to an acceptable extent in most cases. The occasional differences between those data could be explained by model imperfections or neglecting heat storage properties of surface material, which might affect the results even if the thickness of the surface element in the computational model is reasonably small. The differences might also result from changing parameters of the computational model, environmental conditions, or heat flux sensor properties. If substantial differences between q_{IN} and q_{OUT} are detected, such as at the end of the second period (see highlighted part of Figure 9), re-initialization should be done using recent data according to Section 3.1 in order to achieve a higher level of accuracy. In this case, it is obvious that the initialization, which was done between Day 1 and 14, is out-of-date and not effective anymore. The entire process of re-initialization can be easily done by utilizing the principles of the proposed method. The discrepancy detected in this particular case could be attributed to the overall change in environmental conditions (represented by temperature in the first place). However, occupants' behavior associated with an excessive increase of convection or radiation may be a causation as well.

On the other hand, the heat flux data from the FHF01 surface probe (green line) obtained by using default factory calibration proved to be insufficient for a building performance analysis as the probe undervalued the heat flux to a significant extent. This fact, indicating the low sensitivity of the sensor, may bring major uncertainties to any analysis built on those outputs. Therefore, a custom calibration was essential to improve the sensor accuracy and to keep the sensor involved as an integrated part of the experimental arrangement.

The sensor sensitivity adjustment (see Figure 10) was done by means of seeking for such a combination of *A* and *B* coefficients that will produce the least *RMSE* between average value of q_{IN} and q_{OUT}, i.e., q_{REF} (see Section 2.2, step 4), and the heat flux $q_{EXP,\,adj}$ calculated from the raw data as

$$q_{EXP,adj}(t) = A \cdot U(t) + B, \tag{8}$$

where $U(t)$ is the thermoelectric voltage recorded on the thermopile (heat flux probe) in particular time t. Following Equation (2), when the condition $T_i = T_s$ is met, a zero-heat flux should occur on the interior surface of the wall. However, the quick inspection of experimental data revealed that when $T_i = T_s$, a non-zero-heat flux is detected by the surface probe indicating that some kind of systematic error is present in the measurement. Regardless the source of that error, searching for the adjusting function *S* in the linear form should compensate this type of error. Such an approach to the sensor adjustment makes the entire process slightly different from that performed by the manufacturers, which usually tend to determine the sensitivity as a single value, i.e., without the shift of the linear function defined by *B*. The optimized parameters *A* ($W \cdot V^{-1} \cdot m^{-2}$) and *B* ($W \cdot m^{-2}$) are shown in Table 3.

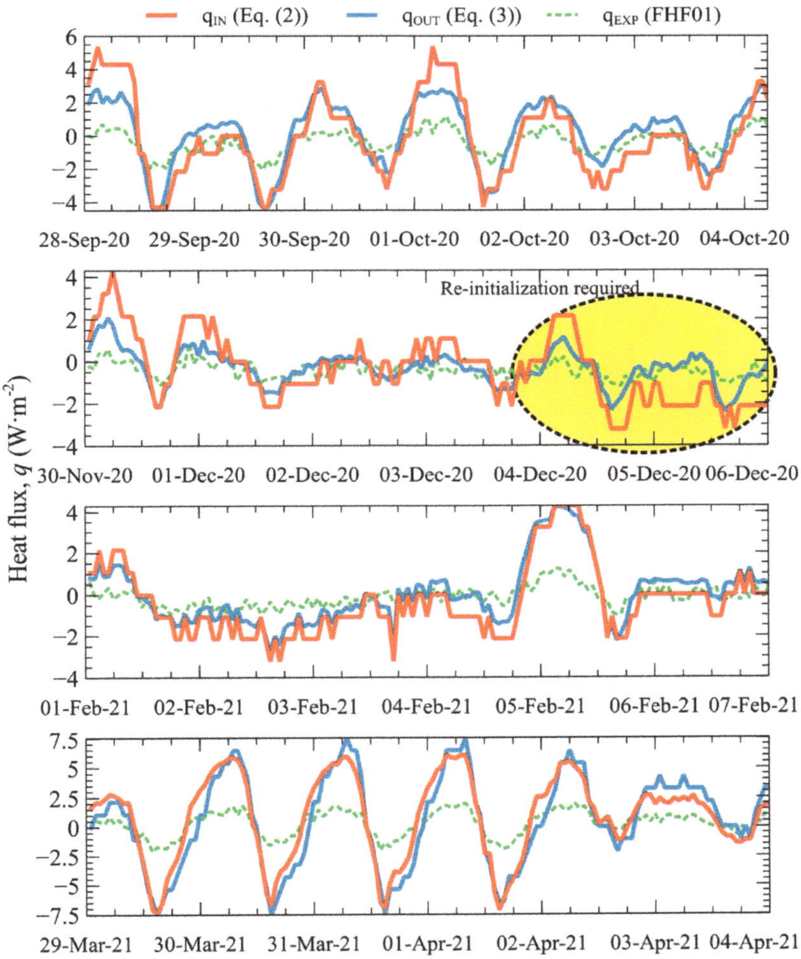

Figure 9. Comparison of surface heat fluxes before calibration (Phase 2).

Table 3. Results of sensor adjustment and optimized parameters.

Period	A (W·V^{-1}·m^{-2})	B (W·m^{-2})	*RMSE*
28 September 2020–4 October 2020	0.4880	0.8915	1.078
30 November 2020–6 December 2020	0.3412	0.2712	0.888
1 February 2021–7 February 2021	0.4570	0.3482	0.889
29 March 2021–4 April 2021	0.6468	−0.0156	1.270

It should be noted that the data presented in Figure 10 are supposed to be carefully analyzed to provide relevant information on the applicability of the designed calibration procedure or to suggest modifications that might improve the approach with respect to the particular areas of its application.

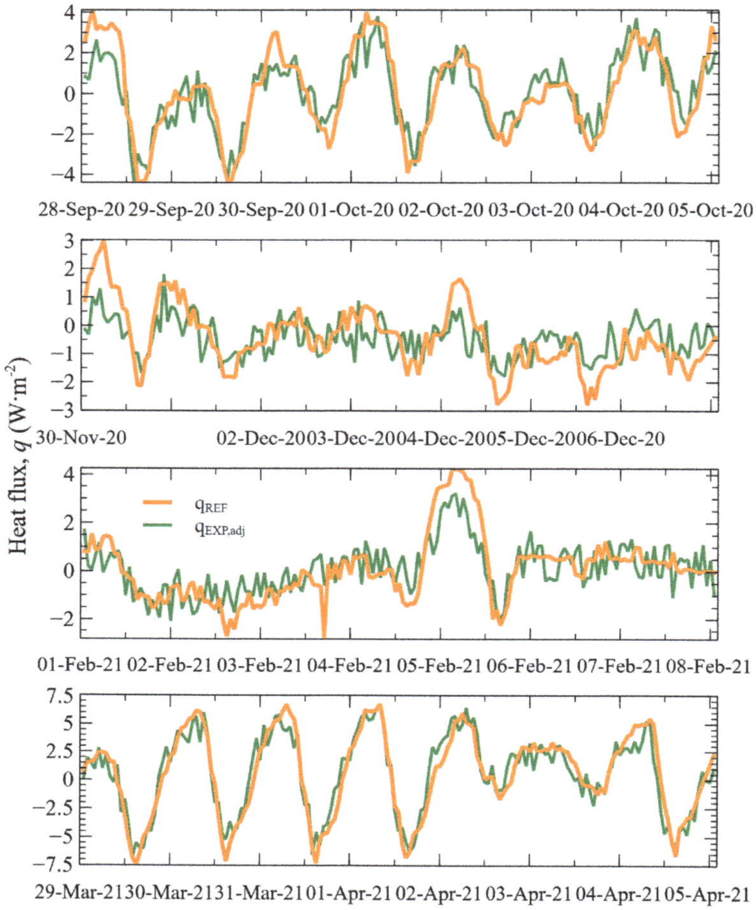

Figure 10. Sensor performance after sensitivity adjustment (Phase 2).

3.3. Phase 3—Sensor Performance Verification

With the adjusted sensitivity of the heat flux sensor, the sensor sensitivity adjustment procedure could proceed to the verification stage using data from Phase 3. For the verification, four testing periods between autumn 2020 and spring 2021 that followed the adjustment periods were selected (see Table 1 and Figure 6) and performance data was obtained for those periods. The idea of the verification was to prove the stability of the performed adjustment on blind data in the periods sufficiently close to the point of readjustment. The performance data included indoor and surface temperature measurements as well as outdoor climatic data for the simulation of the thermal response of the wall. Similarly to the previous steps, the surface heat fluxes q_{IN} and q_{OUT} were derived from experimental-computational data and compared against the output of the adjusted heat flux sensor.

Generally, the most important data reflecting the accuracy of demonstrated application are shown in Figure 11. Here, various accuracies of the adjusted sensor can be observed within the periods of verification. To provide a better insight into the accuracy of the adjusted heat flux sensor, a regression analysis of data was done and the *R*-squared values together with *RMSE* were determined for each period. The results of regression analysis are shown in Figure 12. The discrepancy from the adjustment phase as discussed in Section 3.2

(see Figure 9) has been clearly projected into this verification (as marked in Figure 11). Since the verification was done on the periods that followed the adjustment periods immediately, some performance deviations in the period 7 December 2020–13 December 2020 were anticipated. This observation underlines the need for a proper re-initialization based on detailed surveillance and analysis of the adjusted performance to optimize the computational model and to adapt the proposed procedure to changing environmental conditions.

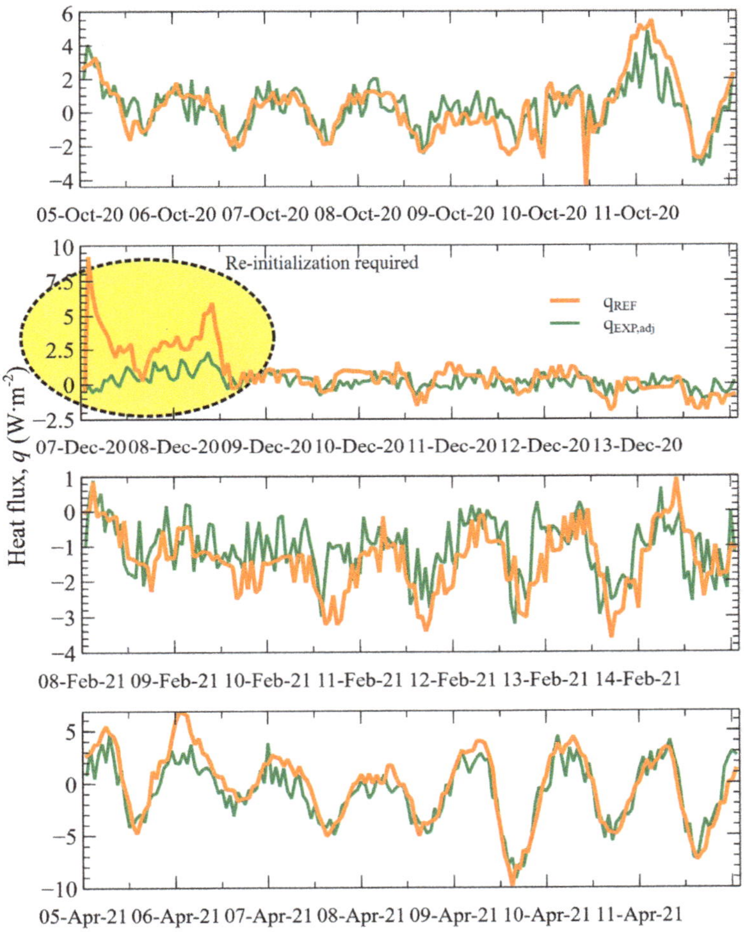

Figure 11. Performance verification of the self-adjustment technique.

It should be noted that the investigated object serves recreational purposes which is typical of a seasonal regime of occupation. As for this, some parts of the object, such as the entry foyer or some technical rooms, are considered "free-running" rooms, i.e., without space heating and/or cooling during the entire year. This stands also for the room, where all the sensors were placed, which is apparent from the interior temperature development during the measuring period (see Figure 6). This fact can seriously affect the results as the operating conditions might dramatically change over time, which can affect the sensor performance and physical behavior of construction materials.

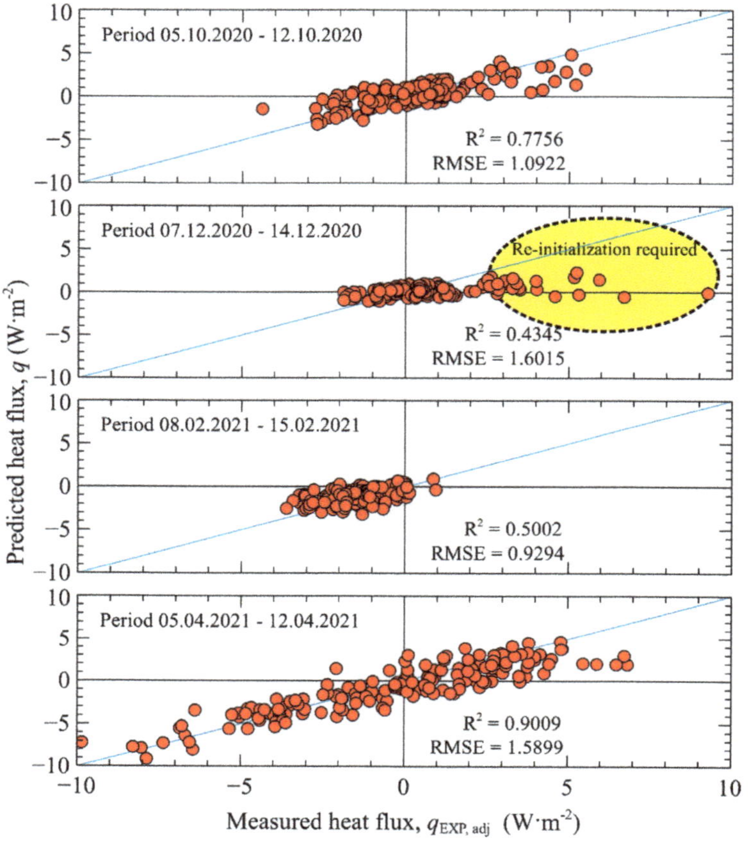

Figure 12. Regression analysis of the verification results.

3.4. Validation of the Self-Adjustment Technique

For the validation, the field data were obtained between 1 March 2022 and 15 March 2022. In reference to the beginning of the experiment, the validation period corresponded to the interval between Days 568 and 582. Right after the validation period, the self-adjustment of the HFP01SC sensor sensitivity was done to update the accuracy and validate the outputs from the HFP01 sensors. The self-adjustment of the sensitivity was done several times, while the interval between individual executions was at least 6 h as recommended by the manufacturer. The summary of experimental data obtained during the validation period is shown in Figure 13. The thermoelectric voltage in this figure refers to the output recorded on the HFP01SC and will be further processed with the adjusted sensitivity determined through the adjusting procedure. The data from the sensor sensitivity self-adjustment procedure are summarized in Table 4.

After performing the sensitivity adjustment, the thermoelectric voltage from Figure 13 was divided by the new sensitivity to obtain the time-resolved heat flux which served as the reference value in the validation procedure. Simultaneously, the heat fluxes $q_{IN}(t)$ and $q_{OUT}(t)$, obtained from Steps 2–4 as described in Section 2.2, were plotted in Figure 14. Prior to the calculation of $q_{IN}(t)$ and $q_{OUT}(t)$, the computational model was adjusted according to Step 1 of the proposed procedure to achieve the best performance of the semi-virtual environment. The heat transfer coefficient α was optimized to 5.0 $W{\cdot}m^{-2}{\cdot}K^{-1}$ producing the least difference ($RMSE$ = 0.2627) between computationally and experimentally determined surface temperatures $T_{s,sim}(t)$ and $T_{s,exp}(t)$, respectively. The validation data presented in Figure 14 represent the key input for the validation of the presented technique itself. Here, the reference heat flux from the high-sensitive probe HFP01SC was compared with the computational-experimental one (q_{REF}), which substitutes the electrically generated flux in the proposed sensor adjustment technique. The correspondence of the heat flux data in Figure 14 is very good, which can be considered a good result of the validation procedure showing that the heat flux from the semi-virtual environment was obtained with reasonable accuracy, and could be used as a target value in any subsequent calibration or sensitivity adjustment.

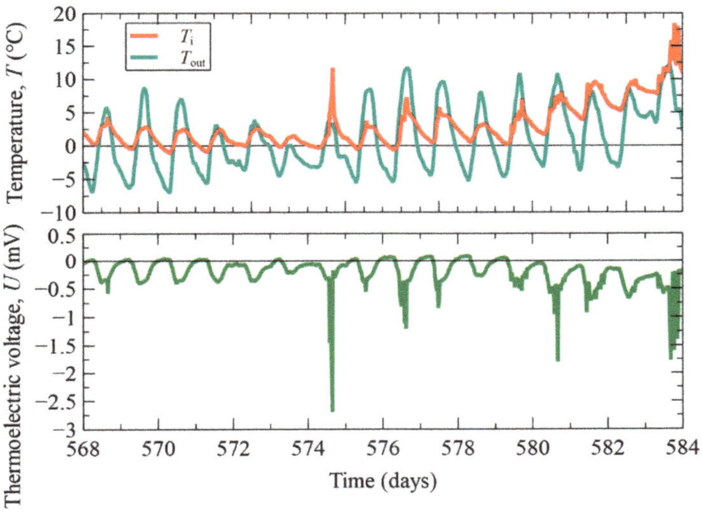

Figure 13. Experimental data for the validation of sensor sensitivity adjustment technique.

Table 4. Results of self-adjustment procedure of HFP01SC sensor sensitivity.

	U_{SC} (mV)	$U_{current}$ (mV)	$R_{current}$ (Ω)	R_{heater} (Ω)	A_{heater} (m^2)	S_{SC} (V·m^2·W^{-1})
1	3.012	0.596	10	99.2 ± 1.0	(3885 ± 136) × 10^{-6}	(60.34 ± 2.41) × 10^{-6}
2	2.958	0.594	10	99.2 ± 1.0	(3885 ± 136) × 10^{-6}	(65.69 ± 2.63) × 10^{-6}
3	2.925	0.592	10	99.2 ± 1.0	(3885 ± 136) × 10^{-6}	(65.42 ± 2.62) × 10^{-6}
4	2.938	0.593	10	99.2 ± 1.0	(3885 ± 136) × 10^{-6}	(65.46 ± 2.62) × 10^{-6}
5	2.859	0.588	10	99.2 ± 1.0	(3885 ± 136) × 10^{-6}	(64.66 ± 2.59) × 10^{-6}
	S_{SC} = (64.31 ± 2.57) × 10^{-6} V·m^2·W^{-1}					

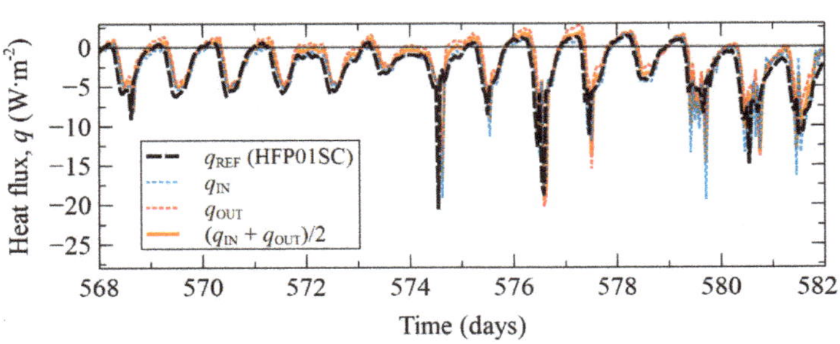

Figure 14. Validation of the proposed sensitivity adjustment technique.

4. Discussion

The results presented in the previous subsections demonstrated that heat flux measurements in buildings need to reflect specific requirements given by the experimental conditions. Besides the changing environment, which might be typical for "free-running" buildings, the selection of appropriate heat flux probes is crucial for obtaining accurate data. However, the results reported in this paper showed that even such issues can be effectively addressed if a proper approach is chosen.

The analysis of the experimental data underlined the necessity of sensor re-calibration as the heat flow rate between the indoor environment and building walls may dramatically change during the measurement as a consequence of changing environmental conditions. The adaptive sensitivity adjustment proved that the heat transfer coefficient α was changed during the measurement which could be attributed to different conditions during model initialization [38,39]. Therefore, careful monitoring of environmental conditions as well as field variables should be done to provide sufficient data for a decision on whether re-initializations and/or re-adjustments are required and what is the recommended time span for performing the individual procedural steps.

A periodic calibration is often required in the building sector to ensure reliable quality of measured quantities. The conventional approach of calibration, on the other hand, may be exceptionally difficult to perform as the sensors might be difficult to reach or can't be removed from the construction due to ongoing measurement. There have been limited studies published in the literature focusing on addressing those issues [40–42]. The novel approaches are mostly called virtual experiments, online calibrations, self-calibrations, or various combinations using these or similar other terms. However, most of the research published in the literature tends to prefer the traditional ways of sensor calibrations, which depend primarily on the type of used sensors [43–45]. The importance of heat flux monitoring and analysis in the field of building and civil engineering can be documented by reference to numerous works that have been published recently [46–55]. Also, the variety of those applications proves the importance of heat flux analysis that spans across multiple disciplines. Heat flux measurement is a key part of practically any thermal performance analysis [46,52,53,55], experimental validation of thermal properties of novel materials or building components [47,49,54], monitoring of thermal performances of buildings or their parts [48,50,51], etc. Therefore, the quality and accuracy of heat flux measurement should be always maintained at a high level, especially when long-term monitoring and observations are required [49–51]. In the field of heat flux monitoring, the producers brought new types of sensors with self-calibration capability, which clearly documents the awareness of the necessity for sensor recalibration during experimental measurement. The proposed method may contribute to the state-of-the art in this respect with the utilization of standard probes without advanced features, thus, providing high-tech solutions at standard costs. Also, the proposed approach can be applied "ex-post" to existing experimental data if some discrepancy is detected and appropriate environmental data are available along with heat flux variations. The accuracy of the proposed technique depends mostly on the selected re-adjustment period. Shorter periods bring higher accuracy to the heat flux measurement as the approach may efficiently reflect and address the changing environmental conditions. Basically, the readjustment can be triggered by user-defined values describing the change of environmental conditions or scheduled in regular periods. Consequently, a novel time-saving solution for maintenance-free application may be proposed as an alternative to traditional laboratory techniques.

5. Conclusions

In this paper, a smart adjustment technique for sensitivity enhancement of the surface heat flux probes was designed and validated in a critical experiment performed in real conditions. The proposed technique was based on combining the experimental and computational data to define a prescribed heat flux, thus omitting the need for traditional hardware equipment that would have to be used to perform the sensor performance check in the standard way. The proposed method allows the heat flux to be determined using raw data from the heat flux sensor without a need to interfere with the ongoing experiment. As such, the sensor sensitivity can be automatically adjusted during the measurement based on the surveillance of experimental variables. The applicability of the proposed method was demonstrated and validated in a real experiment. The following conclusions can be drawn, based on the performance analysis, to summarize the main outcomes of the research:

- Factory calibration of heat flux sensors is not always sufficient as the real experimental conditions may differ from those effective during the calibration. Also, some errors may arise from the type of installation of the heat flux sensors which can subsequently affect the sensitivity of the sensor.
- Continuous monitoring of heat flow sensor performance and silent checking of their accuracy allows the proposed technique to be applied instantaneously during the ongoing experiment without interfering with experimental apparatus or taking technological breaks required for sensor manipulation or replacement.
- The proposed approach is suitable for both short-term and long-term heat flux measurements while bringing an entirely new potential for maintenance-free applications. However, a qualified decision on sensor sensitivity re-adjustment or re-initialization of the computational model running in the background is crucial to ensure high accuracy of the method.
- The principles of the adaptive re-adjustments of heat flux sensor performance allow for their integration into automatic systems and smart HVAC technologies to efficiently control the indoor environment and thermal comfort of residential buildings based on real-time performance data.
- Although the validation was performed with sufficient accuracy, the additional testing would be beneficial to include various conditions, surface materials, or different kinds of probes. Such a complex validation should be considered an essential precondition.

Author Contributions: Conceptualization, J.K.; methodology, J.K. and R.Č.; software, J.M.; validation, J.K. and J.M.; investigation, J.K.; resources, J.K.; data curation, J.K. and J.M.; writing—original draft preparation, J.K.; writing—review and editing, R.Č.; visualization, J.K.; supervision, R.Č.; project administration, J.M.; funding acquisition, J.K. All authors have read and agreed to the published version of the manuscript.

Funding: This research has been supported by the Czech Science Foundation, under project No. 22-08786S.

Institutional Review Board Statement: Not applicable.

Informed Consent Statement: Not applicable.

Data Availability Statement: All data supporting the results presented in this manuscript are stored in the repositories of the Czech Technical University with an authorized access. and will be available upon a reasonable request sent to the corresponding author (J.K.).

Conflicts of Interest: The authors declare no conflict of interest.

References

1. Wang, S.; Chen, Y. Transient Heat Flow Calculation for Multilayer Constructions Using a Frequency-Domain Regression Method. *Build. Environ.* **2003**, *38*, 45–61. [CrossRef]
2. Zhang, Y.; Sun, X.; Medina, M.A. Calculation of Transient Phase Change Heat Transfer through Building Envelopes: An Improved Enthalpy Model and Error Analysis. *Energy Build.* **2020**, *209*, 109673. [CrossRef]
3. Fu, T.; Zong, A.; Zhang, Y.; Wang, H.S. A Method to Measure Heat Flux in Convection Using Gardon Gauge. *Appl. Therm. Eng.* **2016**, *108*, 1357–1361. [CrossRef]
4. Purpura, C.; Trifoni, E.; Petrella, O.; Marciano, L.; Musto, M.; Rotondo, G. Gardon Gauge Heat Flux Sensor Verification by New Working Facility. *Measurement* **2019**, *134*, 245–252. [CrossRef]
5. Gardon, R. An Instrument for the Direct Measurement of Intense Thermal Radiation. *Rev. Sci. Instrum.* **2004**, *24*, 366. [CrossRef]

6. Kidd, C.T.; Nelson, C.G. How the Schmidt-Boelter Gage Really Works. In Proceedings of the 41st International Instrumentation Symposium, Aurora, CO, USA, 7–11 May 1995.
7. Kidd, C.T.; Adams, J.C. Fast-Response Heat-Flux Sensor for Measurement Commonality in Hypersonic Wind Tunnels. *J. Spacecr. Rockets* **2012**, *38*, 719–729. [CrossRef]
8. Lam, C.S.; Weckman, E.J.; Lam, C.S.; Weckman, E.J. Steady-State Heat Flux Measurements in Radiative and Mixed Radiative–Convective Environments. *Fire Mater.* **2009**, *33*, 303–321. [CrossRef]
9. Mocikat, H.; Herwig, H. Heat Transfer Measurements with Surface Mounted Foil-Sensors in an Active Mode: A Comprehensive Review and a New Design. *Sensors* **2009**, *9*, 3011–3032. [CrossRef]
10. Lakatos, Á.; Kovács, Z. Comparison of Thermal Insulation Performance of Vacuum Insulation Panels with EPS Protection Layers Measured with Different Methods. *Energy Build.* **2021**, *236*, 110771. [CrossRef]
11. Ballestrín, J.; Estrada, C.A.; Rodríguez-Alonso, M.; Pérez-Rábago, C.; Langley, L.W.; Barnes, A. Heat Flux Sensors: Calorimeters or Radiometers? *Solar Energy* **2006**, *80*, 1314–1320. [CrossRef]
12. Sharkov, A.V.; Korablev, V.A.; Nekrasov, A.S.; Minkin, D.A.; Fadeeva, S.V. A Radiometer for Measuring High-Intensity Heat Flux Density and a Method of Calibrating It. *Meas. Tech.* **2012**, *54*, 1276–1279. [CrossRef]
13. Childs, P.R.N.; Greenwood, J.R.; Long, C.A. Heat Flux Measurement Techniques. *Proc. Inst. Mech. Eng. C J. Mech. Eng. Sci.* **1999**, *213*, 655–677. [CrossRef]
14. Nakos, J. *Description of Heat Flux Measurement Methods Used in Hydrocarbon and Propellant Fuel Fires at Sandia;* Sandia National Laboratories: Albuquerque, NM, USA; Livermore, CA, USA, 2010.
15. Levinson, R.; Chen, S.; Ferrari, C.; Berdahl, P.; Slack, J. Methods and Instrumentation to Measure the Effective Solar Reflectance of Fluorescent Cool Surfaces. *Energy Build.* **2017**, *152*, 752–765. [CrossRef]
16. *ASTM C177-19;* Standard Test Method for Steady-State Heat Flux Measurements and Thermal Transmission Properties by Means of the Guarded-Hot-Plate Apparatus. ASTM International: West Conshohocken, PA, USA, 2019.
17. *ASTM C518-17;* Standard Test Method for Steady-State Thermal Transmission Properties by Means of the Heat Flow Meter Apparatus. ASTM International: West Conshohocken, PA, USA, 2017.
18. Lakatos, Á.; Csík, A.; Csarnovics, I. Experimental Verification of Thermal Properties of the Aerogel Blanket. *Case Stud. Therm. Eng.* **2021**, *25*, 100966. [CrossRef]
19. *ASTM C1130-21;* Standard Practice for Calibration of Thin Heat Flux Transducers. ASTM International: West Conshohocken, PA, USA, 2021.
20. *ASTM C1363-19;* Standard Test Method for Thermal Performance of Building Materials and Envelope Assemblies by Means of a Hot Box Apparatus. ASTM International: West Conshohocken, PA, USA, 2019.
21. *ASTM C1114-06;* Standard Test Method for Steady-State Thermal Transmission Properties by Means of the Thin-Heater Apparatus. ASTM International: West Conshohocken, PA, USA, 2019.
22. *ISO 8302:1991;* Thermal Insulation—Determination of Steady-State Thermal Resistance and Related Properties—Guarded Hot Plate Apparatus. Technical Committee: ISO/TC 163; International Organization for Standardization: Geneve, Switzerland, 1991.
23. *ASTM C1046-95R21;* Standard Practice for In-Situ Measurement of Heat Flux and Temperature on Building Envelope Components. ASTM International: West Conshohocken, PA, USA, 2021.
24. *ASTM C1155-95R21;* Standard Practice for Determining Thermal Resistance of Building Envelope Components from the In-Situ Data. ASTM International: West Conshohocken, PA, USA, 2021.
25. Zarr, R.R.; Martinez-Fuentes, V.; Filliben, J.J.; Dougherty, B.P. Calibration of Thin Heat Flux Sensors for Building Applications Using ASTM C 1130. *J. Test. Eval.* **2001**, *29*, 293–300. [CrossRef]
26. HFP01SC Brochure HFP01SC Self-Calibrating Heat Flux Sensor Brochure. Available online: www.hukseflux.com (accessed on 17 April 2022).
27. FHF04SC Brochure FHF04SC Self-Calibrating Heat Flux Sensor with Integrated Heater Brochure. Available online: www.hukseflux.com (accessed on 17 April 2022).
28. Info; hukseflux.com FHF05 Series Heat Flux Sensors User Manual. Available online: www.hukseflux.com (accessed on 18 July 2024).
29. Maděra, J.; Kočí, V.; Kočí, J.; Černý, R. Physical and Mathematical Models of Hygrothermal Processes in Historical Building Envelopes. *AIP Conf. Proc.* **2017**, *1906*, 140003. [CrossRef]
30. Maděra, J.; Kočí, J.; Kočí, V.; Kruis, J. Parallel Modeling of Hygrothermal Performance of External Wall Made of Highly Perforated Bricks. *Adv. Eng. Softw.* **2017**, *113*, 47–53. [CrossRef]
31. Kočí, V.; Jerman, M.; Pavlík, Z.; Maděra, J.; Žák, J.; Černý, R. Interior Thermal Insulation Systems Based on Wood Fiberboards: Experimental Analysis and Computational Assessment of Hygrothermal and Energy Performance in the Central European Climate. *Energy Build.* **2020**, *222*, 110093. [CrossRef]
32. Künzel, H.M. *Simultaneous Heat and Moisture Transport in Building Components: One-and Two-Dimensional Calculation Using Simple Parameters;* Fraunhofer IRB Verlag Suttgart; IRB-Verlag: Stuttgart, Germany, 1995.
33. Jerman, M.; Keppert, M.; Výborný, J.; Černý, R. Hygric, Thermal and Durability Properties of Autoclaved Aerated Concrete. *Constr. Build. Mater.* **2013**, *41*, 352–359. [CrossRef]

34. Kočí, V.; Maděra, J.; Jerman, M.; Žumár, J.; Koňáková, D.; Čáchová, M.; Vejmelková, E.; Reiterman, P.; Černý, R. Application of Waste Ceramic Dust as a Ready-to-Use Replacement of Cement in Lime-Cement Plasters: An Environmental-Friendly and Energy-Efficient Solution. *Clean. Technol. Environ. Policy* **2016**, *18*, 1725–1733. [CrossRef]
35. User Manual FHF01, Hukseflux Thermal Sensors. Available online: https://www.hukseflux.com/uploads/product-documents/FHF01_manual_v2006.pdf (accessed on 14 December 2021).
36. User Manual HFP01 & HFP03, Heat Flux Plate/Heat Flux Sensor, Hukseflux Thermal Sensors. Available online: https://www.hukseflux.com/uploads/product-documents/HFP01_HFP03_manual_v1721.pdf (accessed on 14 December 2021).
37. Info; hukseflux.com Self-Calibrating Heat Flux Sensor TM Hukseflux Thermal Sensors. HFP01SC User Manual Self-Calibrating Heat Flux Sensor TM Hukseflux Thermal Sensors. Available online: www.hukseflux.com (accessed on 5 April 2022).
38. Maffulli, R.; He, L. Wall Temperature Effects on Heat Transfer Coefficient for High-Pressure Turbines. *J. Propuls. Power* **2014**, *30*, 1080–1090. [CrossRef]
39. Janssen, H. Dependence of Heat Transfer Coefficients on the Wall Temperature Shape and Consequences for the Calculation of Overall Heat Transfer Coefficients for Heat Exchangers. *Heat Transf. Coeff.* **1996**, *31*, 153–161. [CrossRef]
40. Yu, Y.; Li, H. Virtual In-Situ Calibration Method in Building Systems. *Autom. Constr.* **2015**, *59*, 59–67. [CrossRef]
41. Bychkovskiy, V.; Megerian, S.; Estrin, D.; Potkonjak, M. A Collaborative Approach to In-Place Sensor Calibration. *Lect. Notes Comput. Sci.* **2003**, *2634*, 301–316. [CrossRef]
42. Dulev, V.; Ermishin, S.; Khoteev, N.; Lopatin, A.; Shabanov, P.; Menshikov, A.; Startsev, A. Automated Measuring System Designed to Calibrate Measuring Devices Using Virtual Standards® Technology. In Proceedings of the AUTOTESTCON, San Antonio, TX, USA, 20–23 September 2004; pp. 158–164. [CrossRef]
43. Sapozhnikov, S.Z.; Mityakov, V.Y.; Mityakov, A.V.; Pokhodun, A.I.; Sokolov, N.A.; Matveev, M.S. The Calibration of Gradient Heat Flux Sensors. *Meas. Tech.* **2012**, *54*, 1155–1159. [CrossRef]
44. Sharkov, A.V.; Korablev, V.A.; Makarov, D.S.; Minkin, D.A.; Nekrasov, A.S.; Gordeichik, A.A. Measurement of High-Density Heat Flux Using an Automated Installation. *Meas. Tech.* **2016**, *59*, 67–69. [CrossRef]
45. Zhang, J.; Mei, G.; Bao, T.; Yan, L.E. An Experimental Study on Thermal Analyses of Crushed-Rock Layers with Single-Size Aggregates. *Cold Reg. Sci. Technol.* **2019**, *160*, 110–118. [CrossRef]
46. Ha, T.-T.; Feuillet, V.; Waeytens, J.; Zibouche, K.; Peiffer, L.; Garcia, Y.; le Sant, V.; Bouchie, R.; Koenen, A.; Monchau, J.-P.; et al. Measurement Prototype for Fast Estimation of Building Wall Thermal Resistance under Controlled and Natural Environmental Conditions. *Energy Build.* **2022**, *268*, 112166. [CrossRef]
47. Chihab, Y.; Bouferra, R.; Garoum, M.; Essaleh, M.; Laaroussi, N. Thermal Inertia and Energy Efficiency Enhancements of Hollow Clay Bricks Integrated with Phase Change Materials. *J. Build. Eng.* **2022**, *53*, 104569. [CrossRef]
48. Al-Absi, Z.A.; Hafizal, M.I.M.; Ismail, M. Experimental Study on the Thermal Performance of PCM-Based Panels Developed for Exterior Finishes of Building Walls. *J. Build. Eng.* **2022**, *52*, 104379. [CrossRef]
49. Balocco, C.; Pierucci, G.; de Lucia, M. An Experimental Method for Building Energy Need Evaluation at Real Operative Conditions. A Case Study Validation. *Energy Build.* **2022**, *266*, 112114. [CrossRef]
50. Briga-Sá, A.; Leitão, D.; Boaventura-Cunha, J.; Martins, F.F. Trombe Wall Thermal Performance: Data Mining Techniques for Indoor Temperatures and Heat Flux Forecasting. *Energy Build.* **2021**, *252*, 111407. [CrossRef]
51. Stella, P.; Personne, E. Effects of Conventional, Extensive and Semi-Intensive Green Roofs on Building Conductive Heat Fluxes and Surface Temperatures in Winter in Paris. *Build. Environ.* **2021**, *205*, 108202. [CrossRef]
52. Rincón-Casado, A.; Jara, E.Á.R.; Pardo, A.R.; Lissén, J.M.S.; Flor, F.J.S. de la Reducing the Cooling Energy Demand by Optimizing the Airflow Distribution in a Ventilated Roof: Numerical Study for an Existing Residential Building and Applicability Map. *Appl. Sci.* **2024**, *14*, 6596. [CrossRef]
53. Santos, P.; Abrantes, D.; Lopes, P.; Moga, L. The Relevance of Surface Resistances on the Conductive Thermal Resistance of Lightweight Steel-Framed Walls: A Numerical Simulation Study. *Appl. Sci.* **2024**, *14*, 3748. [CrossRef]
54. Küheylan, C.D.; Özkan, D.B. Experimental and Theoretical Study of Heat Transfer in a Chilled Ceiling System. *Appl. Sci.* **2024**, *14*, 5908. [CrossRef]
55. Harangozó, J.; Tureková, I.; Marková, I.; Hašková, A.; Králik, R. Study of the Influence of Heat Flow on the Time to Ignition of Spruce and Beech Wood. *Appl. Sci.* **2024**, *14*, 4237. [CrossRef]

applied sciences

MDPI

Article

Eularian–Eularian Model for Agglomeration Behavior of Combusted Iron Particles

Warnakulasooriya Dinoja Sammani Fernando and Jamal Naser *

Department of Mechanical Engineering and Product Design Engineering, Swinburne University of Technology, John St, Hawthorn, VIC 3122, Australia; ddfernando@swin.edu.au
* Correspondence: jnaser@swin.edu.au

Abstract: Direct reduction of iron (DRI) technology in fluidized beds has been identified as a promising approach due to its environmental benefits over other methods. Nevertheless, the process of iron particle sintering in the DRI approach poses a significant obstacle to its advancement. The present work investigated the phenomenon of agglomeration in fine iron particles across various temperatures and with multiple sintering force models of different intensities of solid bridge force. The study utilized a simple but comprehensive and cost-effective CFD model developed using the Eularian–Eularian two-fluid model. The model was explicitly incorporated with user-defined subroutines for the solid phase, while the gas phase was modeled with AVL Fire advance simulation software. The solid bridge force between solid particles was modeled as the inter-particle cohesive force. The model was validated with the experimental results and results from another CFD-DEM model for the same experiment. High temperatures with increased sintering forces were observed to have the most impact on the iron particle agglomeration, while the gas's superficial velocity had a minimal effect on it. The predictions of this model closely align with the CFD-DEM model results, providing sufficient reliability to implement this model on a large scale.

Keywords: particle agglomeration; solid bridge force; fluidized bed; two-fluid method; DRI

1. Introduction

The current state of the global energy supply requires cleaner, renewable, and CO_2-neutral energy sources to replace fossil fuels and address the threats of climate change globally. In recent times, the DRI method has been identified as one of the core technologies for the decarbonizing iron process in favor of the road map of climate change mitigation targets [1]. Further, industry metal fuels were identified as a very encouraging and environmentally beneficial energy source due to their higher energy density and zero CO_2 emissions when in the DRI process [1,2]. Iron powders, particularly those in micron size, are initially prioritized because of their high volumetric energy density and widespread availability [2–5]. Iron powders undergo combustion to produce thermal energy. The essential step in achieving a carbon dioxide-free iron fuel cycle is the DRI, utilizing sustainable energy sources such as hydrogen (H_2) after the combustion process [6,7]. It is beneficial to use fluidized bed combustion plants for the DRI process due to their flexibility in using fine iron particles and their cost-effectiveness. It is estimated that a fluidized bed reduces the material cost by half compared with the MIDREX method using a shaft furnace of DRI technology [8].

Nevertheless, a significant issue arises in agglomeration caused by particle sintering; this often occurs when iron oxides are chemically reduced at high temperatures. Agglomeration/sintering has been detected in reactors that promote efficient mass/heat transmission and adequate solid mixing, such as fluidized beds [9]. As a result, when agglomeration is not avoided, fluidization is significantly reduced or even stopped completely, negatively impacting the operation's efficiency. A detailed understanding of the agglomeration, as well as defluidization characteristics of iron/iron oxide fines at high temperatures, is necessary to create an operational design of fluidized bed DRI.

Agglomeration occurs in the fluidized bed combustion due to several types of cohesive forces [10–13]. Those inter-particle cohesive forces fall under several categories according to their natures. They include electrostatic forces, van der Walls forces, solid bridge, or liquid bridge forces. Any of these forces can become dominant according to the specific case. Out of these force categories, it was identified that the liquid bridge force is most influential at high temperatures of the DRI method [14].

Liquid bridges are more versatile than van der Waals forces in practical applications. They can be adjusted. Liquid bridges are more complex as they involve both dynamic and static forces, and they use up energy. The static liquid bridge force arises from the combination of surface tension and pressure deficit within the liquid bridge. This emphasizes how these factors work together to influence liquid bridges in industrial processes. When a particle's surface is coated with a mobile liquid film, liquid bridges will develop at points of contact with other particles and surfaces. Moreover, if the partial pressure of a suitable vapor in the surrounding gas is high enough, condensation will take place at these contact points, leading to the formation of additional liquid bridges [14].

The geometrical/physical parameters of the iron/iron oxide particles and operating conditions play a major role in particle agglomeration behavior [15,16]. The physical parameters include the composition of the particle, size, shape and microstructure, while operational conditions encompass operating temperature, reduction of gas composition, and flow rate of the gas. Hence, it has become very difficult to model the solid bridge force accurately in agglomeration. Although experimental studies can be conducted to understand the agglomeration behavior, the operational costs of these experiments are very high. Therefore, agglomeration formation in the DRI method has become a challenge to be addressed because it mitigates the efficiencies of the process [17]. Therefore, it is crucial to understand the process of inter-particle cohesion by solid bridge force in the DRI method, which causes defluidization of fluidized bed combustion plants.

Thus, different approaches to numerical simulations, such as the two-fluid method and the discrete element method (DEM), provide tools to investigate the fluidization phenomena that take place in fluidized bed combustion. The numerical model simulation strategy developed by Tazleem et al. uses two fluid approaches based on the Eularian–Eulatian technique to predict agglomeration. The established modeling technique is verified using experimental data from the literature for a simplified two-dimensional geometry of a pilot-scale fluidized bed combustor [18]. Another numerical model based on the DEM approach was developed by Kuwagi et al. [12] to investigate metallic solid bridge forces in agglomeration. This model encompasses a surface diffusion mechanism that considers the surface roughness. Further, particle sintering in agglomeration is analyzed by Mansourpour et al. [19,20] with a CFD-DEM simulation. Their study incorporates the effect on the agglomerate sizes by temperature and gas velocity.

There are several studies that have utilized CFD-DEM simulations to investigate the sintering behavior of iron/iron oxide fines in fluidized beds. However, the high computing cost makes it challenging to simulate large-scale experiments using the CFD-DEM approach. The computer simulation cost of the DEM is significantly affected by the enormous number of particles in the micron-sized cohesive iron/iron oxide powder. To overcome this challenge and to expand the modeling capability to a large/industrial scale, the two-fluid modeling approaches can be applied. However, there is very little research work conducted using the two-fluid approach to simulate agglomeration in fluidized beds for the production of iron/iron oxide powder. Even the available literature does not provide proper agglomeration behavior with significant parameters, like solid pressure. They do not provide solid fraction variations with defluidization in the fluidized bed, which is crucial for understanding the agglomeration patterns in the fluidized bed. In this model, solid fraction variation aligned with the solid pressure variation is analyzed in detail and validated with the CFD-DEM study. Moreover, this study provides a cost-effective multiphase CFD model that can be scaled up for industrial-level reactors to predict agglomeration.

Therefore, in this study, a simplified yet comprehensive CFD 3D numerical model using the Eularian–Eularian method was developed using AVL Fire advanced simulation software (2024 R1) and user-defined FORTRAN subroutines. The gas phase of the model was modeled using the inbuilt AVL Fire algorithms, while the solid phase was developed explicitly using user-defined subroutines and coupled with the software AVL Fire.

In this work, defluidization behavior of iron particles were analyzed at various temperatures in a controlled adiabatic and reduced atmosphere. It is crucial to understand the process of sintering iron particles in DRI method. When in the DRI process, sintering can also occur at lower temperatures due to chemical reactions of converting iron oxide to metallic iron [6]. This model includes a variety of temperature settings, from low to high, to analyze the micron-sized iron particle sintering in the fluidized bed.

Further, the paper provides a detailed explanation modeling procedure and the application of solid bridge force into the model in Section 2. Section 3 of this paper includes a detailed description of the geometry of the reactor and other parameters. Section 4 covers the numerical results achieved with this model and validation of the results with CFD-DEM model results produced for the same experiment. Section 5 outlines the conclusions drawn from the previous section of results and discussion and the intended future work of this model.

2. Model Description

Within the framework of this model, the solid phase is explicitly simulated by means of user-defined subroutines, and it is coupled with the AVL fire gas phase simultaneously. Primarily, the model is developed to predict the solid agglomeration based on the sintering force estimations for four different temperatures and three different sintering force factors. Further, the modeling of solid phases is comprised of several other sub-models that have been developed by a number of authors and validated using experimental data [21,22].

2.1. Modeling of Bed Hydrodynamics

The equations of continuity and momentum are solved for both gas and solid phases with the Eulerian–Eulerian modeling approach. Equations (1) and (2) depict the continuity equations for gas and solid, illustrating mass flow from and into the controlled volume, as well as the net mass transfer in and out of the control volume. Equations (3) and (4) depict momentum transfer for gas and solid phases, encompassing all factors associated with momentum transfer into and out of a regulated volume.

$$\frac{\partial(\varepsilon_g \rho_g)}{\partial t} + \nabla \cdot (\varepsilon_g \rho_g u_g) = S_{gs} \tag{1}$$

$$\frac{\partial(\varepsilon_s \rho_s)}{\partial t} + \nabla \cdot (\varepsilon_s \rho_s u_s) = S_{sg} \tag{2}$$

$$\frac{\partial(\varepsilon_g \rho_g u_g)}{\partial t} + \nabla \cdot (\varepsilon_g \rho_g u_g u_g) = -\varepsilon_g \nabla P + \nabla \cdot \overline{\overline{\tau_g}} + S_{mom}(u_g - u_s) + \varepsilon_g \rho_g g + S_{gs} \tag{3}$$

$$\frac{\partial(\varepsilon_s \rho_s u_s)}{\partial t} + \nabla \cdot (\varepsilon_s \rho_s u_s u_s) = -\varepsilon_s \nabla P + \nabla \cdot \overline{\overline{\tau_s}} - S_{mom}(u_g - u_s) + \varepsilon_s \rho_s g + S_{sg} \tag{4}$$

Interphase Momentum Transfer

Drag force calculation is conducted using two approaches in this study. The calculation is based on Equations (5) and (8), utilizing the solid volume percentage in each cell of the domain. The flow resistance is calculated using the empirical pressure drop correlation proposed by Ergun [21] when the solid volume fraction exceeds 0.2 ($\varepsilon > 0.2$, dense regions).

The Gidaspow model [21] is utilized to determine the pressure drop for regions with a solid volume fraction below 0.2 ($\varepsilon < 0.2$, dilute regions). The coefficients of permeability and inertial are denoted by η and γ, computed in Equations (6) and (7), respectively. They are utilized to determine the S_{mom} for the dense phase in Equation (5). The gas phase flow resistance factor is added to the solid phase momentum equation with a sign that is opposite. Equations (9) and (10) determine the drag coefficient of the gas phase and the Reynolds number of the gas flow.

$$S_{mom} = -\left(\frac{\mu_g}{\eta}|v_\infty| + \gamma\frac{1}{2}\rho_g v^2\right) \qquad (\text{for } \varepsilon_s > 0.2) \tag{5}$$

$$\eta = \frac{\psi^2 d_{eq}^2}{150}\frac{(1-\varepsilon_s)^3}{\varepsilon_s^2} \tag{6}$$

$$\gamma = \frac{3.5}{\psi d_{eq}}\frac{\varepsilon_s}{(1-\varepsilon_s)^3} \tag{7}$$

$$S_{mom} = \frac{3}{4}C_d\frac{\varepsilon_g\rho_g|u_g - u_s|\varepsilon_s}{d_s}\varepsilon_g^{-2.65} \qquad (\text{for} \varepsilon_s \leq 0.2) \tag{8}$$

$$C_d = \begin{cases} \frac{24\left(1+0.15Re^{0.687}\right)}{Re}, & Re \leq 1000 \\ 0.44, & Re \geq 1000 \end{cases} \tag{9}$$

$$Re = \frac{\rho_g|\boldsymbol{u}_g - \boldsymbol{u}_s|\varepsilon_g d_s}{\mu_g} \tag{10}$$

2.2. Cohesive Force Model

The particle agglomeration model requires a model of the chosen interparticle cohesive force and adding it to the momentum equation in addition to the other main forces. The contact force can be an electrostatic force, a liquid bridge force, etc. In this model, micron-sized iron powder is combusted for different temperatures. Therefore, solid bridge force is recognized as the major contributor to the particle agglomeration in the DRI method in a fluidized bed. When two or more particles are bonded together at a temperature lower than that particle's melting point, the solid bridge force emerges among those agglomerated particles. The DRI process is carried out at elevated temperatures of more than 500 °C. These temperatures cause the outer layer of the iron oxide particles to become adhesive and malleable. Hence, the environment favorable for the particle sintering is created in the DRI method when in higher temperatures. According to Knight et al., surface diffusion is driven by the atoms of the particles moving to the junction of the solid bridge [23]. Takafumi et al. explain the surface tension mechanism in detail based on the balance between the buoyancy force of the bubbles and the breaking force of the solid bridge [24]. Equation (11) represents the solid bridge force between two particles.

$$F_{sb} = \pi\sigma x^2, \tag{11}$$

where tensile strength and the neck radius are represented by σ and x, respectively. The neck radius is modeled by using Kuczynski's surface diffusion model, which is explained in Equation (12) [25].

$$x = \left(\frac{56\gamma\delta^4}{KT}D_s a^3 t\right)^{1/7}, \tag{12}$$

where the variables $\gamma, \delta, K, T, a,$ and t stand for surface tension, lattice constant, Boltzmann constant, temperature, curvature radius, and contact time, respectively. The surface diffusion coefficient (D_s) in Equation (13) is determined by the frequency factor ($D_{o,s}$) and activation energy (E_s).

$$D_s = D_{o,s}\exp\left(\frac{-E_s}{RT}\right) \tag{13}$$

Moreover, a general equation for solid bridge force modeling can be derived using Equation (14) [12,26].

$$F_{sb} = C\pi\sigma\left(\frac{56\gamma\delta^4}{KT}\,D_s r_p^3 t\right)^{2/7}, \tag{14}$$

where r_p is the minimum radius value of two particles in contact.

In this model, solid bridge force is modeled in three different impact models by varying the C value in Equation (14). The model for value C = 0.417 is called the three-micro contact-point model. Similarly, C = 1 represents the smooth surface model, and C = 1.251 refers to the nine-microcontact-point model. This classification for the model for different intensities of sintering force is similar to the CFD-DEM study conducted by Liu et al. [26].

Another key parameter in Equation (14) is the contact time, t. Initially, the contact time is zero. Then, if the consecutive cells have similar velocities within the (+/−5%) range, solid mass in those cells is said to be in contact, and the contact time is added to the stored contact time of the previous time step. If the velocities fall out of the boundary, the contact time remains the same.

3. Experimental Setup and Grid

3.1. Setup

Figure 1 represents the simulation setup used by Liu et al. for their CDF-DEM modeling [26]. In this model, a grid with the same dimensions is also applied as a cylindrical reactor with a 1.6 cm diameter and 4.8 cm height. The reactor has an inner diameter of 1.6 cm, while the height of the cylinder is maintained at 4.8 cm and 9.6 cm for the simulation. In the experiment, three distinct diameters of particles were used. However, Liu et al.'s coarse-grained method results for particle size, which is 58 μm, were used in this study. The total weight of the 0.01 kg iron particles was used to fill the bed [17].

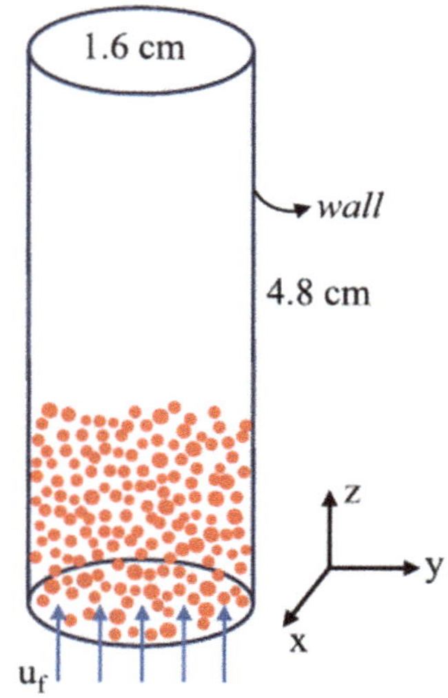

Figure 1. Simulation domain [26].

Grid independence was analyzed for three different grids depending on the number of cells. The number of cells in grid 1, grid 2, and grid 3 are 12,200, 22,000, and 35,000, respectively. The average values of flow velocities taken from cross-sectional cuts at different reactor heights from the bottom of the reactor are compared and presented in Figure 2. The variation of these parameters in the three different meshes was found to be reasonably acceptable and within the expected range. Therefore, grid 2 was selected as the mesh in the simulations of this study.

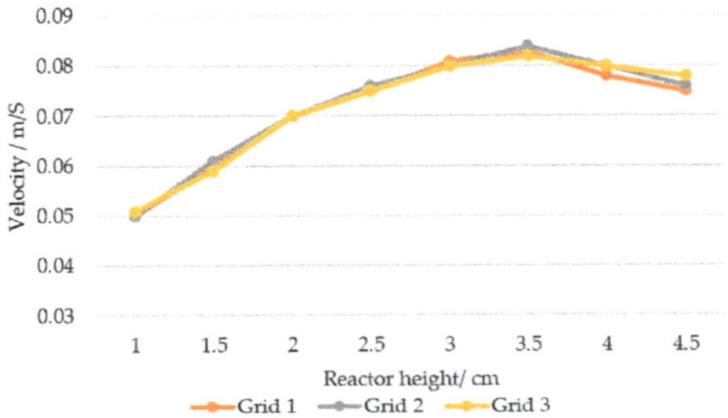

Figure 2. Grid sensitivity.

3.2. Settings and Parameters

The simulations were carried out in a reduced environment by feeding inert gas (N_2) from the bottom of the gasifier uniformly into the bed. Cylinder walls were given zero gradient boundary conditions. Environmental conditions were set to ambient pressure settings. All the simulation settings associated with iron combustion were taken from the CFD-DEM model, which is listed in Table 1 [26]. Tensile strength was taken from an existing analysis performed by Hidaka et al. to determine tensile strength variation with differential temperatures [27]. Gas phase parameters were given according to the AVL Fire in-built algorithms that are specific to temperatures. The simulations were performed for a maximum of 20 s with a time step of 0.001 s. Simulation runs were conducted at four different temperatures, from 500 °C to 1000 °C. Benchmark simulations for zero agglomeration (C = 0) were conducted for all four temperatures for the purpose of comparison. Then, the three-micro contact-point model l (C = 0.417), smooth-surface model (C = 1), and nine-micro contact-point model (C = 1.251) were performed as agglomeration force models.

Table 1. Physical properties and settings used in the simulations [26].

Particle Properties	Value	Source
Coefficient of restitution e	0.9	
Coefficient of friction μ	0.5	
Density ρ_p	5240 kg/m^3	
Lattice constant δ	1.38×10^{-9} (m)	
Surface tension γ	1 N/m	[28]
Frequency factor $D_{o,s}$	2.8×10^{-13} (≤900 °C) 1.6×10^{5} (>900 °C) m^2/s	[29]
Activation energy	1.74×10^{5} (≤900 °C) 5.79×10^{5} (>900 °C) J/mol	[29]

4. Results and Discussion

This section examines the findings of simulated cases that were carried out with varying degrees of sintering force and temperatures during the process. The primary focus of this investigation was on the downward trend in the pressure drop of the solid bed surface with the presence of a sintering force. Further, visual representations of the solid fraction and solid velocity are also presented. The features of (de)fluidization of combusted iron powder at increased gas velocities are presented and evaluated in this section.

4.1. Minimum Fluidized Velocity and Bed Conditions

To validate the CFD model for solid hydrodynamics of combustion iron powder, the minimum fluidization velocity (u_{mf}) must be estimated. Minimum fluidization velocity is the minimum velocity required to make the bed fluidized by keeping the upward drag force of the gas equal to the total solid bed weight. This is one of the most crucial elements in fluidized bed combustion. Numerical simulations were run to estimate the u_{mf} with an N_2 environment by varying the flow rate at a constant temperature of 20 °C.

Minimum fluidized velocity (u_{mf}) for this simulation setup was found to be 0.01 m/s, while 0.15 m/s was found to be excessive based on the range of simulations carried out from 0.001 m/s to 0.15 m/s. Solids climbed to the top of the gasifier at a velocity of 0.15 m/s, suggesting that the velocity was too high. On the other hand, the velocity of 0.001 m/s resulted in non-fluidized behavior because the velocity was not enough to fluidize the bed. This result for minimum fluidization velocity estimation aligns with DEM simulations conducted by Liu et al., indicating that 0.01 m/s is the minimal fluidization velocity for this arrangement, though there are some discrepancies with theoretical calculations derived using Beestra's correlation [26]. The variation of the particle size distribution can be attributed to the slight variation in theoretical and numerical results [17,30].

Moreover, in the CFD-DEM study, 0.15 m/s is used to analyze the effect of fluidization velocity on particle agglomeration. However, 0.15 m/s was found to be excessive for this simulation setup. This distinction can be ascribed to the fact that these simulations were carried out using the Eularian–Eularian two-fluid approach, in which the solid is considered as a fluid, but the CFD-DEM method is based on the Lagrangian approach, which considers solid as discrete particles. However, in general, the simulation outcomes reasonably agree with the outcomes of the other referenced literature mentioned above, providing evidence for the applications of accurate hydrodynamic theories.

4.2. Solid Bed Hydrodynamics

In a proper bubbling fluidized bed, bubbles originate at the bottom of the solid bed and rise to the top, increasing in bubble size until their buoyancy force exceeds the solid drag force. The buoyancy force rises as the bubble diameter increases, causing the bubbles to climb to the surface of the solid substrate and burst. This situation recurs in the fluidized bed reactor to enhance fuel mixing capabilities for improved combustion. As the bubble ascends, solid particles move, resulting in the formation of zones with lower solid fraction. However, when agglomeration occurs, it affects the bubble growth in the fluidized bed, driving the bed toward defludization. This numerical model has the capability to analyze the strength of the solid bridge force for various temperatures and four different agglomeration force models.

Figure 3 illustrates the normalized bed pressure drop profiles over time for each sintering force model temperature varying from 500 °C to 1000 °C. The standard deviation (STD) of the pressure drop curves was calculated from 0.5 s to 15 s. At low temperatures (e.g., 500 °C), the cohesive force impact was observed to be negligible in any sintering force model. This can be identified in the snapshots of Figure 4, as both cohesive and non-cohesive cases have healthy bubbling behavior at 500 °C. Comparatively, the standard deviation was increased in the three-micro contact-point model (C = 0.417) at 650 °C compared with the 500 °C model cases despite there being no agglomeration visible in Figure 4b. This is because even with reduced sintering force, the bed may have difficulty maintaining bubbling at temperatures exceeding 650 °C. Additionally, when C = 0.417 (three-micro contact-point model) at 1000 °C, a consistent reduction in pressure drop is observed after 8 s. It is further demonstrated by the snapshots in Figure 4d. Images captured at 1000 °C with a C value of 0.417 display blue areas indicating minimal solid presence and channeling caused by particle sintering. The void fraction snapshots from the CFD-DEM model performed by Liu et al. provide similar evidence to this discussion. However, there are some distinctions in the three-micro contact-point model (C = 0.417) at 1000 °C in the CFD-DEM study, as it shows good bubbling hydrodynamics in the CFD-DEM study even with a higher STD value. In this study, agglomeration occurred after 8 s, which shows significant pressure reduction. To illustrate that behavior, 15 s of run time was maintained in this study [26].

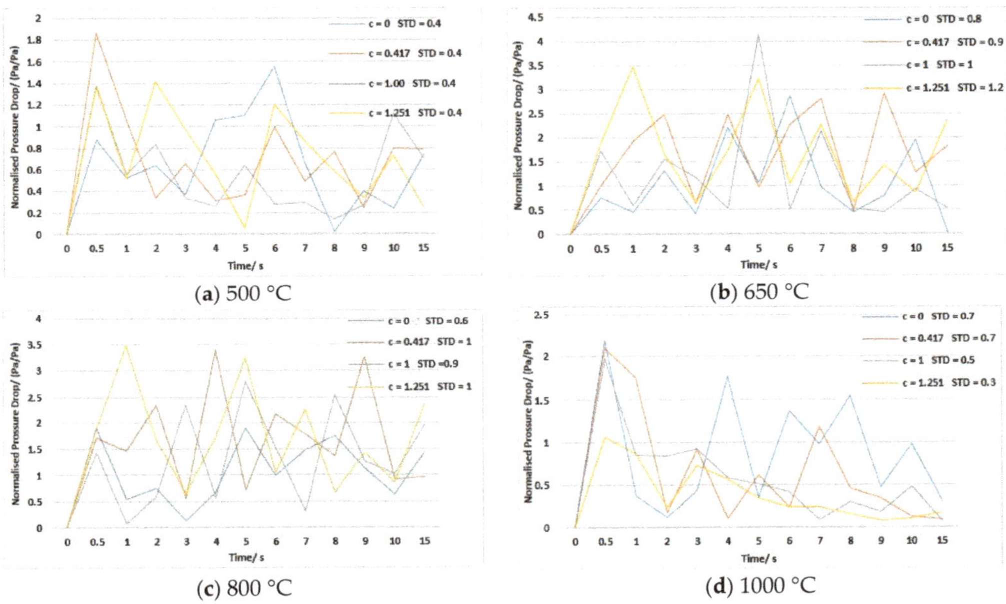

Figure 3. Pressure drop variation with time.

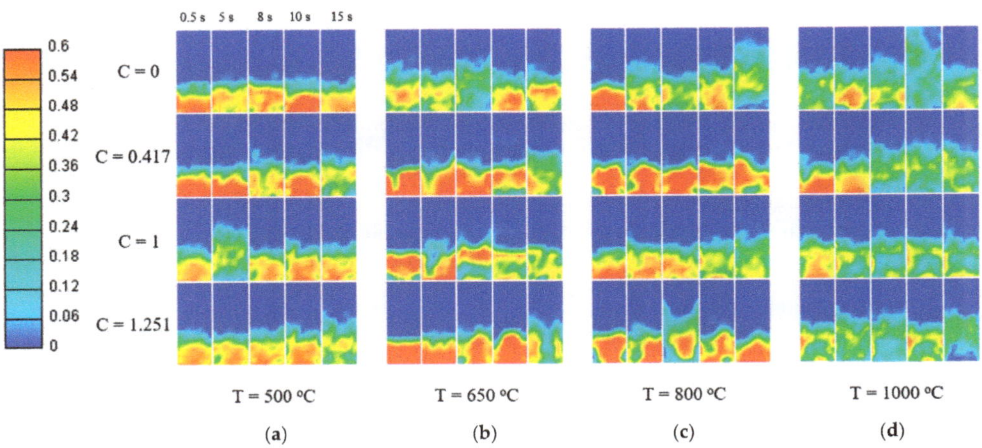

Figure 4. Solid fraction variation.

In sintering force models at higher temperatures (i.e., 800 °C and 1000 °C), the fluctuations of the pressure drop increased gradually, showing higher standard deviation

(STD) values. This notable variation proposes an increase in bubble sizes, resulting in more extreme bed dynamics. The sintering force models of a smooth surface (C = 1) and the nine-microcontact-point model (C = 1.251) had significant differences in terms of the magnitude (approximately 2.4 to 3 times) compared with the three-micro contact-point model (C = 0.417). Hence, the rate of particle clumping was high when in higher sintering force models ($C \geq 1$), leading to inhomogeneity in the bed. This scenario can be well captured in a three-micro contact-point model sintering model at 1000 °C. Then, when the sintering force was further increased by applying the smooth surface model (C = 1) and the nine-microcontact-point model (C = 1.251) at the maximum temperature used in this study (i.e., 1000 °C), the pressure drop and the standard deviation (STD) decreases rapidly, clearly indicating particle agglomeration in Figures 3d and 4d. Despite the similar trend observed in the CFD-DEM study, the sintering bonds remain permeable in the CFD-DEM cases, whereas in this model, some solid movements were still visible. This model demonstrates the occurrence of cohesive bonding, where both the formation and dissolution of bonds take place. However, it is important to note that healthy bubbling is entirely halted at 1000 °C in both the smooth surface and nine-micro contact-point model. These arguments are supported by Figure 5, as solid velocities decrease with the increase of sintering force impact and with the rise of temperature.

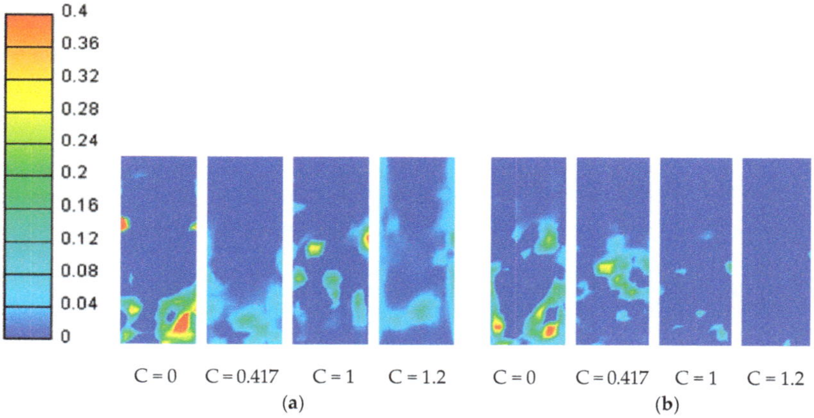

Figure 5. Solid velocity (**a**) 800 °C and (**b**) 1000 °C.

In Figure 3c, a decrease in bed height and flow inhomogeneity can be observed in sintering force models, especially when increasing the impact of sintering force at 800 °C. This is attributed to the impact of agglomeration, which is noticeable even without complete defluidization.

A reduction in the bed expansion can be seen Figure 4d from three-micro contact point model after 8 s. Further, a significant reduction in bed expansion is observed in Figure 4d for the smooth and nine-micro contact-point models. This was also clearly observed in the CFD-DEM study.

The creation of significantly larger agglomerates and the strong adhesion of particles to the bed wall within a few seconds is observed, as shown in Figure 4c,d, which depicts smooth and nine-micro contact-point model situations, respectively. In Figure 4c, in the smooth and nine-micro contact-point models, making and breaking agglomerates can be seen closer to the wall after 10 s. In Figure 4d, for sintering force cases, the formation of agglomerates can be observed. This behavior is indicated in the smooth and nine-micro contact-point models due to the high intensities of the sintering force. Agglomeration can be observed clearly in those images of smooth and nine-micro contact-point model situations at 1000 °C, which suggests that the solid bridge force that was forecasted with this model at this temperature is very high. This observation, similar to the CFD-DEM study results, suggests that this two-fluid model reasonably accurately predicts the particle agglomeration leading to defluidization in fluidized bed combustion.

In the CFD-DEM work, it is demonstrated that there is no movement in any of the particles in the smooth and nine-micro contact-point model situations when the temperature is 1000 °C, which indicates that the defluidization pressure drop remains constant. Nevertheless, this study demonstrates that even in smooth and nine-micro contact-point model cases at a temperature of 1000 °C, there is little motion in solid particles, which indicates that there are slight differences in pressure drops, and certain solid particles have low-velocity values. There is evidence of this in the pressure drop graphs shown in Figure 3d and the solid velocity snapshots shown in Figure 5b. A potential reason for this variance is that the different methods of calculating contact time that were discussed in the previous section can cause it.

The results of this model reasonably reflect the observation of the experiment and CFD-DEM findings. It is noticed that, in general, low-temperature runs (500 °C and 650 °C) could operate continuously for more than two hours without any defluidization appearance. Nonetheless, the experiments conducted at 950 °C were soon found to be non-fluidized behavior within several minutes [17].

Moreover, Figure 5b depicts that the solid toward the bottom of the bed has lower velocities than the top in agglomerated models, whereas when C = 0, the bottom of the bed has higher velocities than the top. According to the experimental findings, non-fluidized zones initially started from the bottom of the bed. Thus, it is fair to say that the smooth-surface model and the nine-micro contact-point model provide better predictions of solid sintering force, which is closely related to the experimental observations. The neck size of the solid bridge grows with the particle temperature and the contact time according to Equation (14). Particle cohesion begins from the lower part of the bed, particularly at the beds' corners, because of the high density, as depicted in Figure 4c,d. Then, with the contact period of the particles rising, more particles tend to accumulate into the existing agglomerates, increasing the size of the agglomerates [17].

4.3. *Effects of the Gas Superficial Velocity*

Defluidization occurs due to the interaction between the sintering force (solid bridge force) and the breakage force in the reactor. Combining forces of drag force and the particle collision force establishes the breakage force needed to break down the agglomerated particles. Increasing the gas's superficial velocities can provide the breakage force to separate particles, reducing the contact time. Thus, high fluidization velocities have the ability to break the particles that stick together. Another set of simulations was performed to study the effect of gas superficial velocity on particle agglomeration by increasing the velocity to 0.12 m/s. The simulations were carried out for a smooth surface (C = 1) and a nine-micro contact-point model (C = 1.251) at 1000 °C.

When the gas superficial velocity is increased to 0.12 m/s, the bed seems more dynamic for both the smooth and nine-micro contact-point cases model compared with the previously stated scenarios. After 13 s, the smooth surface model bed begins to lose its fluid-like properties, whereas the nine-micro contact-point model starts to do so after 9 s.

After 9 s of the nine-micro contact point model, the sintering region has expanded from the bottom of the bed over the whole bed, indicating almost defluidization. The results indicate that higher gas surface velocities can lead to an increased gas flow through the bed, resulting in the formation of big bubbles and higher collision frequencies between particles. Consequently, the decreased contact time between colliding particles weakens the solid bridge, potentially delaying defluidization. However, defluidization could still be observed in those simulations, as indicated in Figure 6, the pressure drop curve, and the snapshots of solid fraction in Figure 7. This phenomenon is further proved by the solid velocity snapshots depicted in Figure 8.

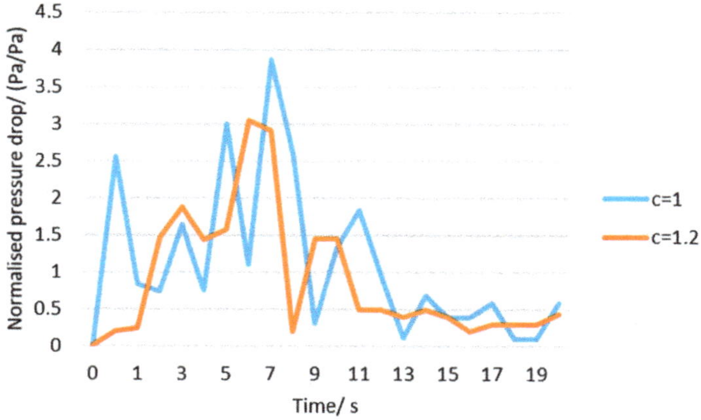

Figure 6. Pressure drop variation for increased velocity V = 0.12 m/s.

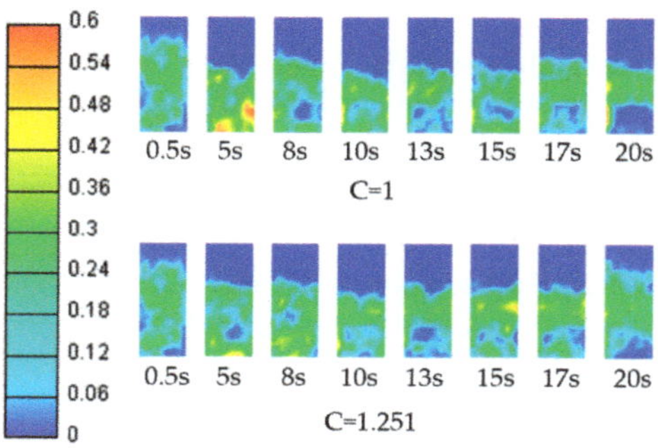

Figure 7. Solid fraction variation for V = 0.12 m/s.

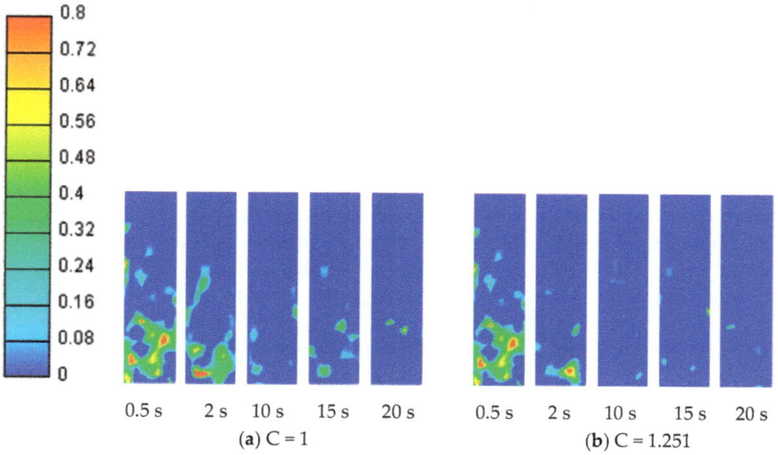

Figure 8. Solid velocities/ms^{-1}.

5. Conclusions

The agglomeration hydrodynamics of the iron powder particles in a fluidized bed were analyzed using a CFD numerical model. The model was developed with a simplified approach of a two-fluid method involving the implementation of the user-defined subroutines written by the authors. The conclusions of the study are as below:

(1) An Eulerian–Eulerian multiphase model, where the solid phase is explicitly simulated, was implemented through user-defined subroutines and is successfully extended to cohesive dense gas-solid flows by including temperature-dependent solid bridge (sintering) force.
(2) The iron fine powder agglomeration due to the sintering force developed in the model qualitatively agrees with the experiment and CFD-DEM studies. The solid temperature of the iron particles was found to be a significant parameter in estimating the sintering force effect. Higher gas superficial velocities can delay the defluidization in the fluidized bed combustion. However, the bed still leads toward the defluidization regardless of the gas's superficial velocity.
(3) It was observed that there is a significant impact on particle cohesion of the smooth surface/nine-micro contact-point model (C = 1/1.251) model (2.4/3 scaling factor) results in higher temperatures, which is closely aligned with the realistic phenomena. This behavior is similar to that of the CFD-DEM study of the same experiment. Therefore, the smooth surface/nine-micro contact-point model sintering force model is more suitable for representing the true phenomena observation in the study of combusted iron powder.
(4) The normalized pressure drop curve follows a similar pattern for each temperature and model type similar to the experiment. With the temperature and sintering force factor increase, the higher pressure drop variation could be observed attributing to higher standard deviations. Nevertheless, the standard deviation of the pressure drop significantly drops in clear agglomeration cases at higher temperatures.
(5) The study proposed a cost-effective, simplified CFD model for predicting particle agglomeration in fine iron particles. The results of the model reasonably agree with the other referenced CFD-DEM work along with the experiment observations. Future work can lead to the implementation of the model on a larger scale to minimize cost.

Author Contributions: Conceptualization, W.D.S.F.; Software, J.N.; Investigation, W.D.S.F.; Resources, J.N.; Data curation, J.N.; Writing—original draft, W.D.S.F.; Writing—review & editing, W.D.S.F. and J.N. All authors have read and agreed to the published version of the manuscript.

Funding: This research received no external funding.

Institutional Review Board Statement: Not applicable.

Informed Consent Statement: Not applicable.

Data Availability Statement: The original contributions presented in the study are included in the article, further inquiries can be directed to the corresponding author.

Conflicts of Interest: The authors declare no conflict of interest.

References

1. Kim, J.; Sovacool, B.K.; Bazilian, M.; Griffiths, S.; Lee, J.; Yang, M.; Lee, J. Decarbonizing the iron and steel industry: A systematic review of sociotechnical systems, technological innovations, and policy options. *Energy Res. Soc. Sci.* **2022**, *89*, 102565. [CrossRef]
2. Julien, P.; Bergthorson, J.M. Enabling the metal fuel economy: Green recycling of metal fuels. *Sustain. Energy Fuels* **2017**, *1*, 615–625. [CrossRef]
3. Ning, D.; Shoshin, Y.; van Stiphout, M.; van Oijen, J.; Finotello, G.; de Goey, P. Temperature and phase transitions of laser-ignited single iron particle. *Combust. Flame* **2022**, *236*, 111801. [CrossRef]
4. Poletaev, N.I.; Khlebnikova, M.Y. Combustion of Iron Particles Suspension in Laminar Premixed and Diffusion Flames. *Combust. Sci. Technol.* **2022**, *194*, 1356–1377. [CrossRef]
5. Guo, Y.; Ren, K.; Wei, A.; Tao, C.; Huang, W.; Zhao, P.; Wu, D. Iron dust explosion characteristics with small amount of nano-sized Fe_2O_3 and Fe_3O_4 particles. *Fuel* **2022**, *324*, 124786. [CrossRef]
6. Liu, X.; Zhang, X.; Li, J.; Zhu, Q.; Deen, N.G.; Tang, Y. Regeneration of iron fuel in fluidized beds Part II: Reduction experiments. *Powder Technol.* **2023**, *420*, 118183. [CrossRef]

7. Hessels, C.J.M.; Homan, T.A.M.; Deen, N.G.; Tang, Y. Reduction kinetics of combusted iron powder using hydrogen. *Powder Technol.* **2022**, *407*, 117540. [CrossRef]
8. Guo, L.; Bao, Q.; Gao, J.; Zhu, Q.; Guo, Z. A Review on Prevention of Sticking during Fluidized Bed Reduction of Fine Iron Ore. *ISIJ Int.* **2020**, *60*, 1–17. [CrossRef]
9. Liu, X.; Deen, N.G.; Tang, Y. On the treatment of bed-to-wall heat transfer in CFD-DEM simulations of gas-fluidized beds. *Chem. Eng. Sci.* **2021**, *236*, 116492. [CrossRef]
10. Hartley, P.A.; Parfitt, G.D.; Pollack, L.B. The role of the van der Waals force in the agglomeration of powders containing submicron particles. *Powder Technol.* **1985**, *42*, 35–46. [CrossRef]
11. Staniforth, J.N. Ordered mixing or spontaneous granulation? *Powder Technol.* **1985**, *45*, 73–77. [CrossRef]
12. Kuwagi, K.; Mikami, T.; Horio, M. Numerical simulation of metallic solid bridging particles in a fluidized bed at high temperature. *Powder Technol.* **2000**, *109*, 27–40. [CrossRef]
13. Seville, J.P.K.; Willett, C.D.; Knight, P.C. Interparticle forces in fluidisation: A review. *Powder Technol.* **2000**, *113*, 261–268. [CrossRef]
14. Seville, J.P.K.; Tüzün, U.; Clift, R. *Processing of Particulate Solids*; Springer: Berlin/Heidelberg, Germany, 1997.
15. Barustan, M.I.A.; Jung, S.-M. Morphology of Iron and Agglomeration Behaviour During Reduction of Iron Oxide Fines. *Met. Mater. Int.* **2019**, *25*, 1083–1097. [CrossRef]
16. Oh, J.; Noh, D. The reduction kinetics of hematite particles in H_2 and CO atmospheres. *Fuel* **2017**, *196*, 144–153. [CrossRef]
17. Liu, X.; Zhang, X.; Li, J.; Zhu, Q.; Deen, N.G.; Tang, Y. Regeneration of iron fuel in fluidized beds, Part I: Defluidization experiments and theoretical prediction model. *Powder Technol.* **2023**, *420*, 118182. [CrossRef]
18. Tasleem, A.; Ullah, A.; Li, F.; Yi, Q.; Nimmo, W.; Daood, S.S. Detection of onset of agglomeration in a bubbling fluidized bed biomass combustor using reactive Eulerian computational fluid dynamics. *Particuology* **2024**, *90*, 504–515. [CrossRef]
19. Mansourpour, Z.; Mostoufi, N.; Sotudeh-Gharebagh, R. Investigating agglomeration phenomena in an air-polyethylene fluidized bed using DEM–CFD approach. *Chem. Eng. Res. Des.* **2014**, *92*, 102–118. [CrossRef]
20. Mansourpour, Z.; Mostoufi, N.; Sotudeh-Gharebagh, R. A numerical study on agglomeration in high temperature fluidized beds. *J. Chem. Pet. Eng.* **2014**, *48*, 15–25.
21. Diba, M.F.; Karim, M.R.; Naser, J. Fluidized bed CFD using simplified solid-phase coupling. *Powder Technol.* **2020**, *375*, 161–173. [CrossRef]
22. Karim, M.R.; Naser, J. Progress in Numerical Modelling of Packed Bed Biomass Combustion. In Proceedings of the 19th Australasian Fluid Mechanics Conference, Melbourne, Australia, 8–11 December 2014.
23. Knight, P.C.; Seville, J.P.K.; Kamiya, H.; Horio, M. Modelling of sintering of iron particles in high-temperature gas fluidisation. *Chem. Eng. Sci.* **2000**, *55*, 4783–4787. [CrossRef]
24. Mikami, T.; Kamiya, H.; Horio, M. The mechanism of defluidization of iron particles in a fluidized bed. *Powder Technol.* **1996**, *89*, 231–238. [CrossRef]
25. Kuczynski, G.C. Self-Diffusion in Sintering of Metallic Particles. In *Sintering Key Papers*; Sōmiya, S., Moriyoshi, Y., Eds.; Springer: Dordrecht, The Netherlands, 1990; pp. 509–527. [CrossRef]
26. Liu, X.; Hessels, C.J.M.; Deen, N.G.; Tang, Y. CFD-DEM investigation on the agglomeration behavior of micron-sized combusted iron fines. *Fuel* **2023**, *346*, 128219. [CrossRef]
27. Hidaka, Y.; Anraku, T.; Otsuka, N. Deformation of iron oxides upon tensile tests at 600–1250 °C. *Oxid. Met.* **2003**, *59*, 97–113. [CrossRef]
28. Hadjipanayis, G.C.; Siegel, R.W. *Nanophase Materials: Synthesis-Properties-Applications*; Springer Science & Business Media: New York, NY, USA, 2012; Volume 260.
29. Atkinson, A.; Taylor, R.I. Diffusion of 55Fe in Fe_2O_3 single crystals. *J. Phys. Chem. Solids* **1985**, *46*, 469–475. [CrossRef]
30. Beetstra, R.; van der Hoef, M.A.; Kuipers, J.A.M. Drag force of intermediate Reynolds number flow past mono- and bidisperse arrays of spheres. *AIChE J.* **2007**, *53*, 489–501. [CrossRef]

applied sciences

MDPI

Article

The Assessment of the Influence of Low-Frequency Electromagnetic Fields Originated from the Power Infrastructure on Humans' Health

Leszek Sławomir Litzbarski [1,2], Marek Olesz [1,2], Grzegorz Redlarski [1,2,3,*], Piotr Mateusz Tojza [1,2], Arkadiusz Żak [1,2], Emanuel Gifuni [1,2], Zuzanna Cieślikowska [1,2] and Mieszko Czapliński [2,3]

1 Department of Electrical and Control Engineering, Gdańsk University of Technology, Narutowicza 11/12, 80-233 Gdańsk, Poland; leszek.litzbarski@pg.edu.pl (L.S.L.); marek.olesz@pg.edu.pl (M.O.); piotr.tojza@pg.edu.pl (P.M.T.); arkadiusz.zak@pg.edu.pl (A.Ż.); s194407@student.pg.edu.pl (E.G.); s194322@student.pg.edu.pl (Z.C.)

2 Fahrenheit Union of Universities, Gdańsk Zwycięstwa Ave. 27, 80-219 Gdańsk, Poland; mieszkocz4@gumed.edu.pl

3 Faculty of Medicine, Medical University of Gdańsk, M. Skłodowskiej-Curie Street 3a, 80-210 Gdańsk, Poland

* Correspondence: grzegorz.redlarski@pg.edu.pl

Abstract: The objective of this study is to assess the impact of low-frequency electromagnetic fields (LF EMFs) generated by power infrastructure on the nearby environment. Measurements of electric (E) and magnetic (H) field intensities were conducted around high-voltage power lines, transformer stations and facilities related to them. Numerical simulations were also performed to model the distribution of the field values around real buildings in close proximity to power delivery systems. Given the ongoing scientific debate regarding the effects of EMFs on living organisms, the current analysis was based on the existing standards—particularly ICNIRP 2010 guidelines, which set the maximum allowable E and magnetic induction (B) values at 5 kV/m and 200 μT, respectively. Stricter national regulations were also examined, such as Poland's 1 kV/m E limit in residential areas and Belgium's 10 μT limit for B. The results showed that while most cases complied with ICNIRP 2010 standards, certain stricter local regulations were exceeded. Specifically, 9 of 14 cases exceeded Poland's E limits, and 8 failed to meet Belgium's B requirements. Only in one place—a warehouse near 110 kV power lines (in a critical case)—the ICNIRP limit B was exceeded. These findings underscore the variability in regulatory standards and highlight the need for localized assessments of EMF exposure.

Keywords: electromagnetic field; health hazards; power infrastructure; electric power distribution; electromagnetic compatibility

1. Introduction

Over the centuries, humans as biological organisms have lived and developed in an environment in which the electric field (EF) and magnetic field (MF), which are part of the electromagnetic field (EMF), have played and still play a very important role, presenting two pillars sustaining life on Earth. However, since the industrial revolution and the emergence of modern electricity, the inevitable technical and technological progress has been gradually polluting the human environment, exposing humans to various artificial sources of EMFs of the intensities unknown in nature. Further discoveries of electromagnetism and electromagnetic waves attributed to such great names as Oersted, Ampere, Faraday and Maxwell and the invention of the radio as a means of distance communication began the era of the omnipresence of EMFs in the human environment. Nowadays, humans as biological organisms are surrounded by different EFs and MFs emitted from numerous sources of various intensities, such as electrical machines, devices or domestic appliances, not to mention computers, mobile phones, etc.

One of the side effects of each working electrical device is EMFs generated near their workplaces. These fields interfere with various elements of the environment, among which all living organisms are of the greatest concern. Therefore, it is important to accurately determine the nature and the side effects of its impact. Every day, living organisms, including humans, are exposed to the influence of different types of fields of different physical parameters, at a wide frequency (f) range. Electrical devices such as mobile phones, microwave ovens, television or radio transmitters emit electromagnetic radiation (EMR) with a frequency f from 300 MHz to 300 GHz. A portion of this radiation is converted into kinetic energy, which then transforms into heat. This heat can negatively affect health in various ways [1]. On the other hand, the lower frequencies from the radio spectrum frequencies from 100 kHz to 300 MHz range do not cause such phenomena. What should be pointed out is that, during the radiation of high-frequency fields, i.e., above 100 kHz, accompanied with the fields of high intensities, non-ionizing radiation phenomena occur [2,3]. Another type of radiation can be found in the case of low-frequency fields caused by, e.g., power lines (f = 50 Hz or 60 Hz). It is believed that such fields have a quasi-stationary character, and the two components of EMFs, i.e., the electric field (EF) and the magnetic field (MF), can be considered separately [2]. Nowadays, the opinions of the research community about the effects of radiation attributed to different fields on living organisms are divided. This is due to the fact that earlier studies have not unambiguously shown a negative impact of EMFs. Already at the beginning, at the turn of the XIX and XX centuries, there was a widespread view (e.g., in medical textbooks) of the positive impact of EFs and MFs on humans [2]. In the 1950s and 1960s, with the development of science trying to describe the radiation phenomena, this view gradually disappeared, and the view about the neutral effect of EMFs on living organisms began to dominate [2]. The first study that reported the potentially harmful effects of MFs on living organisms was an epidemiological study published in 1979 by Wertheimer and Lepper [4]. In this study, children from the city of Denver, CO, USA, who lived in homes exposed to MFs of higher intensities were tested. This intensity was estimated based on the number of transmission lines near homes and the number of power lines transmitting electricity to homes. It was noted that children exposed to MFs had a "slightly higher" risk of developing leukemia compared to other children who were not exposed to such radiation. The authors developed their own code [5] in which they managed to visually determine the level of this exposure (based on the number of lines in the area of residence). The result of the study raised a lot of controversy, mainly because of the method used in the research, which was very subjective. Nevertheless, it caused an increased interest among researchers in the subject of the influence of EMFs on living organisms, including humans. It should be pointed out that further tests confirmed the validity of the charges against Wertheimer and Lepper after examining children of Rhode Island [6], where no connection between the influence of EMFs and the increase in morbidity was observed. However, different results were obtained by researchers in France and Corsica, who found that people exposed to EMF radiation may be affected by an increased incidence of brain tumors [7]. What is more, it is suggested to be more cautious when it comes to children, as it affects the young nervous system in a more sophisticated manner than an adult one [8]. The National Institute of Environmental Health Sciences [9] modified the method of the analysis by considering the effect of MFs on the basis of the health cards of children, instead of the methodology employed by Wertheimer and Lepper in their study [4], which was based on death certificates. In this study, no negative effects of EMF radiation on health were observed.

However, disturbing results were published by researchers from Sweden [10], who, after examining children up to 16 years of age who, in the 1960s and 1985, lived at a distance of no more than 300 m from power lines (220 kV or 400 kV), observed that the incidents of leukemia were 2.4 times greater than among their peers. Very surprisingly, at the same time, quite opposite results were obtained by researchers from Denmark [11] and Finland [12], who under similar assumptions found no such effects.

A study carried out in California concentrated on the correlation between the distance from the power lines and the effect of magnetic fields on children [13]. The researchers found that neither close proximity to high-voltage lines alone nor exposure to high calculated fields was associated with childhood leukemia.

As we can notice, there have been many studies from 1979 to the present reporting a raised risk for childhood leukemia and tumors with exposure to power-frequency magnetic fields. There are also suggestions that the reported risk is decreasing [14], but there have not been consistent data due to the fact that the results may have been affected by the source of funding [15].

This article considers typical values of electric and magnetic field intensities occurring in the vicinity of power infrastructure. Both direct E and H measurements and their values estimated from numerical simulations for real cases are presented. They all are compared with the limits known from standards and legal regulations in force in various countries.

2. Materials and Methods

Experimental measurements of B were taken using the Spectrum Analyzer Aaronia model NF-5035 (Calderara di Reno, Italy), which is equipped with a three-axis sensor and has the following characteristics: measurement range from 1 pT to 0.1 mT (in the f range from 1 Hz to 1 MHz), measurement error 3%. The value of H was estimated from the well-known equation

$$H = \frac{B}{\mu_0}$$

where $\mu_0 = 4 \cdot \pi \cdot 10^{-7}$ H/m is the magnetic permeability of free space, whose value is close to its value for air. In order to determine the value of E, the Spectrum Analyzer Aaronia model NF-5035 was used, which is equipped with a one-axis sensor. This device has a measurement error of 3% in the range from 0.1 V/m to 20 kV/m. All measurements were taken using an instrument placed on a tripod insulated from the ground. The person reading the measurement results stood about 1 m away from the instrument to reduce the impact of their presence on the field distribution. Measurements were conducted at locations reported by employers or investors, which were situated near power lines and electrical installations. The electromagnetic field measurements were performed outdoors or in enclosed spaces—residential buildings, industrial facilities and warehouses—in order to obtain approval for human occupancy for 8 h (industrial areas) or continuous presence (residential locations).

The actual values of E and H were determined using the finite element method by the use of the FEMM 4.2 software for specific boundary conditions. The use of numerical simulations to calculate the distribution of electromagnetic fields has been described in detail, among others, in [16].

3. Results

The results presented in this section originate from expert opinions on potential threats related to the occurrence of EMFs generated by energy infrastructure. The analyzed cases concerned locations in Poland in the case of planned and existing facilities.

3.1. Measurements of the EMF Strength around Energy Infrastructure

In the case of existing energy infrastructure, the simplest solution is often to carry out measurements that show the actual E and B values in a given area. Measurements of actual B values are presented on the example of a 15 kV/0.4 transformer-distribution station located in the basement and on the ground floor of a residential building in the north of Poland. In order to determine the distribution of the MF in the vicinity of the mentioned transformer station, B measurements were carried out, based on which the values of H were estimated. The obtained results are presented in Figure 1. Further investigations focused on the determination of the maximum values of H in the building under consideration as well as in the apartment next to the transformer station. The extremum denoted value of H

was 5.3 A/m and 0.4 A/m, respectively. For the purposes of further analysis, exposure to the highest recorded *H* value was assumed, i.e., 15 A/m (close to the grounding wire of the power transformer).

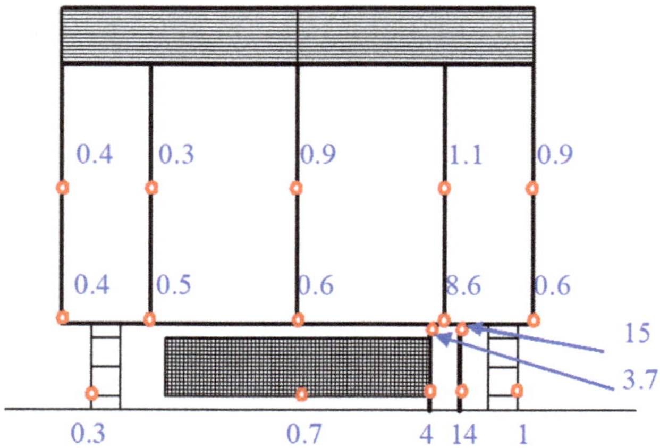

Figure 1. Experimentally measured values of the magnetic field intensity *H* in [A/m] at the 15/0.4 kV/kV transformer station. The red circles in the side view indicate the measurement points.

An example of *E* measurements is a study conducted on a building plot in northern Poland. An overhead 110 kV line of the following parameters runs over this plot: three AFL-6 240 mm^2 working cables (L1–L3) and a shield line (LR) with an OPGW 43/22mm^2/555 optical fiber with a cross-section of 96 mm^2. Knowing the parameters of the poles located in the immediate vicinity of the plot (Type P+5, Series S24, and Type ON120+5, Series S24), the minimum height of the wires above the ground was determined, taking into account the cable sag (see Figure 2).

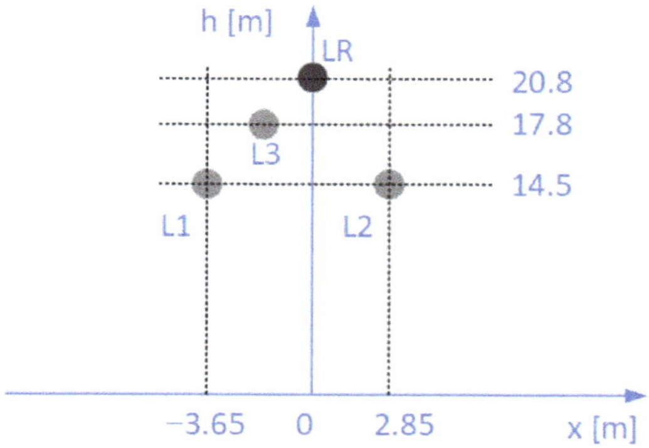

Figure 2. Distribution of cables (working cables L1, L2 and L3 and a grounding wire LR) in the 110 kV line in the middle of the span between the poles. The distances in the drawing (in meters) are given from the base of the power pole.

During measurements at a height of 2 m above the ground in the middle of the span, the results were obtained as shown in Figure 3. The coordinates on the x-axis match the data from Figure 2. The maximum value of *E* is 0.83 kV/m. It should be also mentioned that in a specific position under two line wires placed one above the other, *E* reaches a value of 0.66 kV/m due to the screening properties of EFs emitted by the wires.

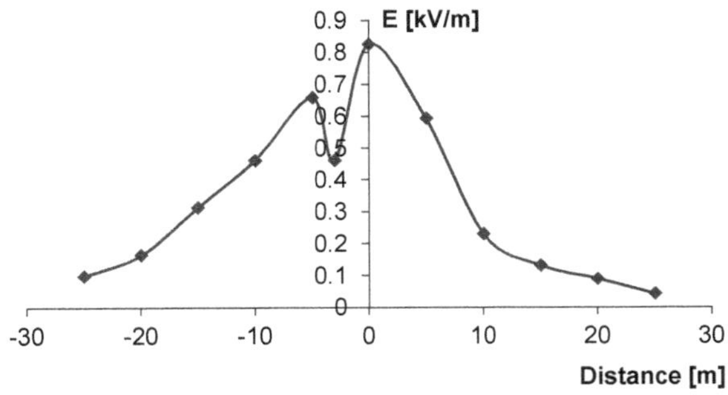

Figure 3. Experimentally measured values of the electric field intensity E under the 110 kV line.

Below is a brief description of other objects examined and the highest measured electric (E_{max}) and magnetic (H_{max}) field intensity values, which are listed in Table 1:

- A multi-family building with balconies near 110 kV overhead line: Measurements of the vertical component of E were taken under a high-voltage line at a height of 2 m above the ground on a straight line connecting the power pole with the corner of the building, near the pole and on the balcony on the fourth floor. The E values at a height of 2 m above the ground do not exceed 0.5 kV/m. Higher field intensities were measured at a distance of 2 m from the balcony grate, i.e., 1.8 kV/m, and above the balcony balustrade, 1 kV/m;
- An office space with inappropriate equipotential bonding: The highest B values were found in the room near the computer monitor screen (0.58 μT). At a distance of 0.1 m from the monitor surface, the value of B is reduced to 0.3 μT. Higher B values were noted in rooms through which the gas connection made of steel pipe passed. The probable cause of the observed phenomenon was the lack of an insulating insert on the gas connection or incorrectly made equalization connections in the building. The MFs measured directly on the pipe surface outside the building reached $B \approx 5$ μT, which is the result of the main frequency current flow;
- A room above the MV/LV indoor station: Measurements were performed on the floor surface above the room where the 15/0.4 kV transformer was installed. The highest H of about 4 A/m was found for f of 50 Hz with a current flow in the order of 200 A in the busbars. The reinforcement elements of the ceiling of the transformer room caused EF to be damped to a level of only 0.005 kV/m;
- An LV switchboard: The MF did not exceed $B \approx 20$ μT, which corresponds to H of 16 A/m. The highest field value was measured outside the switchboard cover. The H values above the top and bottom edge of the cover are 8.8 A/m and 12 A/m, respectively. On the other hand, H in front of the computer screen is 3 A/m, and at the window, it is 6 A/m. For comparison, only 0.3 A/m was measured at the entrance door to the room behind which the switchboard is located. This result clearly locates the source of the MF in the switchboard.

Table 1. The maximum measured values of electric (E_{max}) and magnetic (H_{max}) field intensity.

	Investigated Object	E_{max} [kV/m]	H_{max} [A/m]
1	Transformer-distribution station 15 kV/0.4 kV	<0.05	5.3
2	Building plot under 110 kV overhead line (only E)	0.83	-
3	Multi-family building with balconies near 110 kV overhead line	1.2	<0.2
4	Office space with inappropriate equipotential bonding (only H)	-	4
5	Room above the MV/LV indoor station (only E)	4.01	-
6	LV switchboard (only H)	-	<16

3.2. Numerical Simulations of Electric and Magnetic Field Intensities around Energy Infrastructure

The actual values of E and H depend on the configuration of the lines and objects nearby and are usually determined using finite element method calculations in specialized software. Below is an extreme case of the actual implementation of an industrial warehouse located only 4.5 m from the outermost conductors of a two-track 110 kV high-voltage line. Analysis of the field distribution in the FEMM software was performed for specific boundary conditions, i.e., with phase conductors L1 closest to the warehouse walls reaching the highest peak phase voltage value of 100.3 kV, while the remaining phases L2 and L3 reach values of (−50.1) kV. Therefore, all analysis results are given in maximum voltage values related to 1 m. Furthermore, it was assumed that the lightning rod and the construction outline of the proposed warehouse with an external conducting facade have a ground potential of 0 V. Distances between the conductors were determined according to the data for D2 ON and P series poles (Figure 4A). Due to the close proximity of the outer conductors of the overhead line to the structure of the proposed warehouse, to limit corona discharges, it is recommended to round off the side edges, especially where there is proximity to the 110 kV line. In the calculations, this rounding was set at r = 0.2 m. Figure 4B shows the distribution of E at a height of 2.0 m above ground level (E_{2m}). Because of the shielding of EF by the building structure (Faraday cage), its intensity inside is equal to 0 kV/m.

Figure 5A specifies the values of E on the roof surface and on the side wall of the warehouse. For the analyzed cases, the maximum value of E does not exceed E_{max} = 4.0 kV/m on the side walls and roof (Figure 5B) and E_{2m} = 3.0 kV/m at a height of 2.0 m above ground level.

The H value for the 50 Hz component was calculated assuming a restriction based on the occurrence of maximum current values in phases L1, L3 (track 1) and L1 (track 2) located closest to the side plane of the warehouse. According to the obtained data, instantaneous maximum current values of 888.3 A were assumed for phases L1 and L3 (track 1), corresponding to an effective value of 630 A resulting from the assumed long-term load capacity of the conductors for temperatures below 10 °C. In the remaining phases L2 and L3 (track 2), there is a negative instantaneous current value (−444.15 A). To restrict the simulation results, the current value in phase L3 was set to match that of phase L1 for the line closer to the warehouse wall (assumption of a critical asymmetry of load currents resulting in an increase in the magnetic field level). Aluminum lines were assumed, and a 3 mm thick steel sheet was assumed for the warehouse construction to ensure proper mesh generation in the FEMM program. In practice, with smaller steel thicknesses or for other materials, there the attenuation of the MF is smaller, and the field penetrates into the interior of the warehouse. In Figure 6, the distribution of H at a height of 2 m above the ground surface (H_{2m}) is shown. It is worth noting that, unlike the EF, the MF is able to penetrate the interior of the building under investigation.

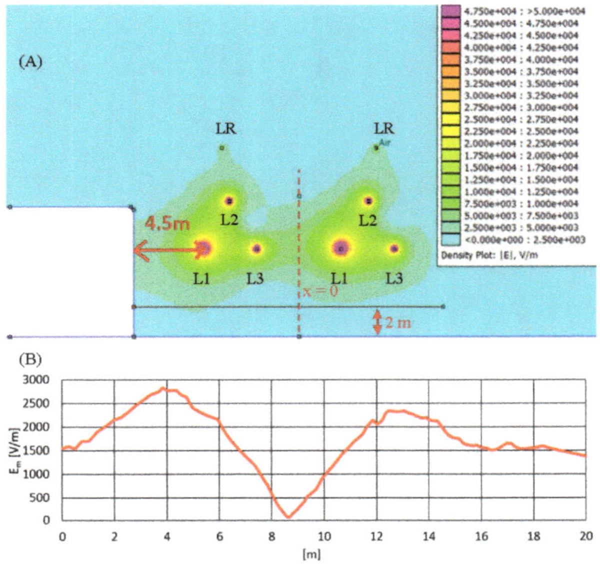

Figure 4. Results of numerical simulations for the distribution of the electric field intensity *E*: (**A**) coordinates of suspension of wires of span 51–52 in the vicinity of the warehouse (phase wires L1–L3, shield line—LR). The earth potential of 0 V was assumed on the walls and roof of the warehouse. The red line marks the height of 2 m above the ground. (**B**) Electric field level at a height of 2.0 m above the ground according to the profile given in Figure 4A) (solid red line) calculated for catastrophic conditions assuming a height of 5.8 m for the lowest conductors above the ground.

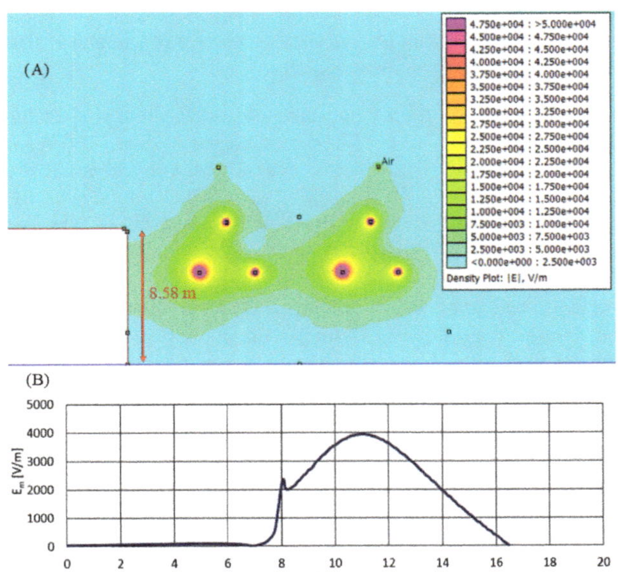

Figure 5. Results of numerical simulations for the distribution of the electric field intensity *E*: (**A**) wire distribution profile, (**B**) electric field level on the roof and side wall of the warehouse according to the profile marked in red in Figure 5A). The local maximum for the abscissa of 8 m results from the field concentration at the assumed radius of rounding the roof edge of r = 0.2 m.

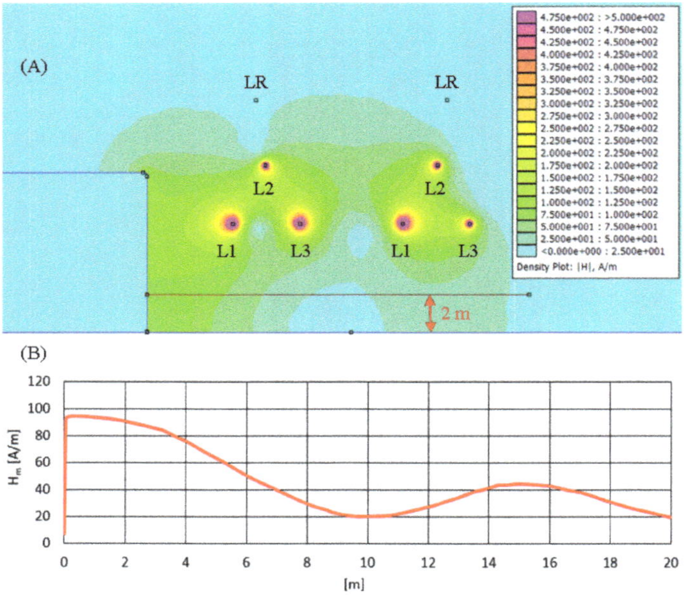

Figure 6. Results of numerical simulations for the distribution of the magnetic field intensity H: (**A**) Magnetic field distribution in the warehouse environment. Coordinates of suspension of cables of span 51–52 in the vicinity of the warehouse (phase cables L1–L3, lightning rod—LR). The walls and roof of the warehouse are made of 3 mm thick steel. The height of 2.0 m above the ground is marked with a red line. (**B**) The level of the magnetic field at a height of 2.0 m above the ground in accordance with the profile given in Figure 6A) (red line) calculated for catastrophic conditions assuming the height of the lowest cables above the ground of 5.8 m. The hall walls are made of structural steel. The maximum value of the field intensity occurs at the warehouse wall.

Figure 7A determines the H value on the roof surface and on the side wall of the warehouse. In each case, the maximum value of H does not exceed H_{max} = 125.0 A/m on the side walls and roof (Figure 7B) and H_{2m} = 96.0 A/m at a height of 2 m above the ground.

In a similar way, numerical simulations were performed for other selected objects. The obtained values of E_{max}, E_{2m}, H_{max} and H_{2m} are listed in Table 2. A short characterization of the investigated cases is described below:

- A multi-family building with balconies near double-track 110 kV overhead lines: The analysis was carried out under the following assumptions: the building on the side of the line has balconies placed, according to the design, at a minimum distance of 6 m from the line (horizontal projection). The double-circuit line is characterized by the following technical parameters of poles and conductors (the heights of suspension are given in relation to the base of the poles, taking into account the dimensions of the insulators): OS24 P+5-type poles, a suspension height of the lowest conductor of 19.4 m and a span length equal to 240 m. The minimum suspension height of the lowest phase wire was estimated at 11.9 m. For the line located closer to the building, the maximum phase current value is 460 A. For the other line, the maximum phase current value is 230.0 A;
- A shopping center under a 110 kV overhead line: A commercial building located partially under a high-voltage line with the following parameters: an SW24 ON120+10 pole with a suspension height of the lowest conductor of 23.9 m and an S24 ON120+10 pole with a suspension height of the lowest conductor of 21.8 m. In the analyzed case,

the span length was 256.1 m. The value of the long-term current-carrying capacity in the phase cables was 634 A. For the simulation purposes, the maximum permissible cable proximity of 6.4 m to the roof of the building was assumed;

- A residential building next to double-track overhead 110 kV lines: A single-family house located in the vicinity of a 110 kV double-circuit overhead line. The minimum distance of the building wall from the extreme high-voltage line wire is 7.4 m. Line parameters: OS24 P-type pole with a suspension height of the lowest wire of 15 m and OS24 ON150-type pole with a suspension height of the lowest wire of 12 m, a span length is 215 m. The minimum suspension height of the lowest phase wire was estimated at 8.05 m. In order to tighten the simulation results, the highest effective value of the phase voltage of 68.4 kV was assumed in the L1-phase wires (based on energy quality measurements, the average value of the phase-to-phase voltage was 118.4 kV). In the remaining L2- and L3-phase wires, the values then occur at (−34.2) kV. The maximum phase current values for the first and the second circuit are 437.1 A and 310.1 A, respectively;
- Parking under a 110 kV overhead line: Parking for cars and trucks located under a 110 kV high-voltage line with the following parameters: a maximum long-term load capacity of 709 A. The minimum suspension height was determined on the basis of geodetic measurements carried out at a temperature of 2 °C. For phase L1 (−3.37 m from the line axis) and L3 (3.13 m from the line axis), the suspension height was 12.28 m, and for phase L2 (2.32 m from the line axis), the suspension height was determined at 15.76 m. The obtained heights were reduced by a distance of 1.3 m, which results from a higher temperature of the conductor or catastrophic conditions;
- Terraced buildings near 110 kV double-track overhead lines: A residential building located 9 m from the extreme conductor (phase L1) of a double-circuit overhead line with a voltage of 110 kV. The analyzed line section is characterized by the following parameters: an AEG ON-type pole with a suspension height of the lowest conductor of 12.65 m and an AEG O-type pole for which the suspension height of the lowest conductor is 14.75 m (the suspension heights are given in relation to the base of the poles, taking into account the dimensions of the insulators), a span length of 131.14 m. The minimum suspension height was estimated as 9.5 m. A maximum current of 475 A was assumed in each of the line circuits;
- A multi-family building with balconies near a 15 kV overhead line: A residential building constructed 3.8 m from the end pole of the 15 kV MV overhead line. A medium-voltage line terminated with a Kgo-13.5/25 MV cable connection. The long-term load capacity of the investigated line was equal to 280 A;
- A multi-family building under a 110 kV overhead line: A residential building for which the smallest distance to the extreme conductors of the 110 kV HV line is 6.12 m. The analyzed line section is characterized by the following parameters: an SW24 ON120+5-type pole with a suspension height of 70.36 m and an SW24 KN60+10-type pole with a suspension height of 43.58 m (the suspension heights are given in relation to the zero reference level defined by sea level). The minimum suspension height over the analyzed building was estimated as 53.3 m. A maximum current of 675 A was assumed in each of the line circuits.

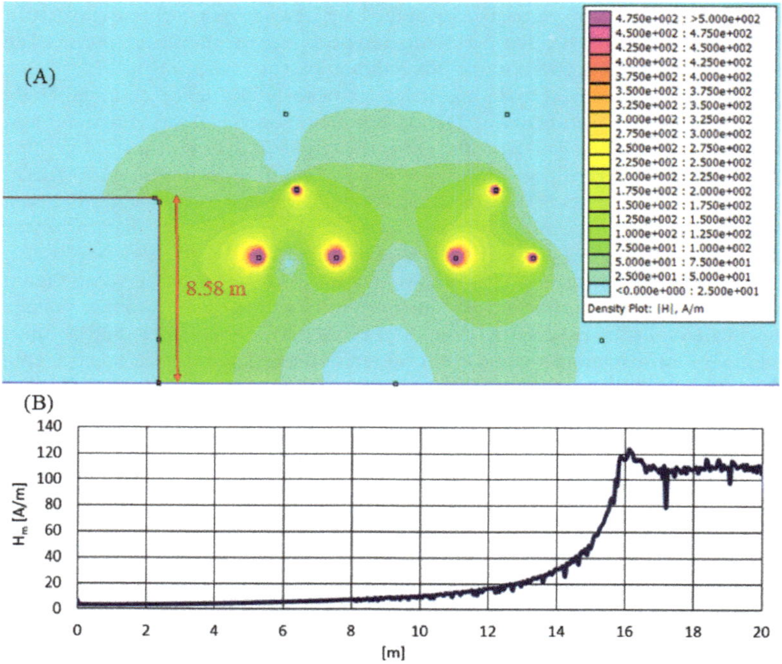

Figure 7. Numerical simulations of the magnetic field distribution. (**A**) Magnetic field distribution in the space above the roof and side wall of the warehouse—roof height—8.58 m. According to the project assumptions, the lowest cables in the conditions of catastrophic soot. (**B**) The level of the magnetic field on the roof and side wall of the warehouse according to the profile marked in red in Figure 7A). The local maximum for the abscissa of 16 m occurs halfway up the side wall.

Table 2. The values of E_{max}, E_{2m}, H_{max} and H_{2m} obtained from numerical simulations in FEMM software.

	Calculated Case	E_{max} [kV/m]	E_{2m} [kV/m]	H_{max} [A/m]	H_{2m} [A/m]
1	Warehouse hall near 110 kV double-track overhead lines	4.0	3.0	125.0	96.0
2	Multi-family building with balconies near double-track 110 kV overhead lines	3.95	0.89	46.1	12.6
3	Shopping center under 110 kV overhead line	2.2	0.3	46.0	27.0
4	Residential building next to double-track overhead 110 kV lines	2.2	1.25	39.8	48.2
5	Parking under 110 kV overhead line	1.4	0.82	65.0	12.0
6	Terraced buildings near 110 kV double-track overhead lines	1.6	0.9	15.0	15.0
7	Multi-family building with balconies near 15 kV overhead line	0.55	1.3	9.2	4.4
8	Multi-family building under 110 kV overhead line	3.5	0.1	26.1	7.5

For the purposes of numerical simulations for the above cases, the following assumptions were made. In the case of 110 kV lines, AFL 6-240-type working conductors with a design diameter of 21.7 mm were used, while for the 15 kV line, the AFL 70 type (diameter equal to 11.3 mm) was employed. Distances between conductors were estimated according to the data for the used poles. The suspension heights of conductors were determined on the basis of the line parameters except in cases where they were measured directly.

4. Discussion

Due to the lack of scientific consensus regarding the impact of EMFs on living organisms, the evaluation of the results described in the previous section was carried out based on the criteria described by the standards. Since there are different standards and local regulations regarding the issue of the allowed values of EMF fields, in order to improve the readability of this article, a subsection relating to this issue is introduced.

4.1. The Overview of Legal Regulations and Standards

The governments of states address the problem of the potential harms of EMFs and introduce some measures of protection in their legislation. One of the legal approaches to ensure no harm to humans happens is the introduction of norms that determine the maximal amount of electric field force or magnetic force that can be emitted by devices that create the field.

There is an independent international body called the International Commission on Non-Ionizing Radiation Protection [17], which was founded in 1992 and issues guidelines regarding the scope of its activities. In 1998, ICNIRP issued a set of guidelines [18] on EMFs and the protection of the general public against their harm. Three physical quantities were analyzed, and their reference levels were established (reference levels are not supposed to be exceeded; sticking to them guarantees no harm to the population); these quantities were E (kV/m), H (A/m) and B (μT); the levels depend on the f (Hz) and are shown in Table 3.

Table 3. The reference values of E, H and B for EMFs with different f following ICNIRP.

	<1 Hz	**1–8 Hz**	**8–25 Hz**	**0.025–0.8 kHz**	**0.8–3 kHz**	**3–150 kHz**	**0.15–1 MHz**	**1–10 MHz**	**10–400 MHz**	**400–2000 MHz**	**2–300 GHz**
E [kV/m]	-	10,000	10,000	250/f	250/f	87	87	$87/(f^{0.5})$	28	$1.375(f^{0.5})$	61
H [A/m]	3.2×10^4	$3.2 \times 10^4/(f^2)$	4000/f	4/f	5	5	0.73/f	0.73/f	0.073	$0.0037/(f^{0.5})$	0.20
B [μT]	4×10^4	$4 \times 10^4/(f^2)$	5000/f	5/f	6.25	6.25	0.92/f	0.92/f	0.092	$0.0046 \cdot f^{0.5}$	0.20

ICNIRP amended its guidelines in 2010 [19] and then further in 2020 [1], allowing for higher reference values. For our further considerations, the f of 50 Hz (standard frequency of electric current) is assumed, and the norms analyzed are for the general public (there also exist norms for professionals working with EMFs, which can be found in the same documents as the ones for the general public). The guidelines' comparison is in Table 4.

Table 4. The comparison of guidelines published by ICNIRP.

	1998	**2010**	**2020**
Scope of regulation (frequencies)	Up to 300 GHz	1 Hz–100 kHz	100 kHz–300 GHz
Physical quantities regulated	Electric field strength, magnetic field strength, magnetic flux density	Electric field strength, magnetic field strength, magnetic flux density	Electric field strength, magnetic field strength, power density S (W/m^2)
E for general public (f = 50 Hz)	5 kV	5 kV	Not affected
B for general public (f = 50Hz)	100 μT	200 μT	Not affected

In Europe, the majority of countries' legislation ought to be consistent with the laws of the European Union. The European Union took a stance on the issue already in 1999 when the recommendation of the European Council on the limitation of exposure of the general public to electromagnetic fields was passed (1999/519/EC [20]). This legal act was inspired by the guidelines of ICNIRP from 1998; it constitutes a framework for all European countries, and among its annexes are the enumeration and definitions of physical quantities that are to be used for EMF evaluation and tables with reference values and formulae for calculations of summary exposure to different EMFs simultaneously. The document itself calls on member states to fulfill important duties concerning the education of the general public about the possible dangers of EMFs, promote research on that topic and ensure the safety of workers exposed to EMFs. The regulations of the EU remain unchanged since their original publication, and the original act is still in force. Different states of the EU may have variations in their legislation (the EU recommendation describes the minimal standards that every country should have implemented, but the member states are free to add additional restrictions or not obey the recommendation). The suggested minimal standards are indicated in Annex B, Table 1 of 1999/519/EC. For the frequency of 50 Hz, the reference values are 5 kV for E and 100 μT for B.

Some countries in Europe have adopted these standards as their own; such examples are Portugal [21], Germany and France [22]. Sometimes, they added some special precautions. For example, in Germany, pursuant to *26.* Verordnung zur Durchführung des Bundes-Immissionsschutzgesetzes [23] is a list of places (residential buildings, schools, hospitals) where special care should be taken not to exceed the permissible limits. In France, the so-called Abeille law [24] makes it compulsory to provide information on options for exposure reduction and prohibits constant use of Wi-Fi (which is also a type of electromagnetic field) in the presence of children up to the age of 3 years.

Yet another group of countries base their policy on the topic on recommendations or have not legally implemented the recommendation of the EU in their legal systems [25]; in some of these countries, a more lenient set of reference values is used. Among these states are the United Kingdom, Scandinavian countries, Austria and the Netherlands. In the United Kingdom, the National Radiological Protection Board recommends sticking to the values from ICNIRP 1998 [26], but it is not the law. Austria uses the ICNIRP 1998 standard, and in some parts of the country (Salzburg province), pursuant to a law from 2009, additional restrictions have been added that described minimal distance between human dwellings and high-voltage installations [27].

Finally, a group of countries decided to implement their own, stricter regulations: Italy, Belgium, Poland and Switzerland. A summary of their legal solutions concerning EMF protection can be found in Table 5.

Table 5. The comparison of EMF protection regulation in selected European countries.

Country	Ref.	EMF Restrictions
Italy	[28]	Basic limit as in 1999/519/EC, but in restricted places (schools, nurseries, residential dwellings), allowed B is only 10 μT, new installations are not to exceed 3 μT, and in 3 regions (Veneto, Tuscany, Emilia-Romagna), allowed B is 0.2 μT [29]
Poland	[30]	E of 1 kV and B of 75 μT allowed in residential buildings, E of 10 kV and B of 75 μT allowed in public places
Belgium	[31]	Differences between provinces; the value of E varies between 5 kV and 10 kV; B is reduced in dwellings or house interiors and is usually 10 μT
Switzerland	[32]	Basic limit as in 1999/519/EC, but in new installations and/or objects of "sensitive use", B allowed is 1 μT

This group of states also includes countries that are not members of the European Union—in Serbia, permissible reference values are E of 2 kV and B of 40 μT [33].

Outside Europe, the topic is regulated independently by the states. In the following part of the review, selected examples outside of Europe will be analyzed.

In Canada, the regulations about protection from EMFs are gathered in Safety Code 6 [34] published by the Radiation Protection Bureau of Canada. It provides the norms of exposure both within the body and (in the form of reference levels) in the environment. The scope of this regulation encompasses f from 3 kHz to 300 GHz. As for the f of 50 Hz, Canada does not have any national regulation, but in practice, ICNIRP guidelines are used [35].

In the United States, there are no federal regulations concerning reference levels or restrictions to EMFs. This matter was left to the states, and some of them introduced their own regulations and limits for EMF exposure—some states chose not to establish any limit value and focused on precautions to avoid excessive exposure, e.g., by establishing minimal distances between installations and human dwellings; in other states, there are limits established, and they range between 20% and 240% of ICNIRP depending on the state [25].

Japan has a standard regulated by the Ordinance of the Ministry of Economy, Trade and Industry, in which the limit of E = 3 kV and of B = 200 μT are established [36].

China has its own reference set of values, which are regulated mostly by Chinese norms GB 8702-88 [37] and GB 9175-88 [38]. They are lower than EU/ICNIRP standards [18,39] with percentage differences differing depending on f [40]—up to 800 Hz, they are mostly concordant with the EU standard; at higher f, they are lower than ICNIRP. Australia also does not have any national standard, but the country's authorities decided to use the guidelines from ICNIRP.

The reference limit values for 50 Hz EMFs defined by different standards and local regulations are summarized in Table 6.

Table 6. The reference limit values for 50 Hz EMF following different standards and local regulations.

	IC-NIRP 1998	ICNIRP 2010	1999/519/EC	Portugal	Germany	France	Belgium	Italy	Poland	The Netherlands	Switzerland	UK	Spain	Canada	USA	Japan	China	Australia
E [kV/m]	5	5	5	5	5	5	5–10 [1]	5	1/10 [2]	5	5	5	5	5	1–12 [1]	3	4	5
B [μT]	100	200	100	100	100	100	10	3–100 [1]	75	100	1/100 [3]	100	100	100	-	200	100	100
H [A/m]	80	160	80	80	80	80	8	2.4–80 [1]	60	80	0.8–80 [3]	80	80	80	-	160	80	80

[1] Differences between provinces/states. [2] Residential buildings/other buildings. [3] New installations and objects of "sensitive use"/basic limit.

4.2. The Impact of LF EMF on Humans

The cases described in Section 3 were assessed for compliance with various EMF limit standards and regulations applicable in selected countries. The obtained results are presented in Table 7. It can be observed that the current version of standard ICNIRP is less restrictive in the context of the maximum permissible values of H than the older one. For this reason, all of the investigated cases obey limitations imposed by ICNIRP 2010. Nevertheless, it should be mentioned that there was only one case (warehouse hall near the 110 kV double-track overhead lines) when the requirements of ICNIRP 1998 [18] and 1999/519/EC standards were not met. Local legal regulations proved to be more demanding in terms of permissible exposure to EMFs. For example, the limit for H in Belgium is one order of magnitude smaller than the value defined by the ICNIRP 2010 standard, which is why most of the investigated cases are not in line with these requirements. In the case of Italy, the failure to meet local EMF requirements concerns in most cases the lowest limits of H, which apply only in selected provinces. The situation is similar in Switzerland, but the restrictions imposed mainly apply to new installations and objects of "sensitive use". Polish regulations are distinguished by a very low permissible E value, which is equal to 1 kV/m for residential areas, and a reduced permissible value of B = 75 μT compared to standard ICNIRP 2010. The above restrictions mean that the construction of residential buildings under high-voltage lines or in their immediate vicinity is significantly limited. The restrictions on the maximum allowable value of E in Japan and China are also more restrictive than those defined by the ICNIRP 2010 standard, which is why (especially in the first case) some of the facilities considered did not meet their requirements. Among the cases analyzed, the most problematic in terms of compliance with the above-mentioned requirements was the case of the warehouse hall near the 110 kV double-track overhead lines.

Table 7. Compliance of the examined cases with various standards and legal regulations regarding electromagnetic field limits.

		Standards			Local Regulations					
		ICNIRP 1998	**ICNIRP 2010**	**1999/519/EC**	**Belgium**	**Italy**	**Switzerland**	**Poland**	**Japan**	**China**
Measurements	Transformer-distribution station 15 kV/0.4 kV	YES	YES	YES	YES	NO [1]	NO [2]	YES	YES	YES
	Multi-family building with balconies near 110 kV overhead line	YES	YES	YES	YES	YES	YES	NO	YES	YES
	Building plot under 110 kV overhead line (only *E*)	YES	YES	YES	YES	YES	YES	YES	YES	YES
	Office space with inappropriate equipotential bonding (only H)	YES	YES	YES	YES	NO [1]	NO [2]	YES	YES	YES
	Room above the MV/LV indoor station (only *E*)	YES	YES	YES	YES	YES	YES	NO	NO	NO
	LV switchboard (only H)	YES	YES	YES	NO	NO [1]	NO [2]	YES	YES	YES
Numerical simulations	Warehouse hall near 110 kV double-track overhead lines	NO	YES	NO	NO	NO	NO	NO	NO	NO
	Multi-family building with balconies near two 110 kV overhead lines	YES	YES	YES	NO	NO [1]	NO [2]	NO	NO	YES
	Shopping center under 110 kV overhead line	YES	YES	YES	NO	NO [1]	NO [2]	YES	YES	YES
	Residential building next to double-track overhead 110 kV lines	YES	YES	YES	NO	NO [1]	NO [2]	NO	YES	YES
	Parking under 110 kV overhead line	YES	YES	YES	NO	NO [1]	NO [2]	NO	YES	YES
	Terraced buildings near 110 kV double-track overhead lines	YES	YES	YES	NO	NO [1]	NO [2]	NO	YES	YES
	Multi-family building with balconies near 15 kV overhead line	YES	YES	YES	NO	NO [1]	NO [2]	NO	YES	YES
	Multi-family building under 110 kV overhead line	YES	YES	YES	NO	NO [1]	NO [2]	NO	NO	YES

[1] Applies to provinces with the most restrictive regulations. [2] Applies to new installations and objects of "sensitive use". In other cases, the requirements are met.

5. Conclusions

The overall objective of this article was to estimate the impact of LF EMFs generated by power infrastructure. For this purpose, there were analyzed values of *E* and *B* obtained during measurements on objects localized in the neighborhood of HV power lines, transformer stations, etc. Furthermore, numerical simulations of the distribution of the EF, as well as MF distributions, around real buildings next to power delivery system elements were conducted. To date, it has not been clearly established to what extent EMF radiation affects the development of cancer in humans. These concerns have prompted calls for the IARC [41] to reassess the risks of RF radiation. The results of animal studies suggest the need to consider upgrading the classification to "probably carcinogenic" or higher [42]. Given the lack of scientific consensus on the impact of EMFs on living organisms, the evaluation of the obtained results was based on the criteria outlined in existing standards. Various standards and local regulations address the permissible levels of EMF exposure. The guidelines included in the ICNIRP 2010 standard were chosen as the basic reference point, defining the maximum permissible values of $E = 5$ kV/m and $B = 200$ μT. Compared to the previous version of this standard from 1998, the permissible value of *B* has been doubled. However, individual countries may introduce stricter limits on these amounts. In terms of permissible limits of *E*, the most restrictive requirements among the analyzed examples were those of Polish regulations (1 kV/m in the case of residential areas). In the matter of permissible values of *B*, Belgium has introduced the most restrictive regulations ($B = 10$ μT). Nevertheless, some countries introduce even lower limits of *E* and *B*, but this only applies to selected aspects, e.g., the *B* limit in some Italian provinces has been reduced to 3 μT. Another example is Switzerland, where new installations and objects of "sensitive use" require *B* smaller than 1 μT. Of the cases examined, only one (warehouse hall near the 110 kV double-track overhead lines) did not meet the requirements set by standard ICNIRP 2010. Due to the lowest permissible *B* values, 8 of the 14 cases analyzed did not meet the requirements set by Belgian regulations. In light of the regulations in force in Poland, and especially the low permissible E values, the established limits were exceeded in nine cases.

Numerical simulations performed using FEMM software showed that the value of the E strongly depends on the distance from the source (in this case, from the 110 kV double-track overhead lines). At distances greater than 30 m from the chosen reference point, this parameter did not exceed the background level. Based on these simulations, it can also be concluded that a building wall made of steel with a thickness of at least 3 mm is sufficient to provide shielding from the EF. In the case of reinforced concrete structures, the so-called Faraday cage effect will also be observed. However, it should be noted that the MF is capable of penetrating through building walls. Actual EMF values depend on many factors, such as the configuration of the high-voltage line wires, the design of the pole, the type of wires used and the maximum voltage and current flowing through them. Therefore, it is advisable to measure the actual EMF levels with appropriate instruments once the construction process is complete.

Author Contributions: Conceptualization and methodology, L.S.L., M.O., A.Ż. and G.R.; formal analysis and investigation, P.M.T., E.G., Z.C., M.C., M.O. and L.S.L.; resources, P.M.T., E.G., Z.C. and M.C.; data curation, L.S.L., M.O., A.Ż. and G.R.; writing—original draft preparation, A.Ż., P.M.T., E.G., Z.C., M.C., M.O. and L.S.L.; writing—review and editing, M.O., A.Ż., L.S.L. and G.R.; visualization, L.S.L., Z.C. and E.G.; supervision, M.O., A.Ż. and G.R. All authors have read and agreed to the published version of the manuscript.

Funding: This research received no external funding.

Institutional Review Board Statement: Not applicable.

Informed Consent Statement: Not applicable.

Data Availability Statement: The original contributions presented in the study are included in the article, further inquiries can be directed to the corresponding author.

Conflicts of Interest: The authors declare no conflicts of interest.

Abbreviations

The following abbreviations are used in the article

B	magnetic induction
B_{2m}	magnetic induction at height of 2 m
B_{max}	the maximum value of magnetic induction
E	intensity of electric field
E_{2m}	intensity of electric field at height of 2 m
E_{max}	the maximum value of intensity of electric field
EF	electric field
EMF	electromagnetic field
EMR	electromagnetic radiation
f	frequency
H	intensity of magnetic field
H_{2m}	intensity of magnetic field at height of 2 m
H_{max}	the maximum value of intensity of magnetic field
LF EMF	low-frequency electromagnetic field
MF	magnetic field

References

1. Guidelines for Limiting Exposure to Electromagnetic Fields (100 kHz to 300 GHz). Available online: https://ntnuopen.ntnu.no/ntnu-xmlui/handle/11250/2737117 (accessed on 14 October 2024).
2. Szuba, M. *Linie i Stacje Elektroenergetyczne w Środowisku Człowieka*; Informator PSE-Operator SA: Warszawa, Poland, 2008.
3. Saliev, T.; Begimbetova, D.; Masoud, A.-R.; Matkarimov, B. Biological effects of non-ionizing electromagnetic fields: Two sides of a coin. *Prog. Biophys. Mol. Biol.* **2018**, *141*, 25–36. [CrossRef] [PubMed]
4. Wertheimer, N.; Savitz, D.A.; Leeper, E. Childhood cancer in relation to indicators of magnetic fields from current sources. *Bioelectromagnetics* **1995**, *16*, 86–96. [CrossRef] [PubMed]

5. Wertheimer, N.; Leeper, E. Electrical wiring configurations and childhood cancer. *Am. J. Epidemiol.* **1979**, *109*, 273–284. [CrossRef] [PubMed]
6. Savitz, D.A.; Wachtel, H.; Barnes, F.A.; John, E.M.; Tvrdik, J.G. Case-control study of childhood cancer and exposure to 60-hz magnetic fields. *Am. J. Epidemiol.* **1988**, *128*, 21–38. [CrossRef] [PubMed]
7. Carles, C.; Esquirol, Y.; Turuban, M.; Piel, C.; Migault, L.; Pouchieu, C.; Bouvier, G.; Fabbro-Peray, P.; Lebailly, P.; Baldi, I. Residential proximity to power lines and risk of brain tumor in the general population. *Environ. Res.* **2020**, *185*, 109473. [CrossRef] [PubMed]
8. Moon, J.-H. Health effects of electromagnetic fields on children. *Clin. Exp. Pediatr.* **2020**, *63*, 422–428. [CrossRef] [PubMed]
9. N. I. of Environmental Health Sciences and others. *Assessment of Health Effects from Exposure to Power-Line Frequency Electric and Magnetic Fields*; National Institutes of Health: Research Triangle Park, NJ, USA, 1998.
10. Feychting, M.; Alhbom, M. Magnetic Fields and Cancer in Children Residing Near Swedish High-voltage Power Lines. *Am. J. Epidemiology* **1993**, *138*, 467–481. [CrossRef] [PubMed]
11. Olsen, J.H.; Nielsen, A.; Schulgen, G. Residence near high voltage facilities and risk of cancer in children. *BMJ* **1993**, *307*, 891–895. [CrossRef] [PubMed]
12. Verkasalo, P.K.; Pukkala, E.; Kaprio, J.; Heikkila, K.V.; Koskenvuo, M. Magnetic fields of high voltage power lines and risk of cancer in Finnish adults: Nationwide cohort study. *BMJ* **1996**, *313*, 1047–1051. [CrossRef] [PubMed]
13. Crespi, C.M.; Swanson, J.; Vergara, X.P.; Kheifets, L. Childhood leukemia risk in the California Power Line Study: Magnetic fields versus distance from power lines. *Environ. Res.* **2019**, *171*, 530–535. [CrossRef] [PubMed]
14. Swanson, J.; Kheifets, L.; Vergara, X. Changes over time in the reported risk for childhood leukaemia and magnetic fields. *J. Radiol. Prot.* **2019**, *39*, 470–488. [CrossRef] [PubMed]
15. Carpenter, D.O. Extremely low frequency electromagnetic fields and cancer: How source of funding affects results. *Environ. Res.* **2019**, *178*, 108688. [CrossRef] [PubMed]
16. Huo, F.; Lu, W.; Qiu, Z.; Huang, D.; Huang, C. Study on electric field distribution characteristics of the maintenance area when UHV AC transmission line crossing 220 kV tower. *J. Eng.* **2019**, *2019*, 2986–2990. [CrossRef]
17. ICNIRP. Aim, Status & History. Available online: www.icnirp.org/en/about-icnirp/aim-status-history/index.html (accessed on 16 March 2024).
18. International Commission on Non-Ionizing Radiation Protection. Guidelines for limiting exposure to time-varying electric, magnetic, and electromagnetic fields (up to 300 GHz). *Health Phys.* **1998**, *74*, 494–522.
19. I. C. on Non-Ionizing Radiation Protection and others. Guidelines for limiting exposure to time-varying electric and magnetic fields (1 Hz to 100 kHz). *Health Phys.* **2010**, *99*, 818–836. [CrossRef] [PubMed]
20. Recommendation, Council. Recommendation and Others, Limitation of Exposure of the General Public to Electromagnetic Fields (0 hz to 300 ghz). Available online: https://osha.europa.eu/pl/legislation/guidelines/council-recommendation-1999519ec-limitation-exposure-general-public-electromagnetic-fields-0-hz-300-ghz (accessed on 14 October 2024).
21. Diário da República n.o 275/2004, Série I-B de 2004-11-23. Available online: https://diariodarepublica.pt/dr/detalhe/diario-republica/275-2004-122583 (accessed on 14 October 2024).
22. Décret n 2002-775 du 3 mai 2002. Available online: https://www.legifrance.gouv.fr/loda/id/JORFTEXT000000226401 (accessed on 14 October 2024).
23. BGBl. I S. 3266. Available online: https://www.gesetze-im-internet.de/bimschv_26/BJNR196600996.html (accessed on 14 October 2024).
24. LOI n° 2015-136 du 9 Février 2015. Available online: https://www.legifrance.gouv.fr/loda/id/JORFTEXT000030212642 (accessed on 14 October 2024).
25. Stam, R. Comparison of International Policies on Electromagnetic Fields: (Power Frequency and Radiofrequency Fields). 2018. Available online: https://www.arpansa.gov.au/comparison-international-policies-electromagnetic-fields-power-frequency-and-radiofrequency-fields (accessed on 14 October 2024).
26. UK Government Web Archive. Available online: https://www.nationalarchives.gov.uk/webarchive/ (accessed on 14 October 2024).
27. Landeselektrizitätsgesetz-Novelle. Available online: www.salzburg.gv.at (accessed on 14 October 2024).
28. La Normativa Italiana: DPCM 8 Luglio 2003. Available online: https://www.repertoriosalute.it/ (accessed on 14 October 2024).
29. EMFS Info. Available online: https://www.emfs.info/limits/world/ (accessed on 14 October 2024).
30. Dz. U. 2019 poz. 1448. Available online: https://isap.sejm.gov.pl/isap.nsf/DocDetails.xsp?id=WDU20190002448 (accessed on 14 October 2024).
31. BBMEG. Available online: https://www.bbemg.uliege.be/belgian-legislation/ (accessed on 14 October 2024).
32. Verordnung über den Schutz vor Nichtionisierender Strahlung (NISV) vom Dezember 1999 (Stand am November 2023). Available online: https://www.fedlex.admin.ch/eli/cc/2000/38/de (accessed on 14 October 2024).
33. Law on Protection against Non-Ionizing Radiation (" ",). Official Gazette of RS. volume No. 36/09. Available online: https://leap.unep.org/en/countries/rs/national-legislation/law-protection-non-ionizing-radiation (accessed on 14 October 2024).
34. Health Canada. Available online: www.canada.ca/en/health-canada/services/publications/health-risks-safety/limits-human-exposure-radiofrequency-electromagnetic-energy-range-3-300.html (accessed on 14 October 2024).

35. Electric and Magnetic Field (EMF) Safety Program. Environmental Health & Safety. Available online: https://ehs.utoronto.ca/our-services/radiation-safety/emf-safety-program/#ICNIRP2010 (accessed on 14 October 2024).
36. T. and I. No. 34 of 2013 Order of the Ministry of Economy. Ministerial Order to Provide Technical Standards for Electrical Appliances and Materials. Available online: https://www.japaneselawtranslation.go.jp/en/laws/view/2961/en (accessed on 14 October 2024).
37. *GB 8702-1988 Standard;* Regulations for Electromagnetic Radiation Protection. State Environmental Protection Agency: Beijing, China, 1988.
38. *GB 9175-88 Standard;* Hygienic Standard for Environmental Electromagnetic Waves. State Environmental Protection Agency: Beijing, China, 1988.
39. International Commission on Non-Ionizing Radiation Protection. Guidelines on limiting exposure to static magnetic fields. *Health Phys.* **2009**, *9*, 504–514.
40. Wu, T.; Shao, Q.; Yang, L.; Qi, D.; Lin, J.; Lin, X.; Yu, Z. A large-scale measurement of electromagnetic fields near GSM base stations in Guangxi, China for risk communication. *Radiat. Prot. Dosim.* **2012**, *155*, 25–31. [CrossRef] [PubMed]
41. IARC. Available online: https://www.iarc.who.int (accessed on 14 October 2024).
42. Microwavenews. Available online: https://microwavenews.com/short-takes-archive/iarc-urged-reassess-rf (accessed on 14 October 2024).

applied sciences

MDPI

Article

Full Road Transport Sector Transition Towards 100% Autonomous Renewable Energy Supply in Isolated Systems: Tenerife Island Test Case

Itziar Santana-Méndez, Óscar García-Afonso * and Benjamín González-Díaz

Departamento de Ingeniería Industrial, Escuela Superior de Ingeniería y Tecnología (ESIT), Universidad de La Laguna (ULL), Camino San Francisco de Paula, nº 19, 38200 San Cristóbal de La Laguna, Spain; isantana@ull.es (I.S.-M.); bgdiaz@ull.es (B.G.-D.)
* Correspondence: ogarciaa@ull.es

Abstract: The transition towards sustainable energy systems is a key challenge faced by society. Among the different sectors, road transport becomes one of the most difficult due to the large energy consumption and infrastructure requirements. In this context, although zero-tailpipe-emission vehicle adoption is seen as a promising route, the energy provision through renewable sources is still uncertain, especially with hydrogen. This paper explores a 100% renewable energy supply scenario for both power-generation and road transport sectors in the isolated system of Tenerife. With this aim, the island's energy system has been modelled in the software EnergyPLAN. Taking as reference the current renewable technology roadmap in the island, the impact of a full deployment of zero-tailpipe-emission vehicles on the energy system has been evaluated, providing the power and energy storage capacity requirements. The obtained results indicate the need for 6 GW of renewable power (nearly 20 times the current figures) and 12 GWh of a yet non-existent storage capacity. This deployment must be accompanied with approximately 1 GW of dispatchable sources and 1.3 GW of electrolysis capacity to carry out a complete decarbonisation of the transport sector in the island. Finally, a series of recommendations to policy makers are suggested to support the definition of future roadmaps.

Keywords: autonomous renewable energy supply; isolated energy system; road transport; electrification; decarbonisation

1. Introduction

Worldwide transport emissions, with nearly 8 Gt CO_2 in 2022, have grown at an annual average rate of 1.7% since 1990 [1]. With an oil dependence of approximately 91%, the transport sector accounts for approximately one-quarter of global anthropogenic CO_2 emissions [2].

In the European Union, road transport represents approximately 70% of the transport sector's CO_2 emissions. Among the different modes, light-duty vehicles (LDVs) are by far the group with largest emissions with 60.6%, followed by heavy-duty vehicles (HDVs) contributing to 38% [3]. To face this situation, the European Union has launched a strong push to adopt more sustainable road transport driven mainly by the European Green Deal and the Sustainable and Smart Mobility strategy [4]. In addition to several aspects that cover the use of low-carbon fuels, the roll out of public charging for light- and heavy-duty vehicles, the EU established that all new cars and vans that come on the market from 2035 must have zero tailpipe emissions. In addition, the EU set the ambitious milestones of at least 30 million of these vehicles operating on European roads by 2030, and nearly all cars, vans, buses, as well as new heavy-duty vehicles being zero-emission vehicles by 2050.

Recently, the Council of the European Union ratified the approved revision of the HDV CO_2 emissions standards [5]. Compared to the 2019 levels, the new standards for almost all new HDVs with certified CO_2 emissions must provide 45%, 65%, and 90% reductions from 2030, 2035, and 2040, respectively. In addition, all new buses providing urban services must be zero-tailpipe-emission vehicles from 2030.

The aforementioned objectives involve a progressive substitution of the fleet vehicle technology over the coming years. This additional energy must be covered by a greater penetration of renewable energy (RE) [6,7], which eventually puts pressure on the grid and requires a massive deployment of energy storage systems to avoid energy spill. Among the different storage technology options, hydrogen-based storage systems and their use to generate synthetic fuels for sectors where direct electrification is challenging could play a key role in increasing the RE share in the primary energy supply [8]. However, further research and development are still needed to reduce electrolysis cost [9], ensuring its viability. In addition, the hydrogen supply chain must be carefully studied and planned [10], as there is no infrastructure in place to support the technology's deployment.

In this context, small-to-medium isolated energy systems are the most vulnerable. Their geographical conditions restrict their willingness to be self-sufficient and also generate significant economic costs [11]. In addition to this, a small power grid size is a barrier for the mass penetration of renewable sources, as large quantities can generate an imbalance in the network, endangering its stability and generating intermittency problems in almost all networks [12].

1.1. Literature Review

Due to the singularity of the isolated systems, the impact of battery electrical vehicles (BEVs) on isolated grids has been studied in the literature. A common characteristics of an isolated energy system is the noticeable dependence of fossil fuels (especially diesel) for power generation [13], therefore decreasing the benefit of EV adoption to reduce the overall CO_2 emissions. Several authors have tried to quantify reduction figures in representative small-to-medium islands. Applying a Well-to-Wheels approach, a deployment of EVs might bring CO_2 reductions with respect to conventional powertrains between 0 and 20% and 40 and 50% for RE penetration levels ranging from limited up to 20% [14?]. In addition, these studies reveal that hybrid powertrain architectures currently provide similar CO_2 reduction figures, requiring a much larger RE penetration to enlarge the abatement potential [16–18].

Many studies have been focused on the EV possibilities to promote the integration of renewable energy and to reinforce the grid stability [19–21]. To face the variability and uncertainty issues of the resource, there are mainly two approaches: (i) by means of controlling methods, where a limitation of the quantity of the renewable energy in the grid is achieved, or (ii) by means of energy storage systems [22], where larger energy penetration could be reached [23]. Therefore, it has been demonstrated that the BEVs can perform a two-way delivery of energy, improving the peak-shaving effect and reducing the need for conventional power plant generation [24].

On the contrary, a large deployment of BEVs entails a remarkable additional load to electrical systems, hindering the optimal management of the grid constraints and making necessary a precise planning of the grid to avoid undesirable electrical problems [25–27]. In this context, previous studies report that uncertainties in EV charging and RE resources could require up to a 30% increase in the installed power capacity from the estimated one to cover certain demands to avoid an energy shortage risk [28].

One relevant aspect is the BEV capability to act as an active element through Vehicle to the Grid (V2G), allowing the interchange of power between the electrical grid and the vehicle. The application of V2G technology can contribute to peak shaving, self-consumption, frequency control, reactive power compensation, and, when introduced in a larger scale, this technology rises the share of renewable energy in the energy mix [29]. In this sense, previous results have revealed a reduction of both grid harmonics and frequency deviation, favouring the integration of variable energy sources into power grids [30]. In addition, V2G penetration can be adequately used to avoid grid overloading and load matching [31–33], reducing the renewable energy surplus and therefore turning into a reduction of the renewable capacity needed. Other studies have highlighted the benefit of V2G penetration to reduce the undesired effects of the reactive power generated by the power electronics in various types of small- and medium-size distribution networks through a proper combination of charging stations location and management algorithms [34].

Several research works have studied decarbonisation scenarios in isolated energy systems, where the EV penetration and its interaction with the grid are not considered in detail and the HD transport demand is dismissed [35–37].

Despite the multiple contributions to the state of the art, the literature survey related to transport decarbonisation in isolated systems reflects that the majority of works have addressed specific topics. However, analysis that covers how the isolated energy system must evolve to accommodate the energy transition of the whole vehicle fleet (light- and heavy-duty) considering energy sovereignty have been dismissed. The present research work is devoted to filling this gap by means of evaluating a full transition scenario of road transport from a renewable energy autonomy standpoint. First, a complete transition of the power-generation sector is studied to later address the impact of the transport sector demand on the energy system response. The provided figures of energy demands, new power and storage capacities, and the efficiency indicators represent a solid contribution to general knowledge. In addition, the obtained results point out the main bottlenecks of this route, which allows policy makers and industry stakeholders to develop future roadmaps.

1.2. Tenerife Energy System

Tenerife, Canary Islands, with an area of 2034.38 km^2 and a population of 944,107 inhabitants is the largest and most populated island in Spain. Located in the Atlantic Ocean and close to the African coast (see Figure 1), Tenerife is one of the European Union's outermost regions.

Figure 1. Geographical location of Tenerife, Canary Islands.

A basic representation of this isolated energy system is depicted in Figure 2, developed with the most recent published data (2022) [38]. Oil (mainly heavy fuel oil, diesel, and petrol) is the dominant primary energy source, exceeding the 22 TWh of energy import. The two main demand sectors are power-generation (29.06%) and transportation (70.94%). Oil consumption in the industrial/service sectors and LPG, although existent, has been omitted as its values are relative low. Renewable sources (mainly WE and PV) only represent 2.8% of the total primary energy consumption, making clear the great dependence on fossil fuels across all sectors.

In 2022, the electricity demand was 3.4 TWh/year, with a peak power of 557 MW. A total share of 84% of this demand was covered with conventional sources, a common factor in other isolated power systems [11,38,39]. The current structure of the power-generation sector is summarised in Table 1, which includes the technology, power capacity, generation share, and fuel type. Combined Cycles and steam turbines supply most of the required power and provide the base load, with a total share of 74%. The rest of the technologies are two-stroke marine-derivative compression ignition engines and gas turbines, the latter being mainly used to support peaks and provide a spinning reserve. With no infrastructure in place to deploy natural gas, gas turbines make use of diesel fuel, whereas heavy fuel oil is the current fuel to feed the two-stroke engines and steam turbines.

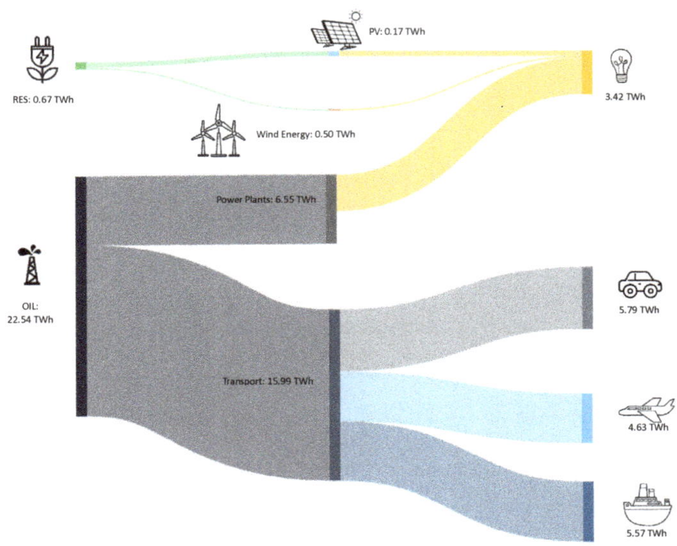

Figure 2. Sankey diagram of current energy system of Tenerife island. Data extracted from [38].

Table 1. Description of the current power-generation system of Tenerife. Data extracted from [38].

Technology	Gross Installed [MW]	Generation [GWh]	Share [%]	Fuel	Average Thermoelectric Efficiency [%]
Combined Cycle	457	1943	54.2	Diesel	47
Steam turbine	160	706	19.7	HFO	33
Gas turbine	248	99	2.8	Diesel	24
Diesel engine	48	149	4.1	HFO	43
Wind energy (WE)	222.6	508	14.2	-	-
Solar photovoltaic (PV)	107.6	168	4.7	-	-
Small hydropower	1.2	3	0.1	-	-
Biogas plant	1.6	9	0.3	Biogas	-

As a result of the technical difficulties to reach large RE penetration percentages [40], the current contribution of RE technologies to the electrical energy produced on the island is 19.2% (17% across this archipelago) [38]. This percentage is considerably far from interconnected systems such as mainland Spain, where the current figure is 48.4% [41].

The road transport sector, with a fleet mainly powered by internal combustion engines (ICEs), has an oil demand of 5.79 TWh (25.68% of the total demand). The breakdown consumption, grouped by vehicle application and fuels, is shown in Table 2. The oil demand of LDVs, including passenger cars and vans, represents 65% of the total road transport demand. The fleet (approximately 800k vehicles) [42] is dominated by petrol engines, as the consumption per fuel type evinces. The rest of the conventional vehicles uses LPG as fuel (energy demand of 7.73 GWh), representing only 0.21% of the total fleet [38]. Plug-in vehicles account for 0.54% of the LDV fleet [42]. Their share in the new registered vehicles has moved from 1.37% in 2019 to 12.28% in 2023 [42], confirming a progressive technology shift due to government incentives.

Table 2. Oil-based annual energy consumption of road transportation.

	LDV Energy Consumption [TWh]	HDV Energy Consumption [TWh]
Petrol	2.52	-
Diesel	1.21	2.03
Total	3.73	2.03

HDVs represent 36% of the road transport energy demand and almost 10% of the total oil imported to the island. In this case, diesel is practically the only source of power, as the petrol HDV share is negligible and battery electric buses for urban services are still being commissioned.

1.3. Decarbonisation Strategy

The regional government has recently published the PTECan-2030 (Plan de Transicion Energetica de Canarias), that establishes a roadmap to decarbonise the Canarian energy system. This main objective is divided into several specific objectives: energy autonomy, decarbonisation of the economy, electrification of heat/cooling and transport demands, as well as progressing in circular economy approaches. Although the PTECan plan envisages a full decarbonisation of the economy in 2040, only domestic-specific goals are set for 2030, described as follows:

- A 62% share of RE in the generated electricity.
- A 29% of RE share in the total final energy consumption.
- A 27% of energy efficiency improvement.

The PTECan plan does rely on a series of specific studies covering aspects such as net metering potential [43], energy storage [44], transport electrification [45], future dispatchable power generation [46], or the deployment of H_2 as fuel [47]. An in-depth analysis of future energy demands and transport evolution was performed as a basis to propose the suitable energy system structure subjected to a series of identified constraints.

With respect to the power-generation sector, the plan for Tenerife proposes a massive deployment of renewable sources in the 2025–2040 timeline. From the current value of 333 MW, the strategy suggests that RE capacity will be increased by approximately 14 times, as Table 3 depicts. In this context, PV, both self-consumption and utility scale, would experience a significant growth, with 2302 MW. The plan also relies on increasing the on-shore WE capacity by 7.5 times, by means of installing new units and re-powering the old ones. Off-shore WE will also play a key role in the future energy mix, as the plan expects 505 MW. To a lesser extent, geothermal and wave energy will also be introduced in the energy system, with 20 MW and 5 MW, respectively. The plan also points out the need for 210 MW of firm generation provided by gas turbines, using internally generated H_2 from renewable sources.

Although the PTECan-2030 lacks clarification regarding the energy storage systems (ESSs) for the island of Tenerife, the specific strategy supporting the plan does rely on a significant deployment of user-based EES (batteries) accompanying the expected massive penetration of self-consumption PV. Up to 2.5 GWh of available EES is expected, allowing higher rates of PV share in the power-generation mix. Distributed EES, located across the different electricity substations, could also reach the significant figure of 1.1 GWh. The space constraints and the orography of the island limit the deployment of large-scale energy storage systems, such as pumped storage hydropower, compressed-air energy storage or Liquid Air Energy Storage. Multiple assessments regarding plant locations have been conducted [44]. Current projections suggest 300 MW/4800 MWh of maximum capacity for this kind of system. Any additional energy storage requirements will be covered by batteries.

Table 3. Current structure and PTECan proposal for renewable energy and energy storage capacities in Tenerife.

	Current System [38]	**PTECan [48]**
Wind On-shore [MW]	222.6	1700
Wind Off-shore [MW]	0.0	505.3
PV On-shore [MW]	107.6	1650
PV Off-shore [MW]	0.0	27
PV Self-Consumption [MW]	38	829
Wave Drive [MW]	0.0	5.0
Small Hydropower[MW]	1.2	2.6
Biogas [MW]	1.6	17.8
Geothermal [MW]	0.0	20.0
H_2 turbines [MW]	-	210
EES–User [GW/GWh]	-	2.6/2.5
EES–Distributed [GW/GWh]	-	1.1/1.1
EES–Large Scale [MW/GWh]	-	313/5.2

From PTECan's perspective, BEVs will be the dominant technology across the light-duty transport sector. From the current low figures of a 12% share in new registrations, BEVs will progressively take the place of ICEs until the total abandonment of fossil fuels. In the context of the heavy-duty transport sector, Fuel Cell Electric Vehicles (FCEVs) will be the major technology to replace the diesel fleet.

Although the work behind the aforementioned strategies has been extensive and covers the entire period until decarbonisation, the final PTECan plan, which is a combination of the most relevant outcomes, lacks clarity defining the future energy demands and technology deployment beyond 2030. The aim of this paper is to provide an insight into this future plan, by complementing the results and extending the analysis to the full decarbonisation of both sectors from an autonomous energy supply perspective. The paper is organised as follows. Section 2 is devoted to explaining the methodology applied within this research work, covering the model configuration in EnergyPLAN and its validation. In Section 3, a description of the problem statement and the required energy system structure to meet the decarbonisation with autonomous energy supply goals are covered in detail. In the discussion, conclusions, and policy implications, the authors provide a set of recommendations based on the data gathered that might support future roadmap revisions.

2. Materials and Methods

2.1. *EnergyPLAN Simulation Tool*

The nature of this work requires the use of a bottom-up approach, as both energy sectors and technologies must be analysed in detail. EnergyPLAN, HOMER or H2RES are examples of the bottom-up model options that can be found in the literature. Among them, the free and user-friendly software EnergyPLAN v16.2 (Aalborg University, Aalborg, Denmark) has been selected as it is one of the most used tools to perform energy planning and support energy decision makers [49].

The EnergyPLAN tool applies a purely deterministic simulation approach with a description of all sectors: electricity, heating, cooling, industry, and transportation. Developed by Aalborg University, EnergyPLAN has been specifically designed to assist in the design of planning strategies under the theory of Choice Awareness [50]. Instead of applying numerical solvers to resolve optimisations, the software uses an heuristic approach, performing hourly energy balances and taking a set of predefined dispatching priorities [51]. This fact, together with the sectorial aggregation principle, significantly reduces the computational effort, being able to run a one-year period in just few seconds.

Figure 3 includes an overview of the software structure. As inputs, the user can define the main aspects of the analysed energy system: the different energy demands, the renewable energy sources, the capacities of the energy stations, the costs, and the regulation strategies. Among the outputs, EnergyPLAN provides the energy balance and the resulting annual production, fuel consumption, electricity import/export, CO_2 emissions, etc. [51]. Additional description of the software can be found in [52].

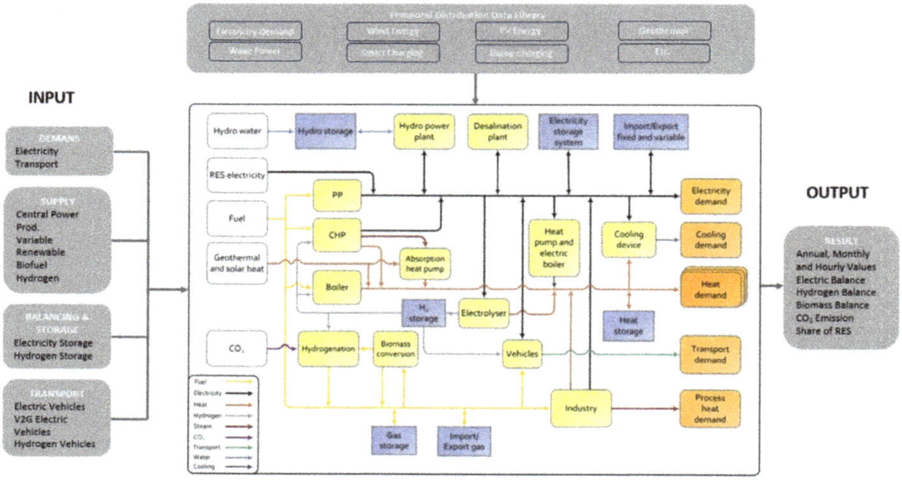

Figure 3. EnergyPLAN model structure. Definition of inputs/outputs. Adopted from [51].

EnergyPLAN has been widely applied in energy planning exercises with a special focus on sustainable energy solutions [51] at national [53], regional [54], and local levels [55]. The literature also reports that EnergyPLAN is one of the most common tools applied to evaluate isolated systems [49]. Although these systems might bring additional constraints such as reliability and robustness of the power grid, EnergyPLAN can successfully address them using alternative methods based on specific indicators. Several application examples taking, as a test case, isolated island systems, like the present research work, can be found in the literature [55,56].

2.2. Baseline Model Configuration and Validation

The energy system of Tenerife (described in Section 1.2) was modelled in EnergyPLAN. The baseline model was configured according to 2022 data, the most recent and representative year with a complete available dataset [38]. The input data applied to the model are described as follows:

- The aggregate demands for electricity (grid) and transport oil demand.
- The hourly distribution of electricity demand and the renewable energy production (WE and PV), extracted from data registered by the system operator [57].
- Grid technical restrictions, incorporated to the model as a minimum baseload. A value of 140 MW was applied based on previous history [57].
- Power-generation installed capacities and the average thermoelectric efficiency of the conventional powerplant [38].
- Heat/Cooling power demands were discarded as they are negligible compared to the rest of demands.

Once the model of the energy system was configured, the next step comprised the validation phase. With this in mind, the years 2019–2022 were simulated, imposing the input parameters described above. The same normalised profiles for electrical demand and

WE/PV generation (2022 data) were used across every test case, therefore only changing the aggregate demands and power capacities. The model was considered valid if the RE energy production and conventional generation/oil consumption were close to the data published.

Table 4 includes the main results of the validation phase. The simulation outputs in terms of annual electricity demand (just for reference, as it was imposed), RE generation, and the conventional park performance are compared against published data. The same values for simulated WE generation are found in the years 2019–2021, as no new capacity was installed in those years and the same normalised generation profiles were used. The same reason applies to simulated PV generation across the four years (no changes in capacity).

Table 4. Energy system model validation.

	2019			2020			2021			2022		
	Model	Data [58]	Error [%]	Model	Data [59]	Error [%]	Model	Data [60]	Error [%]	Model	Data [38]	Error [%]
Demand [TWh]	3.55	3.54	0.06	3.17	3.16	0.09	3.25	3.25	0.15	3.42	3.42	−0.06
WE generation [TWh]	0.48	0.49	3.08	0.48	0.48	0.77	0.48	0.52	−8.40	0.50	0.50	0.00
PV generation [TWh]	0.19	0.19	0.52	0.19	0.17	7.34	0.19	0.18	3.83	0.19	0.18	2.15
Oil demand (generation) [TWh]	7.37	7.57	2.64	6.42	6.37	0.71	6.61	6.57	0.66	7.18	7.16	0.28
Peak grid demand [MW]	571	576	0.87	510	556	−8.27	522	529	−1.32	551	557	−1.08

The low error values in predicting the annual renewable energy production, below 8.4%, evince that the annual energy source variability is not high, and, therefore, the same normalised profiles can be applied without introducing important errors to assess future scenarios. The acceptable estimation of RE generation ensure an accurate representation of the electrical energy generated and oil consumption of conventional generators, even in the event of a drop in demand for both 2020 and 2021 associated with the lockdown during the COVID-19 pandemic. The estimation errors are below 5% in terms of oil consumption with the sources related to the actual gen-sets warm-up/cool-down phases during the operation, as the model does not take this into account. Overall, the acceptable accuracy figures provided by the model confirm that it can be used with confidence to assess future scenarios of the island´s energy system.

3. Results

3.1. Problem Statement: Decarbonisation of Power-Generation and Road Transport Sectors Through Autonomous Energy Supply

The validated EnergyPLAN model of Tenerife's energy system will be used to illustrate the problem of reaching the status of a fully autonomous and decarbonised energy system in terms of both power-generation and road transport sectors. As a starting point, the regional government targets for RE power capacity and EES previously shown in Table 3 would be considered. The main hypotheses applied when simulating this scenario in EnergyPLAN are described as follows:

- The annual electricity demand is increased from the current 3.6 TWh to 5.1 TWh. The latter figure, extracted from the data supporting PTECan [46], was estimated using multiple regression techniques taking the trends for future population and GDP projections as independent variables.
- The non-dispatchable renewable power capacity per technology is described in Table 3. The validated normalised generation profiles are applied in EnergyPLAN to obtain the power generation of on-shore wind energy and PV. With respect to the off-shore WE, a correction factor to cover the additional 20% of energy availability obtained in several resource measurements in the island [61] is applied.
- The dispatchable power injection would be provided by a high-enthalpy geothermal system (see capacities in Table 3) and gas turbines fuelled with H_2. Neither baseload nor grid stability restrictions were considered in this work. The large amount of energy storage capacity, generation unit hybridisation to provide primary and secondary

control reserves [62,63], or the development and deployment of power electronics [64] support this hypothesis.

- Energy storage capacity was considered at user- (self-consumption), distributed- (electricity distribution system), and large-scale levels. The power and storage capacities are also included in Table 3. The energy storage associated with the self-consumption PV is considered centralised; therefore, the charging and discharging process is modelled based on EnergyPLAN dispatch priorities.
- PTECan assumptions indicate that BEVs will practically be the sole technology across the LDV fleet. In this work, the same criteria was assumed. In order to account for a conservative hypothesis, the LDV energy demand was kept constant. Therefore, the demand of fossil fuels (ICEs) shown previously in Table 2 was shifted to electricity, applying the following expression:

$$Elec_{LDV} = (E_P \eta_{ICE_P} + E_D \eta_{ICE_D}) \frac{1}{\eta_{Char} \eta_{EV} \eta_{trans}} \quad (1)$$

where $Elec_{LDV}$ is the electric demand of the LDV fleet, E_P and E_D are the petrol and diesel energy demands and η_{ICE_P} and η_{ICE_D} are average petrol and diesel ICE efficiencies. Finally η_{EV}, $\eta_{Charging}$ and η_{trans} correspond to the average electric motor efficiency, the charging efficiency, and the grid transportation efficiency, respectively. As a result, the total annual electricity demand to propel the LDV fleet is estimated to be 1.71 TWhe (half of the current overall grid demand in the island), similar to the 1.5 TWhe figure projected by the government in [45]. The dump charging profile (applied in previous works [17]) was extracted from the mobility strategy supporting PTECan [45].
- The PTECan hypothesis relies on the H_2 use as a primary fuel for heavy-duty (HD) transportation through fuel cells. Unlike LDVs, the technology selection for HD transportation is challenging due to the longer distance, power demands, and the wide range of business cases. Therefore, current OEMs' roadmaps are considering BEVs, FCEVs, or ICEs fuelled with renewable fuels such as H_2, biofuels, or e-fuels [65]. In this work, a combination of BEV and FCEV technologies to propel the HDV fleet was considered. Different shares were evaluated with the aim of providing trends of the energy system response to cover the resulted demand in terms of electricity and H_2 (produced internally with electrolysers). Similar to the LDV fleet, the fossil fuel demand was shifted to electricity and H_2, applying the following expressions:

$$Elec_{HDV} = (E_D \eta_{ICE_D}) \chi_{EV} \frac{1}{\eta_{Char} \eta_{EV} \eta_{trans}} \quad (2)$$

$$H2_{HDV} = (E_D \eta_{ICE_D}) \chi_{H_2} \frac{1}{\eta_{FCEV}} \quad (3)$$

where $Elec_{HDV}$ and $H2_{HDV}$ are the electric and H_2 demand of the island´s heavy-duty fleet, E_D is the current diesel energy demand, and $\eta_{ICE_{HD}}$ is the average HD diesel ICE efficiency. η_{EV}, $\eta_{Charging}$, η_{trans}, and η_{FCEV} correspond to the average electric motor efficiency, the charging efficiency, the grid transportation efficiency, and fuel cell efficiency, respectively. Finally, χ_{EV} and χ_{H_2} correspond to the demand share of electricity and H_2, respectively.
Due to the nature of the HD service, the majority of the fleet will be charged overnight. Therefore, only a dump charging strategy was used.
- The required H_2 for both power-generation and transport would be generated by electrolysers. Rather than imposing a predefined operation by the user, EnergyPLAN switches on the electrolysers based on the H_2 demand and the availability of renewable energy. The electrolyser capacity required to cover the H_2 demand will be a simulation output rather than imposed by the user.

The previously mentioned hypotheses required several efficiency parameters to perform the different energy conversions that are fed later to EnergyPLAN. Table 5 includes the efficiencies of the H_2 route (Power-to-H_2-to-Power) and the energy storage charging/discharging processes. The information is completed with the efficiency figures applied to shift from the internal combustion engine-based propulsion demand to both electricity (BEVs) and H_2 (FCEVs).

Table 5. Overall efficiency values applied to obtain the LDV electricity demand.

	Efficiency Values
Average H_2 turbine thermoelectric efficiency	0.40
Electrolysis efficiency	0.75
ESS charging/discharging efficiency	0.90
η_{ICE_P}	0.28
η_{ICE_D}	0.32
η_{EV}	0.86
$\eta_{Charging}$	0.90
η_{trans}	0.93
$\eta_{ICE_{HD}}$	0.39
η_{FCEV}	0.54

3.1.1. Power-Generation Sector

Figure 4 exemplifies the operation of the decarbonised power-generation sector through the condition of full energy autonomy. The road transport demand is been included in order to show its impact in a latter stage. Power injection per technology and the operation of the energy storage system (both electricity and H_2) are shown for two representative weeks. Left-side plots show a week with large WE resources whereas the right-side plot corresponds to limited WE resources.

The remarkable PV generation, together with high wind resources, entail a large amount of RE production, as Figure 4a shows. This excess of RE is mostly employed to feed the electrolysers that generate the gas turbine fuel, as Figure 4c depicts. The remaining renewable power that can neither be injected nor stored (see Figure 4e) is computed as surplus RE energy. The 3rd and 4th days show curtailment levels that reach up to 1500 MW.

Despite the large EES capacity considered in this scenario, a short intermittency of WE generation changes the system's operation considerably. In a first instance, the EES injects power until total depletion (see Figure 4c,e). Then, the H_2 storage level is significantly reduced, since turbines must operate to partially cover the grid demand and there are limited opportunities to switch on the electrolysers. This behaviour becomes critical for larger periods with limited or no WE resources, as right-side plots evince. The island energy system responds with backup power fuelled with H_2, depleting the gas reserves until relevant wind resources are available again. In addition, the PV power excess is mainly used to feed the electrolysers, limiting the available energy to be stored.

A summary of the simulation outcome for an entire year is included in the first column of Table 6. The WE intermittency and the limited PV capacity factor implies a large operation of H_2 turbines, with 1665 h of annual working time and almost 1 TWhe of electricity generation. The simulation results suggest that 650 MW of capacity would be required, similar in size to the current conventional park. The H_2 provision (2.1 TWh) would be supplied with 1.1 GW of electrolysis, demanding 2.8 TWhe of electricity and 0.56 hm^3 of water. The estimated capacity factor exhibits the poor value of 2475 heq, which might put the economic viability into question.

Despite the large storage capacity and the use of H_2 as an energy vector, the estimated amount of RE curtailed is 2.6 TWhe (27.3%). This behaviour, which causes inefficiencies and profitability losses, might discourage the implementation of RE and the entry of private investment capital.

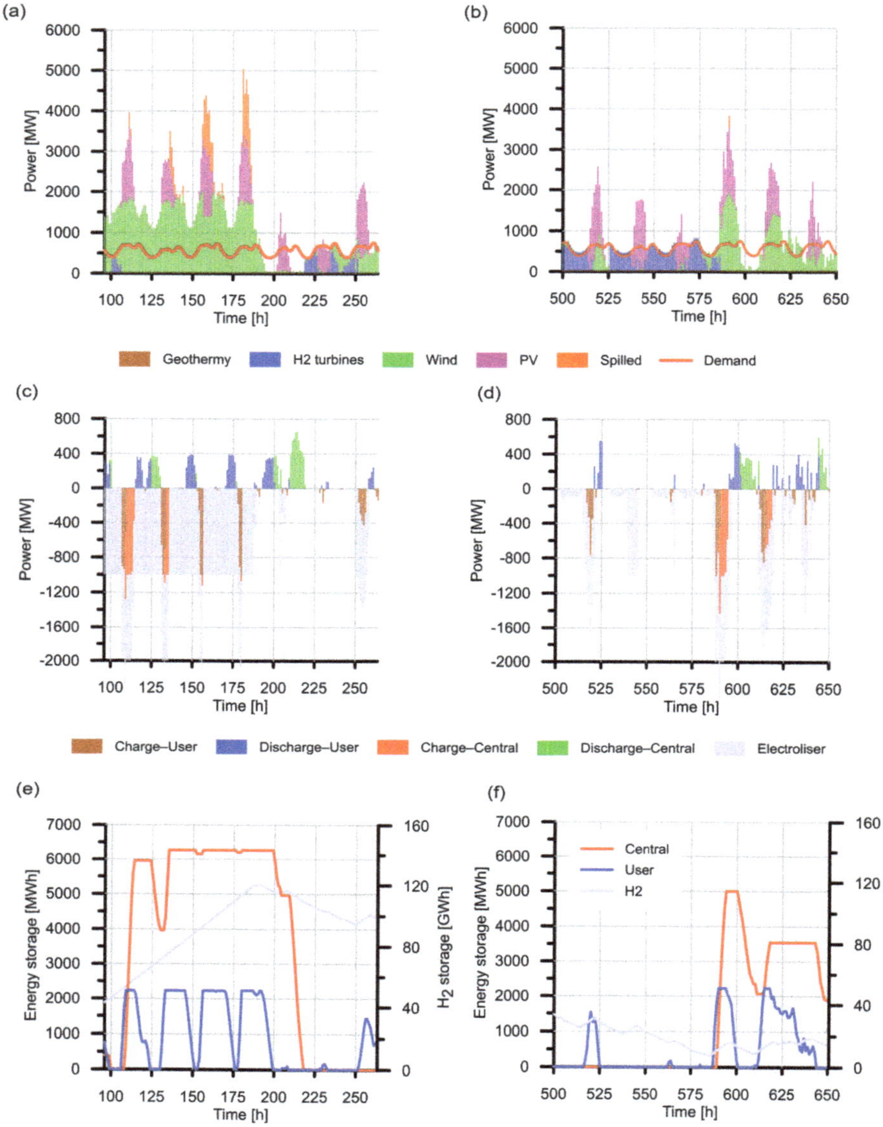

Figure 4. Operation of the decarbonised grid system response with no transport demand considered: representative week with large (**a**,**c**,**e**) and limited (**b**,**d**,**f**) WE resources.

3.1.2. Light-Duty Vehicle Transport Sector

The electrification of the entire LDV fleet through BEVs would result in remarkable changes of the total grid demand (power and profile), as Figure 5a depicts. The expected charging profile in the island, in which overnight charging events prevail over daily and/or opportunistic ones, does modify the grid valley period, increasing the demand from approximately 400 to 600 MW. The average contribution of dispatchable sources (mainly gas turbines fuelled with H_2) during the overnight period is 12%, which means that a considerable amount of the LDV transportation demand would be covered through the Power-to-H_2-to-Power path. Although this route would enable zero CO_2 tailpipe emissions, the low overall round trip efficiency (28% in this work) requires overcapacity and infrastructure over-design. Peak demand also changes considerably. From an initial estimation of 745 MW with no BEVs considered, the grid demand reaches 827 MW, 44% higher than current figures.

Table 6. Main annual simulation results: autonomous renewable energy supply with and without LDV fleet demand.

	W/O LDV	LDV	LDV–Smart Charging
Demand [TWhe]	5.13	6.85	6.85
H_2 Turbines [TWhe]	0.91	1.11	0.99
Total RE [TWhe]	9.51	13.21	13.21
Wind Energy [TWhe]	3.73	4.15	4.15
Wind Energy Off-shore [TWhe]	1.71	3.62	3.62
Photovoltaic Energy [TWhe]	4.05	5,41	5.41
Geothermal [TWhe]	0.17	0.17	0.17
Wave Drive [TWhe]	0.02	0.03	0.03
Biomass [TWhe]	0.17	0.17	0.17
Curtailment [TWhe]	2.60	4.12	4.36
Curtailment [%]	27.34	31.19	33.01
EES Discharge [TWh]	0.66	0.94	0.69
EES Charge [TWh]	0.73	1.04	0.76
Storage Capacity at Maximum Level [% of time]	36.21	46.70	49.77
H_2 produced [TWh]	2.10	2.60	2.30
Electrolysis Capacity [MW]	1131	1206	1225
Electrolysis Electricity Demand [TWhe]	2.80	3.47	3.07
Electrolysis Operation [heq]	2475	2877	2505
Electrolysis Water Consumption [hm^3]	0.56	0.70	0.62
H_2 Turbine Capacity [MW]	653	877	779

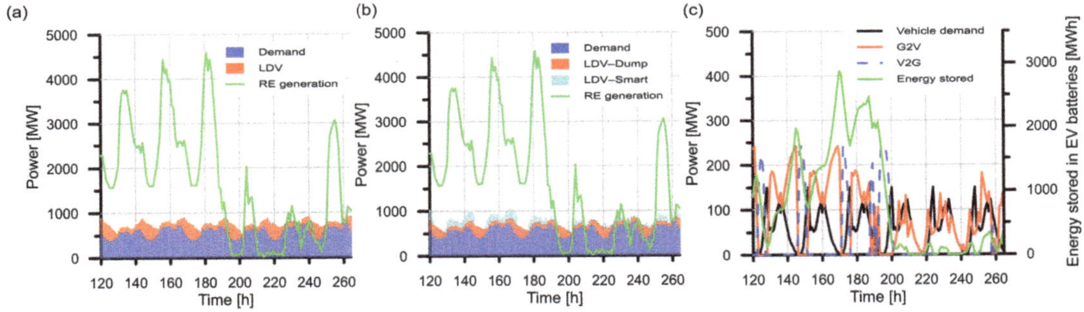

Figure 5. Impact of LDV demand on the grid system: (**a**) only dump charging considered, (**b**) smart charging for 30% of the fleet, and (**c**) example of bidirectional smart charging operation.

Smart charging (G2V/V2G technology) allows moving part of the transport demand to periods with large RE generation, located mainly around midday. In order to assess the impact of this technology on the system response, a sub-scenario was considered where a 30% of the light-duty transport demand was computed as smart charging. EnergyPLAN scheduled vehicle smart charging in periods with excess of RE generation and available battery energy capacity in vehicle batteries. Based on the overall storage capacity and the daily vehicle usage pattern, both imposed by the user, EnergyPLAN computed the hourly overall storage level of the smart fleet. The overall storage capacity value was estimated from the number of vehicles comprising the smart fleet, assuming an average battery capacity of 70 kWh [66] and 80% of the energy available for V2G/G2V functionality. The most recent vehicle usage pattern developed by the local government was imposed [67] to compute the hourly energy consumption and vehicle availability for charging. The charging power was limited by grid/vehicle connection capacity. A value of 351 MW was employed, assuming that 2 kW of charging infrastructure was available per vehicle (the current ratio in Europe is 1 kW/vehicle) [66].

EnergyPLAN allows power injection from vehicles to replace conventional generation if enough energy is available in batteries. The amount of power is, again, capped by the grid/vehicle connection capacity. Table 7 summarises the input parameters employed to simulate the smart charging functionality in EnergyPLAN.

Table 7. Input parameters employed to simulate the smart charging functionality.

	Values
Estimated smart fleet [-]	175.9 k
Average BEV consumption [kWh/km]	0.2
G2V/V2G connection capacity [MW]	352
Unit battery capacity [kWh]	70 [66]
Overall battery capacity [MWh]	985

Figure 5b,c exemplify the operation of the G2V/V2G functionality and the resulting load profile derived from its potential deployment. In the events of large RE generation (first three days), the grid enables significant charging (G2V) loads, increasing the direct contribution of RE to cover the vehicle demand, reducing the energy spilled and the grid stress at peak hours. The bidirectional power flow can also be observed in those days, with electrical vehicles providing power to the grid while they are parked (V2G) at night-time. The opposite behaviour can be observed in the next three days, with limited RE generation and, therefore, G2V capacity practically restricted to cover the daily vehicle demand.

As the reader can expect, the power-generation structure that provided the annual results previously shown in Table 6 is insufficient to cope with the additional LDV fleet demand. Further contribution of dispatchable sources (H_2 turbines) will be required, which means more H_2 production requirements and, therefore, more electricity to feed the electrolysers. This behaviour turns into a snowball effect, preventing both sectors to meet the goals of decarbonisation with an autonomous energy supply. In order to provide an estimation of the additional resources required to cope with the demand, the RE power and energy storage capacities were increased based on the following criteria:

1. For the sake of consistency with local government planning, the PV/WE ratio was kept constant.
2. Due to space constraints, the on-shore WE capacity was capped at 1700 MW, as previous reports suggest [46].
3. User-based EES was increased according to PV self-consumption growth.
4. Large-scale EES limits were set to 5.2 GWh, based on the estimations carried out by the local government [44].
5. A maximum curtailment level of 30% was considered to set the distributed EES capacity growth.

Table 8 includes the RE structure evolution required to extend the decarbonisation and autonomous supply targets to the LDV sector. An additional 1400 MW of RE capacity will be required (33% increase) to support a fully electric LDV fleet. To put these figure into context, the required additional capacity exceeds 3.9 times the current RE installed in the island. The deployment of additional capacity of non-dispatchable energy sources brings along new requirements for EES capacity, as Table 8 illustrate. Simulation results point out that 1GWh of EES, located at user and distributed levels, must be installed to avoid excessive curtailment levels.

The second column in Table 6 shows the annual results of this new scenario, considering the pre-defined dump charging. The problem of covering the additional LDV fleet demand (1.7 TWhe) with non-dispatchable sources characterised by high intermittency and low-capacity factors can be clearly seen in the tabulated data. To avoid a significant contribution of H_2 turbines that creates the aforementioned snowball effect, 1400 MW is required, generating 3.70 TWhe but increasing the curtailment level by 1.5 times despite the larger EES capacity. In addition to this, the H_2 turbine backup capacity must be expanded to 1200 MW and its contribution to support the grid demand increased by 200 GWh, which can be supported with a similar electrolyser capacity.

The annual results of testing the same power-generation structure assuming that 30% of the LDV fleet makes use of intelligently managed EV charging are included in the third column of Table 6. Shifting part of the charging loads to periods with larger RE availability reduces the contribution of H_2 turbines by 10%, as it requires a similar power capacity than the reference case without considering transport demand. The lesser electrolyser demand, which operates in periods with RE availability, explains the slightly higher curtailment values. Although the smart charging functionality seems to exhibit a benefit for the energy system, a much higher percentage of the fleet must be incorporated to reduce the RE and ESSs' capacity required for deployment. In addition, the infrastructure requirements are significant and a robust regulatory framework of this bidirectional functionality must be developed and put into practice.

Table 8. Current structure and RE and EES capacities required to reach autonomous supply in power-generation sector and light-duty vehicle transport.

	Current	Power Generation	LDV
Wind On-shore [MW]	222.60	1530	1700
Wind Off-shore [MW]	-	455	961
PV On-shore [MW]	107.60	1485	1980
PV Off-shore [MW]	-	24	32
PV Self-Consumption [MW]	38	746	995
EES–User [GW/GWh]	-	2.4/2.2	3.2/3.0
EES–Distributed [GW/GWh]	-	1.1/1.1	1.3/1.3
EES–Large-Scale [MW/GWh]	-	313/5.2	313/5.2

3.1.3. Heavy-Duty Transport

Although the use of H_2 as fuel for HD transport overcomes the main drawbacks of BEVs related to limited range and charging infrastructure requirements, the impact of its internal production on the energy system response must be evaluated. A parametric study varying the H_2 share as HD transport fuel was performed to illustrate how the energy system structure must evolve to supply the demand. The same five criteria described in Section 3.1.2 have been followed to carry out the simulations.

Figure 6 shows the additional RE and EES capacity requirements to reach full decarbonisation of the HDV fleet with an internal energy supply as a function of the H_2 penetration as fuel. Readily apparent from Figure 6 is the lesser RE and EES capacity requirements with the H_2 share. The different origin of the additional energy required to charge (BEVs) or fill the tanks (FCEVs) explain this behaviour. With respect to BEVs, large electricity demands located during overnight hours significantly stress the grid system. With no direct support of PV, the additional grid demand must be covered by EES (limiting the energy available to support RE intermittency) and H_2 turbines that must operate for longer periods. This behaviour increases the proportion of the low efficient Power-to-H_2-to-Power path. As a result, the RE capacity must be increased to reach 100% autonomy.

The opposite behaviour is observed by introducing higher shares of H_2 fuel. Although the system must generate the H_2 to feed the HDV fleet, the electrolysers' dispatch approach prioritises an excess of RE, turning into a better use of the available energy. Therefore, for energy systems similar to the one examined in this study, the use of H_2 requires less power-generation infrastructure and improves the overall system efficiency, quite the opposite of only taking the vehicle efficiency into account to carry out an analysis.

Regardless of the type of technology employed to propel the HDV fleet, the predicted additional power and storage capacity to be deployed are significant, with 600–900 MW and 2470–2750 MWh of RE and ESS capacities, respectively.

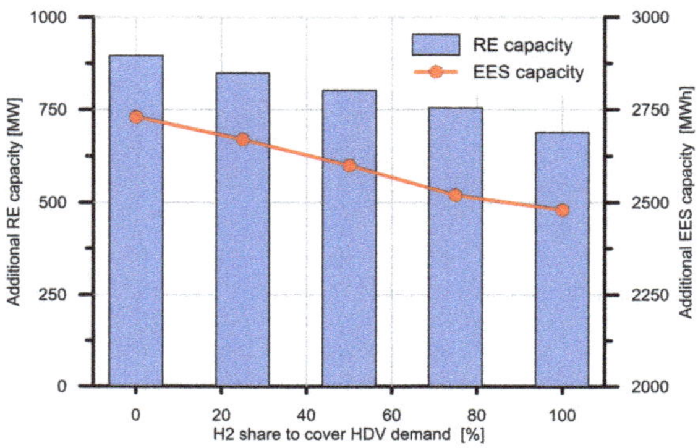

Figure 6. Additional RE and ES storage capacity requirements to reach 100% RE energy supply for the HDV fleet.

4. Discussion

Transitioning to zero-tailpipe-emission propulsion requires a deep analysis, not only from the CO_2 emission reduction potential standpoint, but also in terms of the infrastructure and energy provision. This is critical for isolated systems that are moving towards the objectives of climate neutrality and energy sovereignty from a totally different position. Figure 7 clearly exemplifies this statement, describing the percentage of energy self-sufficiency as function of the RE and EES capacity deployment requirements. Starting from situations with a large fossil fuel dependency, the penetration of non-dispatchable RE sources is very effective to reduce the contribution of fossil fuel-fired power plants and remove ICEV vehicles from the road. However, the entire replacement of technologies or processes that use fossil fuels with electrically-powered equivalents turns into a titanic effort. The significant increase of the energy demand would require up to 6 GW and 12 GWh of RE and EES capacities, respectively.

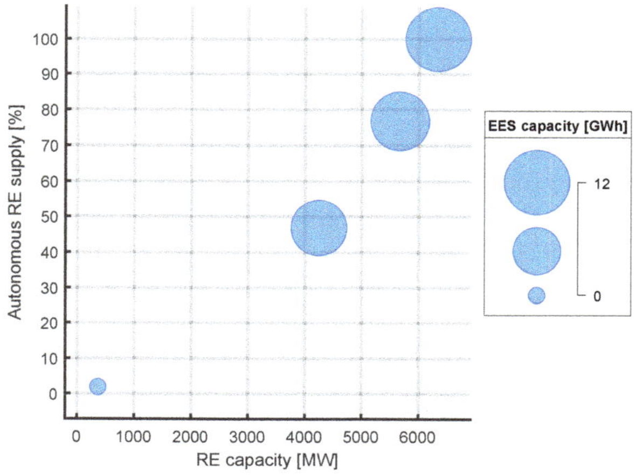

Figure 7. Percentage of energy autonomy as a function of the RE and EES capacity.

Besides the associated costs, deploying this level of infrastructure requires a considerable time frame. If 2050 is set as a reference, aligned with EU goals, if an up-to-date structure is taken as starting point, and, assuming an acceleration of capacity installation in the next decades, the annual penetration level should follow the trends included in Figure 8. However, current access and connection permits, which provide an approximate figure of oncoming new RE capacity, are 330 MW [68]. Although relevant (current RE capacity in the island is 333 MW), this figure is well below the required rate.

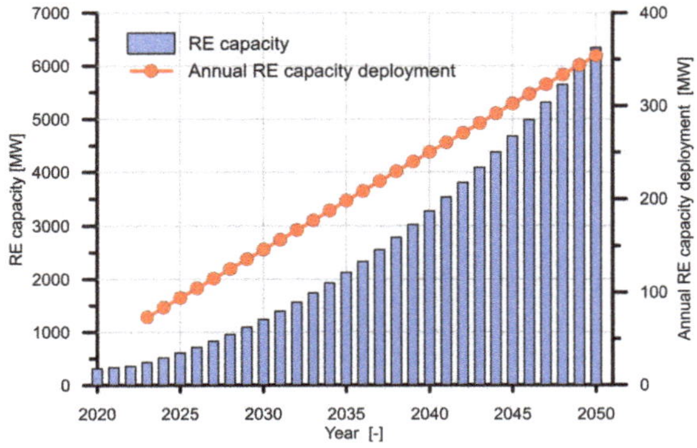

Figure 8. RE capacity deployment rate in the 2025–2050 time frame to reach decarbonisation with a fully autonomous RE supply.

5. Conclusions and Policy Implications

This paper explores a 100% renewable energy supply scenario for both power-generation and road transport sectors in the island of Tenerife. Representative of medium isolated energy systems, Tenerife island represents an excellent test case as both methodology and results can be completely transferable to the rest of isolated systems worldwide.

The island's energy system was modelled in EnergyPLAN, and validated against published data obtained from official governmental sources. Starting from the reference scenario and taking into account the current renewable technology roadmap, the impact of a full deployment of zero-tailpipe-emission vehicles on the energy system was evaluated under the consideration of an autonomous renewable energy supply.

Autonomous decarbonisation of the power-generation sectors implies a large deployment of renewable power capacity. The characteristics of these highly variable technologies, together with the nature of an isolated system, turn into high levels of curtailment (above 27%), even when considering energy storage systems. A significant back-up energy provided by dispatchable gen-sets fed by H_2 will be required to cover demand during renewable intermittency events. The internal H_2 production via electrolysis requires significant capacities that exhibit low equivalent hours of operation.

The impact of a complete technology shift in the transport sector (light- and heavy-duty) was analysed in detail. To accommodate the additional energy demand, the renewable capacity must experience a significant growth (more than 30%), requiring additional storage capacity to avoid excessive curtailment. Although the presence of V2G functionality in 30% of the light-duty fleet allows a slight reduction of the required back-up energy, the same deployment of renewable and energy storage sources is required. The analysis of the heavy-duty fleet reveals that large H_2 shares are beneficial for the energy system, as electrolysers' dispatch approach prioritises events of high availability of renewable energy. This behaviour eventually allows reducing the power and storage capacity to be installed.

In light of this, several recommendations to policy makers could be suggested:

- Energy demand reduction as a first objective. Although the results evince that a full autonomous energy supply is possible, the amount of RE to be deployed is massive. This results in a significant amount of electrical energy spilled and large EES capacities, which might put the profitability of the RE producers at risk.
- Dispatchable RE sources must be prioritised. Given the space constraints of the island, high enthalpy geothermal resources arise as the main opportunity on the island, which

require government boosting in terms of economic support during exploration and future exploitation (if it exists).

- The use of imported energy vectors to support internal demands. In this context, energy carriers such as drop-in fuels could play a key role in transportation. Their high energy density and the fact that can be used with existing technologies and infrastructure make these renewable fuels an excellent option to reduce the RE and ES requirements on the island. Without a doubt, the viability of this route must be studied in depth and it will eventually depend on how the power-to-fuel process maturity evolves.

Future research work includes a deeper analysis of different paths to reach decarbonisation on the island. Energy demand reductions in specific sectors and the comparison of additional scenarios that consider a partial dependence on energy vectors for transportation and/or power-generation will be evaluated. Every solution, to be optimised, must be quantified in terms of economic costs, deployment feasibility, and CO_2 abatement effectiveness.

Author Contributions: Conceptualisation, I.S.-M. and Ó.G.-A.; methodology, I.S.-M. and Ó.G.-A.; software, I.S.-M.; validation, I.S.-M.; formal analysis, I.S.-M. and Ó.G.-A.; investigation, I.S.-M. and Ó.G.-A.; writing—original draft preparation, I.S.-M., Ó.G.-A. and B.G.-D.; writing—review and editing, Ó.G.-A. and B.G.-D. All authors have read and agreed to the published version of the manuscript.

Funding: This research received no external funding.

Data Availability Statement: The data presented in this study are available on request. The data are not publicly available due to the fact that the dataset production is part of the innovative point of this paper.

Conflicts of Interest: The authors declare no conflicts of interest.

Abbreviations

The following abbreviations are used in this manuscript:

BEV	Battery electrical vehicle
E_D	Diesel energy consumption
E_P	Petrol energy consumption
$Elec_{LDV}$	Electric demand of the light-duty vehicle fleet
$Elec_{HDV}$	Electric demand of heavy-duty fleet
EES	Electrical Energy Storage
EV	Electric vehicle
FCEV	Fuel Cell Electric Vehicle
G2V	Grid To Vehicle
HD	Heavy duty
HDV	Heavy-duty vehicle
H_{2HDV}	Hydrogen demand of heavy-duty fleet
HFO	Heavy fuel oil
ICE	Internal combustion engine
ICEV	Internal combustion engine vehicle
LDV	Light-duty vehicle
PV	Photovoltaic
RE	Renewable energy
RES	Renewable energy system
V2G	Vehicle To Grid
WE	Wind energy
η_{ICE_P}	Petrol Internal Combustion Engine Efficiency
η_{ICE_D}	Diesel Internal Combustion Engine Efficiency
$\eta_{Charging}$	Charging efficiency
η_{EV}	Electric Vehicle Efficiency
η_{Trans}	Transport Efficiency

References

1. International Energy Agency. Tracking Transport. Available online: https://www.iea.org/energy-system/transport (accessed on 10 June 2024).
2. The International Council on Clean Transportation. *Vision 2050: A Strategy to Decarbonize the Global Transport Sector by Mid-Century*; The International Council on Clean Transportation: Washington, DC, USA, 2020.
3. European Environment Agency. *Transport and Environment Report 2021: Decarbonising Road Transport—The Role of Vehicles, Fuels and Transport Demand*; European Environment Agency: Copenhagen, Denmark, 2022.
4. European Commision. *Sustainable and Smart Mobility Strategy—Putting European Transport on Track for the Future*; European Commision: Brussels, Belgium, 2020.
5. Regulation (EU) 2024/1610 of the European Parliament and of the Council of 14 May 2024. *Amending Regulation (EU) 2019/1242 as Regards Strengthening the CO_2 Emission Performance Standards for New Heavy-Duty Vehicles and Integrating Reporting Obligations, and Repealing REGULATION (EU) 2018/956*; European Commision: Brussels, Belgium, 2024.
6. Maestre, V.M.; Ortiz, A.; Ortiz, I. Decarbonizing the Spanish transportation sector by 2050: Design and techno-economic assessment of the hydrogen generation and supply chain. *Int. J. Hydrogen Energy* **2023**, *48*, 39514–39530.
7. Alonso-Cepeda, A.; Villena-Ruiz, R.; Honrubia-Escribano, A.; Gómez-Lázaro, E. A Review on Electric Vehicles for Holistic Robust Integration in Cities: History, Legislation, Meta-Analysis of Technology and Grid Impact. *Appl. Sci.* **2024**, *14*, 7147.
8. Aneke, M.; Wang, M. Energy storage technologies and real life applications—A state of the art review. *Appl. Energy* **2016**, *179*, 350–377.
9. Kumar, S.; Lim, H. An overview of water electrolysis technologies for green hydrogen production. *Energy Rep.* **2022**, *8*, 13793–13813.
10. Sgarbossa, F.; Arena, S.; Tang, O.; Peron, M. Renewable hydrogen supply chains: A planning matrix and an agenda for future research. *Int. J. Prod. Econ.* **2023**, *255*, 108674.
11. Alves, M.; Segurado, R.; Costa, M. Increasing the penetration of renewable energy sources in isolated islands through the interconnection of their power systems. The case of Pico and Faial islands, Azores. *Energy* **2019**, *182*, 502–510.
12. Notton, G.; Stoyanov, L.; Ezzat, M.; Lararov, V.; Diaf, S.; Cristofari, C. Integration Limit of Renewable Energy Systems in Small Electrical Grid. *Energy Procedia* **2011**, *6*, 651–665.
13. Mustayen, A.; Rasul, M.; Wang, X.; Negnevitsky, M.; Hamilton, J. Remote areas and islands power generation: A review on diesel engine performance and emission improvement techniques. *Energy Conv. Manag.* **2022**, *160*, 115614.
14. Garcia-Afonso, O.; Santana-Mendez, I.; Delgado-Torres, A.M.; Gonzalez-Diaz, B. On the road to a sustainable transport mobility in isolated power systems: The role of light-duty powertrain electrification. *J. Clean. Prod.* **2021**, *320*, 128646.
15. Ramírez-Díaz, A.J.; Ramos-Real, F.J.; Barrera-Santana, J. Well-to-Wheels for Light-Duty Vehicle Powertrains by Segments in Isolated Systems. *Energies* **2023**, *16*, 1018.
16. Nuez, I.; Ruiz-García, A.; Osorio, J. A comparative evaluation of CO_2 emissions between internal combustion and electric vehicles in small isolated electrical power systems—Case study of the Canary Islands. *J. Clean. Prod.* **2022**, *369*, 133252.
17. García-Afonso, Ó.; González-Díaz, B. Effectiveness of zero tailpipe vehicles to reduce CO_2 emissions in isolated power systems, a realistic perspective: Tenerife Island test case. *Energy* **2023**, *273*, 127211.
18. Garcia-Afonso, O. Impact of powertrain electrification on the overall CO_2 emissions of intercity public bus transport: Tenerife Island test case. *J. Clean. Prod.* **2023**, *412*, 137365.
19. Parazdeh, M.; Kashani, N.; Fateh, D.; Eldoromi, M.; Moti Birjandi, A. EVs vehicle-to-grid implementation through virtual power plants. In *Scheduling and Operation of Virtual Power Plants: Technical Challenges and Electricity Markets*; Zangeneh, A., Moeini-Aghtaie, M., Eds.; Elsevier: Amsterdam, The Netherlands, 2022; pp. 299–324.
20. Clairand, J.-M.; Rodríguez-García, J.; Álvarez-Bel, C. Electric Vehicle Charging Strategy for Isolated Systems with High Penetration of Renewable Generation. *Energies* **2018**, *11*, 3188.
21. Díaz, A.R.; Ramos-Real, F.J.; Marrero, G.A.; Perez, Y. Impact of Electric Vehicles as Distributed Energy Storage in Isolated Systems: The Case of Tenerife. *Sustainability* **2015**, *11*, 15152–15178.
22. Wang, W.; Yuan, B.; Sun, Q.; Wennersten, R. Application of energy storage in integrated energy systems—A solution to fluctuation and uncertainty of renewable energy. *J. Energy Storage* **2022**, *52*, 104812.
23. Sayed, E.T.; Olabi, A.G.; Alami, A.H.; Radwan, A.; Mdallal, A.; Rezk, A.; Abdelkareem, M.A. Renewable Energy and Energy Storage Systems. *Energies* **2023**, *16*, 1415.
24. Shi, R.; Li, S.; Zhang, P.; Lee, K.Y. Integration of renewable energy sources and electric vehicles in V2G network with adjustable robust optimization. *Renew. Energy* **2020**, *153*, 1067–1080.
25. Tasnim, M.; Akter, S.; Shahjalal, M.; Shams, T.; Davari, P.; Iqbal, A. A critical review of the effect of light duty electric vehicle charging on the power grid. *Energy Rep.* **2024**, *89*, 111654.
26. Shukla, R.; Garg, M.; Panda, A. Driving grid stability: Integrating electric vehicles and energy storage devices for efficient load frequency control in isolated hybrid microgrids. *J. Energy Storage* **2023**, *10*, 4126–4147.
27. İnci, M.; Çelik, Ö.; Lashab, A.; Bayındır, K.Ç.; Vasquez, J.C.; Guerrero, J.M. Power System Integration of Electric Vehicles: A Review on Impacts and Contributions to the Smart Grid. *Appl. Sci.* **2024**, *14*, 2246.

28. Álvarez-Piñeiro, L.; Rivera, Y.; Berna-Escriche, C.; Blanco, D. Formulation of best estimate plus uncertainty methodologies for economy decarbonization in high-energy-demand isolated systems: Canary Islands forecasts for 2040. *Energy Conv. Manag.* **2024**, *314*, 118691.
29. Goncearuc, A.; De Cauwer, C.; Sapountzoglou, N.; Van Kriekinge, G.; Huber, D.; Messagie, M.; Coosemans, T. The barriers to widespread adoption of vehicle-to-grid: A comprehensive review. *Energy Rep.* **2024**, *12*, 27–41.
30. Farhoodnea, M.; Mohamed, A.; Shareef, H.; Zayandehroodi, H. Power Quality Impact of Renewable Energy based Generators and Electric Vehicles on Distribution Systems. *Procedia Technol.* **2013**, *11*, 11–13.
31. Bellocchi, S.; Manno, M.; Noussan, M.; Vellini, M. Impact of Grid-Scale Electricity Storage and Electric Vehicles on Renewable Energy Penetration: A Case Study for Italy. *Energies* **2019**, *12*, 1303.
32. Fachrizal, R.; Qian, K.; Lindberg, O.; Shepero, M.; Adam, R.; Widén, J.; Munkhammar, J. Urban-scale energy matching optimization with smart EV charging and V2G in a net-zero energy city powered by wind and solar energy. *eTransportation* **2024**, *20*, 100314.
33. Jiménez, A.; Cabrera, P.; Medina, J.F.; Østergaard, P.A.; Lund, H. Smart energy system approach validated by electrical analysis for electric vehicle integration in islands. *J. Energy Convers. Manag.* **2024**, *302*, 118121.
34. Xu, B.; Zhang, G.; Li, K.; Li, B.; Chi, H.; Yao, Y.; Fan, Z. Reactive power optimization of a distribution network with high-penetration of wind and solar renewable energy and electric vehicles. *Prot. Control Mod. Power Syst.* **2022**, *7*, 51.
35. Qiblawey, Y.; Alassi, A.; Zain ul Abideen, M.; Bañales, S. Techno-economic assessment of increasing the renewable energy supply in the Canary Islands: The case of Tenerife and Gran Canaria. *J. Energy Policy* **2022**, *162*, 112791.
36. Blanco, D.; Rivera, Y.; Berna-Escriche, C.; Muñoz-Cobo, J. Economy decarbonization using green hydrogen and electricity, forecasts and sensitivity analysis for the Canarian Islands in 2040. *J. Energy Storage* **2024**, *80*, 110232.
37. Meschede, H.; Child, M.; Breyer, C. Assessment of sustainable energy system configuration for a small Canary island in 2030. *J. Energy Storage* **2018**, *165*, 363–372.
38. Gobierno de Canarias. *Anuario Energético de Canarias 2022*; Gobierno de Canarias: Canarias, Spain, 2024.
39. Papadopoulos, A.M. Renewable energies and storage in small insular systems: Potential, perspectives and a case study. *Renew. Energy* **2020**, *149*, 103–114.
40. Garcia-Afonso, O.; Delgado-Torres, A.M.; Gonzalez-Diaz, B. Potential of Organic Rankine Cycle for waste heat recovery from piston engine-based power plants in isolated power systems. *Appl. Therm. Eng.* **2022**, *203*, 117815.
41. Red Eléctrica. Available online: https://www.ree.es/es/datos/balance (accessed on 10 July 2024).
42. Instituto Canario de Estadistica. Available online: https://www.gobiernodecanarias.org/istac/estadisticas/sectorservicios/transporte/terrestre (accessed on 19 January 2022).
43. Gobierno de Canarias. *Estrategia para el Autoconsumo Fotovoltaico*; Gobierno de Canarias: Canarias, Spain, 2021.
44. Gobierno de Canarias. *Estrategia del Almacenamiento Energético*; Gobierno de Canarias: Canarias, Spain, 2021.
45. Gobierno de Canarias. *Estrategia del Vehículo Eléctrico*; Gobierno de Canarias: Canarias, Spain, 2021.
46. Gobierno de Canarias. *Estrategia de la Generación Gestionable*; Gobierno de Canarias: Canarias, Spain, 2021.
47. Gobierno de Canarias. *Estrategia Canaria del Hidrógeno Verde*; Gobierno de Canarias: Canarias, Spain, 2022.
48. Gobierno de Canarias. *Plan de Transicion Energetica de Canarias (PTECan-2030)*; Gobierno de Canarias: Canarias, Spain, 2023.
49. Giacomo Prina, M.; Groppi, D.; Nastasi, B.; Astiaso Garcia, D. Bottom-up energy system models applied to sustainable islands. *Renew. Sustain. Energy Rev.* **2021**, *152*, 111625.
50. Lund, H. Chapter 2—Theory: Choice Awareness Theses. In *Renewable Energy Systems*; Academy Press: Boston, MA, USA, 2014; pp. 15–34.
51. Lund, H.; Thellufsen, J.; Østergaard, P.; Sorknæs, P.; Skov, I.; Mathiesen, B. EnergyPLAN—Advanced analysis of smart energy systems. *Smart Energy* **2021**, *1*, 100007.
52. EnergyPLAN. Advanced Energy Systems Analysis Computer Model. Available online: https://www.energyplan.eu/ (accessed on 19 June 2024).
53. Connolly, D.; Lund, H.; Mathiesen, B.V.; Leahy, M. A review of computer tools for analysing the integration of renewable energy into various energy systems. *Appl. Energy* **2010**, *87*, 1059–1082.
54. Yue, C.-D.; Chen, C.-S.; Lee, Y.-C. Integration of optimal combinations of renewable energy sources into the energy supply of Wang-An Island. *Renew. Energy* **2016**, *86*, 930–942.
55. Cabrera, P.; Lund, H.; Carta, J.A. Smart renewable energy penetration strategies on islands: The case of Gran Canaria. *Energy* **2018**, *162*, 421–443.
56. Lund, H.; Duić, N.; Krajačić, G.; da Graça Carvalho, M. Two energy system analysis models: A comparison of methodologies and results. *Energy* **2007**, *32*, 948–954.
57. Red Electrica de España. Canary Electricity Demand in Real-Time. Available online: https://demanda.ree.es/visiona/canarias/canarias/total (accessed on 30 November 2020).
58. Gobierno de Canarias. *Anuario Energético de 2019*; Gobierno de Canarias: Canarias, Spain, 2020.
59. Gobierno de Canarias. *Anuario Energético de Canarias 2020*; Gobierno de Canarias: Canarias, Spain, 2022.
60. Gobierno de Canarias. *Anuario Energético de Canarias 2021*; Gobierno de Canarias: Canarias, Spain, 2023.
61. Sistema de Información Territorial de Canarias. Available online: https://visor.grafcan.es (accessed on 12 January 2024).
62. Monitor Deloitte. *Los Territorios No Peninsulares Descarbonizados en 2040: La Vanguardia de la Transición Energética en España*; Deloitte: Madrid, Spain, 2020.

63. Hybrid Power Plants. Available online: https://www.wartsila.com/energy/solutions/hybrid-power-plants (accessed on 10 July 2023).
64. Curto, D.; Favuzza, S.; Franzitta, V.; Guercio, A.; Navarro Navia, M.A. Grid Stability Improvement Using Synthetic Inertia by Battery Energy Storage Systems in Small Islands. *Energy* **2022**, *254*, 124456.
65. European Road Transport Research Advisor Council (Belgium). *Long Distance Freight Transport: A Roadmap for System Integration of Road Transport*; ERTRAC: Brussels, Belgium, 2019.
66. Global EV Outlook 2020. Available online: https://www.iea.org/reports/global-ev-outlook-2020 (accessed on 1 July 2021).
67. Cabildo de Tenerife. Area de Carreteras, Movilidad e Innovacion. Available online: https://www.tenerife.es/portalcabtfe/es (accessed on 6 June 2024).
68. Red Electrica de España. Access and Connection Permits to Transport and Distribution Nodes. Available online: https://www.ree.es/es/clientes/generador/acceso-conexion (accessed on 20 July 2024).

applied sciences

MDPI

Article

Experimental Verification of a Compressor Drive Simulation Model to Minimize Dangerous Vibrations

Marek Moravič [1], Daniela Marasová [1,*], Peter Kaššay [2], Maksymilian Ozdoba [3], František Lopot [4] and Piotr Bortnowski [3]

[1] Faculty of Mining, Ecology, Process Control and Geotechnology, Technical University of Kosice, Letna 1/9, 042 00 Kosice, Slovakia; marek.moravic@tuke.sk
[2] Faculty of Mechanical Engineering, Technical University of Kosice, Letna 1/9, 042 00 Kosice, Slovakia; peter.kassay@tuke.sk
[3] Faculty of Geoengineering, Mining and Geology, Wroclaw University of Science and Technology, Na Grobli 15, 50-421 Wroclaw, Poland; maksymilian.ozdoba@pwr.edu.pl (M.O.); piotr.bortnowski@pwr.edu.pl (P.B.)
[4] Faculty of Mechanical Engineering, Czech Technical University in Prague, Technicka 4, 160 00 Prague, Czech Republic; frantisek.lopot@fs.cvut.cz
* Correspondence: daniela.marasova@tuke.sk; Tel.: +421-55-602-2591

Abstract: The article highlights the importance of analytical computational models of torsionally oscillating systems and their simulation for estimating the lowest resonance frequencies. It also identifies the pitfalls of the application of these models in terms of the accuracy of their outputs. The aim of the paper is to control the dangerous vibration of a mechanical system actuator using a pneumatic elastic coupling using different approaches such as analytical calculations, experimental measurement results, and simulation models. Based on the known mechanical properties of the laboratory system, its dynamic model in the form of a twelve-mass chain torsionally oscillating mechanical system is developed. Subsequently, the model is reduced to a two-mass system using the method of partial frequencies according to Rivin. The total load torque of the piston compressor under fault-free and fault conditions is simulated to obtain the amplitudes and phases of the harmonic components of the dynamic torque. After calculating the natural frequency and the natural shape of the oscillation, the Campbell diagram is processed to determine the critical revolutions. There is a pneumatic flexible coupling between the rotating masses, which changes the dynamic torsional stiffness. The dynamic torque curves transmitted by the coupling are compared with different dynamic torsional stiffnesses during steady-state operation and one cylinder failure. The monitored values are the position of the critical revolutions, the natural frequency, the natural shape of the oscillation, and the RMS of the dynamic load torque. The experimental model is verified by the simulation model. The accuracy of the developed simulation model with the experimental data are apparently very good (even more than 99% of the critical revolutions value obtained by calculation); however, it depends on the dynamic stiffness of the coupling. In this study, a detailed, comprehensive approach combining analytical procedures with simulation models is presented. Experimental data are verified with simulation results, which show a good agreement in the case of 700 kPa coupling pressure. The inaccuracy of some of the experiments (at 300 and 500 kPa pressures) is due to the interaction of the coupling's apparent stiffness and the level of the damped vibration energy in the coupling, which is manifested by its different heating. Based on further experiments, a solution to these problems will be proposed by introducing this phenomenon effectively into the simulation model.

Keywords: natural frequency; mode shape; piston machines; torsional vibration; pneumatic flexible coupling

1. Introduction

Nowadays, ever growing emphasis is being placed on low levels of vibration and noise generated during the operation of machinery and equipment and transmitted to the surrounding area. They represent a detrimental factor in terms of environmental and general operational safety [1]. Numerical simulation tools such as Fluent software and k–ϵ turbulence models are currently used to assess and control dangerous factors (including vibration) in the workplace [2].

Vibrations have a negative effect on the operation of mechanical systems and on the safety of operation—their reliability and also their service life are reduced [3–5]. It is generally known that various types of sources are involved in the generation of vibrations in the structural parts of the machine, mainly unbalance, misalignment, mechanical clearance of the bearings, resonance of the structure, bent shaft, excessive wear of bearings, damage to the gearing [6,7], lubrication [8–10], etc. These negative phenomena and their causes can be identified during the operation of the equipment by measuring the methods of technical diagnostics and monitoring [11–13]. As a result of inertial forces and variable load torque of individual machines and equipment, torsional oscillations occur [14–17]. Torsional vibration can be dangerous, depending on the level for a particular mechanical system and the place of its occurrence. In torsionally oscillating mechanical systems (mainly mechanical systems with internal combustion engines [18,19], piston compressors, piston pumps, and fans), which include flexible members, it is possible to reduce the dangerous torsional oscillation to an acceptable level by suitable adaptation of the dynamic properties (i.e., stiffness [20–24], damping coefficient [25], and mass moment of inertia [26]) of these members to the dynamics of the system [27,28]. However, this may not only apply to elastic members but also to other members in the mechanical system. Recent developments include so-called active vibration isolation, where an additional torque is generated in real time directly at the driving electric motor or a special active damper [29–32]. Dynamic properties largely affect the operation of the mechanical system, its lifetime, and reliability; therefore, to perform the above-mentioned procedure, the torsionally oscillating mechanical system needs to be appropriately tuned in advance [33]. The tuning of any mechanical system is most often verified on the basis of the results of the so-called Campbell diagram. It is used to determine whether resonance will occur [34]. The research of vibration problems in mechanical systems is not a new phenomenon in the field of engineering sciences. Vibrations represent phenomena occurring in every mechanical system since its manufacture. They are negative phenomena that interfere with the smooth operation of mechanical systems [35] and act as a threatening and risk factor. Therefore, it is both necessary and required to look for ways of their minimization [36,37].

Based on this fact, the main objective of this paper is to control the dangerous vibrations of the drive of a mechanical system by applying a pneumatic flexible coupling based on dynamic simulation models verified by experimental measurements. The main objective of the paper can be achieved by fulfilling the partial objectives, namely: selection of the comparison model, preparation of the mechanical system in the laboratory for experiments, obtaining input data for the model, calculation of natural frequencies and mode shapes, dynamic analysis, creation of the simulation model, realization of the experiments, verification of the model, and design of recommendations for control of dangerous vibrations.

2. Materials and Methods

With the development of new information and simulation technologies, sensors, and more and more methods are being used for the elimination of dangerous vibrations. Current approaches based on theoretical models using parameter-driven mathematical expressions to describe the mechanical system or monitoring-based data acquisition are no longer sufficient to ensure their reliability. Increasingly, these methods are being replaced by new techniques and software such as simulation models (e.g., digital twins) [38], or machine learning approaches [39]. Machine learning algorithms are difficult to interpret, and a large volume and quality of data are required [40]. By integrating analytical models and online data with each other, digital twins will enable real-time control of mechanical systems.

2.1. Research Methodology

The design of the research methodology presented in this paper is based on the integration of analytical models and simulation models. To calculate the dynamic properties of aggregates, it is necessary to replace the aggregate with a suitable model of the mechanical system describing to some extent the real system before constructing the differential equations of motion. When calculating the dynamic phenomena of torsionally oscillating mechanical systems, the subject of the calculation is mainly the stress on the flexible couplings, as the weakest member of the mechanical system [41]. Here, it is appropriate to simplify the complex mechanical system as much as possible, but only enough to determine the stresses on the couplings with sufficient accuracy.

In order to achieve the research objective of controlling the dangerous vibration of the drive of the mechanical system by means of a flexible coupling based on simulation and experimental measurements, the following methodology was proposed (Figure 1).

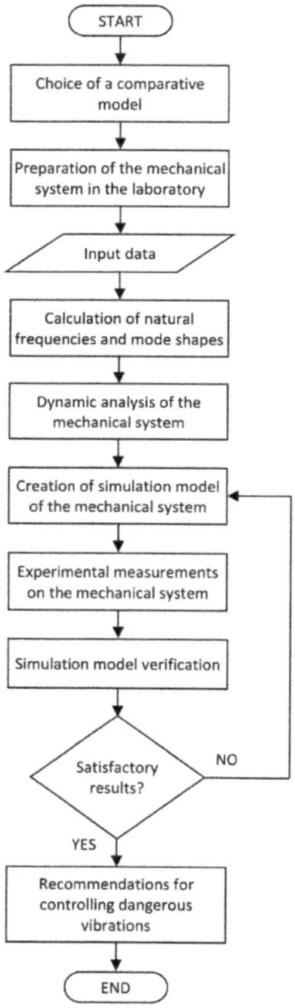

Figure 1. Research methodology.

In the reduction in masses in mechanical systems, the method of partial frequencies according to Rivin [42] was used. The mechanical system was reduced to a two-mass system. After calculating the natural frequency and the natural shape of the oscillation, the Campbell diagram was processed to determine the critical revolutions. The dynamic calculation methodology of the mechanical system was applied in our research.

The dynamic analysis was carried out using a MATLAB program, which outputs the values of harmonic components of the torque transmitted by the flexible coupling. The output of the paper is the results of experimental verification of the simulation model of the mechanical system with variable torque operating at constant load and its comparison with the mathematical model.

2.2. Description of the Mechanical System

The subject of the research is a laboratory mechanical system driving a compressor by an electric motor (Figure 2). It is a mechanical system operating with variable torque at constant load. Between the rotating masses there is a pneumatic flexible coupling whose dynamic torsional stiffness can be varied during operation by changing the air pressure [27].

Figure 2. Experimental model of a mechanical system: 1—electric motor, 2—gearbox, 3—torque sensor, 4—bearing housing, 5—pneumatic flexible coupling, 6—three-cylinder piston compressor, 7—rotation supply.

The revolutions of the electric motor (1) are continuously vector controlled by a frequency converter. The electric motor drives a three-cylinder ORLIK 3JSK-75 piston compressor (6) through a 1:1 gearbox (2) and a 4-1/70-T-C type pneumatic flexible coupling (5). The compressor is mounted on a rubber layer and has no flywheel; therefore, its impact on dynamics is higher. The compressed air from the compressor flows into an air pressure vessel with a capacity of 300 L. The compressed air is fed to the pneumatic coupling via a rotation supply (7). The magnitude of the torsional vibrations is measured by a torque sensor type 7934 MOM KALIBERGYAR (3). Table 1 describes the technical parameters of some members of the mechanical system.

The air pressure in the compression chamber of the pneumatic flexible coupling is measured by a pressure sensor. The signals from both sensors are amplified and processed by a universal eight-channel measuring device MX840 from the manufacturer HBM, and the data are subsequently sent to a PC (Figure 3) [43]. The operating revolutions of the mechanical system are $n_P = 600\ \text{min}^{-1}$.

Sensors used:

1. Torque sensor type 7934, manufacturer MOM KALIBERGYÁR, measuring range 0–500 N·m, accuracy of the torque sensor is 0.1% of its measuring range, i.e., 0.5% (combined failure—nonlinearity, hysteresis, and reproducibility).
2. pressure sensor type MBS 3000, manufacturer Danfoss, overpressure measurement range 0–1 MPa, accuracy of the MBS 3000 sensor with metal diaphragm is 0.5% of its measurement range, i.e., 5 kPa (combined failure—non-linearity, hysteresis, and reproducibility).

Table 1. Basic parameters of some members of the mechanical system.

Elements of the Mechanical System	Parameters
Three-phase asynchronous motor 1LE1001-1DB234AF4-Z SIEMENS 	Nominal revolutions n_N = 1470 min^{-1} pri 50 Hz Mass moment of inertia I_{AM} = 0.065 kg·m^2 Synchronous revolutions n_S = 1500 min^{-1} Nominal power P_N = 11 kW
Compressor 3JSK-75 ORLIK 	Mass moment of inertia I_{PK} = 0.02967 kg·m^2 Number of cylinders i_{VPK} = 3 Cylinder diameter d_{VPK} = 82 mm Piston stroke z_{PPK} = 70 mm
Two-stage spur gearbox OBERAIGNER 	Gear ratio i_P = 1:1 Axial distance a = 250 mm
Pneumatic flexible coupling 4-1/70-T-C 	Nominal torque M_{KN} = 170 N·m Mass moment of inertia of the flange I_{SA} = 0.06 kg·m^2 Mass moment of inertia of the flange I_{SB} = 0.07 kg·m^2 Dynamic torsional stiffness: ▪ k_{DSMAX} = 2250 N·m·rad^{-1} at pressure p_S = 700 kPa in the elastic element, ▪ k_{DSMAX} = 1780 N·m·rad^{-1} at pressure p_S = 500 kPa in the elastic element, ▪ k_{DSMAX} = 1360 N·m·rad^{-1} at pressure p_S = 300 kPa in the elastic element.

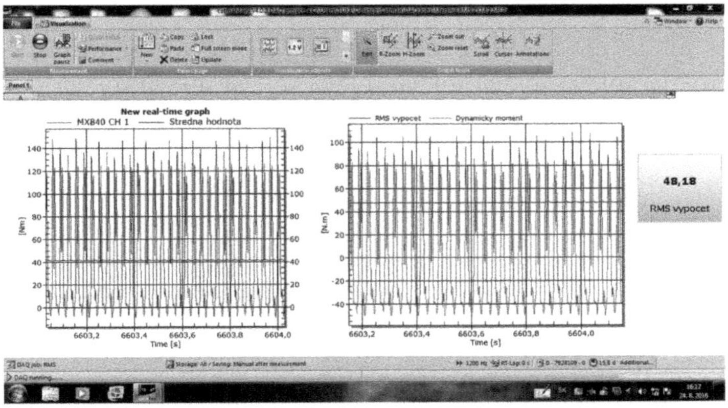

Figure 3. Time record from an experiment measurement in the CatmanEasy program.

3. Results and Discussion

3.1. Determination of the Operating State of the System in Terms of Resonance

The dynamic model (Figure 4) was created based on the known mechanical parameters of the compressor drive system. By means of the model, the values of natural frequencies and mode shapes were calculated as well as the Campbell diagram devised.

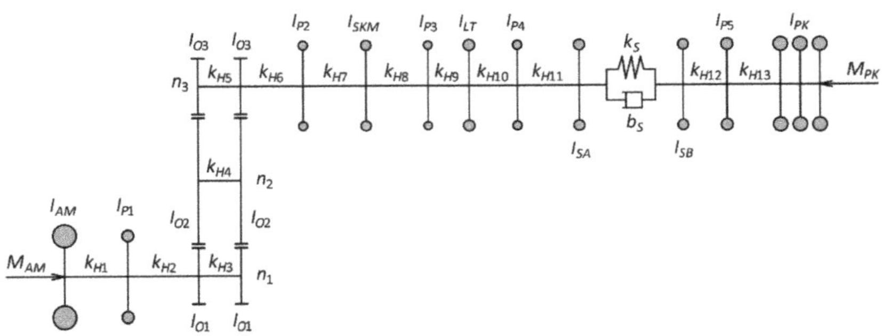

Figure 4. A complete replacement dynamic piston compressor drive model.

The individual elements for the dynamic model are listed in Table 2.

Table 2. Description of parameters of the system shown on Figure 4.

Parameter	Meaning
I_{AM}	Mass moment of inertia of the electric motor
I_{P1}	Mass moment of inertia of the flange behind the motor
I_{O1}, I_{O2}, I_{O3}	Mass moment of inertia of the gear
I_{P2}	Mass moment of inertia of the flange behind the gearbox
I_{SKM}	Mass moment of inertia of the torque sensor
I_{P3}	Mass moment of inertia of the flange behind the torque sensor
I_{LT}	Mass moment of inertia of the bearing housing
I_{P4}	Mass moment of inertia of the flange in front of the pneumatic coupling
I_{SA}	Mass moment of inertia of the flange of the pneumatic coupling
I_{SB}	Mass moment of inertia of the pneumatic coupling flange
I_{P5}	Mass moment of inertia of the flange behind the flexible coupling
I_{PK}	Mass moment of inertia of the piston compressor
$k_{H1}, k_{H2},$	Torsional stiffness of shafts between electric motor and gearbox

Table 2. *Cont.*

Parameter	Meaning
k_{H3}, k_{H4}, k_{H5}	Torsional stiffness of shafts in the gearbox
k_{H6}	Torsional stiffness of the shaft between I_{O3} and I_{P2}
k_{H7}	Torsional stiffness of the shaft between I_{P2} and I_{SKM}
k_{H8}	Torsional stiffness of the shaft between $I_{SKM}$and I_{P3}
k_{H9}	Torsional stiffness of the shaft between I_{P3} and I_{LT}
k_{H10}	Torsional stiffness of the shaft between I_{LT} and I_{P4}
k_{H11}	Torsional stiffness of the shaft between I_{P4} and I_{SA}
k_S	Torsional stiffness of the pneumatic coupling
k_{H12}	Torsional stiffness of the shaft between I_{SB} and I_{P5}
k_{H13}	Torsional stiffness of the shaft between I_{P5} and I_{PK}
b_S	Pneumatic coupling damping coefficient
M_{AM}	Electric motor torque
M_{PK}	Load torque of the piston compressor

The gear ratios between the shafts of the system are calculated as follows:

$$i_{1\text{-}2} = \frac{n_1}{n_2},\ i_{2\text{-}3} = \frac{n_2}{n_3},\ i_{1\text{-}3} = \frac{n_1}{n_3}. \quad (1)$$

Assuming that the shaft stiffnesses k_{H3}, k_{H4}, and k_{H5} (Figure 4) show a higher value than the other stiffnesses and the input shaft revolutions of the gearbox are equal to the output shaft revolutions $n_1 = n_3$, a reduction as shown in Table 3 could be done. This reduction is applied to the high-revolution shaft with stiffness k_{H1}.

Table 3. Reduction in mass moments of inertia and torsional stiffness on the high-revolution shaft.

Mass Moments of Inertia		Element Torsional Stiffnesses	
$I_1 = I_{AM}$	$I_7 = \frac{I_{LT}}{i_{1\text{-}3}^2}$	$k_1 = k_{H1}$	$k_7 = \frac{k_{H10}}{i_{1\text{-}3}^2}$
$I_2 = I_{P1}$	$I_8 = \frac{I_{P4}}{i_{1\text{-}3}^2}$	$k_2 = k_{H2}$	$k_8 = \frac{k_{H11}}{i_{1\text{-}3}^2}$
$I_3 = 2 \cdot I_{O1} + \frac{2 \cdot I_{O2}}{i_{1\text{-}2}^2} + \frac{2 \cdot I_{O3}}{i_{1\text{-}3}^2}$	$I_9 = \frac{I_{SA}}{i_{1\text{-}3}^2}$	$k_3 = \frac{k_{H6}}{i_{1\text{-}3}^2}$	$k_9 = \frac{k_S}{i_{1\text{-}3}^2}$
$I_4 = \frac{I_{P2}}{i_{1\text{-}3}^2}$	$I_{10} = \frac{I_{SB}}{i_{1\text{-}3}^2}$	$k_4 = \frac{k_{H7}}{i_{1\text{-}3}^2}$	$k_{10} = \frac{k_{H12}}{i_{1\text{-}3}^2}$
$I_5 = \frac{I_{SKM}}{i_{1\text{-}3}^2}$	$I_{11} = \frac{I_{P5}}{i_{1\text{-}3}^2}$	$k_5 = \frac{k_{H8}}{i_{1\text{-}3}^2}$	$k_{11} = \frac{k_{H13}}{i_{1\text{-}3}^2}$
$I_6 = \frac{I_{P3}}{i_{1\text{-}3}^2}$	$I_{12} = \frac{I_{PK}}{i_{1\text{-}3}^2}$	$k_6 = \frac{k_{H9}}{i_{1\text{-}3}^2}$	—

A dynamic model with twelve-mass discs after the first reduction is shown in Figure 5.

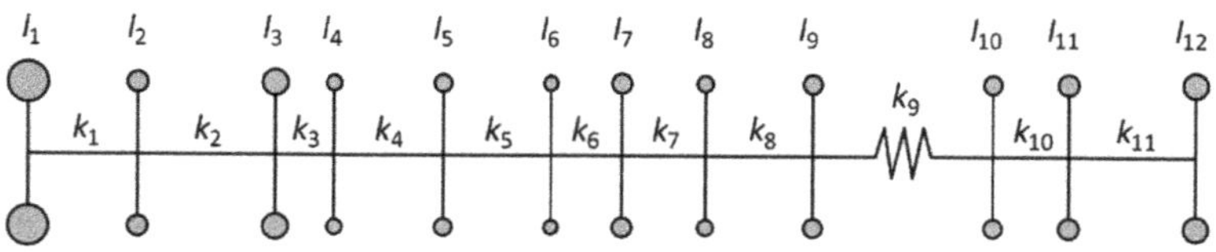

Figure 5. A twelve-mass chain mechanical system.

Using the method of partial frequencies [42], this twelve-mass system was reduced to a two-mass system. The resulting values after reduction by this method usually show an accuracy of 3–5% [42,44].

The dynamical model of this reduced two-mass mechanical system is shown in Figure 6.

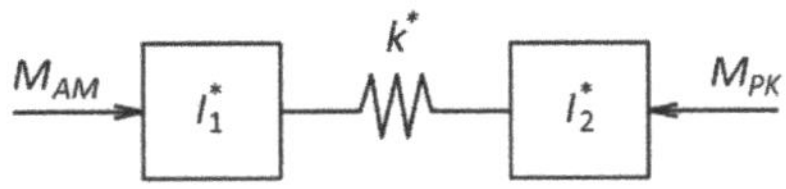

Figure 6. Two-mass reduced mechanical system.

The analysis of torsional vibration of a mechanical system is based on a system of differential equations of motion. The equations describing a two-mass reduced mechanical system without damping (Figure 6) have the following form:

$$I_1^* \cdot \ddot{\varphi}_1 + k^* \cdot (\varphi_1 - \varphi_2) = M_{AM}, \quad (2)$$

$$-I_2^* \cdot \ddot{\varphi}_2 + k^* \cdot (\varphi_1 - \varphi_2) = M_{PK}. \quad (3)$$

The variable load torque of a piston compressor is given by:

$$M_{PK} = M_{NPK} + \sum_{i=1}^{\infty} M_{iPK} \cdot cos(i \cdot \omega_{PK} \cdot t + \gamma_{iPK}). \quad (4)$$

The mode shape of the oscillation at coupling pressure is shown in Figure 7.

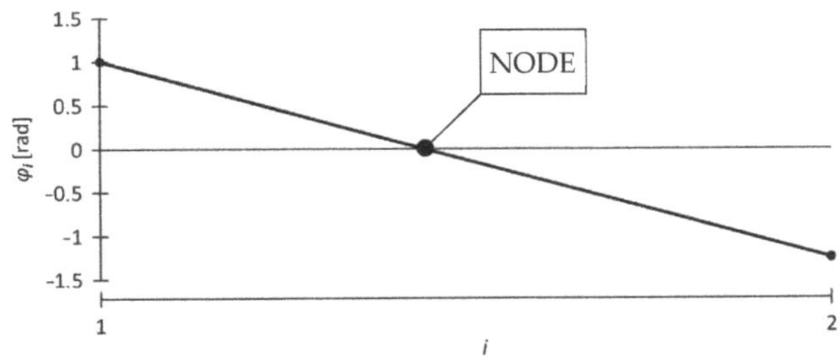

Figure 7. Mode shape of the strain curve for Ω_0 = 201.431 rad·s^{-1}.

A Campbell diagram is an interference diagram that identifies possible matches between the expected excitation frequencies and the calculated natural torsional frequencies over the operating speed range. Each ascending line on the interference diagram represents a single mechanical or electrical excitation frequency or harmonic; when they intersect with the torsional natural frequency (horizontal line), torsional resonance occurs. The intersection of those lines indicates a possible torsional resonance condition. For some systems, an interference diagram may be all that is needed to show that torsional resonances will not be excited [45,46].

Figure 8 shows the Campbell diagram for a system operating at coupling pressure 700 kPa with the following parameters marked in it: the operating rotational frequency n_P = 600 min^{-1}, the critical revolutions from the principal harmonic element n_{K3} = 641.2 min^{-1} and the natural rotational frequency N_0 = 1923.526 min^{-1}.

The detuning coefficient for the *i*-th harmonic component is defined by the formula:

$$\eta_i = \frac{n \cdot i}{N_0}. \quad (5)$$

Table 4 gives the detuning coefficients of the harmonic components operating rotational frequency n_P = 600 min^{-1}. Since this is a three-cylinder piston compressor, the main harmonic components will be integer multiples of the third harmonic component.

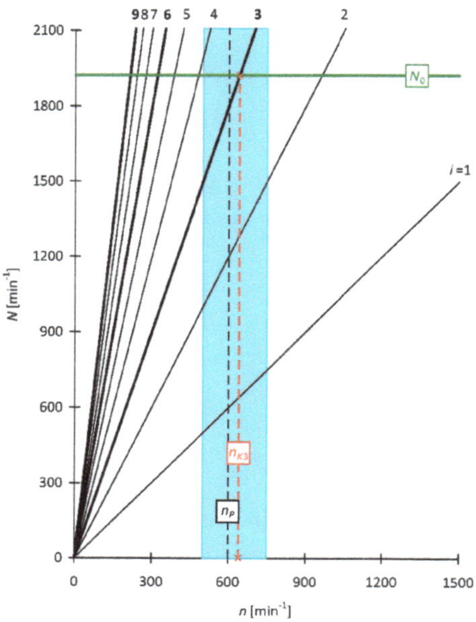

Figure 8. Campbell diagram of the mechanical system studied.

Table 4. The detuning coefficient of the selected harmonic components at coupling pressure 700 kPa.

i	1	2	3	4	5	6
η_i	0.312	0.624	0.936	1.248	1.560	1.872

These will always appear during the load torque; the other harmonic components are the so-called secondary harmonic components, which will only appear if the compressor cylinders do not operate evenly [45–48].

According to the literature [27], the system operates in the resonance region if:

$$0.8 < \eta_i < 1.2. \tag{6}$$

From the results shown in Table 4, we can say that the system at coupling pressure 700 kPa operates in the resonance region.

3.2. *Dynamic Calculation of a Mechanical System Under Forced Oscillation*

To calculate the excitation of the forced oscillation, we based it on the aforementioned compressor parameters, where we first determined the *p-V* diagram of the operating cycle of a single cylinder of a three-cylinder piston compressor, which is contained in Figure 9.

Figure 10 is already a graphical representation of the total load torque M_C and its mean load torque M_S at the crankshaft from all three cylinders of the compressor.

In this case, it is a fault-free condition of the piston compressor. The amplitudes and phases of the harmonic components of the dynamic torque in the fault-free condition are described in Table 5.

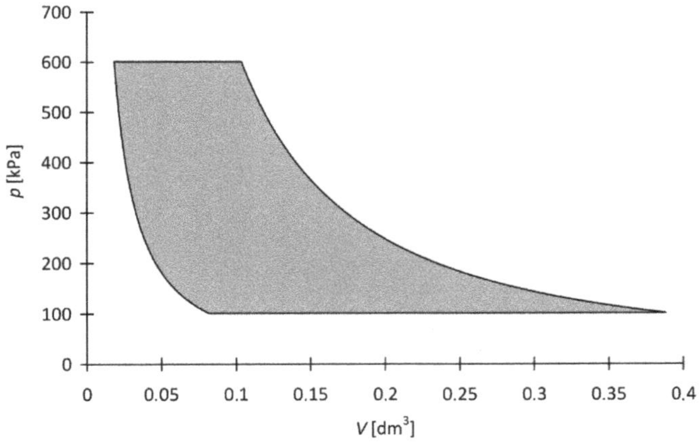

Figure 9. *p-V* diagram of piston compressor cylinder.

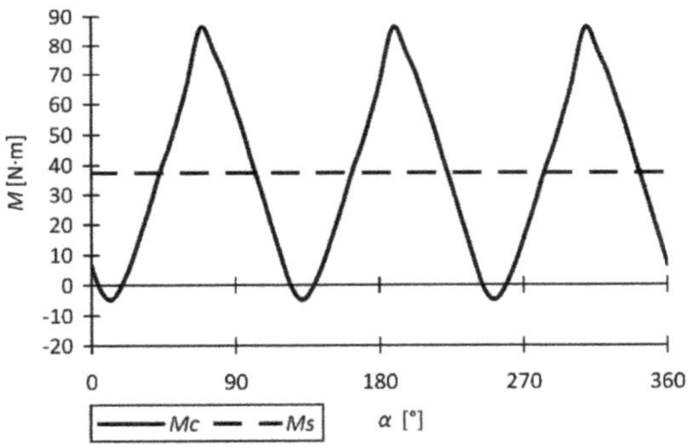

Figure 10. The total load torque M_C vs. the rotation angle of the crankshaft of the piston compressor and its static component M_S.

Table 5. Amplitudes and phases of harmonic components of the dynamic torque of the piston compressor.

i	M_{iPK} [N·m]	γ_{iPK} [rad]	i	M_{iPK} [N·m]	γ_{iPK} [rad]
1	0	0	7	0	0
2	0	0	8	0	0
3	40.381	3.804	9	3.658	4.818
4	0	0	10	0	0
5	0	0	11	0	0
6	1.394	1.286	12	1.233	1.976

3.3. Dynamic Calculation of the Mechanical System at Steady Compressor Operation

Figure 11 plots the calculated time waveform of the dynamic torque transmitted by the coupling for the duration of one operating cycle (one crankshaft revolution) of a piston compressor at its steady-state operation and operating revolutions $n_P = 600\ \text{min}^{-1}$.

The hysteresis loop of the pneumatic flexible coupling (Figure 12) was plotted from the time waveform (Figure 11).

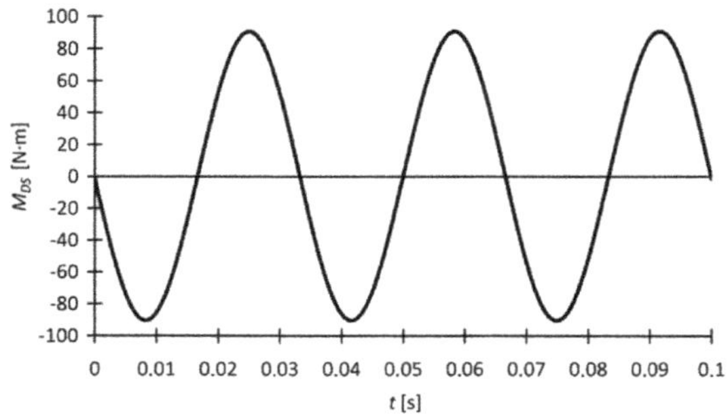

Figure 11. Development over time of dynamic torque transmitted by a pneumatic flexible coupling at pressure 700 kPa.

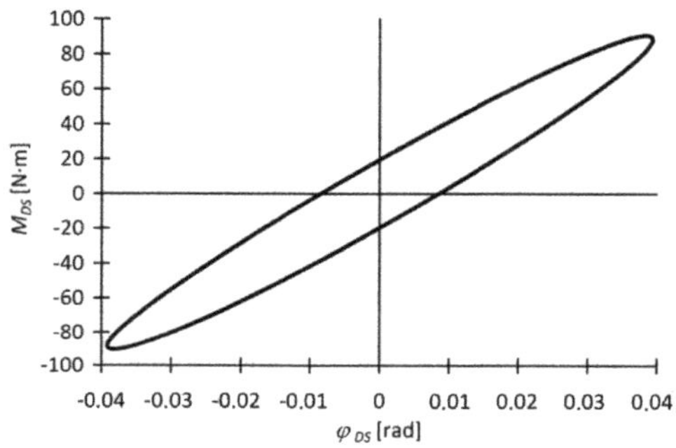

Figure 12. Hysteresis loop of flexible pneumatic coupling at pressure 700 kPa.

The amplitude of the dynamic load torque transmitted by the pneumatic coupling as a function of revolutions is shown in Figure 13.

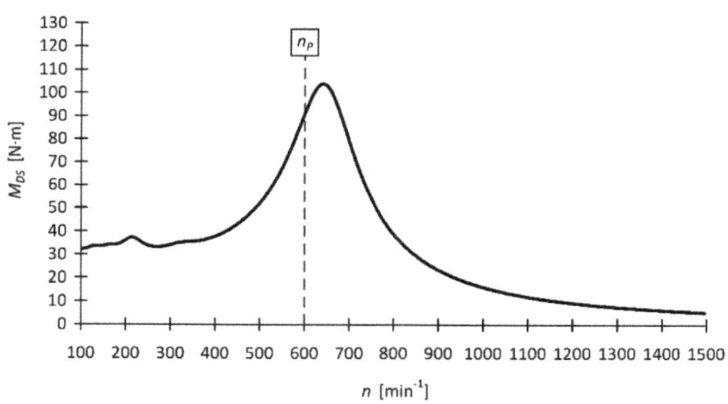

Figure 13. Dynamic load torque transmitted by the flexible pneumatic coupling at pressure 700 kPa.

In order to compare the result of the effective value of the dynamic load torque component of the *RMS* M_{DS} simulation model with the experimental model, we modified the M_{DS} (Figure 13) to the *RMS* M_{DS} (Figure 14).

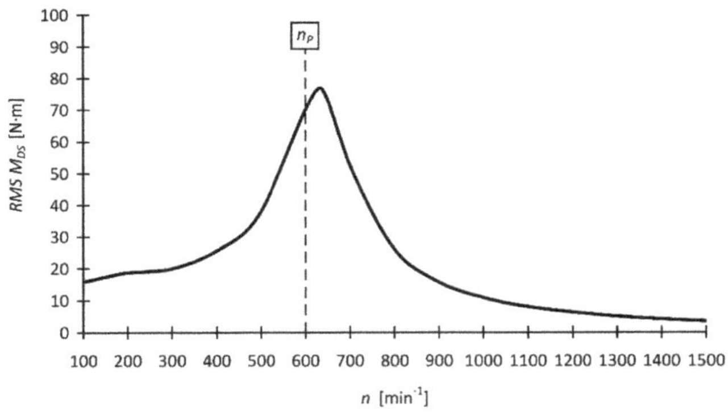

Figure 14. The effective value of the dynamic component of the torque *RMS* M_D transmitted by the pneumatic flexible coupling at pressure 700 kPa.

3.4. Simulation Model for Steady (Even) Compressor Operation

In technical practice, as already mentioned, machine failures often occur, such as cylinders falling out of action. The following section is devoted to this issue. The failure of one cylinder was simulated by setting its overpressure value at displacement against atmospheric pressure to $p = 0$ Pa. Figure 15 shows the calculated waveform of the total load torque M_C and mean load torque M_S at the compressor crankshaft.

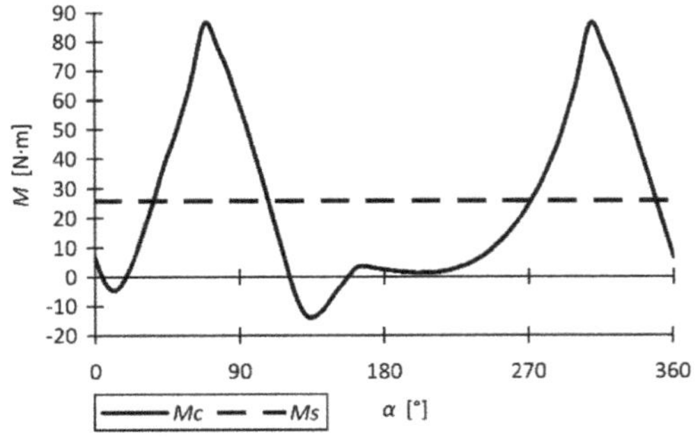

Figure 15. Total load torque M_C depending on the angle of rotation of the α crankshaft when the cylinder of the piston compressor is out of operation and its static component M_S.

Table 6 gives the amplitudes and phases of the harmonic components of the load torque with one cylinder out of operation.

Table 6. Amplitudes and phases of harmonic components of the dynamic torque of the piston compressor with one cylinder out of operation.

i	M_{iPK} [N·m]	γ_{iPK} [rad]	i	M_{iPK} [N·m]	γ_{iPK} [rad]
1	22.804	0.039	7	1.507	0.61
2	19.55	3.4	8	1.676	4.321
3	26.664	3.786	9	2.438	4.817
4	7.84	4.43	10	0.739	5.056
5	3.163	1.884	11	0.526	1.917
6	0.917	1.33	12	0.826	1.989

Figure 16 shows the time waveform of the dynamic load torque transmitted by the pneumatic flexible coupling with one cylinder out of operation.

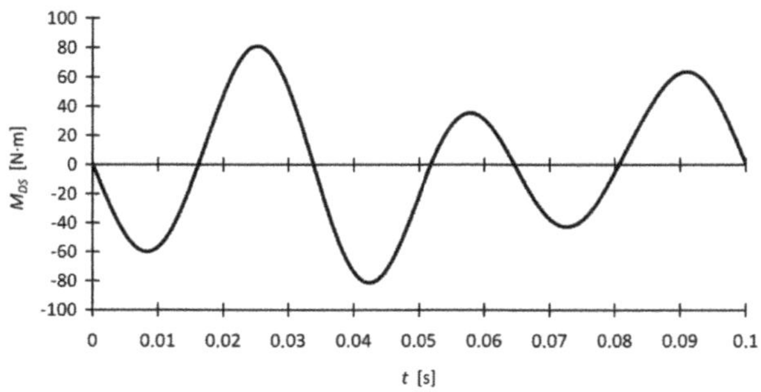

Figure 16. Development over time of the dynamic torque M_{DS} transmitted by a pneumatic flexible coupling at pressure 700 kPa with the cylinder out of operation.

The hysteresis loop of the pneumatic flexible coupling in Figure 17 was derived from Figure 16.

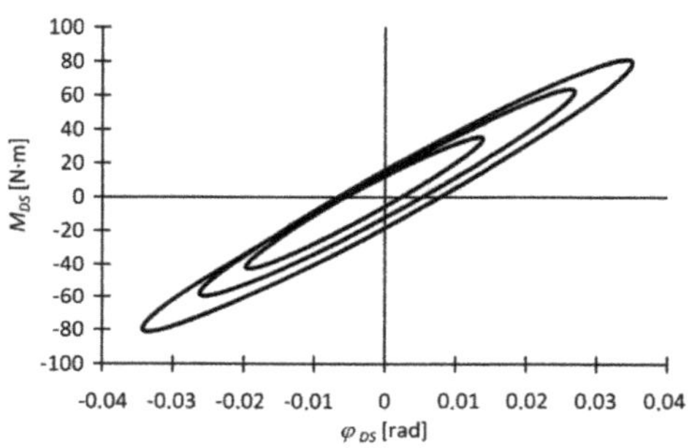

Figure 17. Pneumatic flexible coupling hysteresis loop at pressure 700 kPa with the cylinder out of operation.

Figure 18 describes the waveform of the amplitude of the dynamic load torque transmitted by the pneumatic flexible coupling with one cylinder out of operation as a function of revolutions.

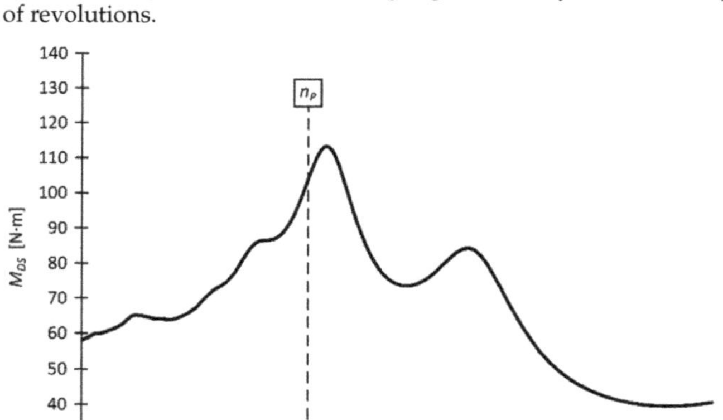

Figure 18. Dynamic torque M_{DS} transmitted by pneumatic flexible coupling at pressure 700 kPa with the cylinder is out of operation.

Figure 19 shows the effective value of the dynamic component of the load torque *RMS* M_{DS} to compare with the experiment.

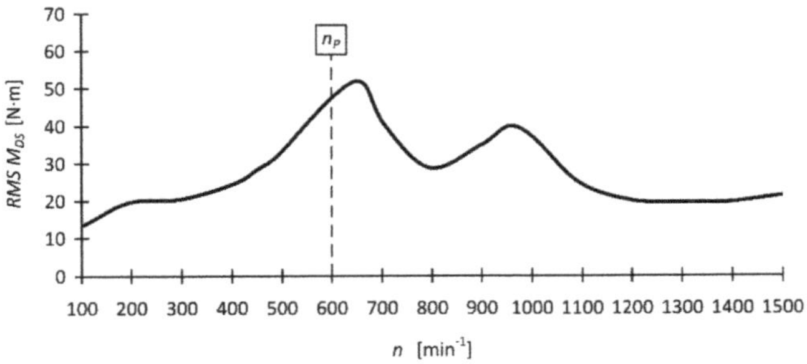

Figure 19. Effective value of the dynamic component of the load torque *RMS* M_{DS} transmitted by a pneumatic flexible coupling at pressure 700 kPa with the cylinder out of operation.

For additional pressures in the elastic element of the coupling, the resulting waveforms are given in Tables 7 and 8.

Table 7. Dynamic torque-time waveform M_{DS} and hysteresis lop at pressure p_S = 500 kPa in the elastic element.

	k_{DS} = 1780 N·m·rad^{-1}	
Steady state	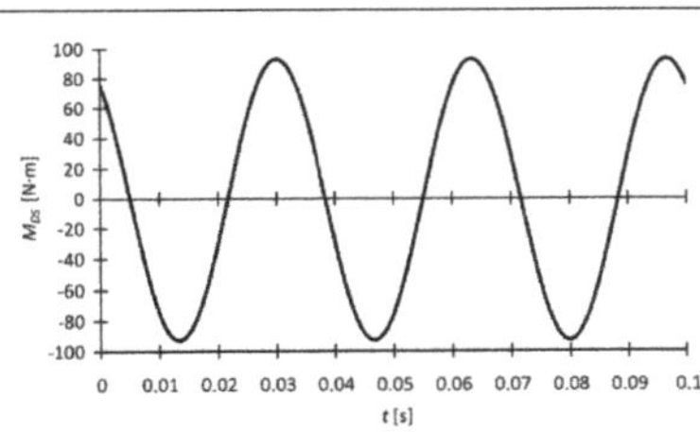	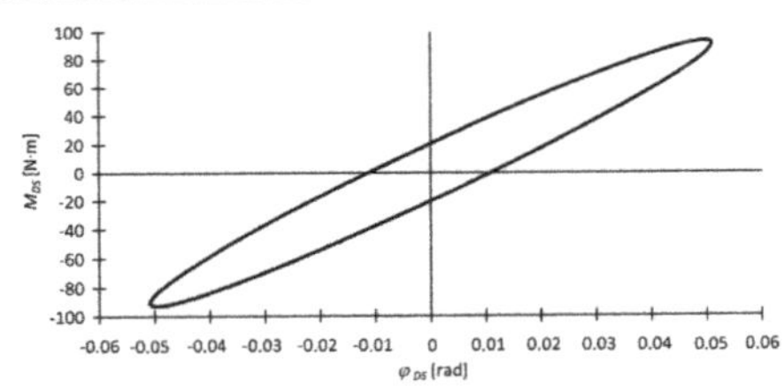
Out of operation cylinder	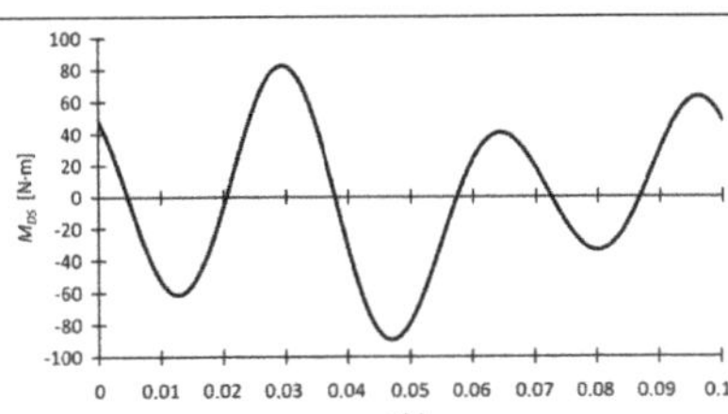	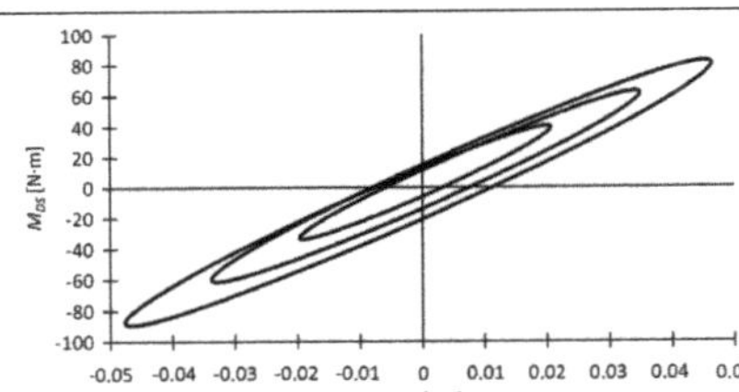

Table 8. Dynamic torque-time waveform M_{DS} and hysteresis lop at pressure p_S = 300 kPa in the elastic element.

	k_{DS} = 1360 N·m·rad^{-1}	
Steady state	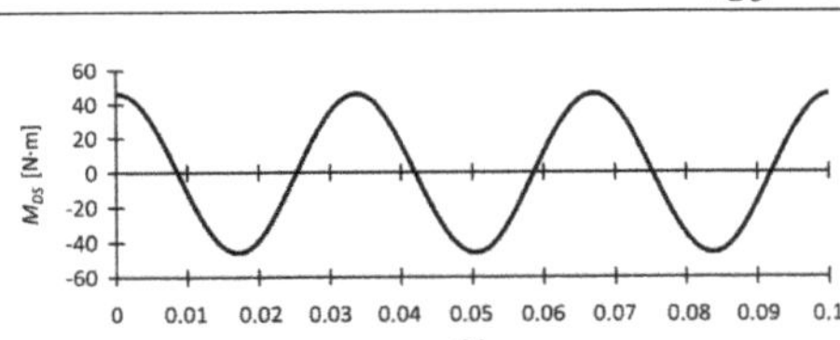	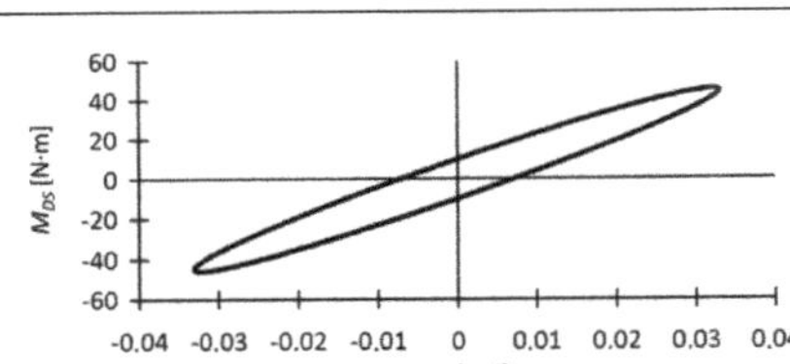

Table 8. *Cont.*

Out of operation cylinder	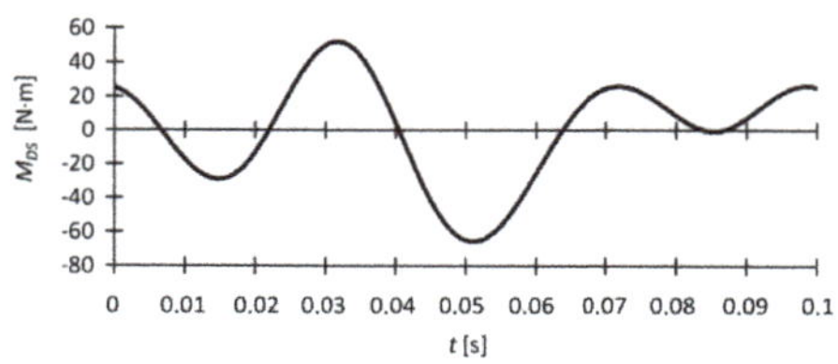	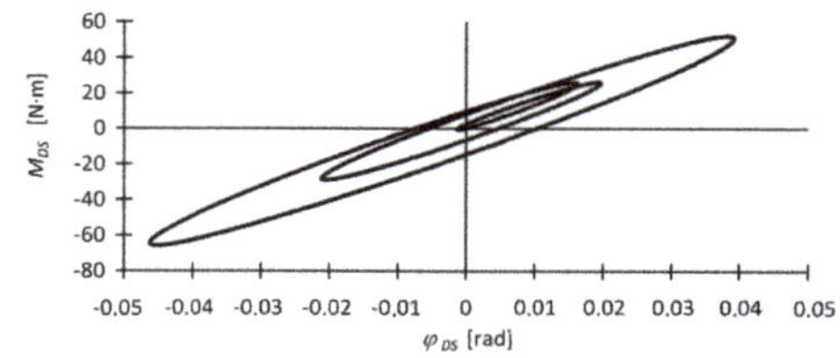

3.5. Simulation of the Output Signals on the Coupling Using Fourier Transformation

For the frequency analysis, a MATLAB simulation program was used to input the amplitudes and phases of the harmonic components of the dynamic torque. The given program also offers the possibility to choose filters (Hamming window and Hann window). The flow and results of the MATLAB analysis for the dynamic torque on the flexible coupling are shown in Figures 20 and 21.

(a)

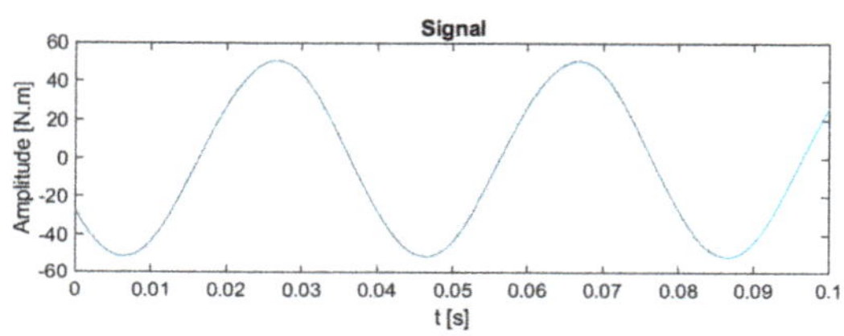

(b)

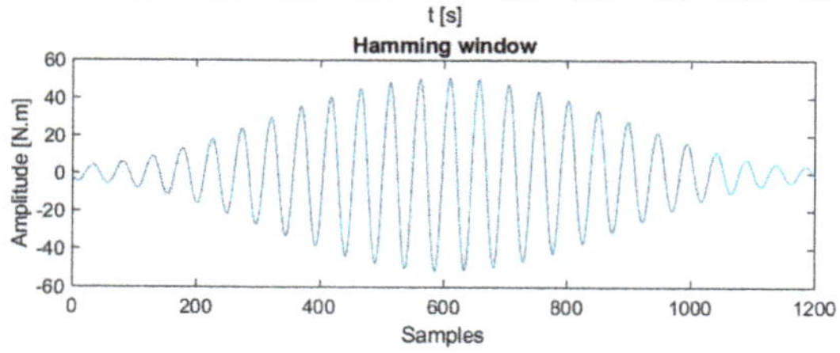

Figure 20. Signal (**a**) and Hamming window (**b**) at coupling pressure p_S = 700 kPa and revolutions n = 500 min^{-1}.

(a)

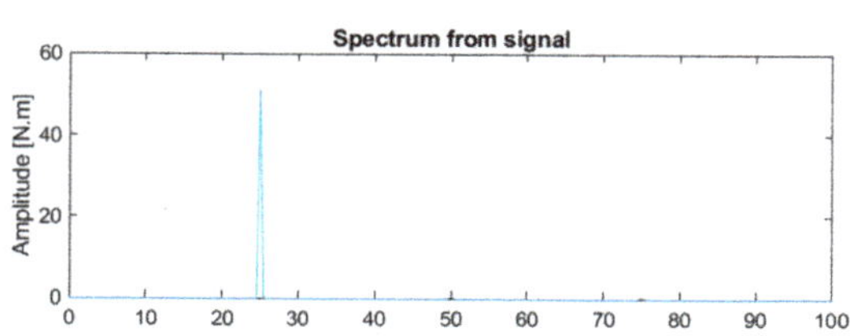

(b)

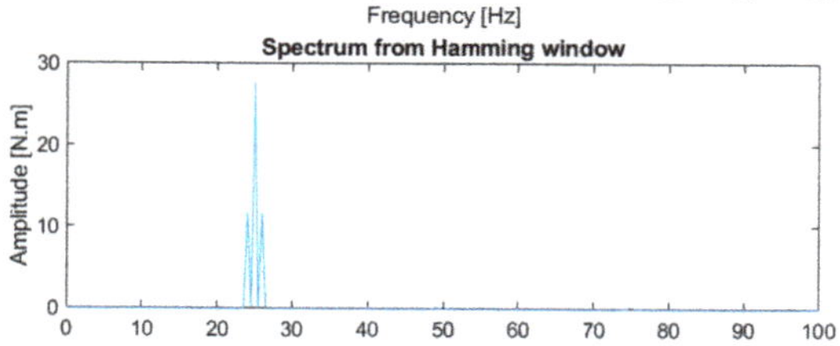

Figure 21. Spectrum from the signal (**a**) and Hamming window (**b**).

3.6. Verification of the Simulation Model Using Experimental Measurement

After the description of the mechanical system of the compressor drive and the dynamic calculations performed, a description of the experimental measurement follows to determine the torsional vibration values in the mechanical system consisting of a three-cylinder air compressor, driven by an electric motor using a pneumatic flexible coupling.

Measuring the Course of the Resonance Curve

The measurement was carried out with stead (even) cylinder operation and then using a rejector so that one cylinder could be disabled by opening the compressor suction valve. The calculation of the effective value of the dynamic component of the load torque $RMS\ M_D$, which is transmitted by the coupling, was the starting point for determining the magnitude of torsional vibration. The formula for its calculation was

$$RMSM_D = \sqrt{\frac{1}{N} \cdot \sum_{i=1}^{N} (M_{Di})^2}, \quad (7)$$

where

$$M_{Di} = M_i - \left(\frac{1}{N} \cdot \sum_{i=1}^{N} M_i\right), \quad (8)$$

where M_i is the i-th signal sample.

The above result from this formula indicates that it was necessary to measure the time course over time of the total load torque, from which the dynamic component M_D was filtered out. In the next calculation of $RMS\ M_D$, a moving average was used. The values were read after a steady state was reached. The measurement was carried out at varying revolutions in the range from n = 105 min^{-1} to n = 1500 min^{-1}, at a pressure in the compression chamber of the pneumatic flexible coupling p_S = 700 kPa, p_S = 500 kPa and p_S = 300 kPa. The compressor output pressure is set to a constant value, p_N = 500 kPa. The time signal (Figure 3) load torque was recorded with a sampling rate of f_V = 1200 Hz (number of samples N = 2400) with steady (even) cylinder operation and then with one compressor cylinder out of operation.

The Campbell diagram (Figure 22) for pressure coupling 700 kPa compiled from the measured values also contains marked, predetermined operating revolutions and critical revolutions from the principal harmonic. Based on the point of intersection of the critical revolutions and the beam (harmonic component) on the diagram, it is possible to determine the value of the rotational natural frequency N_0 and calculate the natural frequency f_0 and the angular natural frequency Ω_0. From what is shown, it follows that resonances at higher values of $RMS\ M_D$ occur at compressor revolutions of approximately n_{K3} = 638 min^{-1}. These are resonances from the main harmonic component of the torque of the three-cylinder compressor i = 3. When the cylinder is out of operation, resonances occur from both the main and the secondary components at critical s n_{K4} = 450 min^{-1}, n_{K3} = 660 min^{-1} a n_{K2} = 956 min^{-1}.

From the measured values, the development of the static (central) load torque is plotted (Figure 23).

Figure 24 shows the resonance curve, i.e., the $RMS\ M_D$ waveform depending on the revolutions n of the mechanical system.

When the cylinder was out of operation (Figure 25), the value of the effective value of the dynamic component of the load torque $RMS\ M_D$ decreased as compared to Figure 24, but secondary harmonic components were manifested.

Shown below is the development of frequency analysis at revolutions n = 500 min^{-1} (Figures 26 and 27). The first representation describes the use of the Hamming window when running evenly, and the second after one cylinder was out of operation.

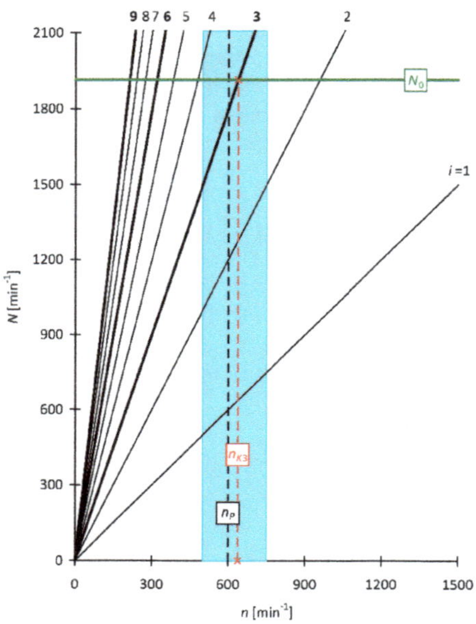

Figure 22. Campbell diagram of the mechanical system from the measured values.

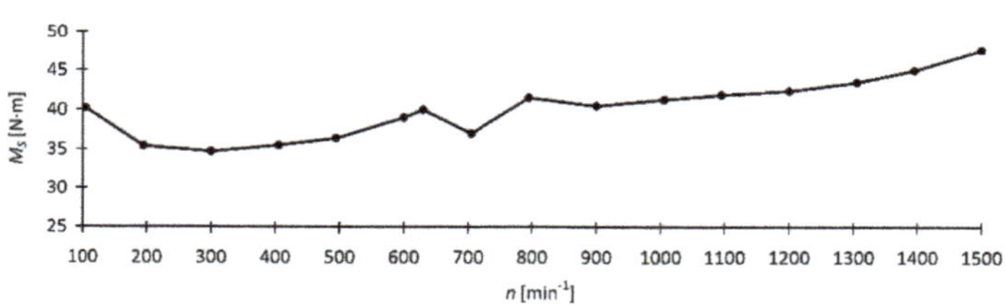

Figure 23. Development of the static component of load torque M_S depending on the engine revolutions n at pressure 700 kPa.

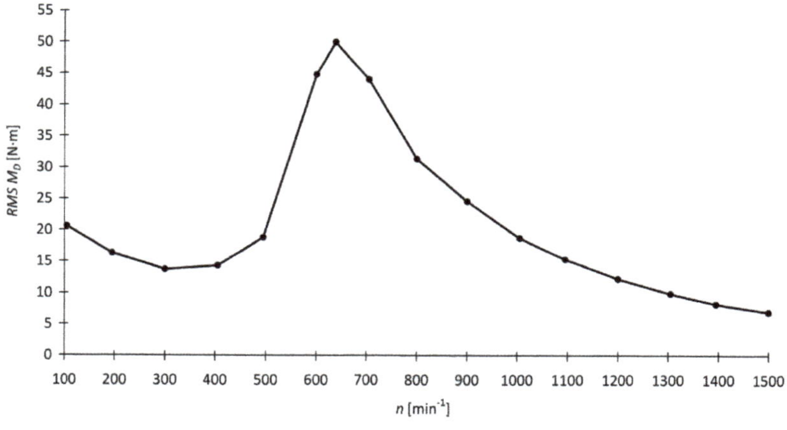

Figure 24. Effective value of the dynamic component of the load torque $RMS\ M_D$ of the pneumatic flexible coupling at pressure 700 kPa vs. the engine revolutions n – steady state.

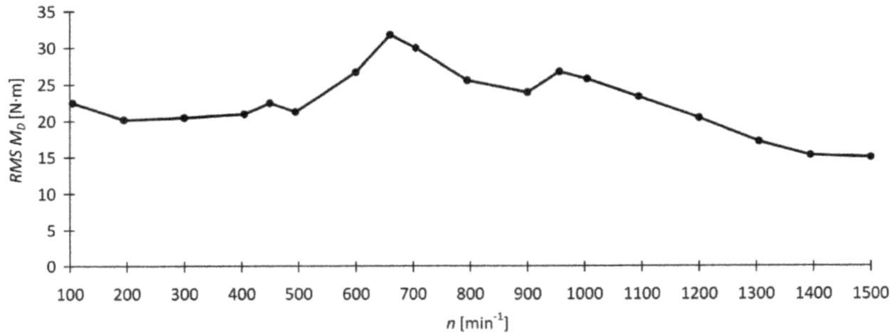

Figure 25. Effective value of the dynamic component of the load torque *RMS* M_D of the pneumatic flexible coupling at pressure 700 kPa vs. the engine revolutions n with the cylinder out of operation.

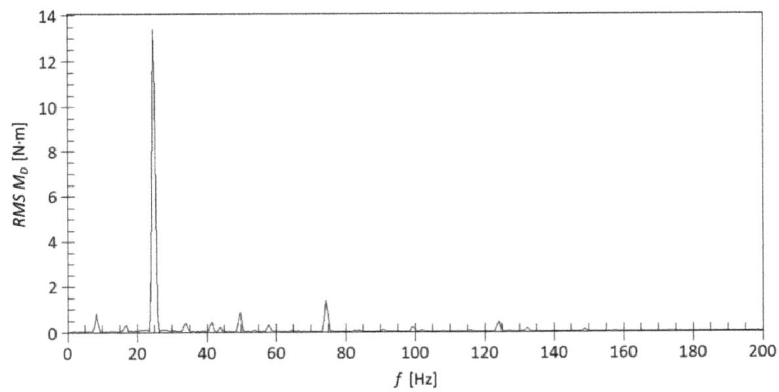

Figure 26. Frequency analysis using a Hamming window at n = 500 min^{-1} at steady operation.

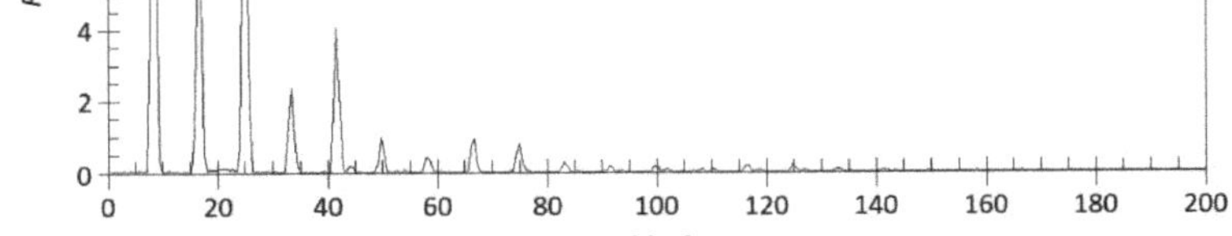

Figure 27. Frequency analysis using a Hamming window at n = 500 min^{-1} and one cylinder out of operation.

4. Comparison of Experimental and Simulation Models

The comparison of the experimental model in the laboratory with the simulation model was mainly aimed at determining the position of the critical revolutions based on the waveform of the effective value of the dynamic component of the *RMS* M_D load torque. The comparison of the experimental and simulation model results is shown in Figures 28–33, where the waveforms of the effective values of the dynamic component of the load torque $RMS\ M_D$ as a function of the steady-state and dropped single cylinder revolutions at pressures of 700, 500, and 300 kPa, respectively, are compared.

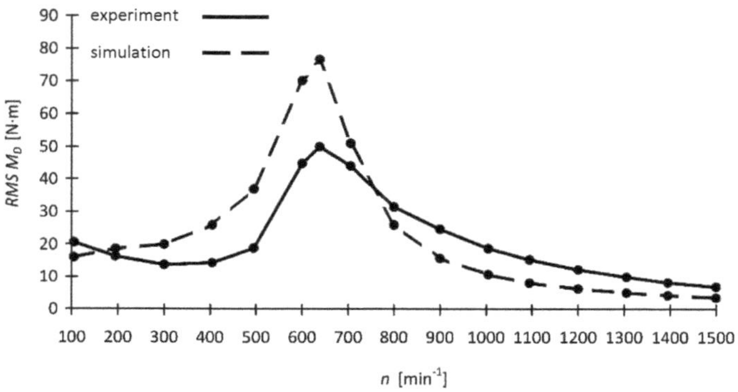

Figure 28. Comparison of the characteristics at steady operation of a piston compressor—pressure in the elastic element of the coupling p_S = 700 kPa.

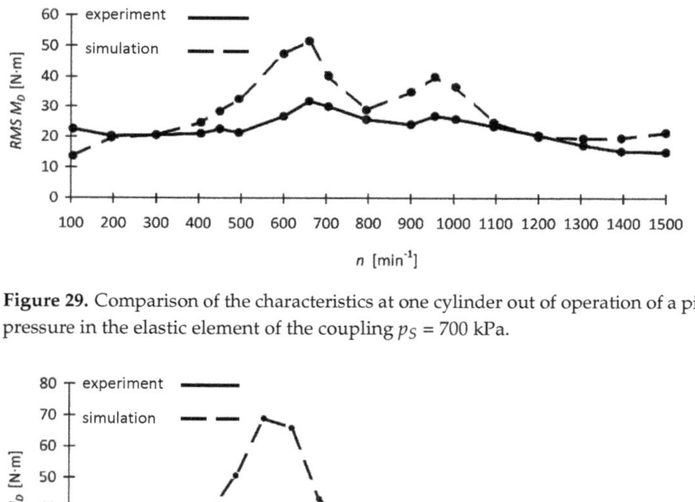

Figure 29. Comparison of the characteristics at one cylinder out of operation of a piston compressor—pressure in the elastic element of the coupling p_S = 700 kPa.

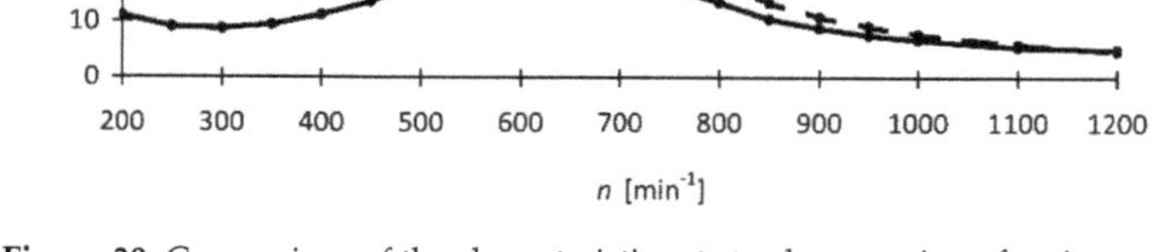

Figure 30. Comparison of the characteristics at steady operation of a piston compressor—pressure in the elastic element of the coupling p_S = 500 kPa.

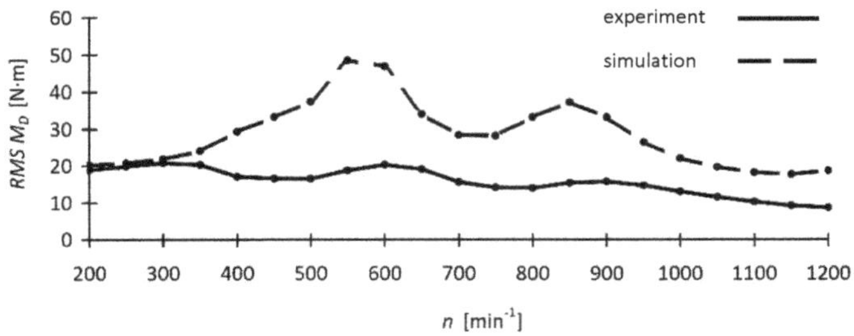

Figure 31. Comparison of the characteristics at one cylinder out of operation of a piston compressor—pressure in the elastic element of the coupling p_S = 500 kPa.

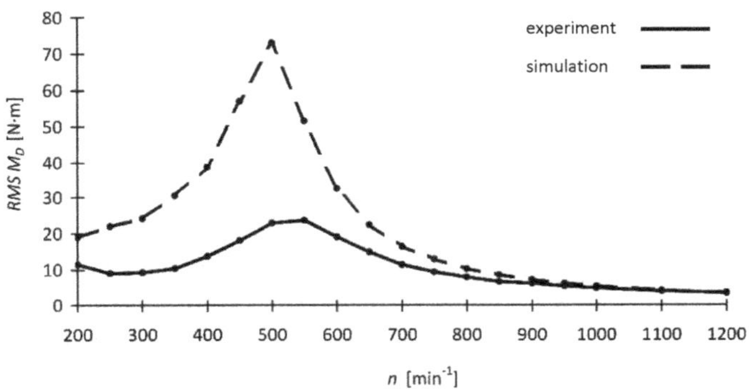

Figure 32. Comparison of the characteristics at steady operation of a piston compressor—pressure in the elastic element of the coupling p_S = 300 kPa.

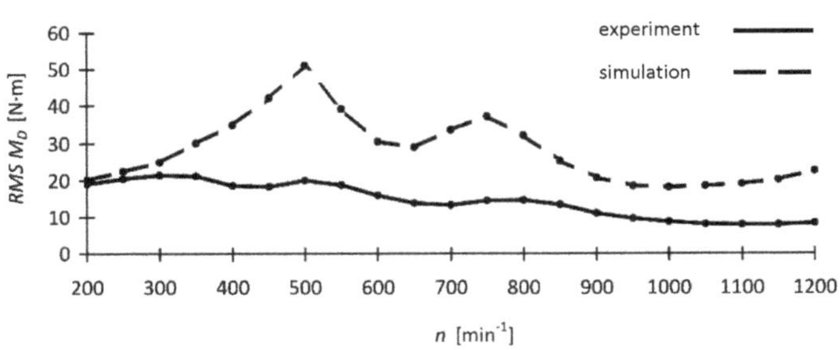

Figure 33. Comparison of the characteristics at one cylinder out of operation of a piston compressor—pressure in the elastic element of the coupling p_S = 300 kPa.

Table 9 shows the values of the natural circular frequencies obtained from the solution of the simulation model and the realization of the experimental measurements for the steady-state operation at pressure p_S = 700 kPa, p_S = 500 kPa, and p_S = 300 kPa in the elastic element of the coupling.

Accuracy in terms of the meaning of correspondence between the calculated and experimentally determined critical revolutions is determined as the ratio of experimentally determined revolutions to the revolutions determined by calculation. The sign (+) in the column indicates an increase, and the sign (−) indicates a decrease in the value of the critical revolutions determined experimentally compared to the value of the critical revolutions determined by calculation.

Table 9. Comparison of simulation and experimental results.

Pressure in Coupling [kPa]	Parameter	Value [rad·s^{-1}]	Parameter	Value [min^{-1}]	Accuracy [%]
700	Ω_{0SIM}	201.431	n_{K3SIM}	641	−0.47
	Ω_{0EXP}	200.434	n_{K3EXP}	638	
500	Ω_{0SIM}	179.167	n_{K3SIM}	570	5.26
	Ω_{0EXP}	188.496	n_{K3EXP}	600	
300	Ω_{0SIM}	156.610	n_{K3SIM}	498	10.44
	Ω_{0EXP}	172.788	n_{K3EXP}	550	

In general, the difference between the results can be caused by a number of different factors and their combination:

- Incomplete information on the input parameters, i.e., torsional shaft stiffness and mass moments of inertia.
- The method used to determine the natural frequencies and mode shapes of oscillations;
- Tooth clearance in the gearbox.
- The effect of frequency on the stiffness of the flexible coupling.
- Influence of the value of the mean torque and amplitude on the equivalent torsional stiffness of the flexible coupling (non-linearity of the loading characteristic of the coupling);
- Torsional vibration in the gearbox.
- Tongues clearance in individual flanges.
- Ambient temperature not included in the calculation.
- Temperature of elastic elements not included in the simulation model.
- Location of the torque meter.
- Insufficient heat dissipation from the piston compressor in operating mode.
- Maintaining and regulating constant pressure by a throttle valve at the outlet of the pressure tank.
- Operating revolutions close to the resonance from the main harmonic component in steady (even) operation.

Because the decrease in accuracy presented in Table 9 corresponds very well with the decrease in coupling stiffness due to temperature, it can be stated that the main cause of the decreasing agreement of the results is precisely this phenomenon. As can be seen, the results differ from each other in the order of units of percent. It can also be stated that the lower stiffness of the connection leads to higher torsional deflections of the oscillating masses, which are manifested by its intense heating. It is therefore very important to capture this aspect and include it in computational models.

5. Conclusions

The main focus of the research was vibrations, or torsional oscillations, often accompanying the operation of mechanical systems. These disturbances, if not controlled, result in machine unreliability in mechanical systems, compromising the safe operation of the plant. Based on the general knowledge of vibration and the known parameters of a real laboratory mechanical system, a simulation model was developed. The original twelve-mass mechanical system was reduced to a two-mass system by the partial frequency method. The MATLAB simulation program was used in the calculations and analyses.

The results obtained by calculation were experimentally verified by measurements on a laboratory mechanical system. After comparing the calculated natural angular frequency with the measured value, we can evaluate that the applied mathematical model is accurate to 99.53% at 700 kPa coupling pressure.

In potential future research, it is possible to include the temperature value of the elastic element of the coupling in the calculations and to determine more accurately the mass moments of inertia of the individual members of the mechanical system.

At the same time, it is planned to include in the simulation the occurrence of non-equilibrium condensation in the compressor, taking into account the research results presented in publications [49–51].

The aim of our paper is to show the importance of analytical calculations in estimating the natural frequencies and the corresponding mode shapes of oscillations in engineering work.

A mathematical model for a torsionally oscillating mechanical system has been developed where the stiffness of the elastic coupling can be continuously varied. Thus, the critical revolutions at different stiffnesses (pressures in the coupling) can be determined.

It was solved in steady state with harmonic excitation, amplitude, and phase over resonance curves for the individual harmonic components, which were then summed. We have simplified the multi-mass system to a two-mass system to analyze the effect of oscillations on the elastic coupling. This is under the condition of accuracy, where the frequencies differed from each other minimally.

Benefits of the simulation model:

- The mass reduction method used allows the use of a two-mass instead of a multi-mass system, which greatly simplifies the calculations.
- We can calculate the critical revolutions as a function of coupling pressure quite accurately. So, with a system operating with a range of operating revolutions, we can avoid resonance without having to directly measure the torque; we just need to know the revolutions, and we can determine the appropriate pressure accordingly.
- Characteristic depending on the ratio of the critical revolutions to the coupling pressures.

Author Contributions: Conceptualization, M.M., P.K. and F.L.; methodology, M.M. and D.M.; software, M.M.; validation, P.K. and F.L.; formal analysis, D.M. and M.M.; investigation, M.M. and P.K.; resources, M.M. and P.K.; data curation, M.M.; writing—original draft preparation, M.M.; writing—review and editing, P.B. and M.O.; visualization, M.M., P.B. and M.O.; supervision, P.K.; project administration, D.M.; funding acquisition, D.M. All authors have read and agreed to the published version of the manuscript.

Funding: This research was funded by the Slovak Research and Development Agency under contract No. APVV-23-0342 "New structural and material elements for sustainable transport of bulk materials" and APVV SK-PL-23-0052 "Modeling of transverse vibrations of belt and diagnostics of pipe conveyors" and from the project KEGA 037TUKE-4/2024 "Creation of interactive tool for increasing of students' skills and competences in teaching of study subjects relating to elaboration of drawing documentation" and VEGA 1/0728/24 "Research on the possibilities of implementing sensory technologies and visualization methods for the needs of digital transformation of pipeline conveyor systems".

Institutional Review Board Statement: Not applicable.

Informed Consent Statement: Not applicable.

Data Availability Statement: Data are contained within the article.

Conflicts of Interest: The authors declare no conflicts of interest.

Abbreviations

I	mass moment of inertia	[kg·m^2]
M	load torque	[N·m]
N_0	rotational frequency	[min^{-1}]
N	number of signal samples	[-]
$RMS\ M_D$	effective value of the dynamic component of the load torque	[N·m]
b	damping coefficient	[N·s·m^{-1}]
f	frequency	[Hz]
i	gear ratio, harmonic component, number of cylinders	[-]
k	torsional stiffness	[N·m·rad^{-1}]

n	RPM	[min^{-1}]
p	pressure	[kPa]
t	time	[s]
Ω_0	natural frequency	[$rad{\cdot}s^{-1}$]
γ	phase shift	[rad]
η	detuning coefficient	[-]
φ	angle of torsion, amplitude of rotation	[rad]
$\dot{\varphi}$	angular velocity	[$rad{\cdot}s^{-1}$]
$\ddot{\varphi}$	angular acceleration	[$rad{\cdot}s^{-1}$]
ω	angular frequency	[$rad{\cdot}s^{-1}$]

References

1. Lu, J.; Zheng, H.; Husnain Haider, M.; Feng, Y.; Zhi, P.; Cheng, J.; Wang, Z. Fracture failure analysis of flywheel hub served in heavy-fuel aviation piston engine. *Eng. Fail. Anal.* **2023**, *151*, 107363. [CrossRef]
2. Amiri, Z.; Bayatian, M.; Mozafari, S. Numerical simulation application in occupational health studies: A review. *Int. J. Occup. Saf. Ergon.* **2024**, *30*, 946–967. [CrossRef]
3. Kassay, P. Effect of torsional vibration on woodchip size distribution. *Sci. J. Silesian Univ. Technol. Ser. Transp.* **2018**, *99*, 95–104. [CrossRef]
4. Stanzl-Tschegg, S.E.; Mayer, H.R.; Tschegg, E.K. High frequency method for torsion fatigue testing. *Ultrasonics* **1993**, *31*, 275–280. [CrossRef]
5. Czech, P. Diagnosing faults in the timing system of a passenger car spark ignition engine using the bayes classifier and entropy of vibration signals. *Sci. J. Silesian Univ. Technol. Ser. Transp.* **2022**, *116*, 83–98. [CrossRef]
6. Henao, H.; Kia, S.H.; Capolino, G.-A. Torsional-Vibration Assessment and Gear-Fault Diagnosis in Railway Traction System. *IEEE Trans. Ind. Electron.* **2011**, *58*, 1707–1717. [CrossRef]
7. Maláková, S.; Sivák, S. GPS Application in the Design of Gearboxes. *Acta Mech. Autom.* **2022**, *16*, 309–315. [CrossRef]
8. Maláková, S.; Sivák, S. Practical Example of Modification of a Gearbox Lubrication System. *Lubricants* **2022**, *10*, 110. [CrossRef]
9. Chmelař, J.; Petr, K.; Mossoczy, P.; Dynybyl, V. Experimental study of lubrication film monitoring in a roller bearing by utilization of surface acoustic waves. *Tribol. Int.* **2020**, *141*, 105908. [CrossRef]
10. Chmelař, J.; Petr, K.; Mikeš, P.; Dynybyl, V. Cylindrical roller bearing lubrication regimes analysis at low speed and pure radial load. *Acta Polytech.* **2019**, *59*, 272–282. [CrossRef]
11. Gładysiewicz, L.; Król, R.; Kisielewski, W. Measurements of loads on belt conveyor idlers operated in real conditions. *Meas. J. Int. Meas. Confed.* **2019**, *134*, 336–344. [CrossRef]
12. dos Santos e Santos, L.; Ribeiro Filho, P.R.C.F.; Macêdo, E.N. Belt rotation in pipe conveyors: Development of an overlap monitoring system using digital twins, industrial Internet of things, and autoregressive language models. *Measurement* **2024**, *230*, 114546. [CrossRef]
13. Urbanský, M.; Grega, R. Contactless Measuring Device for Flexible Shaft Coupling Twist Angle. In *Current Methods of Construction Design*; Lecture Notes in Mechanical Engineering; Springer: Cham, Switzerland, 2020; pp. 193–200, ISBN 9783030331450.
14. Grega, R.; Krajnak, J.; Žul'ová, L.; Kačír, M.; Kaššay, P.; Urbanský, M. Innovative Solution of Torsional Vibration Reduction by Application of Pneumatic Tuner in Shipping Piston Devices. *J. Mar. Sci. Eng.* **2023**, *11*, 261. [CrossRef]
15. Tuplin, W.A. LXV. Torsional vibration in certain mechanical systems. *Lond. Edinb. Dublin Philos. Mag. J. Sci.* **1938**, *26*, 729–752. [CrossRef]
16. Pielorz, A. Vibration problems in selected torsional mechanical systems. *Meccanica* **2003**, *38*, 731–738. [CrossRef]
17. Chen, X.; Chen, R.; Deng, T. An investigation on lateral and torsional coupled vibrations of high power density PMSM rotor caused by electromagnetic excitation. *Nonlinear Dyn.* **2020**, *99*, 1975–1988. [CrossRef]
18. Homik, W.; Mazurkow, A.; Woś, P. Application of a Thermo-Hydrodynamic Model of a Viscous Torsional Vibration Damper to Determining Its Operating Temperature in a Steady State. *Materials* **2021**, *14*, 5234. [CrossRef]
19. Guo, Y.; Li, W.; Yu, S.; Han, X.; Yuan, Y.; Wang, Z.; Ma, X. Diesel engine torsional vibration control coupling with speed control system. *Mech. Syst. Signal Process* **2017**, *94*, 1–13. [CrossRef]
20. Kinnunen, K.; Laine, S.; Tiainen, T.; Viitala, R. Method for Adjusting Torsional Natural Frequencies of Powertrains with Novel Coupling Design. *Machines* **2022**, *10*, 162. [CrossRef]
21. Wieczorek, A.N.; Konieczny, Ł.; Wojnar, G.; Wyroba, R.; Filipowicz, K.; Kuczaj, M. Reduction of dynamic loads in the drive system of mining scraper conveyors through the use of an innovative highly flexible metal coupling. *Eksploat. I Niezawodn. Maint. Reliab.* **2024**, *26*, 181171. [CrossRef]
22. Urbanský, M. Comparison of piston and tangential pneumatic flexible shaft couplings in terms of high flexibility. *Sci. J. Silesian Univ. Technol. Ser. Transp.* **2018**, *99*, 193–203. [CrossRef]
23. Laine, S.; Haikonen, S.; Tiainen, T.; Viitala, R. Rotor resonance avoidance by continuous adjustment of support stiffness. *Int. J. Mech. Sci.* **2024**, *270*, 109092. [CrossRef]

24. Flek, J.; Dub, M.; Kolář, J.; Lopot, F.; Petr, K. Determination of Mesh Stiffness of Gear—Analytical Approach vs. FEM Analysis. *Appl. Sci.* **2021**, *11*, 4960. [CrossRef]
25. Li, W.; Ge, X.; Liang, M.; Deng, X. Magneto-rheological variable damping and variable stiffness torsional vibration control of powertrain transmission. *J. Mech. Sci. Technol.* **2023**, *37*, 3887–3900. [CrossRef]
26. Zhang, X.; He, W.; Hu, J. Impact of Inertia Control of DFIG-Based WT on Torsional Vibration in Drivetrain. *IEEE Trans. Sustain. Energy* **2020**, *11*, 2525–2534. [CrossRef]
27. Homišin, J. *New Types of Flexible Shaft Couplings: Development, Research, Application*; Vienala: Košice, Slovakia, 2002. Available online: https://scholar.google.sk/scholar?cluster=17005307171049548256&hl=en&as_sdt=2005&sciodt=0,5 (accessed on 31 October 2024). (In Slovak)
28. Homišin, J. Optimal tuning of mechanical systems by application of a pneumatic tuner of torsional oscillation. *Sci. J. Silesian Univ. Technol. Ser. Transp.* **2017**, *97*, 47–55. [CrossRef]
29. Bartel, T.; Herold, S.; Infante, F.; Käsgen, J.; Matthias, M.; Millitzer, J.; Perfetto, S. Active Vibration Reduction of Ship Propulsion Systems. In Proceedings of the 2018 Joint Conference—Acoustics, Acoustics 2018, Ustka, Poland, 11–14 September 2018; Polish Acoustical Society: Gdańsk, Poland; Ustka, Poland, 2018; pp. 15–20.
30. Xingyu, L.; Gequn, S.; Lihui, D.; Bin, W.; Kang, Y. Progress and Recent Trends in the Torsional Vibration of Internal Combustion Engine. In *Advances in Vibration Analysis Research*; Ebrahimi, F., Ed.; InTech: Rijeka, Croatia, 2011; pp. 245–272, ISBN 978-953-307-209-8.
31. Kongratana, V.; Tipsuwanporn, V.; Numsomran, A.; Detchrat, A. IMC-based PID controllers design for torsional vibration system. In Proceedings of the 2012 12th International Conference on Control, Automation and Systems, Jeju, Republic of Korea, 17–21 October 2012; pp. 892–895.
32. Zhong, B.; Deng, B.; Zhao, H. Simulation Model and Method for Active Torsional Vibration Control of an HEV. *Appl. Sci.* **2019**, *9*, 34. [CrossRef]
33. Homišin, J.; Kaššay, P.; Urbanský, M. High-flexibility characteristics of pneumatic flexible shaft couplings. *Pneumatyka* **2011**, *79*, 26–29.
34. Kaššay, P.; Urbanský, M. Torsional natural frequency tuning by means of pneumatic flexible shaft couplings. *Sci. J. Silesian Univ. Technol. Ser. Transp.* **2015**, *89*, 57–60. [CrossRef]
35. Jírová, R.; Pešík, L.; Grega, R. An Original Vibrodiagnostic Device to Control Linear Rolling Conveyor Reliability. *J. Mar. Sci. Eng.* **2022**, *10*, 445. [CrossRef]
36. Bruha, M. Importance of control engineering to minimize torsional vibration in variable speed drive systems. In Proceedings of the 2016 Petroleum and Chemical Industry Conference Europe Conference Proceedings, PCIC EUROPE, Berlin, Germany, 14–16 June 2016; pp. 1–8. [CrossRef]
37. Homišin, J. Static optimisation of mechanical systems based on the method of extremal regulation. *Sci. J. Silesian Univ. Technol. Ser. Transp.* **2019**, *103*, 15–29. [CrossRef]
38. Naqvi, S.; Ghufran, M.; Meraghni, S.; Varnier, C.; Nicod, J.; Zerhouni, N. Human knowledge centered maintenance decision support in digital twin environment. *J. Manuf. Syst.* **2022**, *65*, 528–537. [CrossRef]
39. Xia, M.; Shao, H.; Williams, D.; Lu, S.; Shu, L.; Silva, C.W. Intelligent fault diagnosis of machinery using digital twin-assisted deep transfer learning. *Reliab. Eng. Syst. Saf.* **2021**, *215*, 107938. [CrossRef]
40. Wang, H.; Liu, Y.; Mu, Z.; Xiang, J.; Li, J. Real-time precision reliability prediction for the worm drive system supported by digital twins. *Reliab. Eng. Syst. Saf.* **2023**, *240*, 109589. [CrossRef]
41. Zoul, V. Dynamic of propulsion, present situation and trends. *Trans. Univ. Košice* **2014**, *2*, 101–106.
42. Rivin, E.I. *Dinamika Privoda Stankov*; Mašinostrojenije: Moskva, Russia, 1966. (In Russian)
43. Urbanský, M.; Homišin, J.; Kaššay, P.; Moravič, M. Influence of Piston Compressor Inner Failure on Mechanical System Objective Function. *Diagnostyka* **2016**, *17*, 47–52.
44. Chen, M.; Ouyang, H.; Li, W.; Wang, D.; Liu, S. Partial Frequency Assignment for Torsional Vibration Control of Complex Marine Propulsion Shafting Systems. *Appl. Sci.* **2020**, *10*, 147. [CrossRef]
45. Wachel, J.C.; Szenasi, F.R. Analysis of Torsional Vibrations in Rotating Machinery. In Proceedings of the Twenty-Second Turbomachinery Symposium, Dallas, TX, USA, 14–16 September 1993; Texas A&M University: College Station, TX, USA, 1993; pp. 127–151.
46. Harris, C.M.; Piersol, A.G. *Harris' Shock and Vibration Handbook*, 5th ed.; McGraw-Hill Handbooks: New York, NY, USA, 2002.
47. Wilson, W.K. *Practical Solution of Torsional Vibration Problems: Frequency Calculations*, 3rd ed.; Chapman & Hall Ltd.: London, UK, 1967; Volume 1.
48. Wilson, W.K. *Practical Solution of Torsional Vibration Problems: Amplitude Calculations*, 3rd ed.; Chapman & Hall Ltd.: London, UK, 1963; Volume 2.
49. Ding, H.; Dong, Y.; Zhang, Y.; Wen, C.; Yang, Y. Performance of supercritical carbon dioxide (sCO_2) centrifugal compressors in the Brayton cycle considering non-equilibrium condensation and exergy efficiency. *Energy Convers. Manag.* **2024**, *299*, 117849. [CrossRef]

50. Zhang, G.; Yang, Y.; Chen, J.; Jin, Z.; Majkut, M.; Smołka, K.; Dykas, S. Effect of relative humidity on the nozzle performance in non-equilibrium condensing flows for improving the compressed air energy storage technology. *Energy* **2023**, *280*, 128240. [CrossRef]
51. Zhang, G.; Yang, Y.; Chen, J.; Jin, Z.; Dykas, S. Numerical study of heterogeneous condensation in the de Laval nozzle to guide the compressor performance optimization in a compressed air energy storage systém. *Appl. Energy* **2024**, *356*, 122361. [CrossRef]

applied sciences

MDPI

Article

The Integration of Internet of Things and Machine Learning for Energy Prediction of Wind Turbines

Christos Emexidis and Panagiotis Gkonis *

Department of Digital Industry Technologies, National and Kapodistrian University of Athens, 34400 Dirfies Messapies, Greece; emexidis.xristos@gmail.com
* Correspondence: pgkonis@uoa.gr

Abstract: Wind power has emerged as a crucial substitute for conventional fossil fuels. The combination of advanced technologies such as the internet of things (IoT) and machine learning (ML) has given rise to a new generation of energy systems that are intelligent, reliable, and efficient. The wind energy sector utilizes IoT devices to gather vital data, subsequently converting them into practical insights. The aforementioned information aids among others in the enhancement of wind turbine efficiency, precise anticipation of energy production, optimization of maintenance approaches, and detection of potential risks. In this context, the main goal of this work is to combine the IoT with ML in the wind energy sector by processing weather data acquired from sensors to predict wind power generation. To this end, three different regression models are evaluated. The models under comparison include Linear Regression, Random Forest, and Lasso Regression, which were evaluated using metrics such as coefficient of determination (R^2), adjusted R^2, mean squared error (MSE), root mean squared error (RMSE), and mean absolute error (MAE). Moreover, the Akaike Information Criterion (AIC) and Bayesian Information Criterion (BIC) were taken into consideration as well. After examining a dataset from IoT devices that included weather data, the models provided substantial insights regarding their capabilities and responses to preprocessing, as well as each model's reaction in terms of statistical performance deviation indicators. Ultimately, the data analysis and the results from metrics and criteria show that Random Forest regression is more suitable for weather condition datasets than the other two regression models. Both the advantages and shortcomings of the three regression models indicate that their integration with IoT devices will facilitate successful energy prediction.

Keywords: Internet of Things (IoT); Machine Learning (ML); data analysis; data preprocessing; regression analysis

1. Introduction

Given the increasing environmental concerns and the need to shift to sustainable energy sources, wind power has emerged as a significant and feasible alternative to conventional fossil fuels [1]. Wind farms, comprised of a collection of wind turbines, have become essential structures in the worldwide effort to provide environmentally friendly and sustainable energy [2]. The increasing interest in wind energy coincides with a crucial phase in the energy sector, marked by the integration of advanced technologies like the internet of things (IoT) and machine learning (ML). These state-of-the-art technologies are not simply enhancing the abilities of current energy systems; they are completely transforming them. They pledge to introduce a new era marked by intelligent, reliable, and extremely efficient energy systems that not only improve performance but also tackle urgent energy issues worldwide. The wind energy sector is progressively utilizing IoT devices to collect vital data, playing a key role in this technological transformation [3]. The features may encompass parameters like wind speed, wind direction, temperature, humidity, air pressure, and metrics related to the health of the turbine. These variables offer crucial insights into the elements that affect wind power output. Once gathered and examined, these data act as a catalyst for enhancing the effectiveness of wind turbines, precisely predicting energy generation, simplifying maintenance approaches, and identifying possible hazards. Furthermore, the instantaneous transfer of sensor measurements to distant control centers enables uninterrupted surveillance, although it presents certain difficulties, particularly with real-time control at the system and component levels. Currently, supervisory control and data acquisition (SCADA) systems offer advanced services and capabilities for wind energy conversion systems (WECS) that go beyond basic monitoring and management of wind turbines, utilizing meteorological observations and weather data to predict wind power generation [1]. Thoughtless usage of IoT devices can introduce unforeseen costs. A cost-effective strategy can address crucial energy and infrastructure costs incurred by IoT integration [4].

This article provides foundational insights into IoT applications in wind energy systems, incorporating data analysis and ML through three different regression models. The preprocessing phase encompasses outlier removal, data transformation, correlation analysis, cross-validation, standardization and data splitting, all of which prepare the dataset for ML processing. To this end, an innovative approach is presented that involves preprocessing followed by Linear, Random Forest, and Lasso regression, yielding valuable results. These regression models are the basis for ML-assisted prediction. The objective is to demonstrate how the utilization of these technologies can significantly enhance the efficiency, reliability, and ecological sustainability of wind energy systems, thereby facilitating the global transition towards a more sustainable and environmentally conscious energy sector.

Research Context

In recent years, numerous scientific works have dealt with the deployment of IoT in wind energy and the prediction of energy output. The work in [3] examines the implementation of IoT in wind farms and the broader energy sector. It asserts that the incorporation of IoT into wind farms enhances the existing 743 GW worldwide wind power capacity to a sufficient level to provide 20% of the world's electricity by 2030. In the same context, the obstacles related to the implementation of IoT in wind farms and the prospective role of blockchain technology and green IoT in energy systems are discussed as well. In [5], an IoT-based communication architecture is proposed to ensure a reliable connection between wind turbines and the control center, utilizing repeat-accumulate coded communication to improve reliability. The numerical findings indicate that the suggested technique can accurately predict the condition of a wind turbine and greatly surpasses previous estimating methods. In [6], a set of multi-objective predictive models was developed employing various advanced ML algorithms, such as artificial neural networks (ANNs), recurrent neural networks (RNNs), convolutional neural networks (CNNs), and long short-term memory (LSTM) networks. According to the outcomes of this work, the LSTM, RNN, CNN, and ANN algorithms were effective in predicting wind power. The efficacy of these models was assessed by the integration of statistical metrics for performance deviation. In conclusion, the LSTM model is more effective in predicting wind power. In [7], long-term wind power forecasting was conducted utilizing daily wind speed data through five machine learning algorithms: LASSO regression, k-nearest neighbors (kNN) regression, xGBoost regression, random forest regression (RFR), and support vector regression (SVR). The findings of this work indicated that ML algorithms are capable of predicting long-term wind power values based on previous wind speed data and can be utilized in sites distinct from those used for model training. Among these algorithms, RFR had improved performance compared to the other approaches, while LASSO had the worst performance metrics due to its linear basis. In [8], wind speed and solar radiation are the fundamental inputs used in the context of renewable energy sources. To this end, the performance of various artificial neural networks algorithms has taken place to ensure high-quality predictions. These algorithms can accurately predict data like temperature, relative moisture, solar radiation, rain, and wind speed. Overall, the results show that the Levenberg–Marquardt and Bayesian algorithms for regulation are better at predicting properties that are not linear. Components, such as sensors, actuators, protocols, and gateways, enhance energy efficiency, reliability, failure detection, and production optimization. After gathering precise data, IoT emerges as the superior platform for real-time data analysis. In [9], through the analysis of climate change and ground temperature data, Random Forest (RF), LSTM, and XGBoost models were utilised to predict the artificial permafrost table, while grid search and cross-validation techniques were applied to optimise the hyperparameters of each model. Thereafter, the significance of the variables in the ML model was assessed. The results demonstrated a robust association with alterations in the artificial permafrost table. With mean squared error (MSE), mean absolute percentage error (MAPE), root mean squared error (RMSE), mean absolute error (MAE), and R^2 values of 0.003, 0.052, 0.0085, 0.029, and 0.989, respectively, the integrated model was better at making predictions than the individual models. This model provides a swift, precise, and dependable method for predicting the permafrost table and enhancing subgrade stability research under difficult permafrost conditions. In [10], the operational complexities indicate that traditional methods are inadequate, and emerging "big data" techniques, particularly ML and deep learning, are set to significantly influence the design and optimization of offshore wind turbines and farms.

In all the aforementioned studies, the researchers concentrate their efforts either on IoT in wind energy or in wind power prediction. Therefore, a key novel point of this work is the integration of IoT technology with wind energy systems, as well as the comparison of three fundamental ML algorithms to extract valuable information about their strengths and weaknesses. The model evaluation indicators (R^2, adjusted R^2, MSE, RMSE, and MAE) are employed to evaluate the performances of the ML models as well as the Akaike information criterion (AIC) and the Bayesian Information Criterion (BIC) criteria for model selection and comparison [11]. To this end, this work initially examines the design and architecture of WESCS in wind energy, followed by an exploration of ML, particularly focusing on three regression models and their capabilities.

The rest of this paper is organized as follows: In Section 2, the incorporation of IoT in the wind energy sector is analyzed, and in particular the composition of WESC, the cyber-physical integration of a wind turbine, the SCADA systems, and machine to machine (M2M) for internet of everything (IoE)-enabled wind farms. Section 3 is focused on ML in wind energy prediction. To this end, the data and preprocessing steps are described, with an emphasis on regression analysis. In the same context, Linear, Random Forest, and Lasso models are then analyzed, with their basic characteristics and results on various key performance indicators (KPIs), including cross-validation results. In Section 4, KPIs' performance is presented on the test set and cross-validation, concluding with the complexity and interpretability of the models. Finally, concluding remarks are outlined in Section 5.

2. IoT and Wind Energy

The efficacy and magnitude of wind technologies are advancing at a rapid pace. The energy specialists have ambitious goals for the future integration of wind energy into the industry. The primary challenge in the advancement of wind energy lies in the inherent unpredictability of these resources. The enormous amount of data generated by devices, as well as the need for real-time data, has led to the widespread use of IoT devices [12]. Therefore, real-time operation can make the necessary arrangements in the power system to counteract fluctuations without experiencing sudden changes in power output. Furthermore, the availability of real-time data sharing is essential for effective collaboration between energy storage facilities and wind units [12]. Figure 1 illustrates IoT's contribution to the wind energy sector, using weather data as input and providing real-time, accurate information for energy prediction.

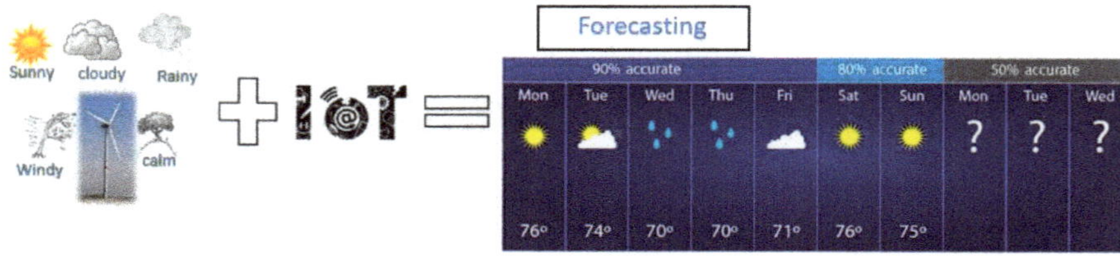

Figure 1. IoT in wind energy sector.

Furthermore, the use of IoT technology in conjunction with information and communications technology (ICT) infrastructures enables wind farm operators to effectively implement precise predictive maintenance schedules, thereby mitigating the risk of incurring substantial losses. A timetable of this nature can be implemented through the utilization of ML and data mining methodologies. Timely maintenance can lower the levelized cost of energy (LCoE) index for wind assets [13]. The LCoE index quantifies the discounted value of the average cost of electricity over the whole operational lifecycle of the turbines. The indispensability of IoT in harnessing wind energy lies in the prompt collection and analysis of data pertaining to wind turbines and wind farms. Currently, the challenges of data transfer latency for offshore wind farms and the restricted capacity to transmit information to distant areas are two significant difficulties that need to be resolved. Hence, by gathering and examining crucial data in real-time, the process of making decisions, such as the prompt shutdown of a turbine to prevent further damages, can be expedited or even automated [5]. The integration of IoT technologies in the wind industry emphasizes the necessity for more holistic approaches to develop economical, secure, and reliable frameworks for the planning, operation, installation, and maintenance of wind farms and turbines.

Typically, wind farms are placed in distant areas, resulting in control centers being situated many hours away from the wind farms. Utilizing an IoT network, remote data transmission connectivity can enable control centers to effectively monitor the condition of a wind turbine and exert control over its operation [5]. Wind turbines, being located in remote areas, necessitate the use of wireless networks like cellular or satellite networks for IoT network access. Figure 2 illustrates a combined system consisting of wind turbines and a wireless IoT network.

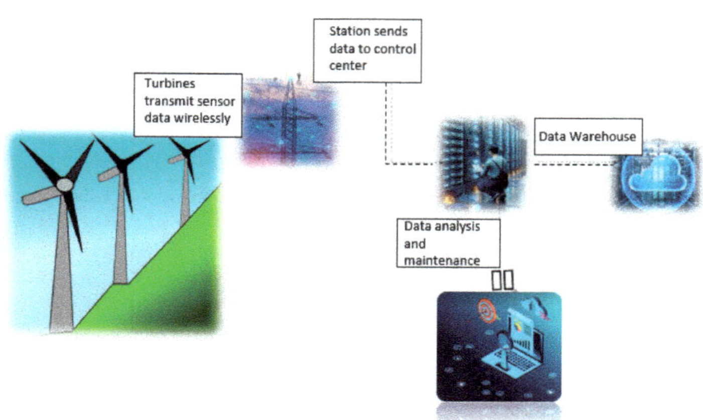

Figure 2. Utilization of an IoT network to transfer data from turbines to the control center.

2.1. Composition of WECS

Physical Layer in WECS: The wind energy conversion system comprises the following physical components: blades, rotor hub, nacelle, and tower foundations, as shown in Figure 3. The nacelle consists of several components, including shafts, a gearbox, a generator, and other electrical and mechanical systems. The various components are continuously monitored and regulated by a multitude of sensors and actuators [14]. The wind turbine is fitted with sensors that quantify multiple characteristics related to the functioning and state of each component. Simultaneously, the control system governs and manages the functioning of the wind turbine through a sequence of actuators.

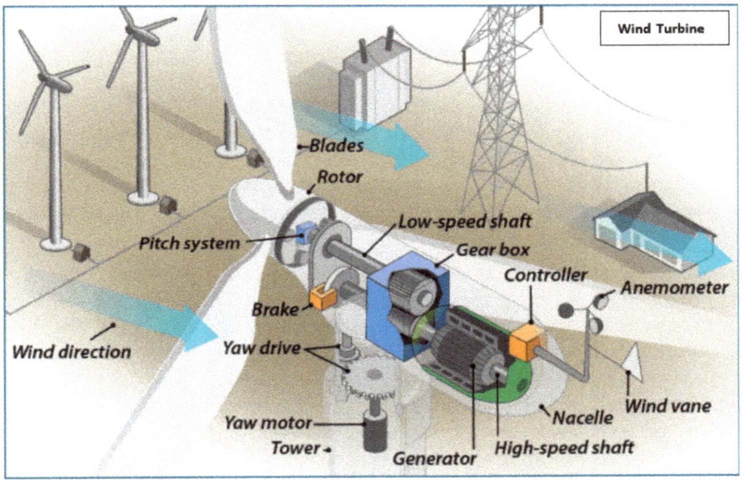

Figure 3. Wind turbine components.

Cyber Layer in WECS: The cyber layer in WECS incorporates a variety of hardware and software technologies that collaborate to accomplish shared goals [14]. The cyber layer typically comprises networking, SCADA, and content management systems (CMS) as illustrated in Figure 4.

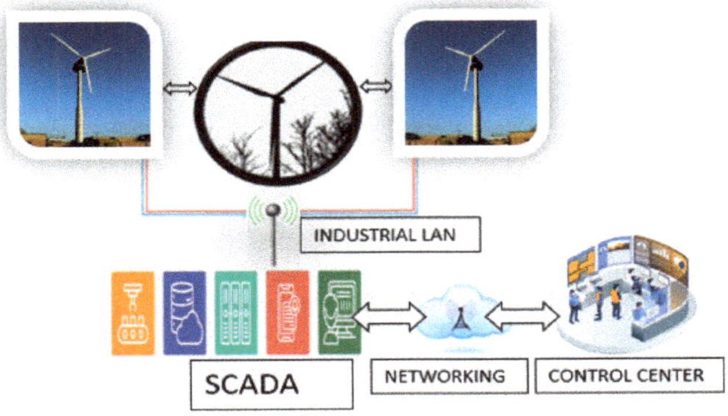

Figure 4. Cyber layer implemented in a wind farm.

Networking: Reliable communication networks between subsystems within a wind turbine are necessary for the successful deployment of WECS. Furthermore, it facilitates the connection of sophisticated machinery and intricately integrated devices across a wind farm. Networking essentially enables the efficient transmission of data and control signals among controllers, actuators, sensors, supervisory centers, and data storage stations [15]. When developing communication networks for wind farms, especially those located offshore, it is crucial to take into account many factors such as data transmission rates and network resilience.

Various sensors are integrated inside a wind turbine to measure its numerous components, such as the generated current, voltage, and rotor speed. Let $O_i(t), i = 1, 2, \ldots N$ be the measured state by the ith sensor of the wind turbine. We define $O_i(t)$ by,

$$O_i(t) = C_i\mathbf{X}(t) + \mathbf{n}_i \tag{1}$$

- $O_i(t)$: A vector of measurements with dimensions $p \times 1$, where p is the number of components or parameters being measured by the ith sensor.
- $C_i(t)$: A matrix that represents the measurement or sensing data from the sensor i with dimensions $p \times n$ where n refers to the number of state variables of the system being observed by the sensors.
- $\mathbf{X}(t)$: A vector that contains the state variables that describe the system's internal dynamics, such as electrical power generation, rotor speed, or internal parameters that are being monitored or controlled.

The matrix $C_i(t)$ represents the measurement or sensing data from sensor i, while $\mathbf{n}_i$ represents the observed noise during the measurement process. This noise is modeled as Gaussian noise with zero-mean and covariance R_i, similar to the process noise. The measured state is transmitted at regular intervals to the control center for the implementation of the necessary actions. As wind turbines are generally located in remote areas, there is frequently no direct communication link between a wind turbine and the control center. Normally, the transmitter of a wind turbine establishes a connection with a nearby base station, which subsequently transmits the message to the control center. The communication link between the base station and control center is assumed to be reliable, as it is part of a solid backbone network. However, the wireless communication link between the wind turbine and the base station encounters challenges in ensuring reliable data transmission [5]. Dependable communication is crucial for precise state estimates and control applications. The observed state is denoted as $O_i(t) = [o_{i1}, o_{i2}, \ldots, o_{ip}]$, where o_{ij} represents the measurement of the jth component of X obtained from the ith sensor. Every element of $O_i(t)$ is transformed and discretized into K bits. The bit block that corresponds to the jth component is represented by $b_{ij}(t)$, where $b_{ij}(t) \in \{0, 1\}^k$. Next, a repeat-accumulate code is used on $b_{ij}(t)$ to produce a code word, where c refers to the encoded bits or codewords. The codewords are organized in a sequential manner to create $m_i(t) = [c_{i1}(t), c_{i2}(t), \ldots, c_{ip}(t)]$. Once the $m_i(t)$ is modulated onto the wireless carrier signal, the resulting carrier signal $s_i(t)$ is transmitted from the wind turbine to the base station:

$$y(t) = h \cdot s_i(t) + n_w \tag{2}$$

where n_w represents the additive white gaussian noise (AWGN) with a mean of zero and a standard deviation of σ_w and h is the fading component. After receiving $y(t)$, the receiver carries out the inverse procedure (such as demodulation, decoding, demapping, etc.) to create the observed state. Figure 5 depicts the communication architecture for wind turbines based on IoT sensors.

SCADA and CMS: Currently, SCADA systems offer advanced services and capabilities for WECS that go beyond basic monitoring and management of wind turbines. CMSs are essential systems that are seamlessly connected with SCADA systems. CMSs utilize a variety of methods to detect defects in wind turbines at an early stage. The incorporation of CMS systems has shown significant improvements in the functioning and upkeep of wind turbines. CMS systems commonly employ a greater number of sensors with higher sample frequencies, as opposed to SCADA systems [14]. CMSs provide significant advantages in data communication, calculation, and storage, in addition to increasing overall costs. Consequently, numerous proposals are put forward to utilize SCADA data for condition monitoring in order to decrease the expenses of wind energy conversion systems [16]. Nevertheless, a thorough examination of specialized CMS in comparison to SCADA-based configuration management (CM) reveals that CMSs are considerably more expensive but possess more diagnostic capabilities due to their higher frequency of information.

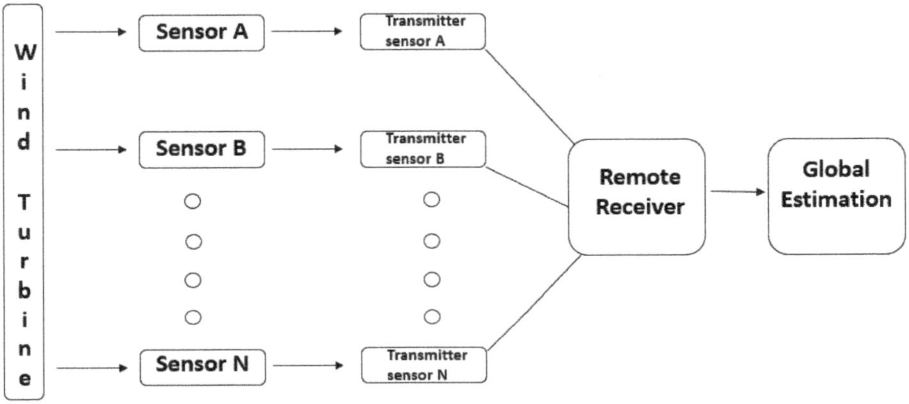

Figure 5. Communication architecture for a wind turbine equipped with multiple sensors.

2.2. Cyber-Physical Integration of a Wind Turbine

Upon conducting a thorough analysis of the many components of WECS, it is evident that they can be classified as intricate technologies that incorporate embedded systems. The cooperation between WECS layers, as depicted in Figure 6 exhibits a significant degree of diversity and constitutes a characteristic cyber-physical system (CPS) [14]. Viewing WECS as CPS introduces additional levels of technology adaption and amplifies the capabilities of WECS to be seamlessly included in intelligent power grids and the IoE.

The advancement of CPS necessitates the creation of novel models and design approaches. The primary objective of these new models and methodologies is to strike a harmonious equilibrium between intricacy and practicality. Integrating all diverse components of the cyber and physical levels to model wind turbine systems poses a significant challenge. Various elements of characteristics must be taken into account when creating CPS compositional models, including functional, non-functional, physical, component interfaces, and interface coordination. Thus, the intricate and diverse nature of components in contemporary wind turbines necessitates the application of similar concerns to be extended to WECS. It is necessary to use comprehensive analysis and verification methods to guarantee the precise operation of control systems for different electrical and mechanical components. Moreover, it is important to furnish comprehensive models of wind resources and electrical loads that influence the operational conditions. In the same context, it is imperative to establish a framework to guarantee the continuous viability, protection, defense, and adaptability of WECS [16].

Verification techniques are necessary to assess the physical requirements, such as size, power, dynamics, and memory, of various computing and networking components used in wind turbines within the context of CPS. Additionally, there is a requirement for tools to verify the compatibility of various interfaces of SCADA, CMS, sensor nodes, and power electronics with control circuits [5]. Future WECS necessitate the implementation of unified techniques to enhance their extensibility, facilitate interaction with smart grids, and improve human–machine interfaces [12]. WECS models are expected to incorporate both continuous physical dynamics and discrete occurrences. There needs to be a consistent understanding of time across sensor nodes, networks, and computing platforms. In addition, contemporary wind turbine technology incorporates computing systems that operate at varying speeds [17]. SCADA systems in wind facilities function at a low frequency to record performance data, whereas CMS operates at a high frequency to monitor components effectively. Moreover, it is imperative to depict the tangible movements of mechanical structures, the study of airflow around objects, the behavior of electrical parts, and the control of electrical power using descriptive programming abstractions. Therefore, it is becoming more and more necessary to combine conceptual computational and information flow models from sensor networks with physical-mechanical and electrical models in wind turbines. Alternatively, black box approaches can be used, relying on system identification using real data obtained from wind facilities.

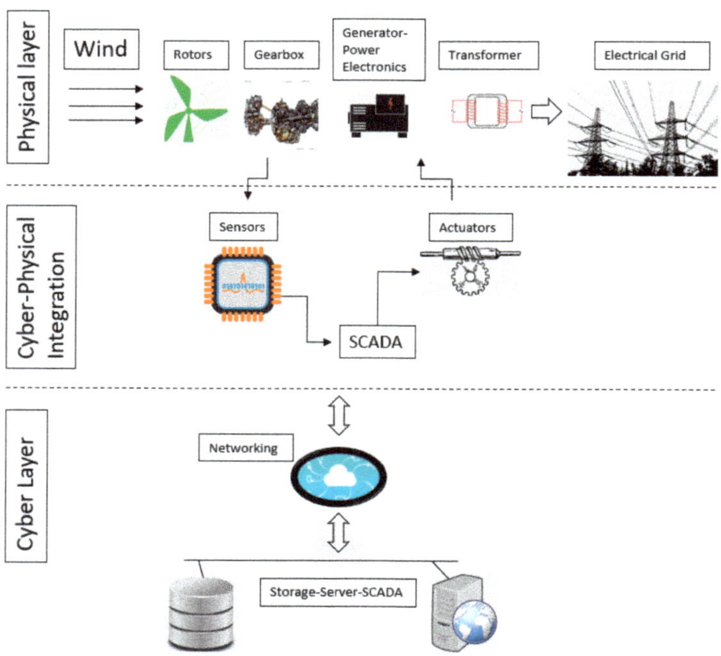

Figure 6. CP integration of WECS.

2.3. SCADA Systems and M2M for IoE-Enabled Wind Farms

Due to the harsh, expansive, and isolated nature of wind farm sites, SCADA systems must possess the ability to efficiently monitor and control operations. Furthermore, the intricate nature of a wind turbine necessitates the perception of wind farms as a collection of interconnected systems. Hence, it is imperative to enhance the existing SCADA systems employed for monitoring and controlling grids and power plants, including wind turbines, by including advanced multilayered interactive sensing, communication, and control functionalities [17]. Presently, energy providers are requiring the incorporation of wind farms SCADA systems into their asset management software, such as enterprise resource planning (ERP) and customer resource management (CRM). Figure 7 presents a concise overview of the responsibilities and objectives that upcoming SCADA systems for wind energy will encompass.

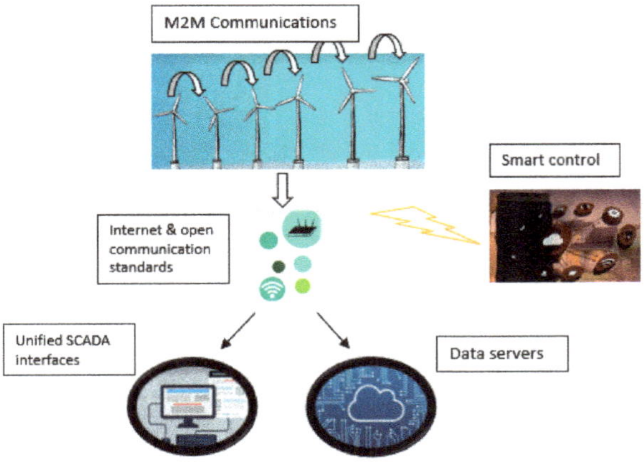

Figure 7. The objectives of upcoming SCADA systems for WECS.

In the future, wind farms will require the use of CP SCADA systems that are implemented using IoT technology for their operation and management. The notion of the IoT is exemplified by the implementation of a wireless sensor network (WSN) for an industrial process, which allows for remote monitoring via the internet [16,18]. A customizable SCADA system with decentralized intelligence and decision-making capabilities is provided based on a CP model of a power system. The system comprises three primary elements: Intelligent machines, analytics, and operators. This concept is employed in a wind control platform to oversee the synchronization of wind turbines in a farm. The SCADA system is anticipated to leverage current advancements in computing and networking to offer monitoring and control services via the internet, aligning with the fundamental essence of the IoT. Efficiently processing large amounts of raw data requires the use of self-organizing CPS networks. Hence, CP wind farms necessitate the implementation of novel network standards, protocols, and infrastructures. M2M communications is an increasingly important component of the IoT [19]. M2M connections enable the transmission of information between intelligent equipment, business applications, and data servers. Anticipated developments in M2M technology will broaden the scope of connections beyond individualized interactions to a model where producers and consumers are interconnected. A planned infrastructure for M2M communications aims to enable smart wind farms to efficiently communicate measured data and enhance their intelligence among wind turbines. A proposed cloud-based M2M telemetry system aims to efficiently handle and visually represent data for suppliers of renewable energy [20]. Furthermore, content management systems are getting more intelligent using M2M technology.

3. ML for Wind Energy Prediction

This section delineates the procedures for data collecting, data preprocessing, and the implementation of ML algorithms.

Wind energy prediction analysis jobs involve the processing and interpretation of large quantities of weather-related data to generate accurate estimates of future energy generation. These tasks employ a diverse range of analytical methodologies and tactics to enhance the precision of forecasts. The meteorological data, consisting of wind speed, wind direction, temperature, humidity, wind gusts, and dewpoint, is now accessible and can be utilized for power generation predictions, as Figure 8 shows [20]. Significant statistical connections can be identified between different meteorological factors and energy production. These correlations can be used to construct models that accurately represent the influence of certain weather conditions on the efficiency of wind turbines. The selection of features for predicting models is determined by this examination. Selecting suitable machine learning techniques, like regression models, for predictive modeling depends on the unique demands of the prediction issue and the features of the data.

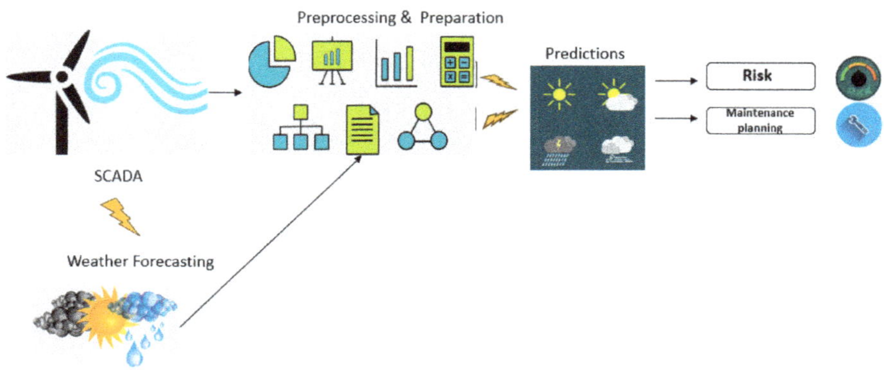

Figure 8. Employing meteorological data to predict electric energy levels.

3.1. Dataset and Preprocessing

The Wind Power Generation Data—Prediction dataset was acquired from Kaggle (https://www.kaggle.com/datasets/mubashirrahim/wind-power-generation-data-forecasting/data, accessed on 7 November 2024) and uploaded to the Kaggle platform by MUBASHIR RAHIM. The meteorological equipment—IoT devices deployed at the site was used to meticulously gather the data. There is no guarantee of data quality, which could affect the accuracy of the results. Furthermore, there is no validation or assurance that the report on time measurements is correct. Deep neural networks could potentially replace regression models in situations where the dataset is significantly larger and includes real-time data. The meteorological apparatus measured temperature, humidity, dew point, and wind properties at predetermined elevations of 2 m, 10 m, and 100 m. Concurrently, sensors were installed on wind turbines to monitor their efficiency and electricity production. The datasets consist of a detailed hourly log obtained from four distinct sites, spanning from 2 January 2017, 00:00:00, to 31 December 2021, 23:00:00. The data underwent rigorous quality checks to detect and rectify any anomalies or inconsistencies, ensuring a high level of data reliability. Regular equipment maintenance has consistently ensured the quality of data over time.

The following are the columns and weather parameters in the data:

- Time: The moment in the day when the measurements were made.
- temperature_2m: The temperature in degrees Fahrenheit at two meters above the surface.
- relativehumidity_2m: The proportion of relative humidity at two meters above the surface.
- dewpoint_2m: Dew point, measured in degrees Fahrenheit at two meters above the surface.
- windspeed_10m: The wind speed, expressed in meters per second, at 10 m above the surface.
- windspeed_100m: The speed of the wind at 100 m above sea level, expressed in meters per second.
- winddirection_10m: The wind direction at 10 m above the surface is represented in degrees. (0–360).
- winddirection_100m: The direction of the wind at 100 m above the surface, expressed in degrees (0–360).
- windgusts_10m: A wind gust is an abrupt, transient increase in wind speed at 10 m.
- Power: The normalized turbine output, expressed as a percentage of the turbine's maximum potential output, and set between 0 and 1.

The normal or Gaussian distribution that is indicated in Figure 9 represents the famous bell-shaped curve, which is characterized by the arithmetic mean μ and the standard deviation σ. The normal distribution is the most frequently employed probability and statistics distribution. Contemporary techniques like linear regression, analysis of variance (ANOVA), and t-tests heavily depend on the assumption that the data follow a normal distribution. Because the dataset contains outliers, which indicate the accuracy of the sensors' measurements, the curves do not have a well-defined shape [21].

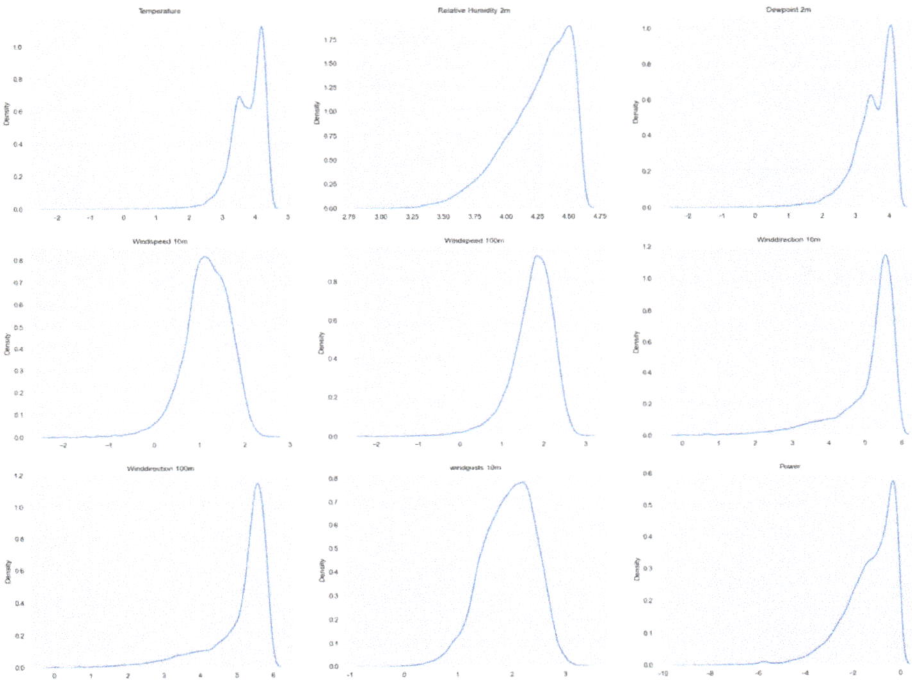

Figure 9. Gaussian distribution including outliers.

Libraries: Python's libraries for data analysis, visualization, and scientific computing are extensively utilized. They provide a comprehensive range of tools and features that make it easier to explore data and generate insights [22]. The libraries to be utilized in the preprocessing stage are as follows:

- Pandas is a robust Python 3.13.0 (https://www.python.org/, accessed on 26 September 2024) package utilized for the manipulation and analysis of data. The software provides data structures such as DataFrames and Series, which facilitate the manipulation and analysis of organized data.
- NumPy is an essential library for scientific computation in Python, commonly referred to as "Numerical Python". The software provides support for large, complex arrays and matrices, together with a collection of mathematical algorithms to effectively handle these arrays.
- Matplotlib is a flexible toolbox that enables the generation of static, interactive, and animated visualizations in the Python computer language. The pyplot module offers a MATLAB-like interface for producing plots and visualizations, simplifying the process of generating charts, histograms, scatter plots, and other graphical representations.
- Seaborn is a data visualization package that enhances the capabilities of matplotlib and provides a more sophisticated interface for creating visually appealing and meaningful statistical graphics. It streamlines the procedure of generating intricate visualizations

and provides pre-installed themes and color palettes to increase the visual appeal of plots.

Dataset: There are four data frames, namely loc1, loc2, loc3, and loc4, that are all of equal size. All data in the datasets originate from the utilization of IoT devices to measure meteorological conditions with consistent precision. The number of rows in each data frame is 43,800, and the number of columns is 10. This study centers on the examination carried out utilizing the loc1 dataset. The columns consist of the following variables: time, temperature, relative humidity, dewpoint, wind direction, wind speed, and wind gusts at 2, 10, and 100 m, as shown in the first five rows of Figure 10. There are six variables of float64 data type, three variables of int64 data type, and one variable of object data type.

Time	temperature_2m	relativehumidity_2m	dewpoint_2m	windspeed_10m	windspeed_100m	winddirection_10m	winddirection_100m	windgusts_10m	Power
2017-01-02 00:00:00	28.5	85	24.5	1.44	1.26	146	162	1.4	0.1635
2017-01-02 01:00:00	28.4	86	24.7	2.06	3.99	151	158	4.4	0.1424
2017-01-02 02:00:00	26.8	91	24.5	1.30	2.78	148	150	3.2	0.1214
2017-01-02 03:00:00	27.4	88	24.3	1.30	2.69	58	105	1.6	0.1003
2017-01-02 04:00:00	27.3	88	24.1	2.47	4.43	58	84	4.0	0.0793

Figure 10. The first 5 rows in dataset of location1.

Since time is an object, it will be transformed to datetime before being used for analysis. The conversion is performed using the function (pd.to_datetime) from the Pandas package [22]. The subsequent tables and figures originate from the exploratory data analysis conducted at loc1. Based on the corresponding time values, in Table 1, organize the power-generating data into separate columns for year, month, and day.

Table 1. Data columns from location 1.

Columns	Null Values
Time	43,800 non-null datetime64
temperature_2m	43,800 non-null float64
relativehumidity_2m	43,800 non-null int64
dewpoint_2m	43,800 non-null float64
windspeed_10m	43,800 non-null float64
windspeed_100m	43,800 non-null float64
winddirection_10m	43,800 non-null int64
winddirection_100m	43,800 non-null int64
windgusts_10m	43,800 non-null float64
Power	43,800 non-null float64
Year	43,800 non-null int32
Month	43,800 non-null int32
Day	43,800 non-null object

Null values: A non-null value refers to any numerical, textual, or other type of value that is not null [23]. The data frame has 43,800 non-null values in each column, corresponding to the index range of 0–43,799. By utilizing Python's function .null().sum(), can determine the number of null values for each variable. Datasets do not contain any null values as described in Table 2. Determining influential and anomalous data points is essential as it will aid in future data collection and the proper utilization of existing knowledge.

Table 2. Null values in the dataset.

Columns	Null Values
temperature_2m	0
relativehumidity_2m	0
dewpoint_2m	0
windspeed_10m	0
windspeed_100m	0
winddirection_10m	0
winddirection_100m	0
windgusts_10m	0
Power	0
Year	0
Month	0

Outliers: The degree to which a data point deviates from the mean in terms of standard deviations is measured statistically by the z-score [23]. The z-score can be calculated using the following formula:

$$z = (x - mean) \;/\; std \tag{3}$$

In this context, x represents a specific data point, *mean* represents the average value of the dataset, and *std* represents the standard deviation of the dataset.

It appears that the dataset contains some outliers, as illustrated in Figure 11. Consequently, Table 3 shows the results of removing outliers to achieve improved outcomes.

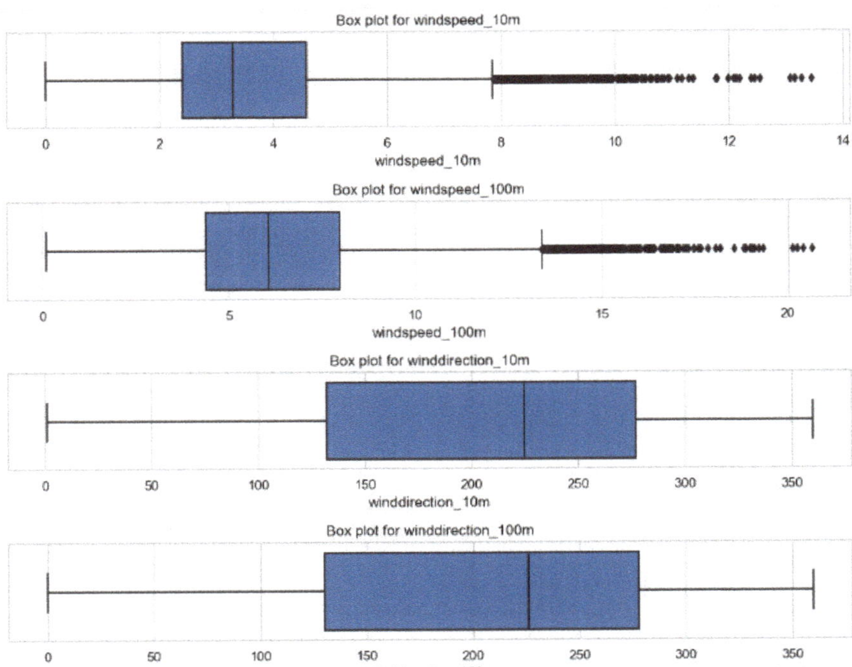

Figure 11. *Cont.*

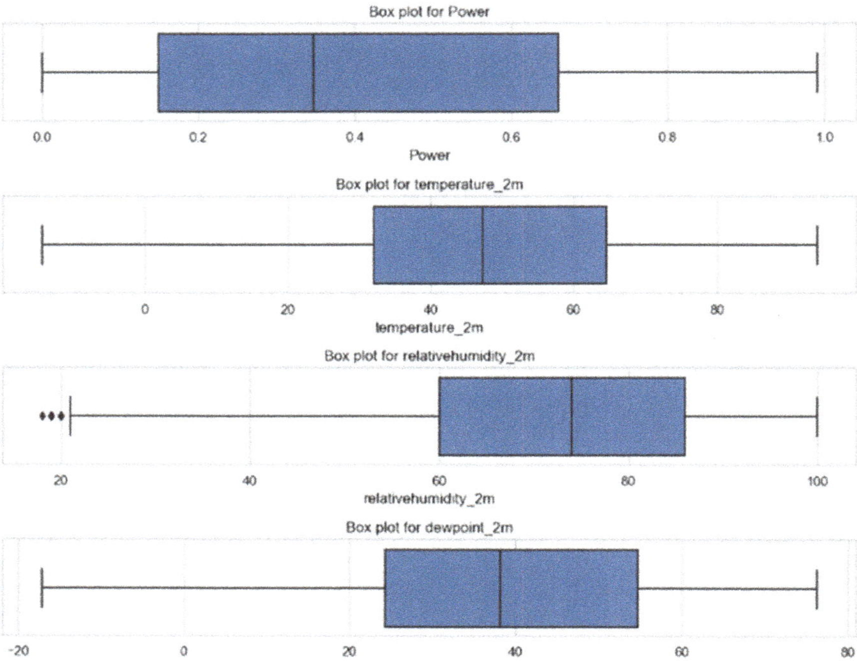

Figure 11. Box plots for numeric columns (time, year, month no needed to plot).

Table 3. Removed outliers from columns.

Columns	Removed Outliers
temperature_2m	5
relativehumidity_2m	11
dewpoint_2m	0
windspeed_10m	318
windspeed_100m	199
winddirection_10m	0
winddirection_100m	0
windgusts_10m	337
Power	Not included
Year	Not included
Month	Not included

The distribution of wind gusts in Figure 12 may show significant skewness due to extreme values, which can obscure the true underlying patterns. After outlier removal, the distribution typically becomes more normal, allowing for clearer insights and more accurate analyses.

Correlation: The dataset is currently accessible and prepared for utilization in deriving significant insights. Correlation study between measurements such as temperature, relative humidity, and wind speed will assist in selecting parameters for prediction [24]. These highly correlated metrics aid in predicting the power generated by wind turbines. A correlation heatmap is a visual representation that presents the correlation between many variables in the form of a matrix, utilizing color codes. By utilizing the Python programming language and the seaborn library, it is possible to generate a helpful heatmap that displays the association between variables. The correlation coefficients for various variables are displayed in a correlation table.

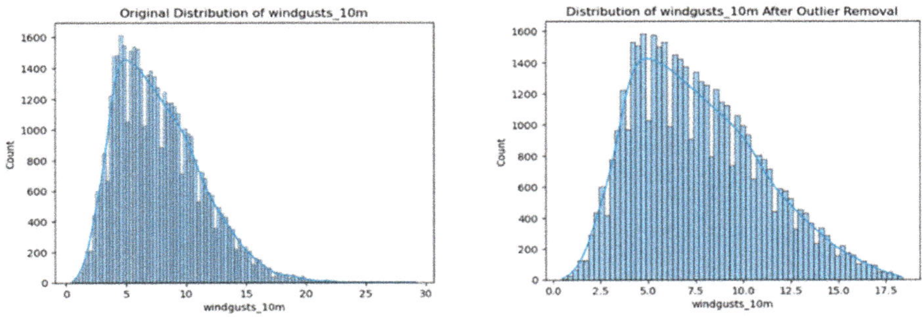

(a) (b)

Figure 12. Distribution of wind gusts (**a**) before and (**b**) after removal of outliers.

Conventional clustering and correlation analysis face difficulties when dealing with the vast amount and low density of valuable information in big data. To improve energy prediction, it is recommended to use big data-driven correlation analysis with clustering. Conducting correlation analysis among several measures such as temperature, dew point, relative humidity, wind direction, wind gusts, and wind speed will aid in the selection of forecast parameters. These strongly connected parameters contribute to the accurate prediction of the electricity produced by wind turbines.

The heatmap in Figure 13 displays the relationships between every conceivable combination of values. It is a potent tool for detecting and visualizing patterns in data, as well as condensing large amounts of data. A Python program, utilizing the Seaborn module, may generate a heatmap that visually represents the association between variables [21].

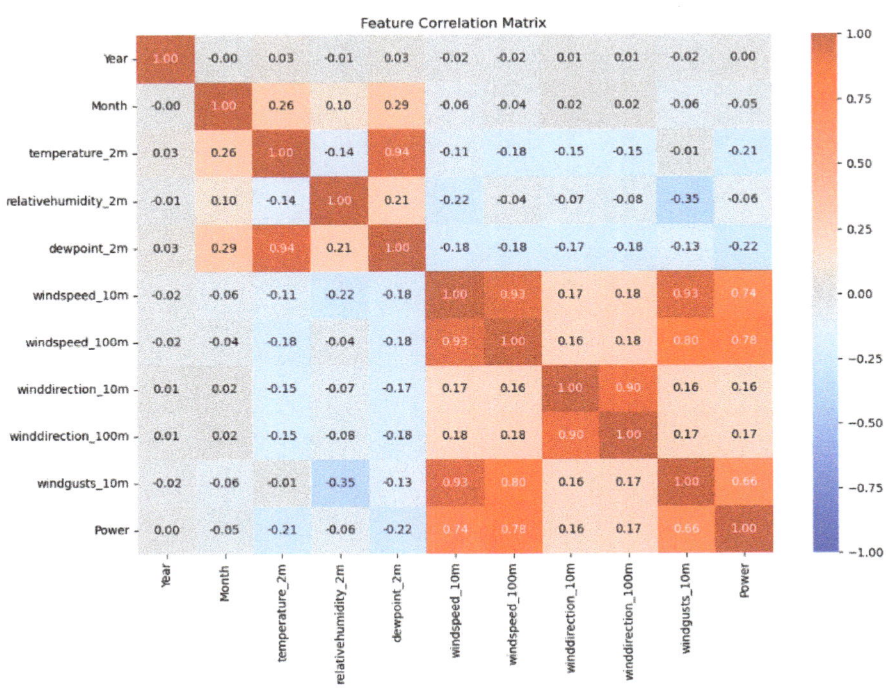

Figure 13. Heatmap—correlation matrix.

When preparing a dataset for ML models, preprocessing stages include data standardization and splitting. Consequently, all three regression models (linear regression, random forest regression, and lasso regression) follow this preprocessing phase.

Standardization: This procedure guarantees that the dataset's features (or variables) have a mean of 0 and a standard deviation of 1 [4,25]. This phase is essential as numerous ML algorithms exhibit enhanced performance when the data is normalized or standardized. The StandardScaler is a prevalent technique that subtracts the mean and normalizes data to unit variance.

The standardization formula for each feature x is as follows:

$$z = (x - \mu) \ / \ \sigma \tag{4}$$

- μ is the mean value of the feature;
- z is the standardized feature;
- σ is the standard deviation of the feature.

Splitting the data: This pertains to partitioning the dataset into training and testing (and occasionally validation) subsets [25]. The training set is utilized to develop the model, whereas the testing set is employed to assess its performance on novel data. The validation set is also employed to optimize the model without affecting the test set, particularly during hyperparameter optimization. This mitigates overfitting and guarantees the model generalizes effectively to novel inputs. The data are typically divided between 70% and 80% for training and 20% and 30% for assessment. The precise ratio is contingent upon the task at hand and the extent of the dataset. After completing the data purification procedure, the dataset (loc1) is now available and ready to be used for extracting meaningful insights.

3.2. Machine Learning and Wind Energy Forecast

ML algorithms have the capability to identify alterations in the surroundings and adjust their actions accordingly. Regression analysis refers to this specific subset of categorization [26,27]. The objective of this part is to structure the prediction in the wind energy domain. Following the technique or data analysis, ML is employed to predict the power energy output. This extensive dataset provides valuable insights on the correlations between various weather patterns and the generation of wind energy. By using predictive models and analyzing meteorological data, it is feasible to forecast power output.

Regression models are utilized to determine the correlation between alterations in one or more explanatory variables and alterations in the dependent variable. To determine the regression model that exhibits the greatest efficiency and the lowest MSE, three regression models will be used and compared [28]. The AIC and BIC criteria will be part of the comparison and selection of the most appropriate model for this dataset, identifying accurately the underlying patterns while minimizing overfitting. The comparison of linear regression, random forest regression, and lasso regression will yield valuable results and provide an opportunity to gain a more comprehensive comprehension of each regression model's capabilities and weaknesses.

Regression is extensively used in the field of big data to build predictive models. These models are designed to forecast certain outcomes for incoming data, rather than interpreting existing data. Regression analysis is a reliable method for identifying the variables that have an impact on a specific topic of interest [27]. Regression analysis allows for the accurate identification of the key aspects, the ones that may be ignored, and the correlations between these elements.

- Dependent Variable: The dependent variable is the primary factor that one seeks to anticipate or comprehend.
- Independent Variables: These variables are postulated to exert an influence on the dependent variable.

The metrics included in the regression models are R^2, Adjusted R^2, MSE, RMSE, and MAE. These indices represent the goodness of fit, with the following definitions for each [29]:

- R^2 (Coefficient of Determination): Assesses the model's efficacy in elucidating the variance of the target variable. Varies from 0 to 1, with proximity to 1 indicating a superior fit [28].
- Adjusted R^2: Analogous to R^2, although modified to account for the quantity of predictors in the model. Addresses overfitting; increases solely if additional predictors enhance the model.
- MSE: The mean of the squared deviations between expected and actual values. Imposes more penalties on larger faults compared to lesser ones.
- RMSE: The square root of the MSE. Denotes the mean error in the identical units as the target variable.
- MAE: The mean of the absolute discrepancies between expected and actual values. More robust to outliers than MSE or RMSE.

The AIC and BIC are two commonly used criteria in statistical model selection and comparison.

- AIC: Assesses a model's quality in comparison to other models, considering both the model's goodness-of-fit and its complexity. The AIC is computed as follows:

$$\mathrm{AIC} = 2k - 2\ln(L) \tag{5}$$

- k is the number of parameters in the model;
- L is the maximum likelihood of the model.

The term $2k$ acts as a penalty for model complexity, reducing the chance of overfitting by favoring simpler models when they achieve a comparable fit.

- BIC: Also evaluates model quality by imposing a more stringent penalty on the number of parameters, particularly as the sample size grows. The BIC is defined as follows:

$$\mathrm{BIC} = \ln(n)k - 2\ln(L) \tag{6}$$

where:

- n is the number of observations;
- k is the number of parameters;
- L represents the likelihood function, which represents the probability of observing the given data under specific model parameters. For a collection of observations x1, x2, associated with a model characterized by parameters θ, the likelihood function is typically expressed as:

$$L(\theta) = \prod_{i=1}^{2} P(x_i|\theta) \tag{7}$$

The $\ln(n)k$ penalty term grows with sample size n, meaning BIC tends to favor simpler models more strongly than AIC, particularly for large datasets [11]. Lower AIC values signify a better model fit. Similar to AIC, lower BIC values indicate a better model fit.

Overfitting in ML occurs when models are selected and hyperparameters are adjusted based on test loss, which challenges the assumption that the model's performance is independent of the test set [30]. The ultimate classifier may exhibit high performance only on a certain sample of examples within the test set, especially when method designers evaluate numerous models on the identical test set [31]. K-fold cross-validation is used to assess the performance of predictive models. The dataset is partitioned into k folds, each of which is a subset. As shown in Figure 14, for each of the k training and assessment cycles, the model uses a distinct fold as the validation set. The model's generalization performance is measured by calculating the average of the performance metrics obtained from each fold.

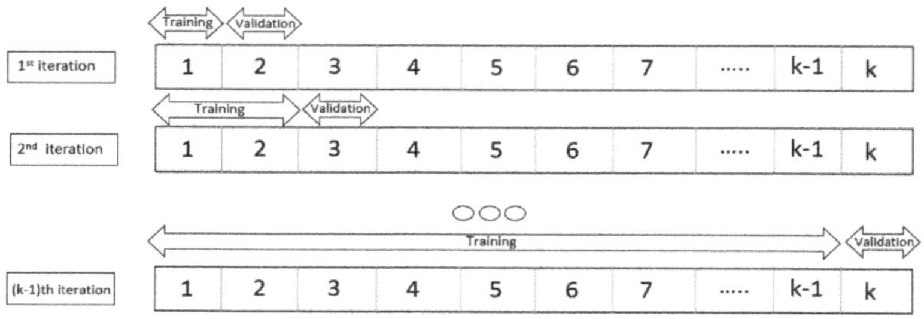

Figure 14. Diagram of K-fold forward cross-validation.

3.2.1. Linear Regression

Linear regression, sometimes known as LR, is a prevalent and extensively utilized modeling technique in the fields of statistics and machine learning. The objective outcome is to establish a linear relationship between the input and target variables. The model postulates a linear amalgamation of the input features to predict the continuous output variable. In order to calculate the coefficients of these input variables, several optimization approaches, such as least squares, are utilized [32]. LR is an excellent option when there are linear relationships between variables due to its simplicity and ease of comprehension. Multiple linear regression (MLR) is an appropriate form of linear regression for this particular situation. MLR produces equations that establish a connection between several input variables (x_n) and a target variable (y).

$$y = w_0 + w_1x_1 + \cdots + w_nx_n \tag{8}$$

Here, n represents the total number of input variables, w_n denotes the coefficient for x_n, and w_o refers to the intercept. Regularization approaches, such as the inclusion of a penalty term on the model's input variables, can restrict the freedom of the input variables during the learning process, hence improving the accuracy of predictions on data that were not used for training.

MLR is a statistical technique that estimates the value of a dependent variable based on multiple independent variables. The objective of MLR is to construct a precise mathematical model that accurately depicts the linear relationship between the independent variables (x_n) and the dependent variable (y) being studied [33]. The primary MLR model is described as:

$$y = \beta_0 + \beta_1x_1 + \cdots + \beta_mx_m + \varepsilon \tag{9}$$

- y is the dependent variable;
- β_0 is the intercept;
- The coefficients $\beta_1, \ldots, \beta_m$ represent the values assigned to the independent variables $x_1, \ldots, x_n$.

The intercept, often known as the "constant", in a regression model signifies the average value of the response variable when all predictor variables in the model are set to zero. The intercept, denoted as β_0, is the estimated value of y when all x_i values are equal to zero. The baseline level of y (dependent variable) is established when the explanatory variables have no influence. The coefficients represent the weights of each independent variable, indicating the extent to which each variable contributed to the prediction.

The assumption of independence in linear regression is crucial. Figure 15 displays the residuals across the order of the test data. There are no discernible patterns or systematic trends, suggesting a random distribution of the residuals around zero. This suggests that the assumption of independent residuals is probably satisfied, indicating there are no strong correlations between consecutive errors. This suggests that the validity of the model was positively affirmed in this regard.

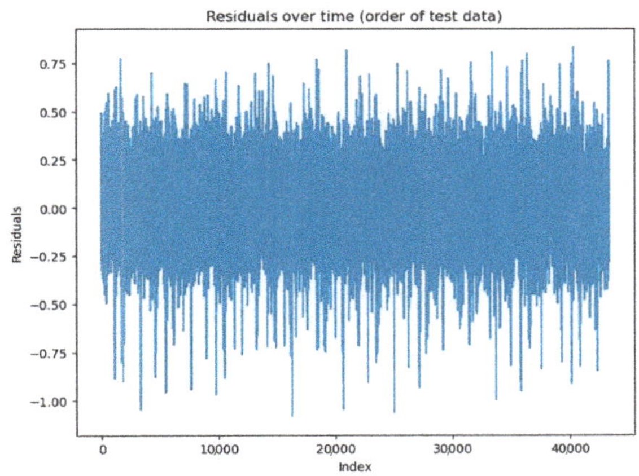

Figure 15. Residuals over time of the test data.

In linear regression, one assumption is that the residuals should have a normal distribution [34]. Figure 16 shows that the residuals are roughly normally distributed, with a peak around zero. It closely approximates a normal distribution, which is suitable for the normality assumption. Minor deviations from normality, especially toward the tails, are present, but these do not seem significant enough to invalidate the assumption of normality.

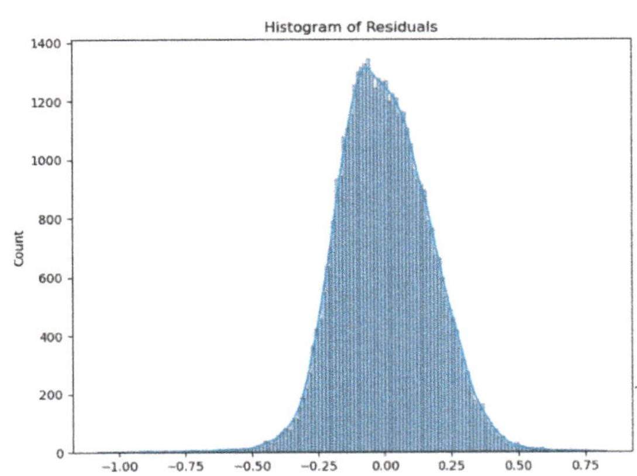

Figure 16. Distribution of residuals using histogram.

The implementation of the linear regression technique, along with cross-validation results, yielded the aforementioned metrics [28]:

- R^2 (0.6199): This means that about 61.99% of the variance in the target variable can be explained by the model's features. This indicates a moderately strong fit, but there is still 38% of variability in the target that the model does not explain.
- Adjusted R^2 (0.6194): The Adjusted R^2 is slightly lower than the R^2 (0.6194 vs. 0.6199), which accounts for the number of predictors. It is close to R^2, suggesting that the added features are useful, but not overfitting.

- MSE (0.0312): The low value of 0.0312 of MSE indicates that the model's predictions are generally close to the actual values, though it is harder to interpret MSE without comparing it to the scale of the data.
- RMSE (0.1767): An RMSE of 0.1767 means that, on average, the model's predictions are off by around 0.18 units from the actual values.
- MAE (0.1389): An MAE of 0.1389 means that, on average, the model is off by 0.14 units, which is slightly lower than the RMSE. This suggests the model is performing well with relatively small errors.

Cross-Validation Results (Mean ± Std): These results give insight into how the model performs across multiple data splits during cross-validation. They help confirm the robustness of the model.

- R^2 (0.6299 ± 0.0082): The average R^2 across cross-validation is 62.99%, slightly higher than the original R^2. The standard deviation (±0.0082) indicates stable performance across different data splits.
- Adjusted R^2 (0.6290 ± 0.0082): The adjusted R^2 is 62.90% with minimal variability, confirming that the model generalizes well without overfitting.
- MSE (0.0303 ± 0.0007): The average error across cross-validation sets is 0.0303 with a small standard deviation (±0.0007), showing that the model is consistent.
- RMSE (0.1741 ± 0.0021): The average RMSE is 0.1741, meaning the average prediction error is about 0.174 units, with slight variability (±0.0021).
- MAE (0.1376 ± 0.0015): The average MAE is 0.1376, indicating that, on average, the model is 0.1376 units off. The small standard deviation (±0.0015) shows good consistency.

The cross-validation results validate the model's stability, as the metrics consistently align across several data splits. Figure 17 shows that there is a clear downward trend, which means that the residuals change systematically as the predicted values increase. Specifically, the variance of the residuals decreases with the increase in the predicted values. The model struggles to predict values consistently across the range of predictions.

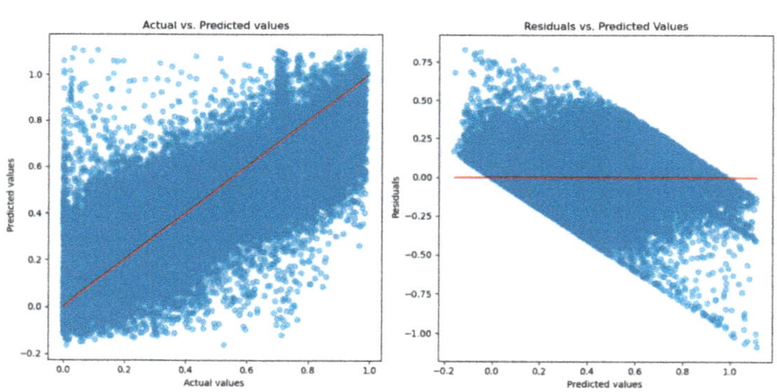

(a) (b)

Figure 17. (**a**) Actual vs. predicted values, (**b**) residual vs. predicted values.

3.2.2. Random Forest Regression

Leo Breiman [35] and Cutler Adele [36] proposed the Random Forest Regression (RFR) algorithm in 2001 as an ML method for both regression and classification tasks. Categorical regression tree (CART) techniques can be classified into two categories depending on the nature of the output variables: regression decision trees and categorical decision trees. It is a flexible ML technique employed for prediction of numerical values. In order to reduce overfitting and improve accuracy, it uses the predictions of many decision trees [37]. Python's machine-learning modules facilitate the efficient optimization and implementation of this method.

Random forest regression involves adjustable parameters, similar to other ML techniques. Some of the factors that influence a regression tree include the minimum number of observations at each terminal node, the fraction of data to sample in each regression tree, the number of trees, and the number of predictor variables randomly picked at each node [38]. Cross-validation is employed to optimize these independent parameters. It is often recommended to set the number of decision trees to a high value in order to achieve a steady minimum for the prediction error, rather than making adjustments.

The equation for the generalization error and margin function in random forest is given as follows:

$$PE^{*} = P_{X,Y}(mg(X,Y) < 0) \tag{10}$$

where

$$mg(X,Y) = av_k I(h_k(X) = Y) - \underset{j \neq Y}{max} av_k I(h_k(X) = j) \tag{11}$$

The equation's relevant term, av_k, specifies the weighting of each tree's vote to determine the final classification or regression output.

Let X and Y represent random vectors. The margin function, mg, determines the average votes for the correct output compared to other outputs. The function $I(.)$ is an indicator function, and h_K represents the classifiers.

In Random Forest regression, RandomizedSearchCV is employed to optimize the model's hyperparameters by exploring a spectrum of potential parameter values [39]. The get_param_grid method produces a dictionary of hyperparameters and their associated values for tuning in the Random Forest model. Each key in the dictionary signifies a hyperparameter (n_estimators: the number of trees in the forest, max_depth: the maximum depth of every tree in the forest, min_samples_split: the minimum number of samples required to split an internal node, min_samples_leaf: the minimum number of samples required to be at a leaf node) and the corresponding list comprises various values that RandomizedSearchCV will investigate.

The implementation of the RFR technique, along with cross-validation results, yielded the aforementioned metrics [40]:

- R^2 (0.76087): An R^2 of 0.76087 signifies that the model accounts for approximately 76.09% of the variance in the target variable, which is commendable. The model effectively catches most patterns within the data.
- Adjusted R^2 (0.76060): Adjusted R^2 is closely aligned with the R^2, indicating that the model is appropriately fitted without superfluous complexity.
- MSE (0.01976): The MSE is notably low, signifying that the model's prediction errors are minimal.
- RMSE (0.14057): An RMSE of 0.14057 implies that, on average, the model's predictions diverge from the actual values by approximately 0.14 units, reflecting commendable performance, particularly relative to the data's scale.
- MAE (0.10439): An MAE of 0.10439 indicates that, on average, the model's predictions deviate by around 0.10 units. Given that MAE exhibits less sensitivity to outliers compared to MSE, it indicates that the model is continuously producing relatively minor mistakes.

Mean Cross-Validation Results: Better knowledge of how the model spans several subsets of the data comes from cross-valuation results. Often used against the single assessment on the test set, the "Mean" values show the average across several folds (splits) of the dataset.

- Mean Cross-Validation R^2 (0.64517): With an average R^2 value of 0.64517—lower than the test set R^2—0.76087—over the cross-valuation folds. This implies that, on average, during cross-valuation, the model explains roughly 64.5% of the variation, whereas on the test set it explains about 76% of the variance. Although this difference suggests some variation in model performance over several data subsets, overall, the finding is still really strong.

- Mean Cross-Validation Adjusted R^2 (0.64509): Considered as lower than the Adjusted R^2 on the test set (0.76060), the average Adjusted R^2 across cross-valuation is 0.64509. Like the R^2 score, this indicates that although the model may be slightly overfitting the test data relative to its performance on several validation sets, it generalizes somewhat reasonably.
- Mean Cross-Validation MSE (0.02943): Higher than the test set MSE (0.01976), the average MSE among several cross-valuation folds is 0.02943. This implies that, on the test data, the model did rather better than on the average validation folds. Still, the variation is not significant, suggesting a rather steady performance.
- Mean Cross-Validation RMSE (0.17155): Higher than the test RMSE (0.14057), the average RMSE for the validation sets is 0.17155. This suggests that, although still within a reasonable range, the model's mistakes during cross-valuation are rather greater than on the test set on average.
- Mean Cross-Validation MAE (0.13200): Higher than the test MAE, 0.10439, the average MAE during cross-valuation is 0.13200. Consequently, the model performs really well over several data splits but makes somewhat more mistakes on the cross-valuation folds.

Cross-validation results reveal that, when tested on several subsets of the data, the model's performance is consistent but rather less. Although the lower cross-valuation R^2 (0.64517) points to some variation in the generalizing capacity of the model, the test and cross-valuation metrics differ only in minor degree.

Currently, the evaluation of the model utilizing the most associated features, namely 'windspeed_10m', 'windspeed_100m', and 'windgusts_10m', yields the following results:

- R^2 Score: 0.67234
- Adjusted R^2 Score: 0.67223
- MSE: 0.02707
- RMSE: 0.16453
- MAE: 0.12603

By picking solely the most correlated features, the model may forfeit significant interactions or information offered by less correlated variables. The optimal parameters chosen yielded a less intricate model (fewer trees, reduced depth), which may not adequately depict the previous degree of intricacy. A strong correlation may not necessarily reflect a feature's complete impact on a model's performance, particularly in non-linear models such as Random Forests.

3.2.3. Lasso Regression

LASSO regression, also known as Least Absolute Shrinkage and Selection Operator regression, is a commonly employed method for reducing the size of coefficients and choosing variables in regression models. The computationally demanding nature of statistical software is no longer concerning due to developments in processing power and integration. The objective of LASSO regression is to identify the variables and corresponding regression coefficients that minimize the prediction error of the model [41]. The parameter λ, often denoted by alpha, serves as the L1 penalty or shrinkage parameter, controlling the degree of regularization applied to the model [42]. Regularization helps prevent overfitting by penalizing large coefficients. A higher value of λ increases the penalty for large coefficients, which causes more coefficients to be shrunk towards zero, effectively removing less important features from the model; a lower value of λ lets the model retain more features, which might lead to overfitting should too many irrelevant features be included. During the configuration of the Lasso model and grid search using GridSearchCV, the algorithmic process selects the best value of λ based on the training process and evaluations from cross-validation. Multicollinearity can affect the selection of the optimal λ during model training. The goal is to find a λ value that maintains the model's performance while simultaneously minimizing overfitting. The λ value found was 0.0001. The results of the variance inflation factor (VIF) calculation show that the variables windspeed_100m, windspeed_10m, and windgusts_10m have high VIF values (49.51, 135.40, and 51.69,). These values indicate that there is significant multicollinearity among these variables. High multicollinearity can affect the reliability of the model parameters, and there may be difficulties in interpreting the effect of individual variables on the target variable. The process of calculating and removing features with a high VIF index did not result in any improvements. LASSO conducts regression analysis using the below equation, where N represents the sample size of a and β, j denotes the parameter coefficients, and $\hat{a}$ represents the prediction.

$$(\hat{a}, \hat{\beta}) = \text{argmin}\left(\frac{1}{N}\sum_{i=1}^{N}\left(y_i - a - \Sigma_{j=1}^{p} x_{i,j} \times \beta_j\right)^2\right) \tag{12}$$

The provided formula can be compacted and represented in Lagrangian form, as illustrated in the equation [43]. The equation below demonstrates that L1 regularization is the preferred method in LASSO. L1 regularization incorporates the absolute value of feature coefficients as a penalty term to regulate the impact of the features.

$$(\hat{a}, \hat{\beta}) = \text{argmin}\left(\frac{1}{N}||y - X\beta||_2^2 + \lambda||\beta||_1\right) \tag{13}$$

The implementation of the Lasso regression, along with cross-validation results, yielded the aforementioned metrics [44]:

- R^2 (0.6110): This means that 61.10% of the variance in the target variable (y) is explained by the Lasso regression model. It is a moderate grade, demonstrating that the model captures a good percentage of the variability; however, there is potential for improvement.
- Adjusted R^2 (0.6108): The Adjusted R^2 value is quite close to the R^2 score (0.6108 vs. 0.6110). This shows that the model's performance does not diminish when accounting for the amount of predictors used. Since the model is not overfitting with irrelevant variables, the adjusted R^2 stays virtually the same as the regular R^2.
- MSE (0.0319): A lower MSE (0.0319) shows the model's predictions are pretty close to the actual values, while there are some inaccuracies.
- RMSE (0.1787): RMSE is 0.1787, suggesting on average, the predictions are wrong by around 0.1787 units of the target variable, which is a substantial amount of inaccuracy.
- MAE (0.1410): With a MAE of 0.1410, the predictions average from the actual values by roughly 0.1410 units. This implies somewhat minimal error, although RMSE (which penalizes more significant errors) indicates somewhat more fluctuation in the errors.

Cross-Validation Results for Lasso Regression:

- Mean R^2 (0.6132 $\pm$ 0.0533): With an average R^2 score of 0.6132—rather close to the test set R^2 of 0.6110—the 10 cross-valuation folds Although the performance of the model fluctuates somewhat throughout the few cross-valuation folds, the standard deviation ($\pm$0.0533) indicates minimal fluctuation that suggests consistency.
- Adjusted R^2 (0.6108 $\pm$ 0.0533): With a mean of 0.6108, the modified R^2 is also rather consistent; it indicates that the model can generalize effectively over several folds and is not overfitting.
- Mean MSE (0.0313 $\pm$ 0.0052): With a tiny standard deviation ($\pm$0.0052), the average MSE over the cross-valuation folds is 0.0313, somewhat near to the test set MSE of 0.0319. This indicates that the model is not unduly sensitive to several subsets of the data and is rather steady in performance.
- Mean RMSE (0.1769 $\pm$ 0.0722): Again, revealing a comparable average prediction error, the RMSE from cross-validation (0.1769) is once more near to the test set RMSE of 0.1787. Though it is still reasonable, the standard deviation ($\pm$0.0722) indicates far more fluctuation in mistakes between folds than in MSE.

- Mean MAE (0.1399 ± 0.0105): With cross-validation, the average MAE (0.1399) is rather close to the test set MAE (0.1410). Furthermore, showing consistency in the prediction accuracy across several subsets is the low standard deviation (±0.0105).

The cross-validation outcomes closely align with the test set findings, indicating that the model generalizes effectively and is not overfitting the data. The minimal standard deviations for all measures indicate the model's stability across various data splits.

4. Results and Discussion

This comparison analysis assesses the efficacy of three regression models: Linear Regression, Lasso Regression and RFR, utilized for wind power predicting data. The performance of each model, the selection, and the comparison are evaluated using standard statistical metrics, such as R^2, Adjusted R^2, MSE, RMSE, and MAE, and criteria, such as AIC and BIC. Furthermore, cross-validation was conducted to assess the models' stability and generalizability. The following is a comprehensive comparison of the three regression techniques based on the data acquired. Initially, the Random Forest Regression and Linear Regression attain exceptionally high R^2 and Adjusted R^2 values, ranging from 0.98 to 0.99. Subsequent to preprocessing, it was observed that the ML algorithms exhibit overfitting issues. In conclusion, preprocessing reduces the risk of overfitting, ensuring accurate predictions on both training and new data. The final results are presented in Table 4.

Table 4. Comparison of metrics of three regression models.

Models	R^2	Adjusted R^2	MSE	RMSE	MAE	AIC	BIC
Linear Regression	0.6199	0.6194	0.0303	0.1741	0.1376	−23,063.5	−22,970.5
Random Forest Regression	0.7608	0.7606	0.0294	0.1715	0.1320	−34,355.5	−34,284.8
Lasso Regression	0.6110	0.6108	0.0313	0.1769	0.1399	−29,847.0	−29,825.7

4.1. Performance Metrics on Test Set

- R^2 and Adjusted R^2: Random Forest exhibits enhanced predictive capability, evidenced by significantly elevated R^2 and Adjusted R^2 values relative to Linear and Lasso models. Both Linear and Lasso regressions exhibit comparable performance; however, Lasso slightly underperforms Linear Regression due to the effects of regularization. On Figure 18, the blue bar represents the RFR, the orange bar represents the LASSO regression, and the green bar represents Linear regression.

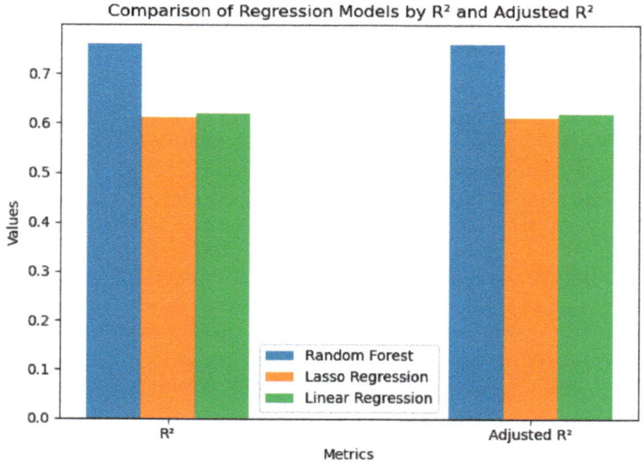

Figure 18. Comparison of regression models by metrics (R^2 and Adjusted R^2).

- MSE, RMSE, and MAE: The Random Forest model exhibits significantly lower MSE, RMSE, and MAE, underscoring its enhanced accuracy and diminished prediction errors. Linear Regression and Lasso have similar performance, while Lasso demonstrates somewhat inferior outcomes due to its penalization of certain characteristics. The differences are minimal, as evidenced by the proximity of each bar's heights in Figure 19, but they have a significant impact.

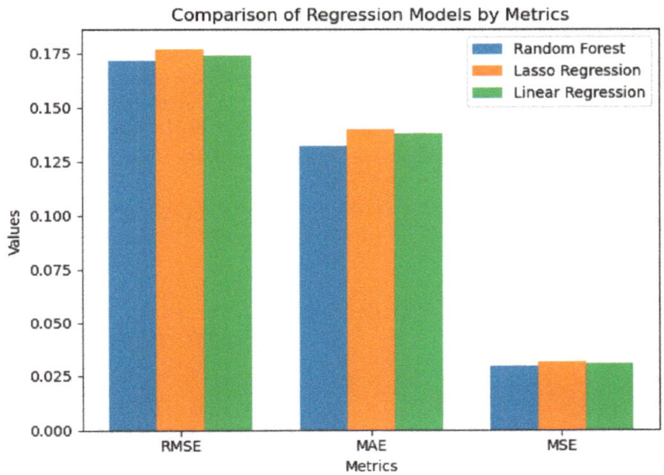

Figure 19. Comparison of regression models by metrics (RMSE, MAE, MSE).

4.2. Cross-Validation Results

- R^2 and Adjusted R^2: The Random Forest algorithm has improved performance on average over cross-validation folds; however, it displays significantly greater variability than on the test set. Linear Regression exhibits marginally superior performance compared to Lasso in cross-validation, although the disparity is negligible.
- MSE, RMSE, and MAE: Random Forest consistently surpasses both linear and Lasso regressions in terms of MSE, RMSE, and MAE throughout cross-validation folds, exhibiting a narrower error range. Linear Regression exhibits marginally superior cross-validation performance compared to Lasso; yet, both models demonstrate considerable stability with minimal discrepancies in error.

4.3. Model Complexity and Interpretability

- Linear Regression: LR is the most elementary of the three models, yielding highly interpretable outcomes with direct coefficients that represent the correlation between features and the target variable. Nonetheless, it may encounter difficulties in capturing intricate, non-linear interactions.
- Random Forest Regression: RFR is an advanced, non-linear model that identifies relationships among variables and accommodates intricate patterns within the data. Nonetheless, it compromises interpretability for enhanced efficiency, as the aggregation of decision trees complicates the understanding of each feature's individual impact.
- Lasso Regression: Employs regularization to penalize insignificant characteristics, hence potentially streamlining the model by removing unimportant variables. This enhances interpretability and mitigates overfitting. Nonetheless, it fails to account for non-linearity in the data.

4.4. Criteria for Model Selection and Comparison

- Random Forest Regression: Exhibits the lowest AIC and BIC values among the three models, indicating that it has the best fit to the data while maintaining a balance between complexity and performance. This indicates that a random forest is the most appropriate model for this dataset, effectively capturing the underlying patterns while minimizing overfitting.
- Lasso Regression: Comes in second place, exhibiting significantly lower AIC and BIC values than linear regression, but outperforming Random Forest. This indicates that Lasso is better than linear regression for fitting the data, likely due to its ability to perform regularization, which mitigates overfitting by penalizing large coefficients.
- Linear regression: Exhibits the highest AIC and BIC values, indicating it is the least effective model for this particular dataset. The model's simplicity may result in underfitting, as it fails to capture the complexities in the data compared to the more sophisticated models.
- Figure 20 illustrates a bar comparison between the two criteria, revealing a significant difference between them. The bars fall below the horizontal axes due to the negative values of the criteria.

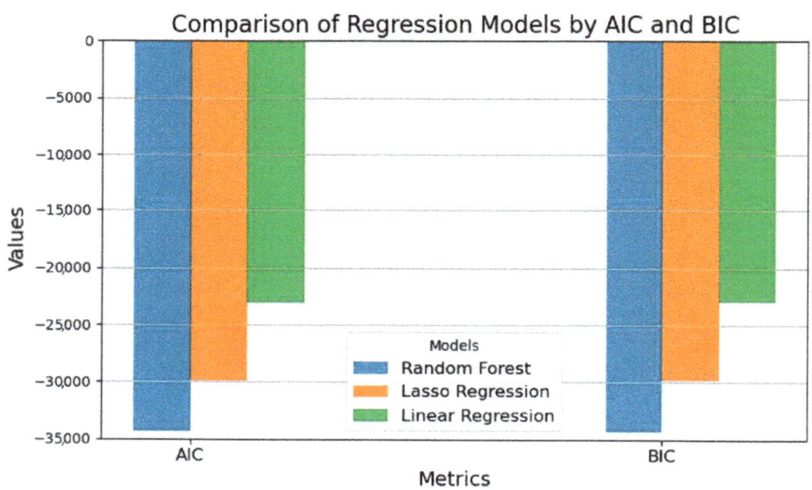

Figure 20. Comparison of regression models by criteria (AIC, BIC).

5. Conclusions

Amidst the worldwide transition to sustainable energy sources, wind power has become a crucial alternative to conventional fossil fuels. Wind farms, consisting of multiple turbines, play a key part in the production of sustainable energy. This article has conducted an in-depth investigation into the possibilities for transformation by combining IoT technology and machine learning in the wind energy industry. The research has demonstrated the capacity to gather crucial data from wind turbines by employing IoT devices in the wind energy industry, which is then converted into practical insights. The successful incorporation of weather data obtained from the IoT into the prediction of wind power generation has effectively connected meteorological observations with data on wind energy production. Moreover, the utilization of diverse machine learning methodologies can facilitate accurate prediction of energy generation. The comparison of Linear, Random Forest and Lasso Regression models provides insight into their distinctions, enhancing the understanding of their respective applications and identifying the strengths, weaknesses and suitability of each model. The assessment of each model's performance is derived from the comparison of R^2, Adjusted R^2, MSE, RMSE, MAE, and criteria AIC, BIC. The results show that the random forest algorithm attains the highest R^2 and Adjusted R^2, along with the lowest MSE, RMSE, and MAE. The criteria display the lowest AIC and BIC values among the three models, indicating optimal data fit while maintaining a balance between complexity and performance. Future development should combine ML techniques with a cost-effective strategy that lowers the costs associated with IoT usage, fully integrating the significance of IoT.

Author Contributions: Conceptualization, C.E. and P.G.; methodology, C.E.; software, C.E.; validation, C.E. and P.G.; formal analysis, P.G.; investigation, C.E.; resources, C.E.; data curation, C.E.; writing—original draft preparation, C.E.; writing—review and editing, P.G.; visualization, C.E.; supervision, P.G. All authors have read and agreed to the published version of the manuscript.

Funding: This research received no external funding.

Institutional Review Board Statement: Not applicable.

Informed Consent Statement: Not applicable.

Data Availability Statement: The original contributions presented in the study are included in the article, further inquiries can be directed to the corresponding author.

Conflicts of Interest: The authors declare no conflicts of interest.

Abbreviations

The following abbreviations are used in this manuscript:

Adjusted R^2	Adjusted Coefficient of Determination
AIC	Akaike Information Criterion
ANOVA	Analysis of Variance
ANNs	Artificial Neural Networks
AWGN	Additive White Gaussian Noise
BIC	Bayesian Information Criterion
CART	Categorical Regression Tree
CM	Configuration Management
CMS	Content Management Systems
CNNs	Convolutional Neural Networks
CPS	Cyber-Physical System
CRM	Customer Resource Management
ERP	Enterprise Resource Planning
IoE	Internet of Everything
IoT	Internet of Things
kNN	k-Nearest Neighbors
KPI	Key Performance Indicator
LCoE	Levelized Cost of Energy
LR	Linear Regression
LSTM	Long Short-Term Memory
MAE	Mean Absolute Error
M2M	Machine to Machine
ML	Machine Learning
MLR	Multilinear Regression
MSE	Mean Squared Error
R^2	Coefficient of Determination
RFR	Random Forest Regression
RMSE	Root Mean Squared Error
RNNs	Recurrent Neural Networks
SCADA	Supervisory Control and Data Acquisition
VIF	Variance Inflation Factor
WECS	Wind Energy Conversion System

References

1. Mohammed, N.Q.; Ahmed, M.S.; Mohammed, M.A.; Hammood, O.A.; Alshara, H.A.N.; Kamil, A.A. Comparative Analysis between Solar and Wind Turbine Energy Sources in IoT Based on Economical and Efficiency Considerations. In Proceedings of the 22nd International Conference on Control Systems and Computer Science (CSCS), Bucharest, Romania, 28–30 May 2019; pp. 448–452.
2. Kaur, N.; Sood, S.K. An Energy-Efficient Architecture for the Internet of Things (IoT). *IEEE Syst. J.* **2017**, *11*, 796–805. [CrossRef]
3. Adekanbi, M.L. Optimization and digitization of wind farms using internet of things: A review. *Internet Things* **2021**, *45*, 15832–15838. [CrossRef]
4. Famoso, F.; Oliveri, L.M.; Brusca, S.; Chiacchio, F. A Dependability Neural Network Approach for Short-Term Production Estimation of a Wind Power Plant. *Energies* **2024**, *17*, 71627. [CrossRef]
5. Noor-A-Rahim, M.; Khyam, M.O.; Li, X.; Pesch, D. Sensor Fusion and State Estimation of IoT Enabled Wind Energy Conversion System. *Sensors* **2019**, *19*, 71566. [CrossRef]
6. Karaman, Ö.A. Prediction of Wind Power with Machine Learning Models. *Appl. Sci.* **2023**, *13*, 11455. [CrossRef]
7. Demolli, H.; Dokuz, A.S.; Ecemis, A.; Gokcek, M. Wind power forecasting based on daily wind speed data using machine learning algorithms. *Energy Convers. Manag.* **2019**, *198*, 111823. [CrossRef]
8. Malik, P.; Gehlot, A.; Singh, R.; Gupta, L.R.; Thakur, A.K. A Review on ANN-Based Model for Solar Radiation and Wind Speed Prediction with Real-Time Data. *Arch. Comput. Methods Eng.* **2022**, *29*, 3183–3201. [CrossRef]
9. Zhang, Y.Z.; Liang, S.J.; Chen, J.B.; Wang, M.; Jia, M.T.; Jiang, Y.T. Enhancing artificial permafrost table predictions using integrated climate and ground temperature data: A case study from the Qinghai-Xizang highway. *Cold Reg. Sci. Technol.* **2024**, *229*, 104341. [CrossRef]
10. Das, P.; Mashiata, M.; Iglesias, G. Big Data Meets Big Wind: A Scientometric Review of Machine Learning Approaches in Offshore Wind Energy. *Energy AI* **2024**, *18*, 100418. [CrossRef]
11. Dziak, J.J.; Coffman, D.L.; Lanza, S.T.; Li, R.; Jermiin, L.S. Sensitivity and specificity of information criteria. *Brief. Bioinform.* **2020**, *21*, 553–565. [CrossRef]
12. Alhmoud, L.; Al-Zoubi, H. IoT Applications in Wind Energy Conversion Systems. *Open Eng.* **2019**, *9*, 490–499. [CrossRef]
13. Shields, M.; Beiter, P.; Nunemaker, J.; Cooperman, A.; Duffy, P. Impacts of Turbine and Plant Upsizing on the Levelized Cost of Energy for Offshore Wind. *Appl. Energy* **2021**, *298*, 117189. [CrossRef]
14. Moness, M.; Moustafa, A.M. A Survey of Cyber-Physical Advances and Challenges of Wind Energy Conversion Systems: Prospects for Internet of Energy. *IEEE Internet Things J.* **2016**, *3*, 134–145. [CrossRef]
15. Ahmed, M.A.; Eltamaly, A.M.; Alotaibi, M.A.; Alolah, A.I.; Kim, Y.C. Wireless Network Architecture for Cyber Physical Wind Energy System. *IEEE Access* **2020**, *8*, 40180–40197. [CrossRef]
16. Maldonado-Correa, J.; Martín-Martínez, S.; Artigao, E.; Gómez-Lázaro, E. Using SCADA Data for Wind Turbine Condition Monitoring: A Systematic Literature Review. *Energies* **2020**, *13*, 3132. [CrossRef]
17. Chen, H.; Chen, J.; Dai, J.; Tao, H.; Wang, X. Early Fault Warning Method of Wind Turbine Main Transmission System Based on SCADA and CMS Data. *Machines* **2022**, *10*, 1018. [CrossRef]
18. Chen, X.; Eder, M.A.; Shihavuddin, A.S.M.; Zheng, D. A Human-Cyber-Physical System toward Intelligent Wind Turbine Operation and Maintenance. *Sustainability* **2021**, *13*, 561. [CrossRef]
19. Win, L.L.; Tonyalı, S. Security and Privacy Challenges, Solutions, and Open Issues in Smart Metering: A Review. In Proceedings of the 2021 6th International Conference on Computer Science and Engineering (UBMK), Ankara, Turkey, 15–17 September 2021; IEEE: New York, NY, USA, 2021.
20. Cox, S.L.; Lopez, A.J.; Watson, A.C.; Grue, N.W.; Leisch, J.E. *Renewable Energy Data, Analysis, and Decisions: A Guide for Practitioners*; National Renewable Energy Lab. (NREL): Golden, CO, USA, 2018.
21. Pontes, E.A.S. A Brief Historical Overview Of the Gaussian Curve: From Abraham De Moivre to Johann Carl Friedrich Gauss. *Int. J. Eng. Sci. Invent.* **2018**, *7*, 28–34.
22. Pedregosa, F.; Varoquaux, G.; Gramfort, A.; Michel, V.; Thirion, B.; Grisel, O.; Blondel, M.; Prettenhofer, P.; Weiss, R.; Dubourg, V.; et al. Scikit-learn: Machine Learning in Python. *J. Mach. Learn. Res.* **2011**, *12*, 2825–2830.
23. Kwak, S.K.; Kim, J.H. Statistical data preparation: Management of missing values and outliers. *Korean J. Anesthesiol.* **2017**, *70*, 407–411. [CrossRef]
24. Gu, Z.; Eils, R.; Schlesner, M. Complex heatmaps reveal patterns and correlations in multidimensional genomic data. *Bioinformatics* **2016**, *32*, 2847–2849. [CrossRef] [PubMed]
25. Ahsan, M.M.; Mahmud, M.P.; Saha, P.K.; Gupta, K.D.; Siddique, Z. Effect of Data Scaling Methods on Machine Learning Algorithms and Model Performance. *Technologies* **2021**, *9*, 52. [CrossRef]
26. Alkesaiberi, A.; Harrou, F.; Sun, Y. Efficient Wind Power Prediction Using Machine Learning Methods: A Comparative Study. *Energies* **2022**, *15*, 72327. [CrossRef]
27. Palmer, P.B.; O'Connell, D.G. Research Corner: Regression Analysis for Prediction: Understanding the Process. *J. Chiropr. Med.* **2009**, *8*, 89–93. [CrossRef]
28. James, G.; Witten, D.; Hastie, T.; Tibshirani, R. *Linear Regression. An Introduction to Statistical Learning*, 2nd ed.; Springer: New York, NY, USA, 2023; pp. 69–134.

29. Li, K.; Horton, R.; He, H. Application of Machine Learning Algorithms to Model Soil Thermal Diffusivity. *Int. Commun. Heat Mass Transf.* **2023**, *149*, 107092. [CrossRef]
30. Roelofs, R.; Shankar, V.; Recht, B.; Fridovich-Keil, S.; Hardt, M.; Miller, J.; Schmidt, L. A Meta-Analysis of Overfitting in Machine Learning. In Proceedings of the 36th Conference on Neural Information Processing Systems (NeurIPS 2019), Vancouver, BC, Canada, 8–14 December 2019; Volume 32.
31. Xiong, Z.; Cui, Y.; Liu, Z.; Zhao, Y.; Hu, M.; Hu, J. Evaluating explorative prediction power of machine learning algorithms for materials discovery using k-fold forward cross-validation. *Comput. Mater. Sci.* **2020**, *171*, 109203. [CrossRef]
32. Maulud, D.; Abdulazeez, A.M. A Review on Linear Regression Comprehensive in Machine Learning. *J. Appl. Sci. Technol. Trends* **2020**, *1*, 140–147. [CrossRef]
33. Filzmoser, P.; Nordhausen, K. Robust linear regression for high-dimensional data: An overview. *Wiley Interdiscip. Rev. Comput. Stat.* **2021**, *13*, e1524. [CrossRef]
34. Alita, D.; Putra, A.D.; Darwis, D. Analysis of Classic Assumption Test and Multiple Linear Regression Coefficient Test for Employee Structural Office Recommendation. *IJCCS Indones. J. Comput. Cybern. Syst.* **2021**, *15*, 295–306. [CrossRef]
35. Breiman, L. Random Forests. *Mach. Learn.* **2001**, *45*, 5–32. [CrossRef]
36. Cutler, A.; Zhao, G. Pert-Perfect Random Tree Ensembles. *Comput. Sci. Stat.* **2001**, *33*, 90–94.
37. Lingjun, H.; Levine, R.A.; Fan, J.; Beemer, J.; Stronach, J. Random Forest as a predictive analytics alternative to regression in institutional research. *Pract. Assess. Res. Eval.* **2018**, *23*, 1–10.
38. Sadorsky, P. A Random Forests Approach to Predicting Clean Energy Stock Prices. *J. Risk Financ. Manag.* **2021**, *14*, 20048. [CrossRef]
39. Aljuboori, A.; Abdulrazzq, M.A. Enhancing Accuracy in Predicting Continuous Values through Regression. *Int. J. Comput. Dig. Syst.* **2024**, *16*, 1–10.
40. Steurer, M.; Hill, R.J.; Pfeifer, N. Metrics for evaluating the performance of machine learning based automated valuation models. *J. Prop. Res.* **2021**, *38*, 99–129. [CrossRef]
41. Ranstam, J.; Cook, J.A. Lasso regression. *Br. J. Surg.* **2018**, *105*, 1348. [CrossRef]
42. Lee, J.H.; Shi, Z.; Gao, Z. On LASSO for Predictive Regression. *J. Econ.* **2022**, *229*, 322–349. [CrossRef]
43. Lind, S.J.; Rogers, B.D.; Stansby, P.K. Review of Smoothed Particle Hydrodynamics: Towards Converged Lagrangian Flow Modelling. *Proc. R. Soc. A* **2020**, *476*, 20190801. [CrossRef]
44. Tatachar, A.V. Comparative Assessment of Regression Models Based on Model Evaluation Metrics. *Int. J. Innov. Technol. Explor. Eng.* **2021**, *8*, 853–860.

applied sciences

MDPI

Article

Dual Control Strategy for Non-Minimum Phase Behavior Mitigation in DC-DC Boost Converters Using Finite Control Set Model Predictive Control and Proportional–Integral Controllers

Alejandra Marmol [1], Elyas Zamiri [1,*], Marziye Purraji [1], Duberney Murillo [2], Jairo Tuñón Díaz [2], Aitor Vazquez [2] and Angel de Castro [1]

1 HCTLab Research Group, Electronics and Communications Technology Department, Universidad Autónoma de Madrid, 28049 Madrid, Spain; alejandra.marmol@estudiante.uam.es (A.M.); marziye.purraji@uam.es (M.P.); angel.decastro@uam.es (A.d.C.)

2 Electrical, Electronic, Communications and Systems Engineering Department, Universidad de Oviedo, 33204 Gijón, Spain; murilloduberney@uniovi.es (D.M.); tunonjairo@uniovi.es (J.T.D.); vazquezaitor@uniovi.es (A.V.)

* Correspondence: elyas.zamiri@uam.es

Featured Application: The proposed control strategy is specifically designed for applications with non-minimum phase behavior. It can be effectively applied to DC-DC boost converters, dual active bridge converters, and other topologies like buck-boost, Cuk, SEPIC, and ZETA converters, all of which commonly exhibit non-minimum phase characteristics.

Abstract: Model Predictive Control (MPC) has emerged as a promising alternative for controlling power converters, offering benefits such as flexibility, simplicity, and rapid control response, particularly when short-horizon algorithms are employed. This paper introduces a system using a short-horizon Finite Control Set MPC (FCS-MPC) strategy to specifically address the challenge of non-minimum phase behavior in boost converters. The non-minimum phase issue, which complicates the control process by introducing an initial inverse response, is effectively mitigated by the proposed method. A Proportional–Integral (PI) controller is integrated to dynamically adjust the reference current based on the output voltage error, thereby enhancing overall system stability and performance. Unlike conventional PI-MPC methods, where the PI controller has an influence on the system dynamics, the PI controller in this approach is solely used for tuning the reference current needed for the FCS-MPC controller. The PI controller addresses small deviations in output voltage, primarily due to model prediction inaccuracies, ensuring steady-state accuracy, while the FCS-MPC handles fast dynamic responses to adapt the controller's behavior based on load conditions. This dual control strategy effectively balances the need for precise voltage regulation and rapid adaptation to varying load conditions. The proposed method's effectiveness is validated through a multi-stage simulation test, demonstrating significant improvements in response time and stability compared to traditional control methods. Hardware-in-the-loop testing further confirms the system's robustness and potential for real-time applications in power electronics.

Keywords: DC-DC converters; finite control set; boost converter; non-minimum phase; hardware-in-the-loop

1. Introduction

The integration of renewable energies, such as solar and wind power, into the electrical grid is revolutionizing the traditional paradigm of electricity generation. Unlike conventional thermal, nuclear, or hydraulic power plants that rely on electrical machines, renewable energy systems primarily use power electronic converters to interface with the grid. Among these, DC-DC converters, particularly boost converters, are crucial for stepping up voltage levels to meet grid requirements [1,2]. Traditionally, these converters have used Proportional–Integral (PI) controllers [3]. However, Model Predictive Control (MPC) is now recognized as a superior strategy due to its ability to handle fast dynamic behaviors in modern power converters. MPC predicts future system states and optimizes control inputs in real time, ensuring a rapid response to changes in operating conditions. This predictive capability allows MPC to manage various performance criteria, such as voltage regulation, current control, and efficiency optimization, simultaneously through a single cost function [4–6].

MPC is implemented in two main forms: Continuous Control Set (CCS) and Finite Control Set (FCS). CCS-MPC operates on a continuous range of control inputs, providing smooth control but requiring high computational power and complex algorithms, making it less suitable for fast dynamics and real-time applications [7,8]. In contrast, FCS-MPC uses a discrete set of control actions, significantly reducing computational demands and enabling quicker decision making [9,10]. However, FCS-MPC faces challenges, such as stability issues when the sampling step is relatively big and the difficulty of managing non-minimum phase behavior in systems like boost converters when the prediction horizon is short [11].

To address the non-minimum phase issue in boost converters, a long-horizon approach is presented by Gardezi et al. [12] that employs machine learning for the adaptive prediction horizon in FCS-MPC, but the method's complexity can be a disadvantage in real-time applications. Among short-horizon solutions, Villarroel et al. [13] propose an input-state linearization method for FCS-MPC, transforming system dynamics to eliminate instability with a short prediction horizon. Another method presented by Villarroel et al. [14] designs a specific cost function within the FCS-MPC framework for stable and precise voltage tracking, minimizing the output voltage error and smoothing control actions. Similarly, Pérez et al. [15] address the issue in active front end rectifiers with a predictive control strategy based on output power and DC link capacitor energy, avoiding additional modulators and achieving a zero steady-state error. Karamanakos et al. [16] present a DMPCC strategy for DC-DC boost converters using a hybrid model for precise current regulation, reducing computational complexity. Firpo et al. [17] use a non-linear MPC method with a partially saturating ferrite core inductor model to manage magnetic saturation and losses, ensuring correct voltage regulation. All these methods require precise parameter estimation, with inaccuracies leading to suboptimal performance, and are affected by load and input condition variations, necessitating robust estimation and adjustment mechanisms.

Another approach is to control the inductor current with an FCS-MPC controller, where the reference for the controller is provided by a PI controller [18–20]. In this method, the PI controller generates the reference current for the FCS-MPC based on the output voltage error. While this hybrid approach aims to combine the strengths of both controllers—using the PI controller for steady-state accuracy and the FCS-MPC for handling fast dynamics—the overall speed is compromised. The PI controller's slower response time can negate the advantage of the FCS-MPC's rapid switching capability, potentially missing out on the primary benefit of using an FCS-MPC. This results in a less responsive system that may not fully leverage the fast dynamic response that FCS-MPC controllers are designed to provide.

This paper presents a significant contribution to the field of power electronics by presenting a novel application of FCS-MPC for DC-DC boost converters. It addresses the non-minimum phase issue, a common challenge in the control of boost converters, by incorporating a PI regulator to adapt the reference current. The proposed method distinguishes itself by employing the Ideal Reference Calculator (IRC) block, which generates the primary reference current for the system based on ideal conditions. This allows the PI controller to focus solely on making small corrections to the reference current in steady-state situations, rather than managing fast dynamic changes. By decoupling the PI controller from fast dynamics, the FCS-MPC, combined with the IRC block, takes full responsibility for real-time adjustments and load changes, ensuring a fast dynamic response. This dual approach leverages the strengths of both controllers—steady-state precision from the PI controller and rapid adaptation from the FCS-MPC combined with the IRC block. The effectiveness of the proposed method is validated through both simulation and experimental results by hardware-in-the-loop (HIL) tests, demonstrating its potential for real-time applications requiring optimal performance and rapid response.

The rest of the paper is organized as follows: In Section 2, the DC-DC boost converter and its discrete model are introduced, establishing the foundation for the control strategy. The FCS-MPC strategy is detailed in Section 3, including its implementation, the choice of prediction horizon, and the formulation of the cost function. Section 4 focuses on the development of the proposed dual control approach, in which a PI controller is integrated to dynamically adjust the reference current, addressing the non-minimum phase issue. In Section 5, the simulation and hardware-in-the-loop (HIL) test results are presented, vali-dating the proposed method, and demonstrating significant improvements in response time and stability compared to traditional control methods. Finally, Section 6 provides the conclusions, summarizing the findings.

2. Application Example

This paper focuses on the application example of a boost converter, although the idea can be adapted to other topologies that face non-minimum phase issues. The proposed method is necessary for boost converters because conventional FCS-MPC strategies struggle to manage the inherent non-minimum phase dynamics, which cause initial voltage undershoots when increasing the output voltage. This topology is also useful for understanding the non-minimum phase phenomena in power converters and FCS-MPC due to its simplicity. In a boost converter, the main objective is to have a higher output voltage than in the input; this means that there is an increment of the DC voltage. Between the elements composing this circuit, there is a DC input voltage source (*Vin*), an inductor to store energy, a power electronic switch (represented by the variable *Q*), a diode (*D*) to transfer the energy stored in the inductor to the load, and a capacitor to smooth the output voltage. While different types of power transistors, such as MOSFETs or IGBTs, can be used as the switch, MOSFETs are more suitable for the mid-high switching frequencies used in the experimental results. IGBTs, on the other hand, may limit the maximum attainable frequency. For this specific design, a MOSFET is used and shown in the figures and schematics.

When the converter works in Continuous Conduction Mode (CCM), there are two equivalent sub-circuits, shown in Figure 1: when the MOSFET is turned *on* and when it is turned *off*. When the MOSFET is *on* (Figure 1a), it creates a low-impedance path for the current, allowing the inductor to store energy from the input voltage. During this state, the diode is reverse-biased, preventing current from flowing to the output. When the MOSFET is turned *off* (Figure 1b), the inductor's stored energy is released. The inductor generates a high voltage, which forward-biases the diode, allowing current to flow to the output and simultaneously charging the output capacitor. This process boosts the output voltage to a level higher than the input voltage, hence the name "boost converter". The rapid switching between these two states ensures a continuous transfer of energy, maintaining a higher steady-state output voltage. In Discontinuous Conduction Mode (DCM), illustrated in Figure 1c, the inductor current falls to zero during the *off* state of the MOSFET. This introduces a third sub-circuit where the diode is *off*, and both the inductor and capacitor are disconnected from the input voltage. During this state, the inductor does not transfer energy, and the output voltage is maintained solely by the charge stored in the capacitor. This mode typically occurs under light load conditions when the energy demand is low, resulting in periods where no current flows through the inductor.

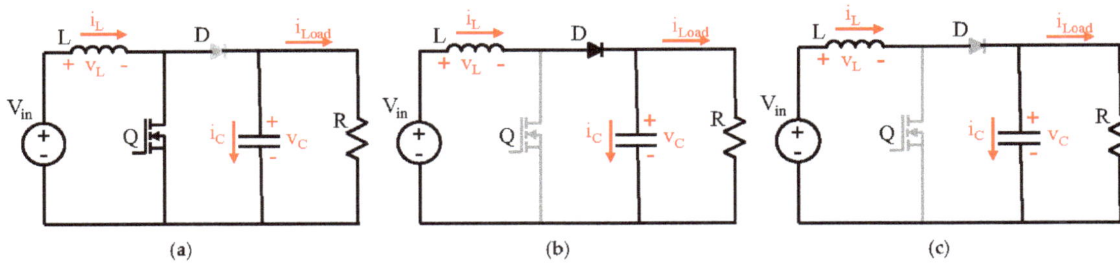

Figure 1. Switching states of a boost topology; (**a**) MOSFET *on* in CCM; (**b**) MOSFET *off* in CCM; and (**c**) MOSFET *off* in DCM.

The discrete differential equations of the boost converter, as represented by Equations (1) and (2), are essential for accurately modeling the system's dynamics and are crucial for the prediction block in FCS-MPC. These equations enable the controller to predict the future inductor current and capacitor voltage, allowing for precise switching decisions.

$$v_C(k+1) = v_C(k) + \frac{\Delta t}{C} \cdot i_C(k) \tag{1}$$

$$i_L(k+1) = i_L(k) + \frac{\Delta t}{L} \cdot v_L(k) \tag{2}$$

For the sake of clarity, $i_L(k)$ and $v_C(k)$ represent the current values of the inductor current and output voltage, respectively, while $i_L(k+1)$ and $v_C(k+1)$ denote their future values. These future values will be the predicted ones in the next section. There are two primary states, as shown in Table 1, representing the possible decisions that the controller can make: switching the MOSFET *on* (situation I) or *off* (situation II). Table 1 also provides the inductor voltage (v_L) and the capacitor current (i_C) for these two states, which are needed for the prediction block Equations (1) and (2).

Table 1. Summary of the state behaviors of a boost converter in different situations.

	Situation I	Situation II
Parameters	Q: *on*	Q: *off*
v_L	V_{in}	$V_{in} - v_c$
i_c	$-v_c/R$	$i_L - v_c/R$

Once the equations are defined, the model can be implemented using different numeric representations, depending on the objectives. The selected data type determines the minimum time step achievable for executing the model equations in an FPGA, as well as the design effort and hardware resources required. The two main possibilities for numerical formats are floating-point and fixed-point formats. The floating-point format, represented by the IEEE-754 standard [21], offers high precision and a wide dynamic range, simplifying the design process but requiring more computational resources and higher power consumption. The fixed-point format, where the number of bits for each variable are specified and remain fixed, provides lower computational costs and reduced hardware complexity due to the smaller size and faster computation times. This makes the fixed-point format more efficient in terms of resource usage and power consumption, ideal for real-time FPGA applications. However, it demands higher design effort to balance accuracy and resource utilization. Given these considerations, the fixed-point numeric type is the best option for implementing a combination of the HIL system and MPC algorithm in an FPGA, offering the necessary balance of efficiency, speed, and resource utilization for real-time control applications.

3. FCS-MPC Algorithm for Boost Converters

The FCS-MPC algorithm simplifies the optimization problem by limiting the controller outputs to a finite set, significantly reducing computational complexity. This characteristic makes it highly suitable for applications requiring fast dynamic responses, such as DC-DC boost converters. The core of FCS-MPC lies in predicting the future behavior of the converter based on a discrete model and optimizing control actions to minimize a cost function that typically includes terms for tracking the reference voltage and minimizing inductor current deviation. The short-horizon FCS-MPC implementation discussed in this paper focuses on predicting future states and evaluating potential switching actions within a single timeframe, enhancing the controller's responsiveness to dynamic changes. However, this approach introduces challenges, particularly in managing the non-minimum phase behavior characteristic of boost converters.

The general block diagram for the short-horizon FCS-MPC implementation is demonstrated in Figure 2. It consists of a capacitor voltage predictor and an inductor current predictor that forecast the future states of these variables based on the discrete Equations (1) and (2). A cost function calculator evaluates the performance of each possible switching state, and an optimum state finder identifies the control action that minimizes the cost function. Finally, a pulse generator translates the optimal control action into switching signals for the power converter without involving any PWM modulator. For each possible switching state, the predicted inductor current (i_{LP}) and capacitor voltage (v_{CP}) can be calculated as follows:

$$i_{LP} = \begin{cases} i_L + \frac{\Delta t}{L}.\, V_{in} & \textit{State } 1\ (n=1) \\ \begin{cases} i_L + \frac{\Delta t}{L}.\, (V_{in} - v_C) & if\ i_{LP} > 0 \\ 0 & if\ i_{LP} \leq 0 \end{cases} & \textit{State } 2\ (n=2) \end{cases} \tag{3}$$

$$v_{CP} = \begin{cases} v_C + \frac{\Delta t}{C}.\, (-\frac{v_C}{R}) & \textit{State } 1\ (n=1) \\ v_C + \frac{\Delta t}{C}.\, (i_L - \frac{v_C}{R}) & \textit{State } 2\ (n=2) \end{cases} \tag{4}$$

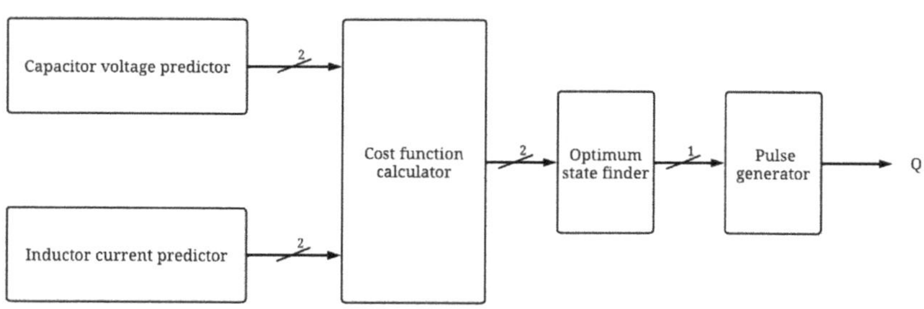

Figure 2. Block diagram of a short-horizon FCS-MPC algorithm for boost converters.

As can be seen in Equation (3), when the predicted inductor current (i_{LP}) becomes negative in State 2, the predictive value is adjusted to zero, which reflects the operation of the boost converter in DCM. The cost function *CF(n)* is designed to minimize the error between the predicted states and the desired reference values. For the boost converter, the cost function is formulated as follows:

$$CF(n) = (i_{L,ref} - i_{LP}(n))^2 + \lambda \cdot (v_{C,ref} - v_{CP}(n))^2 \tag{5}$$

where $v_{C,ref}$ is the reference output voltage, $i_{L,ref}$ is the reference inductor current, and λ is a weighting factor that balances the importance of voltage and current control. The cost function controls both the output voltage and the inductor current simultaneously to ensure optimal performance. The voltage control ensures that the output voltage meets the desired setpoint, providing stability and efficiency in the power supply. Meanwhile, the current control prevents excessive current, which could lead to component stress, thermal issues, and potential damage. By balancing these two aspects, the MPC can achieve a harmonious operation, maintaining precise voltage regulation while protecting the system from overcurrent conditions, thereby enhancing the overall reliability and lifespan of the boost converter system.

In (5), the references can be constant or variable over time. For this study, the output voltage reference is assumed to be constant, while the reference current is dynamically updated by an external circuit that includes a PI controller and a load detector, as detailed in the next section. The discrete model of the converter and the FCS-MPC algorithm are implemented on an FPGA using MATLAB HDL Coder (MATLAB R2024a). This setup enables real-time emulation within a single FPGA, simplifying the system and avoiding clock mismatch issues between the controller and the converter model simulated in real time. Thanks to HDL Coder, the entire system is translated into a VHDL code for FPGA programming via the MATLAB tool without requiring deep knowledge of VHDL programming.

3.1. Non-Minimum Phase Issue in Boost Converters

A significant challenge in controlling boost converters is their inherent non-minimum phase behavior, where the initial response of the system deviates in the opposite direction of the desired outcome. For instance, in a boost converter, attempting to increase the output voltage results in an initial undershoot in the output before the desired voltage level is attained. This is due to the gain sign changing from a negative to a positive value after an initial transient. This non-minimum phase behavior complicates the control process, particularly when using an MPC with a short prediction horizon of one, as it fails to consider the system's long-term behavior. In such cases, controlling the output voltage by FCS-MPC can exacerbate the problem, leading to instability and performance degradation.

In a boost converter, when the controller aims to increase the output voltage from the input voltage to a higher desired level, the non-minimum phase behavior leads to an initial undershoot. The FCS-MPC controller, using a short prediction horizon, evaluates two states: State 1, where the switch is closed and the inductor stores energy, and State 2, where the switch is open and the inductor releases energy to the load. Due to the non-minimum phase behavior, the controller predicts that closing the switch will initially cause the output voltage to drop, while opening the switch will cause a slight immediate increase. To minimize immediate error, the controller opts for State 2 to avoid the undershoot. However, this prevents the inductor from storing sufficient energy to achieve the higher desired output voltage. Consequently, the controller keeps selecting State 2, avoiding the necessary energy buildup, leading to instability and preventing the output voltage from reaching the desired higher level.

Despite these challenges, MPC with a horizon of one step is often preferred due to its simplicity. The short prediction horizon reduces computational complexity, making real-time implementation feasible in mid/high-frequency applications.

3.2. Solutions to the Non-Minimum Phase Issue

To enhance controller performance, additional measures must be incorporated to handle the non-minimum phase dynamics. One approach involves modifying the cost function to include terms that anticipate the initial undershoot and adjust the control actions accordingly. Apart from that, using long-horizon FCS-MPC can solve the non-minimum phase issue; however, this is not the focus of this study for two main reasons. First, it is difficult to determine the number of horizons needed to guarantee the stability of the controller, as it would vary with each application. Gardezi et al. [12] propose a machine learning-based algorithm to provide the needed horizon in each case. The other reason is the complexity of implementing long-horizon FCS-MPC algorithms.

Recognizing the limitations of direct voltage control, the focus shifted toward regulating the inductor current to improve the system's response. A well-known method is to use a double-loop control strategy, integrating auxiliary controllers such as PI controllers to adapt the reference current for the MPC controller based on the error in the output voltage [18–20]. This approach allows the MPC controller to regulate only the inductor current, thus avoiding the non-minimum phase issue.

Figure 3 shows the block diagram of such a system. While this approach addresses the non-minimum phase issue in FCS-MPC, it introduces slowness due to the PI controller's slower timescale for setting the reference current. This lag compromises the fast dynamic response of FCS-MPC, diminishing its quick decision making and adaptability, resulting in suboptimal performance in scenarios requiring rapid adaptation and precise voltage regulation.

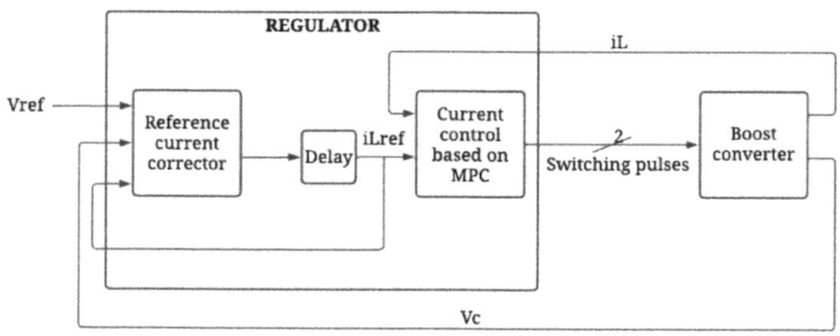

Figure 3. Block diagram of the MPC controller with a PI regulator adapting the current reference based on output voltage error [18–20].

3.3. *Delay Compensation Consideration*

In control applications with larger sampling steps (typically greater than 20 µs), delay compensation methods are often necessary to mitigate the negative impact of delays on controller performance [22,23]. However, in cases where smaller sampling steps are employed, such as the 5 µs step used in this work, the effects of delays are minimal. Introducing delay compensation in such cases would unnecessarily increase the system's complexity without providing significant improvements in response. Therefore, to maintain clarity and simplicity in the control method, no explicit delay compensation techniques are applied in this implementation. The small sampling rate and real-time operation of the FCS-MPC algorithm are sufficient to ensure the system's high performance without the need for additional delay handling mechanisms, as will be shown in Section 5. However, the proposed dual control technique would be applied in the same way if delay compensation were implemented.

4. Proposed Dual Control Strategy Based on Load Detector and Offset Corrector

This section introduces the proposed horizon one FCS-MPC controller strategy based on the load detector and offset corrector. The proposed method consists of an FCS-MPC controller only regulating the inductor current and a reference current calculator, involving two sensors, one for measuring the load current (i_{Load}) and the other one for the output voltage (v_C). The method is summarized in the block diagram shown in Figure 4.

The Ideal Reference Calculator (IRC) block computes the input reference current in an ideal scenario without considering the electrical losses of the switches or other components of the system. Its calculations are based on ensuring that the steady-state power at the input ($P_{in} = V_{in} \cdot i_{L,ref}$) equals the power at the output of the boost converter ($P_{out} = V^2{}_{C,ref}/R$). Consequently, the reference current is calculated as shown in Equation (6), where R is the output load resistance.

$$i_{L,ref} = \frac{(v_{C,ref})^2}{R \cdot V_{in}} \tag{6}$$

By modifying Equation (6) with $R = v_C / i_{Load}$, the final reference current calculation results in Equation (7):

$$i_{L,ref} = \frac{(v_{C,ref})^2 \cdot i_{Load}}{v_C \cdot V_{in}} \tag{7}$$

The calculated reference current is derived from ideal equations, which do not account for electrical losses and variations in component values due to inaccurate measurements or external factors such as temperature. These losses are especially unpredictable in conventional FCS-MPC, which does not control the switching frequency precisely, making real-time loss prediction challenging. To correct minor steady-state errors in the calculated reference current, the proposed dual control strategy for a DC-DC boost converter includes an offset tuning mechanism. The process begins by measuring the actual output voltage and comparing it with the reference output voltage. The difference between these two values represents the voltage error, which is then fed into a low-speed PI controller.

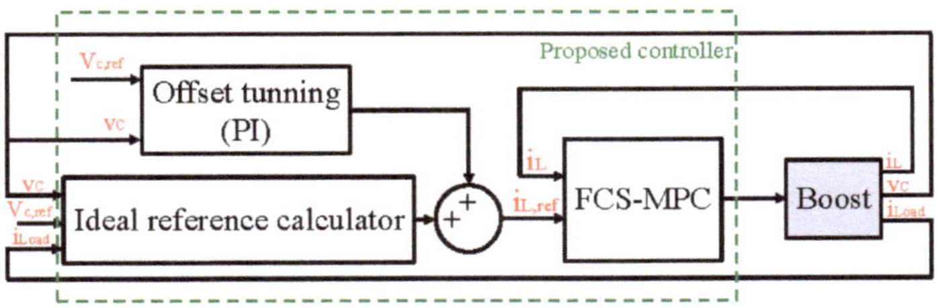

Figure 4. Block diagram of the proposed FCS-MPC controller with load detector and offset corrector.

As shown in the flowchart in Figure 5, the control system is designed using a dual-loop configuration. The outer loop features a PI controller, which adjusts an offset based on the accumulated error between the actual and reference output voltages. This offset is added to the ideal reference current generated by the IRC block, which operates under ideal conditions and without delay. This setup ensures that the system compensates for minor deviations not accounted for by (7). The IRC block provides the primary reference current for the FCS-MPC, which can track this reference quickly due to its fast response capabilities. While the FCS-MPC ensures the rapid tracking of the reference current and real-time system adjustments, the IRC block itself is based on ideal conditions. To account for any steady-state errors in the output voltage caused by non-ideal factors, a slower offset tuner, managed by the PI controller, is used. The FCS-MPC handles the direct control of the inductor current by predicting future system states (using Equations (3) and (4)), evaluating possible switching actions (Equation (5)), and selecting the optimal action to minimize the cost function. This approach allows the FCS-MPC to quickly adapt to load changes detected by the IRC block. By decoupling the dynamic regulation from the PI controller, the FCS-MPC is fully responsible for fast real-time adjustments of the inductor current, without being slowed by the PI loop's response. This combination allows the system to maintain fast dynamic responses while ensuring accurate steady-state performance.

Since the first loop generates the reference current, the second loop (FCS-MPC) only focuses on controlling the current, not the voltage. In fact, the voltage is controlled by the first loop. Therefore, the cost function of the FCS-MPC algorithm is simplified from (5) to the following:

$$CF(n) = (i_{L,ref} - i_{LP}(n))^2 \tag{8}$$

This eliminates the need for tuning the parameter λ, which is removed from the algorithm. Regarding the PI controller in the first loop, it can be designed with any classical techniques, such as Ziegler–Nichols tuning method, since its role is the standard one of adapting the reference current to control the output voltage. Given that the main component of the reference current is calculated by the IRC, which operates with fast dynamics, the

PI controller plays a secondary role, handling only minor steady-state discrepancies. This allows for increasing the phase and gain margins without compromising the system's dynamics. By increasing the phase and gain margins of the PI controller, the system becomes more robust, allowing for greater tolerance to variations in system parameters without compromising performance. Therefore, a fine-tuning method is used in this study to determine the proportional and integral gains. This fine-tuning is justified because the PI controller has a low-speed response, which leaves large phase and gain margins.

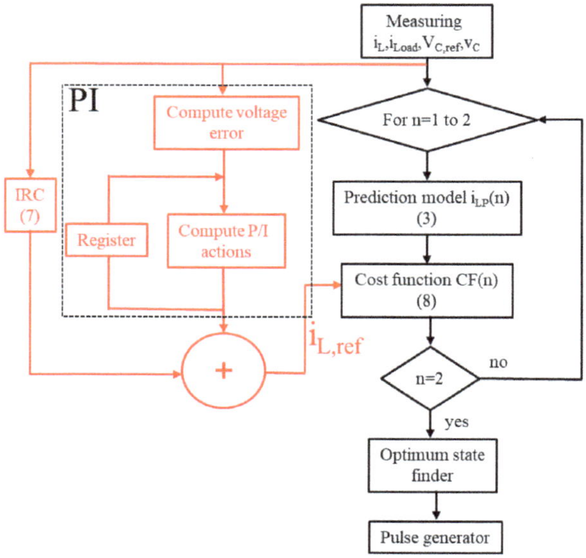

Figure 5. Flowchart of the proposed dual-loop control strategy, integrating PI-based offset tuning for steady-state correction and FCS-MPC for fast dynamic response.

This dual approach optimizes the overall performance of the control system in both transient and steady-state conditions. However, a drawback of this method is the need for an extra sensor to measure the load current when the load is not known, which adds complexity and cost to the system. In the absence of a physical load sensor, observer-based methods offer an alternative for load current estimation, though they may introduce noise-related issues. Talbi et al. [24] demonstrated the effectiveness of an adaptive sliding mode observer (ASMO) for the sensorless control of DC-DC boost converters. The ASMO accurately estimated load resistance and output voltage using the system's mathematical model, maintaining the controller performance even under load variations. However, this method can be sensitive to noise, which might affect the accuracy of the estimations, particularly in systems with a high switching frequency. Similarly, Heydari-doostabad and Ghazi [25] proposed a Luenberger observer for load current estimation in UPS systems, which replaces sensors with model-based estimations of the load current. While the Luenberger observer improves system reliability and reduces complexity, it is also prone to noise issues, especially during sudden load changes or in non-linear conditions. These observer-based approaches could be integrated into the proposed algorithm to estimate the load current without a dedicated load sensor, though noise mitigation strategies should be considered to ensure precise control and system performance, making the whole system more complex.

5. Results

This section presents simulation and experimental results validating the proposed method for addressing the non-minimum phase issue in the short-horizon FCS-MPC applied to a boost converter. The simulation setup is conducted on the MATLAB/Simulink platform, where the boost converter model and FCS-MPC algorithm are implemented and tested under various operating conditions to evaluate the performance of the proposed method. Scenarios with different load changes and input voltage variations are included to assess the controller's ability to handle non-minimum phase behavior and maintain stable output voltage. Apart from the simulation verification, experimental tests are performed by implementing the proposed algorithm inside an FPGA board, specifically the Arty Z7, which is equipped with an XC7Z020-1CLG400C FPGA (AMD, Santa Clara, CA, USA). This FPGA-based implementation enables real-time control and testing, with the controller and the boost converter's discretized model integrated into the same FPGA, ensuring synchronized operation and eliminating clock mismatch issues.

5.1. Simulation Results

Both simulation and experimental tests use the parameters in Table 2 to validate the effectiveness of the proposed method and ensure consistency between the results. A 5 µs sampling step is used, which is smaller than the state-of-the-art sampling steps. Although the proposed method can achieve even smaller sampling steps, down to 50 ns, the 5 µs step is selected due to the limitations in the commutation frequency of power electronic switches. Achieving small sampling steps in FCS-MPC is crucial, as it allows for faster and more precise control, improving the dynamic response and stability of the system. Additionally, in high-frequency applications, reaching small sampling steps is necessary to accurately track and respond to rapid changes in the system.

Table 2. Parameter values used in the simulation and experimental tests.

Parameters	Symbol	Value	Unit
Input voltage	V_{in}	40	V
Output voltage reference	$v_{C,ref}$	60	V
Inductor current reference	$i_{L,ref}$	1.99	A
Load resistance	$R1$	45	Ω
Load resistance	$R2$	22.5	Ω
Capacitance	C	32	µF
Inductance	L	0.5	mH
Sampling time	Δt	5	µs

Initially, the non-minimum phase issue in a traditional FCS-MPC algorithm is illustrated in Figure 6. In this test, the initial capacitor voltage is set to 40 V, i.e., the same as the input voltage, to focus on the behavior of the controller when it starts to reach an output voltage greater than the input. As can be seen, instead of closing the switch, the controller erroneously opens it (switch Q is *off*), preventing the output voltage from reaching the desired 60 V reference. This incorrect action is a direct consequence of the non-minimum phase behavior, where the system's immediate response to the control action is opposite to the desired effect. Consequently, the output voltage fails to reach the target of 60 V, as shown in Figure 6.

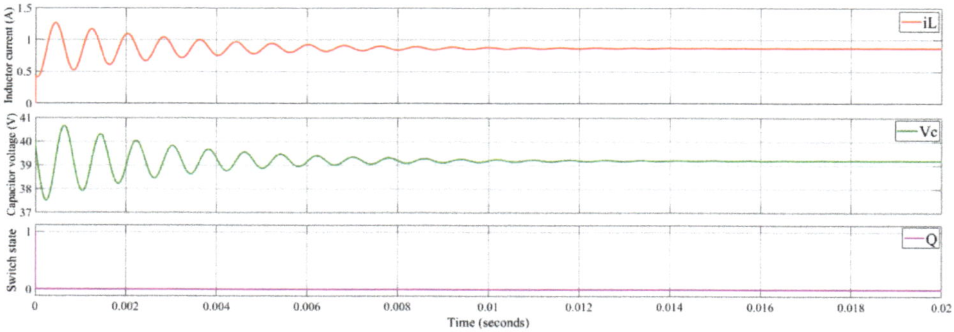

Figure 6. Non-minimum phase issue in conventional FCS-MPC for a boost converter.

Figure 7 illustrates the performance of the control strategy presented in Section 3.1 that uses a PI controller to calculate the reference current [18–20]. This hybrid approach contrasts with the proposal of this paper, highlighting the differences in response and performance. In this approach, the PI controller adjusts the reference current based on the output voltage error, and the FCS-MPC is used to control the inductor current. As shown in the figure, when the load resistance changes from R1 = 45 Ω to R2 = 22.5 Ω at 0.1 s, there is a noticeable voltage drop up to 25% with a settling time of 20 ms. This drop occurs because the PI controller's slower dynamic response introduces a delay in adjusting the reference current to match the new load conditions. Consequently, the system takes longer to stabilize and return the output voltage to the desired level, affecting the overall speed and efficiency of the control system.

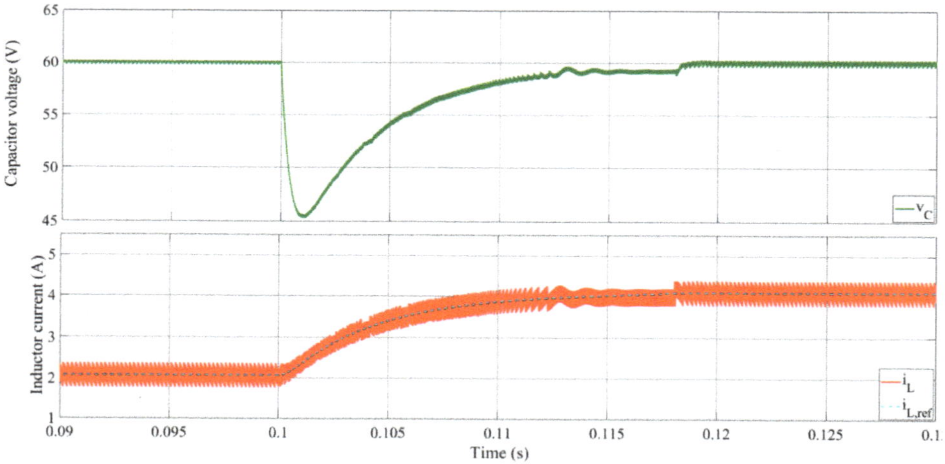

Figure 7. PI-based FCS-MPC response to a step change in load from R1 = 45 Ω to R2 = 22.5 Ω [18–20].

The aim of this study is to integrate a PI controller for correcting offset in steady-state performance, along with a sensor to measure load current and adjust the reference current according to Equation (7). To understand the role of the offset corrector block, the first simulation is conducted without incorporating this block, serving as a proof for verification of the IRC block. In this test, the controller updates the reference current without incorporating the offset corrector block. When subjected to a step change in the load, as shown in Figure 8, this method demonstrates an improved dynamic response. The system exhibits a significantly reduced undershoot in the output voltage and a much faster settling time compared to previous approaches. Specifically, the undershoot is limited to 3%, and the settling time is reduced to 4 ms. However, due to parasitic impedances, the system still shows a slight steady-state error, with the output voltage settling at 59 V instead of the targeted 60 V. In hardware tests, this error can be even more significant due to inaccuracies in the prediction model or discrepancies between the real parameters and those used for calculating the reference current (Equation (7)).

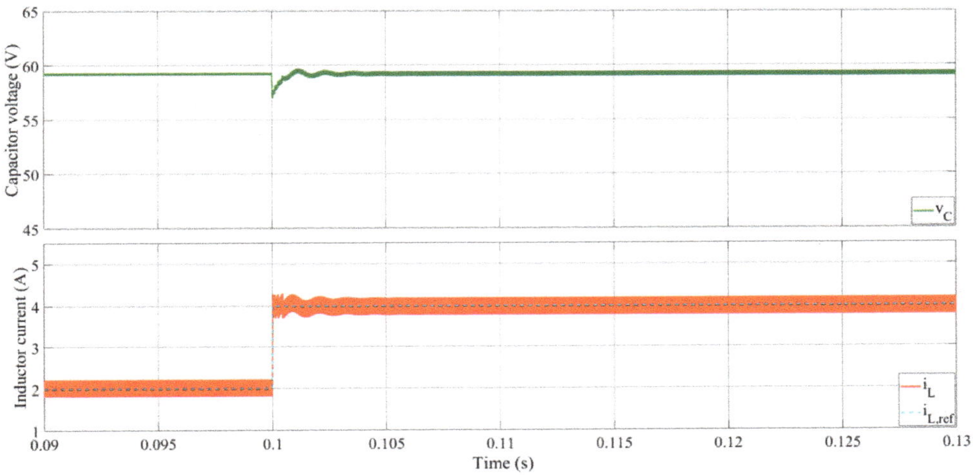

Figure 8. Performance under step change in load with immediate reference current update based on (7), where the IRC block is employed directly without the involvement of the PI controller.

The proposed method combines the advantages of immediate inductor current control using Equation (7) with the steady-state stability provided by the PI controller used for the fine-tuning of the reference current, ensuring accurate and fast reference current adjustments. The IRC block ensures that the power input equals the power output under ideal conditions, while the PI regulator addresses any deviations caused by non-ideal components and external factors.

The simulation results, as depicted in Figure 9, demonstrate that the proposed method is faster than the one presented in [18–20]. The system effectively mitigates the non-minimum phase problem inherent in FCS-MPC applications. It maintains the desired output voltage of 60 V without deviation and quick recovery from load changes. Specifically, the system shows a settling time of 2 ms and an undershoot of 3.2%. Additionally, there is no offset error, and the method remains fast, as the PI controller is only involved for offset correction, not in the reference current calculation block. The presence of sub-harmonic oscillations in the results can be attributed to the low temporal resolution associated with a sampling step of 5 μs. Future work will explore the potential for improvement by reducing the sampling step.

In addition to the load resistor change results, two further simulations were conducted to validate the feasibility and robustness of the proposed converter under different dynamic conditions. The first simulation focuses on the startup behavior of the converter. As shown in Figure 10, the inductor current quickly increases to charge the capacitor to the desired voltage level of 60 V. The output voltage reaches its reference value within 3.5 ms with a small overshoot, demonstrating the effectiveness of the proposed control strategy in managing the initial charging process and achieving fast settling with minimal transient effects.

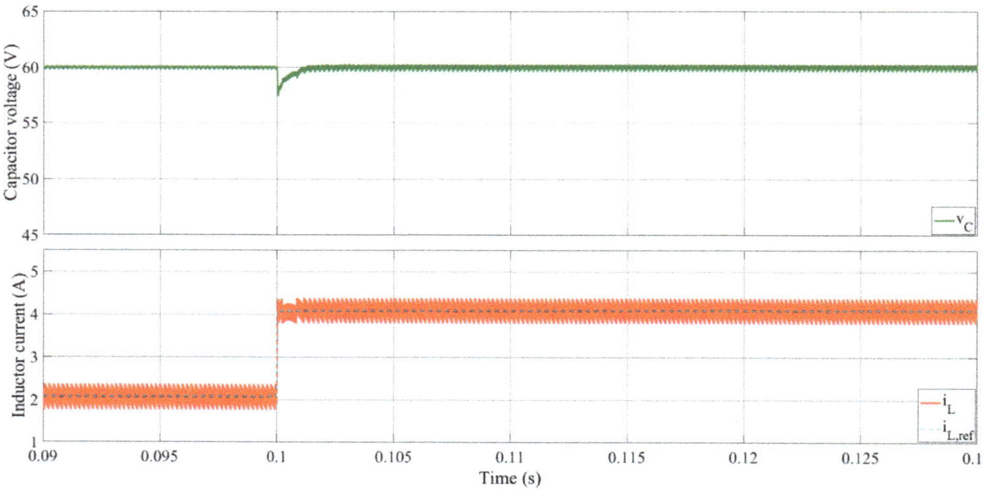

Figure 9. Proposed FCS-MPC method under a step change in load from R1 = 45 Ω to R2 = 22.5 Ω.

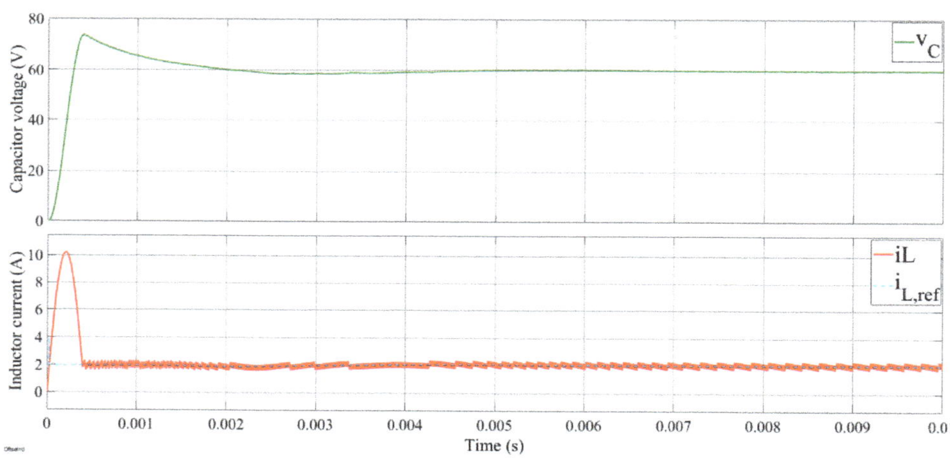

Figure 10. Startup of the boost converter using the proposed method, showing the inductor current charging the capacitor to reach the 60 V output.

The second simulation illustrates the system's response to an output voltage reference change, as depicted in Figure 11. At t = 0.1 s, a step-up change is applied to the output voltage reference, increasing it from 60 V to 80 V. Initially, the output voltage briefly decreases due to the non-minimum phase characteristic of the system, before it increases to reach the new reference value. The inductor current responds to the step change much faster, as seen in the simulation results. However, the dynamics of the boost converter cause the output voltage to respond more slowly. Despite this, the transition occurs without any significant overshoot, and the transient lasts for about 5 ms. This behavior demonstrates the ability of the proposed dual control strategy to handle non-minimum phase behavior while maintaining stable and accurate voltage regulation.

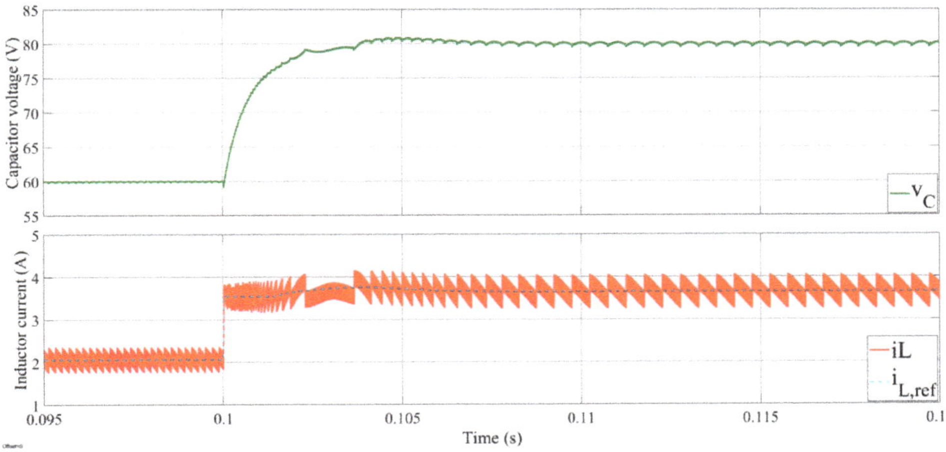

Figure 11. Response of the boost converter with the proposed method to an output voltage step from 60 V to 80 V at t = 0.1 s.

It is important to note that while the FCS-MPC algorithm can operate in both CCM and DCM, it faces specific challenges in DCM. This is because the cost function tends to favor maintaining the zero-current state, particularly under light load conditions, which can lead to the system remaining in this state longer than desired. As a result, the inductor current will not increase when needed, impacting voltage regulation. This issue is unique to conventional FCS-MPC, as other methods like CCS-MPC or those with fixed switching frequencies [16] do not encounter this problem. The focus of the work is on the combination of conventional FCS-MPCs with the IRC block and the offset tuner to mitigate the non-minimum phase issue. However, the limitations in DCM performance related to the FCS-MPC algorithm remain, and addressing these challenges is the focus of ongoing research aimed at improving future implementations. To address this issue, the possible solution under development involves modifying the cost function to focus on average values over a longer period, rather than relying solely on instantaneous values.

5.2. Experimental Results in FPGA

To validate the effectiveness of the proposed control strategy in a practical setting, FPGA-based experimental tests are conducted using the same parameters as in the simulation section. The experimental setup is illustrated in Figure 12, showing the Arty Z7 FPGA board connected to a PC for data acquisition and control implementation. The FPGA is programmed using HDL code generated from MATLAB HDL Coder. The FPGA implementation includes an Integrated Logic Analyzer (ILA) block to monitor internal signals, facilitating debugging and the validation of the control strategy. Notably, the tests utilize hardware-in-the-loop (HIL) techniques, with the entire system implemented within a single FPGA. This approach mitigates communication issues between the controller and converter model, and prevents clock mismatch problems, simplifying the hardware setup and reducing overall system complexity and cost.

The experimental results obtained using the Integrated Logic Analyzer (ILA) are depicted in Figure 13, illustrating the controller's performance. In this test, the performance of the controller is evaluated by implementing a real-time model of the boost converter in the same FPGA. It is notable that the ILA records signals in integer form, necessitating that each signal amplitude be divided by 2^F, where F represents the number of fractional bits defined in MATLAB HDL Coder. Additionally, each sample in the ILA corresponds to 8 ns, reflecting the internal clock period of the Arty Z7 FPGA board.

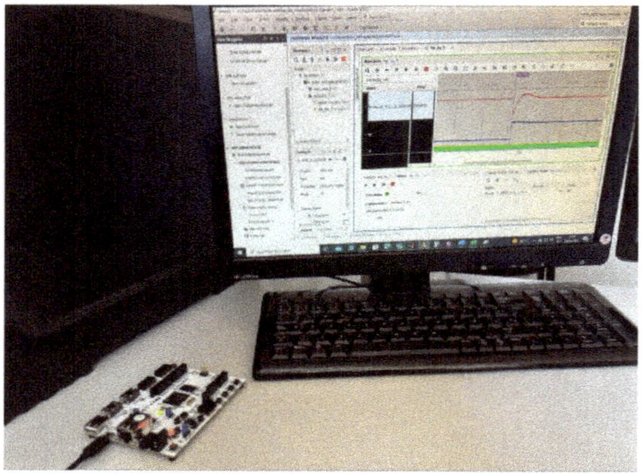

Figure 12. The Arty Z7 FPGA board used for experimental testing of the controller.

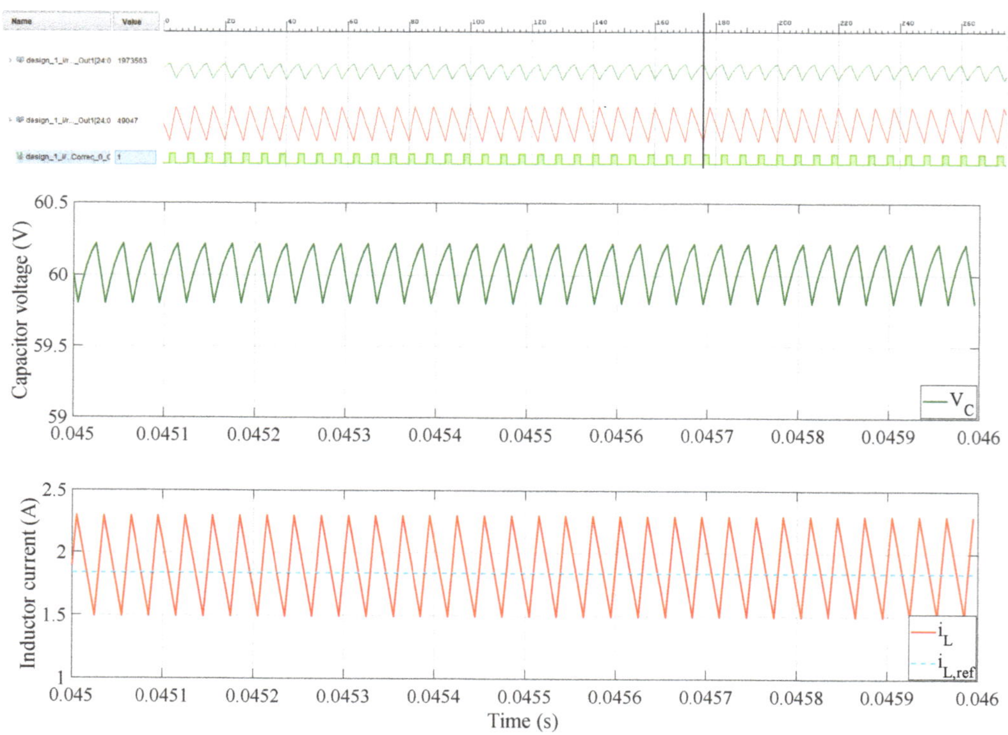

Figure 13. Steady-state performance maintaining 60 V output, obtained from experimental tests in the FPGA, represented by ILA.

Due to the difficulty in discerning the details of the signals directly from the ILA, the data are extracted in .csv format and subsequently plotted in MATLAB, providing a clearer visualization of the results, as seen in Figure 13. Figure 13 demonstrates that the output voltage closely reaches the target of 60 V, exhibiting no deviation and a stable response. The ripple in the output voltage can be decreased by reducing the sampling step of the controller, depending on the practical limitations of the converter and the power rate.

Additionally, the inductor current is maintained around 2 A, ensuring adequate power to sustain the output voltage at the desired 60 V. This precise current control enables the system to effectively adapt to the initial load of 45 Ω. Despite the clarity of steady-state performance, this figure does not capture the dynamic behavior of the controller, as it emphasizes steady-state rather than transient responses.

Figure 14 illustrates the performance of the controller under a step change in load resistance, where the load resistance is varied from 45 Ω to 22.5 Ω. This step change test further validates the proposed method's capability to handle dynamic load variations. The output voltage remains stable at 60 V, with the inductor current adapting smoothly to the new load conditions. Notably, the undershoot in the voltage is about 3.3% similar to the simulation test, and the output voltage is restored to around 60 V in less than 4 ms. These experimental results closely match the simulation results shown in Figure 9, demonstrating the effectiveness of the controller in minimizing transient deviations and quickly stabilizing the output during sudden changes in load.

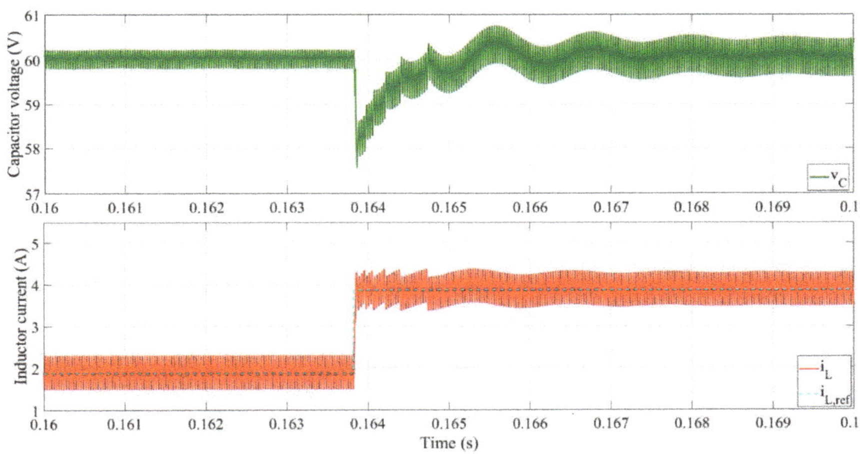

Figure 14. Controller's performance based on the proposed FCS-MPC method under a step change in load from R1 = 45 Ω to R2 = 22.5 Ω, captured by the ILA in the FPGA implementation of the entire system.

As shown in this section, the system response is consistent with simulation results, exhibiting a negligible undershoot and rapid settling time. This approach shows great potential for applications requiring precise voltage regulation and adaptability to changing load conditions. Moreover, it can be applied to any topology where the non-minimum phase issue makes it difficult to use horizon one FCS-MPC, enhancing control performance and stability. Future work will focus on optimizing the control algorithm and exploring its application to more complex power electronic systems.

The experimental results have shown the correct behavior of the controller with a passive load. To further validate the robustness of the proposed controller, additional results with an active load are included. Figure 15 illustrates the behavior of the proposed control method when applied to a boost converter connected to a single-phase H-bridge inverter, which powers a resistive-inductive (R-L) load (20 Ω and 20 mH) on the AC side. The H-bridge is controlled by a sinusoidal PWM modulator to provide a sinusoidal voltage waveform with an amplitude of 30 V. The controller maintains the input DC link voltage of the H-bridge around 60 V while effectively handling active loads, such as the inverter with a DC bus. Unlike passive resistor loads, the system dynamically adjusts the reference current to maintain the desired voltage across the DC link while compensating for the inductive effects on the load side. The 100 Hz component in the inductor current, shown in Figure 15, results from the AC-side power demand, where the H-bridge inverter converts the DC link voltage into a single-phase AC output. Since the output power is sinusoidal, corresponding to the AC-side frequency, the inductor current in the boost converter—responsible for supplying power to the inverter—also contains this 100 Hz frequency component. This occurs because the inductor current fluctuates to meet the alternating power demand of the R-L load on the AC side. While the DC link voltage remains steady at around 60 V, the current drawn by the inductor mirrors the sinusoidal nature of the load's power consumption, oscillating at twice the AC mains frequency (100 Hz).

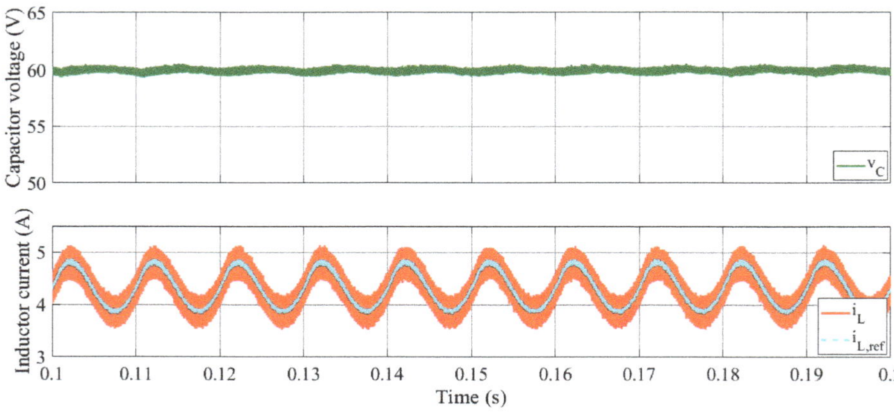

Figure 15. Performance of the proposed FCS-MPC method applied to a boost converter connected to a single-phase H-bridge inverter controlled by sinusoidal PWM.

Figure 16 presents the experimental comparison of the capacitor voltage (a) and inductor current (b) between the proposed method and the PI-based MPC method from references [18–20], with results obtained using the ILA. In both cases, a sampling step of 5 μs is used. The system response to a step load change from 45 Ω to 22.5 Ω is shown. The proposed method demonstrates a significantly faster response, with the capacitor voltage stabilizing more quickly and with smaller overshoot, while the PI-based MPC method shows slower recovery. Similarly, the inductor current in the proposed method settles much faster than in the PI-based MPC approach, highlighting its superior dynamic performance. These experimental results closely match the simulation results previously obtained, confirming the proposed method's performance in achieving faster voltage and current regulation during sudden load changes.

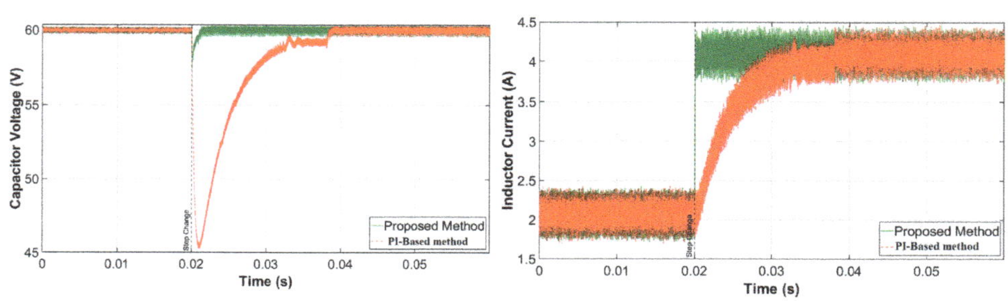

Figure 16. Experimental results comparing (**a**) capacitor voltage and (**b**) inductor current between the proposed method and references [18–20], obtained using the ILA. The step change corresponds to a load transition from 45 Ω to 22.5 Ω.

5.3. Discussion and Potential Applications

The proposed dual control strategy effectively addresses the non-minimum phase behavior in horizon-1 FCS-MPC for DC-DC boost converters by integrating a PI-based reference current offset tuner. In conventional PI-MPC methods, the PI controller is directly involved in generating the reference current and managing both dynamic and steady-state transitions, which often limits the fast response capabilities of the MPC due to the slower reaction time of the PI controller. In contrast, the proposed method decouples the roles of the PI controller and the FCS-MPC. The PI controller is only responsible for correcting small steady-state errors, while the FCS-MPC handles all fast dynamic responses, resulting in improved stability and faster voltage regulation without limiting the use of short-horizon FCS-MPC, even in applications with non-minimum phase issues. Moreover, the flexibility of the MPC control makes it applicable to a wide range of power electronic systems facing similar non-minimum phase challenges.

Table 3 summarizes the key performance characteristics of the proposed dual control strategy (PI + FCS-MPC) compared with other control methods such as Long-Horizon FCS-MPC [26], Adaptive-Horizon FCS-MPC [12], and conventional PI-MPC [18–20]. The comparison includes factors such as response time, the need for sensors or observers, system complexity, and the sampling step. The Long-Horizon FCS-MPC [26] offers a fast response but comes with significantly higher complexity due to the longer prediction horizon. Its ability to handle non-minimum phase behavior depends on the converter dynamics and the number of horizons considered, but it does not require observers or any extra sensors for detecting the load. The Adaptive-Horizon FCS-MPC [12] also provides a fast response. However, it demands very high complexity for implementation without requiring additional sensors. The PI-MPC method shows a slower response due to the nature of the PI control in comparison with Long/Adaptive-Horizon FCS-MPC. This method introduces low algorithm complexity without the need for additional sensors or observers. The proposed dual control strategy (PI + FCS-MPC) achieves a fast response time, similar to both long/adaptive-Horizon FCS-MPC, due to the isolation of the proposed PI controller from the dynamic response of the converter. The steady-state error is similar to other methods thanks to the offset tuning method. It effectively handles non-minimum phase behavior with relatively low complexity and achieves high performance with the additional help of a sensor or observer for correcting the reference current.

Table 3. Comparison of MPC-based control methods for boost converters.

Method	Response Time	Sensor/Observer	Complexity	Sampling Step
Long-Horizon FCS-MPC [26]	Fast	No	High	100 μs
Adaptive-Horizon FCS-MPC [12]	Fast	No	Very high	10 μs
PI-MPC [18–20]	Slow (20 ms)	No	Low	50, 12.5, 10 μs
Proposed Dual Control [PI + FCS]	Fast (2 ms)	Yes	Low	1 and 5 μs

In FCS-MPC, reducing the sampling step has been shown to significantly improve controller performance. The proposed method's 5 μs sampling rate offers precise and rapid control, especially when compared to the 10 μs step used in the adaptive-horizon method by Gardezi et al. [12] and the much longer 100 μs step in the long-horizon approach by Zafra et al. [26]. While the proposed method can achieve sampling steps as low as 100 ns, such high-frequency sampling introduces the risk of damaging power electronic switches due to the limited maximum switching frequency. To prevent this, applying a switching frequency control method, such as the Period Control Approach (PCA), is essential [27,28]. However, the focus of this paper is on the proposed control strategy rather than the impact of calculation resolution on switch performance. As demonstrated in Figure 17, reducing the sampling step from 5 μs to 1 μs leads to greater precision, effectively mitigating the low-frequency subharmonics observed in earlier tests with the sampling step of 5 μs. The smaller sampling period allows for more frequent updates to control inputs, improving prediction accuracy and reducing oscillations in the control response.

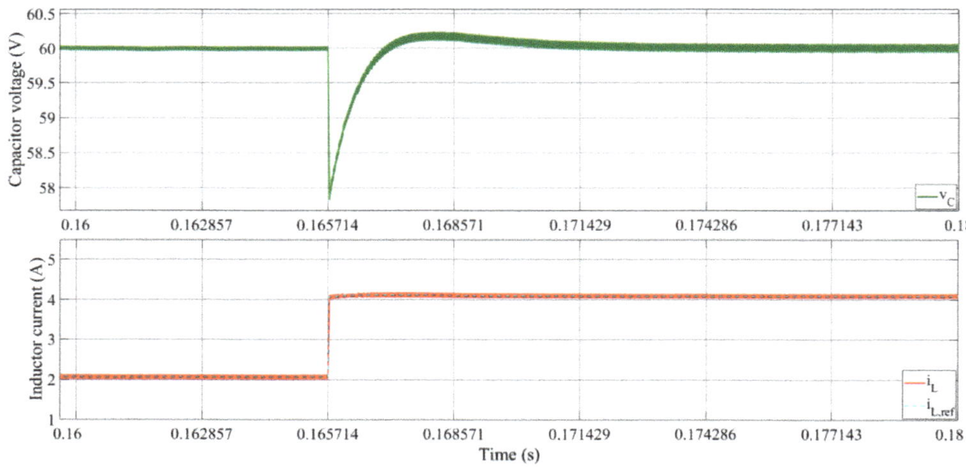

Figure 17. Performance of the proposed FCS-MPC method with a 1 μs sampling step, showing the response to a load change from R1 = 45 Ω to R2 = 22.5 Ω, captured by the ILA.

The correlation between the sampling step and harmonics spectrum is clearly illustrated in the spectra of the inductor current and output voltage in Figure 18. As shown in the subfigures, the amplitude of subharmonics is significantly larger when a sampling step of 5 μs is used compared to 1 μs. Increasing the *Ts* results in additional low-frequency subharmonics that may negatively affect system stability and performance. The maximum achievable switching frequency ($f_{sw,\max}$) is determined by the relationship given by (9); therefore, with *Ts* set to 5 μs, the $f_{sw,\max}$ is limited to 100 kHz. However, in this application, switching generally follows a pattern of one *on*-cycle followed by two *off*-cycles, a behavior attributed to the differing slopes of increasing and decreasing current, as described by (3). This results in a more commonly observed switching frequency of approximately 1/(3·*Ts*). Consequently, with *Ts* = 5 μs, the main harmonic appears at 67 kHz (displayed as 68 kHz due to resolution limitations), while reducing *Ts* to 1 μs shifts this primary harmonic to around 333 kHz (observed as 342 kHz). This shift to higher frequencies with reduced *Ts* allows for more effective filtering by the converter, enhancing performance by mitigating low-frequency subharmonics and stabilizing the system.

$$f_{sw,\max} = \frac{1}{2 \cdot T_s} \tag{9}$$

The integration of the MPC controller with a PI-based reference current offset tuner provides a versatile control mechanism that can be adapted to any non-minimum phase power converter. In such cases, the MPC controller may help simplify the regulator implementation, particularly when compared to linear controllers like PI or PID. This would be the case for boost-derived topologies, such as the buck-boost, four-switch buck-boost, Cuk, SEPIC, or ZETA converters, which are frequently used in low-power battery chargers (below 100 W). These converters, especially the Cuk topology, benefit from inherent current ripple reduction on one of their ports, making them ideal candidates for this control approach.

The extension to AC-DC single-phase active rectifiers would also offer potential advantages over traditional controllers, reducing the complexity of implementation. For example, flyback, flyboost [29], or boost AC-DC power factor correctors [30] could benefit from the MPC strategy. Additionally, some applications impose non-linear behaviors in the control stage, as seen in quadratic boost power converters [31], coupled inductor boost-derived topologies [32,33], Maximum Power Tracking strategies for photovoltaic panels [34], multi-port boost converters [35], or partially saturated inductor boost converters [36,37]. These non-linearities add complexity to the control stage beyond voltage or current regulation. In these cases, the proposed strategy could simplify implementation and reduce computational burdens compared to linear regulators.

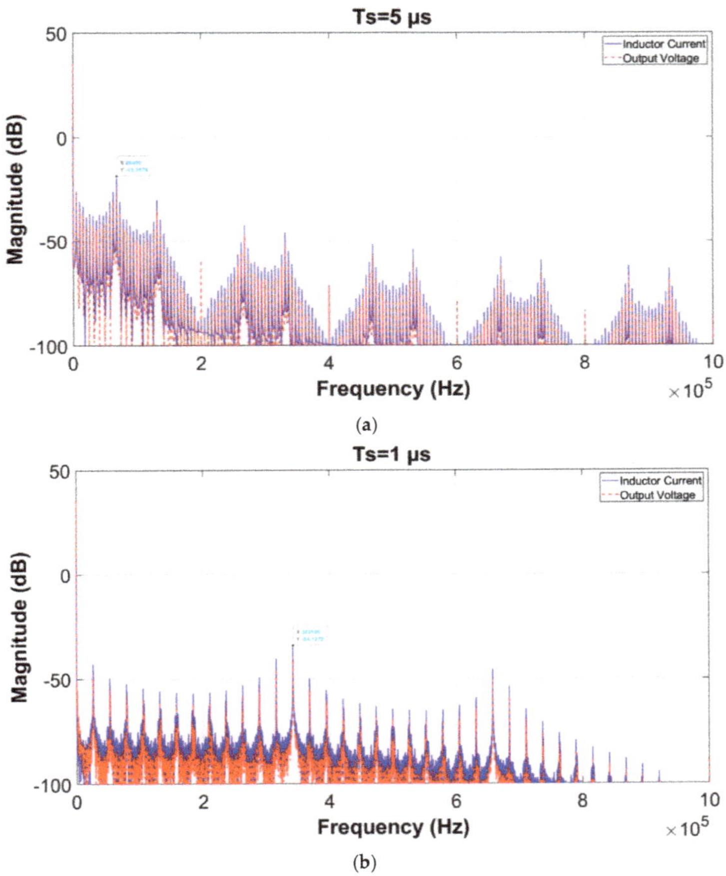

Figure 18. FFT analysis of the output voltage and inductor current with sampling steps of 1 μs and 5 μs.

6. Conclusions

This study presents a significant advancement in the control of DC-DC boost converters by effectively addressing the non-minimum phase behavior, a common challenge in such systems. The main contribution of this work is to examine the limitations of utilizing horizon-one FCS-MPC exhibiting non-minimum phase behavior and to propose a solution that addresses these issues without extending the prediction horizon. Unlike traditional PI-MPC approaches, where the PI controller manages both dynamic and steady-state transitions, the proposed method separates these roles. The FCS-MPC is solely responsible for a fast dynamic response, while the PI controller is only tasked with fine-tuning the reference current calculated by the IRC block, ensuring steady-state accuracy. This decoupling enables the system to handle dynamic load changes more efficiently and maintain high-speed operation, as the PI controller does not interfere with the fast responses of the FCS-MPC.

Simulation results show that the system effectively mitigates the non-minimum phase issue, maintaining the desired output voltage with minimal deviation and quick recovery from load changes. The method's practical viability is demonstrated through comprehensive simulation and HIL experimental tests, showcasing its ability to handle real-world conditions effectively. In particular, the smaller sampling step (5 μs) achieved in this study significantly contributes to the faster response of the system compared to other FCS-MPC methods. Notably, in the HIL tests, both the controller and the converter's discretized model are implemented inside the same FPGA. This integration eliminates clock mismatch issues and negates the need for additional blocks such as ADCs for interfacing between these components, thus simplifying the experimental setup.

The steady-state error is comparable to long-horizon FCS-MPC and conventional PI-MPC methods, but smaller compared to short-horizon FCS-MPC thanks to the offset tuning mechanism employed. With a negligible undershoot and rapid settling time, the system is suitable for industrial applications requiring precise voltage regulation and adaptability to changing load conditions. Furthermore, the proposed dual control strategy can be extended to other power electronic systems facing non-minimum phase issues, broadening its applicability. Future work will focus on optimizing the control algorithm and exploring its application to more complex power electronic systems.

Author Contributions: Conceptualization, A.M. and E.Z.; methodology, A.M., A.V. and A.d.C.; software, A.M. and M.P.; validation, J.T.D., E.Z. and D.M.; formal analysis, A.M. and E.Z.; investigation, E.Z.; resources, A.d.C.; data curation, M.P.; writing—original draft preparation, A.M. and E.Z.; writing—review and editing, A.d.C. and A.V.; visualization, A.M. and E.Z.; supervision, A.d.C.; project administration, A.d.C.; funding acquisition, A.d.C. All authors have read and agreed to the published version of the manuscript.

Funding: This research was partially funded by the project PID2022-137593OBI00, financed by the Spanish Ministry MCIN/AEI/10.13039/501100011033/FEDER, UE, and FSE+.

Institutional Review Board Statement: Not applicable.

Informed Consent Statement: Not applicable.

Data Availability Statement: The original contributions presented in the study are included in the article, further inquiries can be directed to the corresponding author.

Conflicts of Interest: The authors declare no conflicts of interest. The funders had no role in the design of the study; in the collection, analyses, or interpretation of data; in the writing of the manuscript; or in the decision to publish the results.

References

1. Errouissi, R.; Al-Durra, A.; Muyeen, S.M. A Robust Continuous-Time MPC of a DC–DC Boost Converter Interfaced With a Grid-Connected Photovoltaic System. *IEEE J. Photovolt.* **2016**, *6*, 1619–1629. [CrossRef]
2. Bakar Siddique, M.A.; Asad, A.; Asif, R.M.; Rehman, A.U.; Sadiq, M.T.; Ullah, I. Implementation of Incremental Conductance MPPT Algorithm with Integral Regulator by Using Boost Converter in Grid-Connected PV Array. *IETE J. Res.* **2021**, *69*, 3822–3835. [CrossRef]
3. Hidalgo, H.; Orosco, R.; Huerta, H.; Vázquez, N.; Hernández, C.; Pinto, S. A High-Voltage-Gain DC–DC Boost Converter with Zero-Ripple Input Current for Renewable Applications. *Energies* **2023**, *16*, 4860. [CrossRef]
4. Wu, J.; Li, J.; Wang, H.; Li, G.; Ru, Y. Fault-Tolerant Three-Vector Model-Predictive-Control-Based Grid-Connected Control Strategy for Offshore Wind Farms. *Electronics* **2024**, *13*, 2316. [CrossRef]
5. López, M.; Rodriguez, J.; Silva, C.; Rivera, M. Predictive Torque Control of a Multidrive System Fed by a Dual Indirect Matrix Converter. *IEEE Trans. Ind. Electron.* **2015**, *62*, 2731–2741. [CrossRef]
6. Wei, X.; Bi, K.; Lan, G.; Li, W.; Cui, J. Decoupled MPC Power Balancing Strategy for Coupled Inductor Flying Capacitor DC–DC Converter. *Appl. Sci.* **2024**, *14*, 4813. [CrossRef]
7. Dong, Z.; Chen, Q.; Qin, J.; Zhang, Z.; Tse, C.K.; Xu, Y. Noise Tolerance Strategy Based on Virtual Capacitor for DC-DC Converters with Continuous Control Set Model Predictive Control. *IEEE Trans. Power Electron.* **2024**, *39*, 9084–9088. [CrossRef]

8. Nguyen, A.T.; Ryu, S.-W.; Rehman, A.U.; Choi, H.H.; Jung, J.-W. Improved Continuous Control Set Model Predictive Control for Three-Phase CVCF Inverters: Fuzzy Logic Approach. *IEEE Access* **2021**, *9*, 75158–75168. [CrossRef]
9. Gonçalves, P.; Cruz, S.; Mendes, A. Finite Control Set Model Predictive Control of Six-Phase Asymmetrical Machines—An Overview. *Energies* **2019**, *12*, 4693. [CrossRef]
10. Alhasheem, M.; Blaabjerg, F.; Davari, P. Performance Assessment of Grid Forming Converters Using Different Finite Control Set Model Predictive Control (FCS-MPC) Algorithms. *Appl. Sci.* **2019**, *9*, 3513. [CrossRef]
11. Aguilera, R.P.; Quevedo, D.E. On the stability of MPC with a Finite Input Alphabet. *IFAC Proc. Vol.* **2011**, *44*, 7975–7980. [CrossRef]
12. Gardezi, M.S.M.; Hasan, A. Machine Learning Based Adaptive Prediction Horizon in Finite Control Set Model Predictive Control. *IEEE Access* **2018**, *6*, 32392–32400. [CrossRef]
13. Villarroel, F.A.; Espinoza, J.R.; Pérez, M.A.; Ramírez, R.O.; Baier, C.R.; Sbárbaro, D.; Silva, J.J.; Reyes, M.A. Stable Shortest Horizon FCS-MPC Output Voltage Control in Non-Minimum Phase Boost-Type Converters Based on Input-State Linearization. *IEEE Trans. Energy Convers.* **2021**, *36*, 1378–1391. [CrossRef]
14. Villarroel, F.; Espinoza, J.; Pérez, M.; Ramírez, R.; Baier, C.; Morán, L. Shortest horizon FCS-MPC output voltage tracking in non-minimum phase boost-type converters. In Proceedings of the IECON 2019—45th Annual Conference of the IEEE Industrial Electronics Society, Lisbon, Portugal, 14–17 October 2019; pp. 4119–4124. [CrossRef]
15. Pérez, M.A.; Lizana Fuentes, R.; Rodríguez, J. Predictive control of DC-link voltage in an active-front-end rectifier. In Proceedings of the 2011 IEEE International Symposium on Industrial Electronics, Gdansk, Poland, 27–30 June 2011; pp. 1811–1816. [CrossRef]
16. Karamanakos, P.; Geyer, T.; Manias, S. Direct Model Predictive Current Control Strategy of DC–DC Boost Converters. *IEEE J. Emerg. Sel. Top. Power Electron.* **2013**, *1*, 337–346. [CrossRef]
17. Firpo, P.; Ravera, A.; Oliveri, A.; Lodi, M.; Storace, M. Use of a Partially Saturating Inductor in a Boost Converter with Model Predictive Control. *Electronics* **2023**, *12*, 3013. [CrossRef]
18. Li, P.; Li, R.; Shao, T.; Zhang, J.; Fang, Z. Composite adaptive model predictive control for DC–DC boost converters. *IET Power Electron.* **2018**, *11*, 1706–1717. [CrossRef]
19. Chen, F.-Z.; Maksimović, D. Digital Control for Improved Efficiency and Reduced Harmonic Distortion over Wide Load Range in Boost PFC Rectifiers. *IEEE Trans. Power Electron.* **2010**, *25*, 2683–2692. [CrossRef]
20. Nair, H.S.; Lakshminarasamma, N. A Computationally Simple Predictive CCM Average Current Controller With Nearly Zero Tracking Error for Boost PFC Converter. *IEEE Trans. Ind. Appl.* **2020**, *56*, 5083–5094. [CrossRef]
21. IEEE Standard for Floating-Point Arithmetic, in IEEE Std 754-2019 (Revision of IEEE 754-2008), pp.1–84, 22 July 2019. Available online: https://ieeexplore.ieee.org/document/8766229 (accessed on 24 October 2024).
22. Cortes, P.; Rodriguez, J.; Silva, C.; Flores, A. Delay Compensation in Model Predictive Current Control of a Three-Phase Inverter. *IEEE Trans. Ind. Electron.* **2012**, *59*, 1323–1325. [CrossRef]
23. Jin, T.; Shen, X.; Su, T.; Flesch, R.C.C. Model Predictive Voltage Control Based on Finite Control Set With Computation Time Delay Compensation for PV Systems. *IEEE Trans. Energy Convers.* **2019**, *34*, 330–338. [CrossRef]
24. Talbi, B.; Krim, F.; Laib, A.; Sahli, A.; Krama, A. PI-MPC Switching Control for DC-DC Boost Converter using an Adaptive Sliding Mode Observer. In Proceedings of the 2020 International Conference on Electrical Engineering (ICEE), Istanbul, Turkey, 25–27 September 2020; pp. 1–5. [CrossRef]
25. Heydari-doostabad, H.; Ghazi, R. A new approach to design an observer for load current of UPS based on Fourier series theory in model predictive control system. *Int. J. Electr. Power Energy Syst.* **2019**, *104*, 898–909. [CrossRef]
26. Zafra, E.; Vazquez, S.; Geyer, T.; Aguilera, R.P.; Franquelo, L.G. Long Prediction Horizon FCS-MPC for Power Converters and Drives. *IEEE Open J. Ind. Electron. Soc.* **2023**, *4*, 159–175. [CrossRef]
27. Aguirre, M.; Garcia, J.; Perez, M.; Rodriguez, J.; Jha, A.; Romero, F. Period Control Approach Finite Control Set Model Predictive Control Switching Phase Control for Interleaved DC/DC Converters. *IEEE Trans. Ind. Electron.* **2024**, *71*, 8304–8312. [CrossRef]
28. Aguirre, M.; Kouro, S.; Rojas, C.A.; Rodriguez, J.; Leon, J.I. Switching Frequency Regulation for FCS-MPC Based on a Period Control Approach. *IEEE Trans. Ind. Electron.* **2018**, *65*, 5764–5773. [CrossRef]
29. Luo, S.; Qiu, W.; Wu, W.; Batarseh, I. Flyboost power factor correction cell and a new family of single-stage AC/DC converters. *IEEE Trans. Power Electron.* **2005**, *20*, 25–34. [CrossRef]
30. Lazaro, A.; Barrado, A.; Pleite, J.; Vazquez, R.; Olias, E. New family of single-stage PFC converters with series inductance interval. In Proceedings of the 2002 IEEE 33rd Annual Power Electronics Specialists Conference, Cairns, Australia, 23–27 June 2002; pp. 1357–1362. [CrossRef]
31. Monfared, M.T.; Gholizadeh, H.; Ben-Brahim, L. New Enhanced Family of QBC Topologies: Mitigating Capacitor Stress and Increasing Voltage Gain. In Proceedings of the 2024 IEEE 8th Energy Conference (ENERGYCON), Doha, Qatar, 4–7 March 2024; pp. 1–6. [CrossRef]
32. Li, W.; He, X. A Family of Isolated Interleaved Boost and Buck Converters With Winding-Cross-Coupled Inductors. *IEEE Trans. Power Electron.* **2008**, *23*, 3164–3173. [CrossRef]
33. Ambagahawaththa, T.S.; Nayanasiri, D.R.; Jayasinghe, S.G.D. Family of Boost Converters Based on Switched Coupled Inductor and Voltage Lifter Cell. In Proceedings of the 2018 8th International Conference on Power and Energy Systems (ICPES), Colombo, Sri Lanka, 21–22 December 2018; pp. 252–257. [CrossRef]

34. Mohammed, F.A.; Bahgat, M.E.; Elmasry, S.S.; Sharaf, S.M. Design of a Fuzzy Logic Controller for DC Converter of a Stand-Alone PV System Based on Maximum Power Point Tracking. In Proceedings of the 2021 22nd International Middle East Power Systems Conference (MEPCON), Assiut, Egypt, 14–16 December 2021; pp. 7–13. [CrossRef]
35. Meshkati, E.; Farzanehfard, H. Family of High Step-Up Multi-Input Converters with Continuous Battery Current Based on Three Port Boost Topology and Switched Capacitors. In Proceedings of the 2023 14th Power Electronics, Drive Systems, and Technologies Conference (PEDSTC), Babol, Iran, 31 January–2 February 2023; pp. 1–7. [CrossRef]
36. Roberto, S.F.; Sciré, D.; Lullo, G.; Vitale, G. Equivalent Circuit Modelling of Ferrite Inductors Losses. In Proceedings of the 2018 IEEE 4th International Forum on Research and Technology for Society and Industry (RTSI), Palermo, Italy, 10–13 September 2018; pp. 1–4. [CrossRef]
37. Scirè, D.; Lullo, G.; Vitale, G. Assessment of the Current for a Non-Linear Power Inductor Including Temperature in DC-DC Converters. *Electronics* **2023**, *12*, 579. [CrossRef]

applied sciences

MDPI

Article

Green Hydrogen Production—Fidelity in Simulation Models for Technical–Economic Analysis

Adrián Criollo [1,2], Luis I. Minchala-Avila [1], Dario Benavides [3], Danny Ochoa-Correa [1], Marcos Tostado-Véliz [3,*], Wisam Kareem Meteab [3] and Francisco Jurado [3]

[1] Department of Electrical Engineering, Electronics, and Telecommunications (DEET), Universidad de Cuenca, Cuenca 010101, Ecuador; adrian.criollor@ucuenca.edu.ec (A.C.); ismael.minchala@ucuenca.edu.ec (L.I.M.-A.); danny.ochoac@ucuenca.edu.ec (D.O.-C.)

[2] Department of Electrical Maintenance and Industrial Control, Instituto Superior Tecnológico del Azuay, Cuenca 010105, Ecuador

[3] Department of Electrical Engineering, University of Jaén, 23700 Linares, Spain; djbenavi@ujaen.es (D.B.); wkm00001@red.ujaen.es (W.K.M.); fjurado@ujaen.es (F.J.)

* Correspondence: mtostado@ujaen.es

Abstract: Green hydrogen production is a sustainable energy solution with great potential, offering advantages such as adaptability, storage capacity and ease of transport. However, there are challenges such as high energy consumption, production costs, demand and regulation, which hinder its large-scale adoption. This study explores the role of simulation models in optimizing the technical and economic aspects of green hydrogen production. The proposed system, which integrates photovoltaic and energy storage technologies, significantly reduces the grid dependency of the electrolyzer, achieving an energy self-consumption of 64 kWh per kilogram of hydrogen produced. By replacing the high-fidelity model of the electrolyzer with a reduced-order model, it is possible to minimize the computational effort and simulation times for different step configurations. These findings offer relevant information to improve the economic viability and energy efficiency in green hydrogen production. This facilitates decision-making at a local level by implementing strategies to achieve a sustainable energy transition.

Keywords: electrolyzer; proton exchange membrane; fuel consumption; opal-RT; real-time simulation; green hydrogen

1. Introduction

In recent years, hydrogen energy has gradually become a new alternative for clean energy generation. The multiple applications of hydrogen as a means of energy storage in the energy sector, transport, industry, and various other applications have generated great interest in global economic development [1], thus allowing a promising alternative to satisfy and maintain global energy needs with a sustainable tendency [2–5]. As a result, hydrogen generation has sparked great interest in some countries as a fuel of the future [6–8]. However, ensuring green hydrogen production from a renewable energy perspective presents significant challenges. Renewable energy sources (RESs) are naturally variable, requiring energy storage systems (ESSs) to accommodate daily and seasonal changes. Currently, the search for methods and technological development in new energy storage systems is ongoing. One of the processes for generating green hydrogen is through the electrolysis of water using RESs. Recently, there has been significant research and development into this technology, and various methods have been developed to increase efficiency and reduce production costs [9]. Therefore, this article analyzes and studies green hydrogen production in a micro-grid case study based on medium- and high-precision models for a feasible technological and economic analysis. Green hydrogen is generated through an electrolyzer, which is composed of a photovoltaic (PV) system as a renewable resource and an ESS to reduce energy consumption from the electrical grid.

Techno-economic analyses have been fundamental in determining the payback period of investments. However, more studies are currently needed to analyze the return on investment when using excess solar energy to charge energy vectors such as hydrogen. One solution is to produce hydrogen through the electrolysis—splitting with an electric current—of water and to use that hydrogen in a fuel cell (FC) to produce electricity during times of low power production or peak demand or to use the hydrogen in fuel cell vehicles. In addition, hydrogen is a non-toxic and clean energy carrier with a specific energy of 141.9 kJ/g, compared to 46.5 kJ/g of gasoline [10,11]. However, it is important to consider the high cost that its production could generate, so several optimization studies have been carried out. For example, one study has optimized a solar energy system to electrify Kousseri, Cameroon, comparing four algorithms for sizing systems in isolated grids [12]. Similarly, ref. [13] presents the design of a hybrid photovoltaic (PV)/wind turbine (WT)/FC system for three regions in Iran, using the whale optimization algorithm (WOA) to minimize the net present value of the hybrid system while meeting the probability of power generation losses. Furthermore, a study presents a hybrid renewable energy system for domestic and telecommunications loads across India, considering technical, economic, environmental, and social factors [14]. Among the findings, Chennai emerges as the optimal location for domestic and telecommunications loads based on analyses of cost of energy (COE), total net present cost (TNPC), annualized cost of system (ACS), carbon emissions (CEs), and particulate matter (PM), compared to New Delhi and Mumbai.

On the other hand, ref. [15] examines the feasibility of using wind and solar plants for hydrogen generation via electrolysis, focusing on component sizing and the location's impact on the levelized cost of hydrogen. A notable aspect of the research is their finding that hydrogen shows a negative correlation with the amount of electrical production surplus. Similarly, a study analyzed the production of green hydrogen in Paraguay [16]. This study obtained a total production potential of 22.5×10^6 tons/year, which favors the incorporation of the hydrogen economy in the automotive transportation and residential energy sectors. Proton exchange membrane water electrolyzer (PEMWE) technologies boast higher current densities (1–2 Acm^{-2}) than alkaline water electrolyzers (AWEs) and produce a higher purity of hydrogen up to 99.9999% [17]. Barriers associated with those systems have been gathered and recapped to identify key solutions that could aid in their removal. Likewise, in a study, the results found that a 4.2% reduction in the amount of electrical energy needed for electrolysis showed that barriers to solar hydrogen generation are related to the industry of PV cells, atmospheric conditions affecting PV cells' performance, and those related to solar-to-hydrogen (STH) production and storage [18]. According to [19], by connecting a solar module with the voltage at its maximum power point (Vmpp) to a high-efficiency PEM electrolyzer, the group was able to achieve a solar-to-hydrogen efficiency of 12.4%, compared to a past system that only reached 2–6%. This study utilized a 20-cell PEM electrolyzer stack and various PV modules with different output voltages and current levels.

In this context, the modeling of large-sized electrolyzers has also been studied for real-time simulation and the possibility of frequency support by electrolyzers [20]. Also, electrolyzers have been found to support frequency stability more effectively than conventional generators, benefiting the power system's stability. The development of a generic electrolyzer model has allowed simulations to analyze the impact of electrolyzers on power system stability, showing promising results. Field measurements and simulations have validated the model's ability to replicate the behavior of real electrolyzers, highlighting their potential to support frequency and voltage stability in future power systems. In the same line, the reference [21] explores the impact of the operational constraints and the control strategies on the effectiveness of hydrogen electrolyzers in providing frequency control services; the results show that hydrogen electrolyzers can effectively contribute to helping prevent widespread load shedding in power systems. Also, in the studies carried out by [22,23], the results present that proton exchange membrane hydrogen electrolyzer units are a suitable candidate for virtual inertia response, regulation, and contingency frequency control ancillary services, potentially replacing the need for battery installations in future power systems. Electrolyzers can regulate power off-take based on grid frequency. A techno-economic model optimizes revenue by considering power band and ancillary services. Operating the electrolyzer at optimal values maximizes revenue. Offering symmetric primary reserve is a valid option for revenue generation [24].

On the other hand, high-fidelity mathematical models used to perform simulations in the fields of chemical engineering and fluid dynamics can take hours or even days to simulate due to the large amount of data and parameter tuning required in these models. To address these challenges, high-fidelity component-level models can be replaced with reduced-order models that balance accuracy in exchange for lower computational complexity. To create a static reduced-order modeling ROM, curve fitting, look-up tables, and neural networks can be used to create a suitable model [25,26]. This could directly influence the time required for the simulation process of the electrolyzer and the micro-grid system. Therefore, the point of interest is specifically to analyze the energy production and determine the amount of hydrogen that can be produced. Parameters must be taken into account for a technical–economic study.

Consequently, all these factors represent a gap in research for the generation of green hydrogen. Some studies present technical–economic analyses for different countries, which do not allow the approach to be generalized due to the economic power of each one. In addition, there are different conditions for renewable resource generation. This study focuses on a case study to offer alternatives in the regulatory framework and the development and implementation of state support measures for hydrogen energy projects in countries undergoing sustainable development. Many studies in the literature have focused on hydrogen production solely with surplus renewable energy. However, the proposed system allows the generation of green hydrogen continuously throughout the year with minimal consumption from the electrical grid on days of low PV production. Furthermore, the management model achieves an adequate handling of the ESS, which allows for a daily energy balance in an autonomous manner. It also includes data and technical parameters of the equipment used during its experimentation, which brings its effectiveness closer to simulation models for small- and large-scale technical–economic studies. Based on these green hydrogen initiatives, the contributions of this article are as follows:

- A PV energy management and storage model for the uninterrupted production of green hydrogen through self-consumption is proposed.
- The replacement of the high-fidelity model of the electrolyzer with reduced-order modeling is studied based on real experimentation.
- A comparative study is carried out between the energy consumption of methane reforming to generate hydrogen versus the proposal developed with renewable energies.
- A technical–economic analysis is performed for the estimation of local hydrogen production versus the energy cost in kWh/kg.
- Data from experimental tests are used in a real-time simulation environment and micro-grid laboratory tests.

The remainder of this paper is organized as follows: Section 2 presents the details of the modeling of the system components, Section 3 presents the experimental phase with electrolytes, and Section 4 presents the case study, including simulation results and discussion. Finally, Section 5 concludes the study.

2. System Component Modeling

2.1. Production of Green Hydrogen

The production of green hydrogen through the electrolyzer results from the optimal performance of RESs and ESSs with minimal or no consumption from the electrical grid. Therefore, this study proposes the continuous production of hydrogen through electrolyzers. Figure 1 presents a generalized scheme for hydrogen production based on photovoltaics with energy storage support, which covers a pre-established demand profile for the electrolyzer (ETZ). Excess PV energy is stored in the ESS to deliver the energy in case of shortages later. In this context, the PV power injected into the grid P_t^{PV} allows the supply of the demand of the electrolyzer P_t^{ETZ}. In the case of excess ($-$) or the absence ($+$) of generation, the power of the energy storage system P_t^{ESS} allows the energy management to be completed with the objective of reducing the power consumption of the electrical grid P_t^{Gd}. This function is defined in the following Equation (1):

$$min[P_t^{Gd} = P_t^{PV} \pm P_t^{ESS} - P_t^{ETZ}] \quad (1)$$

In this way, self-consumption is achieved from the demand side at the point of common coupling (PCC). Finally, hydrogen can be evaluated as an energy potential in the use of fuel cells.

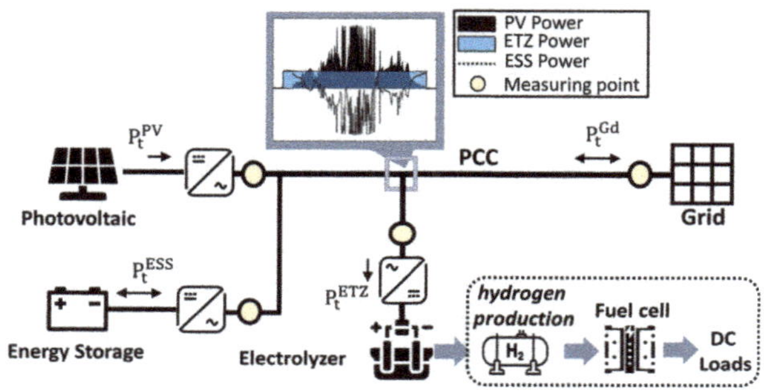

Figure 1. Scheme for hydrogen production based on photovoltaics with energy storage support.

2.2. Photovoltaic Array

The pre-established photovoltaic modules of the system advisor model of the National Renewable Energy Laboratory (NREL) are used. Specifically, it was simulated with the Atersa A-250P model (Elecnor, Valencia, Spain) with a nominal power of 15 kWp based on the 15 × 4 (series–parallel) configuration [27,28]. The PV array is modeled with parameters that use a light-generated current source (I_L), a diode, a series resistor (R_s), and a shunt resistor (R_{sh}) to represent the I–V characteristics of the modules. These values depend on the input variables of irradiance (W/m^2) and temperature (°C) [29]. The equations are detailed below:

$$I_d = I_0\left[exp(\frac{V_d}{V_T}) - 1\right] \quad (2)$$

where I_d is diode current (A), I_0 is diode saturation current (A), V_d is diode voltage (V), and the voltage V_T is calculated as follows:

$$V_T = \frac{kT}{q} \times nI \times Ncell \quad (3)$$

where k is the Boltzmann constant (1.3806×10^{-23} J· K^{-1}), T is the cell temperature (K), q is the electron charge (1.6022×10^{-19} C), nI is the diode ideality factor and $Ncell$ is the number of cells connected in series in a module. Below in Figure 2, the PV power produced $P = IV$ and the characteristic V–I curves are presented.

The operating ranges of the voltage of the PV system are 15 kWp under different levels of solar irradiation G (see Figure 2a). The maximum power tracking for the voltage is around 450 Vdc. Similarly, the nominal current of 36 A with different temperature values established in the model parameters is presented in Figure 2b. Finally, this model is integrated with a DC/DC converter and, subsequently, a DC/AC converter for connection to the PCC [28].

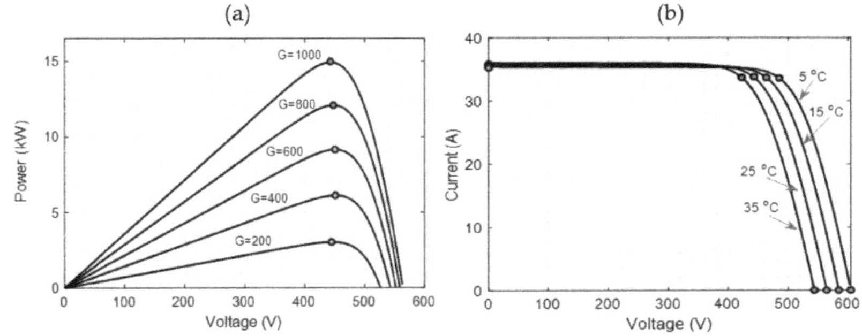

Figure 2. PV system model: (**a**) PV power, (**b**) I-V characteristics.

2.3. Electrolyzer Model

Electrolyzers can be modeled using various approaches. In this case, the chemical–electrical approach is adopted. The electrolyzer block represents the electrical load of the electrolyzer, which consists of an anode and a cathode separated by an electrolyte. This block calculates the hydrogen production based on the electrical energy input and the water temperature in the tank [30]. The electrolyzer block calculates the electrical power using Equation (4):

$$P_E = \frac{v^2}{R_T} \rightarrow P_t^{ETZ} \tag{4}$$

where v is the voltage and R_T is the total resistance given by Equation (5). The R_T, is defined by

$$\begin{aligned} \frac{1}{R_T} &= \frac{1}{R_{el}} + \frac{1}{R} \\ R_{el} &= \frac{\rho(pH) \times x}{A} \end{aligned} \tag{5}$$

where R is the external resistance, A is a resistive term obtained from the cross-sectional transport area, x is the anode–cathode distance, and $\rho(pH)$ is the electrical resistivity of the solution. The energy per mole stored over the electrical load facilitates water electrolysis and can be computed using Equation (6):

$$G_E = \frac{P_E}{\frac{1\ \text{mol}}{\text{s}}} \tag{6}$$

In performing electrolysis, the electrolyzer needs a minimum level of energy according to Equation (7):

$$G_{min} = \Delta H - \frac{T}{T_o} T_{\Delta s} \tag{7}$$

where ΔH is the water enthalpy, $T_{\Delta s}$ is the reaction entropy, T is the water temperature, and T_o is the reference temperature.

If there is not enough electrical energy to split the water molecules, the electrolyzer will not generate hydrogen. Consequently, the electrolyzer block determines Equations (8) and (9), the rates of electron, mole, and hydrogen production, respectively.

$$mol_e = \begin{cases} \frac{\mu(G_E - G_{min})}{e_V N_A} & G_e > G_{min} \\ 0 & G_e \leq G_{min} \end{cases} \tag{8}$$

$$mol_{H_2} = \begin{cases} \frac{N_c mole}{2} & G_e > G_{min} \\ 0 & G_e \leq G_{min} \end{cases} \tag{9}$$

where N_A is the Avogadro constant, e_V is the energy per electron, N_c is the number of cells, and μ is the temperature-dependent efficiency of the electrolysis.

The current passing through the electrolyzer tank, as given by Equation (10), is determined using the following expression, where F represents the Faraday constant (96,485.3321 s·A/mol):

$$i = mol_e \times F \tag{10}$$

The electrolyzer block computes the mass rates of water consumption and hydrogen production using Equations (11) and (12), respectively:

$$\dot{m}_{H_2O} = mol_{H_2} M_{H_2O} \tag{11}$$

$$\dot{m}_{H_2} = mol_{H_2} M_{H_2} \tag{12}$$

where M_{H_2O} is the molar mass of water, and M_{H_2} is the molar mass of hydrogen.

2.4. *Application of Hydrogen as a Fuel Cell*

2.4.1. Simplified Fuel Cell Model

The following model is generic and can emulate the behavior of any fuel cell type fed with hydrogen and air (see Figure 3). The approach involves extracting data from the data sheet to perform simulations without the need for experimental tests on an actual stack [31].

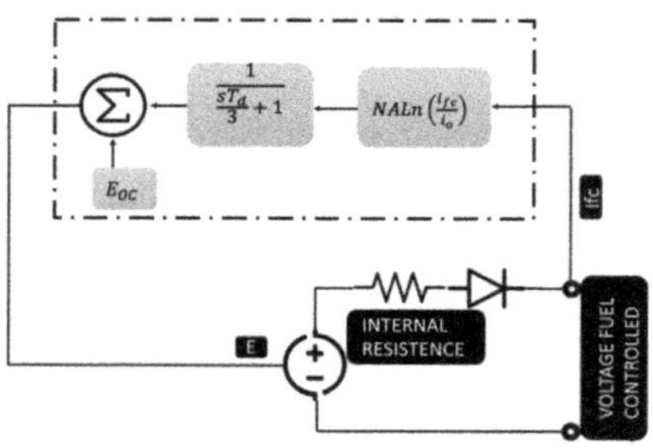

Figure 3. Simplified model representation of fuel cell.

Fuel cell models can be classified into three categories: chemical, experimental, and electrical. Chemical models include complex chemical and thermodynamic phenomena but require excessive parameters and cannot be easily added to electrical simulation programs [32–34]. Experimental models are derived from experiments and represent the fuel cell by look-up tables or empirical expressions [35,36]. Electrical models represent the fuel cell through electrical circuit elements [31]. There, the controlled voltage source E is given by Equation (13). It is important to note that E_{oc} represents open-circuit voltage, N the number of cells, A the Tafel slope, i_o the exchange current, T_d the response time at 95% of the final value in seconds, R_{ohm} the internal resistance, i_{fc} the FC current, and V_{fc} the FC voltage. Equation (14) refers to stack total voltage considering the losses due to electrodes and electrolytes [31].

$$E = E_{oc} - NAln\left[\left(\frac{i_{fc}}{i_o}\right) \times \frac{1}{\frac{sT_d}{3}+1}\right] \tag{13}$$

$$V_{fc} = E - R_{ohm} \times i_{fc} \tag{14}$$

2.4.2. The Detailed Fuel Cell Model

The detailed model accurately simulates a specific fuel cell stack by adjusting for changes in parameters such as pressure, temperature, composition, and flow rates of the fuel and air. These variations influence the Tafel slope (A), the exchange current (i_0), and the open-circuit voltage (E_{oc}). The equivalent circuit of the detailed model (illustrated in Figure 4) is identical to that of the simplified model, except that the parameters (E_{oc}, i_0, A) need to be updated based on the input pressures, flow rates, stack temperature, and gas compositions [31].

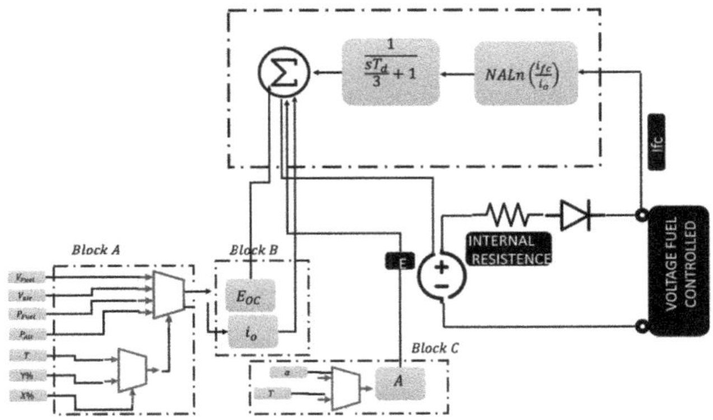

Figure 4. Detailed fuel cell stack model.

According to the detailed model, the next set of equations has been modified:

$$E_{oc} = K_C \times E_n \tag{15}$$

$$i_o = \frac{zFk(P_{H_2} + P_{O_2})}{Rh} \times \exp(\frac{-\Delta G}{RT}) \tag{16}$$

$$A = \frac{RT}{z \times \alpha \times F} \tag{17}$$

where R represents the unit of molar entropy, F the Faraday constant, z the number of moving electrons (two in this case), E_n the Nernst voltage, which is the thermodynamics voltage of the cells and depends on the temperatures and partial pressures of reactants and products inside the stack, α the charge transfer coefficient, P_{H_2} the partial pressure of hydrogen inside the stack, P_{O_2} the partial pressure of oxygen inside the stack, k Boltzmann's constant, h Planck's constant, ΔG the activation energy barrier, T the temperature of operation, and K_C the voltage constant at nominal conditions.

As illustrated in Figure 4, Blocks A, B, and C are used to calculate new values for E_{oc}, i_0, and A. Initially, Block A determines the conversion rates (utilization) of hydrogen (U_{fH_2}) and oxygen (U_{fO_2}) as follows:

$$U_{fH_2} = \frac{60000R \times Ti_{fc}}{z \times F \times P_{fuel} \times V_{fuel} \times x\%} \tag{18}$$

$$U_{fO_2} = \frac{60000R \times Ti_{fc}}{2 \times z \times F \times P_{air} \times V_{air} \times y\%} \tag{19}$$

In Equation (18), P_{fuel} represents the absolute supply pressure of fuel, P_{air} the absolute supply pressure of air, V_{fuel} the fuel flow rate, V_{air} the air flow rate, $x\%$ the percentage of hydrogen in the fuel, and $y\%$ the percentage of hydrogen in the oxidant.

Block B, shown in Figure 4, determines the partial pressures and the Nernst voltage as follows:

$$P_{H_2} = (1 - U_{fH_2}) \times x\% P_{fuel} \tag{20}$$

$$P_{O_2} = (1 - U_{fO_2}) \times y\% P_{air} \tag{21}$$

$$P_{H_2O} = (w + 2y\% \times U_{fO_2}) \times P_{air} \tag{22}$$

$$E_n = 1.299 + (T - 298) \times \frac{-44.43}{zF} + \frac{RT}{zF} \ln\left(P_{H_2} \times P_{O_2}^{1/2}\right) \tag{23}$$

for $T > 373$ K:

$$E_n = 1.299 + (T - 298) \times \frac{-44.43}{zF} + \frac{RT}{zF} \ln\left(\frac{P_{H_2 \times P_{O_2}^{1/2}}}{P_{H_2O}}\right) \tag{24}$$

In Equation (22), P_{H_2O} represents the partial pressure of water vapor and w the percentage of water vapor in the oxidant. The nominal value and Nernst voltage E_n can also be calculated from the Equation (24), where T represents temperature operation, R refers to the amount of energy needed to raise the temperature of one mole of a substance by one kelvin, and finally z is the number of moving electrons.

By knowing the partial pressures of gases and the Nernst voltage, the new values for the open-circuit voltage and the exchange current can be calculated using Equation (13) and (16), respectively. Block C then calculates the new Tafel slope value using Equation (17).

To model the effect of oxygen depletion (due to air compressor delay) on the stack voltage, parameters for flow dynamics such as the peak utilization (Uf_{O_2}(peak)) and the corresponding voltage undershoot (Vu) are needed. Oxygen depletion within the cell increases its utilization above the nominal value, and the Nernst voltage is subsequently adjusted as follows.

2.4.3. Model Assumptions and Limitations

The following assumptions and limitations have been considered in the development of the simulation:

- Gases are assumed to behave as ideal gases.
- The stack is supplied with hydrogen and air.
- A cooling system ensures stable temperatures at the cathode and anode exits, matching the stack temperature.
- A water management system maintains appropriate humidity levels within the cell under all load conditions.
- Pressure drops across flow channels are assumed to be negligible.
- Voltage drops within the cell are primarily due to reaction kinetics and charge transport, as fuel cells typically do not operate in the mass transport region.
- The cell's resistance is assumed to remain constant under all operating conditions.
- Gas or water flow through the membrane is not accounted for.
- The effects of membrane temperature and humidity on stack resistance are not considered.

2.5. *Energy Storage Model*

Energy storage systems have proven to be one of the leading solutions when there is a shortage of renewable energy generation in order to balance demand with generation. Its application is of great importance to achieve the correct performance of all elements in the electrical grid. In this context, the use of batteries is based on the production of green hydrogen with minimal or zero consumption from the grid. Excess PV energy is stored in the batteries during peak power production hours. Then, the accumulated energy is returned in the case of demand from the electrolyzer. This guarantees optimal performance for hydrogen production without interruption. Although each ESS depends on constructive and technological characteristics, the control of the SoC is exclusively detailed and can be applied to any ESS. Therefore, the SoC of the batteries is analyzed through simulation in the following way.

The ESS allows charging or discharging in random sequences with any amplitude between 0 and 100% of its effective capacity. The ESS control algorithm to limit the SoC is shown in Figure 5, where the value of P_t^{ref} is initially assigned as the difference between the photovoltaic power P_t^{PV} and the electrolyzer demand power P_t^{ETZ}, in such a way that being a value greater than zero, it allows the charging of the P_t^{ESS} from a time $(\Delta t_{charge} = t_2 - t_1)$ in the positive cycle and defines $P_t^{charge} = P_t^{ref}$. Subsequently, the charge energy value Ec_t^{ESS} is calculated as the integral of the charge power. The same is applied for the discharge energy, assigned for the negative cycle Ed_t^{ESS}. The total energy of the ESS at E_t^{ESS} during a time t_n and t'_n varies as a function of the difference in the charge and discharge energy, respectively. This procedure is summarized in Equation (25):

$$\begin{cases} P_t^{ref} = P_t^{PV} - P_t^{ETZ} \\ Ec_t^{ESS} = \int_{t_1}^{t_2} P_t^{charge} dt \rightarrow P_t^{ref} \geq 0 \\ Ed_t^{ESS} = \int_{t'_1}^{t'_2} P_t^{discharge} dt \rightarrow P_t^{ref} < 0 \end{cases} \quad (25)$$

Then, the value of the total energy in the storage is calculated as follows:

$$E_t^{ESS} = \eta_c \times Ec_t^{ESS} - \frac{1}{\eta_d} \times Ed_t^{ESS} \quad (26)$$

where η_c is charging efficiency and η_d is discharging efficiency. Then, the calculation to determine the SoC is calculated with the maximum storage energy E_{max}, that is,

$$SoC_t^{ESS} = SoC_{t-1}^{ESS} + \frac{E_t^{ESS}}{E_{max}} \times 100\% \quad (27)$$

The SoC restriction is limited in its maximum SoC_{max} and minimum SoC_{min} values allowed according to the following equation:

$$\begin{cases} P_t^{ESS} = P_t^{ref} \rightarrow (SoC_{min} \leq SoC_t^{ESS} \leq SoC_{max}) \\ P_t^{ESS} = 0 \rightarrow (SoC_{max} < SoC_t^{ESS} < SoC_{min}) \end{cases} \quad (28)$$

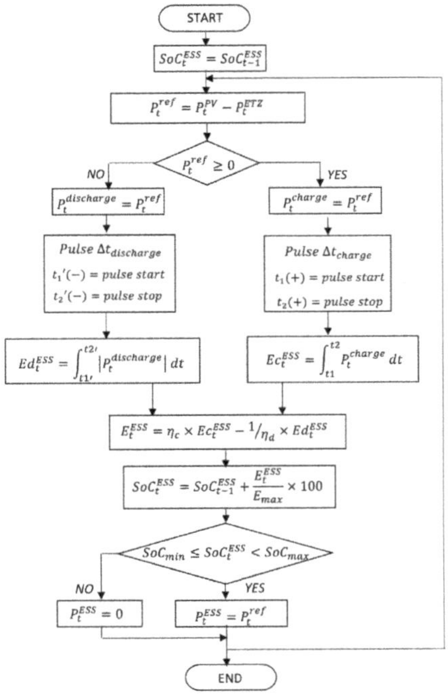

Figure 5. ESS state of charge control algorithm.

3. Experimental Phase with Electrolytes

During the experimental phase for obtaining hydrogen fuel by electrolysis of water, the research was based on the results of a preliminary study [37]. Initially, solutions were prepared with sodium hydroxide (NaOH) at a 1% concentration, sulfuric acid (H_2SO_4) at a 37% concentration, and sodium bicarbonate ($NaHCO_3$) at a 1% concentration. It is worth noting that the electrochemical analysis was carried out in the Laboratory of the Faculty of Chemical Sciences. The results are summarized in Table 1, which presents the outcomes of the experimentation with sulfuric acid for electrodes both with a size of 50 cm^2.

According to Figure 6a for an electrode area of 50 cm^2 and Figure 6b for an electrode area of 100 cm^2, it can be confirmed that at the moment hydrogen is generated, its weight is approximately double that of oxygen, as demonstrated by the graphs. Furthermore, as the distance between electrodes decreases from 39 cm to 1 cm, hydrogen production increases. In Figure 6a, there is a peak of 606 cubic centimeters (c.c.) at a concentration of 26.92%. Meanwhile, for Figure 6b, the maximum limit reaches 1217 c.c. at a concentration of 24.21%.

Based on the study of the ELS500 Hydrogen Generator electrolyzer of the Micro-Grid of the University of Cuenca, to prepare a 1% potassium hydroxide (KOH) solution in 4 L of water, we dissolved 40 g of KOH. This is based on the following Equation (29):

$$Concentration = \frac{\text{mass of solute}}{\text{volume of solution}} \times 100\% \tag{29}$$

For a 1% concentration, the mass of KOH should be 1% of the solution volume in milliliters (4000 mL), resulting in 40 g of KOH required to achieve the desired concentration in 4 L of water.

Table 1. Experimental results with H_2SO_4, electrode area 50 cm^2, electrolyte volume 2 lt, and stainless steel electrodes.

					Distance		cm											
			39	30	20	10	1	39		30		20		10		1	
								Vol (cc)		Vol (cc)		Vol (cc)		Vol (cc)		Vol (cc)	
°C	%	F		Current	mA			H_2	O_2	H_2	O_2	H_2	O_2	H_2	O_2	H_2	O_2
25	37.4	1.19	345	363	470	607	857	182	90	208	105	278	138	366	180	517	255
25	32.16	1.19	333	378	464	619	904	175	86	220	108	276	138	372	182	550	274
25	31.46	1.17	303	332	426	577	822	161	80	189	94	252	125	346	171	496	247
25	29.11	1.16	341	393	486	642	954	178	89	227	113	288	143	385	189	581	291
25	28.14	1.15	345	403	495	644	978	183	91	232	113	294	145	388	191	591	294
25	26.92	1.14	365	422	513	718	1003	194	95	246	122	304	151	432	214	606	303
25	25.35	1.14	364	415	512	682	989	191	95	242	117	302	150	410	202	602	299
25	24.21	1.13	349	405	489	641	968	183	92	236	116	289	141	385	190	584	290
25	23.52	1.11	322	377	461	644	944	170	83	219	107	273	136	386	191	572	286
25	22.20	1.14	331	365	450	621	923	175	88	210	104	267	131	373	183	561	279
25	20.54	1.11	294	333	405	544	821	157	78	190	95	240	119	327	163	498	245
25	18.35	1.13	259	304	377	506	788	136	68	173	87	223	111	304	151	479	237
25	15.64	1.11	233	278	344	472	722	124	62	158	79	203	102	284	141	436	217
25	13.98	1.11	222	261	316	416	633	118	58	151	74	188	93	251	124	384	189
25	12.24	1.21	205	247	295	392	578	108	53	144	71	175	86	235	116	351	174
25	10.49	1.19	179	214	274	357	524	95	48	123	61	162	80	215	107	319	158
25	9.18	1.19	167	190	238	321	454	89	43	108	54	141	70	193	95	275	137
25	7.87	1.19	143	173	208	274	387	75	38	99	49	123	61	165	82	234	116
25	6.29	1.19	119	143	179	232	333	63	30	83	41	106	53	140	69	203	100
25	5.07	1.19	107	129	155	202	262	57	27	75	37	92	46	122	60	159	79
25	3.93	1.16	87	102	125	168	212	45	23	58	30	74	36	101	50	129	64
25	3.32	1.16	70	85	102	133	174	37	19	49	25	60	30	80	39	105	52
25	2.45	1.16	58	67	85	108	143	31	15	38	19	50	25	65	31	87	44
25	1.87	1.16	44	56	67	87	110	23	11	32	16	40	20	52	26	67	33

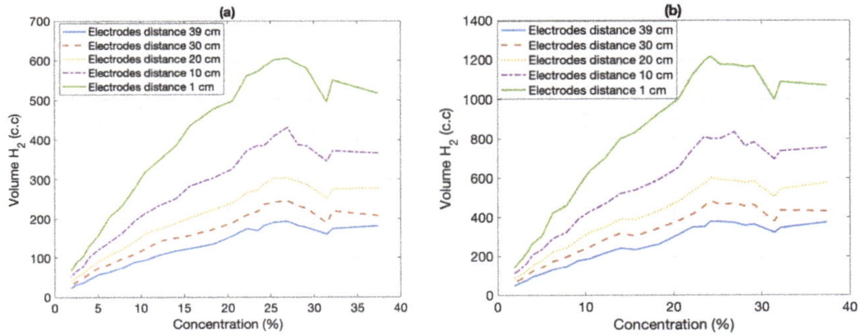

Figure 6. Hydrogen production with H_2SO_4 electrolyte volume 2lt, and stainless steel electrodes: (**a**) electrode area 50 cm^2, (**b**) electrode area 100 cm^2.

Finally, in Figure 7, the three electrolytes under experimentation, $NaHCO_3$, NaOH, and H_2SO_4, are synthesized. The experimental results indicate that the best electrolyte is sulfuric acid, with a 1 cm electrode spacing, an electrode area of 100 cm^2 each, and a hydrogen production of 1217 cc. The lowest quality electrolyte was sodium bicarbonate ($NaHCO_3$). From the practices, it was determined that hydrogen production is also proportional to the electrode area; for example, with NaOH, an electrode area of 50 cm^2 produces 298 cc, while an area of 100 cm^2 produces 596 cc of hydrogen.

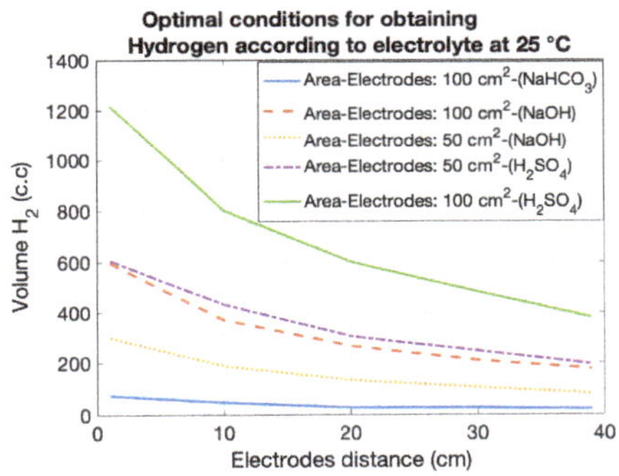

Figure 7. Optimal conditions for obtaining hydrogen with the electrolyte at 25 °C.

4. Simulation Results and Discussion

4.1. Micro-Grid Case Study

The following Figure 8 details the micro-grid equipment used in this study, which is part of the Micro-Grid Laboratory of the Universidad de Cuenca, Ecuador. It consists of a 15 kW PV system with 60 modules from the Atersa brand (Elecnor, Valencia, Spain), Samsung 44 kWh lithium batteries (Samsung, Gyeonggi-do, Republic of Korea) as the ESS, includes a 2-level bi-directional power converter, GPTech (Spain), and the Heliocentris EL500 electrolyzer (Italy) (see Tables 2 and 3), which produces 500 Nl/h at a pressure of 30 bar, with a purity of 99.94% from distilled water and a constant load of 2.39 kW to the PCC with the electrical grid. The hydrogen produced as compressed gas is stored in four tanks (50 L each) at a maximum pressure of 200 bars (Spain) [27].

The data for this simulation are presented in the following table.

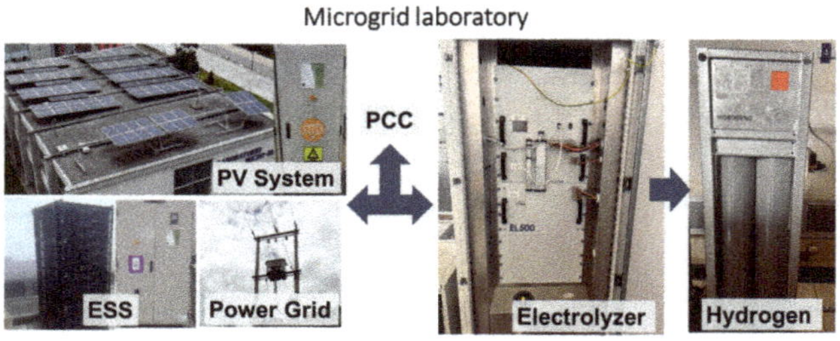

Figure 8. Case study at the Micro-Grid Laboratory, Universidad de Cuenca, Ecuador.

Table 2. Integrated micro-grid equipment.

Description	Parameter Value	Unit
PV System (60 modules 15 × 4 array)	15	kW
Lithium battery bank (11 Cells)	44	kWh
EL500 electrolyzer	2.39	kW
Bi-directional inverter	500 Vdc/230 Vac	V
Power grid	150	kVA
Storage tanks (4 × 50 L)	200	bars

Table 3. Hydrogen production data and equipment parameters.

Description	Parameter Value	Unit
Type of Electrolyzer	1	PEM
H_2 production rate	8.33	L/min
Working pressure	30	bar
Water consumption	0.4	L/h
Water specification	10	μS/cm
Power supply options	AC 220 V–50 Hz DC 40 V–80 V	V
Operative power consumption	2390	W
Universal gas constant R	0.082	L·atm/mol·K
Voltage supply	48	V
Current density	0.2	A/cm^2
Temperature	298	K
Number of cells	72	
Faraday's constant	96,485	C/mol
Electrochemical equivalent	2	cm
Area	76.1362	cm^2
Current I	15.2272	A
Voltage cell	2.4	V
Ohmic resistance of electrolyte R_u	8.05×10^{-5}	Ω·m^2
Ohmic resistance of electrolyte R_d	-2.5×10^{-7}	Ω·m^2
Over temperature shut down	65	°C
V_{ohmic}	0.1486	V
System efficiency	60 to 80	%

The following Algorithm 1 details the steps and configuration for energy control and management. The input parameters to the algorithm required as the measurement points are established according to the scheme in Figure 1. For this, the Modbus communication stations of the equipment must be initialized at a set time. Subsequently, these values are read and processed in real time for the reference values towards the ESS inverter [27,38]. The SoC control limits the energy delivery capacity autonomously, thus offering security during the management system. The system response is practically instantaneous, so the PV energy is stored in the ESS after covering the demand of the electrolyzer.

Algorithm 1 Steps and configuration for energy control and management

Data input: P_t^{PV}, P_t^{ETZ}, P_t^{ESS}, SoC_t^{ESS}
while: $t = 1 : N$ iteration
  Start
  TimerVal = tic
  Start Modbus communication
  m = modbus(Transport, 'Port', Name, Value)
  Read data from a Modbus server
  read(m, target, address, count, serverId, precision)
  $P_t^{ref} = P_t^{PV} - P_t^{ETZ}$ Equation (3)
  If $SoC_{min} \leq SoC_t^{ESS} \leq SoC_{max}$
    $P_t^{ESS} = P_t^{ref}$ Equation (6)
  else
    $P_t^{ESS} = 0$ Equation (6)
  end
  Perform a write operation to the connected
  Modbus server ESS value
  write(m, target, address, values, serverId, 'precision')
end

4.2. Techno-Economic Optimization

In this section, the simulation results are presented according to the scheme in Figure 1 for the production of green hydrogen. In the simulator configuration, steps of minutes, hours, and days are added. The equipment used for the simulation was OPAL-RT Real-Time Simulation (Montreal, QC, Canada), integrating a 4-core, 3.7 GHz, 16 GB RAM, 250 GB SSD Intel® Xeon® E3 computer [39], in which a 2.39 kW electrolyzer profile and a 15 kWp photovoltaic system were assigned. The specific considerations for the technical–economic analysis are as follows:

- Annual input database of solar irradiation, temperature, and maximum storage energy.
- Conducted across a long duration of time at regular intervals ranging from 1 min to 1 h.
- Relies on steady-state operational conditions at each time interval instead of system dynamics.
- Uses a quasi-steady-state simulation.
- Effect of fluctuating PV power as a function of solar irradiation and response of power converters.
- Restriction of battery state of charge based on maximum operating capacity.
- Operating condition of the electrolyzer for 16 h without interruption.

4.2.1. Simulation with Hourly Step Configuration

During the operational simulation of the step-based system configured per hour, the electrical parameters of voltage (V), current (A), and power (kW) were recorded at the different measurement points. Figure 9a,b show the results at the PV system output points and electrolyzer demand, respectively, where a constant voltage can be observed on the 230 Vac in the PCC three-phase system (V_{R-S-T}). In Figure 9a, the PV system current responds almost proportionally to solar irradiation and power as the product of voltage and current. In addition, the electrolyzer current (I_{R-S-T}) remains practically constant due to the consumption of 2.39 kW over 16 continuous hours (see Figure 9b).

On the other hand, Figure 9c,d show the voltage, current, and power at the other measurement points of the energy storage and utility grid, respectively, where a negative power value indicates the support of the electrical grid or the discharge of the ESS; on the contrary, a positive value is the delivery of excess power to the grid or the charge of the ESS (see Figure 9c). In this case, zero consumption from the electrical grid can be seen in Figure 9d. Therefore, the system remains in generation with self-consumption.

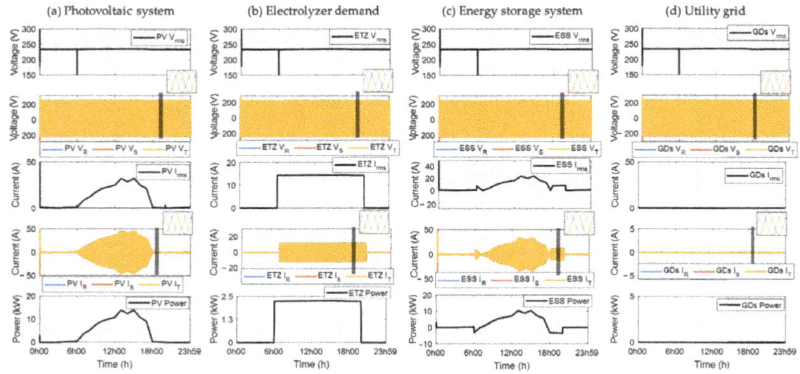

Figure 9. Simulation results: (**a**) PV system, (**b**) electrolyzer demand, (**c**) energy storage, and (**d**) utility grid with step of hours.

4.2.2. Simulation with Steps per Minute Configuration

The simulation results of the system based on steps configured per minute are presented in the following Figure 10. As can be seen, the sampling frequency of the system allows for a detailed analysis of changes in solar irradiation (see Figure 10a), which was not visible in the previous figure. The system response is coupled to more precise and optimal results for controller design and validation. However, the simulation time increases significantly. The parameters in Figure 10b do not show significant changes due to their constant electrolyzer demand value. However, the voltage level generates a variation due to the fluctuating penetration of the PV system. This directly affects all measurement points at PCC.

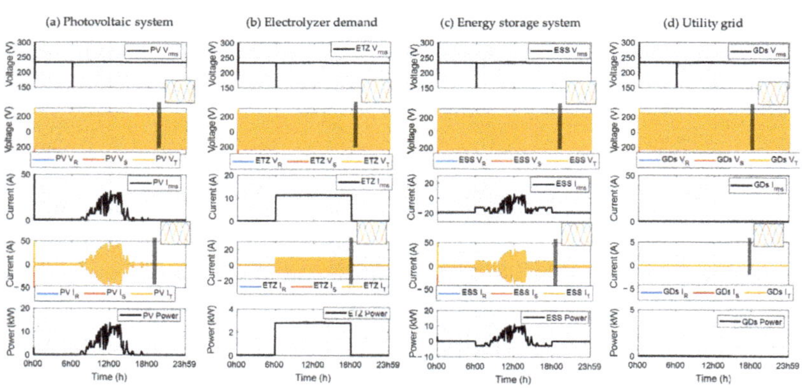

Figure 10. Simulation results: (**a**) PV system, (**b**) electrolyzer demand, (**c**) energy storage, and (**d**) utility grid with step of minutes.

Additionally, Figure 10c shows the simulation results at the ESS output, so it can be seen that the power that exceeds generation allows the charging of the ESS with $P_t^{ESS} < 0$. On the other hand, if the PV power is less than the electrolyzer demand, the batteries are discharged, delivering power to the grid, i.e., $P_t^{ESS} > 0$. In parallel, Figure 10d shows the null response from the utility grid due to the support of the PV system and the ESS. Voltage regulation can also be observed in PCC due to ESS support.

4.2.3. Simulation with Step Configuration by Months

Therefore, to demonstrate the feasibility, an analysis in steps of months is carried out for hydrogen production with the support of PV and ESS systems. In this context, an annual database is required for the technical–economic analysis. The data corresponding to the PV production and electrolyzer demand of one year in steps of minutes are presented in the following Figure 11.

The highest PV energy production is recorded during March, September, and October, and the lowest production months are June and December. Then, considering a constant electrolyzer load of 2.39 kW during the 16 h of operation, it requires an approximate monthly energy of 1147.2 kWh/month. Finally, the viability of the system can be related to PV production and the demand for the electrolyzer during the year. These values are presented in the following Figure 12.

As can be seen, during the ten months, the electrolyzer's demand is fully supplied, and the months of lower production (June and December) require energy from the grid. This results in a model with high expectations for the production of green hydrogen.

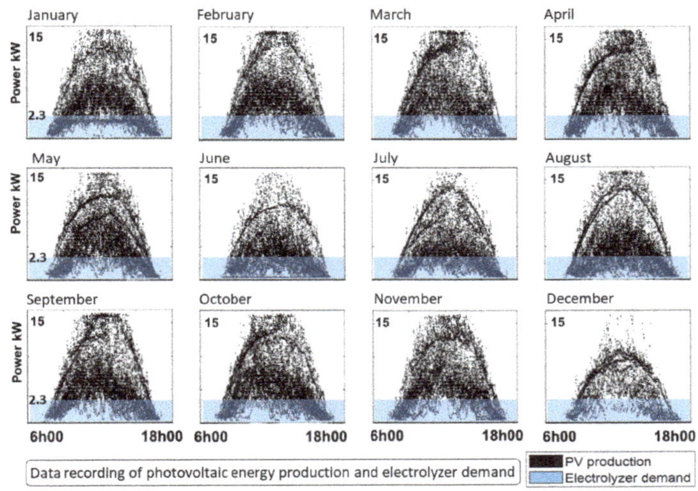

Figure 11. Monthly PV energy production and electrolyzer demand with step of minutes.

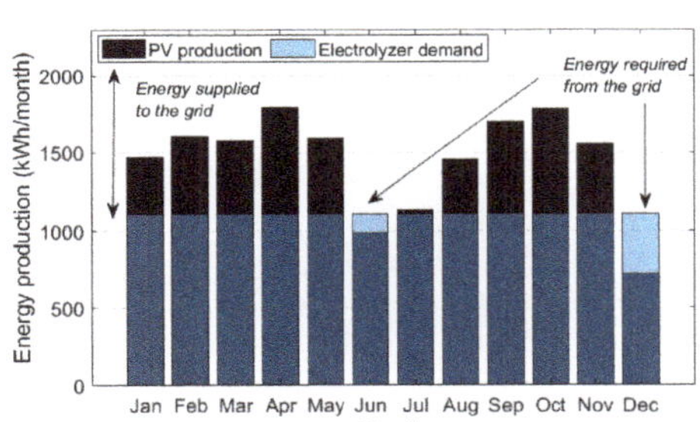

Figure 12. PV energy produced and demand of the electrolyzer monthly.

4.3. Analysis of Seasonal and Diurnal Variations in PV Performance

The variability of PV power is a crucial aspect to consider during solar power generation. This variability is influenced by weather conditions, the geographical location of the installation, and the times of day. The intermittency of solar radiation can cause these fluctuations in energy production. Therefore, in this section, the PV production versus hydrogen production is analyzed under three different climatic conditions. As mentioned above, green hydrogen production is carried out uninterruptedly. Therefore, the energy management system must maintain balance with the demand of the electrolyzer. As can be seen in the following Figure 13, the battery support allows for optimal energy distribution with minimum consumption from the electrical grid.

Figure 13a shows a medium variability in PV production, which has generated around 50 kWh against the 38.24 kWh required by the electrolyzer. The excess value of 23.5% has been supplied to the grid. In this case, the battery SoC has started at its low value, so during the first few hours, it reaches its minimum value and requires energy from the grid. As PV production increases throughout the day, the batteries store the excess energy for discharge during the night hours. On the other hand, in the second case, the variability of the PV production is high (see Figure 13b). The PV energy reaches 65 kWh, which is about 70% of what can be delivered to the grid. Due to the conditions of this day, the energy from the grid is not required and 55% is available in the SoC. Finally, in Figure 13c, a low variability in the PV production is presented, with the energy produced during this day being 62 kWh. Unlike the first case, the state of charge reaches values higher than 60% compared to 35% in the first case. It is worth mentioning that the values are limited by the storage capacity according to Figure 5, so the algorithm must respond to the SoC of the ESS. That is, if the reference power exceeds the SoC values, the power of the ESS must be 0. This can be seen in detail in the enlargement of Figure 13c.

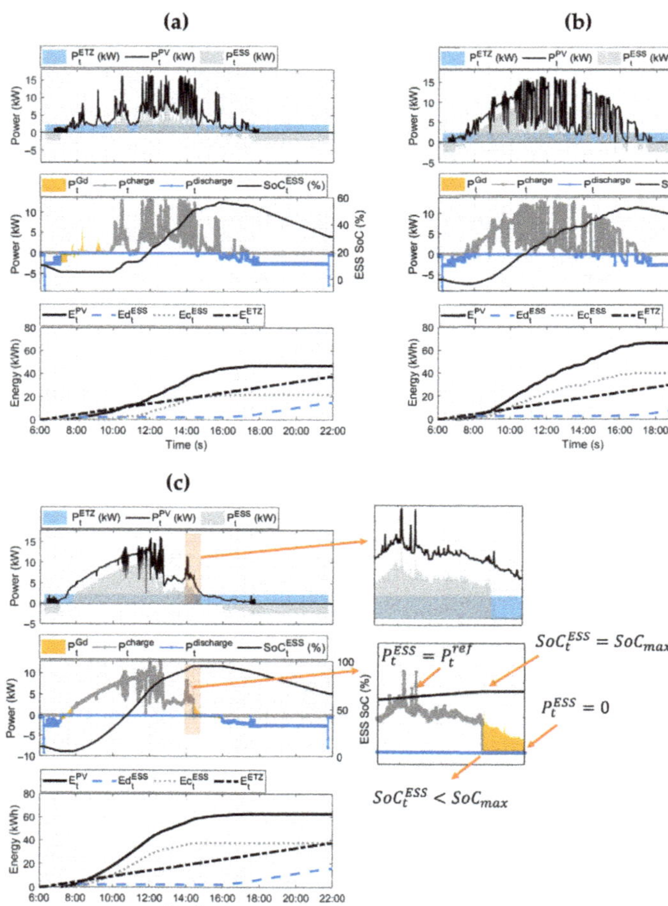

Figure 13. Analysis of seasonal variations in PV system: (**a**) medium variability, (**b**) high variability, (**c**) low variability.

4.4. Validation of Simulation Models Through Experimental Data

The following Figure 14 presents the details of experimental verification for the energy systems as a function of power. As can be observed, the variability of the PV system is due to the fact that the reference values for the model are solar radiation and ambient temperature. A scatter diagram has been included to verify the relationship between the simulation values and the real-time data. Similarly, the response time of the ESS to the variability of the PV power in the charging cycles is observed. The surplus of this energy is stored to be discharged during the night hours. In the latter case, the power response of the electrolyzer is practically a constant value during its operation; however, being a load connection/disconnection device, it causes a very short-duration transient effect. This effect is also absorbed directly by the ESS, which does not cause disturbances in the electrical grid. It should be noted that since these are low rate of change values, the scatter diagram adjusts the largest amount to 2.39 kW.

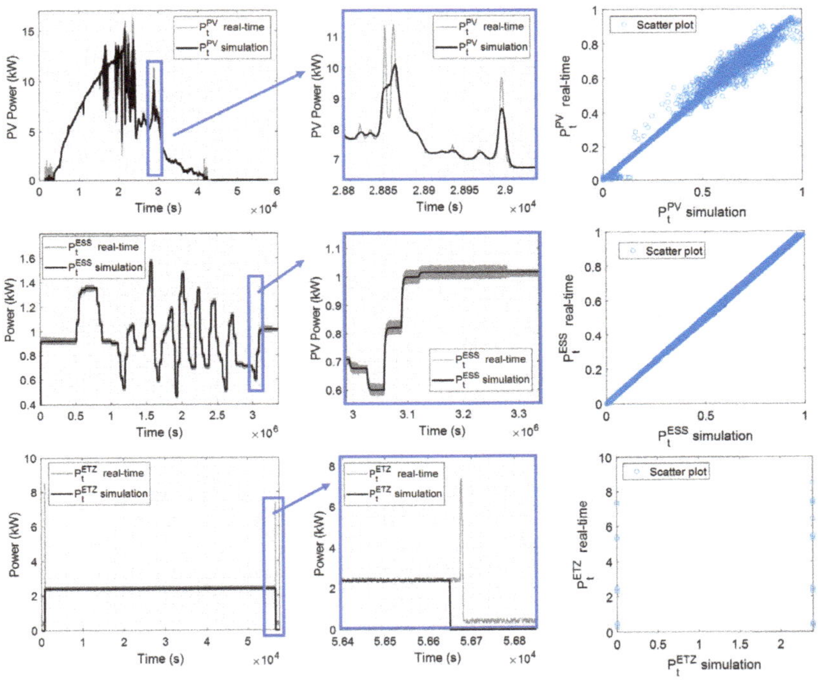

Figure 14. Validation of simulation models through experimental data.

4.5. Cost Analysis and Economic Feasibility

The price of producing green hydrogen is a key aspect for its large-scale adoption. Based on the simulations conducted according to Table 4, the results indicate that the energy consumption to produce one kilogram of hydrogen is 64 kWh. Additionally, the hydrogen production per day and year is 0.287 kg/day and 103.32 kg/year, respectively. On the other hand, a cost of 0.06 USD/day and 36 USD/year was considered as a reference for the price of water in the city of Cuenca, Ecuador. In this case, the required water consumption is 0.4 L per hour.

As shown in Figure 15a, the electrolyzer produces 0.287 kg of hydrogen per day while operating. It is important to note that the electrolyzer runs for 16 h per day and remains off for the remaining 8 h. Therefore, in this case study, the electrolyzer will produce a total of 103.32 kg of hydrogen annually.

Table 4. Costs and results from production.

Description	Parameter Value	Unit
Consumption of energy to produce hydrogen	64	kWh/kg
Production of hydrogen per day	0.287	kg/day
Production of hydrogen per year	103.32	kg/year
Cost of using water	0.06	USD/day
Cost of using water	36	USD/year
Water consumption	0.4	L/h
Water specification	10	uS/cm

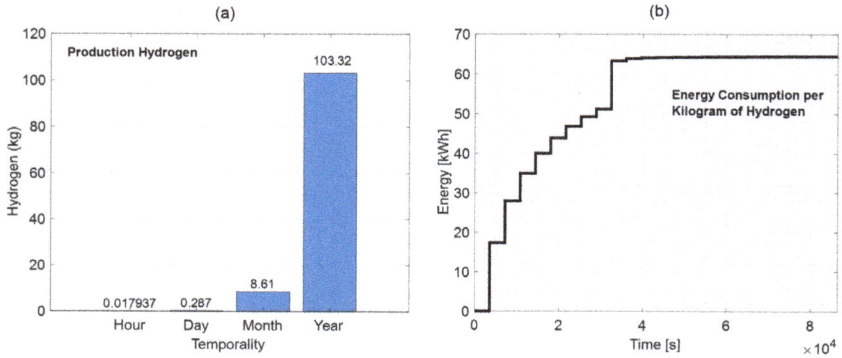

Figure 15. (**a**) Hydrogen production over time: daily to yearly. (**b**) Energy consumption per kilogram of hydrogen (kWh/kg).

In addition, based on Figure 15b, the energy consumption to produce 1 kg of hydrogen is expected to stabilize at 64 kWh after the first 9 h. In this case study, conducted in the city of Cuenca, Ecuador, where the cost of electricity is USD 0.09 per kWh, the production cost for one kilogram of hydrogen would be approximately USD 5.76.

Tables 5 and 6 present the electricity prices per hour in Ecuador for residential, commercial, and industrial consumers [40]. Consequently, the EL-500 electrolyzer consumes 2.382 kWh when operated for 16 h per day, resulting in a total daily demand of 38.08 kWh. Considering the manufacturer's reported efficiency range of 60% to 80%, the net energy consumption is between 22.08 kWh and 30.46 kWh. Additionally, during the experimental phase, it was observed that hydrogen storage at 200 bar in a 50-liter tank provided 8.919 kWh over 3 h, 27 min, and 59 s, When the fuel cell is operating at nominal power with the characteristics of Table 7 of the Ecobox-MR30S1 model (Singapore). This corresponds to a round-trip efficiency of 40.45%, which aligns with values reported in the literature [41].

Table 5. Hourly rate schemes and voltage levels.

Category	Voltage Level	Consumer Group	Demand Registration
Residential	Low Voltage LV <600 V	Residential	No demand
		Commercial	No demand With demand With hourly demand
		Industrial	No demand With demand With hourly demand

Table 6. Electricity prices hour for Ecuador.

Consumption Range (kWh)	Energy Cost (USD/kWh)	Distribution (USD/Costumer)
1 to 50	0.091	
51 to 100	0.0931	
101 to 150	0.095	
151 to 200	0.097	
201 to 250	0.099	
251 to 300	0.101	
301 to 350	0.103	1.414
351 to 500	0.105	
501 to 700	0.1285	
701 to 1000	0.1450	
1001 to 1500	0.1709	
1501 to 2500	0.2752	
2501 to 3500	0.4360	
Greater than 3501	0.6812	

Table 7. Fuel cell Ecobox-MR30S1.

Description	Parameter Value	Unit
Power Rating	0 to 3000	W
Current Rating	0 to 56	A
Voltage Adjustable	43.2 to 57.6	VDC
Voltage Ripple	±1	V
Ambient Temperature	−10 to 45	Centigrates
Relative Humidity	0 to 95%	Non-condensing
Consumption	≤0.78	$\frac{Nm^3}{kWh}$
Efficiency of System	≥42%	N/A

4.6. Analysis Considering CAPEX and OPEX

The analysis of capital expenditure (CAPEX) and operating expenditure (OPEX) is a fundamental practice in the economic evaluation of projects. Table 8 shows all the information required for the calculation of capital and operating costs.

Table 8. Cash flow: CAPEX and OPEX.

System Type	Model	Cells/Modules	Capacity	Unit Price	Capital Cost
PV	A-250P GSE Atersa	60	15 kW	800 USD/kW	12,000 USD
Electrolyzer	EL500 Acta S.p.A	1	2.3 kW	588 USD/kW	1352.4 USD
Battery	Lithium battery bank	11 Cells	44 kWh	290 USD/kWh	12,840 USD
Total Capital Cost					26,192.4 USD
Operating cost	For one year				
Labor	0.3% of total capital cost	1			78.57 USD/year
Distilled water	Ecuador distilled	1	0.4 L/H	0.66 USD/L	1520 USD/year
Maintenance	2% of total capital cost	1			523.84 USD/year
Total operating cost		1			2122.41 USD/year

Based on the values in Table 6 and Figure 12, the profit from the sale of energy to the Ecuadorian electrical system is calculated. These results are presented in the following Figure 16. As can be seen, there is a surplus in energy sales in most months, except in June and December, when additional energy must be purchased from the grid. Annual revenue from electricity sales amounts to $367.63.

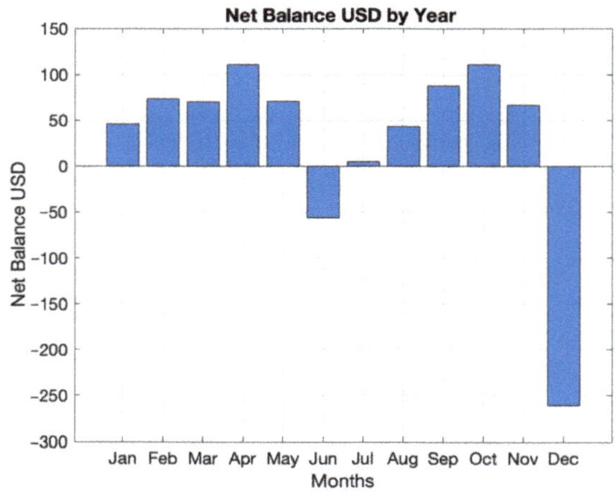

Figure 16. Net cost in the annual energy balance.

Using the cash flow balance and net income from energy sales, the annual cost of the hydrogen production project is calculated according to Equation (30):

$$\begin{aligned} \text{Total Annual Cost (year)} &= \text{Capital Cost} + \text{Annual Operatings Cost} \\ &\quad - \text{Annual Grid Revenue} \\ \text{Total Annual Cost (year)} &= \$\,26192.4 + \$\,2122.41 - \$\,367.62 = \$\,27947.19 \end{aligned} \quad (30)$$

Considering capital costs along with operation and maintenance expenses, the cost per kilogram of hydrogen is approximately

$$\begin{aligned} \text{Unit } H_2 \text{ production Cost} &= \frac{\text{Total anuual Cost}}{\text{Annual } H_2 \text{ Production}} \\ \text{Unit } H_2 \text{ production Cost} &= \frac{27947.19}{103.32\text{ kg}} \\ \text{Unit } H_2 \text{ production Cost} &= \$\,270/\text{kg} \end{aligned} \quad (31)$$

In assessing the economic viability of hydrogen production, the levelized cost of hydrogen (LCOH) based on Equation (32) is used to calculate the average cost of hydrogen production over the entire life of the project. This includes capital, operation, and maintenance costs, as well as the cost of necessary inputs such as electricity and water.

$$\begin{aligned} \text{LCOH} &= \frac{\text{Net Present Value of total Cost (CAPEX plus OPEX) over the lifetime of the plant}}{\text{Net Present Value of total hydrogen production over the lifetime}} \\ &= \frac{27947.19}{103.32\times\ 25} = \frac{\$10.81}{\text{kg}} \end{aligned} \quad (32)$$

Over the 25-year lifetime, the levelized cost of hydrogen under these conditions is compared with international hydrogen production markets. To overcome these limitations, the authors propose the following strategies.

4.6.1. Strategy 1: Offer Subsidy for Green Hydrogen Production

This strategy will enable the Ecuadorian state, through regulatory and control agencies, to analyze the appropriate price to be paid for photovoltaic generation in the production of green hydrogen (Figure 17).

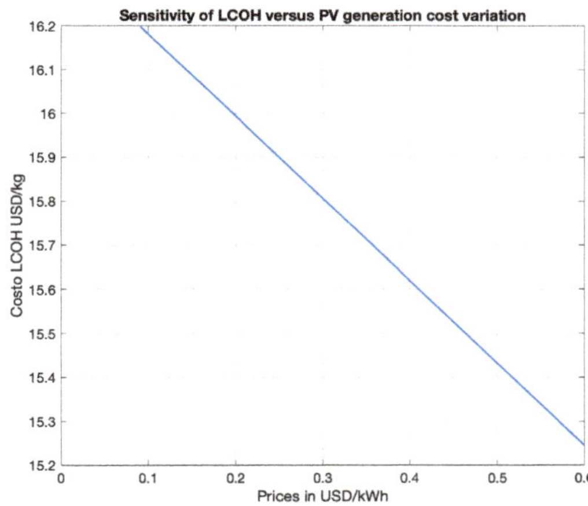

Figure 17. Sensitivity of LCOH versus PV generation cost variation.

4.6.2. Strategy 2: Variable Price per kW for the Electrolyzer

The second strategy involves maintaining an average price of 0.35 cUSD/kWh while varying the price per kW for the electrolyzer (Figure 18).

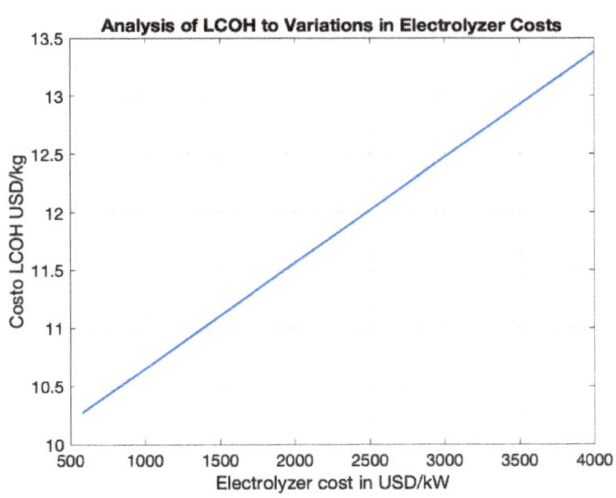

Figure 18. Analysis of LCOH versus variations in electrolyzer costs.

4.7. Comparative of Hydrogen Using Natural Gas

On the other hand, a comparative analysis has been carried out on the production of hydrogen from natural gas. For this purpose, the process was modeled through steam methane reforming, which is the most common method for producing hydrogen from natural gas, which involves two main reactions:

1.—Methane reforming:

$$CH_4 + H_2O \rightarrow CO + 3H_2 \quad (33)$$

2.—Water–gas shift reaction (optional):

$$CO + H_2O \rightarrow CO_2 + H_2 \quad (34)$$

Using the molecular weight of hydrogen $M_{H_2} = 2.016\ \frac{g}{mol}$, we convert 0.287 kg of hydrogen to moles:

$$\text{Moles of } H_2 = \frac{287\ \text{g/day}}{2.016\ \text{g/mol}} = 142.3\ \text{mols/day} \tag{35}$$

The second step was to calculate the required methane and water. Since 1 mole of methane produces 3 moles of hydrogen in the stem methane reforming (SMR) process, the amount of methane required is

$$\text{Moles of } CH_4 = \frac{\text{Moles of } H_2}{3} = \frac{142.3}{3} = 47.3\frac{\text{moles}}{\text{day}} \tag{36}$$

The water requirement is the same as the methane requirement (based on a 1:1 stoichiometric ratio).

Figure 19 represents a simulation of hydrogen production via steam methane reforming (SMR). It can be observed that methane (CH_4), initially at approximately 50 moles, is rapidly consumed, reaching nearly zero within the first 100 s. The water (H_2O), similarly to methane, begins at around 50 moles and is consumed at a similar rate, reaching near-zero. The amount of carbon monoxide (CO) increases rapidly in the beginning but stabilizes at a constant value of approximately 37 moles after a short time. The carbon dioxide (CO_2) is produced in small amounts and stabilizes at a low concentration after an initial peak. Hydrogen is the main product of the reaction, increasing sharply in the first moments and stabilizing at around 142.36 moles, showing a successful hydrogen production via SMR, equivalent to 0.287 kg/day, which closely matches the results in Figure 15.

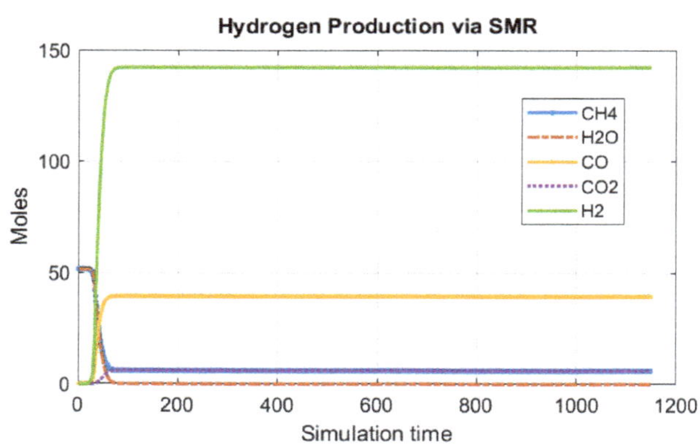

Figure 19. Energy consumption per kilogram of natural hydrogen gas as inlet.

The literature [42] indicates that hydrogen production via steam methane reforming (SMR) of natural gas (CH_4) remains the most cost-effective method for producing hydrogen gas today. Our aim in Table 9 was to make a comparison, and indeed, SMR continues to be the least expensive, even in terms of water consumption. However, it is evident that SMR has a significant environmental impact in terms of carbon emissions, an issue that photovoltaic hydrogen production mitigates entirely. In the context of Ecuador, carbon certificates can be traded, with the current price at USD 50 per ton of CO_2, providing an estimated USD 950 to avoid 19 tons of carbon emissions per year.

Table 9. Comparative hydrogen production with methane reforming SMR vs. PV generation.

Description	SMR Value	Proposal	Literature
Moles (mols/day)	47.3 (CH_4)	143.2 (H_2)	
Water	47.3 mols/day	355.5 mols/day	
(CO_2) emissions in tons in 25 years	19	0	[42]
Heat water to 1000 °C	1.034 kWh	0	
Required energy	18.194 kWh per day	38.9 KWh per day	

However, if we calculate the return on investment through carbon certificates at Ecuador's prices, we obtain an approximate payback period of 1.17 years under this mechanism as shown in the following Equation (37):

$$\begin{aligned} \text{Carbon certificate price (CCP)} &= 50\text{ USD} \\ \text{Total Annual Cost (TAC)} &= 27947.9\text{ USD} \\ \text{Mitigated Carbon Emissions (MCE)} &= 19\text{ Tons} \\ \text{project lifetime (PL)} &= 25\text{ year} \\ \text{Return on Investment} &= \frac{\text{TAC}}{\text{CCP}\times\text{MCE}\times\text{PL}} \\ &= \frac{27947.19}{50\times19\times25} = 1.17\text{ years} \end{aligned} \quad (37)$$

4.8. *Types of Models in Technology Readiness Levels*

Finally, a comparative table of types of models in technology readiness levels applied to reliability in green hydrogen is established. These levels help to identify the development status of different technologies and processes during the simulation. Consequently, they allow the evaluation and planning of the development of green hydrogen technologies for future applications. Table 10 summarizes the relevant points for the selection and type of simulation required under different configurations and analysis of the results of the system modeling.

Table 10. Discussion of the types of models in TRLs.

Types of Models in TRLs	Characteristics
Techno-Economic Optimization	Execution times: 86,400 s/day, 1440 m/day, 24 h/day (monthly analysis). Step configuration: 86,400 s, 1440 m, 24 h. Advantages: Perform feasibility analysis prior to equipment acquisition. Energy feasibility studies based on renewable resources. Low computational performance in simulations. Basic software. Requirement of databases with variable sampling. Ideal for generating budgets. Disadvantages: Results generated are approximate. At the design level, it can generate unforeseen events. Limited simulation events.
Detailed Technical Design	Execution times: ns, µs, ms Step configuration: 50 µs Advantages: Real operating environments. Optimization in control drivers. Power quality analysis. Technology improvement and validation. Disadvantages: High processing capacities. Requires real-time simulators. Increased simulation time.
Operational Optimization	Execution times: 86,400 s/day, 1440 m/day, 24 h/day (monthly analysis). Step configuration: 86,400 s, 1440 m, 24 h. Advantages: Conduct feasibility analysis to expand facilities. Allows for resource optimization studies. Establish economic operations and energy management. Generate long-term studies. Low computational performance in simulations. Programming control algorithms. Disadvantages: Results generated are approximate. Limitations for installed systems. Large-scale data recording requirement.

5. Conclusions

This experimental study demonstrated the impact of different electrolytes on hydrogen production efficiency, focusing particularly on sulfuric acid (H_2SO_4) at 37% concentration. Electrochemical testing, conducted in the Laboratory of the Faculty of Chemical Sciences, provided a comparative analysis using electrodes of 50 cm^2 and 100 cm^2, showing that sulfuric acid significantly enhances current density, thus optimizing hydrogen generation. The results indicate that higher electrode surface areas yield increased hydrogen output.

The production of green hydrogen based on a micro-grid with PV system support and ESS batteries allows for improving sustainability with minimum consumption from the electric grid. This article presents a technical–economic study based on a simulation model approach.

During the electrolyzer's operation, real-time simulation is necessary to de-risk overall technology development as we move closer to system deployment. In this paper, the fidelity of a hydrogen production model is analyzed under different parameterizations in the input variables.

Energy management for green hydrogen production in a micro-grid was presented as a case study. In this case, the electrolyzer's use has been optimized for 16 operating hours per day. Consequently, the micro-grid's 15 kW PV system, together with the ESS batteries, has covered the electrolyzer's annual demand (13,766.4 kWh/year). Despite having high PV generation, energy is required from the electrical grid for two months with low energy demand.

The current simulation fidelity proposal for hydrogen yields a value of 64 kWh/kg of hydrogen. Comparing this with the values reported by IRENA in 2023, which range from 46 to 66 kWh/kg, it is evident that our results fall within this established range.

In this case study conducted in the city of Cuenca, Ecuador, the subsidized electricity cost is approximately 9.20 cUSD/kWh. However, if the actual cost of electricity, which is 16.2 cUSD/kWh, were considered, the cost of producing one kilogram of hydrogen would increase significantly, from 5.76 USD/kg to 10.36 USD/kg.

Finally, using the methane reforming method for hydrogen production, 924 kWh is required to obtain 0.287 kg of hydrogen per day. In contrast, using the PV, ESS, and grid strategy, only 64 kWh is needed, which in Ecuador would represent a cost of USD 10.36/kg in our proposal, compared to USD 134/kg for the traditional method.

Author Contributions: A.C.: Visualization, Supervision, Software, Resources, Writing—Original Draft, Writing—Review and Editing, Methodology. L.I.M.-A.: Resources, Writing—Original Draft, Methodology, Investigation, Formal Analysis, Data Curation, Writing—Review and Editing, Validation. D.B.: Visualization, Supervision, Software, Resources, Writing—Original Draft, Writing—Review and Editing, Project Administration, Methodology. D.O.-C.: Resources, Writing—Original Draft, Methodology, Investigation, Formal Analysis, Data Curation, Writing—Review and Editing, Validation. M.T.-V.: Supervision, Software, Resources, Writing—Original Draft, Project Administration, Methodology, Investigation, Funding Acquisition, Visualization. W.K.M.: Resources, Writing—Original Draft, Methodology, Investigation, Formal Analysis, Data Curation, Writing—Review and Editing, Validation. F.J.: Project Administration, Methodology, Investigation, Funding Acquisition, Formal Analysis, Data Curation, Conceptualization, Visualization. All authors have read and agreed to the published version of the manuscript.

Funding: This research received no external funding.

Institutional Review Board Statement: Not applicable.

Informed Consent Statement: Not applicable.

Data Availability Statement: The original contributions presented in the study are included in the article, further inquiries can be directed to the corresponding author.

Acknowledgments: The authors wish to thank the Spanish Ministry of Science and Innovation and European Union NextGenerationEU/PRTR, under the research Project "Desarrollo de herramientas computacionales para microrredes multi-energía con vectores de electricidad, hidrógeno y gas" (TED2021-129631B-C31). The authors thank the Faculty of Engineering, Universidad de Cuenca, Ecuador, for access to the Micro-Grid Laboratory's facilities, allowing the use of its equipment, and authorizing its staff to provide the technical support necessary to carry out the experiments described in this article. This manuscript is an outcome of the research stay enjoyed by Adrian Criollo in Spain, who gratefully acknowledges Universidad de Cuenca for funding this scholarship and extends his gratitude to the Department of Electrical Engineering at Universidad of Jaén, Spain, for hosting his research stay, which was influential in developing this work.The icons used in this document were developed by Freepik, monkik, Smashicons, and Pixel perfect, from www.flaticon.com (accessed on 30 May 2024).

Conflicts of Interest: The authors declare no conflicts of interest.

References

1. Sharma, G.D.; Verma, M.; Taheri, B.; Chopra, R.; Parihar, J.S. Socio-economic aspects of hydrogen energy: An integrative review. *Technol. Forecast. Soc. Chang.* **2023**, *192*, 122574. [CrossRef]
2. Muhammed, N.S.; Gbadamosi, A.O.; Epelle, E.I.; Abdulrasheed, A.A.; Haq, B.; Patil, S.; Al-Shehri, D.; Kamal, M.S. Hydrogen production, transportation, utilization, and storage: Recent advances towards sustainable energy. *J. Energy Storage* **2023**, *73*, 109207. [CrossRef]
3. Damiri, A.P.; Stamatakis, E.; Bellas, S.; Zoulias, M.; Mitkidis, G.; Anastasiadis, A.G.; Karellas, S.; Tzamalis, G.; Stubos, A.; Tsoutsos, T. A Review of Alternative Processes for Green Hydrogen Production Focused on Generating Hydrogen from Biomass. *Hydrogen* **2024**, *5*, 163–184. [CrossRef]
4. Squadrito, G.; Maggio, G.; Nicita, A. The green hydrogen revolution. *Renew. Energy* **2023**, *216*, 119041. [CrossRef]
5. Ma, N.; Zhao, W.; Wang, W.; Li, X.; Zhou, H. Large scale of green hydrogen storage: Opportunities and challenges. *Int. J. Hydrogen Energy* **2024**, *50*, 379–396. [CrossRef]
6. Bayssi, O.; Nabil, N.; Azaroual, M.; Bousselamti, L.; Boutammachte, N.; Rachidi, S.; Barberis, S. Green hydrogen landscape in North African countries: Strengths, challenges, and future prospects. *Int. J. Hydrogen Energy* **2024**, *84*, 822–839. [CrossRef]
7. Moritz, M.; Schönfisch, M.; Schulte, S. Estimating global production and supply costs for green hydrogen and hydrogen-based green energy commodities. *Int. J. Hydrogen Energy* **2023**, *48*, 9139–9154. [CrossRef]
8. Makepeace, R.W.; Tabandeh, A.; Hossain, M.; Asaduz-Zaman, M. Techno-economic analysis of green hydrogen export. *Int. J. Hydrogen Energy* **2024**, *56*, 1183–1192. [CrossRef]
9. Hassan, Q.; Algburi, S.; Sameen, A.Z.; Salman, H.M.; Jaszczur, M. Green hydrogen: A pathway to a sustainable energy future. *Int. J. Hydrogen Energy* **2024**, *50*, 310–333. [CrossRef]
10. Mirza, U.K.; Ahmad, N.; Harijan, K.; Majeed, T. A vision for hydrogen economy in Pakistan. *Renew. Sustain. Energy Rev.* **2009**, *13*, 1111–1115. [CrossRef]
11. El-Shimy, M.; Afandi, A. Overview of Power-to-Hydrogen-to-Power (P2H2P) systems based on variable renewable sources. In Proceedings of the 5th International Conference on Electrical, Electronics, and Information Engineering, Malang, Indonesia, 6–8 October 2017; Volume 6.
12. Ngouleu, C.A.W.; Koholé, Y.W.; Fohagui, F.C.V.; Tchuen, G. Techno-economic analysis and optimal sizing of a battery-based and hydrogen-based standalone photovoltaic/wind hybrid system for rural electrification in Cameroon based on meta-heuristic techniques. *Energy Convers. Manag.* **2023**, *280*, 116794. [CrossRef]
13. Naderipour, A.; Abdul-Malek, Z.; Nowdeh, S.A.; Kamyab, H.; Ramtin, A.R.; Shahrokhi, S.; Klemeš, J.J. Comparative evaluation of hybrid photovoltaic, wind, tidal and fuel cell clean system design for different regions with remote application considering cost. *J. Clean. Prod.* **2021**, *283*, 124207. [CrossRef]
14. Kushwaha, P.K.; Bhattacharjee, C. Integrated techno-economic-enviro-socio design of the hybrid renewable energy system with suitable dispatch strategy for domestic and telecommunication load across India. *J. Energy Storage* **2022**, *55*, 105340. [CrossRef]
15. Fabianek, P.; Madlener, R. Techno-economic analysis and optimal sizing of hybrid PV-wind systems for hydrogen production by PEM electrolysis in California and Northern Germany. *Int. J. Hydrogen Energy* **2024**, *67*, 1157–1172. [CrossRef]
16. Posso, F.; Galeano, M.; Baranda, C.; Franco, D.; Rincón, A.; Zambrano, J.; Cavaliero, C.; Lópes, D. Towards the Hydrogen Economy in Paraguay: Green hydrogen production potential and end-uses. *Int. J. Hydrogen Energy* **2022**, *47*, 30027–30049. [CrossRef]
17. Ikuerowo, T.; Bade, S.O.; Akinmoladun, A.; Oni, B.A. The integration of wind and solar power to water electrolyzer for green hydrogen production. *Int. J. Hydrogen Energy* **2024**, *76*, 75–96. [CrossRef]
18. Abdelkareem, M.A.; Abdelghafar, A.A.; Mahmoud, M.; Sayed, E.T.; Mahmoud, M.S.; Alami, A.H.; Al Agha, M.M.; Olabi, A.G. Optimized solar photovoltaic-powered green hydrogen: Current status, recent advancements, and barriers. *Sol. Energy* **2023**, *265*, 112072. [CrossRef]
19. Gibson, T.L.; Kelly, N.A. Optimization of solar powered hydrogen production using photovoltaic electrolysis devices. *Int. J. Hydrogen Energy* **2008**, *33*, 5931–5940. [CrossRef]

20. Tuinema, B.W.; Adabi, E.; Ayivor, P.K.; García Suárez, V.; Liu, L.; Perilla, A.; Ahmad, Z.; Rueda Torres, J.L.; van der Meijden, M.A.; Palensky, P. Modelling of large-sized electrolysers for real-time simulation and study of the possibility of frequency support by electrolysers. *IET Gener. Transm. Distrib.* **2020**, *14*, 1985–1992. [CrossRef]
21. Dozein, M.G.; Jalali, A.; Mancarella, P. Fast frequency response from utility-scale hydrogen electrolyzers. *IEEE Trans. Sustain. Energy* **2021**, *12*, 1707–1717. [CrossRef]
22. Dozein, M.G.; De Corato, A.M.; Mancarella, P. Virtual inertia response and frequency control ancillary services from hydrogen electrolyzers. *IEEE Trans. Power Syst.* **2022**, *38*, 2447–2459. [CrossRef]
23. Dozein, M.G.; De Corato, A.M.; Mancarella, P. Fast frequency response provision from large-scale hydrogen electrolyzers considering stack voltage-current nonlinearity. In Proceedings of the 2021 IEEE Madrid PowerTech, Madrid, Spain, 28 June–2 July 2021; IEEE: Piscataway, NJ, USA, 2021; pp. 1–6.
24. Samani, A.E.; D'Amicis, A.; De Kooning, J.D.; Bozalakov, D.; Silva, P.; Vandevelde, L. Grid balancing with a large-scale electrolyser providing primary reserve. *IET Renew. Power Gener.* **2020**, *14*, 3070–3078. [CrossRef]
25. Conn, A.R.; Gould, N.I.; Toint, P.L. *Trust Region Methods*; SIAM: Philadelphia, PA, USA, 2000.
26. Misener, R.; Biegler, L. Formulating data-driven surrogate models for process optimization. *Comput. Chem. Eng.* **2023**, *179*, 108411. [CrossRef]
27. Espinoza, J.; Gonzalez, L.; Sempertegui, R. Micro grid laboratory as a tool for research on non-conventional energy sources in Ecuador. In Proceedings of the 2017 IEEE International Autumn Meeting on Power, Electronics and Computing (ROPEC), Ixtapa, Mexico, 8–10 November 2017; IEEE: Piscataway, NJ, USA, 2017; pp. 1–7.
28. Benavides Padilla, D.J.; Jurado, F.; González, L.G. Data analysis and tools applied to modeling and simulation of a PV system in Ecuador. *Enfoque UTE* **2018**, *9*, 1–12. [CrossRef]
29. Markvart, T.; Castañer, L. Chapter IA-1—Principles of Solar Cell Operation. In *Practical Handbook of Photovoltaics*, 2nd ed.; McEvoy, A., Markvart, T., Castañer, L., Eds.; Academic Press: Boston, MA, USA, 2012; pp. 7–31. [CrossRef]
30. Schalenbach, M.; Zeradjanin, A.R.; Kasian, O.; Cherevko, S.; Mayrhofer, K.J. A Perspective on Low-Temperature Water Electrolysis – Challenges in Alkaline and Acidic Technology. *Int. J. Electrochem. Sci.* **2018**, *13*, 1173–1226. [CrossRef]
31. Tremblay, O.; Dessaint, L.A. A generic fuel cell model for the simulation of fuel cell vehicles. In Proceedings of the 2009 IEEE Vehicle Power and Propulsion Conference, Dearborn, MI, USA, 7–10 September 2009; IEEE: Piscataway, NJ, USA, 2009; pp. 1722–1729.
32. Liu, H.; Zhou, T. CFD based PEM fuelcell models and applications. In Proceedings of the Technical Proceedings of the 2003 Nanotechnology Conference and Trade Show, San Francisco, CA, USA, 23–27 February 2003; Volume 3, pp. 463–466.
33. Yu, Q.; Srivastava, A.K.; Choe, S.Y.; Gao, W. Improved modeling and control of a PEM fuel cell power system for vehicles. In Proceedings of the Proceedings of the IEEE SoutheastCon 2006, Memphis, TN, USA, 31 March–2 April 2006; IEEE: Piscataway, NJ, USA, 2006; pp. 331–336.
34. O'hayre, R.; Cha, S.W.; Colella, W.; Prinz, F.B. *Fuel Cell Fundamentals*; John Wiley & Sons: Hoboken, NJ, USA, 2016.
35. Kim, M.J.; Peng, H.; Lin, C.C.; Stamos, E.; Tran, D. Testing, modeling, and control of a fuel cell hybrid vehicle. In Proceedings of the 2005 American Control Conference, Portland, OR, USA , 8–10 June 2005; IEEE: Piscataway, NJ, USA, 2005; pp. 3859–3864.
36. Acharya, P.; Enjeti, P.; Pitel, I.J. An advanced fuel cell simulator. In Proceedings of the Nineteenth Annual IEEE Applied Power Electronics Conference and Exposition, 2004. APEC'04, Anaheim, CA, USA, 22–26 February 2004; IEEE: Piscataway, NJ, USA, 2004; Volume 3, pp. 1554–1558.
37. Veintimilla, N.A.; Salazar, L.A.G. Obtaining Combustible Hydrogen by Electrolysis of Water, Using Photovoltaic Panels. Bachelor's Thesis, Cuenca, Spain, 2008. Available online: https://dspace.ucuenca.edu.ec/handle/123456789/8378 (accessed on 10 November 2024). (In Spanish)
38. Benavides, D.; Arévalo, P.; Ortega, A.C.; Sánchez-Sutil, F.; Villa-Ávila, E. Energy Management Model for a Remote Microgrid Based on Demand-Side Energy Control. *Energies* **2024**, *17*, 170. [CrossRef]
39. Benavides, D.; Arévalo, P.; Criollo, A.; Tostado-Véliz, M.; Jurado, F. Multi-mode monitoring and energy management for photovoltaic-storage systems. *Renew. Energy* **2024**, *230*, 120820. [CrossRef]
40. *Resolución ARCERNNR-021-2021: Costos SPEE 2022*; Agencia de Regulación y Control de Energía: Quito, Ecuador, 2022.
41. Ho, C.K.; Ambrosini, A. Thermal energy storage technologies. In *The 2020 US Department of Energy (DOE) Energy Storage Handbook (ESHB)*; DOE: Washington, DC, USA, 2020; pp. 162–171.
42. Nyangon, J.; Darekar, A. Advancements in hydrogen energy systems: A review of levelized costs, financial incentives and technological innovations. *Innov. Green Dev.* **2024**, *3*, 100149. [CrossRef]

applied sciences

MDPI

Article

Gradient Descent Algorithm with Greedy Repositioning Using Power Deficit Aggregation of Wakes to Accelerate the Offshore Wind Farm Layout Optimization Problem in Irregular Concession Areas

Angel Gaspar Gonzalez-Rodriguez [1,*], Juan Manuel Roldan-Fernandez [2,†], Javier Serrano-Gonzalez [2,†] and José Vicente Muñoz-Díez [1]

1 Department of Electronic Engineering and Automation, University of Jaen, 23071 Jaén, Spain; jmunoz@ujaen.es
2 Department of Electrical Engineering, Escuela Técnica Superior de Ingeniería, Universidad de Sevilla, 41092 Sevilla, Spain; jmroldan@us.es (J.M.R.-F.); javierserrano@us.es (J.S.-G.)
* Correspondence: agaspar@ujaen.es
† These authors contributed equally to this work.

Abstract: Wind farm layout optimization is essential to maximize the energy production of renewable energy systems. A new layout optimization method for offshore wind farms is proposed to minimize power deficits due to the wake effect without limitation on the number of turbines, the shape, or the extension of the concession area. The main engine of the algorithm is a gradient-descent method in which throughout the optimization process, new turbines are progressively and randomly included within the concession area and quickly expand outward, looking for areas with less perturbation, in turn, pushing the previous ones. When the optimization process ends, to avoid local maxima, it enters into a process of suppression of the set of locations that cause the greatest potential (power deficit). Then, a map of potential for the entire area is created, and a greedy algorithm places new turbines to complete the layout with the final number of turbines. The overall process is completed in 25 s. To drastically speed up the search process and the creation of the potential map, a simplification has been validated and added: for turbines affected by multiple wakes, the resulting power has been calculated by using a linear aggregation of power deficits, instead of the usual linear (or quadratic) aggregation of speed deficits. Owing to this type of aggregation, an analogy is established between power deficit and repulsive non-isotropic electrostatic potential energy, which allows using the properties of conservative fields. Thanks to this, the process is 20 times faster than any other layout optimization algorithm found in the revised literature. Irregular concession areas are easily treated using Stokes' theorem to detect outer points.

Keywords: layout optimization; offshore wind farm; gradient descent; greedy repositioning; power deficit; deficit aggregation; computation time

1. Introduction

The problem of the distribution of turbines in an Offshore wind farm (OWF), also known as Offshore wind farm layout optimization problem (OWFLOP), is the most important factor determining the wind speed deficit, and consequently, it strongly influences the generation of income from the sale of electrical energy. The five most important groups of optimization algorithms used for the OWFLOP are heuristic, pattern search, simulated annealing, evolutionary algorithms, and gradient search [1].

Table 1, taken from [2,3], presents a summary of previous works related to the aforementioned issue of layout optimization, indicating the type of search engine. A more complete review was provided by [4], where most of the cases under study spin around random search algorithms, especially those using a Genetic algorithm (GA). For most of them, the objective function to optimize is the Annual Energy Production (AEP), or correspondingly, the average farm power ($\mathbb{E}[P]$).

Table 1. Distribution of wind farm optimization search papers according to the optimization technique.

Category	Algorithm	Percentage of Articles	Domain
Heuristic	Genetic algorithm	32.82%	Discrete
	Genetic algorithm	6.56%	Continuous
	Evolutionary algorithm	4.65%	Continuous
	Particle swarm	3.28%	Discrete
	Particle swarm	6.55%	Continuous
	Ant colony optimization	1.64%	Continuous
	Monte Carlo simulation	1.64%	Discrete
	Simulated annealing	4.92%	Continuous
	Harmony search	1.63%	Discrete
	Pattern search	4.92%	Continuous
	Random search	3.28%	Continuous
	Greedy heuristic algorithm	3.28%	Discrete
	Greedy heuristic algorithm	1.64%	Continuous
Mathematical programming	Mixed integer programming	13.11%	Discrete
	Gradient-based optimization	13.11%	Continuous
	Total	61	

The review of the literature reveals that most of the research on the OWFLOP starts from a small and uniform rectangular space divided into columns and rows, forming a mesh of cells. Then, the algorithm decides whether to locate a turbine in the center of a cell or not. Obviously, a coarse lattice considerably limits the solution space for the turbine locations, while a sufficiently fine lattice increases the computational cost of the algorithm execution process given that the number of possible solutions is

$$Nsol = \frac{Nc!}{nt!(Nc - nt)!} \tag{1}$$

where Nc is the number of cells and nt is the number of turbines.

Due to the high computational cost of these traditional techniques, most articles are reduced to 10×10 or 20×20 meshes. In [5], a collection of 46 articles with these characteristics is presented, many of them using algorithms based on animal behavior (genetic algorithm, particle swarm, Harris hawk, elephant herding, cuckoo, or algae), and mostly showing a contrived solution to the layout problem. All in all, the general trend is the use of meta-heuristic algorithms where the achievement of the optimum (in relation to the mathematical formulation of the problem) is not guaranteed, especially when the solution space is increased.

Gradient descent (GD) methods are often disregarded because of their poorer performance in terms of accuracy since they are prone to fall in local maxima. Although they perform quite well due to their guaranteed convergence [6], some authors have pointed out that depending on the gradient calculation method, their computational time can be prohibitive for large OWFs [7]. The calculation of the objective function in any type of optimization algorithm requires many operations, and in the case of using its gradient, it requires even more. A finite-difference approximation of gradients (FDAG) with respect to the variables to be optimized (typically the x,y components of each turbine) entails tripling the number of operations necessary to estimate the AEP. Obtaining the gradient using analytical formulas, as in [8], takes an amount of time that would slow down any search method given the high number of trigonometric, quadratic and inverse operations required.

The computation time can increase if, in addition, the Hessian matrix is calculated at each step. Guirguis, in [9], develops differentiable mathematical models to handle the Concession Area (CA) constraints and compares the performance (for an interior point method) using FDAG and the exact gradients, showing, for the former, significantly lower computation times at the expense of a negligible deviation in accuracy. The main problem is that it introduces numerical errors that can affect the convergence of the algorithm, especially for highly constrained problems. This same author, in [10], compares a GD-based multi-objective algorithm to an NSGA-II, claiming the superiority of the former in terms of coverage, spread, and computational cost.

Including spatial constraints in the search space is a design feature addressed in many other works related to layout optimization. A solution for irregularly shaped and discontinuous polygons is also proposed in [11] by identifying the nearest edge and polygon, and in [12] by means of dynamic penalty functions.

In other works where GD methods are used, they are often restricted to very local searches [13], or focused on fine-tuning a heuristic algorithm [14,15].

Furthermore, the optimal location problem is not convex and GD methods may be incapable of finding the global optimum. Consequently, these methods are often combined with some other ones to escape from local maxima. In [7], triangulation is used as the starting point of the gradient method, with the additional idea of maximizing turbine dispersion, thus reducing maximum turbine loads. Quick et al., in [16], use an stochastic gradient descent (SGD) and estimate the gradient of the AEP using the Monte Carlo simulation, achieving lower computation times as long as the farm size is above 225 turbines.

This article presents a method whose search engine is a GD-based optimization. The traditional method has been improved with the following contributions:

Linear aggregation of power deficit (LAPD) as the wake aggregation method. When a turbine is immersed in the wakes of several upstream turbines there are different aggregation methods to obtain the resulting deficit. A new power deficit aggregation method is proposed and used that takes advantages of the linear properties of conservative fields. Based on this linearity, it drastically simplifies and speeds up the computation of the objective function gradient. Two consequences are derived:

- It enables the use of look-up table (LUT).
- The linearization reduces the complexity of the optimization problem, transforming it from an original problem involving 2 *nt* variables, into *nt* optimization problems, each involving two variables.

Greedy reposition of local maxima. Turbines with higher power deficits will be reallocated to avoid local maxima. The linearization in the aggregation method allows the calculation of a map of power deficits in a few seconds (typically 2.5 s for 200,000 locations).

Figure 1 presents the scope of the proposed methods, along with the sequence of operations to be performed during the OWFLOP.

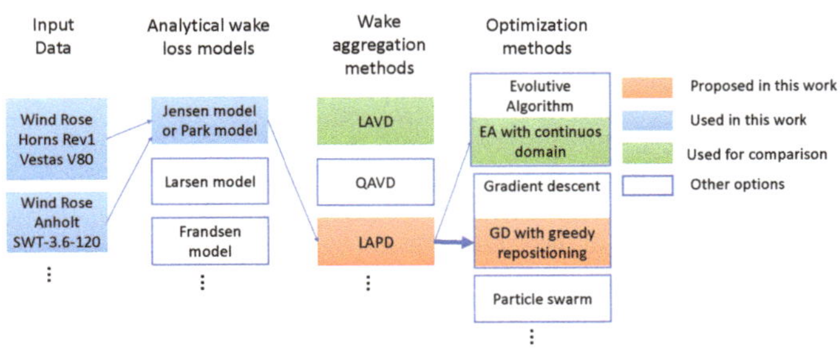

Figure 1. Graphical summary of the proposed method among the existing ones, along with the sequence of the essential operations to be performed.

Other important contributions include the following:

Symmetry. It has been verified that a wind distribution can be replaced by its equivalent using point symmetry (PS). This statement will lead to important simplifications in the evaluation of the power deficit.

Land-use constraints. A fast procedure based on the Theorem of Stokes is used to detect if a turbine location is inside an irregular Concession area (CA). Although not addressed in this work, exclusion areas can also be incorporated.

With these improvements, the resulting optimization method is significantly faster than any other one, while also enabling the handling of a large number of turbines (theoretically unlimited) in CAs of any type.

In order to explain the method used for each tool and its validation, the remainder of the article has been structured as follows:

- Section 2. Brief formulation of the problem and the equations for the AEP deficit used as an objective function. Description and validation of the linear aggregation of the power deficit (LAPD).
- Section 3. Description, validation and implications of the transformation into PS for the wind direction distribution.
- Section 4. Explanation of the iterative algorithm, based on a gradient-descent search method, combined with a Greedy repositioning (GR).
- Section 5. Results and comparisons to the fastest method found in the literature.
- Section 6. Conclusions and limitations of the proposed method.

2. Linearization of the Power Deficit Calculation

2.1. Annual Energy Produced by a Wind Farm

The general form for calculating the energy produced by a wind farm is obtained by adding together the average power, $\mathbb{E}[P_i]$, produced by each of the nt turbines in the wind farm, for every wind direction and for every wind speed. Over the course of a year ($T = 8760$ h), the energy produced by the wind farm, AEP, can be obtained from the following:

$$\text{AEP} = T \sum_{i=1}^{nt} \mathbb{E}[P_i] \tag{2}$$

and

$$\mathbb{E}[P_i] = \int_{u_{ci}}^{u_{co}} \int_0^{2\pi} P_i(u,\theta)\ \text{f}(u,\theta)\text{d}u\text{d}\theta \tag{3}$$

where u is the free-flow wind speed, θ is the wind direction, u_{ci} and u_{co} represent the cut-in speed and cut-out speed, respectively, $P_i(u,\theta)$ is the power produced by turbine i under free-flow wind conditions given by u and θ, and $\text{f}(u,\theta)$ is the frequency with which the wind blows at speed u from direction θ.

More precisely, this summation should be multiplied by the availability and losses factor, k_{AL} ($k_{AL} \leq 1$), which allows the designer to account for the production losses due to the unavailability of turbines and other losses including those occurring in the electrical system, delays and errors in turbine orientation; inaccuracies in the turbine's power curve; or delays in turbine reconnection after a shutdown due to exceeding the maximum speed (hysteresis).

In general, expression (3) is not analytically integrable, regardless of whether an analytic expression for the functions $P_i(u,\theta)$ and $\text{f}(u,\theta)$ can be obtained. Therefore, it must be discretized, transforming the integrals into summations, which yields the following:

$$\mathbb{E}[P] = \frac{\text{AEP}}{T} = \sum_{i=1}^{nt} \sum_{u=u_{ci}}^{u_{co}} \sum_{\theta=1}^{360} P_i(u,\theta)\ \text{f}(u,\theta)\Delta u \Delta\theta \tag{4}$$

Although it is usual to set $\Delta u = 1$ m/s as the discretization step for the wind speed and $\Delta\theta = 1$ deg for its direction, they will be kept for dimensional coherence.

Additionally, the energy deficit and average power deficit due to the wake effect are given by

$$\mathbb{E}[\delta P] = nt P_{1T}^{av} - \mathbb{E}[P] \tag{5}$$

$$\delta\text{AEP} = T\, nt\, P_{1T}^{av} - \text{AEP} \tag{6}$$

where P_{1T}^{av} is the average power produced by an isolated turbine under the same wind conditions:

$$P_{1T}^{av} = \sum_{u=u_{ci}}^{u_{co}} Pc(u)\ \mathrm{f}_{vel}(u)\Delta u \quad \text{with } \mathrm{f}_{vel}(u) = \sum_{\theta=1}^{360} \mathrm{f}(u,\theta)\Delta\theta. \tag{7}$$

where $Pc(u)$ represents the relationship between wind speed and the extracted electrical power (the power curve as given by the turbine manufacturer).

For a wind farm consisting of nt turbines arranged in a specific layout L, each turbine i can suffer the perturbation of another upstream one j which, for certain wind conditions (u, θ), depends on the following factors:

u the free stream wind speed
$s_{j,i}$ the separation between turbines i and j according to the layout L
$\varphi_{j\to i}$ the deviation between the segment linking the turbines (from j to i) and the wind direction θ.

When a turbine is affected by wakes from multiple upstream turbines, it is necessary to combine these wakes in order to obtain the effective power resulting from all the interactions. Therefore, it is necessary to include in (4) the accumulated deficit resulting from these combined perturbations.

The following subsections explain how to calculate $P_i(u,\theta)$ using traditional wake aggregation methods as well as the proposed one.

2.2. *Traditional Aggregation Methods to Obtain the Resulting Average Power P_i*

Various approaches exist for aggregating the wake interactions affecting a single turbine. The traditional ones calculate the wind speed deficit for all the possible combinations between upstream turbines and the turbine i; then aggregate these individual speed deficits and determine the effective wind speed of the airflow reaching the turbine i and finally obtain the power obtained by using the manufacturer power curve. The resulting expression is

$$P_i(u,\theta) = Pc(u - \delta v_i(u,\theta,L)) \quad \text{and } i \leq nt \tag{8}$$

Among these aggregation methods, the Linear aggregation of velocity deficit (LAVD) and mainly the Quadratic aggregation of velocity deficit (QAVD) are preferred [17]. In them, it is assumed that the total wind speed deficit seen by the interaction of several wakes is the sum or, respectively, the mean square sum of each one separately [18].

Linear Aggregation of Velocity Deficit (LAVD)

$$P_i(u,\theta) = Pc(u - \delta v_i(u,\theta)) \quad \text{and} \quad \delta v_i(u,\theta) = \sum_{j\neq i} \frac{4\, A_{j\to i}^2}{\pi D^2}\, \delta v_{j\to i}(u,\theta) \tag{9}$$

where, for clarity, the variable L indicating the farm layout has been omitted, $4\, A_{j\to i}/(\pi D^2)$ is the percentage of the rotor area of the turbine i (with diameter D) that is covered by the wake caused by the turbine j, and $\delta v_{j\to i}$ is the velocity deficit caused by the turbine j on turbine i if only these two exist. $A_{j\to i}$ depends on $s_{j,i}$ and $\varphi_{j\to i}$, while $\delta v_{j\to i}$ depends on $s_{j,i}$. In this work, the popular Jensen model for the wake parametrization has been used. For the sake of brevity, expressions to calculate all the involved variables are not included in this work but can be followed at [17,19]. From (4), the basic equation to calculate $\mathbb{E}[P]$ or AEP is

$$\mathbb{E}[P] = \frac{\text{AEP}}{T} = \sum_{i=1}^{nt} \sum_{u=u_{ci}}^{u_{co}} \sum_{\theta=1}^{360} Pc\left(u - \sum_{j\neq i} \frac{4\, A_{j\to i}^2}{\pi D^2}\, \delta v_{j\to i}(u,\theta)\right) \mathrm{f}(u,\theta)\Delta u \Delta\theta \tag{10}$$

Quadratic Aggregation of Velocity Deficit (QAVD)

$$P_i(u,\theta) = Pc(u - \delta v_i(u,\theta)) \quad \text{and} \quad \delta v_i^2(u,\theta) = \sum_{j\neq i} \frac{4\, A_{j\to i}^2}{\pi D^2}\, \delta v_{j\to i}^2(u,\theta) \tag{11}$$

and

$$\mathbb{E}[P] = \frac{\text{AEP}}{T} = \sum_{i=1}^{nt} \sum_{u=u_{ci}}^{u_{co}} \sum_{\theta=1}^{360} Pc\left(u - \sqrt{\sum_{j\neq i} \frac{4\, A_{j\to i}^2}{\pi D^2}\, \delta v_{j\to i}^2(u,\theta)}\right) \mathrm{f}(u,\theta)\Delta u \Delta\theta \tag{12}$$

The LAVD method results in larger speed deficits, which leads to a lower estimation of the AEP. Although the CAVD method is generally favored [18], there are studies where the linear method provides estimates more closely aligned with those collected by SCADA systems [20].

2.3. *Proposed Aggregation Method for Calculating the Resulting Average Power P_i*

An alternative method for calculating the total energy deficit has been evaluated, where instead of considering the speed deficit for each turbine, the power deficit is considered.

Linear Aggregation of Power Deficits (LAPD)

$$\begin{aligned} P_i(u,\theta) &= Pc(u) - \delta P_i(u,\theta) \quad \text{with} \\ \delta P_i(u,\theta) &= \sum_{j\neq i} \delta P_{j\to i}(u,\theta) \end{aligned} \tag{13}$$

$$\delta P_{j\to i}(u,\theta) = Pc(u) - Pc(u - \delta v_{j\to i}(u,\theta)) \tag{14}$$

and

$$\mathbb{E}[P] = \frac{\text{AEP}}{T} = \sum_{i=1}^{nt} \sum_{u=u_{ci}}^{u_{co}} \sum_{\theta=1}^{360} (Pc(u) - \delta P_i(u,\theta))\, \mathrm{f}(u,\theta)\Delta u \Delta\theta = \tag{15}$$

$$= \sum_{i=1}^{nt} \sum_{u=u_{ci}}^{u_{co}} Pc(u) \sum_{\theta=1}^{360} \mathrm{f}(u,\theta)\Delta u \Delta\theta - \sum_{i=1}^{nt} \sum_{u=u_{ci}}^{u_{co}} \sum_{\theta=1}^{360} \sum_{j\neq i} \delta P_{j\to i}(u,\theta)\, \mathrm{f}(u,\theta)\Delta u \Delta\theta \tag{16}$$

$$= nt P_{1T}^{av} - \mathbb{E}[\delta P] \quad \text{with} \quad \mathbb{E}[\delta P] = \sum_{i=1}^{nt} \sum_{j\neq i} \sum_{u=u_{ci}}^{u_{co}} \sum_{\theta=1}^{360} \delta P_{j\to i}(u,\theta)\, \mathrm{f}(u,\theta)\Delta u \Delta\theta \tag{17}$$

By summing over the range of speeds u and directions θ, the average power deficit for each turbine ($\mathbb{E}[\delta P_i]$) can be easily determined as follows:

$$\mathbb{E}[\delta P_i] = \sum_{j\neq i} \mathbb{E}[\delta P_{j\to i}] = \sum_{j\neq i} \sum_{u=u_{ci}}^{u_{co}} \sum_{\theta=1}^{360} \delta P_{j\to i}(u,\theta)\, \mathrm{f}(u,\theta)\Delta u \Delta\theta \tag{18}$$

yielding

$$\mathbb{E}[P] = \frac{\text{AEP}}{T} = nt\, P_{1T}^{av} - \mathbb{E}[\delta P] \quad \text{with} \quad \mathbb{E}[\delta P] = \sum_{i=1}^{nt} \mathbb{E}[\delta P_i] \tag{19}$$

as in (5).

To sum up, this new aggregation method proposes through (18) that, in the case of multiple wakes affecting a single turbine, the average power deficit experienced by turbine i in the farm can be obtained as the sum of all individual power deficits caused by the interactions between this turbine and each of the others.

This has a significant implication: assuming a certain wind distribution $f(u, \theta)$, for each pairwise interaction, the power deficit $\mathbb{E}[\delta P_{j \to i}]$ only depends on the distance and relative orientation between the two turbines. Therefore, a 2D LUT can be precomputed offline to obtain this deficit for any feasible separation $s_{i,j}$ (e.g., between 4 and 45 diameters) and relative orientation between turbines i and j (between 1 and 360 deg).

2.4. Comparison of Traditional Aggregation Methods and the Proposed Method

The estimates of $\mathbb{E}[P]$ derived from the three methods have been compared for Anholt OWF, based on the data series for the years 2013 and 2014 available in [21], from which the wind rose in Table A1 has been deduced. Using the obtained values, the capacity factor can be calculated as the ratio between the average power of the farm and the sum of the wind turbine-rated capacities P_1^{rated}, i.e.,

$$cf = \frac{\mathbb{E}[P]}{nt\ P_1^{rated}} \quad \text{with } nt = 111 \quad \text{and } P_1^{rated} = 3.6\ MW \tag{20}$$

The results are presented in Table 2 where the second column refers to $\mathbb{E}[P]$; the third one corresponds to this same value multiplied by the availability factor (0.97 [22]); the fourth one is the resulting capacity factor and the fifth one is the observed capacity factor, as given by SCADA. According to [23], this capacity factor is 47.8% (for the year 2018), while [23] indicates that it fluctuates between 40% and 54% from 2013 to 2019. For the period from 2013 to 2014, a value of 47.66% was obtained from [23], which was practically identical to that from 2018.

Table 2. Comparison of capacity factor (cf) obtained from different ways of aggregation.

	Aggregation	$\mathbb{E}[P]$ (MW)	Availability (97%)	cf [1]	SCADA cf
Anholt	LAVD	190.24	184.53	46.2%	47.7%
	QAVD	203.12	197.03	49.3%	
	LAPD	179.42	174.04	43.6%	

[1] Anholt capacity is 399.6 MW.

As expected, the LAPD results in higher deviations from the observed value, while the LAVD and QAVD result in closer estimations. This inaccuracy occurs because the hypothesis of aggregating power deficits instead of wind speed deficits for an accurate estimation of AEP is only valid in the speed range in which the turbine power curve grows linearly with speed. This range extends from about 2 m/s above the u_{ci} to about 3 m/s below the rated wind speed. In general, the wind speed frequently falls outside this range, so a precise estimate of the AEP is not expected. It is worth mentioning that for very high wind speeds, the thrust coefficient is low, and the expected deficit will also be low. Furthermore, if the wind speed value, once disturbed, remains above the rated speed, for most turbines (with variable pitch), the output power will not be affected. Consequently, all aggregation models will yield the same results.

Although the value provided by the LAPD aggregation is not as accurate as the other two methods, it will be shown that it maintains a value distribution very similar to theirs, especially to that of LAVD. This is important because an optimization of the layout based on energy calculations using either method will yield similar solutions regarding the arrangement of the turbines.

This affinity between the three types of aggregation has been tested by creating maps of power deficits in an area of 500 × 428 locations, representing a surface of 3500 × 3000 m under the wind conditions of Horns Rev 1 (HR1) given in Table A1. Figure 2 compares the maps for power deficits using the three types of aggregation, according to the estimation using Equations (9), (11) and (13), respectively, for maps (a), (b), and (c). The evaluation took 29 min 43 s for LAVD, 27 min 20 s for QAVD, and only 2.5 s for LAPD. Although a precise estimate of AEP is not reached using LAPD, a **scaling** effect in energy is detected by comparing maps (a) and (c), in the sense that the distribution of power deficits at all points of the concession zone follows a pattern similar to that obtained with LAVD. This equivalence in the maps of power deficits is translated into an equivalence in the distribution of their gradients, which allows designing an optimization algorithm based on LAVD that is **200 times faster** than on QAVD or LAVD. In summary, by comparing the map for power deficit, as in Figure 2, an equivalence is detected with regard to the distributions of deficits and gradients.

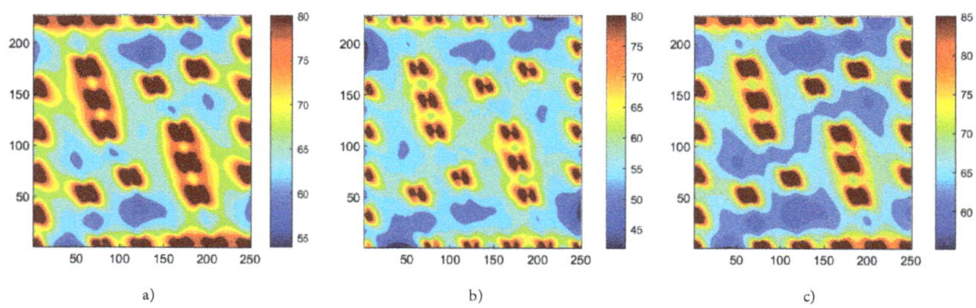

Figure 2. Map of estimated power deficit assuming a certain distribution of turbines: (**a**) estimation using LAVD according to (9); (**b**) estimation using QAVD according to (11); (**c**) estimation using LAPD according to (13).

The reduced time required for the evaluation of AEP using LAPD is due to the fact that by introducing linearity into the calculation of power deficits, this magnitude behaves like the potential energy of a conservative field. Accordingly, when a wind turbine is included or suppressed, or its location modified, it is not necessary to recalculate all the multiple wakes in which the wind turbines are involved (as shadowed or shadower turbines). Instead, taking advantage of linear properties, only the individual interactions between pairs of turbines are considered, which can be reduced to a function of two parameters: the distance between turbines $s_{i,j}$ and the orientation between them, respectively. Consequently, an offline LUT can be created, and bilinear interpolation can be used to derive the power deficit between pairs of turbines, significantly reducing the number of operations. This is not possible if an aggregation (quadratic or linear) of the speed deficits is used.

3. Symmetrization

The traditional method to estimate energy takes into account the interaction between each pair of turbines, considering the entire range of wind speeds and wind directions. However, in this subsection, it will be demonstrated that, if LAPD is used, it is only necessary to calculate the power deficit for half of the duples, assigning to the angle θ the combined probability for θ and $\theta + \pi$. To demonstrate this, we start from the overall power deficit $\mathbb{E}[P]$ calculated from (18) and (19)

$$\mathbb{E}[\delta P] = \sum_{i=1}^{nt} \sum_{j<i} \mathbb{E}[\delta P_{j\to i}] + \mathbb{E}[\delta P_{i\to j}] \tag{21}$$

Since for each pair of turbines i, j the average values of power deficits are as follows:

$$\mathbb{E}[\delta P_{j\to i}] = \sum_{v=u_{ci}}^{u_{co}} \sum_{\theta=1}^{360} \delta P_{j\to i}(\theta,u)\ \mathrm{f}(u,\theta)\Delta u\Delta\theta \tag{22}$$

$$\begin{aligned}\mathbb{E}[\delta P_{i\to j}] &= \sum_{v=u_{ci}}^{u_{co}} \sum_{\theta=1}^{360} \delta P_{i\to j}(\theta,u)\ \mathrm{f}(u,\theta)\Delta u\Delta\theta = \sum_{v=u_{ci}}^{u_{co}} \sum_{\theta=1}^{360} \delta P_{j\to i}(\theta+180,u)\ \mathrm{f}(u,\theta)\Delta u\Delta\theta = \\ &\sum_{v=u_{ci}}^{u_{co}} \sum_{\theta=1}^{360} \delta P_{j\to i}(\theta,u)\ \mathrm{f}(u,\theta+180)\Delta u\Delta\theta\end{aligned} \tag{23}$$

and naming

$$\mathrm{f}^{+\pi}(u,\theta) \equiv \frac{\mathrm{f}(u,\theta)+\mathrm{f}(u,\theta+180)}{2} \tag{24}$$

it yields

$$\mathbb{E}[\delta P] = 2\sum_{i=1}^{nt}\sum_{j<i}\sum_{v=u_{ci}}^{u_{co}} \sum_{\theta=1}^{360} \delta P_{j\to i}(\theta,u)\ \ \mathrm{f}^{+\pi}(u,\theta)\Delta u\Delta\theta \quad \square \tag{25}$$

The previous expressions demonstrate that by using $\mathrm{f}^{+\pi}(u,\theta)$ as the wind probability function, it is possible to obtain the overall power deficit from half of the interactions. This is valid when LAPD is used as an aggregation method. Appendix A shows that, if LAVD is used, this is also valid for OWFs with PS, and in general, for any layout, the deviation in AEP is lower than 0.005%.

For better clarity, and hereafter, a certain magnitude M (power/power deficit) will be overlined when it refers to the result of the following summation for all wind conditions,

$$\overline{M} = \sum_{v=u_{ci}}^{u_{co}} \sum_{\theta=1}^{360} M\ \mathrm{f}^{+\pi}(u,\theta)\Delta u\Delta\theta \tag{26}$$

It is worth noting that $\mathbb{E}[M]$ is really an average, while $\overline{M}$ must be regarded as a mathematical operator, with which important implications will be derived. However, if the farm has PS, then ($\mathrm{f}(u,\theta) = \mathrm{f}(u,\theta+\pi)$), and $\mathbb{E}[M] = \overline{M}$. Even without PS and applied to the overall farm, (25) converts into

$$\mathbb{E}[\delta P] = 2\sum_{i=1}^{nt}\sum_{j<i}\overline{\delta P_{j\to i}} = \overline{\delta P} \tag{27}$$

This means that, although overlined magnitudes do not provide exact average values for individual turbines when summed for all of the OWF turbines, it yields the exact average value. As a consequence, for the purpose of obtaining the overall power deficit, the following expressions are equivalent:

$$\begin{aligned}\mathbb{E}[\delta P_i] &\simeq \overline{\delta P_i} \\ \mathbb{E}[\delta P_{j\to i}] &\simeq \overline{\delta P_{j\to i}}\end{aligned} \tag{28}$$

In the same way as (27), it is easy to demonstrate

$$\mathbb{E}[\delta P] = 2\sum_{i=1}^{nt}\sum_{j<i}\overline{\delta P_{i\to j}} \tag{29}$$

Using LAPD has another implication although, in general, $\mathbb{E}[\delta P_{j\to i}] \neq \mathbb{E}[\delta P_{i\to j}]$, if the turbines are identical in model and height, then the overlined power deficit caused by turbine i on turbine j ($\overline{\delta P_{i\to j}}$) is the same as that caused by turbine j on turbine i ($\overline{\delta P_{j\to i}}$), that is

$$\overline{\delta P_{i\to j}} = \overline{\delta P_{j\to i}} \tag{30}$$

The formal demonstration is given in Appendix B.

Additionally, from (13)

$$\overline{\delta P_i} = \sum_{j \neq i} \overline{\delta P_{j \to i}} = \sum_{j \neq i} \overline{\delta P_{i \to j}} \quad \text{and} \tag{31}$$

$$\sum_{j \neq i} \mathbb{E}[\delta P_{j \to i}] \simeq \sum_{j \neq i} \overline{\delta P_{j \to i}} = \sum_{j \neq i} \overline{\delta P_{i \to j}} \simeq \sum_{j \neq i} \mathbb{E}[\delta P_{i \to j}] \tag{32}$$

As a result, if LAPD is applied, the power deficit in a turbine caused by all others has the same value as the additional power deficit of that turbine on all others. This means that when introducing a new turbine in the wind farm, the increase in power deficit is calculated as twice the power deficit experienced by the introduced turbine.

4. Method: Optimization Using the GD + GR Method

4.1. Overall Algorithm Description

In the micro-siting problem, the location of each turbine must be determined to maximize the AEP, which, assuming a fixed number of turbines nt, is equivalent to minimizing the energy losses δAEP due to the wake effect. For clarity, $\delta AEP/T$, analogous to $\overline{\delta P}$ will be replaced by U, anticipating the analogy of interpreting $\overline{\delta P}$ as **potential energy**. Therefore, the problem is formulated as follows:

$$\begin{aligned} U \equiv \frac{\delta \text{AEP}}{T} = \overline{\delta P} = \sum_{i=1}^{nt} \overline{\delta P_i} = \sum_{i=1}^{nt} \sum_{j \neq i}^{nt} \overline{\delta P_{j \to i}} \\ \min\ U = U(L) \quad \text{with} \quad L = \{\mathbf{t}_1, \mathbf{t}_2, \dots \mathbf{t}_{nt}\} \\ s.t.\ \mathbf{t}_i = (x_i, y_i) \in CA\ \forall i \leq nt \end{aligned} \tag{33}$$

where L for layout refers to a set of the positions $\mathbf{t}_i = (x_i, y_i)$ where the turbines are located, and CA is the concession area, in general, with an irregular shape.

In general terms, this problem will be solved through a sequence of operations by which layouts with decreasing U $= \delta\text{AEP}/T = \overline{\delta P}$ will be obtained. The updating of successive solutions will be carried out using the GD method. However, since the optimization function is non-convex, this method must be accompanied by some modification that allows for the detection of local optima and the ability to skip to another solution with lower potential. Additionally, it does not start from an initial configuration of nt turbines; instead, every certain number of iterations, a set of turbines will be included at random feasible positions, allowing each configuration to evolve to occupy areas with fewer interactions, typically the corners and the contour.

In this work, this combined method will be used and compared to an improved evolutive algorithm. The proposed method is shown in Figure 3 and consists of the following steps, which will be described in the next subsections:

- **Progressive addition by batches with gradient-based repulsion** (see upper frames of Figure 4).
    1. Initial placement of turbines in random but feasible positions within the CA. In Figure 3, six turbines are included in each batch, and four in Figure 4.
    2. The GD method is used to move the turbines so that the overall energy deficit is minimized.
    3. Repeat the previous steps until a total of nt turbines are included.

    Additional details will be presented in Section 4.2.
- **Greedy Repositioning** (see lower frames of Figure 4).
    4. Identification and removal of turbines in areas with high U (i.e., high energy deficit). Symbol □.
    5. Creation of a map to detect positions with lower U that will be used to relocate the removed turbines. Symbol ◇.
    6. Apply GD to move the turbines to a lower potential state. Symbol ○.

7. Repeat steps 4–6 until no new optimal positions are obtained.

Further details will be explained in Section 4.4.

- **Final Tuning**

8. Use the gradient method again to fine-tune the layout.

The GD method is used in steps 2, 6 and 8, with details explained below. In all steps, LAPD will be used, as it is faster than LAVD or QAVD and results in the same U distribution (see Figure 2).

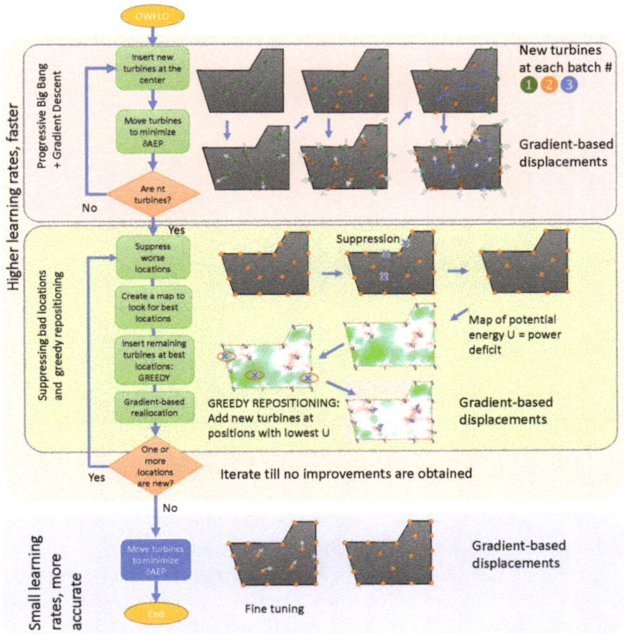

Figure 3. Sequence of operations for the OWFLOP with GD search and GR. Crosses and circles are used for suppressed and added turbines, respectively.

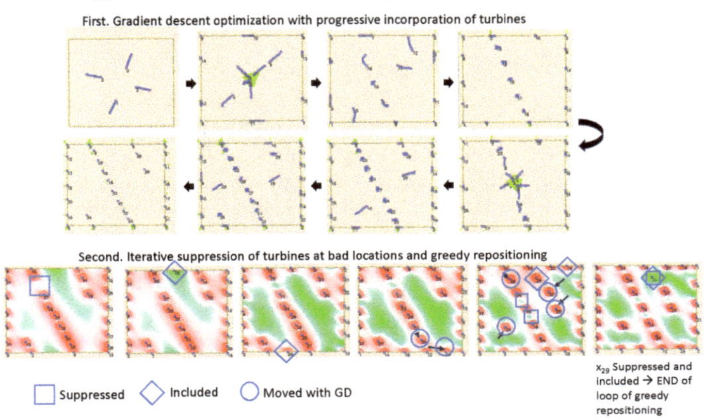

Figure 4. Frames of the process. Top: several batches in which four new turbines are included, followed by the GD algorithm operating to find their optimal locations. Bottom: iterative suppression of turbines at suboptimal locations and repositioning where U is minimum.

Figure 5 shows the evolution of the farm deficit as the algorithm progresses. In magenta, the evolution of the estimated AEP using LAPD is shown, which is the method actually used during the optimization. In green, the evolution of the AEP estimated using LAVD is shown, which is not used during the optimization but is included in the graph to illustrate that both aggregation methods present parallel evolutions. This parallel evolution is coherent with the fact that LAPD can be used during the optimization process to minimize losses due to the wake effect.

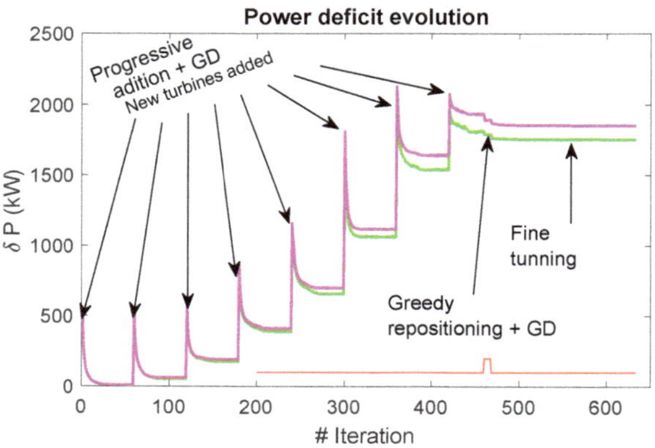

Figure 5. Proposed Algorithm. Evolution of the overall power deficit along the optimization process. In magenta, the $\mathbb{E}[\delta P] = \overline{\delta P}$ estimation using LAPD, which is the aggregation method used during the optimization. In green, the overall power deficit $\overline{\delta P}$ estimation using LAVD. The pulse in red represents the GR of turbines.

Most of the iterations correspond to the initial approach (primary GD optimization with progressive incorporation of turbines), in which turbines are added at random positions. When a batch of turbines is added, there is an abrupt increase in the deficit as represented by the peaks. Just after being included, they repel each other due to the strong interactions they suffer, thus reducing their own potential (power deficit) and that of the farm. The figure illustrates this as a descending trend after each peak. The repulsion mechanism seeking the minimization of the potential presents an analogy with the electrostatic force of repulsion, which turns out to be the potential gradient. Therefore, it seems appropriate to use the gradient as a force that determines the movement of the particles/turbines to minimize their potential.

Once all the turbines have been located, to detect and reallocate the turbines that can cause local optima, the GR mechanism is used to suppress the turbines that cause the greatest power deficit and create a potential map. Then, it tracks the map searching for the optimal location and places a turbine at that particular place. Then, a potential map is recreated to place a new turbine. After completing the repositioning of the suppressed turbines, a second GD is executed. The process (suppression, repositioning and secondary GD) repeats until the best new locations are the same as the previously suppressed ones, which indicates that no further improvement can be obtained.

Finally, a search is run again using the GD method with a smaller learning rate to adjust the final maximum. It should be said that this third GD search does not lead to a significant improvement in the estimated U.

4.2. Formulation of the Gradient Descent Method

Among the variants of the GD method, one of the most used for solving large-scale nonlinear optimization problems is the conjugate gradient method, introduced for nonlinear systems by Fletcher and Reeves in the 1960s. For linear systems, it is faster than the basic GD method. However, in the case under study, obtaining an analytical expression for the function to be minimized (AEP), as in [8–10], increases the computation time. This time increases even more if the inverse of (33) or its Hessian is required to obtain a correct value of the learning rate to meet the inequalities of the strong Wolfe conditions [24].

Traditionally, the GD method is based on the calculation of the gradient of an objective function, in this case $U = \delta P$, which depends on the positions of the nt turbines, giving rise to a gradient vector of 2 nt components.

$$\nabla U = \left[\frac{\partial U}{\partial q_1} \cdots \frac{\partial U}{\partial q_{2nt}}\right]^T \tag{34}$$

$$U(\mathbf{q}) \quad \text{defined and differentiable in } S \subset \mathbb{R}^{2\,nt}$$

where $\mathbf{q}$ is the vector of components (x and y) for all of the nt turbines

$$\begin{aligned} q_{2i-1} &= x_i \\ q_{2i} &= y_i \end{aligned} \qquad \forall\, 0 < i \leq nt \tag{35}$$

and the set S defined as

$$\mathbf{q} \in S \subset \mathbb{R}^{2\,nt} \iff \mathbf{t}_i \equiv (x_i, y_i) = (q_{2i-1}, q_{2i}) \in CA \quad \forall\, 0 < i \leq nt \tag{36}$$

In this work, a method based on the basic GD will be used, but with the proposed modifications, which proves to be faster than any other optimization method based on GD or not. The common expression dictating the evolution of the solution $\mathbf{q}$ at step k is given by the following:

$$\mathbf{q}^{k+1} = \mathbf{q}^k - \gamma^k \nabla U(\mathbf{q}^k) \tag{37}$$

where $\gamma^k \in \mathbb{R}^+$ is the so-called learning rate and establishes the size of the steps taken in the direction of the negative gradient during the optimization process.

4.2.1. Decoupling the Optimization Problem in $\mathbb{R}^{2\,nt}$ into nt Problems in $\mathbb{R}^2$

As discussed in Section 2, when multiple wakes interact with the same downwind turbine, several aggregation methods can be used to estimate the speed deficit experienced by the air stream passing through the downwind turbine. With adequate modeling of wind conditions, the quadratic and linear methods (QAVD and LAVD) yielded results comparable to those obtained directly from SCADA systems. Another method with similar results to LAVD is LAPD; the power deficit generated by a downwind turbine affected by several upstream ones is the sum of the power deficits calculated considering each interaction separately. Despite a higher estimation deviation than QAVD or LAVD, the proposed method, LAPD, produces a similar power deficit map and, therefore, a similar distribution of gradients, as shown by comparing maps (a) and (c) of Figure 2.

Additionally, it will be shown that by using LAPD, the optimization problem involving $2\,nt$ variables shown in (33) is reduced to optimizing nt problems of two variables. Indeed, taking partial derivatives with respect to, for example, component x_i, yields

$$\frac{\partial U}{\partial x_i} = \frac{\partial \sum_k \overline{\delta P}_k}{\partial x_i} = \sum_k \frac{\partial \sum_{j \neq k} \overline{\delta P}_{j \to k}}{\partial x_i} \overset{\text{LAPD}}{=} \sum_k \sum_{j \neq k} \frac{\partial \overline{\delta P}_{j \to k}}{\partial x_i} \tag{38}$$

and since the power deficits $\overline{\delta P}_{j \to k}$ consider their interaction separately

$$\left.\begin{array}{c} j \neq i \\ \wedge \\ k \neq i \end{array}\right\} \Rightarrow \frac{\partial \overline{\delta P}_{j \to k}}{\partial x_i} = 0 \tag{39}$$

it yields

$$\frac{\partial U}{\partial x_i} = \sum_{j\neq i} \frac{\partial \overline{\delta P}_{j\to i}}{\partial x_i} + \sum_{k\neq i} \frac{\partial \overline{\delta P}_{i\to k}}{\partial x_i} = 2\sum_{j\neq i} \frac{\partial \overline{\delta P}_{j\to i}}{\partial x_i} \tag{40}$$

and analogously for y,

$$\frac{\partial U}{\partial y_i} = 2\sum_{j\neq i} \frac{\partial \overline{\delta P}_{j\to i}}{\partial y_i} \tag{41}$$

If $\hat{e}_x$ $\hat{e}_y$ are the unitary Cartesian vectors, then we can define

$$\nabla^i \cdot \equiv \frac{\partial \cdot}{x_i}\hat{e}_x + \frac{\partial \cdot}{y_i}\hat{e}_y \tag{42}$$

and (40) and (41) turn into

$$\nabla^i U = 2\sum_{j\neq i} \nabla^i \overline{\delta P_{j\to i}} \tag{43}$$

From (31), it also results that

$$\nabla^i \overline{\delta P}_i = \sum_{j\neq i} \nabla^i \overline{\delta P_{j\to i}} \quad \Rightarrow \quad \nabla^i U = 2\nabla^i \overline{\delta P}_i. \tag{44}$$

For the coordinates of turbine i, given only by x_i and y_i, it yields

$$\nabla^i U = 2\nabla^i \overline{\delta P}_i = 2\nabla \overline{\delta P}_i \tag{45}$$

Since $\nabla^i U$ is parallel to $\nabla \overline{\delta P}_i$ (in fact, twice), the previous expression together with the method of GD dictates that the displacement direction of turbine i which maximizes the decrease in U and is given by $-\nabla \overline{\delta P}_i$.

This is the key idea of the proposed gradient-based method because it decouples and drastically simplifies the process, turning the problem of optimizing in $\mathbb{R}^{2\,nt}$ into nt problems in $\mathbb{R}^2$. This means that, at each iteration k, the algorithm moves each turbine i to reduce its power deficit $\overline{\delta P}_i$, using the GD method through (45). Applying this for every turbine leads to the overall minimization of the farm power deficit. This way, the method of GD expressed in (37) turns into nt optimization problems in the form

$$\mathbf{t}_i^{k+1} = \mathbf{t}_i^k - \gamma^k \nabla \overline{\delta P_i^k} \tag{46}$$

where $\mathbf{t}_i^k$ is the position of turbine i at step k.

4.2.2. The Gradient in Polar Coordinates

In order to obtain $\nabla \overline{\delta P}_i$, it is necessary to determine the gradient of the power deficit $\overline{\delta P_{j\to i}}$. It is faster to calculate the gradient in polar coordinates:

$$\nabla = \frac{\partial}{\partial r}\hat{e}_r + \frac{1}{r}\frac{\partial}{\partial \vartheta}\hat{e}_\vartheta \Rightarrow \nabla = \frac{\partial}{\partial r}(\cos\vartheta \hat{e}_x + \sin\vartheta \hat{e}_y) + \frac{1}{r}\frac{\partial}{\partial \vartheta}(-\sin\vartheta \hat{e}_x + \cos\vartheta \hat{e}_y) \tag{47}$$

where r and ϑ are the radial and angular components corresponding to the separation between turbines and their relative orientation, and $\hat{e}_r$, $\hat{e}_\vartheta$ are unit vectors in the radial direction and perpendicular direction (counterclockwise). Therefore,

$$\begin{aligned} \frac{\partial \overline{\delta P_{i\to j}}}{\partial x_i} &= \frac{\partial \overline{\delta P_{i\to j}}}{\partial r}\cos\vartheta - \frac{1}{r}\frac{\partial \overline{\delta P_{i\to j}}}{\partial \vartheta}\sin\vartheta \\ \frac{\partial \overline{\delta P_{i\to j}}}{\partial y_i} &= \frac{\partial \overline{\delta P_{i\to j}}}{\partial r}\sin\vartheta + \frac{1}{r}\frac{\partial \overline{\delta P_{i\to j}}}{\partial \vartheta}\cos\vartheta \end{aligned} \tag{48}$$

As previously mentioned, any interaction $\delta P_{i \to j}$ and its gradient components (48) only depend on the distance s between turbines i and j and the relative orientation ϑ of the segment linking them. Therefore, it can be precalculated before execution and accessed during execution time through a 2D LUT.

4.3. Managing Irregular Concession Areas

To ensure that a turbine located at position $\mathbf{t}_i = (x_i, y_i)$ remains within the CA, Stokes' theorem (or more specifically for 2D, Green's theorem) is used. Additional details can be found in Appendix C. This technique allows defining CAs as irregular polygons. This will be used to evaluate whether a turbine is inside, outside, or on the contour of a polygon in the following cases:

- During the execution of the proposed GD +GR algorithm:
  - To progressively add new turbines, to check that the newly added turbine is within the CA.
  - After modifying the turbine components, to determine if the new positions of the turbines are outside the CA.
  - To determine if a turbine is on the boundary, and in that case, apply a modification of the gradient procedure, as explained in Section 4.3.1.
- During the execution of an EA used for comparative purposes, to determine if the position of a new turbine is feasible, either during the generation of individuals in the population or after crossover and mutation operations.

4.3.1. Displacement Along the Contour of the CA

By using the proposed GD method with LAPD, the overall power deficit is minimized by adequately displacing the turbines searching to minimize their individual power deficits. Assuming a polygonal CA, by using the expressions (A6) and (A8) given in Appendix C, it can be determined whether the updated position of turbine i at step $k+1$, given by (46), falls outside the CA (i.e., $\mathbf{t}_i^{k+1} \notin CA$). In that case, the calculated position $\mathbf{t}_i^{k+1}$ must be replaced by the intersection of the contour with the segment $\overline{\mathbf{t}_i^{k+1}\mathbf{t}_i^{k}}$. This intersection point is denoted as $\mathbf{t}_i^{*\,k}$ and the side of the polygon intersected by the mentioned segment, as E_j. Subsequently, the trajectory evolution no longer follows the direction of $\nabla\overline{\delta P_i}$, but a direction p_i^k, whose direction vector is parallel to E_j, and its direction is chosen to satisfy the angle condition as defined in [25].

$$\cos(\theta) = \frac{-\nabla\overline{\delta P_i^{k+1}} \cdot p_i^{k+1}}{\|\nabla\overline{\delta P_i^{k+1}}\| \|p_i^{k+1}\|} > \varepsilon \quad \text{for some } \varepsilon > 0 \tag{49}$$

This ensures global convergence [24]. Checks and, if necessary, saturations must be carried out to prevent the final position from exceeding the endpoint of E_j.

4.3.2. Selecting the Learning Rate

Selecting an appropriate learning rate is essential for achieving optimal optimizer performance. A very small value will slow down the search, while a very large value can cause the search process to oscillate or diverge. There are curvature-based methods, such as the Newton–Raphson method, which uses the Hessian for adjustment, but it is computationally expensive. Another method for determining the learning rate, yielding two values, is the Barzilai–Borwein (BB) method [26], which dynamically adjusts the step size based on information from the previous iterations by approximating the inverse Hessian information without explicitly computing it. These two values are

$$\gamma^k_{LONG} = \frac{(\mathbf{t}_i^k - \mathbf{t}_i^{k-1}) \cdot (\mathbf{t}_i^k - \mathbf{t}_i^{k-1})}{(\mathbf{t}_i^k - \mathbf{t}_i^{k-1}) \cdot (\nabla(\overline{\delta P_i^k}) - \nabla(\overline{\delta P_i^{k-1}}))} \tag{50}$$

$$\gamma^k_{SHORT} = \frac{(\mathbf{t}_i^k - \mathbf{t}_i^{k-1}) \cdot (\nabla(\overline{\delta P_i^k}) - \nabla(\overline{\delta P_i^{k-1}}))}{(\nabla(\overline{\delta P_i^k}) - \nabla(\overline{\delta P_i^{k-1}})) \cdot (\nabla(\overline{\delta P_i^k}) - \nabla(\overline{\delta P_i^{k-1}}))} \tag{51}$$

Experiments have shown that the improvements of adaptive methods like BB vanish when turbines reach the boundaries of the CA. Therefore, a fixed learning rate of 0.3 has been chosen during the initial approximation phase (see Figure 3) and 0.03 during fine-tuning.

4.4. Suppressing Suboptimal Locations and Greedy Repositioning

The GD method may be unable to move a turbine located in the center of three or four other turbines. This situation represents a local optimum that must be avoided. To this end, after using the gradient method with the *nt* turbines in the CA, a potential map is created to detect which turbines have a much higher potential than others. The turbines are ordered by their potential ($\overline{\delta P_i}$), and the differences between consecutive turbines are calculated. This helps to group turbines and identify which cluster should be removed.

Figure 6 illustrates the criteria for selecting the cluster to suppress as follows:

(a) The cluster with the largest gap with the previous one is more likely to be suppressed.
(b) Clusters with higher potential (power deficit) are prioritized for suppression.

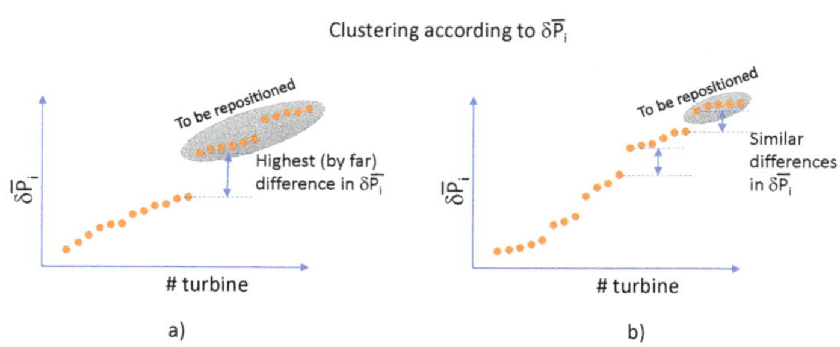

Figure 6. Sorting and clustering the turbines according to their power deficit. (**a**) For a gap significantly larger than the remaining ones, turbines after this gap are suppressed. (**b**) If two or more gaps are similar, the cluster with the worst locations is suppressed.

Once these turbines located in high potential positions have been eliminated, a map of potential is created, the absolute minimum is found, a turbine is placed there, and the potential map is regenerated. Figure 4 shows several frames of this iterative process.

It is worth mentioning that this tool is feasible only if the $U = \overline{\delta P}$ evaluation can be performed in a very short time. As previously mentioned, by using LAPD, each map (500×428 calculations) takes less than 3 s.

4.5. Comparison of Operations Required for Each Type of Aggregation

This subsection demonstrates the drastic reduction in operations achieved by using the LAPD aggregation method compared to the LAVD or QAVD methods. Given that these two methods are very similar in computational terms, the comparison will only be made with the former. A discretization of 1 m/s and 1 deg will be used, as well as PS (see Appendix A) to halve the evaluated directions.

4.5.1. Comparison in the Calculation of AEP or $\mathbb{E}[P]$

In the case of LAVD (or QAVD) from (10), most of the time is spent on obtaining the speed deficits $\delta v_{j\to i}$, which is by far the most computationally expensive task. Therefore, the calculation time is

$$t_{\text{LAVD}}^{\text{AEP}} \simeq 360(v_{co} - v_{ci})\frac{nt(nt-1)}{2}\, t_{\delta v_{j\to i}} \tag{52}$$

where $t_{\delta v_{j\to i}}$ is the time required for each calculation of the speed deficit between two turbines under wind conditions u, θ. In [27], a series of instructions are given to reduce the number of operations by about 15 times.

In the case of LAPD, a direct calculation requires practically the same number of operations. The advantage, however, lies in the possibility of precalculating an LUT with the average power deficit values that a turbine j causes in another turbine i, $(\overline{\delta P_{j\to i}})$ as a function of the distance and relative orientation between them. For a site, this LUT is calculated using (A2) once, taking into account $f(u,\theta)$ and summing for all directions and speeds.

Once the LUT is calculated, for each calculation of AEP or $\overline{P}$, only

$$t_{\text{LAPD}}^{\text{AEP}} \simeq \frac{nt(nt-1)}{2} t_{2D\ int} \tag{53}$$

is required, where $t_{2D\ int}$ is the time needed for a bilinear interpolation, which is also significantly less than $t_{\delta v_{j\to i}}$.

4.5.2. Comparison in the Calculation of the Gradient

In the calculation of the gradient, the difference is even greater. In the case of LAVD (or QAVD), from (34), it is observed that $2nt$ gradient component calculations are required. For each component, the function U (the power deficit in this case) must be calculated after moving each turbine a small amount in X and also in Y. The required time would be

$$t_{\text{LAVD}}^{Grad} \simeq 2\, nt\, 360(v_{co} - v_{ci})\frac{nt(nt-1)}{2} t_{\delta v_{j\to i}}. \tag{54}$$

In the case of LAPD, it is also necessary to first calculate an LUT with the gradient value for each separation distance and relative orientation between turbines, based on (48). Once this is conducted, the time to calculate all the gradient components is reduced to

$$t_{\text{LAPD}}^{Grad} \simeq 2\frac{nt(nt-1)}{2} t_{2D\ int}. \tag{55}$$

4.5.3. Sweep for the Potential Map

In this article, in conjunction with the gradient search, a method is used to obtain the best position to place a new turbine. For this, assuming an incomplete layout with nt_k turbines ($nt_k < nt$), a sweep is performed in the CA, calculating at each point what the increase in the deficit would be for a turbine at that point, as well as the increase in the deficit that this new turbine produces in all the other existing ones. In fact, these two values are identical if LAPD is used, as demonstrated with (32). Since the method is greedy, a partial optimum is obtained at each step of adding a new turbine, but the global optimum as a whole is not guaranteed. This search is performed several times and alternates with gradient searches. It is clear that a finer division of the search space improves the partial result and eventually the global one, although it increases the computation time.

Assuming a grid of the search space with ns cells, the time for LAVD (or QAVD) is as follows:

$$t_{\text{LAVD}}^{Map} = ns\, 360\, (v_{co} - v_{ci}) \sum_{nt_k = nt_0}^{nt} (nt_k - 1) t_{\delta v_{j\to i}}. \tag{56}$$

In the case of LAPD, it is reduced to

$$t_{\text{LAPD}}^{Map} = ns \sum_{nt_k=nt_0}^{nt} (nt_k - 1)t_{2D\ int}. \tag{57}$$

4.5.4. Discussion About Applicability, Scalability, and Used Wake Model

Previous equations show the expected time reduction when using LAPD. This reduction depends on the type of OWFLOP:

- The ratio $t_{\text{LAVD}}^{\text{AEP}}/t_{\text{LAPD}}^{\text{AEP}}$ is of appliance for any optimization algorithm.
- The ratio $t_{\text{LAVD}}^{Grad}/t_{\text{LAPD}}^{Grad}$ is of appliance for Gradient-based OWFLOPs.
- The ratio $t_{\text{LAVD}}^{Map}/t_{\text{LAPD}}^{Map}$ is of appliance when the map of deficits must be calculated.

Expressions (53) and (55) show a clear dependence on nt^2. With regard to (57), the number of suppressed turbines is bounded by the algorithm and, considering that the CA (and in turn ns) can grow linearly with nt, then the time required for the sweep grows with nt^2.

As in Figure 1, the algorithm can be used with another wake model. If a more complex (and slower) wake model is used

- the time required to calculate the look-up tables will increase (with the Park model it takes less than 1 min)
- the computation time required for the optimization search remains constant, unlike that for other algorithms that will increase.

5. Results

A series of experiments have been conducted to validate the following proposals:

- The LAPD aggregation method results in a solution similar to that obtained with the LAVD aggregation method, but 200 times faster. This validates the hypothesis that this method can be used during the optimization process.
- The linear nature of this method mainly improves the performance of gradient-based methods, achieving better performance of the proposed optimization algorithm compared to a genetic algorithm used as a reference.
- The AEP of the layout obtained with the proposed algorithm will be compared with the AEP obtained from the actual layout of the Anholt park, using two types of wake aggregation, LAVD and LAPD.

The farms Horns Rev 1 and Anholt serve as test beds for the experiments. The first testbed is the rhomboid delimited by the Horns Rev 1 wind farm, with dimensions (max_x = 5.5 km, max_y = 3.9 km) in which 80 turbines must be placed. The second test bed is the CA of the Anholt wind farm, which is an irregular polygon (max_x = 10.8 km, max_y = 22 km) in which 111 turbines must be placed. The wind conditions and the model used are shown in Tables A1 and A2.

To compare the proposed algorithm with a traditional one, an EA has been programmed and launched whose fitness is $\overline{\delta P}$. For this algorithm, chromosomes do not contain binary string-type solutions as in most genetic algorithms for OWFLOP; instead, the nt chromosomes of an individual contain the turbine position, that is, its x and y components, coded as real values. Its configuration parameters are offspring size obtained from the crossover, 25%; offspring size obtained from mutation, 25%; individuals disregarded when obtaining the offspring, 20%; population size, 200 individuals; maximum number of generations, 3000; maximum number of stagnant generations, 300.

Figure 7 shows the evolution of the objective function ($\overline{\delta P}$) for both methods: LAPD (green) and LAVD (magenta). Both aggregation methods (LAPD and LAVD) show parallel evolutions in the EA, regardless of which method is used.

The figure also shows that after a certain number of iterations, the EA does not improve its optimum solution, and it yields a higher power deficit than that obtained with the proposed **GD + GR** algorithm. The power deficit estimation for the actual layouts appears as dashed lines, calculated with both aggregation methods and using the same input data. Observing the LAVD curves in green, a traditional EA algorithm does not improve the actual layout. However, the proposed **GD + GR** one reduces the power deficit, especially in the case of HR1. Exact values will be presented in Table 3.

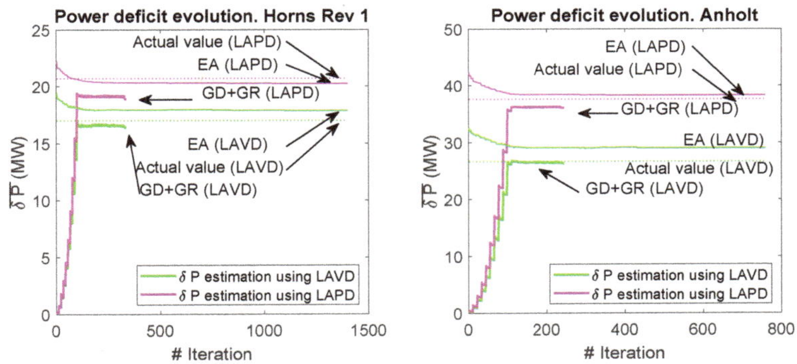

Figure 7. Fitness evolution for the proposed algorithm, compared to the evolution of an EA. In magenta, the power deficit $\mathbb{E}[\delta P] = \overline{\delta P}$ evolution using LAPD, which is the real fitness used in the EA. In green, the same evolution using LAVD.

A set of experiments (ten per situation) have been repeated to compare the performance of both algorithms, both for the test beds of Anholt and HR1. In order to speed up the searches, the LAPD method has been used to estimate $\overline{P}$, and also for the potential maps and gradients. A third set of experiments has been launched using EA implementing the LAVD estimation as fitness. It will be named EA-LAVD and corresponds to a traditional procedure. The original single-thread (ST) code has been modified to allow for multiprocessing (MT) in an Intel® Core™ i7-10750H CPU @ 2.60GHz (6 cores, 12 threads).

Figures 8 and 9, together with Table 3 show the result of comparing the proposed method with an improved EA. The first of the figures represents the computation time, visualizing the superiority of the proposed method in terms of speed. An EA can also benefit from the speed improvement by using LAPD, showing significantly lower times than an EA using LAVD. As seen, the proposed GD + GR algorithm and the improved EA require a few seconds to complete the OWFLOP, in farms with 111 and 80 turbines.

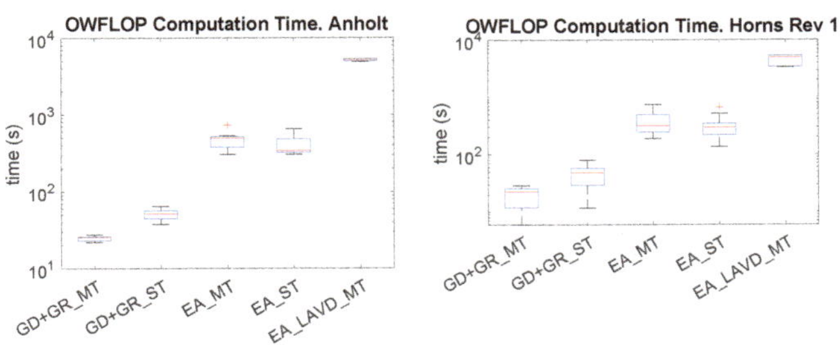

Figure 8. Computation time required to solve the OWFLOP with data from Anholt (**left**) and Horns Rev 1 (**right**). MT and ST refer, respectively, to parallel CPU processing and a single thread. A comparison is presented for the proposed **GD + GR** method (MT and ST) and an improved EA (MT and ST). The four first graphs use LAPD as the wake aggregation method, and the fifth one uses LAVD. Red segments and crosses are the medians and outliers, respectively.

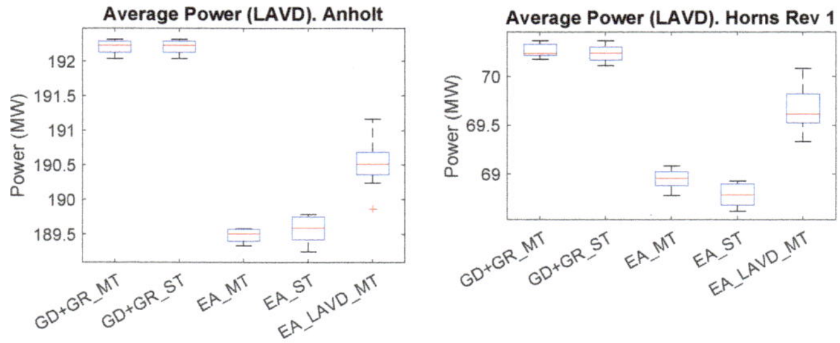

Figure 9. AEP of the solutions found for the test beds with data from Anholt (**left**) and Horns Rev 1 (**right**). MT and ST refer, respectively, to parallel CPU processing and a single thread. A comparison is presented for the proposed **GD + GR** method (MT and ST) and an improved EA (MT and ST). The four first graphs use LAPD as wake aggregation method, and the fifth one uses LAVD. Red segments and crosses are the medians and outliers, respectively.

Table 3. Performance comparison of the proposed GD + GR algorithm, an EA one (both using LAPD for the $\overline{\delta P}$ estimation), and an EA algorithm using LAVD for the $\overline{\delta P}$ estimation. The computation time is presented in the upper table and the obtained power is in the central one. The lower table displays the ratio between the value of δP obtained using the aggregations LAPD y LAVD. For each data the mean value of 10 launches is displayed together with the standard deviation in brackets. Input data from Table A1.

	LAPD				LAVD
Comp. Time	GD + GR		EA		EA
	MT	ST	MT	ST	MT
Anholt	24.78 s (1.90 s)	50.94 s (8.78 s)	477.27 s (132.96 s)	407.20 s (124.44 s)	5168.5 s (194.51 s)
Horns Rev 1	13.44 s (7.73 s)	39.83 s (24.83s)	243.56 s (48.62 s)	218.48 s (50.18 s)	3375.1 s (22.24 s)
	LAPD				LAVD
$\mathbb{E}[P]$ (LAVD used for final solution)	GD + GR		EA		EA
	MT	ST	MT	ST	MT
Anholt	192.20 MW (0.10 MW)	192.20 MW (0.10 MW)	189.48 MW (0.09 MW)	189.57 MW (0.19 MW)	190.55 MW (0.3803 MW)
Horns Rev 1	70.26 MW (0.10 MW)	70.23 MW (0.09 MW)	68.94 MW (0.09 MW)	68.78 MW (0.12 MW)	69.66 MW (0.23 MW)
With the conditions of Table A1 $\mathbb{E}[P]_{Anholt} = 192.0\,\text{MW}$ and $\mathbb{E}[P]_{HR1} = 69.6\,\text{MW}$					
	LAPD				LAVD
$\frac{\mathbb{E}[P]_{\text{LAPD}}}{\mathbb{E}[P]_{\text{LAVD}}}$	GD + GR		EA		EA
	MT	ST	MT	ST	MT
Anholt	0.9499 (0.0005)	0.9499 (0.0005)	0.9518 (0.0009)	0.9514 (0.0009)	0.9317 (0.0081)
Horns Rev 1	0.9636 (0.0007)	0.9640 (0.0009)	0.9642 (0.0008)	0.9639 (0.0010)	0.9043 (0.0187)

The second figure represents the $\mathbb{E}[P] = \overline{P} = \text{AEP}/T$ of the obtained solution after executing the algorithms. Again, the proposed algorithm shows a better performance than EAs. However, using LAPD instead of LAVD can deteriorate the performance of other algorithms (an EA in this case).

These results confirm the two main contributions:

- LAPD can be used as the objective function during the search process, since during the optimization of the power deficit (calculated by means of LAPD), the power deficit (calculated by means of LAVD) is also minimized. Using LAPD, the computation time is substantially reduced.
- The proposed method based on GD together with GR substantially improves the solution obtained with an EA.

In the proposed method (GD-LAPD), the reduction in time is around 1:200 with respect to EA-LAVD, and around 1:20 with respect to EA-LAPD.

Finally, Figure 10 represents the turbine locations for the optimum solutions, both for the set of GD + GR experiments and for the set of EA experiments. In both cases, the turbines are preferentially located at the CA contour in order to maximize the distances between them. However, the inner turbines are more scattered when the EA is used. In the proposed algorithm, the vertices are more likely to be occupied, and a certain trend is observed to dispose of the inner turbines in a more uniform shape; this is similar for all of the algorithm executions.

The results found for Horns Rev 1 show solutions very similar to those found for the same site by the random search algorithm based on continuous formulation [28], which identifies solutions with the majority of turbines packed into the side boundaries and a somewhat regular grid in the interior of the farm. They are also very similar to those found by the SGD method of [16], also on a rectangle but with different wind characteristics.

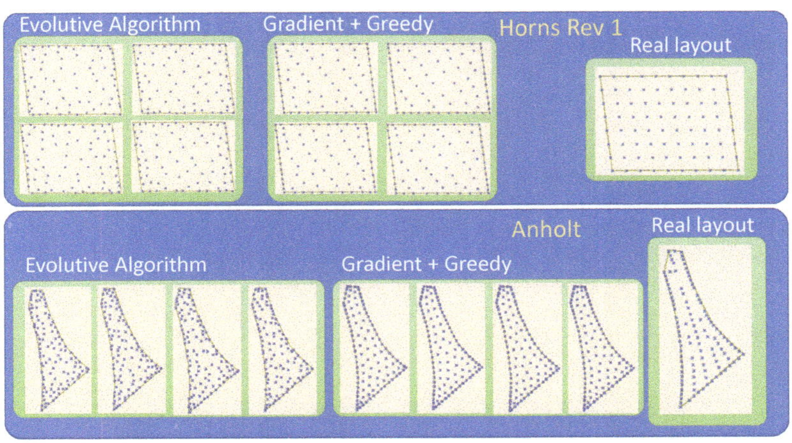

Figure 10. Comparison of different layout solutions as given by the evolutionary algorithm and the proposed method, for both the Horns Rev 1 site and the Anholt site. On the right are the actual layouts.

For comparison, the times achieved in these and other works are listed as follows:

- Ref. [16] 3600 s for SGD nt = 100, using PyWake 2.6.4 and TopFarm 2.3.5. These authors used SLSQP as a benchmark, with a variable computation time depending on the CA shape between 1200 s and 7500 s.
- Ref. [29] 7560 s for a GA and 6480 s for PSO, nt = 18.
- Ref. [28] 3220 s for a random search algorithm, nt = 80.
- Ref. [30] 500 s using TDA 10 K or OpenWind 200 K, nt = 100.
- Ref. [31] 930 s using GA, nt = 40, orientation discretization = 10 deg.

With the method proposed in this article, the time is reduced to around 25 s on average, with 360 samples of a wind direction of 1 deg, 23 samples of speed of 1 m/s, in a regular or irregular CA for 111 turbines. Compared to the fastest of the algorithms found, the proposed method is about 20 times faster.

As a result, it has been seen that the proposed algorithm obtains better solutions than a traditional GA-based one. The required computation time is drastically reduced in both cases if, during the search, the LAPD method is used to evaluate the power deficits. If, as in the case of GA or EA, only the power deficit is estimated, this time reduction is around 1:20. If gradients are also required, it reaches 1:200.

6. Conclusions and Discussion

6.1. Conclusions

This article manages to overcome three inherent problems of the gradient method: finding an adequate initial solution, a high probability of getting stuck in local maxima, and a long execution time.

With regard to the first concern, an inadequate initial solution can create a potential barrier that prevents turbines seeking free spaces from going beyond the barrier [32]. To avoid this, authors often launch a number of experiments, hoping that one starts from an adequate initial solution. This ultimately multiplies the overall time to obtain the optimum. This article presents a progressive method to obtain an initial solution, introducing in batches the turbines within the CA. This way, the turbines are repelled towards the periphery, allowing the turbines to occupy the entire surface (especially the contour), which is usually optimal for wide-spreading purposes.

Regarding the problem of reaching a local maxima that prevents the achievement of a global optimum, after the primary search based on GD, there exists the possibility of falling into local maxima. The GR algorithm is responsible for tackling this issue. This algorithm suppresses the turbines that eventually give rise to the local maxima, and new positions are sought for them. After performing a secondary GD displacement, the process of suppressing and repositioning is repeated.

In order to drastically reduce the execution time, the presented work uses a new aggregation method, LAPD, to calculate the power deficit and its gradient. With this, the problem is partially linearized, and an analogy can be found between turbines and electrostatic charges. Therefore, turbines are moved in the direction of their power deficit gradient in order to obtain an overall optimization.

Many works deal with land-use constraints as concerns that are difficult to handle. In this paper, this issue does not entail a significant problem and is solved in a natural and geometrical way by limiting the movement targets to the concession zone contour. The Theorem of Green (or Stokes) is used to this end.

6.2. Discussion

The power deficit aggregation method can be applied beyond the gradient method. In fact, it has also been used for the EA as a reference for comparison with the proposed gradient method but it can also be used with other heuristic algorithms. However, it is in the gradient method where, taking advantage of its linear superposition character, a problem of $2nt$ variables can be decomposed into nt problems of two variables, making the improvement more evident.

From a purely mathematical point of view, this method does not guarantee the achievement of the global optimum. Given the uncertainties in the modeling of the system, mainly due to inaccurate wind condition forecasting and imprecise wake modeling, deviations in the assumed conditions or models can lead to unrealistic objective function values. Therefore, an exhaustive search for the global optimum is not worthwhile if its neighborhood of monotonicity is very small. In this sense, flatter local optima are preferable, even though they may be of a lower value.

Another problem cited in the literature, which appears for large farms, is the high computing time. Throughout this work, three tools (PS, LAPD and LUT) have been provided to simplify and accelerate the calculation of the objective function and its gradient. Thanks to these tools, the search for the optimum is carried out in a much shorter time compared to previous works. This last tool (LUT) is only possible due to the linearization achieved with LAPD.

Compared to the EA, it is observed that the GD actively seeks the contours as positions for the turbines, noting that these are locations where the potential (i.e., the power deficit) is minimized. On the contrary, for a search engine like an EA, based on crossovers and mutations, it will be more difficult to occupy such a contour, especially when it is irregular.

In summary, this method is by far the fastest of all layout optimization methods found, it does not present limitations regarding the number of turbines or the area of the CA, and it naturally manages land-use constraints for irregular CAs. The solutions offered by the algorithm are consistent, similar to those of other authors who have thoroughly studied layout optimization, and also similar to those of real parks. As a limitation, it is imposed that the turbines must be of the same model and height, and there should not be a nearby farm or obstacle that could cast a shadow on the study area causing heterogeneous wind distribution patterns. It is also restricted to OWFs. For onshore wind farms, several factors generally cause the disturbance exerted by turbine i on turbine j to differ from that exerted by turbine j on turbine i. This symmetry is the basis of the mathematical formulation, and as a result, the proposed reasoning would not be valid. These factors include differences in the rotor height causing different ground interactions; variations in free-stream wind conditions due to a rough orography; different surface roughness and the presence of nearby obstacles.

As a future line of research, disconnected CAs with exclusion zones within them will be studied, as well as the effect of existing parks in the vicinity of the CA.

Author Contributions: Conceptualization, A.G.G.-R.; Data curation, J.S.-G. and J.V.M.-D.; Formal analysis, A.G.G.-R. and J.M.R.-F.; Funding acquisition, J.M.R.-F. and J.S.-G.; Investigation, A.G.G.-R. and J.M.R.-F.; Methodology, A.G.G.-R. and J.M.R.-F.; Project administration, J.M.R.-F. and J.S.-G.; Resources, J.S.-G.; Software, A.G.G.-R. and J.V.M.-D.; Supervision, A.G.G.-R.; Validation, A.G.G.-R. and J.M.R.-F.; Visualization, J.S.-G. and J.V.M.-D.; Writing—original draft, A.G.G.-R.; Writing—review and editing, J.V.M.-D. and J.M.R.-F. All authors have read and agreed to the published version of the manuscript.

Funding: This paper is part of the research project PID2021-127550OA-I00 funded by MICIU/AEI/10.13039/501100011033.

Institutional Review Board Statement: Not applicable.

Informed Consent Statement: Not applicable.

Data Availability Statement: The original contributions presented in the study are included in the article; further inquiries can be directed to the corresponding author.

Conflicts of Interest: The authors declare no conflicts of interest.

Abbreviations

AEP	Annual Energy Production
CA	Concession area
EA	Evolutionary algorithm
GA	Genetic algorithm
GD	Gradient descent
GR	Greedy repositioning
LAPD	Linear aggregation of power deficit
LAVD	Linear aggregation of velocity deficit
LUT	Look-up table
OWF	Offshore wind farm
OWFLOP	Offshore wind farm layout optimization problem
PS	Point symmetry
QAVD	Quadratic aggregation of velocity deficit
SGD	Stochastic gradient descent
nt	Number of turbines
U	Potential energy
$\mathbb{E}$	Average value for all wind conditions
δ	Deficit of

Appendix A. Deviation in AEP Using LAVD Assuming a Symmetrical Wind Distribution

It has been demonstrated that a symmetrical wind distribution can be used without additional errors if LAPD is used. If LAVD is used, this is also valid for wind farms wind PS, meaning that if the entire layout is rotated by 180° about its center, it will result in the same layout. This is, e.g., the case of rhomboidal-shape OWFs, such as HR1. For the sake of space, the demonstration is not included. Nevertheless, it will be seen that, in general, the error considering a symmetrical wind distribution is meaningless.

Several tests have been conducted on various wind farms (see Figure A1) to evaluate the deviation resulting from this simplification:

1. Horns Rev 1 (HR1), characterized by a uniform distribution in rows and columns, thus exhibiting PS.
2. Horns Rev 2 (HR2), featuring a uniform distribution without PS.
3. Anholt is an offshore wind farm with an irregular shape.

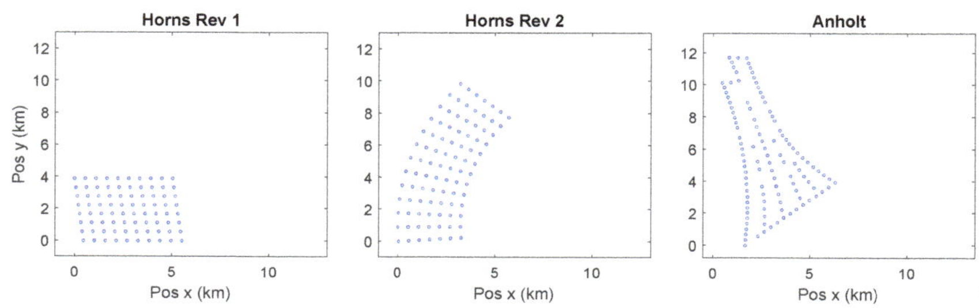

Figure A1. Layout for different OWFs: Horns Rev 1, Horns Rev 2 and Anholt.

Wind conditions at Anholt, HR1, and HR2 (assumed to be equal to HR1), as well as the power curve for the installed wind turbines, are provided in Tables A1 and A2.

Table A1. Wind conditions at the Anholt and HR1 sites. Values for probability and Weibull parameters (scale factor $Weib_A$ and shape factor $Weib_K$) for each sector. $Weib_A$ is given at 135 m for Anholt and at 62 m for HR1.

Anholt	N	NNE	NEE	E	EES	ESS	S	SSW	SWW	W	WWN	WNN
freq (%)	2.38	4.78	6.71	8.76	10.08	8.38	9.87	13.03	10.68	13.33	8.12	3.88
$Weib_A$ (m/s)	8.35	8.07	8.51	10.67	11.16	10.32	11.98	12.56	10.76	11.52	10.42	8.43
$Weib_K$	2.13	2.22	2.26	2.22	2.35	2.17	2.14	2.89	2.79	2.30	1.79	2.17
HR1	N	NNE	NEE	E	EES	ESS	S	SSW	SWW	W	WWN	WNN
freq (%)	3.8	4.3	5.5	8.3	8.7	6.7	8.4	10.5	11.4	12.2	13.9	6.1
$Weib_A$ (m/s)	8.71	9.36	9.29	10.27	10.89	10.49	10.94	11.23	11.93	11.94	12.17	10.31
$Weib_K$	2.08	2.22	2.41	2.37	2.51	2.75	2.61	2.51	2.33	2.35	2.58	2.01

Table A2. Power curve for the turbines installed, respectively, at Horns Rev I, Horns Rev II and Anholt. Speed (first row) in m/s and Power (second to fourth rows) in kW.

Model	3	4	5	6	7	8	9	10	11	12	13	14	15	16	>16
Vestas V80	0	66	154	282	460	696	996	1341	1661	1866	1958	1988	1997	1999	2000
SWT 2.3-93	0	53	163	393	682	1039	1430	1730	2040	2241	2300	2300	2300	2300	2300
SWP 3.6-120	0	161	351	635	1026	1544	2204	2910	3399	3567	3596	3600	3600	3600	3600

The resulting comparison is presented in Table A3. As can be seen, the differences are negligible; therefore, it is not practical to use the original expression (4). Instead, evaluating half of the turbine interactions reduces the computation time by half.

Table A3. Comparison of the $\mathbb{E}[P]$ evaluation without using and using PS for the wind distribution. Estimations made with LAVD.

OWF	Original Expression (4)	Using (24)	Difference
Horns Rev 1	69,602.12 kW	69,602.12 kW	0.0000%
Horns Rev 2	122,353.82 kW	122,353.82 kW	0.0000%
Anholt	191,980.35 kW	191,972.82 kW	0.0039%

Appendix B. Demonstrating $\overline{\delta P_{i\to j}} = \overline{\delta P_{j\to i}}$

To demonstrate this, it can be started from

$$\overline{\delta P_{i\to j}} = \sum_{v=u_{ci}}^{u_{co}} \sum_{\theta=1}^{360} \delta P_{i\to j}(v,\theta)\ \mathrm{f}^{+\pi}(u,\theta)\Delta u \Delta\theta \quad \text{(A1)}$$

$$\overline{\delta P_{j\to i}} = \sum_{v=u_{ci}}^{u_{co}} \sum_{\theta=1}^{360} \delta P_{j\to i}(v,\theta)\ \mathrm{f}^{+\pi}(u,\theta)\Delta u \Delta\theta \quad \text{(A2)}$$

Since $\mathrm{f}^{+\pi}(u,\theta) = f^{+\pi}(u,\theta+\pi)$ and $\delta P_{i\to j}(v,\theta) = \delta P_{i\to j}(v,\theta+\pi)$ the expression (A2) turns into

$$\begin{aligned}\overline{\delta P_{j\to i}} &= \sum_{v=u_{ci}}^{u_{co}} \sum_{\theta=1}^{360} \delta P_{i\to j}(v,\theta+\pi) f^{+\pi}(u,\theta+\pi)\Delta u \Delta\theta = \sum_{v=u_{ci}}^{u_{co}} \sum_{\theta=-179}^{180} \delta P_{i\to j}(v,\theta)\ \mathrm{f}^{+\pi}(u,\theta)\Delta u \Delta\theta \\ &= \sum_{v=u_{ci}}^{u_{co}} \sum_{\theta=1}^{360} \delta P_{i\to j}(v,\theta)\ \mathrm{f}^{+\pi}(u,\theta)\Delta u \Delta\theta = \overline{\delta P_{i\to j}} \quad \square \end{aligned} \quad \text{(A3)}$$

Appendix C. Position of a Point with Regard to a Closed Curve

In order to detect if a wind turbine is inside the CA, it is necessary to program a function to deduce whether a point is inside, outside, or in the contour of a closed polygon. It is convenient to first define the operation $\dot{+}$ as well as its opposite $\dot{-}$ as

$$\begin{aligned} a,b &\in [-\pi,\pi], \\ a \dot{+} b &\Leftrightarrow \begin{cases} a+b & \text{if} \quad a+b \in [-\pi,\pi] \\ a+b+2\pi & \text{if} \quad a+b < -\pi \\ a+b-2\pi & \text{if} \quad a+b > \pi \end{cases} \end{aligned} \quad \text{(A4)}$$

$$a \dot{-} b \Leftrightarrow \begin{cases} a-b & \text{if} \quad a-b \in [-\pi,\pi] \\ a-b+2\pi & \text{if} \quad a-b < -\pi \\ a-b-2\pi & \text{if} \quad a-b > \pi \end{cases} \quad \text{(A5)}$$

The operation $\dot{+}$ together with the interval $[-\pi,\pi]$ constitutes an abelian group satisfying commutativity, closure, associativity, the existence of a neutral element (0), and an opposite element (the opposite of c is $-c$).

With these definitions, and assuming that a closed curve C is composed of a sequence of m points (x_i, y_i) (in which $x_m = x_0$ and $y_m = y_0$), then in order to ascertain whether a point $P(x^p, y^p)$ is inside or outside this curve, it is sufficient to evaluate the summary

$$\Gamma = \sum_{k=1}^{m} \left\{ \arctan(y_{k+1} - y^p, x_{k+1} - x^p) \dot{-} \arctan(y_k - y^p, x_k - x^p) \right\} \quad \text{(A6)}$$

By means of Stokes' theorem, it can be demonstrated that

$$P(x,y) \text{ is inside } C \Leftrightarrow \Gamma = 2\pi \quad \text{or} -2\pi \quad \text{(A7)}$$

$$P(x,y) \text{ is outside } C \Leftrightarrow \Gamma = 0 \quad \text{(A8)}$$

With this formulation, then $\Gamma = \pi$ or $-\pi$ for points belonging to the contour. In order to detect this situation, it is preferable to check each value of the summary in (A6). The point will belong to the contour if and only if any of the addends yield $-\pi$ or π.

References

1. Elkinton, C.; Manwell, J.; Mcgowan, J. Algorithms for Offshore Wind Farm Layout Optimization. *Wind Eng.* **2008**, *32*, 67–84. [CrossRef]
2. Yang, K.; Kwak, G.; Cho, K.; Huh, J. Wind farm layout optimization for wake effect uniformity. *Energy* **2019**, *183*, 983–995. [CrossRef]
3. Cazzaro, D.; Pisinger, D. Variable neighborhood search for large offshore wind farm layout optimization. *Comput. Oper. Res.* **2022**, *138*, 105588. [CrossRef]
4. Herbert-Acero, J.F.; Probst, O.; Réthoré, P.E.; Larsen, G.C.; Castillo-Villar, K.K. A Review of Methodological Approaches for the Design and Optimization of Wind Farms. *Energies* **2014**, *7*, 6930–7016. [CrossRef]
5. Cinar, A.C. A review of 10 × 10 and 20 × 20 grid-type wind turbine placement problems solving by metaheuristics. *Environ. Sci. Pollut. Res.* **2023**, *30*, 11359–11377. [CrossRef]
6. Croonenbroeck, C.; Hennecke, D. A comparison of optimizers in a unified standard for optimization on wind farm layout optimization. *Energy* **2021**, *216*, 119244. [CrossRef]
7. Pérez, B.; Mínguez, R.; Guanche, R. Offshore wind farm layout optimization using mathematical programming techniques. *Renew. Energy* **2013**, *53*, 389–399. [CrossRef]
8. Park, J.; Law, K.H. Layout optimization for maximizing wind farm power production using sequential convex programming. *Appl. Energy* **2015**, *151*, 320–334. [CrossRef]
9. Guirguis, D.; Romero, D.A.; Amon, C.H. Toward efficient optimization of wind farm layouts: Utilizing exact gradient information. *Appl. Energy* **2016**, *179*, 110–123. [CrossRef]
10. Guirguis, D.; Romero, D.A.; Amon, C.H. Gradient-based multidisciplinary design of wind farms with continuous-variable formulations. *Appl. Energy* **2017**, *197*, 279–291. [CrossRef]
11. Risco, J.; Rodrigues, R.; Frederiksen, M.; Quick, J.; Pedersen, M.; Réthoré, P.E. Gradient-based Wind Farm Layout Optimization With Inclusion And Exclusion Zones. *Wind Energy Sci.* **2023**, *9*, 585–600. [CrossRef]
12. Yamani Douzi Sorkhabi, S.; Romero, D.A.; Beck, J.C.; Amon, C.H. Constrained multi-objective wind farm layout optimization: Novel constraint handling approach based on constraint programming. *Renew. Energy* **2018**, *126*, 341–353. [CrossRef]
13. King, R.N.; Dykes, K.; Graf, P.; Hamlington, P.E. Optimization of wind plant layouts using an adjoint approach. *Wind Energy Sci.* **2017**, *2*, 115–131. [CrossRef]
14. Gonzalez-Rodriguez, A.G.; Payan, M.B.; Santos, J.R.; Gonzalez, J.S. Optimization of regular offshore wind-power plants using a non-discrete evolutionary algorithm. *AIMS Energy* **2017**, *5*, 173–192. [CrossRef]
15. Mittal, P.; Kulkarni, K.; Mitra, K. A novel hybrid optimization methodology to optimize the total number and placement of wind turbines. *Renew. Energy* **2016**, *86*, 133–147. [CrossRef]
16. Quick, J.; Rethore, P.E.; Mølgaard Pedersen, M.; Rodrigues, R.V.; Friis-Møller, M. Stochastic gradient descent for wind farm optimization. *Wind Energy Sci.* **2023**, *8*, 1235–1250. [CrossRef]
17. Ma, Y.; Archer, C.L.; Vasel-Be-Hagh, A. The Jensen wind farm parameterization. *Wind Energy Sci.* **2022**, *7*, 2407–2431. [CrossRef]
18. Shao, Z.; Wu, Y.; Li, L.; Han, S.; Liu, Y. Multiple wind turbine wakes modeling considering the faster wake recovery in overlapped wakes. *Energies* **2019**, *12*, 680. [CrossRef]
19. Gonzalez-Rodriguez, A.G.; Serrano-González, J.; Burgos-Payán, M.; Riquelme-Santos, J.M. Realistic Optimization of Parallelogram-Shaped Offshore Wind Farms Considering Continuously Distributed Wind Resources. *Energies* **2021**, *14*, 2895. [CrossRef]
20. Peña, A.; Schaldemose Hansen, K.; Ott, S.; van der Laan, M.P. On wake modeling, wind-farm gradients, and AEP predictions at the Anholt wind farm. *Wind Energy Sci.* **2018**, *3*, 191–202. [CrossRef]
21. Ørsted. Offshore Wind Measurement and Operational Data. Available online: https://orsted.com/en/what-we-do/renewable-energy-solutions/offshore-wind/offshore-wind-data#offshore-meteorological-data (accessed on 1 August 2023).
22. Anholt Offshore Wind Farm Reaches Milestone. Available online: https://orsted.com/en/media/newsroom/news/2014/09/anholt-offshore-wind-farm-reaches-milestone (accessed on 1 August 2023).
23. Energy Numbers. Capacity Factors at Danish Offshore Wind Farms. Available online: https://energynumbers.info/capacity-factors-at-danish-offshore-wind-farms (accessed on 1 August 2023).
24. Nocedal, J.; Wright, S. Conjugate Gradient Methods. In *Numerical Optimization*; Springer: New York, NY, USA, 2006; pp. 101–134. [CrossRef]
25. Kolda, T.G.; Lewis, R.M.; Torczon, V. Optimization by Direct Search: New Perspectives on Some Classical and Modern Methods. *SIAM Rev.* **2003**, *45*, 385–482. [CrossRef]
26. Fletcher, R. On the Barzilai-Borwein Method. In *Optimization and Control with Applications*; Qi, L., Teo, K., Yang, X., Eds.; Springer Nature: Boston, MA, USA, 2005; pp. 235–256.
27. Gonzalez-Rodriguez, A.G.; Burgos-Payan, M.; Riquelme-Santos, J.; Serrano-Gonzalez, J. Reducing computational effort in the calculation of annual energy produced in wind farms. *Renew. Sustain. Energy Rev.* **2015**, *43*, 656–665. [CrossRef]

28. Feng, J.; Shen, W.Z. Solving the wind farm layout optimization problem using random search algorithm. *Renew. Energy* **2015**, *78*, 182–192. [CrossRef]
29. Amaral, L.; Castro, R. Offshore wind farm layout optimization regarding wake effects and electrical losses. *Eng. Appl. Artif. Intell.* **2017**, *60*, 26–34. [CrossRef]
30. Wagner, M.; Day, J.; Neumann, F. A fast and effective local search algorithm for optimizing the placement of wind turbines. *Renew. Energy* **2013**, *51*, 64–70. [CrossRef]
31. Biswas, P.P.; Suganthan, P.; Amaratunga, G.A.J. Optimal placement of wind turbines in a windfarm using L-SHADE algorithm. In Proceedings of the 2017 IEEE Congress on Evolutionary Computation (CEC), San Sebastián, Spain, 5–8 June 2017; pp. 83–88. [CrossRef]
32. Valotta Rodrigues, R.; Pedersen, M.M.; Schøler, J.P.; Quick, J.; Réthoré, P.E. Speeding up large-wind-farm layout optimization using gradients, parallelization, and a heuristic algorithm for the initial layout. *Wind Energy Sci.* **2024**, *9*, 321–341. [CrossRef]

applied sciences

MDPI

Article

Parameter Correction on Waste Heat Recovery System of a Gas Turbine Using Supercritical CO_2 Based on Data Reconciliation

Liang Xu, Jiarui You, Jiahua Wu, Yikang Liu and Di Zhang *

School of Energy and Power Engineering, Xi'an Jiaotong University, Xi'an 710049, China;
xu_liang@stu.xjtu.edu.cn (L.X.); youjiarui189@stu.xjtu.edu.cn (J.Y.); jiahuawu@stu.xjtu.edu.cn (J.W.);
yk_liu@stu.xjtu.edu.cn (Y.L.)
* Correspondence: zhang_di@mail.xjtu.edu.cn

Abstract: At present, the research significance of a waste heat recovery system of a gas turbine using CO_2 is gradually becoming prominent, and accurate parameter measurement and performance monitoring are very necessary in the process of establishing coupled circulation system and actual operation. In this paper, a data reconciliation model is established based on the waste heat recovery system of a gas turbine using supercritical CO_2 (S-CO_2) cycle with two S-CO_2 turbines, and several random errors and gross errors are added to verify the data reconciliation ability of the model. The calculation shows that the data reconciliation model established in this paper can obviously reduce the overall deviation level of each parameter in the thermal system, and can also identify and eliminate gross errors in the system. For some of the key parameters such as the total mass flow of CO_2, data deviation is reduced to less than 1%, and high-precision power values of the equipment are calculated. This means that the measurement accuracy is effectively improved. In general, this paper makes a new attempt to use data reconciliation in parameter measurement and correction of a simple coupled cyclic system, and provides a certain reference for the subsequent application.

Keywords: waste heat recovery; supercritical CO_2 cycle; data reconciliation; gross error identification; measurement data accuracy

1. Introduction

The overall structure of the global energy consumption market is constantly changing and the proportion of clean energy is increasing due to climate change and air pollution control in recent years. However, clean environmental energy, such as wind energy, tidal energy, water energy, solar energy and others, has a very large limitation due to unstable collection [1], so the supporting effect of thermal power generation for the new power system is still significant. In the global energy consumption market in 2023, the proportion of fossil fuel power generation in the field of power generation still reached about 60%, of which the share of gas turbine power generation with natural gas as fuel was also stable at about 23% [2], and its importance is obvious. The gas turbine has the advantages of environmental friendliness, quick start and stop, easy adjustment, etc. At the same time, the gas turbine power generation cycle is also a relatively clean power generation technology which has a wide range of application scenarios.

The relatively high temperature of the gas turbine exhaust is usually around 500 °C [3], so there is still a wide range of waste heat that is worth recycling. For waste heat recovery system design of a gas turbine, the prevailing coupling cycles are steam Rankine cycle, organic Rankine cycle and S-CO_2 power cycle. Steam Rankine cycle is suitable at higher permissible temperatures. In Europe, the United States and Japan, studies have been carried out to increase the efficiency of a gas–steam combined cycle as early as the end of the 20th century [4]. The organic Rankine cycle is suitable under lower temperatures due to the low decomposition temperature of the working medium [5]. However, the decomposition of organic working medium at high temperature may affect the safety of the whole system. The S-CO_2 power cycle has recently aroused the interest of researchers because of its larger recovery temperature range, smaller size and lightweight system design [6]. For the coupling system of gas turbine power generation and S-CO_2 cycle, most of the researches are still at the stage of system design and analysis. Due to the influence of measurement deviation and data transmission error during actual production and operation, it is difficult to obtain accurate measurement data and make it reflect the real operating state of the system. Therefore, further research is needed to perform accurate performance parameter analysis and error diagnosis identification in the simple coupling system.

To solve the problems mentioned above, fuzzy logic [7,8], genetic algorithms [9,10], neural networks [11,12], data reconciliation and other common methods may be used to monitor the performance of thermal system. Among these, data reconciliation technology is very effective in improving the measurement accuracy. Data reconciliation is a method that realizes the coordination and optimization of the input data of the system, reducing various random errors through the model principle and mechanism and various redundant information from the measurement process [13,14]. Data reconciliation technology can reduce the deviation of measurement data and identify gross errors, which can even obtain complete and accurate system performance parameters. Taking the optimization results after data reconciliation as system input data, the calculation accuracy of system performance analysis and error identification can be improved, and it also lays the foundation for the subsequent improvement of system cycle performance and thermal efficiency.

The development of data reconciliation technology has also been through the unremitting efforts of researchers. In theoretical researches, Kretsovalis et al. [15] used graph theory method for data reconciliation calculation and revealed the consistency between this method and the traditional algebraic method in solving data reconciliation problems. Crowe et al. [16,17] proposed a projection matrix to solve the problem of matrix singularity caused by unmeasured variables, and it is generalized to bilinear data reconciliation. Then, Swartz [18] eliminated the unmeasured variables in the constraint equations using the QR matrix orthogonal transformation and improved the solving speed. In addition, the global test method [19], the constraint equation test method [20] and the measurement test method are all suitable for identifying instrument deviation type of gross error. The generalized likelihood ratio test method [21,22] is suitable for identifying the equipment fault type of gross error. In industrial applications, data reconciliation technology is usually used for parameter correction and significant error monitoring in chemical process flow in the field of industrial chemistry [23–26]. By contrast, nuclear power units usually use data reconciliation technology for thermal reactor balance calculations and optimizing reactor power calculations [27–30], while thermal power units usually use data reconciliation technology to improve the measurement accuracy of feed water flow, electric power and other parameters, and to improve the recognition ability of gross errors [31–34].

It is very effective to rectify the parameters of thermal system by using data reconciliation technology, which greatly improves the measurement accuracy of mass flow and power, and it is also the case for the waste heat recovery system of a gas turbine. In the field of gas turbines, the researches on data reconciliation mainly focus on equipment breakdown identification and gas–steam combined cycle system at present, and there are no further researches on data reconciliation in the relatively new system configuration. The data reconciliation model for a waste heat recovery system of a gas turbine using an S-CO_2 cycle established in this paper aims to solve the data reconciliation and suspect measurement identification problems in the relatively new waste heat recovery system of a gas turbine. It lays a foundation for the new system performance monitoring of a gas turbine coupled with S-CO_2.

In this paper, the model is firstly built based on the heat balance equations of the system, and the accuracy of the model is tested using the original calculated design values of the system as the verification benchmark. Then, the optimization results of the model and the recognition of gross errors under 10 different working conditions are calculated, and the availability of the method is analyzed and evaluated. The results show that the model built in this paper improves the measurement data accuracy and the calculation accuracy of the component performance, and provides a certain reference for the model establishment and application of the simple coupled system based on the data reconciliation.

2. System Configuration

The thermal system used for data reconciliation calculation in this paper adopts the waste heat recovery system of a gas turbine proposed by Sun et al. [35], which uses an S-CO_2 cycle with two S-CO_2 turbines. The system layout diagram is shown in Figure 1. The S-CO_2 bottom cycle consists of three heat exchangers, two S-CO_2 turbines, one S-CO_2 compressor, and one cooler. The technical data is listed in Table 1.

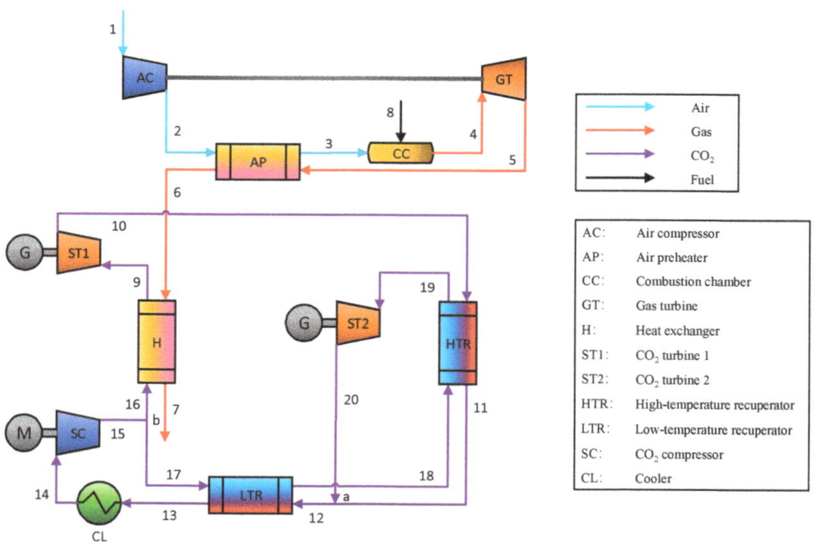

Figure 1. Diagram of coupled system layout.

In the gas turbine cycle, after the air enters the cycle through stream 1, it first enters the air compressor and compressed. The compressed air enters the cold side of the air preheater to be heated by the gas turbine exhaust, then it enters the combustion chamber. In the combustion chamber, the fuel entering the circulation from stream 8 and preheated air are mixed for combustion, and the generated high-temperature gas makes the gas turbine generate power. After the work is done, the gas turbine exhaust enters the hot side of the air preheater for heating, then enters the heat exchanger to continue heating CO_2. The heated exhaust of gas turbine is stream 7.

Table 1. Different equipment characteristic parameters of waste heat recovery system.

Component	Parameter	Value	Unit
Air compressor	Pressure ratio	15.0	--
	Inlet pressure	101.0	kPa
	Inlet temperature	25.0	°C
	Mass flow of air	91.22	kg/s
Air preheater	Outlet exhaust pressure	109.6	kPa
	Outlet exhaust temperature	527.6	°C
Heat exchanger	Terminal temperature difference	10	°C
High-temperature recuperator	Terminal temperature difference	10	°C
Low-temperature recuperator	Terminal temperature difference	10	°C
S-CO_2 compressor	Inlet pressure	8000	kPa
	Outlet pressure	20,000	kPa

In the S-CO_2 bottom cycle, the CO_2 which is heated by the gas turbine exhaust enters the S-CO_2 turbine 1 from stream 9 to do work. After the work is done, the CO_2 successively enters a high-temperature recuperator and a low-temperature recuperator to heat another stream of CO_2 from stream 17 to stream 19, then enters the cooler for cooling and enters the S-CO_2 compressor for recompression. After being compressed by the S-CO_2 compressor, the CO_2 from stream 15 is divided into two streams. Stream 16 enters the heat exchanger to exchange heat with the gas turbine exhaust for continuous circulation. Stream 17 is heated by the two recuperators respectively, then enters the S-CO_2 turbine 2 to generate power. After the work is done, it is merged with another stream of CO_2 at stream 12 between the two recuperators. In this simple coupled circulation system, two S-CO_2 turbines use heat exchangers to exchange heat with the gas turbine exhaust to generate electricity, which solves the recovery of gas turbine waste heat well.

In this paper, a gas turbine with exhaust temperature of 527.60 °C and exhaust mass flow of 92.90 kg/s was used for system construction and data reconciliation calculation. The following assumptions were made to simplify the calculation conditions:

1. The waste heat recovery system is in a stable state, and the CO_2 is always in a supercritical state;
2. The external environment and the cooling water used by the cooler are both at 25 °C, 101 kPa;
3. The pressure loss and heat loss of circulating equipment and pipelines can be ignored;
4. The isentropic efficiencies of the air compressor and the S-CO_2 compressor are constants, which are 80% and 85%, respectively;
5. The isentropic efficiency of the gas turbine and the two S-CO_2 turbines are also constants, which are all 90%.

3. Research Method

3.1. Heat Balance Model

The heat balance model of the waste heat recovery system is established according to the law of thermodynamics. For the system used in this paper, the following parameters are mostly concerned in the data reconciliation calculation process.

The total carbon dioxide mass flow in S-CO_2 cycle expresses as follows:

$$\dot{m}_{CO_2 0} = \dot{m}_{CO_2 1} + \dot{m}_{CO_2 2} \tag{1}$$

where $\dot{m}_{CO_2 1}$ and $\dot{m}_{CO_2 2}$ are the mass flow of CO_2 into ST1 and ST2 branches respectively.

The net output work of the coupled system expresses as follows:

$$W_{net} = W_{GT} + W_{ST1} + W_{ST2} - W_{AC} - W_{SC} \tag{2}$$

where W_{GT} is the output power of the gas turbine, W_{ST1} and W_{ST2} are the output power of the S-CO_2 turbines 1 and 2, respectively, and W_{AC} and W_{SC} are the consumption power of the air compressor and the S-CO_2 compressor, respectively.

The heat balance equation models of the components in the system are shown in Table 2, where $h_i(i = 1, 2, 3, \ldots)$ is the enthalpy value of stream i, $h_{i,s}(i = 1, 2, 3, \ldots)$ is the ideal enthalpy value after an isentropic process, $m_i(i = 1, 2, 3, \ldots)$ is the mass flow of stream i, Q is the heat exchange amount of a heat exchanger, and η is the isentropic efficiency of the equipment.

Table 2. Heat balance equations of each component.

Component		Heat Balance Equation	Description
AC	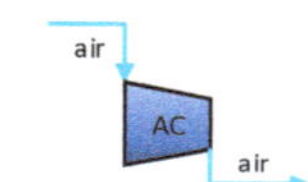	$W_{AC} = m_1(h_2 - h_1)$	Load equation of air compressor
		$\eta_{AC} = (h_{2,s} - h_1)/(h_2 - h_1)$	Efficiency equation of air compressor
AP	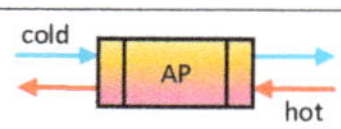	$Q_{AP} = m_2(h_3 - h_2) = m_5(h_5 - h_6)$	Heat load equation of air preheater
CC	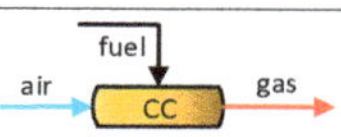	$m_4 h_4 = m_3 h_3 + m_8 h_8$	Conservation equation of combustion chamber
GT	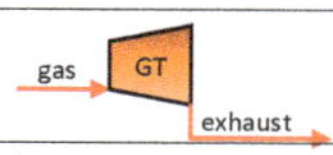	$W_{GT} = m_4(h_4 - h_5)$	Power equation of gas turbine
		$\eta_{GT} = (h_4 - h_5)/(h_4 - h_{5,s})$	Efficiency equation of gas turbine
H	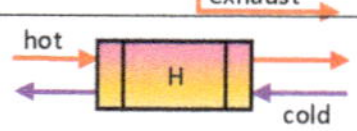	$Q_H = m_6(h_6 - h_7) = m_{16}(h_9 - h_{16})$	Heat load equation of heat exchanger
ST1	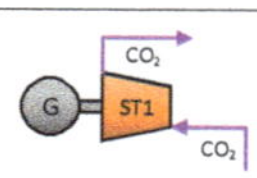	$W_{ST1} = m_9(h_9 - h_{10})$	Power equation of S-CO_2 turbine 1
		$\eta_{ST1} = (h_9 - h_{10})/(h_9 - h_{10,s})$	Efficiency equation of S-CO_2 turbine 1
ST2	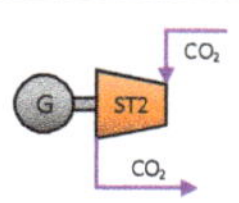	$W_{ST2} = m_{19}(h_{19} - h_{20})$	Power equation of S-CO_2 turbine 2
		$\eta_{ST2} = (h_{19} - h_{20})/(h_{19} - h_{20,s})$	Efficiency equation of S-CO_2 turbine 2
HTR	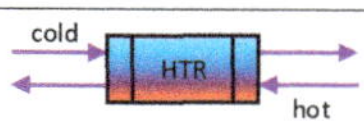	$Q_{HTR} = m_{10}(h_{10} - h_{11}) = m_{18}(h_{19} - h_{18})$	Heat load equation of high-temperature recuperator
LTR	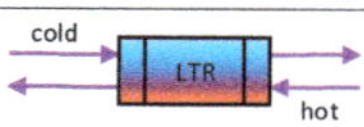	$Q_{LTR} = m_{12}(h_{12} - h_{13}) = m_{17}(h_{18} - h_{17})$	Heat load equation of low-temperature recuperator
SC	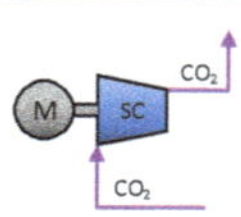	$W_{SC} = m_{14}(h_{15} - h_{14})$	Load equation of S-CO_2 compressor
		$\eta_{SC} = (h_{15,s} - h_{14})/(h_{15} - h_{14})$	Efficiency equation of S-CO_2 compressor
CL	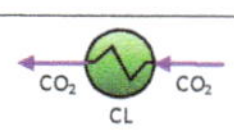	$Q_{CL} = m_{13}(h_{13} - h_{14})$	Heat load equation of cooler

3.2. Data Reconciliation Principle

In order to monitor the operating parameters in different position of the system, it is necessary to set sensors to obtain the measured values, and the data collected by these sensors is merged into the measurement data base as the input to data reconciliation. However, deviations in these measured values are unavoidable. Data reconciliation is a method to modify the measurement data under the premise of minimum total correction, so that all the measurement data can meet the constraint equations of the system. It can also solve the unknown data that is difficult to measure. When the whole thermal system is in a steady state, the vector composed of all known variables including characteristic data and measurement data with only random errors is defined as:

$$\boldsymbol{x} = \boldsymbol{x_0} + \boldsymbol{\delta} \tag{3}$$

where $\boldsymbol{x_0}$ is the true value vector of the known variable, and $\boldsymbol{\delta}$ is the random error vector.

For the known variables in the thermal system, the problem that data reconciliation can solve is to minimize the sum of squares of the differences between the reconciliated values and the initial value under the constraint equations of energy and mass balance. Mathematically, this optimization problem can be converted to solving the least square solution that satisfies multiple constrained equations:

$$\begin{cases} \min\limits_{\hat{x}} \xi(\hat{\boldsymbol{x}}) = \sum\limits_{i=1}^{n} \dfrac{(\hat{x}_i - x_i)^2}{\sigma_i^2} \\ \text{s.t. } \boldsymbol{F}(\hat{\boldsymbol{x}}, \boldsymbol{u}) = \boldsymbol{0} \end{cases} \tag{4}$$

where x_i is the component of known variable $\boldsymbol{x}$ in Equation (3), $\hat{x}_i$ is the reconciliated value of the known variable in group i corresponding to x_i (which compose the reconciled vector $\hat{\boldsymbol{x}}$), σ_i is the mean square deviation (standard deviation) corresponding to x_i, $\xi(\hat{\boldsymbol{x}})$ is the object function of the optimization problem, $\boldsymbol{F}$ is a equation vector representing a group of algebraic equations, and $\boldsymbol{u}$ is the vector of the unknown variable.

If $\boldsymbol{F}$ has a linear function to $\hat{\boldsymbol{x}}$, then there is an analytical solution to this linear constrained problem after using the method of the unknown variable mentioned in Section 3.5. However, for general thermal systems, due to the existence of complex functions such as the property function of the working medium and the characteristic function of the equipment components, $\hat{\boldsymbol{x}}$ to $\boldsymbol{u}$ should be nonlinear, so this problem is a quadratic solution problem with nonlinear constraints. The value of the unknown variable cannot be determined by the optimization problem and must be eliminated in the process of solving. In this paper, the improved Lagrange multiplier method proposed by VDI-2048 [36] is used for the calculation. This method converts constrained optimization into unconstrained optimization and nonlinear constraints into linear constraints in this problem in essence. After eliminating the effects of the unknown variable, the correction vector of the initial value vector is given as follows [36]:

$$\boldsymbol{v} = \boldsymbol{S_x J_{F,x} S_F^{-1} F^{(0)}} \tag{5}$$

where $\boldsymbol{S_x}$ is the covariance matrix of the known variable vector, $\boldsymbol{J_{F,x}}$ is the Jacobian matrix of equation vector $\boldsymbol{F}$ to $\boldsymbol{x}$, $\boldsymbol{S_F^{-1}}$ is the inverse matrix of the covariance matrix of equation vector $\boldsymbol{F}$, which can be calculated as follows:

$$\boldsymbol{S_F} = \boldsymbol{J_{F,x}^T S_x J_{F,x}} \tag{6}$$

and $\boldsymbol{F^{(0)}}$ presents the initial residual vector of $\boldsymbol{F}$ whose notation in the upper right corner indicates that it has not been iterated. Here, the correction matrix of the covariance matrix is also given as follows:

$$S_v = S_x J_{F,x}^T S_F^{-1} S_x J_{F,x} \tag{7}$$

which should be used to modify the covariance matrix S_x.

The linearization method usually expands the nonlinear constrained equation according to Taylor's formula and retains the first derivative term. However, in practical engineering applications, the expanded higher-order derivative term cannot be simply ignored by Jacobian matrix linearization, and the residual of constraint equation cannot be eliminated by a simple correction. Thus, the data reconciliation model established in this paper adopts the iterative algorithm framework developed by You et al. [37]. The correction vector iteration format is constructed as follows:

$$v^{(k)} = \begin{cases} 0, & k = 0 \\ v^{(k-1)} + \Delta v^{(k)}, & k \geq 1 \end{cases} \tag{8}$$

$$\Delta v^{(k)} = S_x J_{F,x}^{(k)} \left[S_F^{(k)} \right]^{-1} F^{(k)} \tag{9}$$

where k is the number of iterations. Before the iterative computation begins, the updated Jacobian matrix $J_{F,x}^{(k)}$ is recalculated by numerical differentiation, and the covariance matrix of the equation residual vector $S_F^{(k)}$ is recalculated by Equation (6), then the iterative process continues. The covariance matrix of the correction vector for the iterative calculation process can be solved as follows:

$$S_v^{(k)} = \begin{cases} 0, & k = 0 \\ S_v^{(k-1)} + \Delta S_v^{(k)}, & k \geq 1 \end{cases} \tag{10}$$

$$\Delta S_v^{(k)} = S_v^{(k)} \left[J_{F,x}^{(k)} \right]^T \left[S_F^{(k)} \right]^{-1} S_v^{(k)} J_{F,x}^{(k)} \tag{11}$$

And the iterative process is terminated when the incremental norm of the correction vector is too small:

$$\| \Delta v^{(k)} \| < \varepsilon \tag{12}$$

where ε is the convergence limit. In this way, the optimizing calculation of data reconciliation for the initial value of variables is completed.

3.3. Gross Error Identification

Under normal circumstances, data reconciliation has a good coordination and optimization effect for random errors obeying statistical distribution law, but it may have a poor effect for gross errors disobeying statistical distribution law. The reason is that gross errors will be dispersed to all related data in the system during the process of data reconciliation, which may lead to a greater degree of overall deviation in the reconciliated data.

Gross error identification usually adopts the statistical hypothesis test method, which is only based on the operation data, has the characteristics of strong applicability, economy and simple operation, and has been widely studied and applied. Different test statistics correspond to the hypothesis statistical test methods, which commonly include the global test method, the constraint equation test method and the measurement test method. The data reconciliation model established in this paper adopts the global test method and the measurement test method.

The global test method is a method to judge whether a gross error in the whole system calculation model exists. In the global test method, the global test statistic can be defined as follows [38]:

$$K = \boldsymbol{v}^T \mathbf{S}_x^{-1} \boldsymbol{v} \quad (13)$$

If the gross error does not exist in the whole system, the test statistic K following the chi-square distribution should meet the following requirements:

$$K < \chi^2_{1-\alpha,n} \quad (14)$$

where $\chi^2_{1-\alpha,n}$ is the chi-square distribution critical value. Normalization of this formula which is easy to use results in the following [38]:

$$Q = \frac{K}{\chi^2_{1-\alpha,n}} < 1 \quad (15)$$

where Q means quality factor is used to characterize the gross error of the whole system.

Even if the whole system satisfies Equation (15), there may still be gross errors in some parameters, and the measurement test method is used to identify the gross errors of parameters. In the measurement test method, in order to realize the identification of suspicious data with gross errors, the test statistics of variable x_i based on the correction vector $\boldsymbol{v}$ can be constructed as follows [36]:

$$r_i = \frac{|\nu_i|}{\sqrt{max\left(\sigma^2_{\nu,i}, \frac{\sigma^2_{x,i}}{10}\right)}} \quad (16)$$

where ν_i is the correction value of variable x_i. For the right side of Equation (16), when $|\nu_i|$ reaches the allowable error maximum ξ_{max}, there is the following equation:

$$r_i < \mathcal{Z}_{1-\alpha/2} \quad (17)$$

where $\mathcal{Z}_{1-\alpha/2}$ is the lower α fraction of the normal distribution, which can be obtained by looking up the statistics table. In the process of systematic data reconciliation calculation, given $\alpha = 0.05$ which means the confidence level is 95%, it can be seen from the table that $\mathcal{Z}_{1-\alpha/2} = 1.96$. Under normal circumstances, this should be:

$$r_i < 1.96 \quad (18)$$

This means that there are no anomalies in this set of data. If Equation (18) is not satisfied, it means that the confidence probability of the reconciliated value is less than 5%, and it can be considered that the parameter is suspicious with gross error. At this time, it is reasonable to think that an instrument error or equipment breakdown may have occurred, and further inspection must be carried out.

Therefore, how do we use practical identification methodology in data reconciliation process? First, we can build the overall model of heat balance equations and redundant constraint equations after determining known and unknown variables. At this time, we can give the global test statistic K of the model and the test statistics r_i of each variable respectively according to above formulas, which can be used as criteria for testing gross errors. Later, as the calculation of iterative process gradually converges, K can be used to test whether the reconciliated system model has gross errors and r_i can be used to test whether the reconciliated single variable contains gross errors.

3.4. Sensor Arrangement

The variables in the system can be divided into the known variables and the unknown variables. The known variables used as input for data reconciliation include two types: characteristic data and measurement data.

Characteristic data are generally based on characteristic parameters of the equipment or design values of system operation because of their relative stability in most situation. The characteristic data used in this paper are shown in Table 1.

For the measurement data, it is necessary to arrange sensors in each part of the whole system. There are three types of sensors: pressure, temperature and mass flow. The arrangement of sensors in this paper is shown in Table 3. It can be seen that three types of sensors are mainly arranged around the S-CO_2 bottom cycle, while some sensors in the gas turbine cycle mainly supplement some necessary data and provide little redundancy for data reconciliation. There are more temperature sensors than sensors for pressure and mass flow, and several temperature sensors are arranged in one stream of waste heat recovery system in some cases.

Table 3. Arrangement of sensors in different stream of waste heat recovery system.

Stream Number	Pressure Number	Temperature Number	Mass Flow Number
Stream 1		TS01	
Stream 2		TS02	MFS01
Stream 3	PS01	TS03	
Stream 4	PS02	TS04	
Stream 5	PS03	TS05	MFS02
Stream 6		TS06	
Stream 7		TS07	
		TS08	
Stream 9		TS09	
		TS10	
Stream 10			MFS03
Stream 11		TS11	
Stream 13		TS12	MFS04
		TS13	
Stream 14	PS04	TS14	
		TS15	
Stream 15	PS05		
		TS16	
Stream 19		TS17	
		TS18	
Stream 20			MFS05

The measurement uncertainty represents the reasonable distribution of the measured value under a specific probability. It can quantify the possible deviation of the measured value. It should be noted that the uncertainty here is not equivalent to the uncertainty level of the measurement instrument, but is only used in data reconciliation calculations. It is similar to a weight coefficient, reflecting the uncertainty of the data before and after the reconciliation.

For each pressure, temperature and mass flow sensor, there is a certain uncertainty. Even small fluctuations in environmental conditions will affect the measurement performance and lead to deviations. In general, the measurement uncertainty can be empirically assumed as either the instrument accuracy or the maximum permissible deviation in practice [28]. In this paper, the specific assumptions of uncertainty are as follows: the absolute uncertainty is taken as 2 °C for temperature sensors and parameters, the relative uncertainty is taken as 2% for pressure sensors and parameters, the relative uncertainty is taken as 5% for the mass flow sensors and parameters and the relative uncertainty is taken as 10% for the characteristic parameters such as ratio and efficiency.

3.5. Unknown Variables and Redundant Equations

For unknown variables in the system, because the value of unknown variables cannot be determined by the optimization problem, it is necessary to eliminate their influence through a reasonable method. The solution process needs to be added after the introduction of the unknown variables. In view of this, this paper adopts the "Intermediate variable method" [37]. By establishing the thermal balance model of the system, the unknown variable is expressed through the known variable and the uncertainty is also derived, so that the unknown variable can be better obtained in the reconciliation process. This method does not need to specify the initial value, and directly establishes the relationship between the known variables in the previous section and the residual vector of the equation. This method of eliminating unknown variables can make the convergence performance of the system more stable.

For example, for the HTR, our known variable is its terminal temperature difference, but we do not know the inlet or outlet temperature of the CO_2 stream on both sides. Then, in the data reconciliation process, these temperature parameters are defined as unknown variables, which need to be expressed by the heat balance of other known variables in the system. After that, these unknown variables and all known variables participate in the data reconciliation process of the model.

After the heat balance model is established, the role of several heat balance equations is to solve the above unknown data using the measured or designed values of some sensors. The heat balance equations used are independent from the equations involved in the optimization problem, and cannot be reused for the subsequent iterative solution of the residual vector of the equation. Among the remaining sensors, there are multiple sensors in the same stream or sensors with known equipment characteristic parameters. These sensors can be involved in the redundant constraint equations as well, which are used to solve the subsequent optimization problem.

4. Model Verification

According to the above principles, we write a computer program of data reconciliation with Python 3.9.0 language, introduced REFPROP physical property library to calculate, so that we can establish heat balance model and data reconciliation model respectively.

The model established in this paper is verified by using the system model established by Sun et al. [35]. The specific verification results are shown in Table 4, in which the parameters used for comparison in this paper are the parameters of the design working condition after heat balance calculation under the same initial conditions. It can be seen from the table that the thermodynamic model established in this paper is in good agreement and has high credibility with the model of Sun L's work.

Table 4. Comparison of model validation.

	Parameter	Reference Model	Present Model	Deviation
Input value	AC pressure ratio	15.0	15.0	--
	AC inlet pressure (kPa)	101.0	101.0	--
	AC inlet temperature (°C)	25.0	25.0	--
	AC mass flow of air (kg/s)	91.22	91.22	--
	AP outlet exhaust pressure (kPa)	109.6	109.6	--
	AP outlet exhaust temperature (°C)	527.6	527.6	--
	SC inlet pressure (kPa)	8000	8000	--
	SC outlet pressure (kPa)	20,000	20,000	--

Table 4. *Cont.*

	Parameter	Reference Model	Present Model	Deviation
Output Value	SC mass flow of CO_2 (kg/s)	132.53	133.420	0.67%
	AC consumption power (kW)	40,014	39,817	0.49%
	GT output power (kW)	70,997	71,129	0.19%
	ST1 output power (kW)	9490	9553	0.66%
	ST2 output power (kW)	6733	6777	0.65%
	SC consumption power (kW)	3457	3480	0.67%

Since the design value parameters of the system under design conditions strictly meet the constraint of energy and mass balance, theoretically there should be no coordination and correction process for data reconciliation calculation using the design value as the initial measured value without adding random errors. To verify the reliability of the data reconciliation algorithm, the data reconciliation calculation results of the model established in this paper are shown in Table 5. The reconciliated value and the design value have a good consistency, which proves that our model is dependable.

Table 5. Verification results of data reconciliation model under the design condition.

Stream Number	Parameter	Design Value	Reconciliated Value	Deviation
Stream 9	H cold side output temperature (°C)	517.600	517.600	0
Stream 10	ST1 output temperature (°C)	390.330	389.086	0.31%
Stream 12	LTR hot side inlet mass flow (kg/s)	132.530	133.419	0.67%
Stream 15	SC output temperature (°C)	62.690	62.690	0
Stream 16	H cold side inlet mass flow (kg/s)	69.890	70.357	0.67%
Stream 17	LTR cold side inlet mass flow (kg/s)	62.640	63.062	0.67%
Stream 18	HTR cold side inlet temperature (°C)	255.470	255.470	0
Stream 19	HTR cold side output temperature (°C)	380.330	380.325	0

5. Results and Discussion

5.1. General Computing Overview

The data reconciliation model used in this paper is based on design values of the system. The known variables of data reconciliation are given by adding certain random errors to the characteristic data and measurement data under the design condition, and the subsequent iterative calculation process of data reconciliation is carried out. The known variables with random errors can be represented here as follows:

$$x_i = x_{i,0} + \sigma_{x,i} \cdot \varphi \tag{19}$$

where $x_{i,0}$ is the design value corresponding to the known variable x_i and φ is the coefficient of random errors. Here, for the random error coefficient, the uniform distribution in $[-1, 1]$ is taken to enhance the convergence of the data reconciliation calculation process. For gross errors, it is necessary to add a gross error much larger than $\sigma_{x,i}$ on the basis of the corresponding design value of the known variable.

The data reconciliation model in this paper calculates ten groups of working conditions. Among them, the first two working conditions do not contain gross errors, and the last eight working conditions all contain one to four gross errors. The specific working conditions are set as shown in Table 6. Among them, Conditions 1 and 2 only contain automatically generated random errors, the purpose of which is to study the reconciliation ability and effect of the data reconciliation model without gross errors. Condition 3 to Condition 6 add one gross error of temperature, pressure or mass flow sensor, respectively, and Condition 7 to Condition 10 add two to four grossa errors, respectively, to study the model ability of identifying suspicious sensors with gross errors and reconciliating of the overall system parameters under the condition of gross errors.

Table 6. Setting of sensors with gross errors under different working conditions.

Working Condition	Sensors	Design Value	Measured Value	Unit
Condition 1	--			
Condition 2	--			
Condition 3 (1 gross error)	TS17	380.330	383.021	°C
Condition 4 (1 gross error)	PS05	24.000	23.282	MPa
Condition 5 (1 gross error)	MFS05	62.640	73.651	kg/s
Condition 6 (1 gross error)	MFS04	132.530	149.173	kg/s
Condition 7 (2 gross errors)	TS17	380.330	382.612	°C
	PS05	24.000	23.195	MPa
Condition 8 (2 gross errors)	MFS05	62.640	73.067	kg/s
	MFS04	132.530	149.819	kg/s
Condition 9 (3 gross errors)	TS17	380.330	383.332	°C
	PS05	24.000	23.208	MPa
	MFS05	62.640	72.613	kg/s
Condition 10 (4 gross errors)	TS17	380.330	382.842	°C
	PS05	24.000	23.211	MPa
	MFS05	62.640	72.865	kg/s
	MFS04	132.530	150.383	kg/s

Under the above ten working conditions, the quality factors of the system in the data reconciliation calculation process which is defined in Equation (15) are shown in Figure 2. The quality factors of ten working conditions are all less than 1, which meets the requirements of global gross error inspection. Quality factor is a test statistic that characterizes whether there are serious errors within the whole system [37]. Only when it is less than 1 it can be indicated that it has passed the global inspection, which means the calculation of the system model does not contain serious errors at the system level. In addition, the quality factors of Conditions 1 and 2 which do not include gross errors are about 0.2. It can also be seen in subsequent working conditions that with the addition and setting of sensors with gross errors, the quality factor is gradually increased. Especially when adding four gross errors in Condition 10, its quality factor has been as high as 0.6 or so.

As we mentioned earlier, the established data reconciliation model requires an iterative process, and its convergence criteria are determined by the residual of redundant equations. Since there are many redundant equations, the residual of equations is introduced here as the determination of iterative convergence, which is calculated as follows [37]:

$$\| \boldsymbol{F} \| = \sqrt{\frac{1}{m}\sum_{j=1}^{m} \boldsymbol{F}_j^2} \tag{20}$$

where m represents the number of equations. It is only used to judge the iteration in the whole reconciliation calculation process, and has no physical significance. The iteration of residual of redundant equations under ten working conditions is shown in Figure 3. The iterative calculation of data reconciliation can converge within three steps, and the residual of redundant equations has been reduced to the magnitude order of 10^{-4} or below, which indicates that the data reconciliation model has a good convergence for the reconciliation process of system parameters, and has a good effect in reducing the residual of the redundant equations.

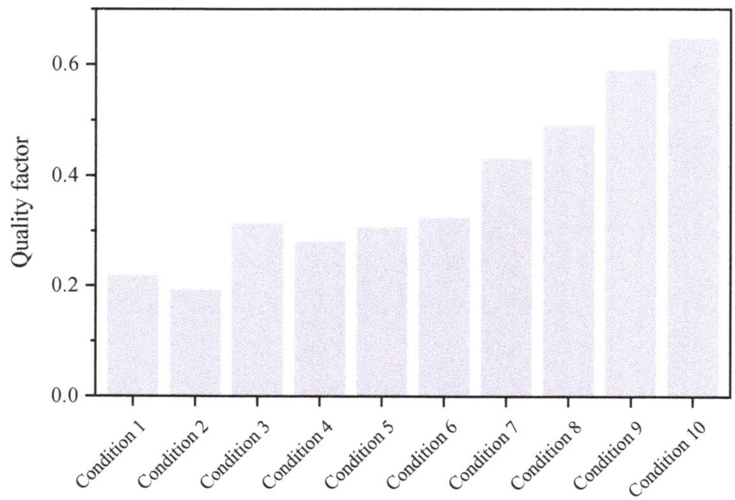

Figure 2. Quality factor of parameters under different working conditions.

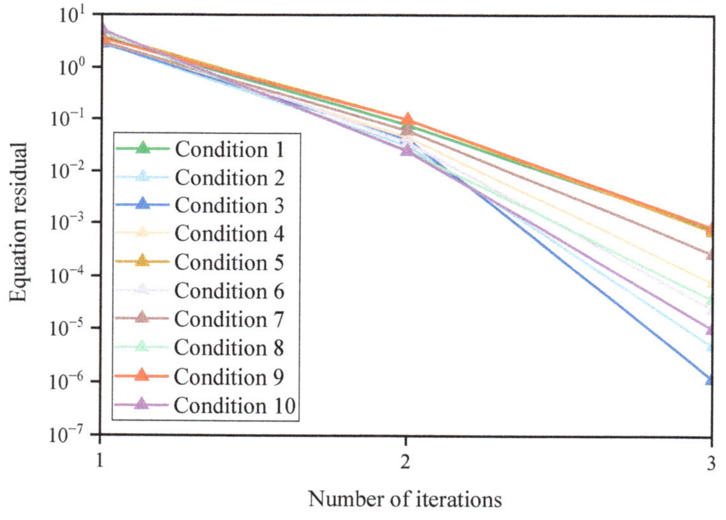

Figure 3. Iterative calculation of equation residual under different working conditions.

5.2. Data Reconciliation Without Gross Errors

This section is an overview of two calculation conditions that do not include gross errors. The reconciliation results of some parameters under Conditions 1 and 2 are shown in Tables 7 and 8, where Val. represents value, De. represents deviation, and Un. represents uncertainty. As can be seen from the tables, since Conditions 1 and 2 do not have gross error sensors, the overall reconciliation effect of the system is relatively good, and the overall reconciliation trend of the two is consistent. The level of reconciliated deviation is small. It can be seen deviations of most parameters under two working conditions are less than 1, and the deviations of very few parameters are also fluctuating around 1%. Measured values of known variables or calculated values of unknown variables still have large deviations from the design values within the allowable deviation range even if they do not contain gross errors. Most of them are mass flow sensors such as mass flow of gas and mass flow of CO_2 in ST1, etc. The deviations of these sensors have been significantly reduced after reconciliation. However, in fact, the deviations of a very small number of parameters are increased, such as SC output pressure under Condition 1 and AC power under Condition 2. This is because data reconciliation is a reconciliation process for parameters of the whole thermal system, and its essence is to disperse the large deviations of some parameters into other parameters to improve the quality of all parameters. In addition, compared with the change of uncertainty before and after reconciliation, the uncertainties of parameters also have a very obvious decline, and those after reconciliation are below 3%. It can be shown that the data reconciliation greatly reduces the uncertainty of the parameter whether known variables or unknown variables, and effectively improve the accuracy of measurement data.

Table 7. Reconciliation results of some parameters in Condition 1.

Parameter	Design Val.	Measured			Reconciliated			Unit
		Val.	De.	Un.	Val.	De.	Un.	
Mass flow of gas	92.900	91.140	1.89%	5.00%	92.476	0.46%	2.54%	kg/s
H hot side output temperature	72.690	72.459	0.32%	0.58%	72.689	0.001%	0.45%	°C
H cold side output temperature	517.600	516.618	0.19%	0.25%	517.585	0.003%	0.12%	°C
Mass flow of CO_2 in ST1	69.890	71.942	2.94%	10.00%	70.085	0.28%	2.55%	kg/s
SC inlet pressure	8.000	7.928	0.90%	2.00%	7.971	0.36%	1.37%	MPa
SC output pressure	24.000	23.981	0.08%	2.00%	23.976	0.10%	1.39%	MPa
SC inlet temperature	32.000	31.288	2.22%	0.66%	31.955	0.14%	0.20%	°C
HTR cold side output temperature	380.330	379.476	0.22%	0.31%	380.278	0.01%	0.13%	°C
Mass flow of CO_2	132.530	132.529	0.001%	10.00%	132.950	0.32%	2.58%	kg/s
Mass flow of CO_2 in ST2	62.640	65.759	4.98%	10.00%	62.865	0.36%	2.74%	kg/s
AC power	40,014	39,817	0.49%	12.38%	39,868	0.36%	2.86%	kW
GT power	70,997	70,949	0.07%	11.21%	71,332	0.47%	2.64%	kW
ST1 power	9490	9480	0.10%	11.73%	9511	0.23%	2.88%	kW
ST2 power	6733	6713	0.29%	15.92%	6752	0.29%	2.85%	kW
SC power	3457	3433	0.69%	11.93%	3456	0.01%	9.37%	kW

Table 8. Reconciliation results of some parameters in Condition 2.

Parameter	Design Val.	Measured			Reconciliated			Unit
		Val.	De.	Un.	Val.	De.	Un.	
Mass flow of gas	92.900	94.240	1.44%	5.00%	92.910	0.01%	2.54%	kg/s
H hot side output temperature	72.690	73.379	0.95%	0.58%	72.821	0.18%	0.46%	°C
H cold side output temperature	517.600	516.579	−0.20%	0.25%	517.768	0.03%	0.12%	°C
Mass flow of CO_2 in ST1	69.890	67.033	−4.09%	10.00%	70.358	0.67%	2.56%	kg/s
SC inlet pressure	8.000	8.041	0.51%	2.00%	8.009	0.12%	1.36%	MPa
SC output pressure	24.000	24.154	0.64%	2.00%	24.081	0.34%	1.39%	MPa
SC inlet temperature	32.000	31.799	−0.63%	0.66%	31.973	−0.08%	0.19%	°C
HTR cold side output temperature	380.330	380.323	−0.002%	0.31%	379.920	−0.11%	0.13%	°C
Mass flow of CO_2	132.530	136.822	3.24%	10.00%	133.456	0.70%	2.60%	kg/s
Mass flow of CO_2 in ST2	62.640	63.773	1.81%	10.00%	63.098	0.73%	2.75%	kg/s

Table 8. *Cont.*

Parameter	Design Val.	Measured Val.	Measured De.	Measured Un.	Reconciliated Val.	Reconciliated De.	Reconciliated Un.	Unit
AC power	40,014	39,817	−0.49%	12.38%	39,692	−0.80%	2.87%	kW
GT power	70,997	71,330	0.47%	11.21%	71,707	1.00%	2.64%	kW
ST1 power	9490	9516	0.28%	11.80%	9594	1.10%	2.87%	kW
ST2 power	6733	6804	1.07%	15.80%	6802	1.03%	2.83%	kW
SC power	3457	3448	−0.26%	12.09%	3512	1.60%	9.54%	kW

The distribution of standard deviation in Conditions 1 and 2 are shown in Figures 4 and 5. The data reconciliation has a good effect on the deviation aggregation of parameters of such a thermal system, and the overall distribution curve of the deviation in Conditions 1 and 2 after reconciliation has an obvious effect of concentration towards the origin than that before reconciliation.

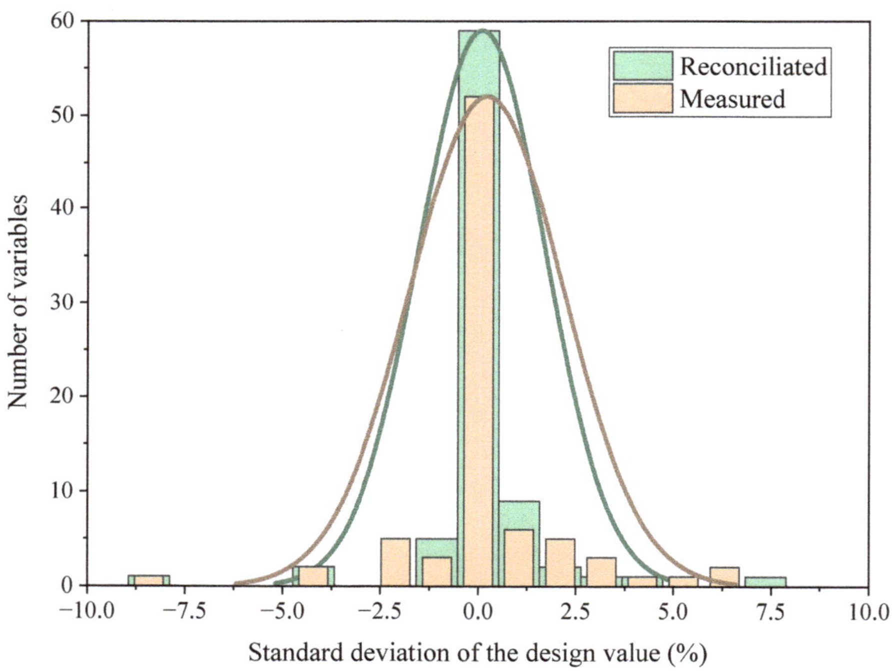

Figure 4. Distribution of standard deviation in Condition 1.

The distribution of uncertainty in Conditions 1 and 2 are shown in Figures 6 and 7. The uncertainty distributions of the two are basically the same, which indicates that the data reconciliation has a good correction for random errors and always converges on the data results with higher precision. At the same time, similar to the deviation distribution, the distribution curve of the uncertainty after reconciliation also has an obvious effect on the origin concentration before reconciliation, which also explains the improvement of the overall data accuracy.

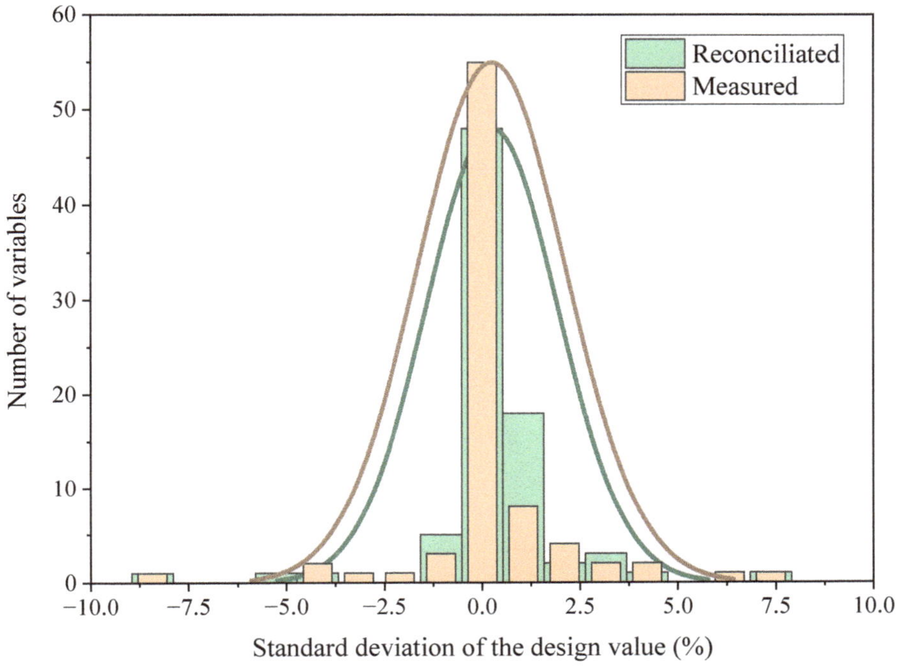

Figure 5. Distribution of standard deviation in Condition 2.

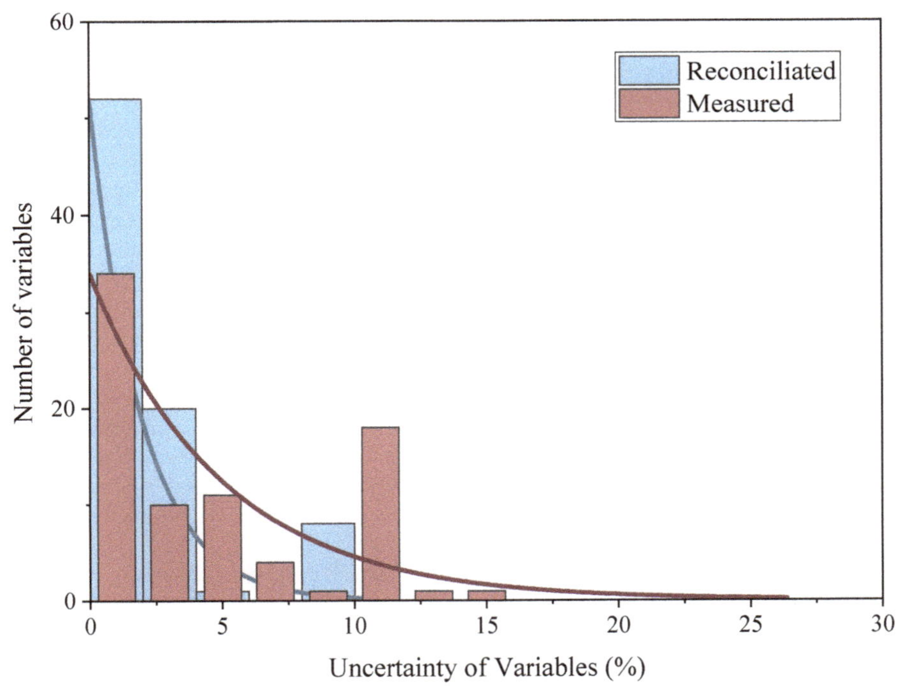

Figure 6. Distribution of uncertainty in Condition 1.

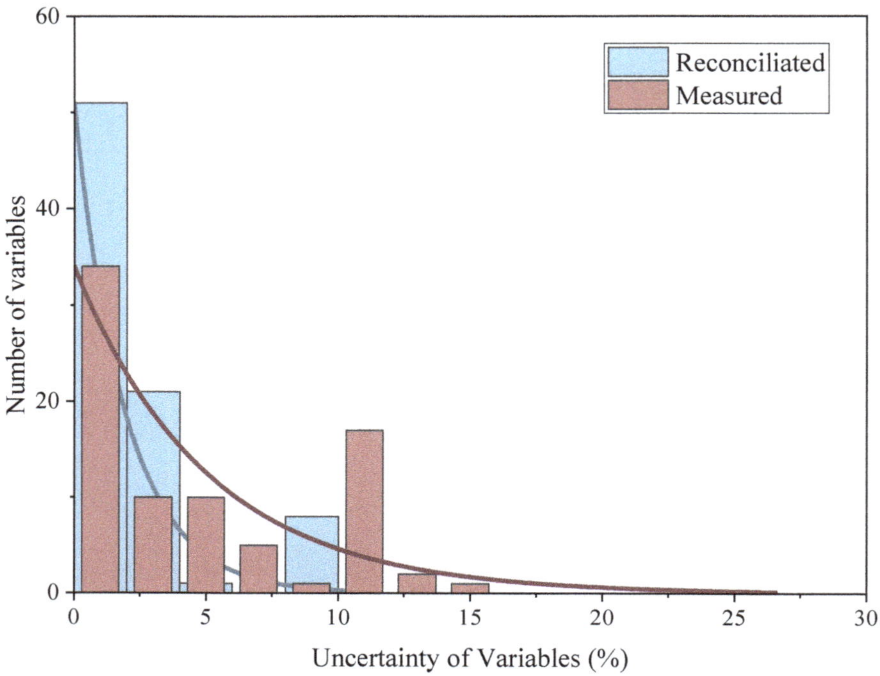

Figure 7. Distribution of uncertainty in Condition 2.

5.3. Data Reconciliation with Gross Errors

This section is an overview of eight calculation conditions with gross errors. For each working condition, the recognition of sensors with gross errors is shown in Table 9. On the basis of passing the global inspection, the sensors with gross errors under each working condition are well identified, which is based on Equation (16) because the calculated criterion value of r is greater than 1.96. Compared with the gross errors setting in Table 6, it is found that the two have a good consistency, which also indicates that the data reconciliation model established in this paper is relatively effective in identifying the gross errors of some sensors in the thermal system.

The reconciliation results of other sensors in Condition 10 except the sensors with gross errors is shown in Table 10. The deviation of sensors with gross errors is dispersed among all parameters, resulting in an increase in the deviation of some parameters from the design value standard, but the overall deviation level is still low. The highest deviation appears in unknown variables of power and fluctuates around 2%. This result can also show that the gross errors have an obvious influence on the data reconciliation effect, because these errors will not be eliminated, but only dispersed. Even so, it can be seen from the change of uncertainty that the data reconciliation still improves the data accuracy significantly in this case.

The distribution of standard deviation in Condition 10 is shown in Figure 8. Due to the addition of gross errors and more random errors, the overall distribution curve has a wider range. However, after data reconciliation, all deviations can be concentrated to the origin. Although the number of variable deviations near the origin is less than before reconciliation due to the effect of large error dispersion, the quality of parameters has been significantly improved. If the gross errors identified can be eliminated in the process of practical application, the reconciliation effect will be better.

Table 9. Identification with gross errors of the working conditions.

Working Condition	Gross Error Detection	Criterion Value of r	Design Value	Measured Value	Reconciliated Value
Condition 3	TS17 (Temperature, °C)	2.559	380.330	383.021	380.658
Condition 4	PS05 (Pressure, MPa)	2.198	24.000	23.282	23.653
Condition 5	MFS05 (Mass flow, kg/s)	2.713	62.640	73.651	63.759
Condition 6	MFS04 (Mass flow, kg/s)	2.345	132.530	149.173	131.808
Condition 7	TS17 (Temperature, °C)	2.376	380.330	382.612	380.418
	PS05 (Pressure, MPa)	2.097	24.000	23.195	23.651
Condition 8	MFS05 (Mass flow, kg/s)	2.688	62.640	73.067	63.347
	MFS04 (Mass flow, kg/s)	2.147	132.530	149.819	133.867
Condition 9	MFS05 (Mass flow, kg/s)	2.599	62.640	72.613	63.273
	TS17 (Temperature, °C)	2.586	380.330	383.332	380.946
	PS05 (Pressure, MPa)	2.042	24.000	23.208	23.653
Condition 10	MFS05 (Mass flow, kg/s)	2.512	62.640	72.865	63.803
	TS17 (Temperature, °C)	2.206	380.330	382.842	380.806
	MFS04 (Mass flow, kg/s)	2.094	132.530	150.383	134.755
	PS05 (Pressure, MPa)	2.004	24.000	23.211	23.651

Table 10. Reconciliation results of some parameters in Condition 10.

Parameter	Design	Measured			Reconciliated			Unit
	Val.	Val.	De.	Un.	Val.	De.	Un.	
Mass flow of gas	92.900	92.650	−0.27%	5.00%	93.495	0.64%	2.54%	kg/s
H hot side output temperature	72.690	73.416	0.10%	0.58%	73.512	1.13%	0.46%	°C
H cold side output temperature	517.600	518.318	0.14%	0.25%	517.932	0.06%	0.12%	°C
Mass flow of CO_2 in ST1	69.890	70.056	0.24%	10.00%	70.952	1.52%	2.56%	kg/s
SC inlet pressure	8.000	8.033	0.42%	2.00%	8.013	0.17%	1.36%	MPa
SC inlet temperature	32.000	32.086	0.27%	0.66%	32.396	1.24%	0.18%	°C
AC power	40,014	39,817	−0.49%	12.38%	39,908	−0.26%	2.85%	kW
GT power	70,997	71,131	0.19%	11.21%	72,305	1.84%	2.64%	kW
ST1 power	9490	9549	0.63%	11.86%	9648	1.67%	2.86%	kW
ST2 power	6733	6789	0.84%	15.91%	6873	2.09%	2.87%	kW
SC power	3457	3502	1.30%	12.23%	3479	0.64%	9.47%	kW

The uncertainty distribution of working condition 10 is shown in Figure 9. There is no significant difference between the distribution trend of uncertainty after adding coarse errors and the previous working conditions. After coordination, the degree of uncertainty reduction is more obvious, the homogeneity is concentrated at the origin and the accuracy of measurement data is improved.

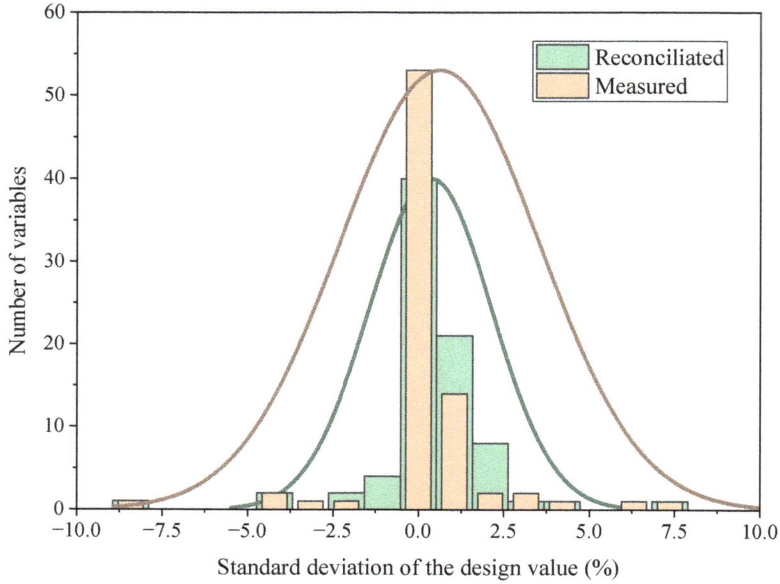

Figure 8. Distribution of standard deviation in Condition 10.

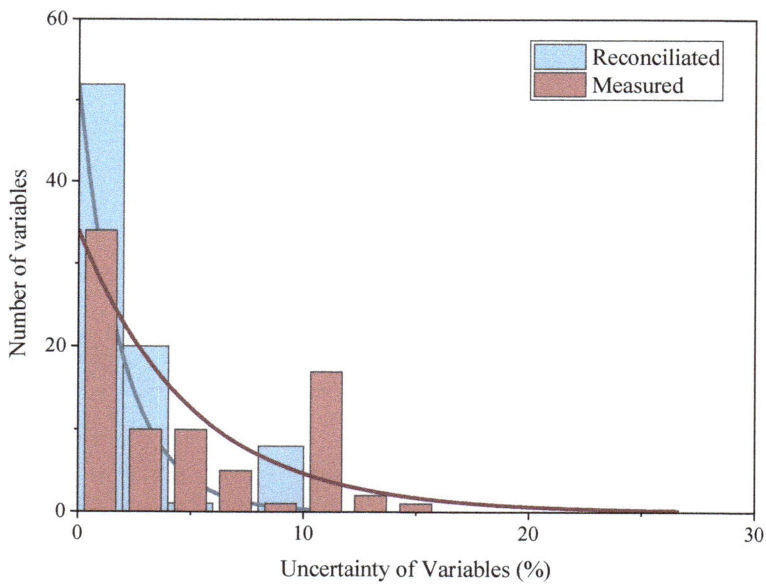

Figure 9. Distribution of uncertainty in Condition 10.

6. Conclusions

In this paper, a data reconciliation model is established based on a waste heat recovery system of a gas turbine using an S-CO_2 cycle with two S-CO_2 turbines. The thermal system and the data reconciliation model are well fitted with the reference values. The data reconciliation calculation shows that the overall deviation level of each parameter of the thermal system after reconciliation is obviously reduced, and the uncertainty is also reduced, which reflects that the data reconciliation technology can improve the measurement accuracy of thermal system. In the case of adding random errors, the measurement deviation of key parameters in the system, such as the total mass flow of CO_2, can be reduced to less than 1%. The power values, such as ST and SC, can be obtained with high data accuracy, and the deviation is also about 1%. In the conditions of adding gross errors, all kinds of sensors containing gross errors can be well identified and eliminated through the data reconciliation model, and the accuracy of the overall measurement data is still improved to a certain extent even under the influence of gross errors.

In summary, the data reconciliation model established in this paper can provide a certain reference for the accuracy of thermal system parameter measurement, and can be applied to scenarios such as parameter correction of thermal system test bench and equipment error detection, which has deep research significance. Building a system test bench can be used to validate the concrete results of the data reconciliation method and bring them into real-world environments. At the same time, it may be necessary to increase the research topics, such as variable operating conditions and heat transfer efficiency, to ensure the rigor of the test. In addition, further exploration of the performance in data reconciliation model in actual advanced power cycle is needed.

Author Contributions: Conceptualization, L.X. and J.Y.; Methodology, L.X. and J.Y.; Validation, J.Y., J.W. and Y.L.; Formal analysis, L.X.; Investigation, J.W., Y.L. and D.Z.; Writing—original draft, L.X.; Writing—review and editing, L.X., J.Y., J.W., Y.L. and D.Z.; Supervision, D.Z. All authors have read and agreed to the published version of the manuscript.

Funding: The work was fully supported by the University Joint Program of Shaanxi Province Key Research Project—Major Project (2022GXLH-01-17) and the Major Project of Shaanxi Province Key Research Project (2024PT-ZCK-47).

Institutional Review Board Statement: Not applicable.

Informed Consent Statement: Not applicable.

Data Availability Statement: Data will be made available on request.

Conflicts of Interest: The authors declare no conflicts of interest.

Nomenclature

Symbol	Meaning
AC	Air compressor
AP	Air preheater
CC	Combustion chamber
CL	Cooler
GT	Gas turbine
H	Heat exchanger
HTR	High-temperature recuperator
LTR	Low-temperature recuperator
SC	S-CO_2 compressor
ST1	S-CO_2 turbine 1
ST2	S-CO_2 turbine 2
h	Enthalpy
m	Mass flow
Q	Heat exchange amount

W	Power
η	Efficiency
$\boldsymbol{F}$	The equation vector representing a set of algebraic equations
$\boldsymbol{F}^{(0)}$	The initial residual vector of the equation vector
$\boldsymbol{F}^{(k)}$	The initial residual vector after k iterations
$J_{\boldsymbol{F},\boldsymbol{x}}$	The Jacobian matrix of equation vector $\boldsymbol{F}$ versus $\boldsymbol{x}$
r_i	The test statistics of variable x_i based on the correction vector ν
$\boldsymbol{S_F}$	The covariance matrix of the equation vector
$\boldsymbol{S_x}$	The covariance matrix of the known variable vector
$\boldsymbol{u}$	The vector of the unknown variable.
x_i	The component of $\boldsymbol{x}$
$\hat{x}_i$	The reconciliated value of the known variable
$\boldsymbol{x}$	The vector composed of all known variables
$\hat{\boldsymbol{x}}$	The reconciliated value vector of the known variable
$\boldsymbol{x_0}$	The true value vector of the known variable
$\mathcal{Z}_{1-\alpha/2}$	The lower α fraction of the normal distribution
δ	The random error vector
ε	The convergence limit
ν	The correction vector of the initial value vector
ν_i	The correction value
$\boldsymbol{\nu}^{(k)}$	The correction vector of the initial value vector after k iterations
ξ_{max}	The maximum allowable error
σ_i	The mean square deviation (standard deviation)
φ	The coefficient of random errors

References

1. Tian, H.; Ding, H.; Bin, W. A Summary of wearable textiles power generation. In *IOP Conference Series: Earth and Environmental Science*; IOP Publishing: Bristol, UK, 2021; Volume 772, p. 012037.
2. *EI Statistical Review of World Energy 2024*; Energy Institute: London, UK, 2024.
3. Heppenstall, T. Advanced gas turbine cycles for power generation: A critical review. *Appl. Therm. Eng.* **1998**, *18*, 837–846. [CrossRef]
4. Can Gülen, S. Étude on gas turbine combined cycle power plant—Next 20 years. *J. Eng. Gas Turbines Power* **2016**, *138*, 051701. [CrossRef]
5. Zhang, C.; Shu, G.; Tian, H.; Wei, H.; Liang, X. Comparative study of alternative ORC-based combined power systems to exploit high temperature waste heat. *Energy Convers. Manag.* **2015**, *89*, 541–554. [CrossRef]
6. Rajinesh, S.; Miller, S.A.; Rowlands, A.S.; Jacobs, P.A. Dynamic characteristics of a direct-heated supercritical carbon-dioxide Brayton cycle in a solar thermal power plant. *Energy* **2013**, *50*, 194–204.
7. Ruby Meena, A.; Senthil Kumar, S. Genetically tuned fuzzy PID controller in two area reheat thermal power system. *Russ. Electr. Eng.* **2016**, *87*, 579–587. [CrossRef]
8. Chowdhury, J.I.; Thornhill, D.; Soulatiantork, P.; Hu, Y.; Balta-Ozkan, N.; Varga, L.; Nguyen, B.K. Control of supercritical organic Rankine cycle based waste heat recovery system using conventional and fuzzy self-tuned PID controllers. *Int. J. Control Autom. Syst.* **2019**, *17*, 2969–2981. [CrossRef]
9. Christoph, S.; Dimache, L.; Lohan, J. Operational Optimisation of a Heat Pump System with Sensible Thermal Energy Storage using Genetic Algorithm: A Case Study. *Therm. Sci.* **2018**, *22*, 2189–2202.
10. Qian, Y.; Du, W.-J.; Cheng, L. Optimization design of heat recovery systems on rotary kilns using genetic algorithms. *Appl. Energy* **2017**, *202*, 153–168.
11. Reza, B.; Eiamsa-ard, S. Genetic algorithm multiobjective optimization of a thermal system with three heat transfer enhancement characteristics. *J. Enhanc. Heat Transf.* **2020**, *27*, 123–141.
12. Lin, Y.; Wang, H.; Hu, P.; Yang, W.; Hu, Q.; Zhu, N.; Lei, F. A study on the optimal air, load and source side temperature combination for a variable air and water volume ground source heat pump system. *Appl. Therm. Eng.* **2020**, *178*, 115595. [CrossRef]
13. Shankar, N.; Jordache, C. *Data Reconciliation and Gross Error Detection: An Intelligent Use of Process Data*; Elsevier: Amsterdam, The Netherlands, 1999.

14. Romagnoli, J.A.; Sánchez, M.C. *Data Processing and Reconciliation for Chemical Process Operations*; Elsevier: Amsterdam, The Netherlands, 1999.
15. Kretsovalis, A.; Mah, R.S.H. Observability and redundancy classification in multicomponent process networks. *AIChE J.* **1987**, *33*, 70–82. [CrossRef]
16. Crowe, C.M.; Campos, Y.G.; Hrymak, A. Reconciliation of process flow rates by matrix projection. Part I: Linear case. *AIChE J.* **1983**, *29*, 881–888. [CrossRef]
17. Crowe, C.M. Reconciliation of process flow rates by matrix projection. Part II: The nonlinear case. *AIChE J.* **1986**, *32*, 616–623. [CrossRef]
18. Swartz, C.L.E. Data reconciliation for generalized flowsheet applications. In Proceedings of the ACS National Meeting, Dallas, TX, USA, 9–14 April 1989.
19. Reilly, P.M.; Carpani, R.E. Application of statistical theory of adjustment to material balances. In Proceedings of the 13th Canadian Chemical Engineering Congress, Montreal, QC, Canada, 23 October 1963.
20. Mah, R.S.H.; Tamhane, A.C. Detection of gross errors in process data. *AIChE J.* **1982**, *28*, 828–830. [CrossRef]
21. Narasimhan, S.; Mah, R.S. Generalized likelihood ratio method for gross error identification. *AIChE J.* **1987**, *33*, 1514–1521. [CrossRef]
22. Narasimhan, S.; Mah, R.S. Generalized likelihood ratios for gross error identification in dynamic processes. *AIChE J.* **1988**, *34*, 1321–1331. [CrossRef]
23. Holly, W.; Cook, R.; Crowe, C.M. Reconciliation of mass flow rate measurements in a chemical extraction plant. *Can. J. Chem. Eng.* **1989**, *67*, 595–601. [CrossRef]
24. Sriniketh, S.; Billeter, J.; Narasimhan, S.; Bonvin, D. Data reconciliation for chemical reaction systems using vessel extents and shape constraints. *Comput. Chem. Eng.* **2017**, *101*, 44–58.
25. Humaad, I.; Ati, U.M.K.; Mahalec, V. Heat exchanger network simulation, data reconciliation & optimization. *Appl. Therm. Eng.* **2013**, *52*, 328–335.
26. Martínez-Maradiaga, D.; Bruno, J.C.; Coronas, A. Steady-state data reconciliation for absorption refrigeration systems. *Appl. Therm. Eng.* **2013**, *51*, 1170–1180. [CrossRef]
27. Magnus, L.; Jansky, J. Process data reconciliation in nuclear power plants. In Proceedings of the EPRI Nuclear Power Performance Improvement Seminar, Saratoga Springs, NY, USA, 15–16 July 2002.
28. Magnus, L. Power recapture and power uprate in NPPS with process data reconciliation in accordance with VDI 2048. *Int. Conf. Nucl. Eng.* **2006**, *42428*, 7–14.
29. Svein, S.; Berg, Ö. Data reconciliation and fault detection by means of plant-wide mass and energy balances. *Prog. Nucl. Energy* **2003**, *43*, 97–104.
30. Damianik, V.E.; Schirru, R. Simultaneous model selection, robust data reconciliation and outlier detection with swarm intelligence in a thermal reactor power calculation. *Ann. Nucl. Energy* **2011**, *38*, 1820–1832.
31. Hartner, P.; Petek, J.; Pechtl, P.; Hamilton, P. Model-based data reconciliation to improve accuracy and reliability of performance evaluation of thermal power plants. *Turbo Expo Power Land Sea Air* **2005**, *47276*, 195–200.
32. Jiang, X.; Liu, P.; Li, Z. Data reconciliation and gross error detection for operational data in power plants. *Energy* **2014**, *75*, 14–23. [CrossRef]
33. Chen, P.-C.; Andersen, H. The implementation of the data validation process in a gas turbine performance monitoring system. *Turbo Expo Power Land Sea Air* **2005**, *46997*, 609–616.
34. Can, G.S.; Smith, R.W. A simple mathematical approach to data reconciliation in a single-shaft combined cycle system. *J. Eng. Gas Turbines Power. Mar* **2009**, *131*, 021601.
35. Sun, L.; Wang, D.; Xie, Y. Energy, exergy and exergoeconomic analysis of two supercritical CO_2 cycles for waste heat recovery of gas turbine. *Appl. Therm. Eng.* **2021**, *196*, 117337. [CrossRef]
36. *VDI-2048: Control and Quality Improvement of Process Data and Their Uncertainties by Means of Correction Calculation for Operation and Acceptance Tests*; Association of German Engineers; VDI Verein Deutscher Ingenieure: Düsseldorf, Germany, 2017.
37. You, J.; Wu, J.; Xu, L.; Xie, Y.; Qiu, J.; Wan, L.; Qin, Q. Exploration on the comprehensive data reconciliation framework for unknown parameter inference in the nuclear power plant system. *Appl. Therm. Eng.* **2024**, *247*, 123138. [CrossRef]
38. Langenstein, M.; Jansky, J.; Laipple, B.; Eitschberger, H.; Grauf, E.; Schalk, H. Finding megawatts in nuclear power plants with process data reconciliation. *Int. Conf. Nucl. Eng.* **2004**, *46881*, 43–52.

applied sciences

MDPI

Article

A Comparative Science-Based Viability Assessment Among Current and Emerging Hydrogen Production Technologies

Yehia F. Khalil

Chemical & Environmental Engineering Department, School of Engineering & Applied Science (SEAS), Yale University, New Haven, CT 06520, USA; yehia.khalil@yale.edu

Featured Application: Advancing environmental sustainability and the circular economy through the production of clean and green hydrogen as a promising energy carrier of the future.

Abstract: This research undertakes a comparative analysis of current and emerging hydrogen (H_2) production technologies, evaluating them based on quantitative and qualitative decision criteria. The quantitative criteria include cost of H_2 production (USD/kg H_2), energy consumption (MJ/kg H_2), global warming potential (kg CO_2-eq/kg H_2), and technology energy efficiency (%). The qualitative criteria encompass technology readiness level (TRL) and availability of supply chain materials (classified as low, medium, or high). To achieve these objectives, an extensive literature review has been conducted, systematically assessing the selected H_2 production technologies against the aforementioned criteria. The insights synthesized from the literature provide a foundation for an informed, science-based evaluation of the potentials and techno-economic challenges that these technologies face in achieving the 1-1-1 goal set by the U.S. Department of Energy (DOE) in 2021. This target aims for a H_2 production cost of USD 1/kg H_2 within one decade (by 2031), including costs associated with production, delivery, and dispensing at H_2 fueling stations (HRSs). Also, the DOE established an interim goal of USD 2/kg H_2 by 2026. This research concludes that among the examined H_2 production technologies, water electrolysis and biomass waste valorization emerge as the most promising near-term solutions to meet the DOE's goal.

Keywords: hydrogen production; clean hydrogen; sustainable hydrogen; waste valorization; bio-based waste upcycling; DOE 1-1-1 goal; clean hydrogen; circular economy

1. Introduction

Hydrogen is a major contributor to the global effort to achieve the net-positive environmental goal that goes beyond the net-zero goal that focuses only on offsetting CO_2 emissions. The net-positive goal reenforces the overarching intent of circular economy bio-based waste valorization for H_2 production as one of the key technology pathways for sustainable H_2 production [1]. However, the energy transition from the current fossil-based technologies to sustainable H_2 production is a physical transformation in its early stages, and there are many technical, economic, and supply-chain logistics challenges that need to be sorted before at-scale adoption of H_2-based technologies can be realized [2–4]. Accordingly, the present research aims to improve the current state of knowledge about these challenges by conducting a comparative science-based and data-driven viability assessment among current and emerging hydrogen production technologies. The assessment evaluates the selected technologies against specific quantitative and qualitative decision criteria. Based on the insights generated from the comparative assessment, conclusions are made about the most promising H_2 production technologies that could meet the U.S. DOE 1-1-1 clean energy goal described in its 2021 "Hydrogen Shot" conference. In addition, techno-economic challenges of the selected H_2 production pathways are highlighted.

The academic merit of this contribution lies in advancing our understanding of the present and future cost competitiveness of key hydrogen production technologies. Additionally, it elucidates the anticipated role of hydrogen as a promising energy carrier and a crucial enabler in achieving net-zero emissions.

The remainder of this manuscript is organized as follows: Section 2 outlines the research methodology employed, Section 3 presents the results, Section 4 discusses the findings, and Section 5 summarizes the conclusions of the comparative assessment. Additionally, Appendix A provides the list of acronyms used in this manuscript.

2. Research Method

To achieve the objectives of this research, extensive literature reviews [*using the following research query: Hydrogen production technologies (All Fields) OR Green hydrogen production (All Fields) AND Hydrogen production cost (All Fields) OR Hydrogen carbon footprint (All Fields) AND Technoeconomic Analysis OR Environmental Impact Assessment AND 2000–2024 (Publication Years)*] were conducted to assemble the relevant quantitative and qualitative data. Section 2.1 highlights the key findings of the literature reviews. The collated information was mapped to specific decision criteria that support the comparative assessment. In addition, the steam-methane reforming (SMR) technology, with and without carbon capture and storage (CCS), was used as the benchmark H_2 production technology.

2.1. Literature Reviews and Key Findings

As an introduction to this subsection, it is worth noting that by 2050, the global energy demand is anticipated to be around 30 terawatts (TW), which is double that of 2011 [5]. Hydrogen is specifically being considered as an alternative fuel in the transportation sector to power fuel cell electric vehicles [6] and for electrifying the commercial aviation (viz., hybrid-electric and all-electric aircraft). The U.S. DOE projected that clean H_2 could be produced for USD 2/kg H_2 by 2025 (*H_2New*, 2021), and in its 2021 "Hydrogen Shot" conference, the DOE drafted its long-term goal to reduce the cost of clean H_2 production to USD 1 per kg H_2 in one decade.

Herein, a literature review of 75 published sources was conducted on various H_2 production technologies, specifically the most prominent conventional technologies and the most promising greener technologies. With a couple exceptions, only papers and reports published in the last decade (2013–2024) were consulted, with an effort to rely on publications from the past 5 years whenever possible, in order to ensure the comparative analysis contained the most updated information. From these sources, both quantitative and qualitative data were extracted to support the comparative assessment of H_2 production technologies.

The conventional methods of H_2 production provide either "gray" or "blue" hydrogen. Steam-methane reforming (SMR) is widely acknowledged as the most well-established method [3], and it is often used as a point of reference to analyze and compare the performance and feasibility of non-conventional methods. Because SMR is a very mature technology, it is thoroughly studied and produces low-cost hydrogen, but it also has low potential for technical improvements that will improve efficiency or performance [7,8]. Several papers studying the conventional methods of producing blue H_2 discuss how CCS reduces emissions significantly but also increases production cost and decreases energy efficiency [8–12].

Waste valorization was evaluated as a more sustainable H_2 production technology. Utilizing waste streams eliminates both the polluting effects of waste build-up and the need for energy-intensive and high-emitting waste incineration. Biomass is a broad term and can be prepared as a feedstock from organic sources, such as grass, wood, agricultural products, animal waste, food scraps, municipal solid waste, and algae, and it can be valorized by methods such as gasification, pyrolysis, supercritical water gasification, and dark fermentation [11,13]. One study noted that food waste is specifically suitable for H_2 production for a number of reasons, such as its carbohydrate richness, wide availability, and low cost [14]. Another study explained that converting food waste to H_2 via dark fermentation is energetically advantageous because it can be achieved under ambient temperature and pressure with low chemical energy requirements, but it must be performed onsite to be feasible due to the high water content and biodegradability of food waste [15]. Even more so than food waste fermentation, gasification of biomass is a highly studied process due to its convenient utilization of diverse waste streams and production of clean biohydrogen. Biomass gasification is currently faced with low conversion and thermal efficiencies, but they can be improved with the optimization of the process temperature, the catalyst used, and the biomass content [16].

Glycerol is another promising waste stream to utilize for H_2 generation, as it is produced abundantly as a byproduct of the transesterification of biomass to biodiesel. By strategically incorporating H_2 production plants into existing biodiesel production plants, glycerol could even become a free material stream. A simulation-based life cycle assessment (LCA) compared the environmental and health impacts of several glycerol-based hydrogen production technologies—ATR, APR, and SCWR—versus those of the conventional method of SMR [17,18]. A techno-economic analysis (TEA) compared the cost of hydrogen produced from APR and steam reforming of glycerol, revealing that APR is slightly less expensive but still not competitive with SMR [18]. Another report on pathways to producing H_2 from biomass provided H_2 production costs and TRLs of SCWR, APR, and SR of glycerol [16].

Water electrolysis is widely considered a promising method of sustainably producing H_2 due to fairly negligent carbon emissions [19–21]. A number of sources compared powering electrolyzers with the local electricity grid mix versus renewables. Because the grid mix varies so widely, a wide range of H_2 cost values were reported, but in general, it is agreed that costs could be reduced by strategically utilizing renewable power sources rather than relying on the grid mix [22–26]. Research has recently explored utilizing wind power and solar PV power to produce green H_2 by electrolysis [20]. While discussed to a lesser extent, other renewables like hydropower and nuclear power are also contenders for more sustainable electrolysis power sources [27–29]. While hydropower is a well-established renewable electricity source, one report noted that it will be overtaken by solar PV power due to its lower levelized cost of electricity (LCOE) and higher energy capacity [30]. Nuclear energy is also a firmly established renewable energy source, but still, setbacks such as social apprehension keep it from becoming the focus of green H_2 research [28]. Different renewable energy sources make more sense for different areas, depending on which natural resources are most abundant and consistent in a given region [25]. Therefore, cost and energy efficiency analyses [31] are further complicated with the additional variable of energy source.

The TRL of water electrolysis varies depending on the technology and the energy source utilized. However, as explored in a variety of reports, electrolysis is a fairly well-developed technology, particularly AKE, PEM, and SOE—more so than AEM [9,16,22,32–35]. In addition to its nearly negligible global warming impact, its high TRL makes electrolysis an attractive green H_2 production technology. However, high H_2 production costs and high energy requirements are setbacks that are keeping electrolysis from replacing the conventional methods. One paper noted that despite its low GWP and high technological readiness, solar PV-powered electrolysis will only be able to take the forefront as a potential option for H_2 production when its energy efficiency is improved, which will also subsequently lower its high production costs [5,36].

Electrolyzers are oftentimes divided into two classes: low-temperature electrolyzers (LTEs) and high-temperature electrolyzers (HTEs). The U.S. DOE has invested R&D funding into these two types of electrolysis in order to promote the most durable, effective, and efficient system of green H_2 production. The characteristics of LTE, such as PEM, is that it is commercially available, but it is not durable, efficient, or affordable enough, while the characteristics of HTE, such as SOE, are that it is less mature but has potential for higher efficiency and a longer lifespan [37]. According to one paper, SOEs are advantageous because of their ability to co-electrolyze steam and carbon dioxide in a single step and, thus, eliminate the need for expensive and maintenance-intensive water–gas shift reactors [38]. SOEs operate at high temperatures, meaning that they require greater energy input, but also that they can utilize low-cost active metals rather than expensive noble metals [38]. The Global Hydrogen Review claims that AEME electrolysis combines the benefits of PEME and AKE, since it does not require expensive platinum utilized by PEME nor the corrosive electrolytes used in AKE, but it is a less mature technology [22].

Emerging technologies for clean H_2 production are also evaluated in this research. Because of the novelty of these technologies, they will not have very high TRLs, with most still being in the research and early development stages [16,22,39]. Several studies of thermochemical decomposition via both solar and nuclear power demonstrate that these technologies are far from being commercially mature [9,34]. However, even though the technologies are immature, their TEA reveals that several of the emerging technologies, like bio-photolysis, biomethane pyrolysis, and ESMR, are potentially able to produce H_2 for competitive prices [11,28,40,41]. ESMR is especially attractive when considered as an easily implemented technology to help aid in the transition from gray to cleaner H_2. It utilizes the same natural gas feedstock but incorporates renewable electricity to lower emissions, resulting in a system that is more sustainable than the conventional methods and more economically and technically feasible than the other emerging technologies [41]. Among the emerging H_2 production technologies is the so-called white color, of natural origin; however, due to its rare occurrence in the Earth's rock, there is no commercial interest at this point in time [42]. This geologic H_2 (aka., white, gold, or natural H_2) is found beneath the Earth's surface and is postulated to be produced by high-temperature reactions between water and iron-rich minerals.

Most of the other emerging technologies still require more technological development to compete with standard hydrogen prices [24,43]. Strategic plant optimization can help lower production costs. For example, H_2 costs from ammonia thermal decomposition via fixed-bed reactors and PSA membranes are expected to decrease if plants are built to be larger and centralized, which would lead to economies of scale and less expensive ammonia costs [44,45]. One article discussed how the fluctuations in H_2 production costs via thermochemical processes are mainly due to variable electricity prices, but the utilization of reliable green electricity from renewable sources like solar PV and wind turbines can greatly reduce production costs [11]. The lack of comprehensive studies of promising emerging technology, like biomethane pyrolysis and SMC, has been called out and discussed. These turquoise H_2 production technologies [46] are preferable to SMR due to their lower emissions and energy requirements, but the information on them is still sparse and the data are often conflicting and inconsistent [47]. There tends to be a disproportionate focus on green H_2 production technologies, particularly water electrolysis, despite other promising candidates. Further analysis of these emerging technologies will open up a slew of diverse avenues for clean hydrogen production.

Another reason to invest more resources in a wider range of promising clean H_2 production technologies is their potential to be combined and integrated into hybrid systems. This integration process can strategically lower H_2 production cost, decrease energy requirements and carbon emissions, increase efficiency, and utilize waste to make useful products. A well-established example of technology integration is the addition of CCS to gray H_2 production technology, a method that succeeds in reducing emissions but also tends to increase production cost and energy consumption while reducing process efficiency [7,8,10,17,47]. More technically advanced and economically and energetically efficient combinations can be devised. One paper performed an energy analysis of an integrated system with SMC and SOE for the production of turquoise H_2 and clean methanol [38]. Another paper examined hybrid wind and solar PV energy systems, finding that the combination of power sources meets the load demand and minimizes emissions and system costs [48]. A study reported that the energy output of FWF can be increased by 7–9 times with a 2-stage fermentation process that results in H_2 produced via dark fermentation and methane produced via anaerobic digestion [15]. This methane could then be utilized as renewable natural gas to be injected into the existing natural grid infrastructure.

The major challenges for H_2 production include high costs, infrastructure development, carbon sequestration, and efficiency issues. Key research priorities should focus on improving technological efficiency, developing cost-effective distributed generation systems, enhancing CCS, and exploring novel catalytic processes. The potential prospects are substantial, with H_2 poised to power vehicles, balance energy grids, decarbonize industries, and meet a significant portion of global energy needs by 2050, thereby aiding in GHG reduction.

2.2. *Quantitative and Qualitative Decision Criteria*

The following set of decision criteria were used as the framework by which the selected H_2 production technologies were compared, as discussed in the results in Section 3.

2.2.1. Quantitative Decision Criteria

- Cost of hydrogen production (USD/kg H_2)

This criterion represents the cost to produce one kilogram of H_2 from the given technology. It is an indicator of the economic viability of each technology. To account for the time–value of money, H_2 production costs that are collected from the published literature are adjusted to the 2024 USD using the Chemical Engineering Plant Cost Index, or CEPCI [11]. To adjust H_2 product cost from a previous year to 2024, Equation (1) is applied:

$$H_2\ Cost_{2024} = H_2\ Cost_{YOP}\left(\frac{CEPCI_{2024}}{CEPCI_{YOP}}\right) \quad (1)$$

where the subscript YOP signifies the year of publication of the H_2 production cost.

- Energy consumption (MJ/kg H_2)

This quantitative criterion represents the amount of energy (MJ) required to produce one kilogram of H_2 from the given technology.

- Global warming impact (Kg CO_2-eq/kg H_2)

This quantitative criterion represents the global warming potential (GWP) of the H_2 production technology and shows how many kilograms of CO_2 are emitted per kilogram of H_2 produced.

- Technology energy efficiency

The energy efficiency of a technology is defined as the ratio of the lower heating value (LHV) of produced H_2 to total energy supplied to the H_2 production process. This efficiency is expressed as a percentage value. Some publications reported this process energy efficiency in terms of high heating value (HHV), and in order to keep comparisons between the production technologies consistent, these values are converted herein to be in terms of LHV by multiplying the reported value by the ratio of H_2 LHV to HHV, as shown in Equation (2).

$$\frac{(\mathrm{LHV})_{\mathrm{H2}}}{(\mathrm{HHV})_{\mathrm{H2}}} = \frac{120\ \mathrm{MJ/kg}}{142\ \mathrm{MJ/kg}} = 0.8451 \tag{2}$$

In addition to creating consistency in this comparative assessment, evaluating the technology energy efficiency in terms of LHV rather than HHV is more conservative, as it gives the lower bound of the efficiency and, therefore, does not overestimate this efficiency value.

2.2.2. Qualitative Decision Criteria

- Technology readiness level (TRL)

Technology readiness level is a qualitative criterion that reflects the maturity of a given technology on a nine-point scale (1–9). A higher number indicates greater maturity and readiness for commercialization. The following color-coded chart depicts the TRL scale as technologies advance from research to development (R&D) to deployment (Figure 1).

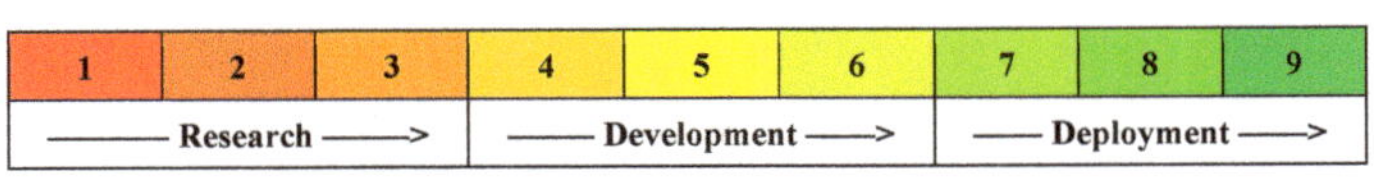

Figure 1. Technology readiness levels (TRL).

- Supply chain material availability

Supply chain material availability is reflective of the availability and ease of accessibility of raw materials, energy, and equipment for the given H_2 production technology. This criterion assesses the technologies on a three-point scale: low, medium, and high. Due to the qualitative and broad nature of this criterion, a simple three-point scale prevents the need for tedious analysis and difficult rankings. However, in order to avoid oversimplification, a comment section is provided to provide clarity on ratings if deemed necessary.

2.3. Hydrogen Production Technologies Being Evaluated

The production technologies being evaluated are as follows:

- Conventional via steam-methane reforming (SMR)

The conventional method for H_2 production is steam-methane reforming (SMR), which accounts for around 50% of H_2 production [5]. This method converts natural gas (primarily methane) and steam to H_2 and CO_2, a byproduct that contributes to global warming. The second conventional method of producing gray hydrogen is coal gasification, but it is not included in this research because it is less common than SMR. Autothermal reforming (ATR) of natural gas is yet another conventional technology for H_2 production. ATR is similar to SMR in that it produces H_2 and CO_2, but it has the advantage of producing more exothermic reactions from the partial oxidation that occurs [8]. The H_2 produced from SMR, coal gasification, and ATR is denoted as "gray" H_2. The integration of carbon capture and storage (CCS) with any of these conventional technologies produces the so-called "blue" H_2. While integration with CCS results in CO_2 emissions reduction, both production cost and energy consumption are shown to increase compared to the case without CCS.

- Hydrogen production by waste valorization

The utilization of various waste streams for hydrogen production is examined in food waste fermentation (FWF), waste plastics gasification (wPG), and biomass gasification with and without CCS. The valorization of biodegradable food waste by dark fermentation can help alleviate the waste management crisis and reduce the production of carbon emissions and toxic pollutants in overcrowded landfills, while also producing more eco-friendly H_2 [49–51]. Data-driven interpretation, comparison, and optimization of H_2 production from supercritical water gasification of biomass has been carried out. Gasification is the high-temperature partial oxidation of waste feedstocks, like plastics or biomass, to produces syngas, which is a mixture of CO and H_2. Syngas can be further converted to make other useful chemicals using the Fischer–Tropsch (FT) process or combusted to generate steam for electricity generation [52].

- Waste byproduct reforming (glycerol from biodiesel production)

A promising waste stream to utilize for H_2 production is the glycerol byproduct that results from biodiesel production via the transesterification of agricultural crops [17]. The technologies studied that convert glycerol into hydrogen are supercritical water reforming (SCWR), aqueous-phase reforming (APR), autothermal reforming, and steam reforming (SR).

- Greener hydrogen production via water electrolysis

The water electrolysis technologies considered herein are alkaline electrolyzer (AKE), proton exchange membrane electrolyzer (PEME), anion exchange membrane electrolyzer (AEME), and solid oxide electrolyzer (SOE) via solar photovoltaic (PV) power, wind power, hydropower, grid mix electricity, and nuclear power. Hydrogen produced via water electrolysis with renewable energy is called green H_2 and is characterized by its low global warming impact. Electrolysis powered by grid mix electricity does not produce green H_2, but it benefits from utilizing the current electricity grid. The US grid electricity mix is roughly as follows: 38% natural gas, 17% coal, 20% nuclear, 11% wind, 7% hydropower, 5% solar, and 2% other [53].

- Emerging hydrogen production technologies

Various emerging clean H_2 production technologies were evaluated in this comparative analysis. The high-temperature processes of biomethane pyrolysis and solar methane cracking (SMC) produce the so-called "turquoise" H_2 and carbon black. SMC utilizes concentrated solar energy to power the high-temperature process, resulting in even cleaner hydrogen [38]. Biophotolysis, electrified steam-methane reforming (ESMR), and ammonia (NH3) thermal decomposition via fixed-bed reactor and via PSA-membrane technology were also evaluated in this research. Biophotolysis utilizes sunlight energy to convert water to hydrogen through biological systems [40]. ESMR combines traditional SMR with renewable electricity. Both solar thermal-powered thermochemical decomposition of water and nuclear-powered thermochemical decomposition of water were researched. There are many variations in nuclear-powered thermochemical decomposition, but only four of the most promising cycles were included in this study. These are the three-step sulfur–iodine (S-I) cycle, the two-step hybrid sulfur (HyS) cycle, the four-step copper–chlorine (Cu-Cl) cycle, and the three-step magnesium–chlorine (Mg-Cl) cycle.

3. Results

The sources introduced in the literature review in Section 2.1 were examined to perform a comparative analysis among the H_2 production technologies for the decision criteria introduced in Section 2.2. The quantitative and qualitative results of the comparative analysis are organized in Tables 1–9. As shown below, Tables 1–3 give the cost of H_2, energy consumption, and global warming impact of the technologies. Tables 4–6 provide the energy efficiencies in terms of the $(LHV)_{H_2}$ of the technologies. Tables 7–9 show the TRL and supply chain material availability of the technologies.

It should be noted that for most decision criteria, the results are recorded as a range of values. This is because multiple sources were cross-referenced in order to provide the most accurate data. For each technology, there are a variety of conditions, such as plant size and local electricity cost, that can vary widely. Therefore, a technology's performance for a given criterion cannot be given in a single value without being very specific about plant specifications (viz., technology type, production capacity, etc.). Providing a range of values from multiple sources is less limiting to the overarching objective of the comparative assessment. It is important to note that for waste plastic gasification [54] with CCS and biomass gasification with CCS (Table 1), as well as biomethane pyrolysis (Table 3), certain papers reported their global warming impact as a negative value. These scenarios imply that net carbon emissions are negative based on utilizing waste streams that would have emitted more carbon otherwise and also due to capturing the carbon emissions associated with gasification.

Table 1. Cost of H_2 (USD/kg H_2), energy consumption (MJ/kg H_2), and GWP (kg CO_2/kg H_2) of conventional production technologies and waste valorization technologies.

Hydrogen Production Technology	Cost of Hydrogen (\$/kg H_2) (All in 2024 U.S. \$)	Energy Consumption (MJ/kg H_2)	Global Warming Impact (kg CO_2/kg H_2)
• **Conventional H_2 production with natural gas**			
• Gray H_2 Production (SMR without carbon capture)	1.19–2.40 [8,19]	183.2 [18]	9.89–11.9 [18,48]
• Blue H_2 Production (SMR with carbon capture)	1.64–2.29 [8,12]	224.75–265.91 [8]	3.32–8.20 [8,48]
• Blue H_2 Production (ATR with carbon capture)	1.44–1.7 [8,9,12]	162.92 [8]	2–4 [8,9]
• **H_2 production by waste valorization (cleaner & sustainable)**			
• Waste plastics gasification (wPG)	2.23–3.41 [53,55]	20.8–25.3 [53]	12.8–15.6 [53], 8.46–5.94(CCS) [9]
• Food waste fermentation (FWF)	2.57–2.88 [9,17]	120–180 [56]	6.6–9.5 [57]
• Biomass gasification	1.72–4.1 [9,12]	115–180 [9,58]	1.9–18 [9]
• Biomass gasification w/CCS	3.7 [9]	180–240 [9]	−15–13 [9]
• Waste byproduct reforming (glycerol from biodiesel production):			
➢ Supercritical water reforming (SCWR) of glycerol	1.69–4.36 [17]	33.17 [18]	51.9 [18]
➢ Aqueous-phase reforming (APR) of glycerol	4.48–9.92 [17,19]	23.76 [18]	4.11 [18]
➢ Autothermal reforming (ATR) of glycerol	1.20 [8]	121.15 [18]	87.2 [18]
➢ Steam reforming (SR) of glycerol	2.03–2.63 [17], 4.86–9.67 [19]	40–60 [19]	4–18 [9]

The key insights to be drawn from the data reported in Table 1 are as follows:

(a) For conventional H_2 production from natural gas: (a.1) Due to the exothermic heat generation, the ATR technology has the lowest energy consumption (MJ/kg H_2); (a.2) without CCS, SMR technology has the highest CO_2 emissions (kg CO_2/kg H_2); and (a.3) as expected, integrating CCS with the conventional H_2 production technology increases the cost of H_2 production (USD/kg H_2). However, this cost increase is impacted by other location-dependent externalities, such as cost of energy, cost of electricity, labor cost, etc.

(b) For H_2 production via waste valorization: (b.1) Biomass gasification with CCS technology has the highest energy consumption (MJ/kg H_2), (b.2) aqueous-phase reforming (APR) of glycerol has the lowest CO_2 emissions (kg CO_2/kg H_2), and (b.3) cost of H_2 production (USD/kg H_2) using ATR of glycerol seems to be the lowest compared to the other waste valorization technologies.

Table 2. Cost of H_2 (USD/kg H_2), energy consumption (MJ/kg H_2), and GWP (kg CO_2/kg H_2) of various water electrolysis technologies with different energy sources.

Hydrogen Production Technology	Cost of Hydrogen ($/kg H_2) (All in 2024 U.S. $)	Energy Consumption (MJ/kg H_2)	Global Warming Impact (Kg CO_2/Kg H_2)
• **Greener H_2 production via water electrolysis**			
• Wind + Alkaline Electrolysers (AKE)	5.25–11.92 [29,33,59]	180–280 [60]	0.0325–4.45 [9,61]
• Wind + Proton Exchange Membrane Electrolysers (PEME)	4–12.27 [26,29,62]	180–280 [60]	0.0325–3.69 [9,61]
• Wind + Solix oxide Electrolysers (SOE)	6.72–7.77 [60,63]	165.6–184 [60,64]	0.0325 [61]
• Wind + Anion Exchange Membrane (AEM) Electrolyzer	4.48–8.96 [23]	185–237 [60]	0.0325 [61]
• Solar PV + Alkaline Electrolysers (AKE)	7.00–9.86 [33]	180–280 [60]	0.37–8.78 [9,61]
• Solar PV + Proton Exchange Membrane Electrolysers (PEME)	10.13–16.51 [29,33]	184–280 [60]	0.37–7.79 [9,61]
• Solar PV + Solix oxide Electrolysers (SOE)	3.36–8.96 [23]	165.6–174.2 [64]	0.37–2.3 [9,61]
• Solar PV + Anion Exchange Membrane Electrolyzer (AEME)	3.36–8.96 [23]	185–237 [60]	0.37 [61]
• Hydro Power + Electrolyzers (AKE, PEME, OXE, AEME)	2.70 [65]	180-216 [48]	0.72-1.75 [9,28]
• Grid Mix + Electrolysers	4.00–8.81 [23–26,66]	180–216 [48]	1–30 [9,25]
• Nuclear power + PEM	4.66–7.15 [29]	180–280 [60]	1.1 [9]
• Nuclear power + SOE	3.79–5 [9,29]	165–180 [64]	0.67–2.1 [9]

The data summarized in Table 2 show that:

- PEME technology that uses solar PV electricity tends to have the highest cost of H_2 production (USD/kg H_2) compared to the other technologies.
- Water electrolyzers that use electricity from the grid mix tend to have the highest CO_2 emissions (kg CO_2/kg H_2).

For the data shown in Table 3 for the emerging H_2 production technologies, the following observations can be derived:

- Thermochemical decomposition integrated with solar thermal as well as ammonia thermal decomposition with PSA membrane separation seem to have the highest cost of H_2 production ($/kg H_2).
- Electrified steam-methane reforming (ESMR) has the lowest energy consumption (MJ/kg H_2).
- CO_2 emissions (kg CO_2/kg H_2) from the evaluated emerging H_2 production technologies seem to be low, with biomethane pyrolysis emerging as the most environmentally friendly, followed by thermochemical decomposition integrated with solar thermal technology.

Table 3. Cost of H_2 production (USD/kg H_2), energy consumption (MJ/kg H_2), and GWP (kg CO_2/kg H_2) of emerging clean H_2 technologies.

Hydrogen Production Technology	Cost of Hydrogen ($/kg H_2) (All in 2024 U.S. $)	Energy Consumption (MJ/kg H_2)	Global Warming Impact (Kg CO_2/Kg H_2)
• **Emerging H_2 production technologies**			
• Turquoise H_2 produced from biomethane pyrolysis	1.36–3.37 [12,17,29]	36–108 [48]	−10.4 to −4.09 [48]
• Electrified steam methane reforming (ESMR)	1.53 [42]	28.8 [23]	5–6 [23]
• Solar methane cracking (SMC)	2.9 [9]	180 [9]	1.9 [9]
• Solar thermal + thermochemical decomposition	7.74–12.22 [12,25,29]	110–150 [29]	0 [25]
• Nuclear + thermochemical decomposition			
• S-I	2.66–7.5 [44,67]	159–223 [67]	0.0–0.5 [25,44]
• HyS	2.39–8.34 [44,67]	116–128 [67]	0.0–0.48 [25,44]
• Cu-Cl	2.24–4.79 [44,64,67]	111 [67]	0.0–1 [25,64]
• Mg-Cl	4.11–5.15 [44,67]	169 [67]	0.0–1 [25,44]
• Bio-photolysis	1.42–7.27 [57]	110–160 [67]	7.5 [57]
• Ammonia thermal decomposition via fixed bed reactor	4.24–11.36 [45]	100–150 [45]	0.0–1 [46]
• Ammonia thermal decomposition with PSA-membrane separation	5.98–11 [45]	100–150 [45]	0.0–1 [46]
• Integration of biomass gasification (BG) with chemical looping hydrogen production (CLHP). • Source: Wu, D. et al. (2024) [68]	For the Air case and O_2 case, the costs are 3.05 and 2.82 $/kg H_2, respectively	n/a	−15.13 to −17.00

Table 4. Energy efficiencies (%) of conventional H_2 production technologies and waste valorization H_2 production technologies.

Hydrogen Production Technology	Energy Efficiency (%) *
• **Conventional with natural gas (NG)**	
• SMR	65–75 [7]
• SMR w/CCS	54 [10]
• ATR w/NG & CCS	45–56 [8,10]
• **Waste valorization**	
• wPG	40–50 [69]
• FWF	57–80 [17,50]
• Biomass gasification	40–50 [29,69]
• Biomass gasification w/CCS	40 [29]
• Glycerol-based:	
➢ SCWR	55–60 [17]
➢ ATR	71–74 [69]
➢ APR	45–55 [69]
➢ Steam reforming (SR)	60–70 [69]

* Efficiency is defined as the ratio of lower heating value (LHV) of hydrogen in product gas to total energy supplied to the process.

Table 5 shows that SMR and steam reforming (SR) of glycerol feedstock have the highest energy efficiency. Biomass gasification technology integrated with CCS has the lowest energy efficiency. Carbon capture and storage (CCS) tends to increase the energy consumption and, hence, lowers the energy efficiency of the technology being evaluated.

Table 5. Energy efficiencies (%) of H_2 production via various water electrolysis technologies with different energy sources.

Hydrogen Production Technology	Energy Efficiency (%) *
▪ **Greener H_2 (water electrolysis)**	
• Wind + AKE	59–70 [24,33,70,71]
• Wind + PEME	58–74 [24,33,70]
• Wind + SOE	71–84 [23,24,72]
• Wind + AEME	69 [9]
• Solar PV + AKE	59–70 [24,33,71]
• Solar PV + PEME	60–68 [24,33,73]
• Solar PV + SOE	69–84 [23,24,26,72]
• Solar PV + AEME	69 [9]
• Hydropower + electrolyzers	61.6–66.10 [34,74]
• Grid mix + electrolyzers	60–84 [23,75]
• Nuclear + SOE	71–81 [24,29,72]
• Nuclear power + PEME	60–65 [24]

* Efficiency is defined as the ratio of lower heating value (LHV) of hydrogen in product gas to total energy supplied to the process.

Table 5 shows that solid oxide electrolyzers (SOEs) using electricity from wind turbines, solar PV, and nuclear power seem to have the highest energy efficiency.

Table 6. Energy efficiencies (%) of emerging clean H_2 production technologies.

Hydrogen Production Technology	Energy Efficiency (%) *
▪ **Emerging technologies**	
• Biomethane pyrolysis	40–56 [23,69]
• ESMR	70 [42]
• SMC	49.4–82.2 [26]
• Solar thermal + thermochemical decomposition	20 [29]
• Nuclear + thermochemical decomposition	
➢ S-I	42–52 [25,44]
➢ HyS	49 [25]
➢ Cu-Cl	43 [44]
➢ Mg-Cl	45 [25]
• Bio-photolysis	10–15 [41,56,57]
• Ammonia thermal decomposition via fixed bed reactor	78–80 [45]
• Ammonia thermal decomposition with PSA-membrane separation	87–90 [45]

* Efficiency is defined as the ratio of lower heating value (LHV) of hydrogen in product gas to total energy supplied to the process.

Table 6 shows that, among the evaluated emerging H_2 production technologies, ammonia thermal decomposition integrated with PSA membrane separation is the most energy-efficient technology and bio-photolysis is the least energy-efficient technology.

Table 7. Color-coded technology readiness level (TRL) and supply chain material availability of conventional and waste valorization H_2 production technologies.

Hydrogen Production Technology	TRL	Supply Chain Material Availability			Feedstock Type
		L	M	H	
■ **Conventional using NG:**					
• **SMR**	9 [17]		✔		methane
• SMR w/CCS	8 [9]		✔		methane
• ATR w/CCS	5 [9]		✔		methane
■ **Waste valorization**					
• wPG	7 [17]			✔	Waste plastics
• FWF	2–5 [9,17]			✔	Food waste
• Biomass gasification	7–9 [9,17]			✔	biomass
• Biomass gasification w/CCS	5 [9]			✔	biomass
• Glycerol-based					
➢ SCWR	4 [17]			✔	Glycerol waste
➢ ATR	7 [9]			✔	Glycerol waste
➢ APR	4–5 [17]			✔	Glycerol waste
➢ SR	8 [75]			✔	Glycerol waste

1	2	3	4	5	6	7	8	9

As shown in Table 7, SMR is the most mature and commercial-scale technology, and thus has a TRL of 9. Food waste fermentation (FWF) technology is the least mature and thus has the lowest TRL, in the 2 to 5 range. Steam reforming of glycerol has a high TRL of 8. Glycerol supercritical water reforming (SCWR) and aqueous-phase reforming (APR) have a lower TRL compared to more mature ATR technology. With respect to supply chain feedstock availability, glycerol waste, biomass, waste plastic, and food waste are more available compared to natural gas (primarily methane).

As displayed in Table 8, the alkaline electrolyzer (AKE) is the most mature technology, with a TRL of 9, and the least mature electrolyzer technology is the anode exchange membrane electrolyzer type, with a TRL as low as 2 or 3. Supply chain availability of metals used in the electrolyzer electrodes are, for the most part, location-dependent.

From Table 9, it can be seen that ammonia thermal decomposition via fixed-bed reactors seems to have the highest TRL of 9 tables. All other H_2 production technologies have lower TRLs, which are indicative that these technologies did not achieve the required maturity level to be commercialized at scale. Once again, supply chain material availability is highly location-dependent, among other externalities.

Table 8. Color-coded technology readiness level (TRL) and supply chain material availability of H_2 production via various water electrolysis technologies with different energy sources.

Hydrogen Production Technology	TRL	Supply Chain Material Availability			Comments
		L	M	H	
▪ **Greener H_2 (water electrolysis):**					
• **Wind + AKE**	9 [9,35]			✔	Nickel, Cupper
• Wind + PEME	6–8 [9,33], 9 [17,35]	✔			Platinum, Iridium, titanium
• Wind + SOE	5 [9], 6–7 [23], 8 [35]			✔	Scandium
• Wind + AEME	2–3 [9], 4–5 [23], 6 [35]			✔	Platinum, Nickel, Transition Metal Catalysts (CeO_2–La_2O)*
• Solar PV + AKE	9 [9,35]			✔	Nickel, Cupper
• Solar PV + PEME	6–8 [9,33], 9 [17,35]	✔			Platinum, Iridium, titanium
• Solar PV + SOE	5 [9], 6–7 [23], 8 [35]			✔	Scandium
• Solar PV + AEME	2–3 [9], 4–5 [23], 6 [35]		✔		Platinum, Nickel, Transition Metal Catalysts (CeO_2–La_2O)*
• Hydropower + electrolyzers	9 [34]		✔		
• Grid mix + electrolyzers	9 [17]		✔		
• Nuclear + SOE	5 [9], 8 [35]			✔	Scandium
• Nuclear power + PEME	6–8 [9], 9 [35]	✔			Platinum, Iridium, titanium

TRL color scale: 1 | 2 | 3 | 4 | 5 | 6 | 7 | 8 | 9

Table 9. Color-coded technology readiness level (TRL) and supply chain material availability of emerging clean H_2 production technologies.

Hydrogen Production Technology	TRL	Supply Chain Material Availability			Comments
		L	M	H	
▪ **Emerging technologies**					
• Biomethane pyrolysis	3–7 [17,71]			✔	
• ESMR	4 [23]		✔		
• SMC	3–5 [9]			✔	
• Solar thermal + thermochemical decomposition	3–4 [9,35]		✔		ZnO/Zn cycle
• Nuclear + thermochemical decomposition					
➢ S-I	3–4 [9,35]		✔		
➢ HyS	3–4 [9,35]		✔		
➢ Cu-Cl	3–4 [9,35]		✔		
➢ Mg-Cl	3–4 [9,35]		✔		
• Bio-photolysis	4 [57]		✔		
• Ammonia thermal decomposition via fixed bed reactors	9 [45]		✔		Precious metal catalyst
• Ammonia thermal decomposition with PSA-membrane separation	4 [45]		✔		

TRL color scale: 1 | 2 | 3 | 4 | 5 | 6 | 7 | 8 | 9

4. Discussion

Based on the results presented (Tables 1–9) in Section 3, the following discission and insights can be synthesized:

- Conventional H_2 production via SMR has the advantage of being low-cost (around USD 1–2/kg H_2) and having a high TRL. However, SMR technology without CCS suffers from a high global warming impact of around 11 kg CO_2/kg H_2, as well as its nonrenewable feedstock requirement of natural gas.
- SMR with CCS and ATR with CCS result in lower global warming impacts, but they still produce around 2–8 kg CO_2/kg H_2, which is higher than the emissions of the majority of H_2 production via water electrolysis. Also, the energy requirements (MJ/kg H_2) of blue H_2 production technologies are generally higher than those of gray H_2 production technologies. The emerging carbon capture, utilization, and storage (CCUS) is expected to add additional environmental and economic values that go beyond CCS only.
- Assessment of waste valorization as an alternative to the conventional methods for H_2 production reveals that its most robust advantage is its high supply chain material availability due to using waste as the feedstock. The various methods utilize different waste streams that would otherwise contribute to costly waste management and environmental issues like toxic leaching and emission of air pollutants from landfills and waste incineration.
- In terms of GWP, waste valorization methods that are integrated with CCS can produce negative carbon emissions with the utilization of waste that otherwise would have been incinerated. Otherwise, when only considering carbon emissions, these technologies are not necessarily more sustainable than the conventional methods, with the exception of APR, with a promising GWP of 4.11 kg CO_2/kg H_2 [18]. Therefore, incorporating CCS technology into waste valorization technologies will enhance their environmental benefits.
- Waste plastics gasification (wPG), food waste fermentation (FWF), and biomass gasification with and without CCS are somewhat more expensive than the conventional methods (around USD 2–4/kg H_2), but not so expensive as to be uncompetitive, particularly when potential for technological improvements and economies of scale are taken into consideration. The energy requirements of the processes range significantly, from 20 to 240 MJ/kg H_2, but as noted earlier, these values vary widely with plant location, energy sources, and many other externalities. Therefore, it is perhaps more meaningful to examine energy efficiency, which is fairly average for wPG and biomass gasification (40–50%) and slightly higher for FWF. Improvements in technology will improve energy efficiency and subsequently further decrease production costs.
- Examining the four methods of glycerol conversion (data in Table 1) reveals that APR is more attractive than both ATR and SCWR, not only due to its lower carbon emissions but also due to its lower energy consumption [17]. Therefore, APR outperforms its glycerol-based counterparts in sustainability and energy consumption, but with a TRL of 4–5, APR is not a mature technology yet and is therefore currently the most expensive, followed by steam reforming (SR), then SCWR, and finally ATR [17]. Also, the energy efficiencies of the four technologies range from 45 to 74, with APR performing the least efficiently. Considering all factors, APR is the most worthwhile technology of the four glycerol-based hydrogen production processes to improve upon.
- The compiled literature data about water electrolysis reveal that it is a very promising technology for green H_2 production, producing close-to-zero carbon emissions. The performance of electrolysis varies depending on the specific technology and electricity source. First examining costs in Table 2, it can be seen that hydropower and nuclear power currently produce the least expensive H_2, at around USD 2.7/kg H_2. The grid mix, wind power, and solar PV power are slightly more expensive, with wind generally being less expensive than solar PV. In terms of the types of electrolyzer, SOE

and AEME are more cost-effective than PEME and AKE. For wind and solar PV power, the most optimal costs for SOE and AEME are USD 3.36–6.72/kg H_2, and for PEME and AKE, the most optimal costs are USD 4–10.13/kg H_2. The energy requirements for the various technologies and power sources are comparable to one another, ranging from 165 to 280 MJ/kg H_2. However, energy efficiency does differ, with wind-powered electrolysis being slightly more efficient than solar PV, and with SOE being the most efficient type of electrolyzer.

- Examining the TRLs of water electrolysis technologies in Table 8, it can be seen that PEME and AKE are the most technologically mature, but their material supply chain availabilities are not very high in terms of availability. PEME specifically has a lower material supply chain availability score because it requires a rare metal (platinum) catalyst. Taking all of the criteria into consideration, a case could be made for any of the water electrolysis processes, but SOE powered by renewable electricity appears to be an especially worthwhile technology to focus on in the future. The source of renewable electricity that should be utilized depends on the characteristics of the region in which the electrolyzer is built. While wind energy showed greater energy efficiency and lower production cost in the literature review, this does not necessarily mean it is the most optimal option everywhere. Also, an important alternative that was not in the scope of this research is the possibility of utilizing multiple sources of renewable electricity in conjunction with, or integrating them into, the electricity grid mix. A hybrid configuration can ensure a more stable and reliable electricity supply, particularly in the case of wind and solar PV, which are known to be intermittent sources.
- The literature review also revealed some encouraging data about the emerging H_2 production technologies. Biomethane pyrolysis, ESMR, supercritical water gasification of sewage sludge [76], industrial CO-to-H_2 via water–gas shift reaction [77], solar thermal thermochemical decomposition, nuclear thermochemical decomposition, bio-photolysis, and ammonia thermal decomposition were herein investigated, all of which are in the late research and early development stages, except for ammonia thermal decomposition via fixed-bed reactors, which exhibits a TRL of 9. Because these technologies are mostly laboratory-scale or small-scale, data rely heavily on simulations and models. Out of the emerging technologies, currently biomethane pyrolysis, ESMR, and bio-photolysis predict the lowest hydrogen costs, with minimum costs of USD 1.36–1.53 per kg H_2. Nuclear thermochemical decomposition is not far behind, at USD 2.24–4.11/kg H_2, with the price varying by almost USD 2 depending on the cycle type. Solar thermal thermochemical decomposition and ammonia thermal decomposition are currently not competitive and will need a lot of technological development and improvements to become viable options for hydrogen production. However, ammonia decomposition via fixed-bed reactors or PSA membranes can produce H_2 at a high energy efficiency with close-to-zero carbon emissions while utilizing ammonia as a renewable feedstock. Therefore, more so than solar thermal decomposition, which has a low energy efficiency, the strategic integration of ammonia thermal decomposition technologies has fairly high prospects as long as production costs are lowered [56–58,60,62,64,67,70,71,73,74,76,77].
- Out of the rest of the H_2 production technologies, nuclear thermal thermochemical decomposition and biomethane pyrolysis appear to be the most promising. The production of "turquoise" H_2 via biomethane pyrolysis has low energy requirements and has a negative global warming potential due to utilizing an otherwise high-emitting feedstock. Thermochemical decomposition with nuclear power has an energy demand similar to that of SMR and produces zero carbon emissions. The most promising cycles

are S-I, HyS, and Cu-Cl. The energy efficiencies of the two technologies are 40–56%, which could be improved upon with more research and design optimization. The other technologies exhibit some advantages, such as ESMR's high energy efficiency and the potentially low production cost of bio-photolysis, but due to their higher GWPs, they are more practical as transitional technologies rather than as long-term solutions [75,78].

5. Conclusions and Recommendations

This research evaluated various H_2 production technologies based on cost (USD/kg H_2), energy consumption (MJ/kg H_2), technology efficiency (%), and carbon emissions (kg CO_2-eq/kg H_2). Steam-methane reforming (SMR) with and without carbon capture and storage (CCS) served as the benchmark H_2 production technology. Despite its low cost, SMR without CCS has high carbon emissions. Water electrolysis, hindered by high energy demands (kWh/kg H_2) and high costs (USD/kg H_2), shows promise, with advancements in renewable energy sources for green electricity production. Waste valorization methods such as biomass gasification or reforming and food waste fermentation offer dual benefits of waste management and sustainable H_2 production, albeit at higher costs at this early stage of development. Emerging technologies like biomethane pyrolysis and nuclear-powered thermochemical cycles also present promising avenues for achieving clean and sustainable H_2 production with minimal environmental impact.

Future research should focus on enhancing the technological maturity, scalability, and cost-efficiency of promising yet underdeveloped H_2 production technologies, such as biomass waste valorization via gasification (or reforming) and water electrolysis using low-cost green electricity, to achieve low carbon emissions and competitive production costs. Additionally, the impact of governmental policies and incentives on H_2 economics should be rigorously analyzed to provide a comprehensive techno-economic framework for sustainable H_2 production. Emphasizing the environmental impacts of these technologies and conducting comparative life cycle assessments along their supply chains can further elucidate the global benefits of green H_2 initiatives.

Funding: This research received no external funding.

Institutional Review Board Statement: Not relevant to this study.

Informed Consent Statement: Not applicable.

Data Availability Statement: The original contributions presented in the research are included in the article; further inquiries can be directed to the corresponding author.

Acknowledgments: The author of this research is grateful for the valuable comments and reviews provided by his colleagues from the University of Oxford, University of Cambridge, and Imperial College of London, in the United Kingdom, as well as by Harvard University, in the U.S.

Conflicts of Interest: The author declares no conflicts of interest.

Appendix A. Acronyms

AEME	Anion exchange membrane electrolyzer
AKE	Alkaline electrolyzer
APR	Aqueous-phase reforming
ATR	Autothermal reforming
CCS	Carbon capture and storage
CCUS	Carbon capture, utilization, and storage
Cu-Cl	Copper–chlorine

ESMR	Electrified steam-methane reforming
FWF	Food waste fermentation
HRS	Hydrogen refueling station
HyS	Hybrid sulfur
Mg-Cl	Magnesium–chloride
PEM	Proton exchange membrane electrolyzer
SCWR	Super-critical water reforming
S-I	Sulfur–iodine
SMR	Steam-methane reforming
SOE	Solid oxide electrolyzer
TEA	Techno-economic analysis
TRL	Technology readiness level
wPG	Waste plastic gasification

References

1. Lui, J.; Paul, M.C.; Sloan, W.; You, S. Techno-economic feasibility of distributed waste-to-hydrogen systems to support green transport in Glasgow. *Int. J. Hydrogen Energy* **2022**, *47*, 13532–13551. [CrossRef]
2. The Paris Agreement | UNFCCC. 2020. Available online: https://unfccc.int/process-and-meetings/the-paris-agreement (accessed on 9 March 2024).
3. Frueh, S. Can Hydrogen Help the Energy Transition? Panel Explores the Benefits and Challenges. National Academies of Science, Engineering, and Medicine. 23 August 2024. Available online: https://www.nationalacademies.org/news/2024/08/can-hydrogen-help-the-energy-transition-panel-explores-the-benefits-and-challenges-of-developing-more-clean-hydrogen-for-fuel (accessed on 25 December 2024).
4. Ghasemi, A.; Rad, H.N.; Akrami, M. Biomass-to-Green Hydrogen: A Review of Techno-Economic-Enviro Assessment of Various Production Methods. *Hydrogen* **2024**, *5*, 474–493. [CrossRef]
5. Dincer, I.; Acar, C. Review and evaluation of hydrogen production methods for better sustainability. *Int. J. Hydrogen Energy* **2015**, *40*, 11094–11111. [CrossRef]
6. Alternative Fuels Data Center: Hydrogen. Energy.gov. 2023. Available online: https://afdc.energy.gov/fuels/hydrogen_basics.html#:~:text=The%20energy%20in%202.2%20pounds,driving%20range%20of%20conventional%20vehicles (accessed on 9 March 2024).
7. IRJET Journal. IRJET-review paper on hydrogen: The fuel of future. *IRJET* **2020**, *7*, 2978–2981. Available online: https://www.academia.edu/44307163/IRJET_REVIEW_PAPER_ON_HYDROGEN_THE_FUEL_OF_FUTURE?amp;swp=rr-rw-wc-57873971&nav_from=7992dc07-7621-4e2d-a8a1-1ed8ca920f22&rw_pos=17 (accessed on 11 March 2024).
8. Oni, A.O.; Anaya, K.; Giwa, T.; Di Lullo, G.; Kumar, A. Comparative assessment of blue hydrogen from steam methane reforming, autothermal reforming, and natural gas decomposition technologies for natural gas-producing regions. *Energy Convers. Manag.* **2022**, *254*, 115245. [CrossRef]
9. Wilkinson, J.; Mays, T.J.; McManus, M. Review and meta-analysis of recent life cycle assessments of hydrogen production. *Clean. Environ. Syst.* **2023**, *9*, 100116. [CrossRef]
10. Martínez-Rodríguez, A.; Abánades, A. Comparative Analysis of Energy and Exergy Performance of Hydrogen Production Methods. *Entropy* **2020**, *22*, 1286. [CrossRef]
11. Maxwell, C. Cost Indices—Towering Skills. Toweringskills.com. 2020. Available online: https://toweringskills.com/financial-analysis/cost-indices/ (accessed on 25 December 2024).
12. Yousefi Rizi, H.A.; Shin, D. Green Hydrogen Production Technologies from Ammonia Cracking. *Energies* **2022**, *15*, 8246. [CrossRef]
13. Patel, S.K.S.; Gupta, R.K.; Rohit, M.V.; Lee, J.-K. Recent developments in hydrogen production, storage, and transportation: Challenges, opportunities, and perspectives. *Fire* **2024**, *7*, 233. [CrossRef]
14. Hosseinzadeh, A.; Zhou, J.L.; Li, X.; Afsari, M.; Altaee, A. Techno-economic and environmental impact assessment of hydrogen production processes using bio-waste as renewable energy resource. *Renew. Sustain. Energy Rev.* **2022**, *156*, 111991. [CrossRef]
15. Cudjoe, D.; Chen, W.M.; Zhu, B. Valorization of food waste into hydrogen: Energy potential, economic feasibility and environmental impact analysis. *Fuel* **2022**, *324*, 124476. [CrossRef]
16. Jarunglumlert, T.; Prommuak, C.; Putmai, N.; Pavasant, P. Scaling-up bio-hydrogen production from food waste: Feasibilities and challenges. *Int. J. Hydrogen Energy* **2018**, *43*, 634–648. [CrossRef]
17. Lepage, T.; Kammoun, M.; Schmetz, Q.; Richel, A. Biomass-to-hydrogen: A review of main routes production, processes evaluation and techno-economical assessment. *Biomass Bioenergy* **2021**, *144*, 105920. [CrossRef]

18. Khalil, Y. Simulation-based environmental-impact assessment of glycerol-to-hydrogen conversion technologies. *Clean Energy* **2021**, *5*, 387–402. [CrossRef]
19. Khodabandehloo, M.; Larimi, A.; Khorasheh, F. Comparative process modeling and techno-economic evaluation of renewable hydrogen production by glycerol reforming in aqueous and gaseous phases. *Energy Convers. Manag.* **2020**, *225*, 113483. [CrossRef]
20. Ashwath, J.; Kanishka, S.; Jannan, B.; Janardhan, A.; Shailesh, A.; Rathore, S. Life Cycle Analysis of Hydrogen. *AIP Conf. Proc.* **2021**, *2396*, 020008. [CrossRef]
21. Sapnken, F.E.; Posso, F.; Kibong, M.T.; Noumo, P.G.; Fantah, A.C.; Tamba, J.G. Green hydrogen demand in Cameroon's energy sectors by 2040. *Renew. Sustain. Energy Rev.* **2024**, *205*, 114834. [CrossRef]
22. Ajeeb, A.; Neto, R.C.; Baptista, P. Life cycle assessment of green hydrogen production through electrolysis: A literature review. *Sustain. Energy Technol. Assess.* **2024**, *69*, 103923. [CrossRef]
23. IEA. *Global Hydrogen Review 2021*; IEA: Paris, France, 2021. Available online: https://www.iea.org/reports/global-hydrogen-review-2021 (accessed on 17 April 2024).
24. Christensen, A. *Assessment of Hydrogen Production Costs from Electrolysis: United States and Europe*; International Council on Clean Transport: Washington, DC, USA, 2020. Available online: https://theicct.org/wp-content/uploads/2021/06/final_icct2020_assessment_of-_hydrogen_production_costs-v2.pdf (accessed on 15 April 2024).
25. Edvan, F. Multiple criteria evaluation of hydrogen production processes for use in automotive sector. *Int. J. Hydrogen Energy* **2023**, *49*, 843–861. [CrossRef]
26. Clark, C.E.; Barker, A.; Brunik, K.; Kotarbinski, M.; Grant, E.; Roberts, O.; King, J.; Andrew Bhaskar, P.; Bay, C.J. Opportunities for green hydrogen production with land-based wind in the United States. *Energy Convers. Manag.* **2023**, *296*, 117595. [CrossRef]
27. Lima, T.M.; Lima, C.G.S.; da Costa Madriaga, V.G.; Vasconcelos, S.C.; Quatrocciochi, D.G.; Ferreira, P.G.; de C. da Silva, F.; da S. M. Forezi, L.; Ferreira, V.F. An Overview of Hydrogen Production from Renewable Sources and Its Main Applications. *ChemistrySelect* **2024**, *9*, e202401237. [CrossRef]
28. Koroneos, C.; Dompros, A.; Roumbas, G.; Moussiopoulos, N. Life cycle assessment of hydrogen fuel production processes. *Int. J. Hydrogen Energy* **2004**, *29*, 1443–1450. [CrossRef]
29. El-Emam, R.S.; Özcan, H. Comprehensive review on the techno-economics of sustainable large-scale clean hydrogen production. *J. Clean. Prod.* **2019**, *220*, 593–609. [CrossRef]
30. Gerhardt-Morsdorf, J.; Incer-Valverde, J.; Morosuk, T.; Minke, C. Exergetic life cycle assessment for green hydrogen production. *Energy* **2024**, *299*, 131553. [CrossRef]
31. Lee, B.; Choe, C.; Kim, H.; Kim, A.; Shin, Y.-U.; Haider, J.; Lim, H. Economic Parity Analysis of Green Methanol Synthesis Using Water Electrolysis Based on Renewable Energy. *ACS Sustain. Chem. Eng.* **2021**, *9*, 15807–15818. [CrossRef]
32. Sharma, M.; Pramanik, A.; Bhowmick, G.D.; Tripathi, A.; Ghangrekar, M.M.; Pandey, C.; Kim, B.-S. Premier, Progress and Prospects in Renewable Hydrogen Generation: A Review. *Fermentation* **2023**, *9*, 537. [CrossRef]
33. Schnuelle, C.; Wassermann, T.; Fuhrlaender, D.; Zondervan, E. Dynamic hydrogen production from PV & wind direct electricity supply—Modeling and techno-economic assessment. *Int. J. Hydrogen Energy* **2020**, *45*, 29938–29952. [CrossRef]
34. Altayib, K.; Dinçer, I. Development of an integrated hydropower system with hydrogen and methanol production. *Energy* **2022**, *240*, 122780. [CrossRef]
35. Guide, T. *ETP Clean Energy Technology Guide—Data Tools—IEA*; IEA: Paris, France, 2023. Available online: https://www.iea.org/data-and-statistics/data-tools/etp-clean-energy-technology-guide (accessed on 10 February 2024).
36. Li, J.; Xie, W.; Zhou, H.; Li, Z.; Shao, M. Techno-economic analysis of electrochemical hydrogen production coupled with alternative oxidation. *Chem. Eng. Sci.* **2024**, *298*, 120322. [CrossRef]
37. Naqvi, S.R.; Kazmi, B.; Taqvi, S.A.A.; Chen, W.-H.; Juchelková, D. Techno economic analysis for advanced methods of green hydrogen production. *Curr. Opin. Green Sustain. Chem.* **2024**, *48*, 100939. [CrossRef]
38. H_2NEW: Hydrogen (H_2) from Next-Generation Electrolyzers of Water LTE Task 3c: System and Technoeconomic Analysis. DOE Hydrogen Program. 2021. Available online: https://www.nrel.gov/docs/fy21osti/79835.pdf (accessed on 25 December 2024).
39. Banu, A.; Biçer, Y. Energy and exergy analysis of an integrated system with solar methane cracking and co-electrolysis of CO_2/H_2O for efficient carbon management. *Int. J. Hydrogen Energy* **2023**, *52*, 580–593. [CrossRef]
40. Mersbet, N.H.; Kerboua, K.; Hoinkis, J. Hydrogen production from wastewater: A comprehensive review of conventional and solar powered technologies. *Renew. Energy* **2024**, *226*, 120412. [CrossRef]
41. Nagarajan, D.; Lee, D.H.; Kondo, A.; Chang, J.S. Recent insights into biohydrogen production by microalgae—From biophotolysis to dark fermentation. *Bioresour. Technol.* **2017**, *227*, 373–387. [CrossRef]
42. Mehanovic, D.; Al-Haiek, A.; Leclerc, P.; Rancourt, D.; Fréchette, L.; Picard, M. Energetic, GHG, and economic analyses of electrified steam methane reforming using conventional reformer tubes. *Energy Convers. Manag.* **2023**, *276*, 116549. [CrossRef]

43. Germscheidt, R.L.; Moreira, D.E.B.; Yoshimura, R.G.; Gasbarro, N.P.; Datti, E.; dos Santos, P.L.; Bonacin, J.A. Hydrogen Environmental Benefits Depend on the Way of Production: An Overview of the Main Processes Production and Challenges by 2050. *Adv. Energy Sustain. Res.* **2021**, *2*, 2100093. [CrossRef]
44. Safari, F.; Dinçer, I. A review and comparative evaluation of thermochemical water splitting cycles for hydrogen production. *Energy Convers. Manag.* **2020**, *205*, 112182. [CrossRef]
45. Kanaan, R.; Henrique, P.; Achard, P.; Beauger, C. Economical assessment comparison for hydrogen reconversion from ammonia using thermal decomposition and electrolysis. *Renew. Sustain. Energy Rev.* **2023**, *188*, 113784. [CrossRef]
46. Cha, J.; Park, Y.; Brigljević, B.; Lee, B.; Lim, D.; Lee, T.; Jeong, H.; Kim, Y.; Sohn, H.; Mikulčić, H.; et al. An efficient process for sustainable and scalable hydrogen production from green ammonia. *Renew. Sustain. Energy Rev.* **2021**, *152*, 111562. [CrossRef]
47. Pashchenko, D. Green hydrogen as a power plant fuel: What is energy efficiency from production to utilization? *Renew. Energy* **2024**, *223*, 120033. [CrossRef]
48. Diab, J.; Fulcheri, L.; Hessel, V.; Rohani, V.; Frenklach, M. Why turquoise hydrogen will Be a game changer for the energy transition. *Int. J. Hydrogen Energy* **2022**, *47*, 25831–25848. [CrossRef]
49. Abdin, Z.; Mérida, W. Hybrid energy systems for off-grid power supply and hydrogen production based on renewable energy: A techno-economic analysis. *Energy Convers. Manag.* **2019**, *196*, 1068–1079. [CrossRef]
50. Mohanakrishna, G.; Sneha, N.P.; Rafi, S.M.; Sarkar, O. Dark fermentative hydrogen production: Potential of food waste as future energy needs. *Sci. Total Environ.* **2023**, *888*, 163801. [CrossRef]
51. Chen, Z.; Wei, W.; Chen, X.; Liu, Y.; Shen, Y.; Ni, B.-J. Upcycling of plastic wastes for hydrogen production: Advances and perspectives. *Renew. Sustain. Energy Rev.* **2024**, *195*, 114333. [CrossRef]
52. Azadvar, S.; Tavakoli, O. Data-driven interpretation, comparison and optimization of hydrogen production from supercritical water gasification of biomass and polymer waste: Applying ensemble and differential evolution in machine learning algorithms. *Int. J. Hydrogen Energy* **2024**, *85*, 511–525. [CrossRef]
53. Afzal, S.; Singh, A.; Nicholson, S.; Uekert, T.; DesVeaux, J.S.; Eric Dutta, A.; Carpenter, A.; Baldwin, R.M.; Beckham, G.T. Techno-economic analysis and life cycle assessment of mixed plastic waste gasification for production of methanol and hydrogen. *Green Chem.* **2023**, *25*, 5068–5085. [CrossRef]
54. Short-Term Energy, Outlook—U.S. Energy Information Administration (EIA). Eia.gov. 2023. Available online: https://www.eia.gov/outlooks/steo/report/BTL/2023/02-genmix/article.php (accessed on 11 January 2024).
55. Lan, K.; Yao, Y. Feasibility of gasifying mixed plastic waste for hydrogen production and carbon capture and storage. *Commun. Earth Environ.* **2022**, *3*, 300. [CrossRef]
56. Nikolaidis, P.; Poullikkas, A. A comparative overview of hydrogen production processes. *Renew. Sustain. Energy Rev.* **2017**, *67*, 597–611. [CrossRef]
57. Acar, C.; Dincer, I. Review and evaluation of hydrogen production options for better environment. *J. Clean. Prod.* **2019**, *218*, 835–849. [CrossRef]
58. Yakaboylu, O.; Harinck, J.; Smit, K.G.; De Jong, W. Supercritical Water Gasification of Biomass: A Literature and Technology Overview. *Energies* **2015**, *8*, 859–894. [CrossRef]
59. Webinar: Wind-to-Hydrogen Cost Modeling and Project Findings. Energy.gov. 2013. Available online: https://www.energy.gov/eere/fuelcells/webinar-wind-hydrogen-cost-modeling-and-project-findings (accessed on 25 December 2024).
60. IRENA. *Making the Breakthrough: Green Hydrogen Policies and Technology Costs*; International Renewable Energy Agency: Masdar City, Abu Dhabi, 2021. Available online: https://www.irena.org/-/media/Files/IRENA/Agency/Publication/2020/Nov/IRENA_Green_Hydrogen_breakthrough_2021.pdf?la=en&hash=40FA5B8AD7AB1666EECBDE30EF458C45EE5A0AA6 (accessed on 25 December 2024).
61. Suleman, F.; Dinçer, I.; Agelin-Chaab, M. Comparative impact assessment study of various hydrogen production methods in terms of emissions. *Int. J. Hydrogen Energy* **2016**, *41*, 8364–8375. [CrossRef]
62. Jang, D.; Kim, K.; Kim, K.H.; Kang, S. Techno-economic analysis and Monte Carlo simulation for green hydrogen production using offshore wind power plant. *Energy Convers. Manag.* **2022**, *263*, 115695. [CrossRef]
63. Lamagna, M.; Ferrario, A.M.; Garcia, D.A.; McPhail, S.J.; Comodi, G. Reversible solid oxide cell coupled to an offshore wind turbine as a poly-generation energy system for auxiliary backup generation and hydrogen production. *Energy Rep.* **2022**, *8*, 14259–14273. [CrossRef]
64. Saeedmanesh, A. Solid Oxide Electrolysis Cell (SOEC) and System Technology for Widespread Use with Renewable Energy. Escholarship.org. 2021. Available online: https://escholarship.org/uc/item/6x19c8wf (accessed on 25 December 2024).
65. Karayel, G.K.; Javanı, N.; Dinçer, I. Hydropower for green hydrogen production in Turkey. *Int. J. Hydrogen Energy* **2023**, *48*, 22806–22817. [CrossRef]

66. IEA. *Energy Technology Perspectives 2023;* IEA: Paris, France, 2023. Available online: https://www.iea.org/reports/energy-technology-perspectives-2023 (accessed on 25 December 2024).
67. El-Emam, R.S.; Özcan, H.; Zamfirescu, C. Updates on promising thermochemical cycles for clean hydrogen production using nuclear energy. *J. Clean. Prod.* **2020**, *262*, 121424. [CrossRef]
68. Wu, D.; Gao, Z.; Wu, S.; Xiao, R. Negative net global warming potential hydrogen production through biomass gasification combined with chemical looping: Environmental and economic assessments. *Int. J. Hydrogen Energy* **2024**, *66*, 24–32. [CrossRef]
69. Parthasarathy, P.; Narayanan, K.S. Hydrogen production from steam gasification of biomass: Influence of process parameters on hydrogen yield—A review. *Renew. Energy* **2014**, *66*, 570–579. [CrossRef]
70. Tebibel, H. Methodology for multi-objective optimization of wind turbine/battery/electrolyzer system for decentralized clean hydrogen production using an adapted power management strategy for low wind speed conditions. *Energy Convers. Manag.* **2021**, *238*, 114125. [CrossRef]
71. Ursúa, A.; Marroyo, L.; Gubía, E.; Gandía, L.; Diéguez, P.M.; Sanchis, P. Influence of the power supply on the energy efficiency of an alkaline water electrolyser. *Int. J. Hydrogen Energy* **2009**, *34*, 3221–3233. [CrossRef]
72. Technical Targets for High Temperature Electrolysis. 2022. Available online: https://www.energy.gov/eere/fuelcells/technical-targets-high-temperature-electrolysis (accessed on 25 December 2024).
73. Li, G.; Cui, P.; Wang, Y.; Liu, Z.; Zhu, Z.; Yang, S. Life cycle energy consumption and GHG emissions of biomass-to-hydrogen process in comparison with coal-to-hydrogen process. *Energy* **2020**, *191*, 116588. [CrossRef]
74. Nnabuife, S.G.; Ugbeh-Johnson, J.; Okeke, N.E.; Ogbonnaya, C. Present and Projected Developments in Hydrogen Production: A Technological Review. *Carbon Capture Sci. Technol.* **2022**, *3*, 100042. [CrossRef]
75. Li, Z.; Brouwer, J. Dynamic operation and feasibility study of a self-sustainable hydrogen fueling station using renewable energy sources. *Int. J. Hydrogen Energy* **2015**, *40*, 3822–3837. [CrossRef]
76. Fatimah, M.; Qyyum, M.A.; Lee, M.; Alshareef, R.S.; Aslam, M.; Saeed, B.; Dai, L.; Gilani, M.A.; Bazmi, A.A.; Chang, I.S.; et al. Industrial waste gases as a resource for sustainable hydrogen production—Resource availability, production potential, challenges, and prospects. *Carbon Capture Sci. Technol.* **2024**, *12*, 100228. [CrossRef]
77. Hu, D.; Ren, C.; Zhang, S.; Ma, M.; Chen, Y.; Chen, B.; Guo, L. Thermodynamic and environmental analysis of integrated supercritical water gasification of sewage sludge for power and hydrogen production. *Energy* **2024**, *299*, 131568. [CrossRef]
78. Frowijn, L.S.F.; van Sark, W.G.J.H.M. Analysis of photon-driven solar-to-hydrogen production methods in the Netherlands. *Sustain. Energy Technol. Assess.* **2021**, *48*, 101631. [CrossRef]

applied sciences

MDPI

Article

Performance of Granulated Ti-Dopped Manganese Oxide Oxygen Carriers in Chemical Looping Processes

Teresa Mendiara *, Amirhossein Filsouf, Iñaki Adánez-Rubio, María Teresa Izquierdo and Alberto Abad

Instituto de Carboquímica, ICB-CSIC, 50018 Zaragoza, Spain
* Correspondence: tmendiara@icb.csic.es

Abstract: The cornerstone in the development of different applications of the Chemical Looping (CL) technology is the oxygen carrier material that transfers the oxygen. We evaluate the performance of manganese oxide doped with TiO_2 to reinforce the release of molecular oxygen and regeneration at high temperatures (850–950 °C). Different manganese-titanium mixed oxides with mass fractions of titanium oxide from 0 to 50 wt.% were considered and prepared by granulation. The sample with 20 wt.%. TiO_2 (MnTi_20) showed better oxygen release/regeneration capabilities than pure manganese oxide under similar conditions and was selected for further development. To evaluate the performance of MnTi_20, long tests (200 cycles) were carried out in a TGA to evaluate the chemical and mechanical stability of the sample. Additionally, MnTi_20 was tested in a batch fluidized bed reactor to evaluate its oxygen uncoupling capability and its reactivity with the main combustion gases (H_2, CO, and CH_4) as well as its fluidization properties. MnTi_20 is capable of regenerating Mn_3O_4 to Mn_2O_3 at 950 °C and maintains its reactivity with the redox cycles. No fluidization problems were encountered during almost 60 h of continuous fluidization, and a lifetime of 2500 h was estimated for this oxygen carrier.

Keywords: chemical looping combustion; chemical looping air separation; oxygen uncoupling; manganese oxide; titanium oxide

1. Introduction

In order to reach the Paris Agreement goals, a clean energy transition from massive use of fossil fuels to sustainable energy sources must be undertaken. To navigate this transition successfully, it is required to focus beyond a single solution or technology. Low-carbon energy sources considered in this transition are solar, wind, and energy storage systems, together with hydrogen and other sustainable fuels, and carbon capture, utilization, and storage (CCUS).

In recent decades, the Chemical Looping (CL) technology has positioned itself as one of the most flexible alternatives to facilitate this transition to a decarbonized economy [1]. CL processes can be integrated into combustion, gasification, and reforming schemes [2]. They can be the basis of oxygen generation production processes and be used in the production of chemicals [3]. They can also be used as energy storage technologies or in combination with the use of solar energy [4]. A recent classification distributed this versatility of CL process in three categories [5]:

(i) systems for fuel full oxidation (Chemical Looping Combustion, CLC; *in situ* Gasification-CLC, iG-CLC; chemical looping with oxygen uncoupling, CLOU)

(ii) systems for fuel partial oxidation (Chemical Looping Reforming, CLR; Chemical Looping Gasification, CLG; Chemical Looping Oxidative Dehydrogenation, CL-ODH)
(iii) systems without fuel (Chemical Air Separation, CLAS; thermal chemical energy storage, TCES)

The basic principle behind Chemical Looping is the use of a solid carrier material to cyclically transport oxygen from one part of a chemical process to another. One of the first configurations to be investigated was that corresponding to combustion. Chemical Looping Combustion (CLC) is classified as a Carbon Capture and Storage (CCS) technology. It has attracted attention due to its potential to reduce the cost associated with CO_2 capture [6]. In CLC, the oxygen carrier is normally a metal oxide (M_xO_y) [7]. It circulates between two interconnected reactors, commonly fluidized beds, as it is shown in Figure 1. First, the oxygen in the oxygen carrier is transferred to the fuel while the oxygen carrier is reduced (fuel reactor). Subsequently, the reduced oxygen carrier reoxidizes with air (air reactor). To develop an adequate oxygen carrier is the cornerstone of the CLC technology.

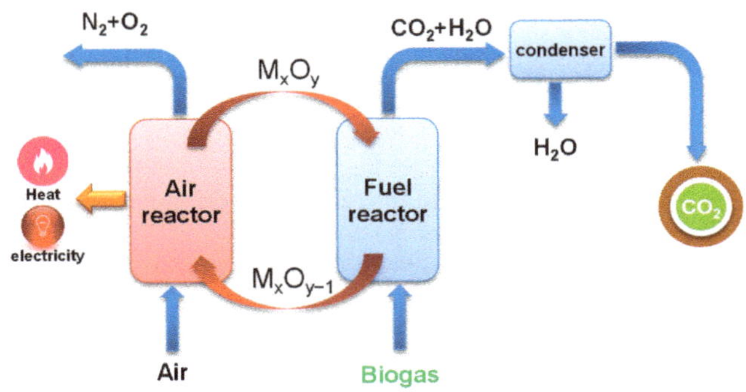

Figure 1. Scheme for the Chemical Looping Combustion (CLC) process.

Commonly, the mechanism for oxygen transfer from the oxygen carrier to the fuel is based on the lattice oxygen of the metal oxide. However, additionally to this mechanism, some oxygen carriers exhibit oxygen uncoupling capability (r1), i.e., the capacity to release molecular oxygen while they are reduced. This is known as chemical looping with oxygen uncoupling (CLOU). Under CLOU combustion, molecular oxygen released by the metal oxide in reaction (r1) can also be used in the combustion of the fuel, improving the combustion efficiency.

$$2\ M_xO_y \leftrightarrow 2\ M_xO_{y-1} + O_2 \quad \text{(r1)}$$

The possibility of generating O_2 during reduction and subsequently regenerating in air is also the basis for the development of other CL processes such as Chemical Looping Air Separation (CLAS) and thermochemical energy storage (TCES) [8]. Since the oxygen carrier is decomposed in steam, the CLAS process can provide a pure oxygen stream once water is condensed at the outlet of the reduction reactor. After that, the reduced oxygen carrier is regenerated with air in an exothermic reaction in the oxidation reactor. Comparing conventional cryogenic air separation to CLAS, the operation cost can be reduced by 40–60% [9]. Based on the CLAS process, the concept of a solar chemical looping air separation (Sol-CLAS) system (Figure 2) was introduced for TCES application [10]. In this process, the energy required for the heating and reduction of the oxygen carrier particles in the reduction reactor (charging step) is provided by concentrated solar thermal radiation. The reduced particles are stored in a reservoir so that the flow of particles to the oxidation reactor (discharging step) is controlled and maintained constant for heat/electricity production despite variation in the solar energy input. The oxygen carrier is then regenerated in air and then stored in another reservoir after being cooled.

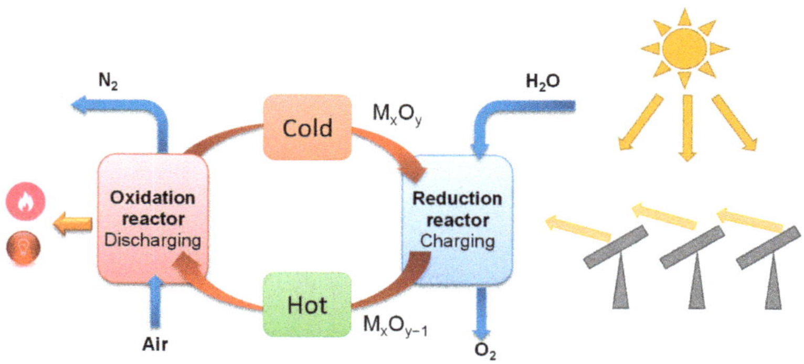

Figure 2. Scheme for the solar Chemical Looping Air Separation (Sol-CLAS) system.

Among the metal oxides with oxygen uncoupling capability, cobalt, manganese, and copper oxides are the most promising redox systems. Ca–Mn perovskites with Mg doping followed by La-based perovskites have also emerged as an adequate option [11]. Compared to cobalt- and copper-based systems, manganese-based oxide systems present the advantage of being inexpensive and harmless. Manganese oxide has several stable phases: MnO_2, Mn_2O_3, Mn_3O_4, and MnO. Reduction of MnO_2 to Mn_2O_3 is an irreversible reaction. Mn_2O_3/Mn_3O_4 transformation releases oxygen according to reaction 2 (r2):

$$6\ Mn_2O_3 \leftrightarrow 4\ Mn_3O_4 + O_2 \quad \text{(r2)}$$

While Mn_2O_3 reduction takes place at high temperatures, Mn_3O_4 exhibited limited re-oxidation at temperatures higher than 800 °C, resulting in a dense and highly sintered structure. In order to narrow this hysteresis loop, manganese oxides have been combined to produce mixed oxides, mainly with iron, copper, magnesium, and silicon oxides [12].

The addition of iron oxide to manganese oxide has proven to be one of the most effective methods for improving the reoxidation of manganese oxide. The addition of iron weakens the Mn-O bond, thus promoting an increase in the reactivity of the material [8]. Through the different studies performed, it can be concluded that the addition of iron oxide (10 to 40 wt.%) reduced the thermal hysteresis, allowed for a lower sintering degree, and also improved the cycling response of the mixed oxides [13,14]. The addition of copper oxide to Mn-Fe materials can further improve the thermal hysteresis since the incorporation of Cu decreases the reduction temperature [15]. Manganese and iron mixed oxides have also been tested in CLOU applications with promising results [16,17].

Magnesium addition to manganese oxide has also demonstrated a good potential for thermochemical energy storage [18]. However, the temperature window for reduction–oxidation (1000–1500 °C) is considered excessively high for solar concentrated applications [19]. Compared to mixed manganese-iron oxides, the addition of SiO_2 to manganese oxide results in materials with more stable redox behavior, also increasing the temperature at which the material can be reoxidized [20] and the reoxidation rate [21]. The combined system Mn/Mg/Si has also recently been tested as a potential oxygen carrier for CLOU with promising results [22]. Other combinations, such as Al_2O_3 [23,24] or ZrO_2 [24]

with manganese oxide, did not improve the thermochemical properties of the original manganese oxide and therefore were discarded.

Recently, the effect of adding TiO_2 to manganese oxide was also considered by the authors. Manganese oxide particles including different mass fractions of titanium oxide ranging from 0 to 50 wt.% were prepared by pelletization [25]. After calcining under certain temperature conditions, the presence of a new solid phase (pyrophanite, $MnTiO_3$) was observed together with the Mn_2O_3-Mn_3O_4 redox pair. Reduction–oxidation cycles were performed with the different samples, showing an improvement with respect to the pure manganese oxide in the reoxidation capacity in air and in the reoxidation temperature, which was increased up to 950 °C. After analyzing the composition of the samples, it was observed that the redox pair responsible for the release of oxygen was Mn_2O_3-Mn_3O_4. The pyrophanite remained stable in the successive redox cycles of reduction–reoxidation and did not intervene in the release of oxygen. However, everything seems to indicate that its presence favored the reoxidation of Mn_3O_4. Similar conclusions were obtained in a more recent study, where manganese oxides from different origins (both commercial and synthesized) were evaluated [26].

In the case of the Mn-Ti samples, besides the oxygen uncoupling capability of the manganese oxygen carriers, lattice oxygen in the manganese oxide can also be transferred to burn the main combustion gases in a conventional CLC process through the following redox reactions involving the redox pair Mn_3O_4/MnO:

$$Mn_3O_4 + H_2 \rightarrow 3\ MnO + H_2O \quad \text{(r3)}$$

$$Mn_3O_4 + CO \rightarrow 3\ MnO + CO_2 \quad \text{(r4)}$$

$$4\ Mn_3O_4 + CH_4 \rightarrow 12\ MnO + 2\ H_2O + CO_2 \quad \text{(r5)}$$

Considering all the above, the objective of the present work is to further investigate the effects of TiO_2 addition to manganese oxide in order to assess the viability of these materials for the CLOU/CLAS or TCES processes as well as for CLC. The novelty of this work in relation to previous studies lies in the method of preparation of the mixed Mn-Ti materials. In this work, the possibility of producing them by granulation of the corresponding commercial powders will be studied. In turn, the durability and the evolution of the reactivity of these materials under several redox cycles will be studied for the first time, as well as their fluidization properties for applications that require the use of fluidized bed reactors. This way, the assessment of their potential as oxygen carriers for different Chemical Looping processes can be done.

2. Materials and Methods

2.1. Materials

Various samples of manganese oxide, including different amounts of titanium oxide ranging from 0 to 50 wt.%, were prepared by mechanical mixing through a high shear mechanical mixing granulator (Eirich laboratory mixer EL1). The raw materials used to prepare these oxygen carriers were Mn_3O_4 (Micromax, Elkem-Norway) and TiO_2 (Acros-Belgium) supplied as powders. After granulation, particles were calcined at the same conditions as those used for particles produced by pelletization [25], i.e., 1030 °C for 2 h in air atmosphere. Then, the calcined granules were sieved in the particle size range of 0.1 to

0.3 mm. This size range was selected to allow adequate particle fluidization in the batch fluidized bed reactor.

2.2. Characterization of Oxygen Carriers

The main phases of the manganese-titanium (Mn-Ti) mixed oxides were identified by X-ray diffraction (XRD). For that purpose, a Bruker D8 Advance X-ray powder diffractometer with an X-ray source (Cu anode) working at 40 kV and 40 mA was used. It included an energy-dispersive one-dimensional detector. In order to obtain the diffractogram of each sample, the scanning step selected was 0.019° over the 2θ range of 10–80°. DIFFRAC.EVA.V5.2 and TOPAS6 software, on the basis of the Joint Committee on Powder Diffraction Standards, were used in the assignation and quantification of crystalline phases.

Crushing strength determination of the different oxygen carriers was carried out in a Shimpo FGN-5 digital force gauge apparatus as the force required to fracture a particle. According to the procedure, the value of crushing strength presented in this work was the average value of 20 measurements of randomly chosen particles within the desired range of 0.1 and 0.3 mm.

Other material properties determined were the specific surface area in ASAP 2020 from Micromeritics, applying the BET method to the isotherm obtained according to ISO 9277. The porosity of the particles was determined by Hg intrusion in a Quantachrome PoreMaster 33, and their real density was measured with a Micromeritics AccuPycc II 1340 helium pycnometer.

A scanning electron microscope SEM EDX Hitachi S-3400N equipped with an EDX analyzer Röntec XFlash of Si(Li) was used to investigate the microstructure of the oxygen carrier materials. Both the morphology of the particles and their cross section were observed after cutting once embedded in resin.

2.3. Thermogravimetric Analyzer

Five redox cycles were carried out in a thermogravimetric analyzer (TGA), CI Electronics type. 50 mg of each sample were loaded in a platinum-wired mesh basket and heated up in air atmosphere to the desired temperature (850 °C or 940 °C). The selection of this temperature range is related to the study of the phase diagram of the Mn-Ti-O system presented in the previous work from the authors [25]. In this work, the theoretical composition of the Mn-Ti samples prepared was analyzed according to this phase diagram, and it was concluded that these temperatures were adequate to carry out the oxidation of the samples in the presence of the Mn_2O_3 or $Mn_3O_4/MnTiO_3$ mixture. All the tests were done with a total flow of 25 L_N/h. More detailed descriptions of the TGA and the procedure can be found elsewhere [25].

First, the capacity to release gaseous oxygen and regenerate in air of the different Mn-Ti mixed oxides at a determined temperature was evaluated. In the reduction step, highly pure N_2 (<2 ppm O_2) was supplied. Then, the reactivity of the Mn-Ti samples with the main combustion gases (CO, H_2, and CH_4) was evaluated, considering that, in this case, reduction to MnO was reached. In the reduction step, the composition of the gaseous reactants was 15% H_2, 15% CO + 20% CO_2, and 15% CH_4 + 20% H_2O, respectively.

Some tests with 200 redox cycles were also performed to evaluate the mechanical and chemical stability of selected Mn-Ti oxides. In this case, the temperature was set to 950 °C. The first experiment was performed using N_2 in the reduction (2 min) and air in the oxidation (8 min), so that the CLOU capacity of the material could be evaluated, i.e., redox transformation Mn_2O_3/Mn_3O_4. In the second experiment, the redox transformation

Mn_2O_3/MnO was evaluated. Thus, reduction conditions were 2% H_2 during 55 s and oxidation with 10% O_2 during 1 min.

The oxygen transport capacity, R_{OC}, was defined as the mass fraction of the oxygen carrier that can be transferred as oxygen:

$$R_{OC} = \frac{m_o - m_r}{m_{oxid}} \quad (1)$$

where m_o is the initial mass of the sample in the oxidized state, m_r the mass of the sample in the reduced state, and m_{oxid} the mass of the fully oxidized sample.

Another parameter to be calculated from the TGA results is the reduction conversion (X_r) and the oxidation conversion (X_o):

$$X_r = \frac{m_0 - m}{R_{OC} m_{oxid}} \quad (2)$$

$$X_O = \frac{m - m_r}{R_{OC} m_{oxid}} \quad (3)$$

Finally, the rate index for the reaction considered can also be calculated as:

$$Rate\ Index\left(\frac{\%}{min}\right) = 60 \cdot 100 \cdot R_{OC} \cdot \left(\frac{dX_i}{dt}\right)\Big|_{norm} \quad (4)$$

where $\left(\frac{dX_i}{dt}\right)\Big|_{norm}$ is the normalized reaction rate expressed as s^{-1}. In the reaction with the main combustion gases (H_2, CO, CH_4), the normalized rate at a fuel gas concentration of 15 vol.% is considered, while in the reaction with O_2, gas concentration is normalized to 10 vol.%.

2.4. Batch Fluidized Bed Reactor

The oxygen release behavior of the oxygen carriers and the reactivity with the main combustion gases, (CO, H_2, and CH_4), were investigated in a batch fluidized bed reactor at different temperatures from 850 to 950 °C (see Figure 3). A detailed description of the batch fluidized bed reactor can be found elsewhere [27].

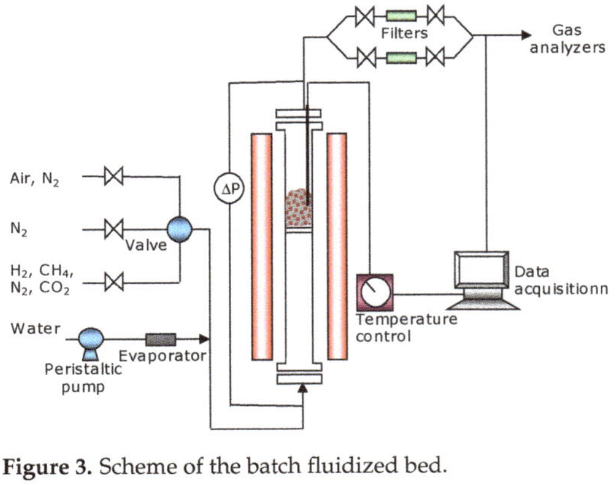

Figure 3. Scheme of the batch fluidized bed.

Initially, 300 g of the oxygen carrier were loaded in the batch fluidized reactor. The sample was heated in air to the desired temperature, and then successive redox cycles were performed. In the oxygen release experiments, pure nitrogen and air were used during reduction and oxidation periods at 850 °C and 950 °C, respectively. In the experiments with combustion gases, the reaction temperature was set in all cases to 950 °C. The reacting mixture used during the reducing periods was 50% H_2 + 20% H_2O, 50% CO + 20% CO_2, and 12.5% CH_4 + 20% H_2O, respectively, while 10 vol.% O_2 was used in all cases during oxidation. The gas velocity in the experiments was set to 12–15 cm/s. Reduction was limited to 2 or 5 min while complete oxidation was always reached. To avoid mixing atmospheres, pure nitrogen was used as a purge between reduction/oxidation periods during 2 min. A total of 10 reduction–oxidation cycles were carried out with each gas. Gas product distribution as well as solid conversion during reduction and oxidation steps were obtained for both types of experiments. The oxygen carrier conversion (X_s) was defined as:

$$X_s = \frac{|m - m_r|}{m_{oxid-m_r}} \tag{5}$$

Details of the calculation of oxygen carrier conversion in the batch fluidized bed can be found elsewhere [27]. From the results, it is also possible to calculate a normalized rate index for the experiments with different gases in the batch fluidized bed reactor: The normalized rate index for the batch experiments is defined as:

$$Rate\ Index\left(\frac{\%}{min}\right) = 60{\cdot}100{\cdot}R_{OC}{\cdot}\left(\frac{dX_i}{dt}\right)\bigg|_{norm} = 60{\cdot}100{\cdot}\frac{P_{ref}}{p_m} \tag{6}$$

where P_{ref} is the reference partial pressure (0.15 atm), which would approximately correspond to a full conversion of the gas, P_m refers to the average partial pressure of the gaseous fuel in the reactor being calculated by the following equation through an integration process, assuming a first-order reaction [27].

$$p_m = \frac{(p_{in} - p_{out}){\cdot}p_{in}}{(p_{out} - p_{in}){\cdot}\varepsilon_g + (p_{in} + \varepsilon_g{\cdot}p_{out}){\cdot}(1 + \varepsilon_g)\ln\left(\frac{p_{in} + \varepsilon_g{\cdot}p_{out}}{(1 + \varepsilon_g){\cdot}p_{out}}\right)} \tag{7}$$

where p_{in} and p_{out} represent the partial pressures of the gaseous fuel at the inlet and outlet of the batch fluidized bed reactor, respectively, and ε_g is the expansion factor, i.e., ε_g = 0 for CO and H_2 and ε_g = 0.25 for reduction with 12.5% CH_4.

The test in the batch fluidized bed also allowed for the analysis of the fluidization behavior with respect to attrition and the agglomeration phenomena of the oxygen carriers. The reactor was equipped with two pressure taps to continuously check the differential pressure between the inlet and outlet streams of the batch reactor and therefore detect agglomeration. In the downstream of the reactor, there were two parallel hot filters to recover elutriated solid during the consecutive redox cycles. The attrition rate was obtained by the following equation:

$$v_{attrition} = \frac{(m_{elut}/\ m_{OC}) \times 100}{t} \times 3600 \quad [\%/\ h] \tag{8}$$

where $v_{attrition}$ is the attrition rate, m_{elut} is the oxygen carrier elutriated from the batch fluidized bed reactor, which is recovered in hot parallel filters and pipes, and m_{oc} is the total mass of oxygen carrier in the experiments.

3. Results

3.1. Granulated Particles: Comparison with Results from Particles Produced by Pelletization

In the first study from the authors on the TiO_2 doping of manganese oxides [25], different Mn-Ti mixtures with Ti/(Mn+Ti) molar ratios between 0 and 0.49 were produced by pelletization. Mixtures of the corresponding powders of Mn_3O_4 and TiO_2 were mixed in a ball and then pelletized by pressure in a hydraulic press. Then they were crushed and sieved to 100–300 μm. Previous studies in literature demonstrated the influence of the particle size of original particles and the preparation method on the redox behavior of manganese oxides [28]. Thus, the objective of the present work was to produce Mn-Ti oxides by granulation following the same Ti/(Mn+Ti) molar ratios previously considered and compare their performance with that observed with particles prepared from pellets. The interest in producing particles by granulation lies in the fact that it is an easily scalable method for considering the commercial production of oxygen carriers. In this regard, a recent study also proposed the granulation method as a preparation procedure to up-scale a newly developed Si-doped manganese oxide [29].

In the present work, granulation parameters were optimized to obtain particles also within the 100–300 μm range. Once produced, particles were calcined in air at 1030 °C during 2 h. Crushing strength tests and XRD analyses were done on the calcined granulated particles and compared to the previous values obtained with pellets. Table 1 and Figure 4 show a comparison of the properties and XRD phases of selected calcined samples regarding pellet-produced and granulated particles.

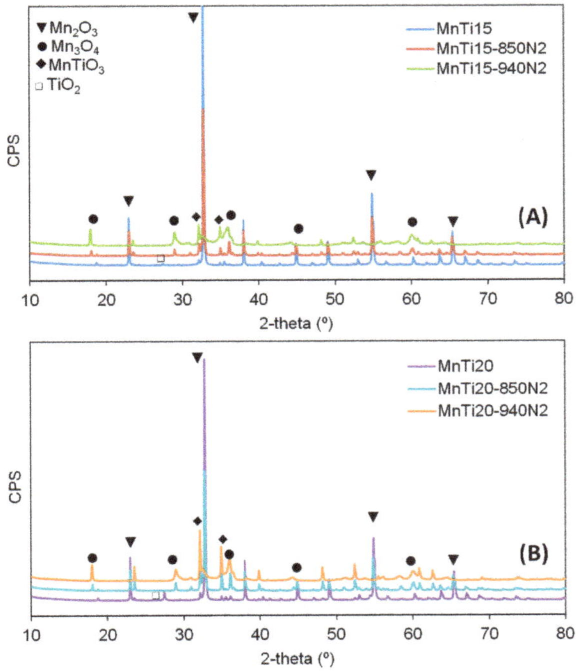

Figure 4. XRD solid phases of (**A**) MnTi_15 and (**B**) MnTi_20 for calcined granulated particles and particles reduced in nitrogen at 850 °C and 940 °C.

Table 1. Main properties of Mn-Ti samples after calcination in air at 1030 °C for 2 h.

Sample	Mn_3O_4 (wt.%)	TiO_2 (wt.%)	Ti/(Ti + Mn) [a]	Crushing Strength (N)	XRD Main Phases (wt.%)		
					$Mn_2O_3 + Mn_3O_4$	$MnTiO_3$	TiO_2
Pelletized samples							
MnTi_15	85	15	0.14	4.6	79.7	19.4	0.9
MnTi_20	80	20	0.19	3.7	75.7	23.6	0.7
Granulated samples							
MnTi_15	85	15	0.14	1.6	95.3	2.5	2.2
MnTi_20	80	20	0.19	2.6	87.9	3.8	8.3

[a] Molar ratio.

Crushing strength values between 1.5 and 2.6 N were obtained for the granulated particles, while higher values were reached when the particles were produced from pellets (3.3–5.5 N). XRD quantitative analyses of the granulated particles showed similar composition to that of pellet-produced particles.

In both cases, the samples were mostly crystalline. The major phase observed was Mn_2O_3, although some Mn_3O_4 was detected in some cases, probably produced during the cooling-down of the sample in the muffle after calcination. Free TiO_2 was detected. In the case of pellets, the amount of free TiO_2 was not significant. Most of it was combined with Mn_2O_3 to form pyrophanite ($MnTiO_3$). In the granulated particles, more free TiO_2 was observed, and the formation of pyrophanite was more limited than in the case of the particles from pellets. This could probably be due to the cooling conditions in the muffle for the granulated particles, which may have favored the decomposition of $MnTiO_3$ to Mn_2O_3 and TiO_2 compared to the case of the pellet-produced particles. Finally, it should be said that in both granulated particles and those obtained from pellets, in the sample with Ti/(Mn+Ti) = 0.49, pyrophanite was the only phase identified.

Manganese-titanium particles produced from pellets showed an improvement with respect to pure manganese oxide regarding oxygen release and regeneration in air at temperatures higher than 800 °C. This behavior was also investigated with the particles produced by granulation by performing successive N_2-air cycles in the TGA at 850 °C and 940 °C to determine the oxygen transport capacity (R_{OC}^{ou}) for oxygen release. Figure 5 shows an example of a typical TGA experiment for determination of R_{OC}^{ou}.

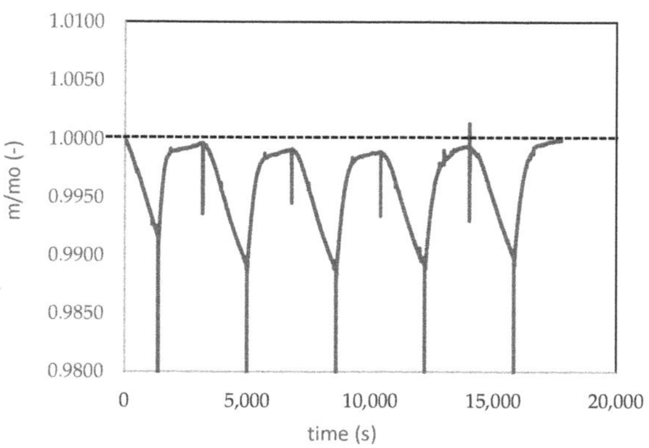

Figure 5. Reduction and oxidation cycles in TGA at 850 °C for the MnTi_20 sample. Reduction was performed in N_2 and oxidation in air during 30 min in both cases.

Figure 6 compares the evolution with the Ti/(Mn+Ti) molar ratio of the values of oxygen transport capacity (R_{OC}^{ou}) for oxygen release obtained for pellet-produced [25] and granulated particles at both 850 and 940 °C obtained from TGA curves similar to that in Figure 5 using Equation (1). In all the calculations, the fifth cycle was considered. The values of R_{OC}^{ou} in the case of particles produced from pellets were similar for the two temperatures tested, i.e., 850 °C and 940 °C. A maximum in the oxygen transport capacity for the oxygen release (R_{OC}^{ou}) was reached at both temperatures for Ti/(Mn+Ti) molar ratios between 0.15 and 0.2. The same behavior is observed for particles produced by granulation, and a maximum in (R_{OC}^{ou}) is detected for similar Mn-Ti compositions. However, there are some differences between particles obtained from pellets and granulated particles. At 850 °C, similar values of R_{OC}^{ou} were obtained for the two types of particles (maximum value close to 1.4%), but at 940 °C, the R_{OC}^{ou} values observed with the granulated particles were lower than those corresponding to particles prepared by pelletization. At 940 °C, the R_{OC}^{ou} values for the MnTi_15 granulated samples were 1.0%, and for MnTi_20, they were 0.8%, while for the particles obtained from pellets, the corresponding values were 1.4% and 1.2%, respectively. In the previous work from the authors [25], it was found that pyrophanite ($MnTiO_3$) was formed but it was not involved in the chemical reactions for oxygen release, which was exclusively attributed to Mn_2O_3 reduction to Mn_3O_4. It was speculated that the role of the $MnTiO_3$ formed was to facilitate the oxidation from Mn_3O_4 to Mn_2O_3 at higher temperatures compared to pure Mn_3O_4. Due to this fact, Mn-Ti pellets were capable of releasing oxygen at 940 °C and being oxidized back. The same behavior seems to be taking place in the particles produced by granulation. But maybe the lower amount of pyrophanite in the granulated particles compared to the pellets affected the Mn_3O_4 reoxidation capacity at 940 °C, making it lower. Thus, calcination conditions should be carefully controlled and optimized in order to favor that most of the TiO_2 added is included in the manganese oxide matrix in the form of pyrophanite.

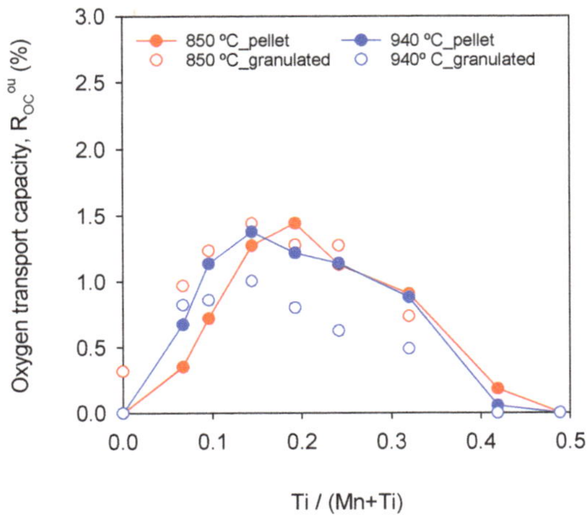

Figure 6. Oxygen uncoupling capability, R_{OC}^{ou}, versus Ti/(Mn+Ti) ratio for manganese-titanium particles produced from pellets and granulated particles (results corresponding to the fifth cycle in TGA at 850 and 940 °C).

The experiments with the samples with 15% and 20% TiO_2, both at 850 and 940 °C, were repeated for further analysis, and after the first reduction step, the samples were not regenerated but cooled down in nitrogen. The reduced samples were also taken to XRD analysis in order to determine the evolution of the solid phases during the reduction step. The corresponding XRD patterns are also shown in Figure 4. As it was already found in similar previous experiments with the pellet-produced particles, the pyrophanite phase remained stable, and only Mn_3O_4 was identified as a reduced species. This corroborates the hypothesis that only the redox pair Mn_2O_3/Mn_3O_4 plays a role in the release of gaseous oxygen. At the lowest temperature (850 °C), both Mn_2O_3 and Mn_3O_4 were found together with $MnTiO_3$. However, at 940 °C the reduction degree was higher, and only Mn_3O_4 and $MnTiO_3$ were identified in the reduced samples.

Following these results, it was decided to select one Mn-Ti mixed oxide for further testing. The sample with Ti/(Mn+Ti) equal to 0.19 (20 wt.% TiO_2) was selected, and from now on it will be denoted as MnTi_20. Table 2 summarizes the main properties of the MnTi_20 oxygen carrier produced by granulation that will be used in the present work.

Table 2. Main properties of the MnTi_20 granulated particles.

	Calcined	Used (Batch FB)
Crushing strength (N)	2.6	1
Porosity (%)	31.2	34.4
Skeletal density (kg/m^3)	4818	4739
BET surface area (m^2/g)	0.6	0.7

Together with the oxygen transport capacity for the gaseous oxygen release (R_{OC}^{ou}) of the oxygen carriers, another parameter to be considered regarding the use of these materials in CLOU/CLAS applications is the rate for this oxygen release. In order to compare the reactivity of different samples, the rate index parameter defined in Equation (4) was used. Table 3 compiles results of the normalized rate index at 950 °C obtained for the granulated Mn-Ti samples both in oxygen release cycles and in the reaction with the main combustion gases (H_2, CO, CH_4) and with O_2. In the table, the corresponding values for the MnTi_20 particles obtained from pellets and granulated particles were included, as well as values for manganese-iron mixed oxides (Mn77Fe and Mn55FeTi7) and Mn-based oxygen carriers (MnSA, MnGBHNE, Mn-ZrSG) previously prepared and evaluated by the authors. Mn77Fe and Mn55FeTi7 particles are Fe-doped manganese oxides [30,31]. MnSA and MnGBHNE are two manganese-based minerals with a high Mn_3O_4 content [27]. Mn-ZrSG is a synthetic oxygen carrier prepared by incipient wetness impregnation of the manganese oxide on ZrO_2 [32].

Table 3. Normalized rate index (%/min) obtained for pelletized and granulated MnTi_20 and other Mn-based oxygen carriers (obtained at TGA, T = 950 °C).

Sample	CLOU		CLC			
	N_2	O_2	H_2	CO	CH_4	O_2
MnTi_20_pelletization	0.56	0.02	15.4	1.9	2.3	9.9
MnTi_20_granulation	0.38	0.16	18.6	2.5	2.8	10.8
MnTi_20_used_TGA	-	-	19.7	-	8.2	10.1
MnTi_20_used_batch	0.26	0.22	18.9	-	2.7	9.4
Mn77Fe[L4] [30]	0.30 *	1.10 *	-	-	-	-
	-	-	10.4	3.3	2.5	8.4
Mn55FeTi7 [31]	0.10	0.45	10.3	0.3	1.0	7.4
MnSA [27]	-	-	14.2	5.1	5.0	11.3
MnGBHNE [27]	-	-	19.2	6.4	9.2	8.6
Mn-ZrSG [32]	-	-	8.8	5.5	13.1	-

* obtained at 870 °C.

As it can be observed in Table 3, only the Ti- or Fe-doped manganese oxides were capable of releasing oxygen and regenerating in air at 950 °C. Granulated MnTi_20 showed a similar rate index for oxygen release to that of the material obtained by pelletization and a better index for the subsequent oxidation. Moreover, the values of the rate index obtained for the main combustion gases are even slightly higher than those found for the material prepared from pellets.

On the other hand, granulated MnTi_20 showed better values for the rate index in the oxygen release than some Mn-Fe mixed oxides [30,31]. The values of the rate index with the main combustion gases were also lower for the Mn-Fe mixed oxides at similar temperatures (950 °C). Thus, MnTi_20 could be considered as an alternative to manganese-iron oxygen carriers. The rate indexes in Table 3 were also compared to those obtained with manganese oxide-based oxygen carriers but with mineral origin or prepared by the impregnation method. In the case of manganese-based minerals, it has been observed that in any case the minerals showed no sufficient CLOU properties. Depending on their initial oxidation state, they could release some molecular oxygen, but once Mn_3O_4 was formed, it could not be regenerated. Then, they can only be used in CLC applications involving the redox pairs Mn_3O_4/MnO [27], even more so considering the high rate indexes for CO and CH_4 found compared to synthetic materials. In the case of the Mn-ZrSG material obtained by impregnation, no oxygen release was observed since the original solid phase was Mn_3O_4 and it was not capable of being oxidized to Mn_2O_3. However, the rate indexes obtained in the reaction with CO and CH_4 with Mn-ZrSG were higher than those found for MnTi_20.

3.2. Stability with TGA Cycles

The behavior of the Mn-Ti mixed oxides under several redox cycles was not studied in the previous work with particles obtained from pellets. Thus, in this work, the chemical and mechanical stability of granulated MnTi_20 was evaluated in the TGA in 200-redox-cycle tests at 950 °C. Figure 7 shows the solid conversion during reduction (Xr) versus time for the successive redox cycles performed. Figure 7A shows the corresponding results for N_2-air cycles, and Figure 7B for hydrogen-diluted air cycles.

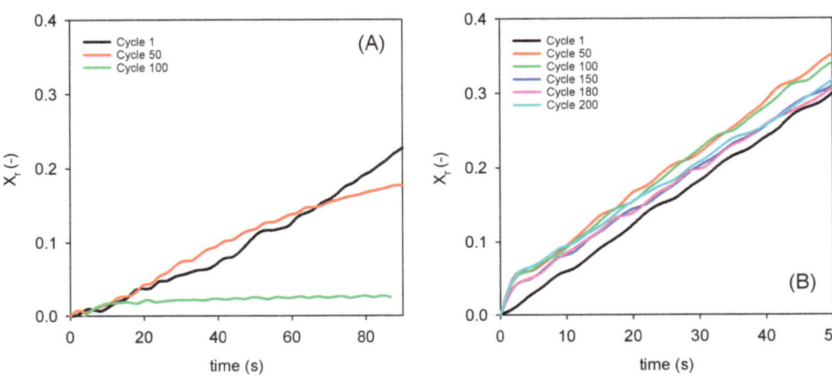

Figure 7. Solid conversion during reduction (*Xr*) versus time through 200 redox cycles in TGA at T = 950 °C (**A**) Reduction = 100% N_2; Oxidation = air (**B**) Reduction = 2% H_2; Purge = 100 vol.% N_2; Oxidation = 10 vol.% O_2.

During the first 50 cycles in Figure 7A, the solid conversion does not experience significant changes; however, from that cycle on, it seems that the complete regeneration of the oxygen carrier during oxidation is not reached, and therefore the reduction conversion in subsequent cycles progressively decreases. Figure 7B shows a slight activation of the sample in the reaction with hydrogen during the first 50 cycles. After this point, there are no significant changes in the reactivity of the material. After the 200-redox-cycle test shown in Figure 7B, an additional reactivity test was done with 15% H_2 and another with 15% CH_4 + 20% H_2O at 950 °C in order to calculate the rate index of the used sample. Results are shown in Table 3. An increase in the rate index for both gases is observed, especially significant in the case of methane, which changes from 2.8 to 8.2%/min.

The solid phases in the reacted TGA samples were also investigated after the 200-redox-cycle tests by XRD, showing again the unchanged permanence of the pyrophanite phase, both in reduced and oxidized samples. Nevertheless, changes in the crushing strength for the calcined and used material were detected. The crushing strength value decreased from 2.6 N in the calcined MnTi_20 to 0.4 N after 200 redox hydrogen-diluted air cycles. This is a low crushing strength value to operate in a fluidized bed system, since the crushing strength limit for a correct fluidization was set at 1 N. Figure 8 shows images obtained in a Scanning Electron Microscope (SEM) of both calcined particles (A and D) and used in the 200-redox-cycle tests in TGA (B–F).

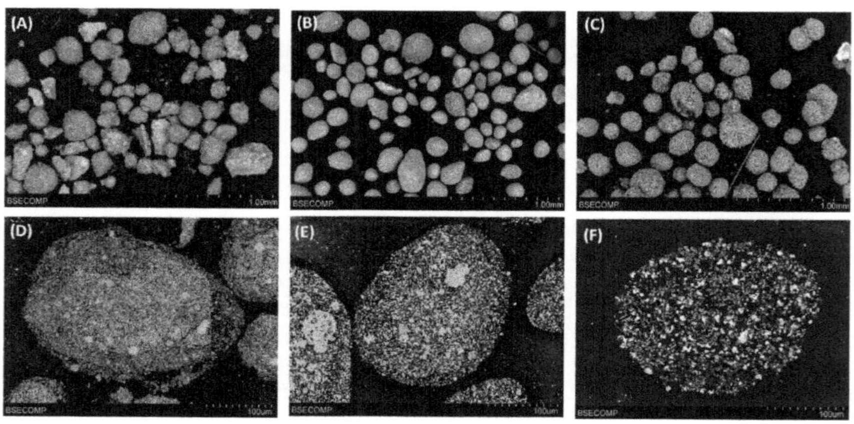

Figure 8. SEM images of MnTi_20 (**A,D**) calcined and after 200 redox TGA cycles at 950 °C (**B,E**) with N_2-air and (**C,F**) with 2% H_2—10% O_2.

Images A to C show a general overview of the external morphology of the (A) calcined and (B-C) used particles. No significant differences are observed. However, differences can be appreciated in images (D to F), which correspond to particles embedded in a resin and then cut so that the internal morphology of the particles can be analyzed. Compared to the calcined particles (D), the images for the used particles (E and F) reveal an increase in the porosity, especially in the case of the samples reacted under hydrogen-diluted air cycles. This increase in porosity may be linked to the decrease in the crushing strength previously mentioned. It seems that the larger the reduction degree of the sample, the higher the increase in porosity. Thus, the degree of reduction in the particle should be controlled to avoid deterioration of the mechanical properties with the cycles.

3.3. Batch Fluidized Bed Experiments

Many of the CL technology-based applications in which the MnTi_20 oxygen carrier may be promising use interconnected fluidized beds as the preferred configuration. For this reason, it is especially important to understand the behavior of the material under fluidization conditions as well as to determine its reactivity in fluidized beds. Thus, oxygen-released experiments and redox cycles with the main combustion gases were carried out in the batch fluidized bed previously shown in Figure 3.

3.3.1. Experiments with N_2-Air Cycles

Three oxygen release-oxidation tests were performed at both 850 °C and 950 °C with the MnTi_20 oxygen carrier. The most significant fact is that the MnTi_20 was capable of reoxidizing in air even at 950 °C, although the oxidation rate was low. Figure 9 shows the oxygen concentration profiles obtained in the third cycle at the two temperatures mentioned. During reduction, the oxygen concentration decreased as the oxygen carrier was more reduced, to values close to the oxygen concentration expected at equilibrium at the reactor temperature. This type of behavior in the oxygen evolution was previously observed in oxygen carriers based on mixed oxides [33].

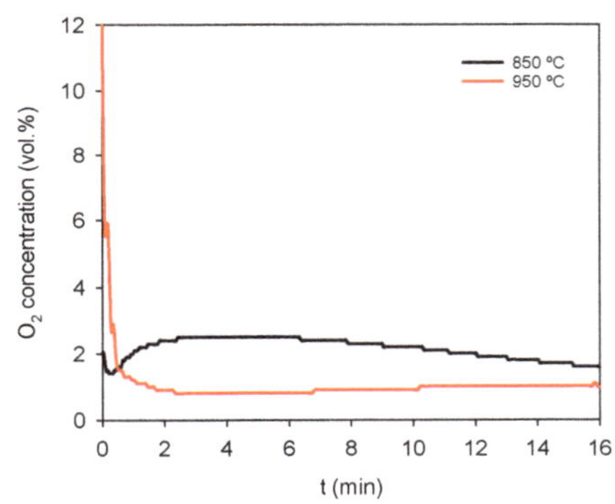

Figure 9. Oxygen concentration versus time during the third cycle at 850 and 950 °C with MnTi_20.

It is interesting to note the behavior of the oxygen carrier at 850 °C. The oxygen concentration initially experiences a minimum, then goes through a plateau, and then slowly decreases. This initial behavior could indicate the existence of what can be called an "induction period" in the oxygen release at this temperature. This phenomenon will be more deeply considered in future research works. At 950 °C, the oxygen release rate was faster, and the maximum in the oxygen release rate was not observed. After 15 min of reduction experiment, a final value of 1.1% O_2 was reached at 950 °C.

3.3.2. Experiments with Combustion Gases

During these tests, ten reduction–oxidation cycles were performed with each reducing gas at 950 °C. Figure 10 shows the gas product profile and solid conversion in the fifth cycle in the batch fluidized bed experiments with H_2, CO, and CH_4.

Reactions with H_2 and CO in Figure 10A,B showed complete combustion of the gas in both cases, since no H_2 or negligible CO was detected at the batch reactor outlet. The solid conversion decreased during reduction. After that, it remained constant during the purge period and increased through MnTi_20 reoxidation using diluted air (10 vol.% O_2). It can be observed that all of the oxygen introduced was consumed during the first stage of the oxidation reaction. During this step, the oxidation was limited by the flow of O_2 fed to the reactor. At the end of the oxidation period, the oxygen concentration rapidly increased to the inlet value. The highest value of solid conversion (*Xs*) was obtained in the case of the experiments with H_2 (~50%). Slightly lower conversion was reached in the case of CO

(33.7%). The lowest solid conversion was reached in experiments with methane (22.6%), in Figure 10C. In this case, not all the methane was converted (35–40% methane conversion), but all the methane reacted was converted to CO_2 since no H_2 or CO were observed as combustion products during reduction. This fact suggests that neither methane reforming nor partial oxidation reactions with H_2O or CO_2 were of significance. However, some thermal decomposition of methane could take place due to the presence of a slight amount of CO_2 at the beginning of the oxidation period. Carbon formation on oxygen carriers depends on the fuel used, the amount of oxygen added with the oxygen carrier, and the operating conditions of pressure and temperature.

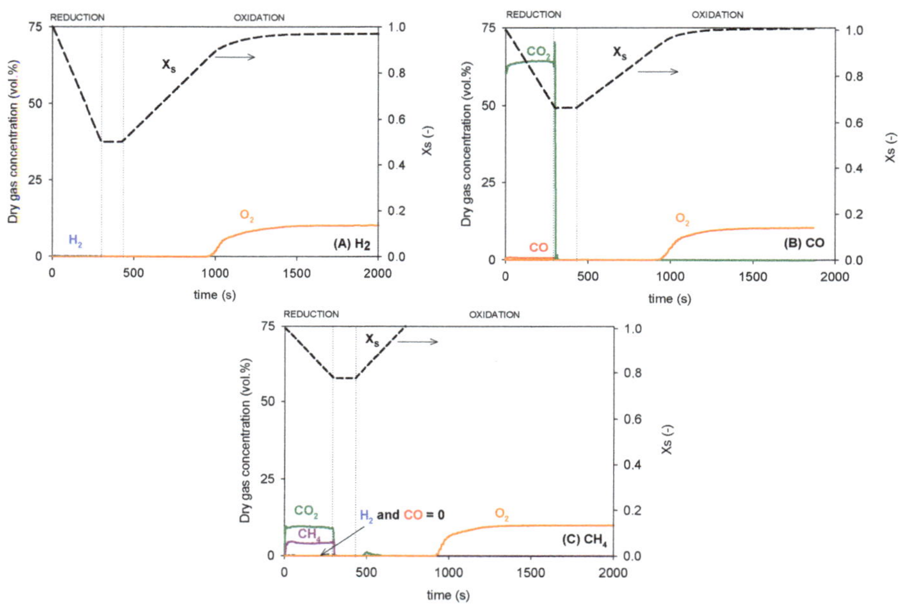

Figure 10. Gas product profile and solid conversion (*Xs*) in the fifth cycle in the batch fluidized bed experiments with (**A**) 50% H_2 + 20% H_2O, (**B**) 50% CO + 20% CO_2, and (**C**) 12.5% CH_4 + 20% H_2O. Oxidation in all cases with 10 vol.% O_2. T = 950 °C; Reduction time = 5 min.

It is also interesting to evaluate the reaction rate of the MnTi_20 in the batch fluidized bed experiments through the calculation of the normalized rate index in the batch fluidized bed following Equations (6) and (7). Table 4 shows the values obtained for the different combustion gases, namely H_2, CO, and CH_4, together with values previously reported for manganese minerals (MnSA and MnHNE) and a manganese oxide oxygen carrier impregnated on ZrO_2 (Mn-ZrSG) [27,32].

Table 4. Normalized rate index (%/min) obtained in the batch fluidized bed reactor for granulated MnTi_20 and other Mn-based oxygen carriers (T = 950 °C).

Sample	H_2	CO	CH_4
MnTi_20_granulation	0.9	0.4	0.5
MnSA [27]	1.6	0.7	0.5
MnGBHNE [27]	1.5	0.4	0.5
Mn-ZrSG [32]	-	-	5.2

Values in Table 4 are calculated for the same materials and reacting gases than those shown in Table 3 but they are lower than in the TGA experiments. This fact can be attributed to mass transfer issues in the bubbles formed in the fluidized bed of the batch reactor. Reacting gases inside the bubbles have to diffuse to the emulsion, thus reducing the reaction rate with the oxygen carrier particles. The MnTi_20 oxygen carrier presents rate indexes in the same order of magnitude as the manganese minerals MnSA and MnGBHNE. Values are similar for CO and CH_4, although a bit lower in the case of H_2. The impregnated material Mn-ZrSG presented a very high rate index for CH_4, more than 10 times that found for MnTi_20.

A total of almost 60 h of continuous fluidization were performed with the MnTi_20 sample in the batch fluidized bed reactor. No fluidization or agglomeration problems were observed during this time. After the experimental campaign, the used sample was characterized. Tables 2 and 3 present the main properties of the used sample and the corresponding rate indexes calculated in TGA experiments at 950 °C in experiments with N_2-air cycles and H_2 or CH_4. In Table 2, it can be observed that the crushing strength value of used particles decreased to values close to 1 N, the limit considered to operate in a fluidized bed. This corresponds with a slight increase in porosity. The H_2 and CH_4 rate indexes calculated in TGA for the used MnTi_20 oxygen carrier in Table 3 remain similar to those found for the calcined granulated particles, while the rate index for oxygen release seems to have slightly decreased. Figure 11 shows the SEM images of the MnTi_20 particles after their use in the batch reactor together with the fines collected in the hot filters after it. Similarly to Figure 8, Figure 11A–C show a general overview of the external morphology, while Figure 11D–F correspond to particles embedded in a resin and then cut.

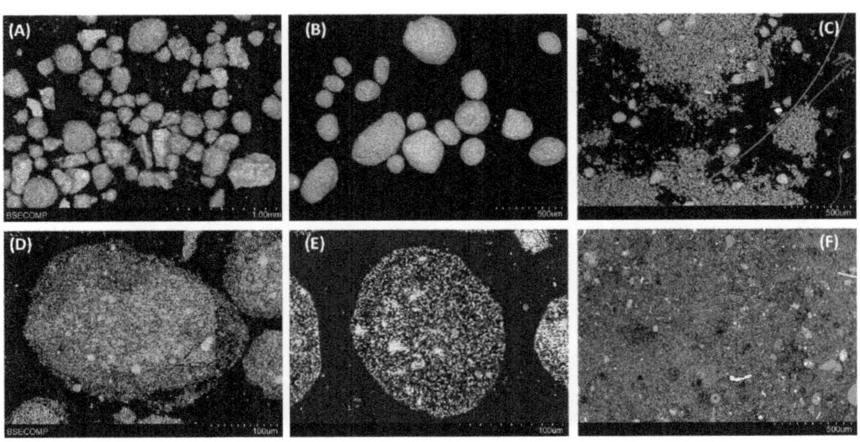

Figure 11. SEM images of MnTi_20 (**A**,**D**) calcined and (**B**,**E**) after experiments in the batch fluidized bed reactor at 950 °C (**C**,**F**) fines collected in the hot filters after the batch fluidized reactor.

As it was the case after the TGA experiments, an increase in the porosity of the samples from the bed can be observed in Figure 11E compared to the calcined particles Figure 11D. Fines collected in the filters after the batch fluidized bed reactor are also shown in Figure 11C,F. Different types of fines are observed. On one hand, those with the smallest size are powder that could be attached to the granulated particles and be lost during the first hours of fluidization of the particles in the batch reactor. The fines with the largest size may correspond to the breakage of the particles during the reaction with the different gases.

Based on the amount of fines collected from the bed, Figure 12 shows the time evolution of the attrition calculated for these experiments following Equation (8).

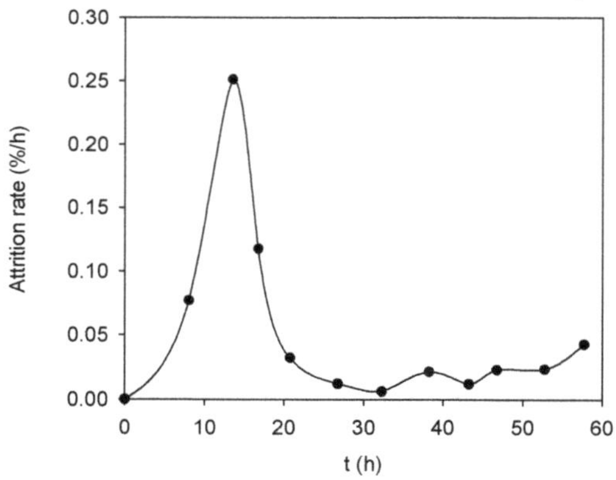

Figure 12. Time evolution of the attrition rate in the batch fluidized bed experiments with granulated MnTi_20 (T = 950 °C).

In Figure 12, the attrition rate increases at the beginning, probably due to the previously mentioned loss of fines attached to the fresh particles during the preparation of the granulated material. After then, the attrition rate decreases to low values, although an upward trend seems to initiate after 50 h of continuous fluidization. At this point, the estimated lifetime from the final value of the attrition rate measured would be around 2500 h. This value is better than those previously reported for manganese-based perovskites and minerals, which were frequently discarded for operation due to the low lifetime they presented [27,34]. Nevertheless, to be confident in the durability of the particles, tens or hundreds of redox cycles would be required under a relevant environment and constant operating conditions, preferably in a CLC continuous unit.

4. Discussion

In this work, a manganese oxide-based material with TiO_2 addition has been produced by granulation and tested under conditions relevant for different Chemical Looping processes. Ti-doped manganese oxides with Ti/(Mn+Ti) molar ratios between 0 and 0.49 were previously identified as promising due to their capacity to release gaseous oxygen and regenerate in air under conditions where pure manganese oxide does not. In a previous work by the authors, these materials were prepared in the form of mixtures of the corresponding oxides followed by pelletization. The present work aimed to reproduce one of the identified materials (MnTi_20) through the granulation method, which allows for easier scale-up in a hypothetical commercial production of the oxygen carrier. Moreover, the preparation method can influence the reactivity of the particles; thus, this aspect was also evaluated.

It has been possible to granulate MnTi_20 particles with adequate mechanical properties to be used in processes based on fluidized beds, such as the Chemical Looping processes. However, control of the calcination conditions of the material was revealed as necessary, especially the cooling conditions after calcination to ensure that most of the introduced TiO_2 is stabilized as $MnTiO_3$ and no free TiO_2 is found. Experimental evidence, both with pellet-produced and granulated particles, seems to indicate that the presence of this phase allows the Mn_3O_4 phase to be reoxidized to Mn_2O_3 under conditions in which pure manganese oxide cannot do. XRD analysis of the granulated particles showed a higher fraction of free TiO_2 and lower fraction of $MnTiO_3$ compared to those found in the corresponding particles prepared from pellets, which would also explain the differences in the oxygen transport capacity for the oxygen release observed between the two types of particles at the highest temperature tested, i.e., 940 °C.

In order to be considered for further scale-up, the evolution of the mechanical properties of MnTi_20 along several redox cycles was studied for the first time. Long TGA tests (200 redox cycles) were performed, and in the experiments in the batch fluidized bed, the attrition rate of the granulated particles was estimated. In both types of experiments, an increase in the porosity of the samples was observed, leading to a decrease in the mechanical strength of the particles that may make them unsuitable for use in a fluidized bed. Despite the deterioration of their mechanical properties with the number of cycles, the granulated MnTi_20 particles maintained their reactivity, which was similar to or even better in some cases than Fe-doped manganese oxides previously reported in the literature.

Another aspect to consider is the rate at which manganese-titanium mixed materials can release oxygen or react with typical combustion gases, i.e., H_2, CO, and CH_4. In this study, we have worked mainly in the 850–950 °C temperature window, which is a temperature range adequate for CLOU, TCES, and CLC processes based on oxygen carriers. It has been observed that although Mn_3O_4 is capable of regenerating to Mn_2O_3 under these conditions, the oxidation rate is slow compared to other non-manganese-based materials, such as copper oxygen carriers. Depending on the Chemical Looping application to which this material is intended, this oxygen release rate should be improved, such as in CLOU or TCES applications, in order to guarantee high combustion efficiencies and the thermochemical storage capability, respectively, of the Mn-Ti-based materials. As it was mentioned in the introduction section, recent works propose the addition of MgO [19], SiO_2 [35], or even CeO_2 [36]. In future research with the MnTi_20 material, the effect of the addition of these oxides can be considered. Furthermore, the addition may not only lead to an improvement in reactivity but also in mechanical properties, which should also be considered for future applications in packed or fluidized beds [37]. In this sense, other preparation methods, also scalable, such as granulation, should also be investigated.

5. Conclusions

Manganese-titanium mixed oxides with different Ti/(Mn+Ti) molar ratios in the 0–0.49 range (corresponding to 0–50 wt.% TiO_2) were prepared by granulation in a high shear mechanical mixing granulator and calcined in air under the same conditions as those used with particles of identical composition but prepared by pelletization. The granulation method can improve the scalability of the production of these oxygen carrier particles as well as influence their reactivity. Once calcined, granulated particles presented similar composition obtained by XRD: mostly Mn_2O_3, with low amounts of Mn_3O_4 and an important part of TiO_2 in the form of pyrophanite ($MnTiO_3$). However, the crushing strength of granulated particles was lower, although higher than 1 N. Both particles prepared by granulation and pelletization showed a maximum in the oxygen transport capacity for oxygen release in the Ti/(Mn+Ti) range of 0.15–0.2 at the two temperatures considered, i.e., 850 and 940 °C. Thus, the mixed oxide with 20 wt.% TiO_2 (MnTi_20) was tested further.

MnTi_20 showed high reactivity to H_2 and maintained it for almost 200 redox cycles in TGA. However, after this long testing, mechanical properties were worsened compared to the calcined material: mechanical strength decreased to values lower than 1 N, and porosity significantly increased, making particles more fragile for breakage.

Regarding the performance of MnTi_20 under fluidized conditions, the material also showed the ability to release oxygen and regenerate at temperatures higher than 850 °C, although the oxidation rate was low at 950 °C. Reactivity with H_2 and CO at 950 °C was also high, comparable to that previously found with manganese-based minerals, although not as high as that reported for an impregnated Mn-based oxygen carrier. Nevertheless, neither manganese minerals nor the impregnated oxygen carrier showed CLOU properties. No agglomeration or fluidization problems were encountered, and the lifetime of the material after 60 h of continuous fluidization was estimated as roughly 2500 h. This value is one of the highest found in literature for manganese-based oxygen carriers.

This behavior in the batch fluidized bed makes this material a promising oxygen carrier to be considered in different CL processes. Its improved CLOU properties enable it to compete against manganese-based minerals, and its reactivity is of the same order and similar to that of other synthetic materials, such as manganese-iron mixed oxides.

Author Contributions: Conceptualization, T.M. and A.A.; methodology, T.M., A.F. and A.A.; validation, T.M. and A.F.; formal analysis, T.M., A.F. and M.T.I.; investigation, T.M., A.F. and M.T.I.; resources, T.M. and A.A.; data curation, T.M. and M.T.I.; writing—original draft preparation, T.M.; writing—review and editing, T.M. and I.A.-R.; visualization, T.M. and I.A.-R.; supervision, T.M. and I.A.-R.; project administration, T.M. and A.A.; funding acquisition, T.M. and A.A. All authors have read and agreed to the published version of the manuscript.

Funding: This research was funded by the PDC2021-121190-I00/AEI/10.13039/501100011033 financed by MCIN/AEI/10.13039/501100011033 and European Union NextGenerationEU/PRTR. The authors also thank the funding received from Gobierno de Aragón—Dpto. de Ciencia, Universidad y Sociedad del Conocimiento—Project LMP166_21.

Institutional Review Board Statement: Not applicable.

Informed Consent Statement: Not applicable.

Data Availability Statement: Data will be available on request.

Acknowledgments: I. Adánez-Rubio acknowledges for "Ramón y Cajal" Program (Grant RYC2022-035841-I funded by MCIU/AEI/10.13039/501100011033 and FSE+).

Conflicts of Interest: The authors declare no conflicts of interest. The funders had no role in the design of the study; in the collection, analyses, or interpretation of data; in the writing of the manuscript; or in the decision to publish the results.

References

1. Joshi, A.; Shah, V.; Mohapatra, P.; Kumar, S.; Joshi, R.K.; Kathe, M.; Qin, L.; Tong, A.; Fan, L.-S. Chemical looping—A perspective on the next-gen technology for efficient fossil fuel utilization. *Adv. Appl. Energy* **2021**, *3*, 100044. [CrossRef]
2. Palone, O.; Cedola, L.; Borello, D.; Markides, C.N. Decarbonizing power and fuels production by chemical looping processes: Systematic review and future perspective. *Appl. Therm. Eng.* **2024**, *254*, 123844. [CrossRef]
3. Zhu, X.; Imtiaz, Q.; Donat, F.; Müller, C.R.; Li, F. Chemical looping beyond combustion—A perspective. *Energy Environ. Sci.* **2020**, *13*, 772–804. [CrossRef]
4. Abanades, S. A Review of Oxygen Carrier Materials and Related Thermochemical Redox Processes for Concentrating Solar Thermal Applications. *Materials* **2023**, *16*, 3582. [CrossRef]
5. Liu, G.; Lisak, G. Cu-based oxygen carriers for chemical looping processes: Opportunities and challenges. *Fuel* **2023**, *342*, 127828. [CrossRef]
6. Bui, M.; Adjiman, C.S.; Bardow, A.; Anthony, E.J.; Boston, A.; Brown, S.; Fennell, P.S.; Fuss, S.; Galindo, A.; Hackett, L.A.; et al. Carbon capture and storage (CCS): The way forward. *Energy Environ. Sci.* **2018**, *11*, 1062–1176. [CrossRef]
7. Lyngfelt, A.; Leckner, B.; Mattisson, T. A fluidized-bed combustion process with inherent CO_2 separation: Application of chemical-looping combustion. *Chem. Eng. Sci.* **2001**, *56*, 3101–3113. [CrossRef]

8. Barrasso, M.; Langella, G.; Amoresano, A.; Iodice, P. Latest Advances in Thermal Energy Storage for Solar Plants. *Processes* **2023**, *11*, 1832. [CrossRef]
9. Klara, J.M.; Plunkett, J.E. The potential of advanced technologies to reduce carbon capture costs in future IGCC power plants. *Int. J. Greenh. Gas Control* **2010**, *4*, 112–118. [CrossRef]
10. Haseli, P.; Jafarian, M.; Nathan, G.J. High temperature solar thermochemical process for production of stored energy and oxygen based on CuO/Cu_2O redox reactions. *Sol. Energy* **2017**, *153*, 1–10. [CrossRef]
11. Dawa, T.; Sajjadi, B. Exploring the potential of perovskite structures for chemical looping technology: A state-of-the-art review. *Fuel Process. Technol.* **2024**, *253*, 108022. [CrossRef]
12. Han, X.; Wang, L.; Ling, H.; Ge, Z.; Lin, X.; Dai, X.; Chen, H. Critical review of thermochemical energy storage systems based on cobalt, manganese, and copper oxides. *Renew. Sustain. Energy Rev.* **2022**, *158*, 112076. [CrossRef]
13. Carrillo, A.J.; Serrano, D.P.; Pizarro, P.G.; Coronado, J.M. Understanding Redox Kinetics of Iron-Doped Manganese Oxides for High Temperature Thermochemical Energy Storage. *J. Phys. Chem. C* **2016**, *120*, 27800–27812. [CrossRef]
14. Carrillo, A.J.; Serrano, D.P.; Pizarro, P.; Coronado, J.M. Improving the thermochemical energy storage performance of the Mn_2O_3/Mn_3O_4 redox couple by the incorporation of iron. *ChemSusChem* **2015**, *8*, 1947–1954. [CrossRef]
15. Carrillo, A.J.; Serrano, D.P.; Pizarro, P.; Coronado, J.M. Manganese oxide-based thermochemical energy storage: Modulating temperatures of redox cycles by Fe–Cu co-doping. *J. Energy Storage* **2016**, *5*, 169–176. [CrossRef]
16. Azimi, G.; Rydén, M.; Leion, H.; Mattisson, T.; Lyngfelt, A. $(Mn_zFe_{1-z})_yO_x$ Combined Oxides as Oxygen Carrier for Chemical-Looping with Oxygen Uncoupling. *AIChE J.* **2013**, *59*, 582–588. [CrossRef]
17. Pérez-Vega, R.; Abad, A.; García-Labiano, F.; Gayán, P.; de Diego, L.F.; Izquierdo, M.T.; Adánez, J. Chemical Looping Combustion of gaseous and solid fuels with manganese-iron mixed oxide as oxygen carrier. *Energy Convers. Manag.* **2018**, *159*, 221–231. [CrossRef]
18. Bo, A.; Randhir, K.; Rahmatian, N.; Klausner, J.; Petrasch, J. Chemical equilibrium of the magnesium manganese oxide redox system for thermochemical energy storage. *Chem. Eng. Sci.* **2022**, *259*, 117750. [CrossRef]
19. Randhir, K.; King, K.; Rhodes, N.; Li, L.; Hahn, D.; Mei, R.; AuYeung, N.; Klausner, J. Magnesium-manganese oxides for high temperature thermochemical energy storage. *J. Energy Storage* **2019**, *21*, 599–610. [CrossRef]
20. Yilmaz, D.; Darwish, E.; Leion, H. Investigation of the combined Mn-Si oxide system for thermochemical energy storage applications. *J. Energy Storage* **2020**, *28*, 101180. [CrossRef]
21. Bielsa, D.; Zaki, A.; Arias, P.L.; Faik, A. Improving the redox performance of Mn_2O_3/Mn_3O_4 pair by Si doping to be used as thermochemical energy storage for concentrated solar power plants. *Sol. Energy* **2020**, *204*, 144–154. [CrossRef]
22. Adánez-Rubio, I.; Mattisson, T.; Jacobs, M.; Adánez, J. Development of new Mn-based oxygen carriers using MgO and SiO_2 as supports for Chemical Looping with Oxygen Uncoupling (CLOU). *Fuel* **2023**, *337*, 127177. [CrossRef]
23. Block, T.; Schmücker, M. Metal oxides for thermochemical energy storage: A comparison of several metal oxide systems. *Sol. Energy* **2016**, *126*, 195–207. [CrossRef]
24. Al-Shankiti, I.A.; Ehrhart, B.D.; Ward, B.J.; Bayon, A.; Wallace, M.A.; Bader, R.; Kreider, P.; Weimer, A. Particle design and oxidation kinetics of iron-manganese oxide redox materials for thermochemical energy storage. *Sol. Energy* **2019**, *183*, 17–29. [CrossRef]
25. Abad, A.; Mendiara, T.; Izquierdo, M.T.; de Diego, L.F.; García-Labiano, F.; Gayán, P.; Adánez, J. Evaluation of the redox capability of manganese-titanium mixed oxides for thermochemical energy storage and chemical looping processes. *Fuel Process. Technol.* **2021**, *211*, 106579. [CrossRef]
26. Wang, B.; Wang, Z.; Dou, B.; Ma, Y.; Liang, Y. Effects of TiO_2 doping on the performance of thermochemical energy storage based on Mn_2O_3/Mn_3O_4 redox materials. *RSC Adv.* **2021**, *11*, 33744–33758. [CrossRef]
27. Mei, D.; Mendiara, T.; Abad, A.; de Diego, L.F.; García-Labiano, F.; Gayán, P.; Adánez, J.; Zhao, H. Evaluation of Manganese Minerals for Chemical Looping Combustion. *Energy Fuels* **2015**, *29*, 6605–6615. [CrossRef]
28. Carrillo, A.J.; Serrano, D.P.; Pizarro, P.; Coronado, J.M. Thermochemical heat storage based on the Mn_2O_3/Mn_3O_4 redox couple: Influence of the initial particle size on the morphological evolution and cyclability. *J. Mater. Chem. A* **2014**, *2*, 19435–19443. [CrossRef]
29. Bielsa, D.; Oregui, M.; Arias, P.L. New insights into Mn_2O_3 based metal oxide granulation technique with enhanced chemical and mechanical stability for thermochemical energy storage in packed bed reactors. *Sol. Energy* **2022**, *241*, 248–261. [CrossRef]
30. Pérez-Vega, R.; Abad, A.; Gayán, P.; de Diego, L.F.; García-Labiano, F.; Adánez, J. Development of $(Mn_{0.77}Fe_{0.23})_2O_3$ particles as an oxygen carrier for coal combustion with CO_2 capture via in-situ gasification chemical looping combustion (*i*G-CLC) aided by oxygen uncoupling (CLOU). *Fuel Process. Technol.* **2017**, *164*, 69–79. [CrossRef]

31. Abián, M.; Abad, A.; Izquierdo, M.T.; Gayán, P.; de Diego, L.F.; García-Labiano, F.; Adánez, J. Titanium substituted manganese-ferrite as an oxygen carrier with permanent magnetic properties for Chemical Looping Combustion of solid fuels. *Fuel* **2017**, *195*, 38–48. [CrossRef]
32. Costa, T.R.; Gayán, P.; Abad, A.; García-Labiano, F.; de Diego, L.F.; Melo, D.M.A.; Adánez, J. Mn-based oxygen carriers prepared by impregnation for Chemical Looping Combustion with diverse fuels. *Fuel Process. Technol.* **2018**, *178*, 236–250. [CrossRef]
33. Adánez-Rubio, I.; Nilsson, A.; Izquierdo, M.T.; Mendiara, T.; Abad, A.; Adánez, J. Cu-Mn oxygen carrier with improved mechanical resistance: Analyzing performance under CLC and CLOU environments. *Fuel Process. Technol.* **2021**, *430*, 106819. [CrossRef]
34. Li, X.; Faust, R.; Purnomo, V.; Mei, D.; Linderholm, C.; Lyngfelt, A.; Mattisson, T. Performance of a perovskite-structured calcium manganite oxygen carrier produced from natural ores in a batch reactor and in operation of a chemical-looping combustion reactor system. *Chem. Eng. J.* **2024**, *497*, 154516. [CrossRef]
35. Huang, Y.; Zhu, P.; Gu, C.; Xu, H.; Zhou, J.; Xiao, G. Investigations on substituted Si-doped Mn_2O_3/Mn_3O_4 redox pair for thermochemical energy storage. *J. Energy Storage* **2023**, *73*, 108804. [CrossRef]
36. Preisner, N.C.; Block, T.; Linder, M.; Leion, H. Stabilizing Particles of Manganese-Iron Oxide with Additives for Thermochemical Energy Storage. *Energy Technol.* **2018**, *6*, 2154–2165. [CrossRef]
37. Almendros-Ibáñez, J.A.; Fernández-Torrijos, M.; Díaz-Heras, M.; Belmonte, J.F.; Sobrino, C. A review of solar thermal energy storage in beds of particles: Packed and fluidized beds. *Sol. Energy* **2019**, *192*, 193–237. [CrossRef]

applied sciences

Article

Numerical Modelling Assessment of the Impact of Hydrogen on the Energy and Environmental Performance of a Car Using Dual Fuel (Gasoline–Hydrogen)

Saugirdas Pukalskas *, Tadas Vipartas, Alfredas Rimkus, Donatas Kriaučiūnas, Justas Žaglinskis, Saulius Stravinskas, Andrius Ušinskas, Romualdas Juknelevičius, Gabrielius Mejeras, Vidas Žuraulis, Vilius Mejeras and Aleksas Narkevičius

Department of Automobile Engineering, Faculty of Transport Engineering, Vilnius Gediminas Technical University, Plytinės Str. 25, 10105 Vilnius, Lithuania; tadas.vipartas@vilniustech.lt (T.V.); alfredas.rimkus@vilniustech.lt (A.R.); donatas.kriauciunas@vilniustech.lt (D.K.); justas.zaglinskis@vilniustech.lt (J.Ž.); saulius.stravinskas@vilniustech.lt (S.S.); andrius.usinskas@vilniustech.lt (A.U.); romualdas.juknelevicius@vilniustech.lt (R.J.); gabrielius.mejeras@vilniustech.lt (G.M.); vidas.zuraulis@vilniustech.lt (V.Ž.); vilius.mejeras@vilniustech.lt (V.M.); aleksas.narkevicius@vilniustech.lt (A.N.)

* Correspondence: saugirdas.pukalskas@vilniustech.lt; Tel.: +370-655-29-573

Abstract: The utilization of "green" hydrogen in transportation areas gives rise to production- and supply infrastructure-related challenges; therefore, its wider application in automotive transport would lead to higher demand with cost reduction and a faster expansion of the hydrogen refuelling network. This study presents energy and environmental performance indicators analyses of a Nissan Qashqai J10 engine during the Worldwide Harmonised Light Vehicles Test Cycle (WLTC), replacing conventional fossil gasoline with dual-fuel (D-F) gasoline and hydrogen. Numerical modelling was conducted using AVL Cruise™ (Version R2022.2) software, utilizing the torque, fuel consumption, and environmental performance data of the HR16DE engine obtained through experimental testing across a wide range of loads and speeds on an engine test bench. The experimental investigation was carried out in two stages: using pure gasoline (G100); injecting a hydrogen additive into the intake air, constituting 5% of the gasoline mass (G95H5). Following similar stages, numerical modelling was conducted using the vehicle's technical specifications to calculate engine load and speed throughout the WLTC range. Instant fuel consumption and pollutant emissions (CO, CH, NO_x) were determined for various driving modes using experimental data maps. CO_2 emissions were calculated considering fuel composition and consumption. By integrating the instant values, the total and specific fuel consumption and emissions were calculated. As a result, this study identified the effect of a 5% hydrogen additive in improving engine energy efficiency, reducing incomplete combustion products and lowering greenhouse gas (CO_2) emissions under various driving modes. Finally, the results were compared with the requirements of EU standards.

Keywords: spark ignition engine; hydrogen; dual fuel; WLTC; AVL Cruise™ specific emissions

1. Introduction

Greenhouse gas (GHGs: carbon dioxide (CO_2), nitrous oxide (N_2O), methane—CH_4, F-gasses, and water vapour (H_2O)) emissions increase across various sectors from three different base years—1990, 2005, and 2021—to 2022 as analyzed in the JRC (Joint Research

Centre—European Commission) science for policy report on GHG emissions of all world countries [1]. The power industry has seen a significant rise in emissions, with a 92% increase from 1990 and a 34% increase from 2005, alongside a modest 1% increase from 2021 to 2022. Similarly, the industrial combustion and processes sector experienced the largest surge, with a 95% increase from 1990 and a 43% increase from 2005. In the buildings sector, emissions had a slight increase of 3% from 2005. Agriculture emissions have steadily increased by 21% from 1990, 15% from 2005, and 1% from 2021 to 2022. The transport sector shows notable emission increases across all periods, with a 72% rise from 1990, a 22% increase from 2005, and a significant 5% rise from 2021 to 2022. The fuel exploitation sector also recorded substantial increases, with a 56% rise from 1990, a 22% increase from 2005, and a 3% rise from 2021.

Mathematical prediction models show that GHG emissions from the transport sector will tend to increase for most European countries for the next several years [2], while China's case shows a rise for both energy consumption and transport CO_2 emissions by 2050 [3]. Modern prediction models are mostly based on neural networks, metaheuristics, prediction factor selection, or hybrid technology, and some regions like the USA show inconsistent results for future emissions [4,5]. In contrast, low-income countries with fast GDP (gross domestic product) growth are facing a GHG emissions increase, and a decrease in energy intensity is one of this problem's solution [6].

Most nations globally are developing strategies to address climate change. The European Union (EU) is particularly proactive, with ambitious goals outlined in the European Green Deal and European Climate Law [7,8]. The EU aims to cut its net domestic GHG emissions by a minimum of 55% by 2030 compared to 1990 levels and achieve climate neutrality (net-zero GHG emissions) by 2050 [1].

Due to the reliance on fossil fuels, the transport sector is a significant contributor to GHG emissions, while the electric vehicle's part is still increasing [9]. Domestic transportation in EU 27 (27 member states of the European Union) follows closely, contributing 23.1%, which reflects the substantial impact of vehicles and transport systems on the environment [10]. Here, emissions initially increased significantly from approximately 672 $MtCO_2e$ (million metric tons of carbon dioxide equivalent) in 1990 to a peak of around 863 $MtCO_2e$ in 2007. This represents a 28.4% increase over 17 years, driven by growing vehicle usage and economic activity. After 2007, emissions from domestic transportation began to decline; however, by 2022, emissions increased up to ~803 $MtCO_2e$, reflecting a 3.8% reduction since 2014. Overall, from 1990 to 2022, there was a ~19.5% increase in emissions, despite the recent declines.

Hydrogen has emerged as a promising solution for reducing GHG emissions due to its potential for zero carbon emissions [11–19]. When burned, hydrogen produces only water as a by-product, thereby eliminating carbon dioxide emissions entirely [20–22]. Hydrogen's wide flammability limits and high flame propagation rates enable stable combustion, even with lean mixtures, which can significantly improve engine efficiency and reduce nitrogen oxide (NO_x) emissions [23]. Moreover, hydrogen has a higher energy content per unit mass compared to petrol, which can result in higher thermal efficiency for engines designed to utilize hydrogen effectively. Additionally, hydrogen's capability to produce power without emitting carbon compounds positions it as a superior alternative in the quest for cleaner automotive technologies; however, it depends on the type of hydrogen production, as so-called black, brown, grey, and blue hydrogen is indicated [24]. When used in spark ignition engines, hydrogen combustion produces no carbon monoxide (CO), CO_2, hydrocarbons (HC), or particulate matter (PM) emissions, making it an exceptionally clean fuel choice [25,26].

Hydrogen's high burning velocity and broad flammability range, in contrast to traditional fossil fuels, elevate the risk of flashbacks, pre-ignition, and accidental explosions [27–29]. Moreover, the combustion of pure hydrogen produces a high combustion temperature, leading to the formation of nitrogen oxides [30–32]. Using hydrogen as fuel at excess air ratios higher than 1.3 reduces NO_x emissions by up to six times compared to emissions at a stoichiometric mixture. Lean air–hydrogen mixtures combust at lower temperatures, which are not enough to support NO_x formation [33,34]. Adjusting spark timing can also help reduce NO_x emissions. At an excess air ratio of 1.4, retarding the spark timing by just 4 CAD (crank angle degrees) results in a 17% reduction in NO_x emissions. For instance, at an excess air ratio of 1.6, retarding the spark timing leads to a 70% reduction in NO_x emissions [35]. Selecting the specific mixture formation method like 'spark-plug mildly stratified charge', NOx emissions are limited to less than 10 ppm [36].

Due to its low energy density under standard temperature and pressure, hydrogen requires specialized storage solutions. For instance, a fuel cell vehicle requires approximately 5–6 kg of hydrogen, compressed to 700 bar, to achieve a driving range of 500 km [37]. Combusting hydrogen in an internal combustion engine (ICE) requires an even greater amount of hydrogen to be stored on board the vehicle. According to modelling results from Chalmers University of Technology, hydrogen fuel cell vehicles demonstrate nearly twice the efficiency of ICEs, with a lower fuel consumption of 1.05 kg-H_2 per 100 km under the Worldwide Harmonised Light Vehicles Test Procedure driving cycle (WLTC), compared to approximately 1.79 kg-H_2 per 100 km for hydrogen combustion vehicles [38]. Replacing petrol with hydrogen to achieve a driving range of at least 500 km presents significant challenges, primarily due to the increased amount of hydrogen that needs to be stored in the vehicle. Also, hydrogen is highly flammable, and special composite H_2 fuel tanks must be used for hydrogen storage. For fuelling systems, stainless steel lines are required, along with a special pressure reducer equipped with a flame arrestor. A hydrogen leak detection system should also be installed, incorporating hydrogen sensors that provide warnings and can automatically shut off the tank's main stop valves in the event of a leak. In case of a fire, a pressure relief system should vent hydrogen away from the vehicle and passengers to minimize the risk of explosions. Additionally, automatic shut-off valves should be installed to automatically close solenoid valves on the hydrogen tanks during a crash, isolating the fuel supply. Even with the installation of safety systems, dual-fuel technology remains a potential alternative for reducing greenhouse gas emissions and fossil fuel consumption.

One potential solution to the storage challenge is implementing a dual-fuel system. In this approach, hydrogen can be used as an additive or substitute for a small portion (up to 15% by mass) of conventional fuel [39]. This strategy reduces the required hydrogen storage, lowering the overall vehicle mass while simultaneously contributing to a reduction in tailpipe emissions [40]. Hydrogen enables the increase of air excess ratios up to two, without substantially prolonging the combustion duration and reducing NO_x emissions. For instance, using a petrol–hydrogen mixture containing 10% hydrogen by mass and increasing the air excess ratio from 1.4 to 1.6 results in only a minor increase in combustion duration, approximately 1–2 degrees of crank angle (CA) [41]. Hydrogen also changes the fraction of carbon-based fuel used, leading to lower emissions of CO, CO_2, and hydrocarbons. Studies have shown that even a small fraction of hydrogen can reduce unburned HC emissions by up to 9% across all load points, including tests conducted at full load, with an additional hydrogen flow of 0.25 kg/h. This reduction is due to the increased combustion speed and improved fuel oxidation by hydrogen, and a rise in combustion temperature [42]. However, while the addition of hydrogen enhances the oxidation of carbon-based fuels, it also influences thermal NO_x formation. The increased combustion temperature caused by hydrogen leads to higher NO_x emissions. Studies have shown that reducing the hydrogen fraction in the fuel mixture from 15% to 10% decreases NO emissions by 14%, with a further 17% reduction observed when the hydrogen fraction is lowered to 5% [43].

To achieve a more comprehensive evaluation of hydrogen's impact—considering the additional mass of H_2 tanks and driving at different speeds—the WLTP driving simulation will be performed [44,45]. This study is designed to assess the effect of hydrogen as an additive in ICE under energy and performance indicators through experimental investigation and numerical modelling. The rest of this paper is organized as follows: Section 2 describes the materials and methods used for experimental research and numerical simulation. Section 3 discusses the results and comparative analysis based on energy and environmental performance indicators. In Section 4, the conclusions and further research are summarized.

2. Materials and Methods

In the experimental part, engine performance and environmental indicators were analyzed using an engine test bench with different fuels: operating with pure gasoline (labelled G100) in dual-fuel mode (D-F), comprising 95% gasoline by mass and 5% hydrogen by mass (labelled G95H5). In the numerical modelling part, the AVL Cruise™ (Version R2022.2) software was utilized, incorporating data from the experimental tests to perform a numerical simulation WLTC. Fuel consumption and emissions were calculated for both G100 and G95H5 cases.

2.1. Fuel Properties

The main properties of the fuels are presented in Table 1, where the properties of the dual-fuel (D-F) mixture are calculated considering the H_2 mass fraction in the fuel. The lower heating value (LHV) of H_2 is 2.76 times higher compared to G100. Consequently, when 5% H_2 by mass is added to the G95H5 fuel mixture, the energy share of H_2 reaches 12.7%, and the fuel's LHV increases by 6.8%. This indicates that under the same engine load, the cyclic mass of fuel supplied is lower in the G95H5 case. Changes in the elemental composition of the fuel resulted in a 27.9% reduction in the carbon/hydrogen (C/H) ratio and an 8.8% increase in the stoichiometric air/fuel (A/F) mass ratio (which is higher than the increase in the LHV). Consequently, for the same engine load, maintaining a stoichiometric mixture (excess air ratio $\lambda = 1$) with G95H5 required a greater throttle opening due to the increased air demand. The low density and high volumetric displacement of hydrogen in the cylinder further reduce the air intake, necessitating additional throttle opening to maintain $\lambda = 1$. During the WLTC, the required engine power does not reach the maximum specified in the engine design. When operating with G100, the throttle is partially opened, whereas with G95H5, there is room to open the throttle further. However, at maximum throttle opening, the volumetric displacement of hydrogen and the increased air demand for combustion with G95H5 resulted in a smaller cyclic fuel mass compared to G100, leading to a reduction in the engine's maximum brake torque. The significantly reduced C/H ratio of G95H5 suggests lower emissions of CO and CH, as well as reduced greenhouse gas CO_2 during the driving cycle.

Table 1. Key fuel properties [46–48].

Parameter		Gasoline (G100)	Hydrogen ($H_2$100)	D-F (G95H5)	Difference Between D-F G95H5 and G100, %
Density at 15 °C, kg/m^3		736.0	0.09	-	-
Lower heating value *LHV*, MJ/kg		43.53	120.00	47.35	+6.8
Elemental composition, %	C	86.42	0.00	82.10	−5.0
	H	13.58	100.00	17.90	+31.8
C/H		6.36	-	4.59	−27.9
Stoichiometric air–fuel ratio A/F, kg air/1 kg fuel		14.7	35.0	15.7	+8.8

2.2. Experimental Engine Test Bench Equipment

In the first stage of this study, the energy and environmental indicators of a naturally aspirated spark-ignition engine (SIE), the Nissan HR16DE (Nissan Motor Co., Ltd., Yokohama, Japan), were investigated using experimental equipment (Figure 1). According to the engine specifications, it delivers a maximum power output of 84 kW at 6000 rpm. The naturally aspirated engine was equipped with a multi-point fuel injection (MPFI) system operating at 3.5 bar fuel pressure, along with a Continuously Variable Valve Timing Control System (CVTCS). The engine was loaded using the AMX200/100 (Automex, Gdansk, Poland) eddy current engine load stand, which has a torque measurement accuracy of ±0.9 Nm and is capable of applying a maximum torque load of 250 Nm (Table 2).

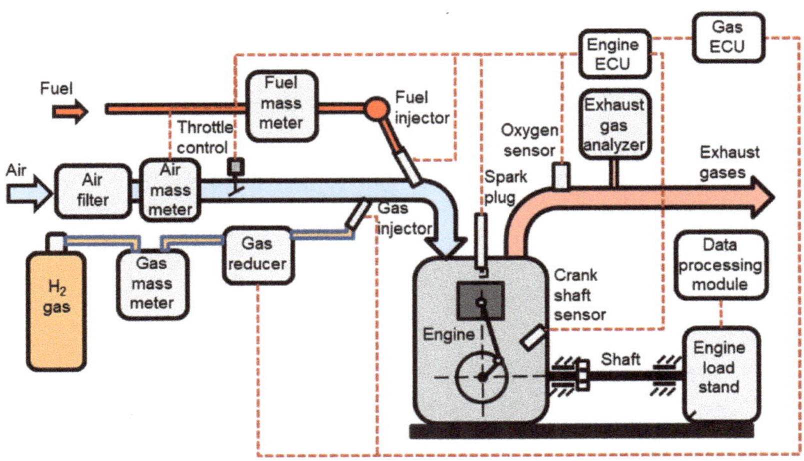

Figure 1. The schematic diagram of the test bench for an SIE under load.

Gasoline consumption was measured using AMX240 (Automex, Gdansk, Poland) gravimetric scales with an accuracy of ±0.10%. Hydrogen consumption was measured using a high-pressure Coriolis mass flow metre RHEONIK RHM 015 (RHEONIK, Odelzhausen, Germany) with an accuracy of ±0.10%. Air mass flow was measured at the intake system using a BOSCH HFM 5 (BOSCH, Stuttgart, Germany) air flow metre, which provides an accuracy of 3%. The composition of the exhaust gasses was analyzed using the AVL DiCOM 4000 (AVL DiTEST, Graz, Austria) gas analyser with the following specifications:

- CO: Measuring range, 0–10% (vol); resolution ±0.01%.

- CO_2: Measuring range, 0–20% (vol); resolution ±0.10%.
- HC: Measuring range, 0–20,000 ppm (vol); resolution ±1 ppm.
- NO_x: Measuring range, 0–4000 ppm (vol); resolution ±1 ppm.
- Excess air ratio (λ): Measuring range, 0–9.999; resolution ±0.001.

In the experimental setup, liquid (G100) and gaseous (H_2) fuels were injected through separate injectors installed in the intake manifold. G100 was injected using the original multi-point injection system of the HR16DE HR16DE (Nissan Motor Co., Ltd., Yokohama, Japan) engine. For injecting hydrogen, a separate multi-point injection system was installed in the intake manifold. To maintain the desired gasoline/hydrogen ratio, the MoTeC M800 (MoTeC, Melbourne, Australia) system was used to adjust the operating ratio of the primary and secondary injectors. The ratio was recalculated based on the mass fuel consumption of G100 and $H_2$100. Engine operating parameters—including throttle position, engine speed, excess air ratio (λ), and ignition timing—were controlled using a programmable electronic engine control unit (ECU), MoTeC M800. A lambda control system maintained a stoichiometric mixture (λ = 1.0) over a wide range of operating conditions. Exceptions occurred at maximum throttle opening, where the mixture was enriched to an excess air ratio of λ = 0.9 to achieve maximum torque. At each measurement point, data were collected five times, and the standard deviation was calculated. If any measurement point showed significant variation, the measurement was repeated.

Table 2. Main parameters of the tested engine.

Parameter	Specification
Engine Type	4 cylinder, DOHC
Displacement	1.6 L (1598 cm^3)
Fuel System	Multi-point fuel injection (MPFI)
Valve Timing	Continuously Variable Valve Timing Control System (CVTCS)
Bore	78.0 mm
Stroke	83.6 mm
Compression Ratio	10.7:1
Block and Head Material	Aluminum alloy
Ignition System	Distributorless ignition system (DIS)
Maximum Power	84 kW (6000 rpm)
Maximum Torque	156 Nm (4400 rpm)
Emissions Compliance	Euro 4

2.3. The Methodology of the Experimental Study on the Engine Load Test Bench

The first stage of the experiments involved testing the engine operation with G100 fuel. Fuel consumption and pollutant concentrations were measured across the full range of engine speeds and loads (Figure 2). The maximum brake torque achieved was 156 Nm. A fuel consumption map (B_{f_G100}, kg/h) was created based on the measured data, illustrating fuel consumption across the operating conditions (Figure 3).

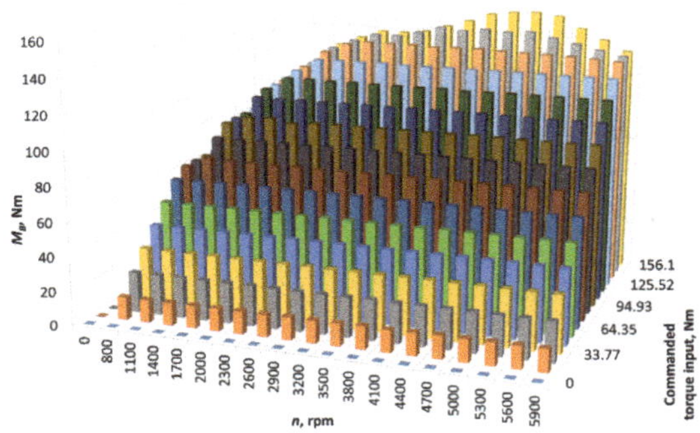

Figure 2. Engine load map.

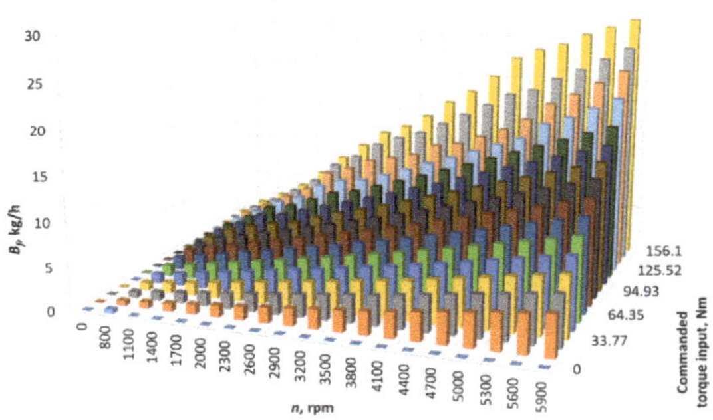

Figure 3. Fuel consumption map for G100.

Before conducting tests with the dual-fuel G95H5 mixture, the gasoline consumption map from the initial G100 testing (B_{f_G100} map, Figure 3) was used to preliminarily calculate the required hydrogen mass flow (5% of G100 by mass). Based on these calculations, a hydrogen consumption map (B_{f_H2}, Figure 4) was created to guide the hydrogen supply during dual-fuel operation. During the dual-fuel experiments, the hydrogen flow specified in the B_{f_H2} map was supplied to the intake manifold through additional gas injectors controlled by the electronically managed OSCAR-N gas injection system. This system received real-time data from the MoTeC M800 ECU, enabling precise synchronization with the gasoline injection system. Gasoline was simultaneously injected via the standard fuel system, with the injector opening duration managed by the MoTeC M800 system. To maintain a stoichiometric mixture ($\lambda = 1.0$), the control system adjusted the gasoline injection volume based on signals from an oxygen sensor installed in the exhaust system. By supplementing the air–gasoline mixture with hydrogen gas (5% of the gasoline mass), while maintaining the same excess air ratio as in the G100 tests, the maximum engine brake torque decreased to 137 Nm, approximately a 12% reduction compared to G100. A dual-fuel consumption map (B_{f_G95H5}, kg/h) was developed (Figure 5). Fuel consumption with G95H5 decreased by approximately 18% compared to G100. This reduction resulted from several complex factors:

1. The low density of hydrogen increases its volumetric displacement, reducing the air intake and thus volumetric efficiency, which decreases torque.
2. G95H5 requires 8.8% more air for combustion compared to G100.
3. The lower heating value (LHV) of the fuel blend increased by 6.8% (Table 1).

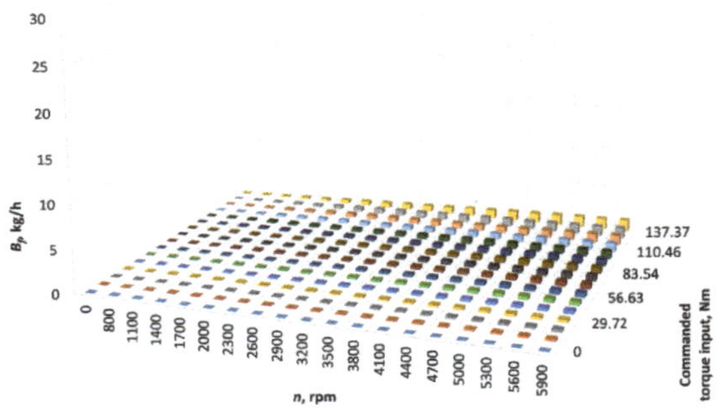

Figure 4. H_2 consumption map.

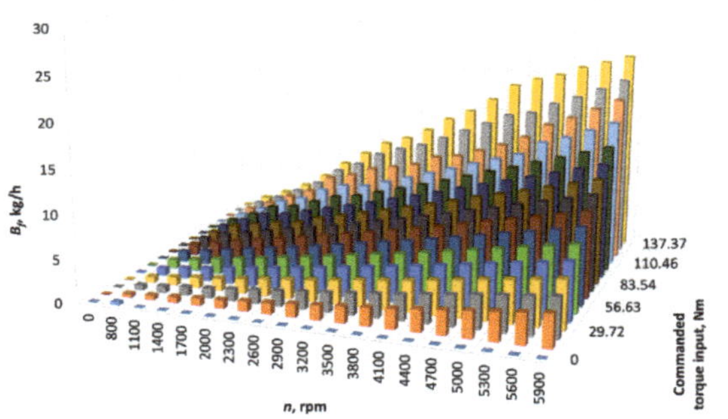

Figure 5. D-F G95H5 consumption map.

The composition of engine exhaust gasses (pollutant concentration) was measured using a stationary gas analyser. Results are given after measuring the pollution before the catalytic converter. The emissions of individual exhaust components were calculated using the specified formulas:

$$E_{CO} = \frac{CO \cdot M_{CO} \cdot m_{ex}}{M_{ex} \cdot 100}, \text{kg/h}; \quad (1)$$

$$E_{CH} = \frac{CH \cdot M_{CH} \cdot m_{ex}}{M_{ex} \cdot 10^6}, \text{kg/h}; \quad (2)$$

$$E_{NO_x} = \frac{NO_x \cdot M_{NO_x} \cdot m_{ex}}{M_{ex} \cdot 10^6}, \text{kg/h}; \quad (3)$$

here, M_{CO}—the molar mass of carbon monoxide, M_{CO} = 28.01 g/mol; CO—the concentration of carbon monoxide in exhaust gasses, %; M_{CH}—the molar mass of unburned hydrocarbons, M_{CH} = 42.5 g/mol; CH—the concentration of unburned hydrocarbons in exhaust gasses, ppm; M_{NO_x}—the molar mass of nitrogen oxides, M_{NO_x} = 30.01 g/mol; NO_x—the concentration of nitrogen oxides in exhaust gasses, ppm; M_{ex}—the molar mass of the exhaust gas mixture, calculated by considering the molar mass of individual exhaust gas components and their concentration in the exhaust gasses, g/mol; m_{ex}—the engine exhaust gas mass emission:

$$m_{ex} = B_f + B_{air}, \text{kg/h}. \quad (4)$$

where B_f—the total mass consumption of gasoline and hydrogen; B_{air}—the air mass consumption.

The load of the SIE was gradually increased to its maximum at various fixed engine speeds. For each measurement point, pollutant emissions were calculated, and pollutant emission maps were created for the engine operating on gasoline and dual fuel. The visualization of CO, CH, and NO_x emissions for the engine operating on D-F G95H5 is presented in Section 3, Results and Discussion.

2.4. The Methodology of WLTC Numerical Simulation

In the second part of this study, to comprehensively evaluate the impact of a hydrogen additive on the performance parameters of a spark-ignition engine operating on gasoline under simulated real-world driving conditions, a numerical model of a Nissan Qashqai J10 (Nissan Motor Co., Ltd., Yokohama, Japan) passenger car was developed using the AVL Cruise™ simulation software (Figure 6). The vehicle is equipped with a four-cylinder spark-ignition engine (HR16DE) with a displacement of 1.598 dm^3, a maximum power output of 84 kW at 6000 rpm, and a maximum brake torque of 156 Nm at 4400 rpm. The vehicle's wheelbase length is 2630 mm, the frontal surface area is 2.32 m^2, the aerodynamic drag coefficient is 0.32, the curb weight is 1272 kg, and the maximum allowable weight is 1830 kg. The car features a five-speed manual transmission with the following gear ratios: 3.73, 2.1, 1.39, 1.09, and 0.89. The capacity of the fuel tank is 65 L, and the dimensions of the tires of 215/65 R16 are used in this model.

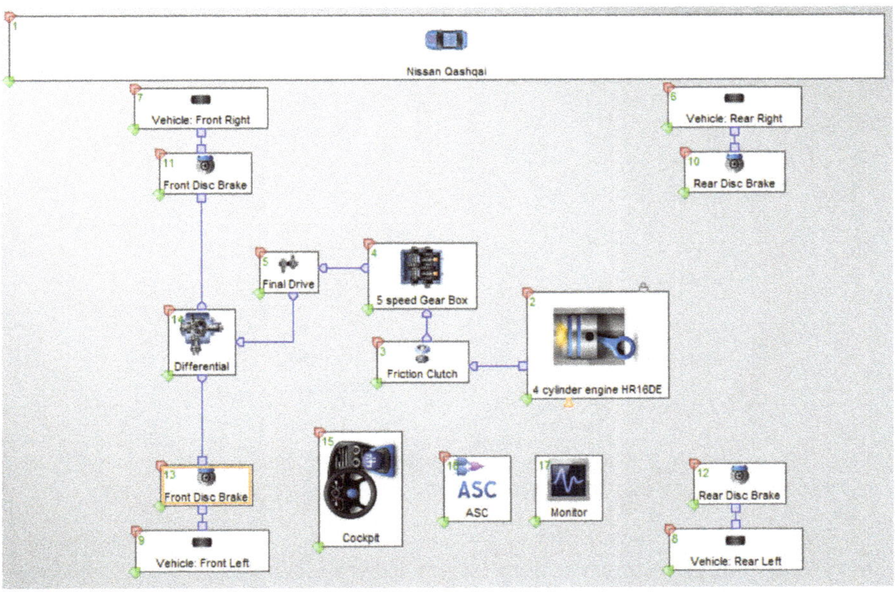

Figure 6. The numerical model of the Nissan Qashqai J10 vehicle in AVL Cruise™ simulation software: 1—the main vehicle component Nissan Qashqai, 2—the naturally aspirated internal combustion engine HR16DE, 3—friction clutch, 4—manual transmission, 5—final drive, 6—rear right wheel, 7—front right wheel, 8—rear left wheel, 9—front left wheel, 10—rear right brake disc, 11—front right brake disc, 12—rear left brake disc, 13—front left brake disc, 14—differential, 15—cockpit, 16—anti-slip control (ASC) system, and 17—monitor.

According to the Worldwide Harmonised Light Vehicles Test Procedure (WLTP), based on the engine power-to-mass ratio (PMR) and the manufacturer-declared maximum vehicle speed, the WLTC driving cycle was assigned to "Class 3b". Vehicles in "Class 3b" have an engine power-to-mass ratio greater than 34 W/kg and a manufacturer-declared maximum speed of at least 120 km/h. The tested vehicle's maximum speed is 175 km/h, and its PMR is 66.04 W/kg.

The WLTC simulation task for the vehicle was created in a dedicated folder within the simulation software by inputting the driving cycle data (time–speed–load). During the numerical simulation, real-world driving conditions were modelled for 30 min across four distinct modes:

1. Urban cycle: a maximum speed of up to 56.5 km/h, average speed of 18.9 km/h, duration of 589 s, and distance of 3.09 km.
2. Medium-speed cycle: a maximum speed up of to 76.6 km/h, average speed of 39.5 km/h, duration of 433 s, and distance of 4.76 km.
3. High-speed cycle: a maximum speed of up to 97.4 km/h, average speed of 56.7 km/h, duration of 455 s, and distance of 7.16 km.
4. Very high-speed cycle: a maximum speed of up to 131.3 km/h, average speed of 92 km/h, duration of 323 s, and distance of 8.25 km.

Environmental conditions during the simulation were set as follows: an ambient temperature of 23 °C, relative humidity of 50%, and atmospheric pressure of 1.013 bar.

As shown in Figure 7, during the WLTC simulation conducted using AVL Cruise™, the tested vehicle travelled a total distance of 23.27 km in 1800 s while maintaining the prescribed speeds. An analogous driving cycle was simulated with the vehicle operating on both G100 and G95H5 fuels.

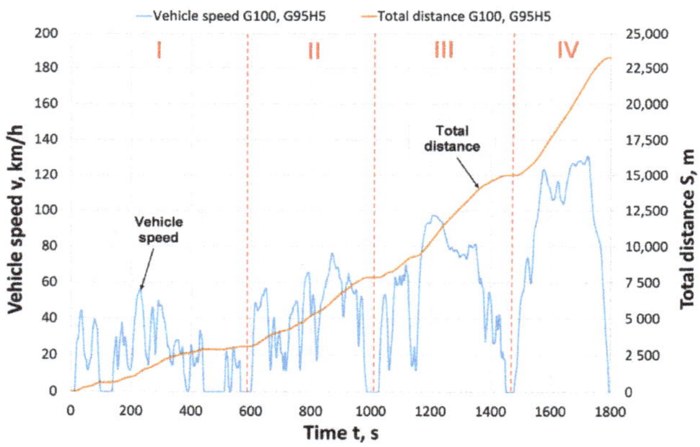

Figure 7. The vehicle speed v and distance S travelled in the WLTC simulation: I—Urban cycle; II—Medium-speed cycle; III—High-speed cycle; IV—Very high-speed cycle.

In this study, the vehicle's WLTC was simulated using the AVL Cruise™ software. In the near future, after installing an additional hydrogen fuelling system in the vehicle, real-world driving tests are planned. The experimental investigation of the WLTC cycle will involve the use of a chassis dynamometer, fuel (gasoline and hydrogen) mass metre, air mass metre, gas analyser, and other equipment. Using data obtained from the experimental tests, the vehicle's energy and environmental performance indicators will be calculated [49], and the validation of the numerical simulation results presented in this article will be conducted. During the WLTC experimental tests on the chassis dynamometer, the spark-ignition engine (SIE) will operate under various transient modes. These operational parameters may not fully match the results obtained during the simulation, as the latter relied on experimental data measured under stable engine operating conditions.

3. Results and Discussion

In the second stage of this study, using data from the Nissan Qashqai J10 vehicle, the AVL Cruise™ program was used to calculate the engine speed (Figure 8), brake torque (Figure 9), and power (Figure 10) at all WLTC driving points. The engine speed was calculated while driving in the following modes: urban cycle (I), average speed (II), and high speed (III) range from 1700 to 3800 rpm, depending on the vehicle speed and the selected gear. When driving at very high speeds (IV), the engine speed increases to 2500–4500 rpm, as the highest gear in the gearbox is five, and this gear is engaged when the vehicle speed reaches 80–90 km/h. Increasing the vehicle speed from 80 km/h to 130 km/h does not allow for a reduction in the gear ratio, and the engine speed increases by approximately 1.6 times. Replacing fuel G100 with G95H5 causes only a slight change in engine speed, as the gears are shifted identically in both cases. Engine speed is one of the coordinates in the fuel consumption and pollutant emission maps used to select the necessary parameters for WLTC indicator calculations.

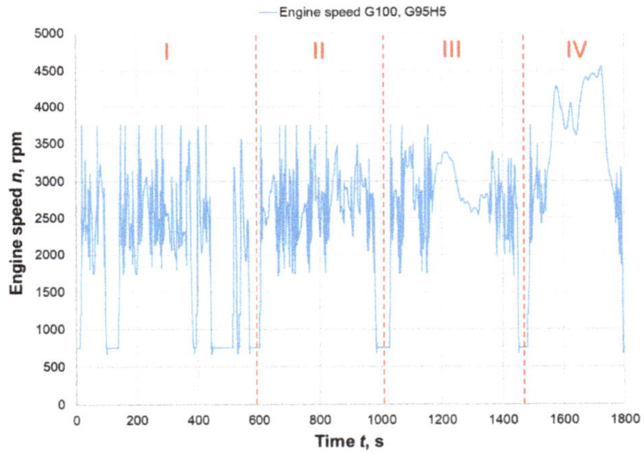

Figure 8. Internal combustion engine speed n during WLTC simulation: I—Urban cycle; II—Medium-speed cycle; III—High-speed cycle; IV—Very high-speed cycle.

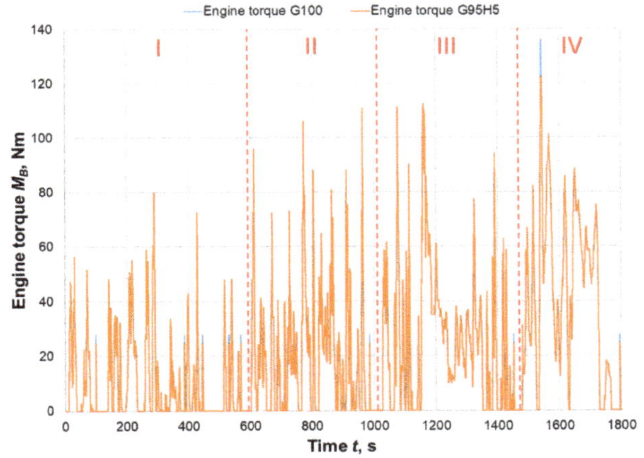

Figure 9. The brake torque M_B of the internal combustion engine during WLTC simulation: I—Urban cycle; II—Medium-speed cycle; III—High-speed cycle; IV—Very high-speed cycle.

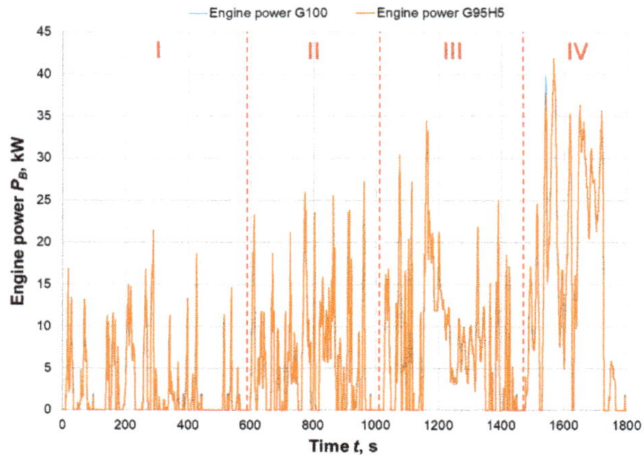

Figure 10. The brake power P_B of the internal combustion engine during WLTC simulation: I—Urban cycle; II—Medium-speed cycle; III—High-speed cycle; IV—Very high-speed cycle.

The SIE brake torque dynamically varies at all WLTC points (Figure 9), which confirms that the engine operates in transitional modes during real driving. The load, in particular, increases as the vehicle accelerates and overcomes inertia, with the maximum value when driving at high and very high speeds approaching 140 Nm. Engine load is the second coordinate used for selecting the necessary parameters for the WLTC indicator calculations.

The engine brake power directly depends on the engine speed and brake torque, and throughout the entire WLTC period, it fluctuates (Figure 10). The engine's effective power is zero when the engine is idling, or engine braking is occurring. When the vehicle is travelling at a high speed of 130 km/h, the engine load power ranges between 20 and 35 kW. During rapid acceleration, the power demand reaches 45 kW, and the engine still has a ~45 kW reserve. Whether the vehicle is powered by G100 or G95H5 fuel, the engine brake power values at all WLTC points are similar, as the fuel type has minimal impact on engine load. The only significant influence is the increased vehicle mass due to the additional hydrogen supply system.

In the WLTC test stage, the fuel consumption B_f data (maps) for the engine operating on G100 and G95H5 fuels (Figures 3 and 5) were uploaded into the AVL Cruise™ numerical simulation program. These maps were created based on the authors' preliminary experimental studies and the results of other studies [50–52]. When analyzing the instantaneous fuel consumption determined through numerical modelling, lower consumption values for G95H5 are visually observed (Figure 11). After calculating the total fuel consumption during WLTC, the fuel consumption for G100 and G95H5 was found to be 1.03 kg (44.92 MJ) and 0.92 kg (43.76 MJ), respectively. The reduction of approximately 10.4% in dual-fuel G95H5 consumption compared to G100 was due to a 6.8% higher D-F LHV (Table 1) and increased brake thermal efficiency due to the hydrogen addition. An evaluation of fuel consumption during WLTC and LHV showed that replacing G100 with G95H5 reduced fuel energy consumption by about 2.57%. The thermal efficiency of combustion increased, as hydrogen accelerated and improved the combustion process, which reduced energy loss through exhaust gasses and increased energy release due to more complete combustion. These results are consistent with other researchers [50,53,54].

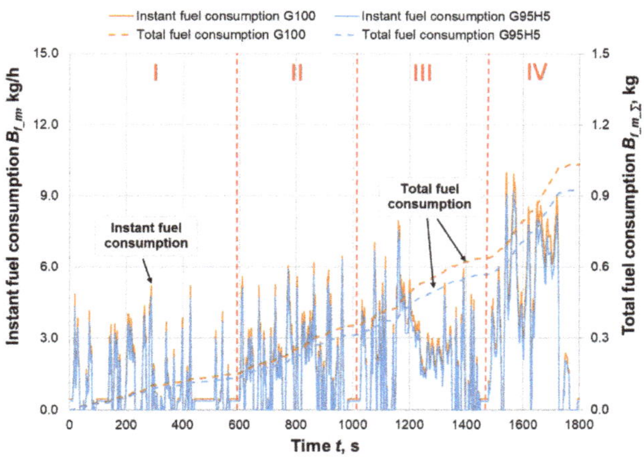

Figure 11. Fuel consumption during the WLTC test using pure gasoline G100 and dual-fuel G95H5: I—Urban cycle; II—Medium-speed cycle; III—High-speed cycle; IV—Very high-speed cycle.

The emission indicators of an SIE (E_{CO}, E_{HC}, and E_{NO_x}) using G100 and G95H5 fuels (Figures 12–14) across a wide range of engine speeds and loads were determined in the first stage of the study through experimental engine testing on a dynamometer. The emission trends of incomplete combustion products (CO and HC) and nitrogen oxides varied with changes in engine load and speed, but certain similarities were also observed. At low and moderate loads and speeds, CO emissions were relatively low due to the maintenance of stoichiometric air excess, ensuring efficient combustion (Figure 12). As engine loads approached maximum levels, the mixture was enriched, leading to a significant increase in CO emissions, especially at maximum speeds, as combustion time decreased. HC emissions consistently increased with rising engine speed and load (Figure 13) due to the greater amount of combustible mixture in the cylinder and the reduced combustion duration. Similarly to the case of CO, a sharp increase in HC emissions was observed when approaching maximum torque, and the air–fuel mixture was enriched [55].

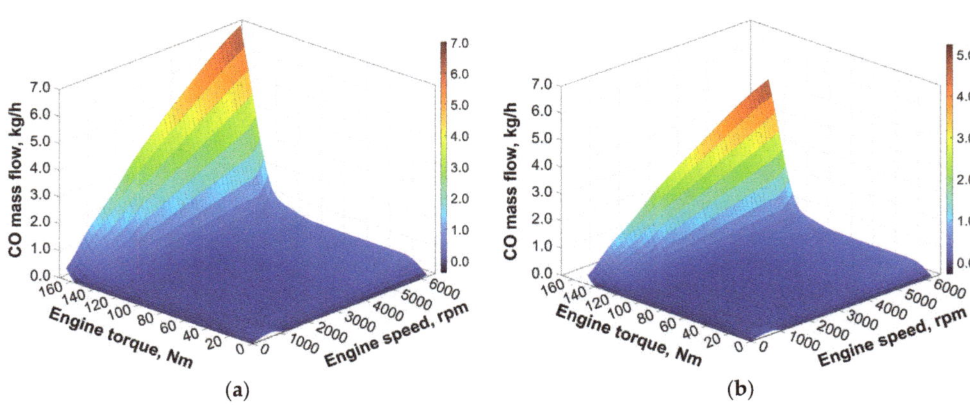

Figure 12. CO emissions map: (**a**) G100; (**b**) G95H5.

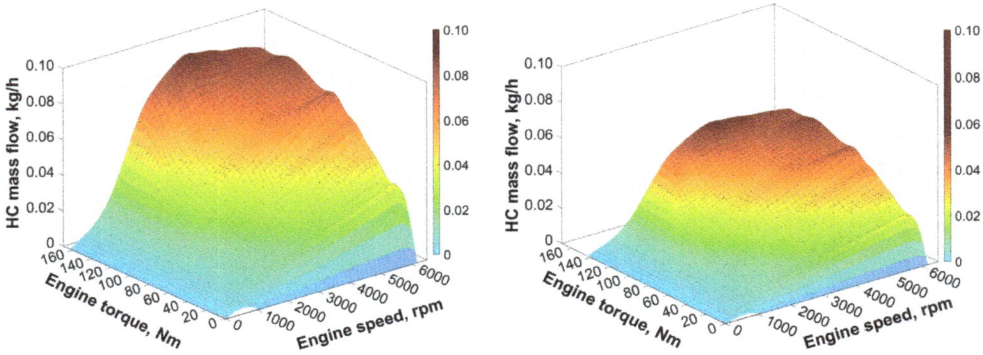

(a) (b)

Figure 13. HC emissions map: (**a**) G100; (**b**) G95H5.

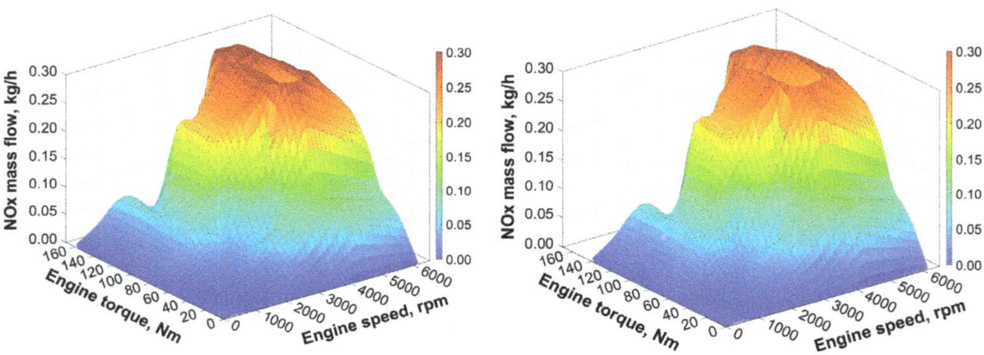

(a) (b)

Figure 14. NO_x emissions map: (**a**) G100; (**b**) G95H5.

The NO_x emissions were determined by the combustion temperature in the cylinder and the air excess coefficient. It was observed that increasing the engine load resulted in a rise in NO_x emissions (Figure 14). This was due to the greater volume of the air–fuel mixture in the cylinder (volumetric efficiency) caused by a wider throttle opening, which increased the combustion temperature in the cylinder. However, as engine loads approached maximum levels, a significant increase in NO_x emissions (as seen in the cases of CO and HC) did not occur due to the richer fuel mixture and a lack of oxygen [54,56].

In the second stage of the research, the engine emission data (E_{CO}, E_{HC}, and E_{NO_x}) were uploaded into the AVL Cruise™ numerical simulation software and used to determine the vehicle's emissions at individual WLTC points (instant emissions), based on specific engine speeds and loads. These point-specific emission data were then used to calculate the total emissions over the entire test cycle. Several researchers have published similar studies, but they used gas rather than petrol [57,58].

Figures 15 and 16 show quite different trends in the emissions of incomplete combustion products (CO and HC). This is due to the different tendency observed in the CO and HC emission maps (Figures 12 and 13). The total emissions (kg) calculated using the AVL Cruise software were used to analyze the specific emissions (g/km) of the different driving phases (cycles) of the WLTC by dividing the total emissions by the distance travelled during driving mode. During the urban cycle (I), the CO specific emissions were low (0.38 g/km for G100 and 0.30 g/km for G95H5) but increased significantly (~8 times) with increased engine load and the enrichment of the air–fuel mixture (3.0 g/km for G100 and 2.3 g/km for G95H5) at very high vehicle speeds (IV). The HC specific emission intensity at low load (I) is even slightly higher (0.85 g/km for G100 and 0.54 g/km for G95H5) compared to very high vehicle speeds (IV) (0.46 g/km for G100 and 0.30 g/km for G95H5), due to the fact that combustion is carried out at a lower temperature at low load, and therefore, the HC emissions are still high, even at idling speed.

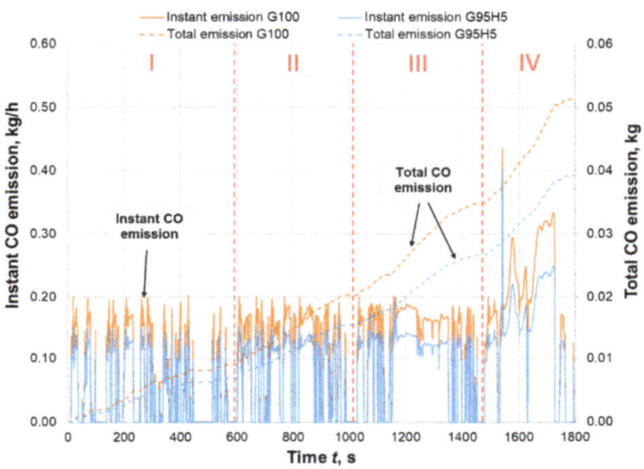

Figure 15. Instant and total CO emissions during the WLTC test: I—Urban cycle; II—Medium-speed cycle; III—High-speed cycle; IV—Very high-speed cycle.

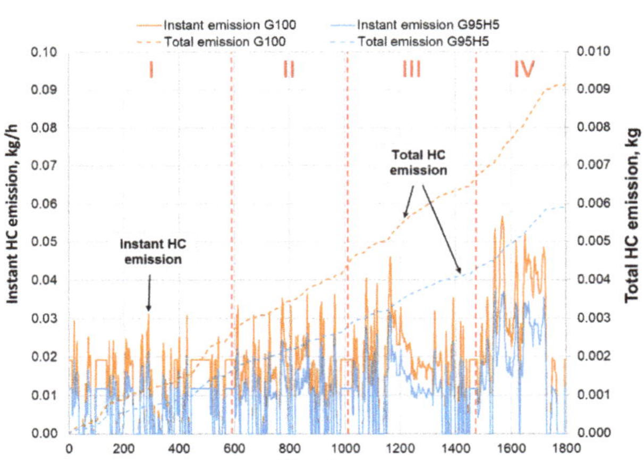

Figure 16. Instant and total HC emissions during the WLTC test: I—Urban cycle; II—Medium-speed cycle; III—High-speed cycle; IV—Very high-speed cycle.

When analyzing carbon monoxide (CO) emission trends over the WLTC using G100 and G95H5 fuels, the addition of 5% hydrogen by mass to gasoline significantly reduced cumulative CO emissions. This reduction was attributed to the altered fuel composition and more efficient combustion process. The hydrogen additive decreased the C/H ratio in the fuel mixture by 27.9% and increased the combustion rate. In city driving conditions (Phase I), the total CO emissions reached 8.9 g for G100 and 6.9 g for G95H5, representing a 22.9% reduction in total emissions. As the vehicle transitioned to suburban conditions (medium-speed regime) in Phase II, hydrogen reduced the total CO emissions by 23.6%. After the first two WLTC phases, the total CO emissions reached 20.3 g for G100 and 15.6 g for G95H5. During rural conditions (high-speed regime) in Phase III, numerical simulation determined that the hydrogen additive reduced CO emissions by 23.4%, with cumulative CO emissions after the first three phases reaching 34.9 g for G100 and 26.7 g for G95H5. In Phase IV, during highway conditions (very high-speed regime), aerodynamic drag increased fuel consumption, leading to the most intense growth in CO emissions. In this phase, a 23.9% reduction in CO emissions due to hydrogen was observed. The total CO emissions after all four WLTC driving phases amounted to 51.3 g for G100 and 39.2 g for G95H5, reflecting a 23.6% decrease. The greater effect of hydrogen in reducing CO emissions was achieved when the engine operated dynamically, under higher loads, and with a richer air–fuel mixture. The specific CO emissions for individual WLTC driving phases using G100 and G95H5 fuels were as follows: Phase I: 0.384 g/km (G100) vs. 0.296 g/km (G95H5), Phase II: 2.57 g/km (G100) vs. 1.97 g/km (G95H5), Phase III: 2.34 g/km (G100) vs. 1.79 g/km (G95H5), and Phase IV: 3.04 g/km (G100) vs. 2.32 g/km (G95H5). Considering the total WLTC driving distance of 23.27 km, the calculated specific CO emissions were 2.21 g/km for G100 and 1.69 g/km for G95H5. The tested vehicle must comply with the Euro 4 emission standard, which permits a maximum CO emission level of 1.0 g/km. This study evaluated emissions before the catalytic converter. With the use of a catalytic converter, CO emissions would be significantly reduced, and for both fuels, they would meet the standard requirements.

The addition of 5% hydrogen to gasoline during the WLTC significantly reduced total HC emissions in all driving conditions. In city driving conditions (Phase I), the hydrogen additive reduced total HC emissions by 37.1%, from 2.65 g (G100) to 1.66 g (G95H5) (Figure 16). During suburban conditions (Phase II), the reduction in cumulative HC emissions with hydrogen was 36.5%, with total HC emissions after the first two phases reaching 4.54 g for G100 and 2.89 g for G95H5. In rural conditions (Phase III), hydrogen reduced HC emissions by 35.2%, with cumulative HC emissions after three phases totaling 6.75 g for G100 and 4.32 g for G95H5. During highway conditions (Phase IV), where increased aerodynamic drag raised fuel consumption, the hydrogen additive reduced HC emissions by 34.1%. After all four WLTC phases, the total HC emissions were 9.19 g for G100 and 5.95 g for G95H5, resulting in a 35.2% decrease in HC emissions across the full driving cycle. The specific HC emissions for individual WLTC phases using G100 and G95H5 fuels were as follows: Phase I: 0.853 g/km (G100) vs. 0.537 g/km (G95H5), Phase II: 0.617 g/km (G100) vs. 0.391 g/km (G95H5), Phase III: 0.405 g/km (G100) vs. 0.262 g/km (G95H5), and Phase IV: 0.455 g/km (G100) vs. 0.300 g/km (G95H5). This analysis shows that HC emissions were of similar intensity across all driving phases. The positive effect of the hydrogen additive was more pronounced in the initial WLTC driving phases, where engine loads were lower. This improved HC combustion at lower combustion temperatures, differing from the trend observed in CO emission reductions with hydrogen. The specific HC emissions over the full WLTC were calculated at 0.395 g/km for G100 and 0.256 g/km for G95H5. The tested vehicle is certified to meet the Euro 4 emission standard, which allows a maximum HC emission level of 0.1 g/km. With the use of a catalytic converter, HC emissions would be significantly reduced and would meet the ecological requirements for both fuels.

Numerical modelling using the AVL Cruise™ software shows that a 5% hydrogen additive in gasoline has a varying impact on NO_x emissions across different WLTC phases. At the end of the city driving cycle (Phase I), the total NO_x emissions were 2.0 g for G100 and 2.46 g (22.7% higher) for G95H5 (Figure 17), with specific emissions of 0.647 g/km and 0.794 g/km, respectively. In the medium-speed regime (Phase II), NO_x emissions increased due to higher engine loads, with hydrogen further increasing NO_x emissions by 25.9%. Specific emissions rose to 0.787 g/km for G100 and 0.991 g/km for G95H5, resulting in total emissions of 5.6 g for G100 and 7.0 g for G95H5 after two phases. In the high-speed regime (Phase III), the impact of hydrogen on NO_x emissions remained significant, increasing emissions by 23.5%. Specific emissions reached 0.897 g/km for G100 and 1.108 g/km for G95H5, with total NO_x emissions after three phases amounting to 11.3 g for G100 and 14.0 g for G95H5. During the very high-speed regime (Phase IV), higher engine loads and elevated combustion temperatures in the cylinder resulted in substantially higher NO_x emissions. Specific emissions in this phase reached 1.714 g/km for G100, while the hydrogen additive increased NO_x emissions by 19.7%, raising specific emissions to 2.051 g/km for G95H5. At the end of the WLTC, the total NO_x emissions were 22.7 g for G100 and 27.6 g for G95H5, representing a 21.4% increase due to hydrogen. An analysis of individual cycle phases indicates that the hydrogen additive had the greatest impact on NO_x emissions during Phases II and III, which involved more accelerations and higher engine loads. Over the full WLTC, the specific NO_x emissions were 0.978 g/km for G100 and 1.187 g/km for G95H5. The Euro 4 emission standard allows a maximum NO_x emission of 0.08 g/km. It is expected that with the use of a catalytic converter, this standard will be met.

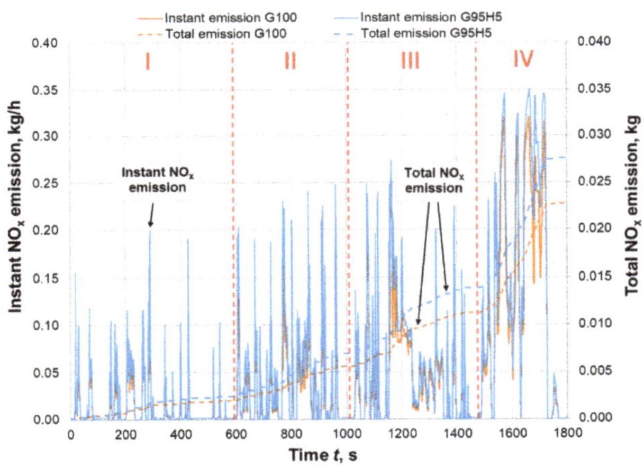

Figure 17. Instant and total NO_x emissions during the WLTC test: I—Urban cycle; II—Medium-speed cycle; III—High-speed cycle; IV—Very high-speed cycle.

The AVL Cruise™ software calculated CO_2 emissions over the WLTC by evaluating the elemental fuel composition and fuel consumption (Figure 18). The addition of 5% hydrogen by mass to gasoline reduced fuel consumption by approximately 10.4%, due to increased engine brake thermal efficiency and a 6.8% higher lower heating value of the fuel (Table 1). This hydrogen additive led to a 5% reduction in the elemental carbon content of the fuel (Table 1), which collectively contributed to a decrease in greenhouse gas CO_2 emissions. In the city driving mode (Phase I), total CO_2 emissions were 476 g for G100 and 399 g for G95H5 (16.1% less), with specific CO_2 emissions of 154 g/km for G100 and 129 g/km for G95H5, respectively. In the medium-speed mode (Phase II), specific CO_2 emissions decreased to 135 g/km for G100 and 115 g/km for G95H5 (15.3% less), and at the end of this phase, total CO_2 emissions were 1.119 kg for G100 and 0.994 kg for G95H5. In the high-speed mode (Phase III), fuel consumption per kilometre and specific CO_2 emissions were at their lowest—126 g/km for G100 and 107 g/km for G95H5 (14.9% less), with total CO_2 emissions at the end of this phase reaching 2.017 kg for G100 and 1.708 kg for G95H5. In the very high-speed mode (Phase IV), specific CO_2 emissions increased to 150 g/km for G100 and 128 g/km for G95H5 (14.4% less), and by the end of the WLTC, total CO_2 emissions were 3.254 kg for G100 and 2.768 kg for G95H5.

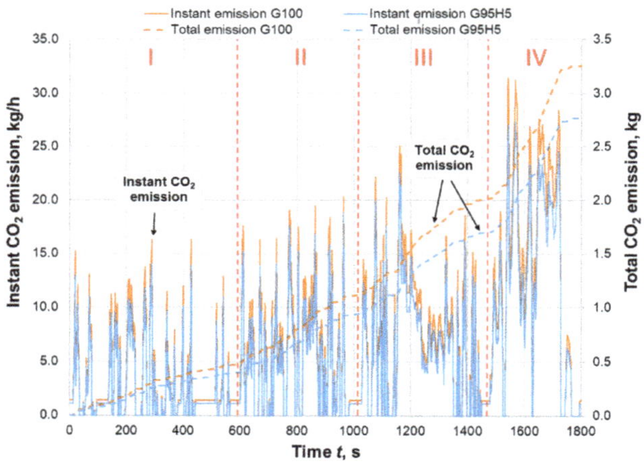

Figure 18. Instant and total CO_2 emissions during the WLTC test: I—Urban cycle; II—Medium-speed cycle; III—High-speed cycle; IV—Very high-speed cycle.

It was calculated that the specific emissions over the WLTC were 140 g/km for G100 and 119 g/km for G95H5 (14.9% lower). According to the European Parliament and Council regulation, each car manufacturer is required to achieve a CO_2 specific emission target of 95 g/km by 2021 [59,60]. The AVL Cruise™ software's results for the CO_2 emissions of the Nissan Qashqai J10 show that with pure gasoline, this target cannot be reached. The 5% hydrogen additive significantly reduced CO_2 emissions, but additional measures are required to meet the target. These could include replacing gasoline with renewable fuels that have a lower C/H ratio, as well as increasing the hydrogen concentration in the dual-fuel mixture. To determine the pollutant and CO_2 emission results using new fuels, further detailed experimental and numerical modelling studies will be needed.

4. Conclusions

The numerical WLTC simulation using the AVL Cruise™ software was conducted for a Nissan Qashqai J10 passenger car equipped with a 1.6 L spark-ignition engine operating on pure gasoline (G100) or dual fuel (D-F) consisting of 95% (by mass) gasoline and 5% hydrogen (G95H5). Real vehicle data, fuel properties, and engine experimental research data (fuel consumption, CO, HC, and NO_x emissions maps) were utilized for the WLTC simulation. Through numerical simulation, the indicators of fuel consumption and pollutant emissions were determined for the vehicle operating under four driving modes: urban cycle (I), medium speed (II), high speed (III), and very high speed (IV).

1. The addition of 5% hydrogen reduced overall fuel mass consumption by about 10% (primarily due to a higher LHV and a roughly 2.57% increase in thermal efficiency).
2. CO and HC emissions decreased significantly (up to 23.9% and 37.1%, respectively, depending on the driving mode), owing to faster combustion and better efficiency. However, NO_x emissions increased (by 19.7% in mode IV and by 25.9% in mode II) due to higher combustion temperatures and more intense heat release. Reduced carbon content (by 5%) and lower fuel consumption cut CO_2 emissions by 14.9% compared to pure gasoline. Despite this reduction, the WLTC simulation showed CO_2 levels remain above the EU's 95 g/km target.

3. Blending gasoline with biofuels (e.g., bioethanol, biobutanol) and/or increasing hydrogen concentration in the fuel are potential pathways to further emission reductions and compliance with stricter EU standards.

To meet EU emission standards, further research is planned, focusing on the use of gasoline and biofuel blends (such as bioethanol, biobutanol, etc.) and increasing the hydrogen concentration in dual fuels.

Author Contributions: Conceptualization, S.P. and A.R.; methodology, A.R. and G.M.; software, T.V. and A.U.; validation, T.V., A.U. and G.M.; formal analysis, S.S. and R.J.; investigation, A.R., T.V. and G.M.; resources, J.Ž., A.U., V.M. and V.Ž.; data curation, T.V., V.Ž. and A.N.; writing—original draft preparation, A.R. and T.V.; writing—review and editing, D.K. and S.S.; visualization, A.R., V.M. and T.V.; supervision, A.R.; project administration, S.P. All authors have read and agreed to the published version of the manuscript.

Funding: Research was conducted as part of the execution of Project "Mission-driven Implementation of Science and Innovation Programmes" (No. 02-002-P-0001), funded by the Economic Revitalization and Resilience Enhancement Plan "New Generation Lithuania".

Institutional Review Board Statement: Not applicable.

Informed Consent Statement: Not applicable.

Data Availability Statement: The data presented in this study are available on request from the corresponding author.

Acknowledgments: A part of the results included in this study have been obtained using internal engine simulation software AVL Cruise™ acquired by signing a Cooperation Agreement between AVL Advanced Simulation Technologies and the Faculty of Transport Engineering of Vilnius Gediminas Technical University.

Conflicts of Interest: The authors declare no conflicts of interest.

Abbreviations

A/F	air/fuel mass ratio
B_f	hourly fuel consumption
CA	crank angle
CAD	crank angle degrees
CH_4	methane
C/H	carbon/hydrogen ratio
CO	carbon monoxide
CO_2	carbon dioxide
D-F	dual fuel
ECU	engine control unit
EU 27	27 member states of the European Union
ICE	internal combustion engine
GDP	gross domestic product
GHGs	greenhouse gasses
G100	pure gasoline
G95H5	hydrogen additive, constituting 5% of the gasoline mass
HC	hydrocarbons
H_2	hydrogen
H_2O	water vapour
H_2100	pure hydrogen
SIE	spark-ignition engine
LHV	lower heating value

$MtCO_2e$	million metric tons of carbon dioxide equivalent
NO_x	nitrogen oxides
N_2O	nitrous oxide
PM	particulate matter
PMR	power-to-mass ratio
WLTC	Worldwide Harmonised Light Vehicles Test Cycle
λ	excess air ratio

References

1. Crippa, M.; Guizzardi, D.; Schaaf, E.; Monforti-Ferrario, F.; Quadrelli, R.; Risquez Martin, A.; Rossi, S.; Vignati, E.; Muntean, M.; Brandao De Melo, J.; et al. *GHG Emissions of All World Countries: 2023*; Publications Office of the European Union: Luxembourg, 2023; ISBN 978-92-68-07550-0.
2. Kazancoglu, Y.; Ozbiltekin-Pala, M.; Ozkan-Ozen, Y.D. Prediction and Evaluation of Greenhouse Gas Emissions for Sustainable Road Transport within Europe. *Sustain. Cities Soc.* **2021**, *70*, 102924. [CrossRef]
3. Ji, T.; Li, K.; Sun, Q.; Duan, Z. Urban Transport Emission Prediction Analysis through Machine Learning and Deep Learning Techniques. *Transp. Res. Part Transp. Environ.* **2024**, *135*, 104389. [CrossRef]
4. Jin, Y.; Sharifi, A.; Li, Z.; Chen, S.; Zeng, S.; Zhao, S. Carbon Emission Prediction Models: A Review. *Sci. Total Environ.* **2024**, *927*, 172319. [CrossRef]
5. Khalil, M.; AlSayed, A.; Elsayed, A.; Sherif Zaghloul, M.; Bell, K.Y.; Al-Omari, A.; Laqa Kakar, F.; Houweling, D.; Santoro, D.; Porro, J.; et al. Advances in GHG Emissions Modelling for WRRFs: From State-of-the-Art Methods to Full-Scale Applications. *Chem. Eng. J.* **2024**, *494*, 153053. [CrossRef]
6. Dolge, K.; Blumberga, D. Economic Growth in Contrast to GHG Emission Reduction Measures in Green Deal Context. *Ecol. Indic.* **2021**, *130*, 108153. [CrossRef]
7. European Climate Law—European Commission. Available online: https://climate.ec.europa.eu/eu-action/european-climate-law_en (accessed on 21 December 2024).
8. State of the Union: Commission Raises Climate Ambition. Available online: https://ec.europa.eu/commission/presscorner/detail/en/ip_20_1599 (accessed on 21 December 2024).
9. Abdul-Manan, A.F.N. Uncertainty and Differences in GHG Emissions between Electric and Conventional Gasoline Vehicles with Implications for Transport Policy Making. *Energy Policy* **2015**, *87*, 1–7. [CrossRef]
10. Statista EU-27: GHG Emissions Breakdown by Sector. Available online: https://www.statista.com/statistics/1325132/ghg-emissions-shares-sector-european-union-eu/ (accessed on 17 December 2024).
11. Foretich, A.; Zaimes, G.G.; Hawkins, T.R.; Newes, E. Challenges and Opportunities for Alternative Fuels in the Maritime Sector. *Marit. Transp. Res.* **2021**, *2*, 100033. [CrossRef]
12. IRENA Hydrogen A Renewable Energy Perspective. Available online: https://www.irena.org/publications/2019/Sep/Hydrogen-A-renewable-energy-perspective (accessed on 17 December 2024).
13. Gao, W.; Fu, Z.; Li, Y.; Li, Y.; Zou, J. Progress of Performance, Emission, and Technical Measures of Hydrogen Fuel Internal-Combustion Engines. *Energies* **2022**, *15*, 7401. [CrossRef]
14. Stępień, Z. A Comprehensive Overview of Hydrogen-Fueled Internal Combustion Engines: Achievements and Future Challenges. *Energies* **2021**, *14*, 6504. [CrossRef]
15. Jacquemin, J.; Thiran, P.; Quoilin, S. Prioritizing the Role of Renewable Fuels and Hydrogen Networks in the Transition towards Net Zero Emissions in Western Europe. *Energy* **2025**, *314*, 134069. [CrossRef]
16. Batgi, S.U.; Dincer, I. A Study on Comparative Environmental Impact Assessment of Thermochemical Cycles and Steam Methane Reforming Processes for Hydrogen Production Processes. *Comput. Chem. Eng.* **2024**, *180*, 108514. [CrossRef]
17. Aydin, M.I.; Dincer, I. Comparative Modeling and Assessment of Renewable Hydrogen Production and Utilization in Remote Communities. *Comput. Chem. Eng.* **2025**, *194*, 108995. [CrossRef]
18. Karczewski, M.; Chojnowski, J.; Szamrej, G. A Review of Low-CO2 Emission Fuels for a Dual-Fuel RCCI Engine. *Energies* **2021**, *14*, 5067. [CrossRef]
19. Szamrej, G.; Karczewski, M. Exploring Hydrogen-Enriched Fuels and the Promise of HCNG in Industrial Dual-Fuel Engines. *Energies* **2024**, *17*, 1525. [CrossRef]
20. Falfari, S.; Cazzoli, G.; Mariani, V.; Bianchi, G.M. Hydrogen Application as a Fuel in Internal Combustion Engines. *Energies* **2023**, *16*, 2545. [CrossRef]
21. Beccari, S.; Pipitone, E.; Caltabellotta, S. Analysis of the Combustion Process in a Hydrogen-Fueled CFR Engine. *Energies* **2023**, *16*, 2351. [CrossRef]

22. Kumaraswamy, A.; Garud, S.S.; Karimi, I.A.; Farooq, S. Hydrogen for Net-Zero Emissions in ASEAN by 2050. *Int. J. Hydrogen Energy* **2024**, *90*, 575–587. [CrossRef]
23. Aghahasani, M.; Gharehghani, A.; Mahmoudzadeh Andwari, A.; Mikulski, M.; Pesyridis, A.; Megaritis, T.; Könnö, J. Numerical Study on Hydrogen–Gasoline Dual-Fuel Spark Ignition Engine. *Processes* **2022**, *10*, 2249. [CrossRef]
24. Nunes, F.C.; Vara Prasad, M.N.; Alves, L.D.J. Contribution of Hydrogen Economy toward Net Zero Emission and Decarbonization. In *Accelerating the Transition to a Hydrogen Economy*; Elsevier: Amsterdam, The Netherlands, 2025; pp. 59–76, ISBN 978-0-443-14039-6.
25. Molina, S.; Ruiz, S.; Gomez-Soriano, J.; Olcina-Girona, M. Impact of Hydrogen Substitution for Stable Lean Operation on Spark Ignition Engines Fueled by Compressed Natural Gas. *Results Eng.* **2023**, *17*, 100799. [CrossRef]
26. Sterlepper, S.; Fischer, M.; Claßen, J.; Huth, V.; Pischinger, S. Concepts for Hydrogen Internal Combustion Engines and Their Implications on the Exhaust Gas Aftertreatment System. *Energies* **2021**, *14*, 8166. [CrossRef]
27. Le, D.K.; Kwon, H.; Lee, M.J. Numerical and Experimental Investigation of the Correlation between Inlet Flow Rates, Temperature, and NOx Emissions in Pure Hydrogen Combustion. *Appl. Therm. Eng.* **2025**, *259*, 124748. [CrossRef]
28. Caricato, A.; Carlucci, A.P.; Potenza, M.E.C.; Laforgia, D.; Torresi, M.; Strafella, L. Autoignition Characterization of Hydrogen Directly Injected into a Constant-Volume Combustion Chamber through a Heavy-Duty Injector. *Energies* **2023**, *16*, 6823. [CrossRef]
29. Giacomazzi, E.; Troiani, G.; Di Nardo, A.; Calchetti, G.; Cecere, D.; Messina, G.; Carpenella, S. Hydrogen Combustion: Features and Barriers to Its Exploitation in the Energy Transition. *Energies* **2023**, *16*, 7174. [CrossRef]
30. Al-Janabi, H.A.-K.S.; Al-Baghdadi, M.A.-R.S. A Prediction Study of the Effect of Hydrogen Blending on the Performance and Pollutants Emission of a Four Stroke Spark Ignition Engine. *Int. J. Hydrogen Energy* **1999**, *24*, 363–375. [CrossRef]
31. Akif Ceviz, M.; Sen, A.K.; Küleri, A.K.; Volkan Öner, İ. Engine Performance, Exhaust Emissions, and Cyclic Variations in a Lean-Burn SI Engine Fueled by Gasoline–Hydrogen Blends. *Appl. Therm. Eng.* **2012**, *36*, 314–324. [CrossRef]
32. Ji, C.; Cong, X.; Wang, S.; Shi, L.; Su, T.; Wang, D. Performance of a Hydrogen-Blended Gasoline Direct Injection Engine under Various Second Gasoline Direct Injection Timings. *Energy Convers. Manag.* **2018**, *171*, 1704–1711. [CrossRef]
33. Meng, H.; Ji, C.; Yang, J.; Wang, S.; Chang, K.; Xin, G. Experimental Study of the Effects of Excess Air Ratio on Combustion and Emission Characteristics of the Hydrogen-Fueled Rotary Engine. *Int. J. Hydrogen Energy* **2021**, *46*, 32261–32272. [CrossRef]
34. Qiang, Y.; Zhao, S.; Yang, J.; Cai, J.; Su, F.; Wang, S.; Ji, C. Effect of Excess Air Ratio and Spark Timing on the Combustion and Emission Characteristics of Turbulent Jet Ignition Direct Injection Hydrogen Engine. *Int. J. Hydrogen Energy* **2024**, *93*, 1166–1178. [CrossRef]
35. Lee, J.; Park, C.; Bae, J.; Kim, Y.; Choi, Y.; Lim, B. Effect of Different Excess Air Ratio Values and Spark Advance Timing on Combustion and Emission Characteristics of Hydrogen-Fueled Spark Ignition Engine. *Int. J. Hydrogen Energy* **2019**, *44*, 25021–25030. [CrossRef]
36. Liang, Z.; Xie, F.; Jiang, B.; Li, X.; Su, Y.; Wang, Z. Evaluating the Potential of Mixture Formation Methods to Achieve Efficient Combustion and Near-Zero Emissions on a Hydrogen Direct Injection Engine. *J. Clean. Prod.* **2024**, *439*, 140930. [CrossRef]
37. Di Profio, P.; Canale, V.; Germani, R.; Arca, S.; Fontana, A. Reverse Micelles Enhance the Formation of Clathrate Hydrates of Hydrogen. *J. Colloid Interface Sci.* **2018**, *516*, 224–231. [CrossRef] [PubMed]
38. Durkin, K.; Khanafer, A.; Liseau, P.; Stjernström-Eriksson, A.; Svahn, A.; Tobiasson, L.; Andrade, T.S.; Ehnberg, J. Hydrogen-Powered Vehicles: Comparing the Powertrain Efficiency and Sustainability of Fuel Cell versus Internal Combustion Engine Cars. *Energies* **2024**, *17*, 1085. [CrossRef]
39. Gültekin, N.; Ciniviz, M. Experimental Investigation of the Effect of Hydrogen Ratio on Engine Performance and Emissions in a Compression Ignition Single Cylinder Engine with Electronically Controlled Hydrogen-Diesel Dual Fuel System. *Int. J. Hydrogen Energy* **2023**, *48*, 25984–25999. [CrossRef]
40. Lin, Y.-C.; Lin, S.-L.; Jhang, S.-R.; Chen, K.-S.; Su, C.-W.; Huang, C.-E. Energy Saving and Pollutant Emission Reduction by Adding Hydrogen in a Gasoline-Fueled Engine. *Aerosol Air Qual. Res.* **2022**, *22*, 220259. [CrossRef]
41. Gabana, P.; Giménez, B.; Martín Herreros, J.; Tsolakis, A. Analysis of the Combustion Speed in a Spark Ignition Engine Fuelled with Hydrogen and Gasoline Blends at Different Air Fuel Ratios. *Fuel* **2025**, *381*, 133563. [CrossRef]
42. Padmanabhan, S.; Punitha, N.; Poyyamozhi, N.; Senthil, S.; Damodharan, D.; Sakunthala, K. Sustainability Improvement of Spark-Ignition Engine Performance Enhanced with Nanoparticles and Hydroxy Gas. *Int. J. Hydrogen Energy* **2024**, *96*, 1171–1185. [CrossRef]
43. Zhang, R.; Shu, G.; Zhao, H.; Chen, L.; Wei, H.; Pan, J. A Comparative Study on NH3/H2 and NH3/CH3OH Combustion and Emission in an Optical SI Engine. *Fuel* **2024**, *369*, 131731. [CrossRef]
44. Fontaras, G.; Ciuffo, B.; Zacharof, N.; Tsiakmakis, S.; Marotta, A.; Pavlovic, J.; Anagnostopoulos, K. The Difference between Reported and Real-World CO_2 Emissions: How Much Improvement Can Be Expected by WLTP Introduction? *Transp. Res. Procedia* **2017**, *25*, 3933–3943. [CrossRef]
45. Anticaglia, A.; Balduzzi, F.; Ferrara, G.; De Luca, M.; Carpentiero, D.; Fabbri, A.; Fazzini, L. Feasibility Analysis of a Direct Injection H2 Internal Combustion Engine: Numerical Assessment and Proof-of-Concept. *Int. J. Hydrogen Energy* **2023**, *48*, 32553–32571. [CrossRef]

46. Li, Y.; Gong, J.; Deng, Y.; Yuan, W.; Fu, J.; Zhang, B. Experimental Comparative Study on Combustion, Performance and Emissions Characteristics of Methanol, Ethanol and Butanol in a Spark Ignition Engine. *Appl. Therm. Eng.* **2017**, *115*, 53–63. [CrossRef]
47. Liu, X.; Wang, H.; Zheng, Z.; Liu, J.; Reitz, R.D.; Yao, M. Development of a Combined Reduced Primary Reference Fuel-Alcohols (Methanol/Ethanol/Propanols/Butanols/n-Pentanol) Mechanism for Engine Applications. *Energy* **2016**, *114*, 542–558. [CrossRef]
48. Rimkus, A.; Matijošius, J.; Bogdevičius, M.; Bereczky, Á.; Török, Á. An Investigation of the Efficiency of Using O_2 and H_2 (Hydrooxile Gas -HHO) Gas Additives in a Ci Engine Operating on Diesel Fuel and Biodiesel. *Energy* **2018**, *152*, 640–651. [CrossRef]
49. Rimkus, A.; Mejeras, G.; Dittrich, A.; Pukalskas, S.; Barta, D. Effect of the Concentration of Bioethanol Mixed with Gasoline on the Energy and Environmental Performance of a Hybrid Vehicle in the Worldwide Harmonized Light Vehicles Test Cycle (WLTC). *Appl. Sci.* **2024**, *14*, 10858. [CrossRef]
50. Duan, X.; Feng, L.; Chu, X.; Sun, Z. The Performance of a Spark Ignition Gasoline Engine with Hydrogen Addition under Low-Load Conditions. *Fuel* **2025**, *379*, 133091. [CrossRef]
51. Purayil, S.T.P.; Hamdan, M.O.; Al-Omari, S.; Selim, M.; Elnajjar, E. Influence of Ethanol–Gasoline–Hydrogen and Methanol–Gasoline–Hydrogen Blends on the Performance and Hydrogen Knock Limit of a Lean-Burn Spark Ignition Engine. *Fuel* **2024**, *377*, 132825. [CrossRef]
52. Manzoor, M.U.; Dou, X.; Yosri, M.; Talei, M.; Yang, Y. Capturing Different Modes of Hydrogen Combustion in a Spark-Ignition Engine Using Numerical Simulations. *Fuel* **2024**, *375*, 132343. [CrossRef]
53. Duan, X.; Chu, X.; Wang, R.; Chen, Z.; Zhou, F.; Abdellatief, T.M.M. The Performance and Emissions Characteristics of the Gasoline Spark Ignition Engine Fuelled with Green and Renewable Methanol and Hydrogen. *Renew. Energy* **2025**, *240*, 122184. [CrossRef]
54. Purayil, S.T.P.; Hamdan, M.O.; Al-Omari, S.A.B.; Selim, M.Y.E.; Elnajjar, E. Impact of Intake Air Pressure and Temperature on Performance and Hydrogen Knock Limit in a Spark-Ignition Engine Using Hydrogen-Gasoline Blends. *Appl. Therm. Eng.* **2025**, *263*, 125331. [CrossRef]
55. Marchitto, L.; Costagliola, M.A.; Berra, A. Influence of Performance Packages on Fuel Consumption and Exhaust Emissions of Passenger Cars and Commercial Vehicles under WLTP. *Energies* **2024**, *17*, 3356. [CrossRef]
56. Purayil, S.T.P.; Martini, E.A.; Elsaid, A.; Khalil, M.; Zoghbour, T.; Seyam, M.; Elnajjar, E. Influence of Steam Induction on the Performance and Hydrogen Knock Limit of a Hydrogen-Gasoline Spark Ignition Engine. *Int. J. Thermofluids* **2024**, *24*, 100933. [CrossRef]
57. Karagöz, Y. Analysis of the Impact of Gasoline, Biogas and Biogas + Hydrogen Fuels on Emissions and Vehicle Performance in the WLTC and NEDC. *Int. J. Hydrogen Energy* **2019**, *44*, 31621–31632. [CrossRef]
58. Barbu, M.C.; Birtaş, A.; Chiriac, R. On the Improvement of Performance and Pollutant Emissions of a Spark Ignition Engine Fuelled by Compressed Natural Gas and Hydrogen. *Energy Rep.* **2022**, *8*, 978–991. [CrossRef]
59. DIESELNET Emission Standards: Europe: Cars GHG. Available online: https://dieselnet.com/standards/eu/ghg.php (accessed on 17 December 2024).
60. European Parliament and of the Council. *Regulation (EU) 2019/631*; European Parliament and of the Council: Brussels, Belgium, 2019; Volume 111.

applied sciences

MDPI

Article

Demand Management in Hybrid Locomotives Through Aggregated Models of Supercapacitors and Railway Units

Antonio Gabaldón [1,*], María Carmen Ruiz-Abellón [2], Francisco Martínez [2] and Antonio Guillamón [2]

1 Power Systems Group, Technical University of Cartagena, 30202 Cartagena, Spain
2 Department of Applied Mathematics and Statistics, Technical University of Cartagena, 30202 Cartagena, Spain; maricarmen.ruiz@upct.es (M.C.R.-A.); f.martinez@upct.es (F.M.); antonio.guillamon@upct.es (A.G.)
* Correspondence: antonio.gabaldon@upct.es; Tel.: +34-968-338944

Abstract: Most European Union governments and numerous railway operators have announced plans to replace most of their diesel units by 2030–2040. However, a significant portion of the rail network remains non-electrified. In some cases, the proposed solution has been to close certain tracks, but this approach entails considerable societal costs for small cities and represents a loss of prior railway investments. Consequently, hybrid locomotives and multiple units (either new or refurbished) emerge as a viable solution during this transitional period to enhance energy efficiency and preserve services on these lines, particularly for freight operations. These hybrid units can operate on both electrified and non-electrified tracks and can also serve as "railway prosumers", contributing to both storage and generation in fully or partially electrified areas. However, implementing these "prosumer tasks" faces challenges, such as the rapid power demand fluctuations during acceleration and the loss of energy recovery potential during braking in hybrid or fully electric units. These losses may also impact the overall power system. This paper presents an alternative approach to modeling double-layer capacitors (supercapacitors) combined with electrical equivalent models for lithium-ion batteries. The Differential Transformation Method (DTM) is used to solve the non-linear ordinary differential equations governing the supercapacitor model, while parameter optimization is achieved through a grid search approach, demonstrating high accuracy compared with laboratory trials. This framework highlights the potential of hybrid units, as illustrated through simulations that analyze storage sizing, energy management, increased energy recovery, and changes in unit performance. These models facilitate a pre-feasibility evaluation of energy storage systems for hybrid railway applications.

Keywords: demand response; physical-based modelling; energy storage systems; energy efficiency; hybrid railway units; non-linear ordinary differential equations; differential transformation method

1. Introduction

Decarbonization is a key political objective for the coming decades. The transportation sector, accounting for a significant share of greenhouse gas (GHG) emissions (approximately 25% in the European Union during this decade), has become a primary target for electrification efforts. Within the railway sector, achieving electrification necessitates the complete electrification of railway infrastructure. However, this process is far from complete and can be described as progressing rather slowly. Table 1, built from data available in [1,2], illustrates the percentage of non-electrified railway lines in the EU-27

https://doi.org/10.3390/app15052412

overall, as well as in selected countries under varying scenarios, highlighting the changes in electrification rates from 2013 to 2022. While it is true that approximately 55% of the rail network in Europe is electrified, and this figure continues to increase, it is not advancing as rapidly as anticipated. In contrast, electrified tracks in the United States and Canada represent only around 1% of their respective networks. This can be justified by reports [3] that estimated costs of electrification are about twice as expensive in the USA compared with the European Union.

Table 1. Percentage of non-electrified lines and percentage of railway in freight traffic (Europe, a perspective from 2013 to 2022).

Country	Percentage of Non-Electrified Lines		Percentage [2] of Railway in Freight Traffic	
	2013	2022	2013	2022
Austria	30.32	27.98	32.1	30.2
Denmark	76.25	67.24	10.9	9.6
France	48.09	39.74	10.8	10.6
Germany	47.37	45.16	19.3	19.8
Poland	38.59	37.28	27.6	23.2
Slovakia	56.32	56.31	36.5	30.4
Spain	48.09	36.16	5.3	4.4
Sweden	25.03	25.00	33.7	28.7
UE 27 [1]	46.19	43.19	21.0	16.0

[1] European Union without UK (2013). [2] Only land transport methods.

The primary factors contributing to this situation are the high cost and technical challenges associated with full electrification. For example, electrifying infrastructure in tunnels or above road bridges poses significant technical and economic difficulties, making it financially impractical, even when considering only operational costs (excluding capital expenses), particularly for low-traffic railway lines.

Moreover, according to European research objectives, freight transport "still no shift from road to rail" [4], but especially in Spain (Table 1). According to statistics [5], Spain lags behind most European countries in terms of freight transport by rail, performing well below the average. Additionally, the trends are negative, as can be observed in Table 1.

Following the deregulation of the railway sector a decade ago, private operators now account for more than 45% of the market share (and over 50% in terms of ton–kilometers) [6]. This shift has led to a surplus of freight locomotives, particularly electric ones, within the fleet of Spain's traditional national railway operator, RENFE. Many of these surplus locomotives are either leased to private operators or decommissioned well before reaching their projected design lifespan (e.g., 4–5 Mkm).

Overview of Energy Storage in Railway Power Systems (RPS)

The technical feasibility of On-Board Energy Storage Devices (OESD) for trains is not a new concept and has been demonstrated by numerous passenger units and some shunting locomotives, which operated successfully for extended periods (e.g., in Germany from 1955 to 1995). However, these early implementations faced significant limitations, including restricted operating range, challenging track profiles, and the substantial weight of storage devices. While advances in storage technologies (such as reduced battery charging times and extended lifespans) have addressed some of these issues, a major challenge persists: the demanding operational conditions of railways, which often differ significantly from the standard applications of energy storage devices, as noted in the specialized literature [7].

Fuel cells have emerged as a popular alternative to electrification, though they still present notable disadvantages, such as lower efficiency compared with batteries and the need for dedicated refueling infrastructure. Nevertheless, fuel cell technology offers distinct advantages, enabling the operation of intercity and regional services (with ranges of approximately 800 km) at low-traffic service routes and with shorter refueling times.

The functions of OESD have also been analyzed in [8] and can be grouped into five main categories. It is worth noting that these functions closely resemble the conventional operation of Distributed Energy Resources (DER) in Public Power Systems (PPS):

- Power assist (peak clipping): OESD can supply power during acceleration phases, partially offsetting the demand on the primary energy source. This functionality enables a reduction in the rated power needed for hybrid units.
- Regenerative braking: Energy recovered during braking can be used to power auxiliary loads or other units. However, it is often challenging (or even impossible) to recover all this energy in real time, making energy storage essential. A similar situation occurs with prosumers in renewable energy systems when excess generation exceeds demand, and injecting energy back into the grid is either technically unfeasible or economically inefficient.
- Operation on non-electrified tracks: OESD facilitates operations on these segments, such as "last-mile" applications near cities or industrial tracks. This reduces the need for shunting locomotives, lowering costs and minimizing environmental impact.
- Assistance in emergencies: OESD can provide power in the event of catenary (Overhead Line Equipment, OLE) failures caused by meteorological factors or mechanical faults, such as issues with the pantograph–catenary contact. This functionality is analogous to a "black start" service in PPS. Unfortunately, some railway operators have overlooked the potential of this capability.
- Load leveling service: By balancing energy demands, OESD can enhance the efficiency of the railway power system (RPS). This approach is closely aligned with PPS strategies for optimizing energy generation and reducing peak energy prices.

OESD, and specifically supercapacitors (Electric Double-Layer Capacitors, EDLC), offer significant advantages in the operation of locomotives and Diesel Multiple Units (DMUs), particularly when hybridized with batteries. These advantages include:

- Improved fuel efficiency during idle operations: EDLCs enable multiple restarts of diesel units without consuming fuel during idle periods. This is particularly beneficial for commuter locomotives and DMU services. For instance, in low-traffic lines, passenger locomotives and DMUs often spend 40–60 min idling on platforms while waiting to begin a new service, as observed in the selected scenario.
- Support for peak current demands: EDLCs can deliver fast peak currents during the initial acceleration phase, which is a significant challenge for batteries and often necessitates oversizing them. This advantage arises from the considerably lower internal impedance of EDLCs compared with batteries (see Section 2).
- Environmental and operational benefits: EDLCs have a lower environmental impact due to easier recycling processes and greater adaptability to varying weather conditions. Additionally, their simplified engineering models reduce design complexity. When used in parallel with batteries, EDLCs enhance load balancing due to their rapid response times [9].

The application of OESD requires detailed models, such as models of railway units, but also a more detailed model for EDLC and other devices to take advantage of its advantages.

The main contributions of this paper can be summarized as follows:

- It presents a non-linear model aimed at enhancing the understanding of EDLC dynamics to balance energy requirements using Physical-Based methodologies (PBLM), in this case, an electrical circuit model. In terms of accuracy and parameter evaluation, this approach surpasses other practical models, particularly due to its simplicity in evaluation and its straightforward physical interpretation.
- The paper also proposes an alternative for repurposing underutilized diesel-electric and electric locomotives on low-traffic lines, leveraging the fact that energy storage and partial electrification are both cost-effective and technically feasible in certain scenarios. This hybridization is particularly applicable to freight services and serves as a viable alternative during the necessary transition period to meet decarbonization targets for 2030–2050. To support this, a specific model for railway units has been calibrated for operations on partially electrified tracks. Furthermore, this approach sustains railway freight operations while avoiding the substantial investments required for full electrification, a significant challenge for several countries like Spain.
- Additionally, the paper illustrates how freight traffic, which typically operates at low speeds and experiences long standby periods, can function as a distributed energy resource. This enables the optimization of energy management (e.g., energy recovery and peak demand management) in electrified routes, making the system more efficient overall.

The rest of the paper is organized as follows. Section 2 depicts the characteristics of the selected track and potential locomotives for hybridization, as well as the literature review of energy storage models for supercapacitors and batteries. Additionally, it develops the proposed model for EDLC and the different laboratory tests. Section 3 deals with the calibration of the model parameters and the analysis of the experimental and simulation results. Finally, the main findings and recommendations of the study are presented in Section 4.

2. Materials and Methods

This section consists of 4 subsections. In Section 2.1, the railway track used in the simulations is depicted. In Section 2.2, we describe some of the energy storage vehicles and how they operate on partially electrified tracks, including a description of the train simulator (PBLM-Rail) used in this work. Section 2.3 deals with the most common energy storage models for EDLC and batteries, and finally, Section 2.4 provides the laboratory trials and the new model for the EDLC proposed in this paper.

2.1. Track Characteristics

The railway track being used for simulation purposes suits well the objectives of this paper for several reasons. First, it has a large track area without electrification (due to its low freight traffic, around 4 trains/day [6], but also has regional and intercity passenger traffic). Second, the elevation of the itinerary presents interesting requirements for tractive power and possibilities for regenerative braking. Also, the track is partially electrified in a zone where there is a common platform for conventional and high-speed trains. Finally, there is a dual-gauge track (European standard, 1435 mm, and Iberian gauge, 1668 mm) in the electrified zone (around 8 km) at the junction of the Alicante route. This track is electrified at 25 kV AC for the passenger high-speed traffic (gauge 1435 and 1668 mm), but it also is electrified in the Iberian gauge track (that usually works at 3 kV DC but is not energized yet). Almost all possibilities in Spanish railways can theoretically play in this route.

The track database consists of six main data points: the elevation profile, the curve characteristics, the maximum speed, the zones with double/triplex track, the gauges of the track, and the electrification of each zone. Data have been partially obtained from the Infrastructure Operator (ADIF) and from on-site measurements through video and inspection near the infrastructure. Figure 1 shows the main characteristics of this track section.

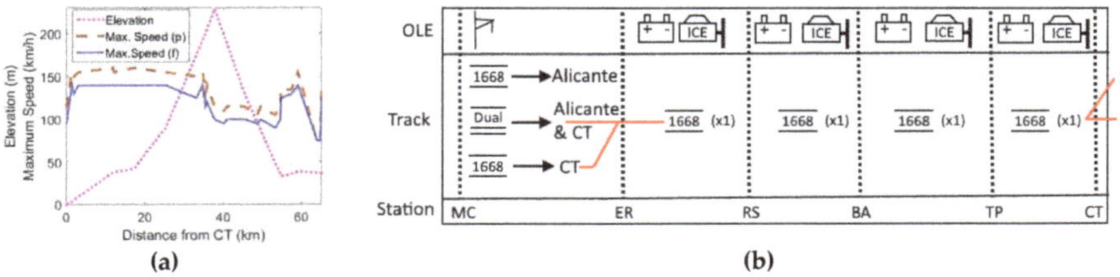

Figure 1. Characteristics of the track: (**a**) Elevation profile and maximum speed on the Cartagena–Murcia track; (**b**) Electrified and track gauge and zones (single-double-triplex). Acronyms for main train stops: MC (Murcia del Carmen); ER (El Regüeron); RS (Riquelme–Sucina); BA (Balsicas); TP (Torre–Pacheco); and CT (Cartagena, Passenger PS, and freight FS station). The main line (MC–ER) has two branches (CT and Alicante; see figure arrows and links in red).

2.2. Vehicles

2.2.1. Energy Storage Vehicles

As stated before, energy storage locomotives and multiple units are not a new concept. First attempts are far from being considered "prototypes" due to the number of units, lifetime, and services that they developed during their careers. These precedents are interesting from an engineering point of view. In the 20th century, some manufacturers started to apply energy storage in railways. A good example is German Railways and their battery multiple units (BEMU) called "ETA" (i.e., Elektrischer Triebwagen mit Akkumulatoren, in German [10]) being developed by Prussian and German railways. The series are 177, 178, 179, or 180 (manufactured from 1907 to 1937) and, later, series ETA 150.0 and 150.5 (around 100 units, further renamed as series 515, built from 1953 to 1957). The target for their services reached 100–150 km with the same battery for older units, and it was extended to 300–350 km with ETA 150.5 (storage capacity of batteries around 520 kWh). Some of those units remained operative until the early 90s. Another interesting example of freight units is Fireless Steam Locomotives. Their duties usually were switching on industrial (chemical) plants. In this case, these locomotives avoid the danger of fire and explosion of conventional steam (notice that early designs started 100 years ago). These locomotives stored steam in a large boiler, and this steam came from steam power plants needed in this industry sector. The last class of those units was built by Henschel in the early 50s. Some units remained in service until 1995 and can be considered, to some extent, as a more successful ancestor of compressed air storage systems (CAES) in PPS. Figure 2 shows these interesting and representative units.

Figure 2. Some examples of first energy storage railway units (recreation). Fireless locomotive Henschel D-fl/25/3,5 (1); ETA 177 (2), ETA 515 (3).

2.2.2. Description of Vehicles Used on the Route

At present, railway operators of this track use diesel-electric locomotives (for freight and passenger services) and diesel-hydraulic DMUs for regional services. The power of these units is around 500 kW for DMUs and 2000–2300 kW for diesel-electric locomotives. Freight locomotives are variants of the Electro-Motive Division (EMD) of General Motors [11] international locomotive platforms, but passenger units too. Usually, the units used for freight are double traction for gas trains. Figure 3 shows some pictures of these units. Note that all of them are refurbished units manufactured in the 80s and rebuilt from 2002 to 2010. These units remain with the original and performant diesel engine and AC generators because they are far from their useful life. Note that the average mileage for these units was around 60.000 km/year [6].

(**a**) (**b**)

Figure 3. Some examples of locomotives on the Murcia and Cartagena route: (**a**) Freight locomotive series RENFE 333.3 (multiple units with tank wagons, used as an example for simulations); (**b**) Freight locomotive series 319 (Continental Rail, former RENFE 319.2).

2.2.3. Vehicle Operation in Non-Electrified and Partially Electrified Tracks

Operation scenarios of railway units are services in non-electrified tracks and in partially electrified tracks. Partial electrification can provide interesting advantages because it reduces costs in infrastructure but increases the costs of railway units. Some reports consider that batteries cannot be applied as a solution for stand-alone (full battery locomotives BEL) heavy-duty freight due to space constraints [12], and other reports that they are only effective for relatively short-range segments [3]. Nevertheless, catenary systems and BEL had the largest annual fuel savings in the USA (2022, [3]), and this supports the interest in hybridized units and electrification. Other literature reports deal with these problems by charging batteries through a specific range of wayside options selected to meet energy requirements [13] or through stationary pantographs, but, especially for non-electrified tracks, the number of studies is relatively low. One of the first proposals of hybrid units to reduce the impact of DMU can be found in [14]. In this work, the authors propose short sections of overhead lines in stations to achieve better acceleration and reduce the size of batteries similar to electrical buses and tramways at some stops. The authors in [15] deal with a Battery Hybrid Electric Multiple Unit (BHEMU) with overhead equipment and a battery. They find an optimal operation of a 2 MW BEMU through novel drivetrain and battery models, considering ancillary (hotel) loads. The main problem is that the battery model is too simple, and in this way, the main problem becomes an optimization one. For the considered itinerary (152 km), they found that the optimization involves around 530 kWh of storage in batteries and three electrified zones. Authors in [16] present the advantages of hybridization by the implementation of a regenerative braking strategy, increasing the efficiency of a DMU by 18%. Reference [17] presents a comparison with two partial electrification options: an optimal strategy and a "common-sense" approach (i.e., high-power sections are electrified first). These strategies condition the size of energy storage devices (only batteries in this work). The work is of interest from the point of view of defining the size of batteries according to itineraries and the degree of electrification

(20–80%); the model of the vehicle is complex, but the model of the battery is very simple again (a voltage source with a resistor). Finally, ref. [18] simulates heavy haul operation by a hybrid, multiple units that integrate two diesel-electric "A units" (3100 kW) and the interesting option of a "B unit" storage locomotive (equipped with a flywheel or Li-ion batteries of 5 MWh capacity and 50 t weight). The problem is that the information provided about storage models is scarce.

2.2.4. Train Simulator (PBLM-Rail)

The simulation is based on several sub-models that exchange information during the simulation. The model for vehicles is physical-based, and it has been developed in detail in previous papers [19]. It applies a scalar equation to calculate the acceleration of the train (DMU, EMU, DHEMU, or locomotives with wagons or coaches). The equation expresses the equilibrium between external and internal forces:

$$\left(m + \frac{J}{R^2}\right)a = \sum_{k=1}^{n} (F_k) \tag{1}$$

$$m(1 + ERM)a = \sum_{k=1}^{n} (F_{tk}) - \sum_{k=1}^{n} (F_{rk}) - F_c - F_g - F_s - F_b \tag{2}$$

where

n: overall number of effects being considered.

m: mass of the train

a: acceleration of the train

J: moment of inertia of the different rotating masses (wheels, motor rotors...) in the transmission, which causes an apparent increase of mass.

R: wheel radius.

F_{tk}: propulsive force, tractive effort of the k locomotive, DMU, EMU, ...

F_{rk}: resistance forces due to the element r of the train, locomotive, coach, generator...

u: number of units in the train

F_c: curve resistance

F_g: resistance due to the slope of the track (positive or negative)

F_s: starting resistance

F_b: resistance due to braking, in our case dynamic braking [20] and pneumatic braking [21].

In this model, ERM is the Equivalent Rotational Mass, which ranges from 0 to 0.30, depending on the type of vehicle. Other values can be found in [22,23].

Resistance forces are the sum of all forces acting on the railway unit at a given time or place. These resistance forces can be examined in detail in the literature by different models [20,23], but they are not the main objective of this paper. To approximate rail vehicle resistance in PBLM-Rail, this paper applies the equation:

$$F_r(v) = A + Bv + Cv^2 \tag{3}$$

where v is the speed of the vehicle (m/s), and A (N), B (Ns/m), and C (Ns^2/m^2) are regression coefficients obtained by fitting the run-down test of passenger and freight units to the previous equation.

This paper uses different data from manufacturers such as Bombardier, Alstom, Patentes Talgo, Stadler Rail, Siemens, Electro-Motive Division (EMD), Mitsubishi, and railway operators (SNCF, DB, or SBB).

With respect to grade resistance with slope β, the force is

$$F_g = mgsin(\beta) \tag{4}$$

With respect to curve resistance, literature approaches use different empirical equations for different curve radii. For example, in reference [24]:

$$F_c = 0.001mg * 198/r_c \tag{5}$$

where r_c is the curve radius.

With respect to starting force F_s, there are few studies on this topic. An accurate approximation of the main interest for haulage (and specifically for freight) is the approach proposed in reference [25]:

$$F_s(v) = \begin{cases} 3F_r(10);\ v = 0\ \text{km/h} \\ 2F_r(10)(1-v) + F_r(1);\ v \in (0,\ 1]\ \text{km/h} \\ F_r(10);\ v \in (1,\ 10]\ \text{km/h} \end{cases} \tag{6}$$

As seen in Equation (6), the approach is that the value of F_S is related to F_r in Equation (3).

2.3. Energy Storage Models (PBLM-EDLC and PBLM-Battery)

Developing accurate models for both trains and OESD is essential for assessing the pre-feasibility of hybrid locomotives in each scenario and set of duties. At this stage, defining the size of each component is crucial, as it directly impacts volume constraints, the overall weight of the railway units, and axle load limits. These models would also enable the optimization of the design of the hybrid system, but optimization of OESD is not the objective of the present work.

2.3.1. Distributed Energy Resources Laboratory Facilities

The OESD analyzed in this work includes Li-ion batteries ($LiFePO_4$) and EDLC. In the first case, Pylontech and Victron [26] (high discharge current) batteries have been tested. EDLC devices (five BMOD0500 P016 units) are from Maxwell Capacitors [27]. The main characteristics of EDLC and batteries are shown in Tables 2 and 3.

Table 2. Characteristics of EDLC.

Characteristic	Value
Rated/Max. capacitance	500/600 F
ESR (equivalent series resistance)	2.1 mΩ
Rated/Maximum Voltage	16/17 V
Maximum/continuous current [1]	1900/100 A
Maximum stored energy [2]	18 Wh
Specific energy (specific power)	3.2 Wh/kg (2700 kW/kg)

[1] Short-circuit current 7600 A; [2] at rated voltage, i.e., 16 V.

Table 3. Characteristics of Li-ion batteries.

Characteristic	Victron Super Pack	Pylontech US2000C
	Value	Value
Chemistry	$LiFePO_4$	$LiFePO_4$
Nominal voltage	12.8 V	48 V
Voltage limits	10–14.2 V	46–54 V
Nominal capacity	100 Ah (1.3 kWh)	500 Ah (2.4 kWh)
Max. continuous current	100 A	25 [2]–50 A

Table 3. *Cont.*

Characteristic	Victron Super Pack	Pylontech US2000C
	Value	Value
Peak discharge current [1]	150 A	90 A
Max. charge current	100 A	25–50 [2] A
Specific energy	91.2 kWh/t	100 kWh/t

[1] 10–15 s; [2] Recommended.

To determine and tune parameters of OESD models, a broad intrusive system (both for monitoring and control) has been designed and installed in the storage devices and their environment. The demonstrator system in the laboratory (also used for load modeling [28]) measures electrical magnitudes but also some thermal and weather variables with relevance to consider device limits and overloads. These are especially important to test the performance of batteries.

To obtain all these variables, multi-sensor devices (temperature and humidity) and thermocouples (through acquisition data cards) on the surface of OESD were installed. In addition, two specific non-commercial devices integrate protections and several switches to manage the operation of contactors and relays to accomplish charging and discharge tests and to perform measurement conditioning. They are also used to equilibrate the voltage (or discharge) of EDLCs when they are switched into series or parallel for the validation of their behavior.

Charge and discharge can be managed manually or remotely through the IP-Symcon platform [29]. Temperatures and weather variables are monitored through this platform too. This system includes the management of several protocols (e.g., EnOcean, Zwave, and especially KNX) for the application and management of energy storage systems in PPS. With respect to voltage and current, they are monitored through several signal conditioning boards and a data acquisition card linked to the workstation.

Figure 4 shows a picture of control devices, power sources (each one from 0 to 60 V, currents between 12 and 25 A, overall current 50 A, 3 kW), electronic loads (overall current 80–100 A, 2 kW), a signal conditioning board, and some OESD devices being used in tests.

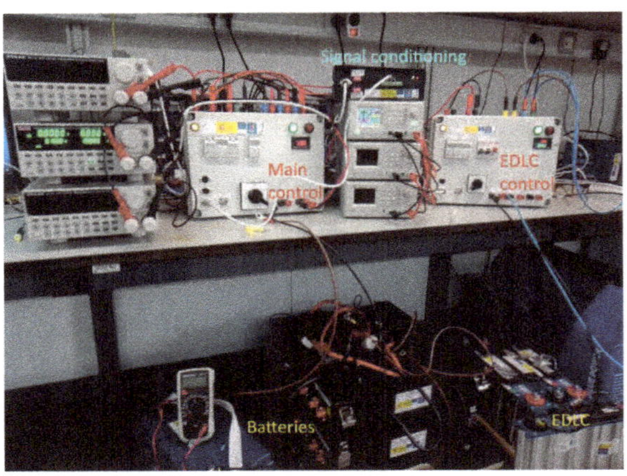

Figure 4. Laboratory test facilities.

2.3.2. Models for EDLC

Several approaches for EDLC models have been developed in the literature. A good review can be found in [30]. These methodologies can be classified as electrochemical, equivalent circuit, intelligent, fractional-order, self-discharge, and thermal models. Electrochemical and fractional-order models are considered the most used to characterize the electrical behavior of EDLC. Electrochemical models provide high precision but poor computational power, according to some authors [31]. Alternatively, equivalent circuit models are based on test measurements and justified by the ability of these approaches to reflect the element dynamics under a wide range of applications. The problem is that these models sometimes lack a physical explanation of values and perhaps can become an adjustment method of measurements through RC networks (i.e., a piecewise exponential approach to voltage or current curves). For instance, the model depicted in Figure 5c uses each RC branch to represent different time constants of charge and discharge (R_0C_0 from the initial time to some seconds; R_1C_1 for some seconds up to 10 min, and the last branch, R_2C_2, for times over 10 min). The main advantages of the electric equivalent approach are the straightforward identification of parameters for basic or classical models and the availability of similar models for batteries and train hotel (thermal) loads in the literature. This facilitates the development of a unified and comprehensive model for all these resources (both loads and storage devices) in the context of railway management. Additionally, this approach enables a further analysis of their interactions with RPS and PPS.

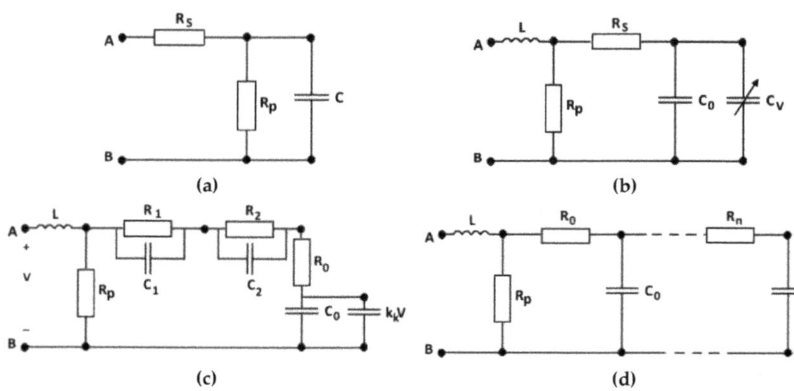

Figure 5. Electric equivalent models for EDLC: (**a**) Classical (real linear capacitor); (**b**) Classical (real non-linear capacitor); (**c**) Series RC branches (or parallel) model [32]; (**d**) Transmission line model (RC branches).

Equivalent circuit models are classified into classical models, Thevenin models, dual polarization models, second and third-order models, or lumped RLC models. Figure 5 depicts the equivalent circuit for some of these models.

2.3.3. Models for Batteries

Proposed methodologies for battery modeling, in the literature are closely related to EDLC methods that have been previously revisited. According to the literature [33], four categories can be used for modeling:

- Empirical model: The voltage is given by a mathematical function of SoC, internal impedances (coefficients), and current, for instance [34]. The main problem is that these models have limited capacity to fit voltage.
- Electric Equivalent model: It is the choice of this paper. The model is understandable (physical-based), but the identification of parameters can be quite complex [33] and sometimes does not make much sense (a table with different values for different scenarios), and it becomes a fitting regression alternative. Some examples (Thevenin and two RC networks in series) are depicted in Figure 6.

- Electrochemical model: It intends to describe the chemical–electrical reactions inside the battery. For example, Faraday and Ohm's laws and laws of diffusion. They have high accuracy, but experimental tests are expensive and complex.
- Data driven models: They are basically black boxes that obtain through machine learning models the relation between the main electrical variables in the battery (voltage, current, SoC...). They do not need knowledge of battery physics, but they need a large training to obtain a database to apply methods for each battery.

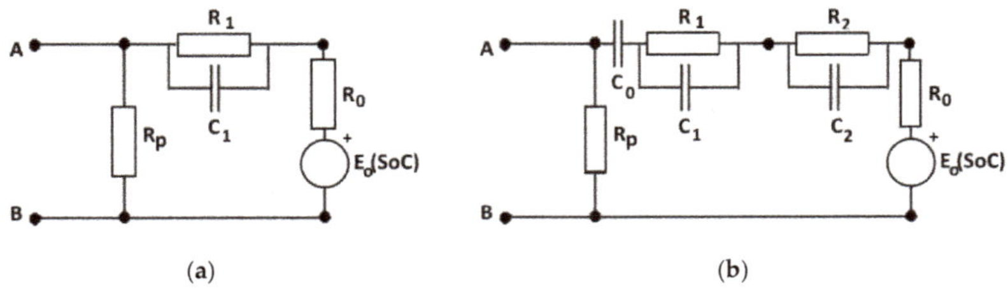

Figure 6. Some examples of electric equivalent models for batteries: (**a**) Thevenin model; (**b**) Series RC branches.

With respect to the electrical equivalent, the main difference between the models in Figure 6 is the number of RC networks (i.e., the order of the circuit). A higher order reflects better the fast and slow dynamics of the battery, but RC parameters are different in charge and discharge processes, which makes the model more complex. Also, the model can include (Figure 6b) a capacitor C_0 that describes the variation of open circuit voltage due to currents. Finally, the open circuit voltage is dependent on SoC, and the dependence of E_0 must be obtained through tests, which makes the evaluation of parameters more complex if temperature is considered (i.e., the operation of ESS in cold climates). Because this paper is focused on EDLC models, and batteries will perform with constant charges and discharges, we will assume a simpler version of the model presented in Figure 6b that has been previously used in railway literature (see Section 2.2). For this task, several charge and discharge tests have been developed with different currents (from 5 to 40 A) to obtain E_0 (SoC) and the internal impedance. Figure 7a shows one of these tests, whereas Figure 7b shows the change in the open voltage of a 100 Ah battery. Table 4 depicts the values used in the simulation.

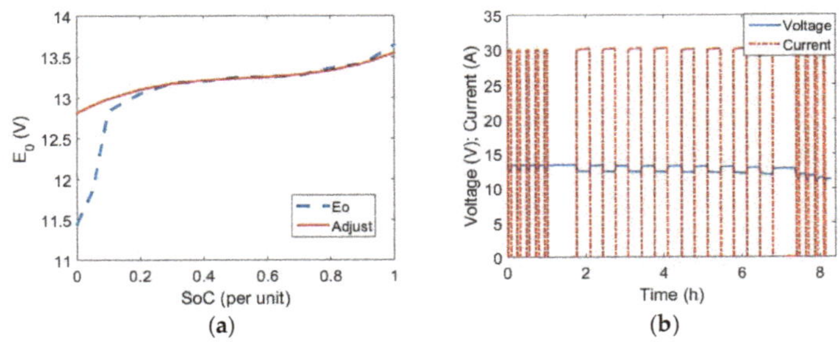

Figure 7. Characteristics of 100 Ah battery (testing at 30 A): (**a**) Open voltage E_0 vs. So12-C in a laboratory test and its adjustment; (**b**) Battery constant current discharge test (30 A).

Table 4. Parameters of 1.28 kWh Li-ion battery (for SoC foreseen operation).

SoC	Open Voltage	Internal Impedance
	Value (V)	Value (mΩ)
0.9	13.41	24.2
0.8	13.36	24.3
0.6	13.25	24.3
0.4	13.20	24.6
0.2	13.05	25.6

As an example, E_0 can be adjusted to SoC (between 0.2 and 0.9) according to:

$$E_0(SoC) = 12.82 + 2.02SoC - 3.45SoC^2 + 2.154\ SoC^3\ (V) \tag{7}$$

2.4. Laboratory Tests—The Proposed Model for EDLC

Our five laboratory EDLCs (90 Wh, 16 V) have undergone testing. Each device was subjected to a duty cycle of six constant–current charge and discharge cycles at 5, 10, 15, 20, 25, and 33 A (see Figure 8a). This testing approach reflects locomotive start-up demands, as simulations indicate that OESD supplies power for 2–4 min, limiting the tested EDLC current to approximately 30–40 A.

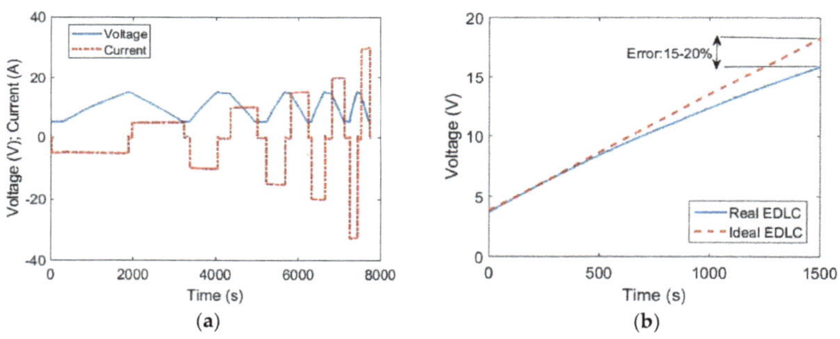

Figure 8. Tests for EDLC: (**a**) An example of laboratory tests; (**b**) Real measurements (charging at 5 A) and the response of the ideal/linear capacitor model.

According to tests and literature, the ideal model does not work and gives significant and cumulative errors of up to 20% (Figure 8b). If the rated capacity is C_0, the voltage decreases from the ideal voltage line defined by slope Ig_0/C_0, and it decreases more and more with the rise of voltage (see Figure 8b). This fact can be "explained" by several RC networks that suit the voltage trend according to their time constants, but it is stated in the literature that capacitors are not linear, and many electrolytes are not linear, too, which changes charge and discharge periods. Consequently, a non-linear model seems the best option, but it is usually considered more complex to solve it from a mathematical point of view, but this is not right, as will be stated later in this sub-section.

Our proposal in this paper is to implement a simple model with good accuracy that reflects the physical base of EDLC and that enables engineers to solve and implement it easily. Such a model can help to size and test the performance of the OESD in railway units. Specifically, the model is presented in Figure 9, connected to an ideal current source (power source and electrical load units in mode CC presented in Figure 5).

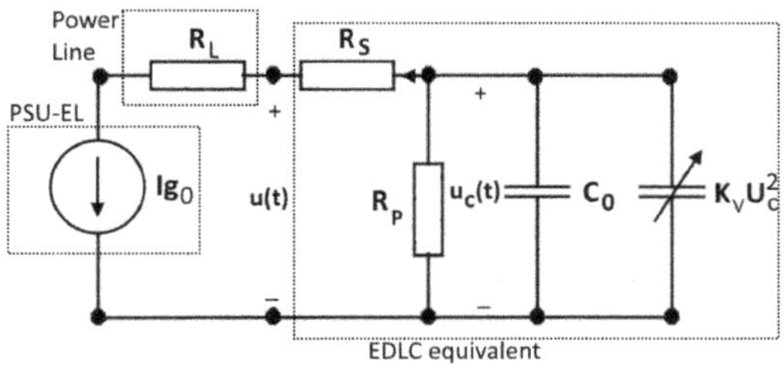

Figure 9. Equivalent circuit proposed for PBLM-EDLC.

Parameters in Figure 9 are the following:

R_P: is the parallel resistance of the capacitor (Ω).

C_0: is the constant capacity of the capacitor, which changes depending on charge and discharge duties due to the consideration of hysteresis on the electrolyte (F).

K_V: is the coefficient of the non-linear term (quadratic) of the capacitor (F/V^2).

R_S: is the series resistance (ESR) given by the manufacturer (Ω).

R_L: the resistance of the test measurement circuit (Ω).

Ig_0: current in A of Power Source Unit (PSU) or Electrical Load (EL).

$u(t)$: is the output voltage in V of EDLC.

$u_C(t)$: is the voltage in V of the ideal capacitor.

Kirchhoff's relations should consider that capacitance is variable, and the current is conducted by the change of charge q(t), $\frac{dq(t)}{dt}$. In this way, the mathematical model is the initial value problem (IVP) involving a non-linear first-order equation given by

$$\frac{d\left[\left(K_V u_C^2(t)\right)(u_c(t))\right]}{dt} + \frac{d[(C_0)(u_c(t))]}{dt} + \frac{u_C(t)}{R_P} = -Ig_0 \tag{8}$$

or

$$3K_V u_C^2(t)\frac{du_C(t)}{dt} + C_0\frac{du_C(t)}{dt} + \frac{u_C(t)}{R_P} = -Ig_0, \tag{9}$$

With the initial condition $u_c(t) = u_0$, u_0 is a real constant. To simplify the formulation of the solution and the search for parameters from measurements, the non-linear part of capacity ($3K_V$) is redefined as K_P, that is, $K_P = 3K_V$:

$$K_P u_C^2(t)\frac{du_C(t)}{dt} + C_0\frac{du_C(t)}{dt} + \frac{u_C(t)}{R_P} = -Ig_0 \tag{10}$$

From the point of view of measurements, $u_C(t)$ can be easily obtained considering R_S (and R_L in some cases, depending on measurement nodes):

$$u_C(t) = u(t) + R_s Ig_0 \tag{11}$$

From Equation (9), it can be considered that in the non-linear capacitor, the user sees at each voltage level an apparent capacitance C_{ap} given by

$$C_{ap} = C_0 + 3K_V u_C^2(t) = C_0 + K_P u_C^2(t) \tag{12}$$

In this paper, to solve the above IVP, we are going to use the Differential Transformation Method (DTM) proposed in [35]. DTM constitutes an important numerical technique used to solve linear and non-linear initial value problems in the analysis of electric circuits [35]. The DT method is an iterative procedure to obtain the analytical solutions of the Taylor series of differential equations, which presents an important advantage in that it can greatly reduce the size of the computational work while still accurately providing the series solution with a fast convergence rate [36]. Moreover, this technique can be directly applied to non-linear ordinary differential equations without requiring linearization, discretization, or perturbation. For this reason, this technique has been of interest to many researchers, where the DT method has been employed to obtain the approximate analytical solution of first, second, and third-order linear differential equations [37]. Furthermore, in [38–40], the DT method is used to solve differential equations of the Lane–Emden type.

Next, we recall the notion of differential transformation (DT) and some of its main properties.

Definition 1. *Assume that f(t) is an infinitely differentiable function. DT of f(t) is expressed as*

$$F(k) = \frac{1}{k!}\left[\frac{d^k f(t)}{dt^k}\right]_{t=0}. \tag{13}$$

Remark 1. *Inverse DT of $F(k)$ is defined as*

$$f(t) = \sum_{k=0}^{\infty} F(k)t^k, \tag{14}$$

By substituting Equation (13) in Equation (14), we obtain

$$f(t) = \sum\nolimits_{k=0}^{\infty} \frac{t^k}{k!}\left[\frac{d^k f(t)}{dt^k}\right]_{t=0} \tag{15}$$

Note that the above equation shows that DT originates from the Maclaurin series expansion.

Now, we present theorems 1–4 that collect some essential properties of DT [36].

Theorem 1. *If $f(t) = \lambda g(t) + \mu h(t)$, then $F(k) = \lambda G(k) + \mu H(k)$, where λ and μ are real constants.*

Theorem 2. *If $f(t) = g(t)h(t)$, then $F(k) = \sum_{l=0}^{k} G(l)H(k-l)$.*

Theorem 3. *If $f(t) = t^n$ with $n \in N$, then $F(k) = \delta(k-n) = \begin{cases} 1 & k = n \\ 0 & k \neq n \end{cases}$.*

Theorem 4. *If $f(t) = \frac{d^n g(t)}{dt^n}$, then $F(k) = (k+1)(k+2)\ldots(k+n)G(k+n)$.*

Having briefly laid the foundations of DTM, we are able to solve our IVP. Thus, taking DT in Equation (10) gives

$$K_P \sum_{l=0}^{k} P(l)D(k-l) + C_0(k+1)U_c(k+1) + \frac{1}{R_P}U_c(k) = -Ig_0\delta(k), \tag{16}$$

where $P(l) = \sum_{r=0}^{l} U_c(r)u_c(l-r)$, $D(k-l) = (k-l+1)U_c(k-l+1)$, and

$$u_c(0) = u_0. \tag{17}$$

By substituting Equation (17) into Equation (16) and using the recurrence relation, we have

$$U_c(1) = -\frac{Ig_0 + \frac{1}{R_P}U_c(0)}{K_P U_c(0)^2 + C_0};\ U_c(2) = -\frac{\left(2K_P U_c(0)U_c(1) + \frac{1}{R_P}\right)U_c(1)}{2\left(K_P U_c(0)^2 + C_0\right)},$$
$$U_c(3) = -\frac{\left(6K_P U_c(0)U_c(1) + \frac{1}{R_P}\right)u_c(2) + K_P U_c(1)^3}{3\left(K_P U_c(0)^2 + C_0\right)}, \tag{18}$$

and so on. Finally,

$$u_c(t) \approx \sum_{k=0}^{3} U_c(k)t^k = U_c(0) + U_c(1)t + U_c(2)t^2 + U_c(3)t^3. \tag{19}$$

3. Results

This section deals first with the practical estimation of EDLC parameters, including the justification of the physical meaning and relevance of these parameters of PBLM-EDLC. Then, hybridization possibilities of an electrical locomotive are presented from the point of view of locomotive layout and weight. Based on PBLM-Rail simulations, the first blueprint of OESD is explained. Finally, power and energy requirements are tested for multiple locomotive hauling (diesel-electric and hybrid-electric), and an approximation of energy management is stated, underlining the necessary accuracy and feedback between PBLM scripts. The feasibility of the proposed hybridization approach is discussed, and the possibility of improving in future research is discussed.

3.1. *Evaluation of Parameters for the EDLC Model (PBLM-EDLC)*

The coefficients involved in the analytical solution of Equation (10) depend on three physical parameters: K_P (recall that $K_P = 3K_V$), C_0, and R_P. The optimization of those parameters was carried out considering a grid and focusing on minimizing the error of the fitted model. Below, we describe the process in detail.

Firstly, a grid was established for the three physical parameters for optimizing (K_P, C_0, and R_P); see Table 5.

Table 5. Grid of parameters to optimize the EDLC model.

Parameter	From	To	Step
K_P (F/V^2)	0.2	1.5	0.1
C_0 (F)	400	600	10
R_P (Ω)	20	40	5

Secondly, for each combination of parameters in the grid, several goodness-of-fit measurements were computed, such as the MAPE (Mean Absolute Percentage Error), RMSE (Root Mean Squared Error), and RRMSE (Relative Root Mean Squared Error). The errors always refer to the difference between the real voltage of the trials, $u_c(t)$, and the estimated voltage by the DTM solution given in Equations (18) and (19).

The measurements MAPE and RMSE are well known, but it is worth noting that RRMSE can be defined as a percentage dividing the RMSE by the absolute value of the arithmetic mean:

$$RRMSE = 100\,\frac{RMSE}{|\overline{y}|} = 100\,\frac{\sqrt{\frac{1}{N}\sum_{t=1}^{t=N}(y_t - \hat{y}_t)^2}}{\left|\frac{1}{N}\sum_{t=1}^{t=N} y_t\right|} \tag{20}$$

Here, y_t is the real value, $\hat{y}_t$ is the point forecast, $\overline{y}$ is the mean of the real data, and N is the length of the dataset under evaluation.

On the other hand, note that the trials have different lengths; therefore, the global measurements (named by global MAPE, global RMSE, and global RRMSE) will give more weight to those trials with a higher amount of data. For this reason, we have also considered the average of the goodness-of-fit measurements in each trial (named by average MAPE, average RMSE, and average RRMSE).

The second stage was developed in two different scenarios. Scenario_1: considering all the trials. Scenario_2: splitting trials of charges and discharges. Note that the dielectric behavior can be different for charges and discharges, so Scenario_2 makes sense.

Results showed some of the following interesting conclusions:

(1) Behavior of the error measurements is similar for different values of parameter R_P, and that is true for Scenarios 1 and 2 as well (see Figure 10 for Scenario_1). Therefore, the accuracy of the fitted model seems to depend on the combination of parameters K_P and C_0.

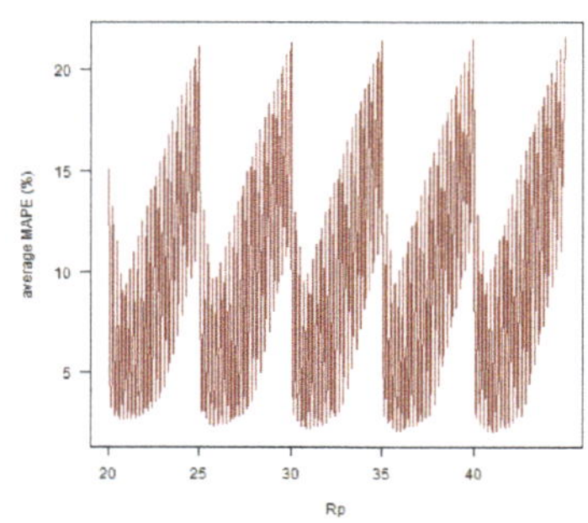

(a)

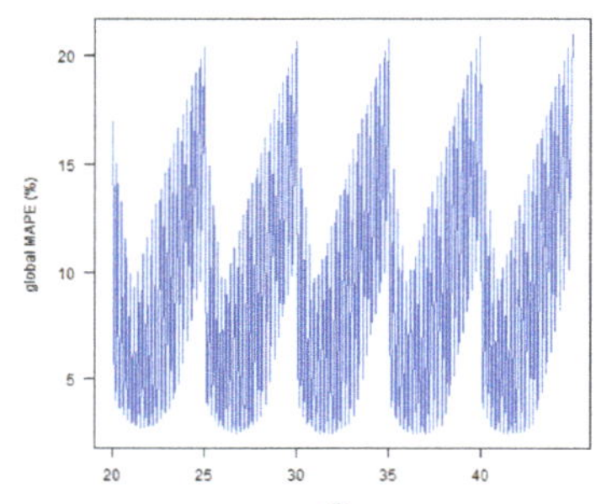

(b)

Figure 10. Plot of the MAPEs against the parameter R_p in Scenario_1: (**a**) Average MAPE; (**b**) Global MAPE.

(2) Checking the results, lower error measurements were found for specific combinations of parameters (K_P and C_0). In particular, the plot of the MAPEs (RMSEs and RRMSEs) against the quantity $M = 100{\cdot}K_P + C_0$ reveals a clear quadratic pattern that allows us to determine the optimum combinations (see Figure 11 for Scenario_1). Similar conclusions were obtained for Scenario_2.

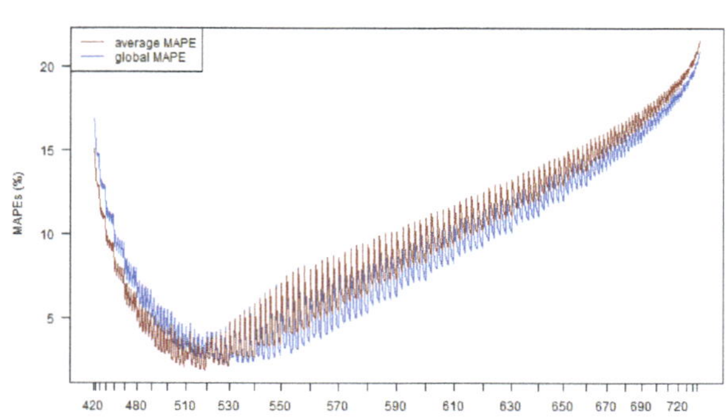

Figure 11. Plot of the average MAPE (brown) and global MAPE (blue) against the quantity $M = 100{\cdot}K_P + C_0$ in Scenario_1.

(3) It is preferred that the quantity $M = 100{\cdot}K_P + C_0$ takes a value in the interval [510, 520] (see Figure 11 and comments in the next section). Indeed, for Scenario_1, the optimal

parameters that minimize the average MAPE are ($K_P = 0.7$, $C_0 = 440$, $R_P = 40$) with average_MAPE = 1.93% (note that the optimal parameters can be different depending on what error measurement is minimized). Anyway, if the combination of parameters verifies that quantity M is in [510, 520], we can ensure that MAPEs and RRMSEs are low enough. For instance, when ($K_P = 0.4$, $C_0 = 480$, $R_P = 40$), we obtain average_MAPE = 2.23%, global_MAPE = 2.57%, average_RRMSE = 2.46%, and global_RRMSE = 3.68%. Similar conclusions can be obtained for Scenario_2, although the optimal range of the quantity M is shifted for charges and discharges. Table 6 shows the optimal parameters (that minimize the average MAPE) and the remaining error measurements for each scenario.

Table 6. Optimal parameters (minimize the average MAPE) and error measurements for each scenario.

Parameter	Scenario_1	Scenario_2 (Only Charges)	Scenario_2 (Only Discharges)
K_P	0.7	0.6	0.2
C_0	440	430	510
R_P	40	20	40
average_MAPE (%)	1.93	1.37	0.82
global_MAPE (%)	2.6	2.43	0.99
average_RMSE	0.24	0.20	0.08
global_RMSE	0.34	0.37	0.1
average_RRMSE (%)	2.34	1.95	0.8
global_RRMSE (%)	3.43	3.59	1

(4) The fitted model provides high accuracy in general, a little better for discharges than for charge trials in Scenario_1. Figure 12 shows the comparison between the actual values of $u_c(t)$ and the fitted ones for different trials. Similar conclusions were obtained for Scenario_2.

Finally, Table 7 shows the MAPE for the optimal parameters (given in Table 6) in both scenarios, distinguished by trials. Note that there is a slight improvement if we consider two EDLC models: one for charges and another for discharges.

Table 7. MAPE by scenarios and trials for optimal parameters in Table 6.

Trial	Scenario_1 (K_P,C_0,R_P)= (0.7, 440, 40)	Scenario_2 (Charges) (K_P,C_0,R_P)= (0.6, 430, 20)	Scenario_2 (Discharges) (K_P,C_0,R_P)= (0.2, 510, 40)
5 A (charge)	4.37	4.1	-
10 A (charge)	1.11	0.83	-
15 A (charge)	1.95	0.85	-
20 A (charge)	1.74	0.63	-
33 A (charge)	1.72	0.45	-
5 A (discharge)	3.9	-	1.44
10 A (discharge)	1.75	-	0.55
15 A (discharge)	1.28	-	0.50
20 A (discharge)	1.13	-	0.66
30 A (discharge)	1.20	-	0.95

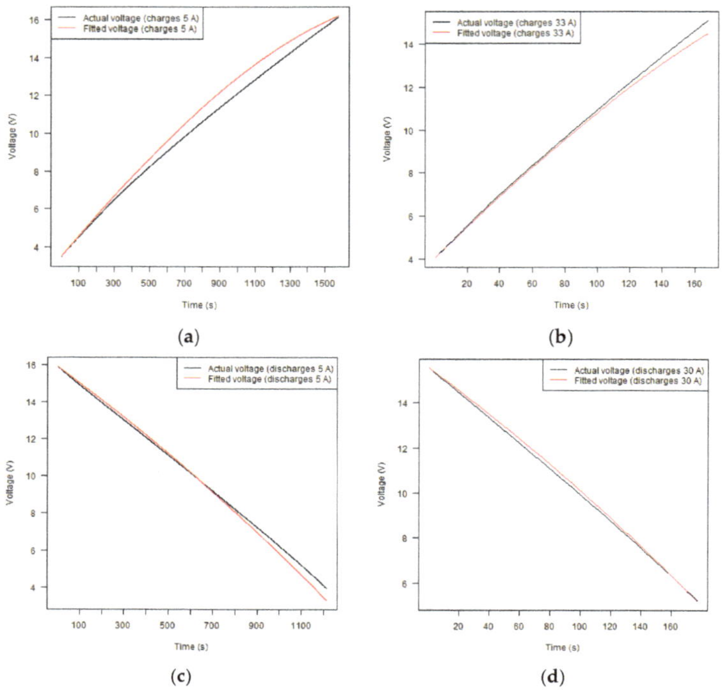

Figure 12. Comparisons between the actual values (in black) of the voltage $u_c(t)$ and the fitted ones (in red) for Scenario_1: (**a**) Trial of charges at 5 A; (**b**) trial of charges at 33 A; (**c**) trial of discharges at 5 A; (**d**) trial of discharges at 30 A.

3.2. Physical Meaning of PBLM-EDLC Parameters

The suitability of a simulation model depends on its performance and the degree of difficulty when the user evaluates the main parameters of the model. The value of R_s is easy to find from the manufacturer's data ("ESR" in Table 3). In other cases, the measurement nodes for voltage are outside EDLC (the overall device is composed of several commercial banks, and cable links have some resistance, R_L in Figure 9). In this case, the value $(R_S + R_L)$ is very easy to find when EDLC changes from charge to discharge because the change in voltage is due to the value $(R_S + R_L)Ig_0$.

The manufacturer also gives a range of values for the EDLC capacitance (Table 2), so if the model is physically based, the capacitance seen in tests should be close to these values. Neglecting R_P in the first round, because the energy losses in EDLC are very low, we can define an apparent (Equation (12)), or instantaneous capacitance at each time t, $C_{ap}(u_C)$, of the parallel branch in Figure 8 as

$$C_{ap}(u_C) = \frac{Ig_0}{dU_C/dt} \tag{21}$$

Values of CC can be obtained from the EDLC differential Equation (10) as

$$C_{ap}(u_C) = C_0 + K_P u_C^2 \tag{22}$$

If a single EDLC module (e.g., the 500 F module tested in the laboratory, see Table 2) should perform in the voltage range from 0 to 16 V (17 V maximum, Table 2). The average capacitance (C_{av}) at the range of voltage is performed by

$$C_{av} = \frac{1}{17}\int_0^{17} CC(u_C)du_C = C_0 + \frac{1}{3}K_P 17^2 = C_0 + 96.3K_P \tag{23}$$

This relation is very close to the quantity $M = 100 \cdot K_P + C_0$ that provides the optimal combination of parameters for the grid searching (see Section 3.1).

3.3. Hybridization Possibilities

Hybrid units can be new, but sometimes it seems reasonable to rebuild "middle-aged" units (<20 years) or units with a low number of kilometers (<2 million), which can increase their operational life and reduce capital costs of new units [41]. Fighting against programmed obsolescence is a challenge for many operators, such as the French operator SNCF [42], to reduce capital costs and support environmental concerns.

As previously mentioned, Spanish operators typically use two locomotive series for the selected track: the Series 319.2 and Series 333.3 diesel-electric locomotives, with rated powers of 1450 kW and 2240 kW, respectively (both based on EMD locomotive platforms). As the authors have discussed in previous works [43], the average power output of these units is approximately 800–1300 kW, meaning that the full rated power is not applied if a storage device can handle peak demand periods.

However, implementing freight hybrid trains presents challenges, particularly regarding the space needed to accommodate OESD. One potential solution is the use of a dedicated Battery-Electric Locomotive (BEL) alongside diesel units operating in a single consist, an approach tested by U.S. operators, e.g., [13]. This seems practical given the heavy weight of American freight trains, which often exceed 2500 tons. A second option is the use of a service vehicle, which has proven effective for passenger trains [44] but is more challenging for freight operations since a dedicated wagon for each train would be costly and complex because locomotives change from head to rear at CT and MU stations.

Finally, the third option involves finding space for OESD integration within a conventional locomotive unit. This last approach has been considered in the present study. Reasons for this choice are discussed in the following paragraphs.

At present, the main Spanish railway operator has decreased its freight activity [6,45], and some new electric locomotives manufactured by Bombardier in 2010–2015 (units belonging to the TRAXX family) are surplus. This represents that older series 269.2 (built in 1985, designed by Mitsubishi Electric Co. [46]) are available for location (12 + 20 units of another 269 subseries). Moreover, around 18 units of the 319.2 and 333.3 series are still available for renting [47].

Class S269 weighs 88 tons (its subseries are S269.2 and S269.6). They are 3 kV DC locomotives that develop 2 × 1550 kW with two motor groups (mono-motor group in each bogie) with independent control and electronics (1–3 and 2–4 in Figure 13). The hypothesis is that one of these mono-motor groups and the 380 V-AC motor–generator group (covered by power electronics in modern diesel units [48]) can be replaced by OESD. The result is a locomotive with 1550 kW comparable to a diesel-electric unit series 319.2. As stated before, this reconstruction procedure is not new and has been applied by some manufacturers in DMU and EMU to convert them into hybrid EMUs [49]. Sometimes, hybrid units such as the Talgo S-730 Diesel Hybrid Electric Multiple Unit (DHEMU) have removed one of the two diesel generators located in a technical car to include a hydrogen fuel cell for tests of a research project dealing with green hydrogen [44]. It is important to note that a new active cooling system is necessary to control the temperature of OESD (15–20 °C in the case of batteries). Figure 13 shows the layout and dimensions of standard class S269.

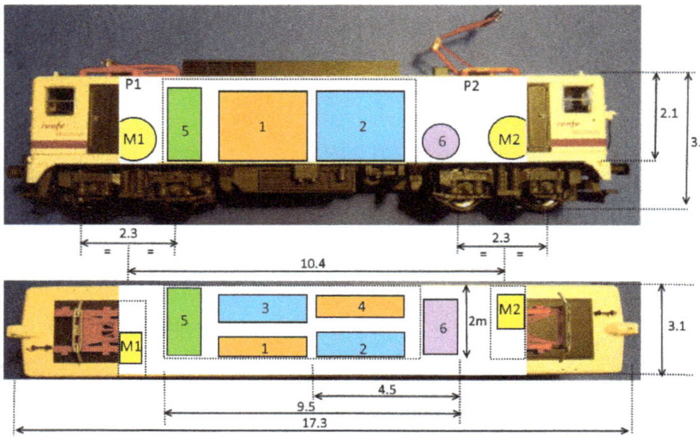

Figure 13. Simplified layout for locomotive subseries 269.6 according to Mitsubishi Electric specifications for Spanish manufacturers [50]. M1 and M2 (DC motors); 1 (3 kV rack for M1); 2 (chopper for M2); 3 (chopper for M1); 4 (3 kV rack for M2); 5 (filter); 6 (ancillary motor–generator group, 380 V); P1 and P2 (3 kV pantographs).

Tractive effort curves of diesel-electric and electric locomotives are depicted in Figure 14 and elaborated from public manufacturers' data. Figure 14b shows the joint tractive effort of the hybrid multiple units. For the sake of comparison, the tractive effort of a modern TRAXX series is shown in Figure 14b. Notice that TRAXX develops more power, but hybrid units develop high tractive effort at low speeds. All curves have been used in the simulation model for railway units (PBLM-Rail).

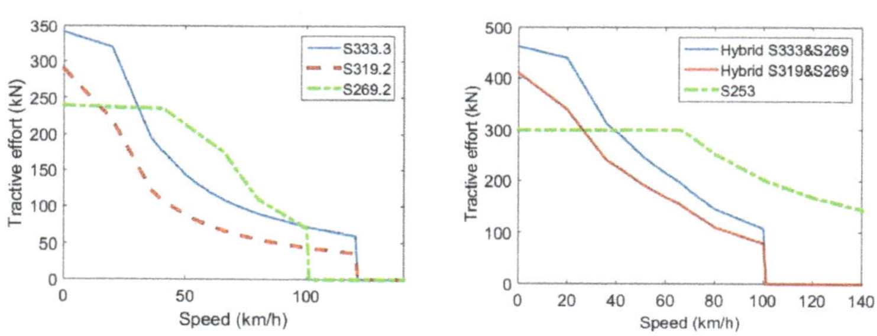

Figure 14. Tractive effort of single units: (**a**) Series 333.3, 319.2, and 269.2; (**b**) Hybridization of 333.3 and 269.2 (rebuilt to a single-motor unit), 319.2, and 269.2 (single-motor) and Bombardier 3 kV DC TRAXX family (e.g., class S253 of RENFE in Spain).

3.4. Energy and Power Requirements

To evaluate power and energy requirements, a double traction 1300 t train (1 Hbis and 12 tank wagons, e.g., four-axle Zags 117 m^3 transporting liquified petroleum gas type 2 [51], and two locomotives 333.3, see Figure 3a) has been simulated in the example track. Another possibility in this route is a train with 12 intermodal-container wagons class Sgnss 60′ to link Cartagena harbor, but the weight should be very close to Zags. The justification of this choice is based on the methodology used in some freight transport statistics [6] based on public European databases. According to these reports and the proposed methodology for data analysis, the average weight of goods loaded onto trains in Spain was 350 t with an average distance of 390 km in 2005 (and an average weight of around 500 t). Nevertheless, heavy trains could transport twice or threefold these numbers. Secondly, the admissible length of trains (according to the infrastructure operator, ADIF) is 320 m on this route. Third, the city of Cartagena has developed refining activities since the 50s, and most gas products are delivered by train. This justified the weight and class of wagons.

Figure 15 shows the main results from our PBLM-Rail simulator in the Cartagena–Murcia itinerary. From Cartagena (CT) to El Regüeron (ER), there are four stations with a secondary track to make a crossing possible. In the simulation, two stops have been considered because some passenger services would require these stops (this is a single-track section; see Figure 1a). Stop times are not very significant from the point of view of energy requirements, and three minutes are considered for simulation. The second stop is just before the train climbs the highest slope on the track (10–12%), which corresponds to the highest power requirements for locomotives.

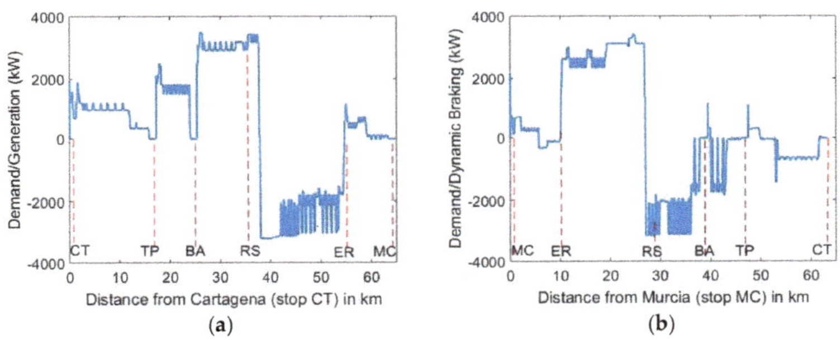

Figure 15. Demand requirements and maximum potential of regenerative braking for a 1300 t freight train: (**a**) From Cartagena to Murcia; (**b**) From Murcia to Cartagena.

Results show that the freight train demands around 900 kWh, and braking accounts for 590 kWh (around 500 kWh go to braking resistors). Figure 15a,b clearly show that the main braking zone corresponds to the braking effort developed to maintain speed limitations when the train goes down the hill after the station RS; see the elevation profile, Figure 1a. The reader should also consider that Spain is a mountainous country (the second-highest country in Europe, surpassed only by Switzerland) and that conventional routes (1668 mm gauge) have important slopes and frequent curves that make haulage difficult but also benefit regenerative braking. Also, it is interesting to note that power demand does not require 4450 kW from the two 333.3 locomotives except when the train starts after the train stops and when the train climbs the hill in the route section from BS to RS stations. A cumulative demand/generation curve for this route (e.g., from CT to MC train stops) is depicted in Figure 16a.

Cumulative load-duration curves (LDCs, as shown in Figure 16a) are commonly used by system operators in PPS to determine the utilization of generation resources. In our case, the analysis focuses on a small Power System (RPS) composed of two diesel-electric units. A preliminary examination of Figure 16a suggests that a single S333.3 unit could meet hauling requirements 85% of the time, including braking (albeit with reduced performance and increased reliance on pneumatic braking). The dotted green line in the figure indicates the power requirements exceeding the maximum output of a single S333.3 unit, highlighting that only ten minutes of operation in non-consecutive intervals would necessitate a second coupled unit.

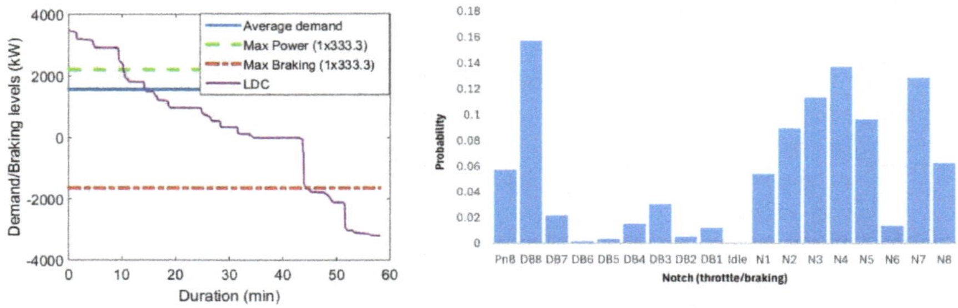

(**a**) (**b**)

Figure 16. Demand/braking characteristics from the simulation of a 1300-ton freight train from Cartagena to Murcia: (**a**) Load Duration Curve (LDC, for double S333.3 hauling) (**b**) Bar plot of throttle (N) and braking (DB). Notes: PnB means that pneumatic braking was applied. Idle position has been omitted because of its large difference in magnitude in actual freight duties.

Figure 16b illustrates the throttle and dynamic braking notch positions. The data reveal that the use of a double unit is only required for short, specific periods and that the N8 notch (maximum output power) is not the most frequently used. This insight provides a foundation for defining a blueprint for the hauling power requirements.

Based on this analysis, it can be hypothesized that a second locomotive unit with lower traction power (in the range of 1100–1400 kW) could adequately fulfill operational requirements. While this configuration might slightly increase trip duration, it could significantly reduce energy demands, thereby improving overall efficiency.

It is worth noting again that S269 locomotives, although providing lower tractive effort than diesel units at low speeds (<30 km/h), exhibit superior tractive effort at medium speeds (30–50 km/h). Additionally, 269 locomotives weigh 32 tons less than the S333.3 series, which could further contribute to operational efficiency (as illustrated in Figure 14a).

3.5. Sizing of OESD

The sizing of On-Board Energy Storage Devices (OESD) depends on several factors, including the vehicle type, itinerary, degree of electrification, and the operational duties of the railway unit. Preliminary values can be inferred from past experiences, scientific literature, and manufacturer data, although specific details are often unavailable.

As mentioned in Section 2, the German ETA series (EMU) required 528 kWh of energy to cover a 300 km trip using lead–acid batteries. Reference [17] suggests that for routes with 20% electrification, approximately 2100 kWh of battery storage is needed, while the minimum storage size for routes with 36% electrification is 544 kWh. The system must also handle peak currents of around 6000 A, achieved using Lithium Titanate Oxide (LTO) batteries. The DC bus operates at a rated voltage of 750 V, which is consistent with the voltage used in third-rail electrification in the UK.

In [14], the authors propose limiting charge and discharge rates to 0.5 C and 1 C, respectively, to optimize battery capacity, resulting in a required storage range of 2200 to 4800 kWh for a six-car Battery Electric Multiple Unit (BEMU) intercity train. They emphasize that lower C-rates reduce battery degradation and extend their operational lifetime, as supported by the literature [52]. SoC is also limited to a range from 0.9 to 0.2.

Partial storage systems were also analyzed in [19] to manage RPS peak loads caused by the synchronization of train timetables. In this context, a combination of batteries (approximately 100 kWh) and EDLCs was proposed. However, these systems primarily operate under electrified tracks, and the battery sizing aligns more with last-mile applications than with the hybridization goals of this study. Table 8 summarizes the main OESD options for some representative railway systems.

Table 8. Battery and diesel options and characteristics for EMU in Europe and USA.

Operator, Country, Year	Rail System	Vehicle, Manufacturer	ESS	Energy (MWh)	Power (kW)	Other	Ref
DB, Germany, 1959	BMU	ETA 515, Orenstein and Koppel, W & M Donauwörth, MAN	Lead acid	0.528	200	100 km/h 300 km	[10]
DB Regio, Germany, 2022	EMU, regional	Flirt Akku, Stadler Rail	LTO	0.18	N/A	150 km/h 224 km [1]	[53]
SNCF, France, 2019–2023 [3]	DBEMU, regional	TER Regiolis, Alstom-SNCF	LTO	0.14	600	80 km w/o catenary	[54]
ÖBB, Austria, 2019, 2023 [3]	EMU, regional	Cityjet Desiro, Siemens	LTO	0.53	1300	100 km/h [2]	[55]
Rumo Logistics, Brazil, 2023	Freight	GT38H Joule (hybrid), Progress Rail	$LiFePO_4$	0.55	2200	60 km/h	[56]
Prototype tests, USA, 2020	Freight	SD70J EMD Joule (BEL), Progress Rail	$LiFePO_4$	2.4–8	3200	120 km/h	[57]
Zanzeff project, USA, 2021	Freight	ET44C4, BNSF	N/A	2.4	3000	<100 km/h	[13]

[1] 2022 attributed "record" for a BEMU; [2] in battery mode; [3] commercial service.

Based on manufacturer proposals and on the requirements of our freight route, a 550 kWh capacity for OESD is assumed. As the reader can see in Table 8, European manufacturers prefer LTO batteries due to their higher current, but [57] is confident in $LiFePO_4$ technology for its Joule BEL due to three reasons: its thermal stability, inherent safety, and energy density. Moreover, this manufacturer (a Caterpillar company expert in diesel-electric engines, formerly EMD) also rebuilds middle-aged locos. These improvements on middle-aged series have usually been used by operators, and it has also been applied in Spain for S333.3. Especially both the original AC generators and the diesel engine remain in refurbished units. For this reason, it is interesting to depict some characteristics of this generator (GT26 according to the EMD catalog; see Figure 17).

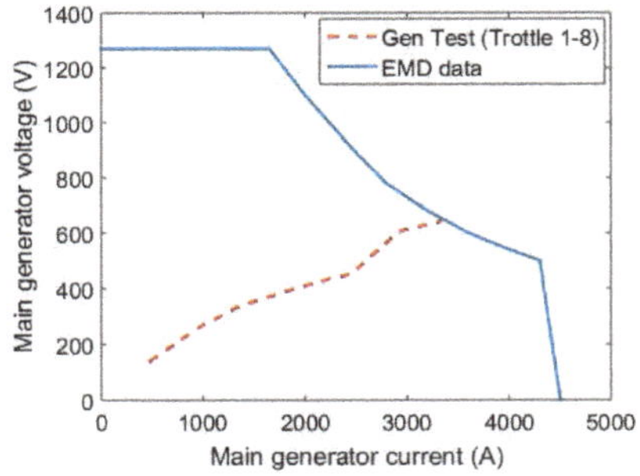

Figure 17. Main Generator GT26 voltage vs. current curve and data from a real acceleration test of the generator of the original locomotive of 333 series (an international platform by EMD).

The current requirements for the six traction motors are approximately 3600 A at 650 V. For the hybrid system, a DC bus operating at around 750 V is assumed, this being a typical rated voltage for trams and third-rail railway systems [17]. This voltage facilitates the availability of DC–DC converters, such as 3 kV-to-750 V converters [58], which can be used for supply and braking energy transfer to and from the OLE and locomotives. It also enables the potential recharging of batteries from the diesel-electric unit without requiring dedicated charging infrastructure at wayside locations or train stops.

Using $LiFePO_4$ batteries (commonly applied in prosumer energy management) and considering a maximum current of approximately 3600 A (for a diesel unit alone), the OESD would require between 36 and 72 rows of batteries. For instance, each row would consist of 62 Victron 100 A batteries to reach 750 V. However, this configuration presents significant challenges for a fully battery-electric locomotive (BEL). Achieving adequate storage capacity would require 2200–4000 batteries (e.g., 62 $\times$ 36 elements, each providing 1.28 kWh) and a total storage capacity of approximately 2700 MWh. Although this sizing aligns with data from BEL manufacturers, such as the SD70J by Progress Rail (see Table 8), it exceeds the storage constraints for a hybrid unit. Consequently, European reports like [12] conclude that batteries alone are not a feasible solution for stand-alone locomotives in heavy freight services.

Given the need for fast response from the OESD, a hybrid approach combining batteries with Electric Double-Layer Capacitors (EDLCs) is proposed. The primary hypotheses used for sizing this hybrid solution are as follows:

- Track Characteristics and Urban Mobility Trends: In the scenario analyzed, the Murcia (MC) station will integrate a new transportation hub, including a bus station, parking facilities, and a tramway line. This infrastructure development suggests the need for fast charging points for electric buses. Additionally, passenger train traffic (especially high-speed trains) requires rapid adjustments in demand and generation, which EDLCs are well-suited to handle. The hybrid locomotive could serve as a mobile generation resource for both Railway Power Systems (RPS) and Public Power Systems (PPS). According to the authors' previous work [17], passenger trains generate 20–45 kWh during braking, while acceleration demands 40–50 kWh—levels difficult to achieve and store using only batteries.
- Demand and braking profiles: The track's demand profiles, as shown in Figure 15, exhibit both base and fluctuating demand patterns. Since batteries have higher internal impedance and limited instantaneous power delivery, the RPS must handle high peak loads. Additionally, frequent braking and acceleration cycles require high energy efficiency. EDLCs are better suited to handle these rapid demand fluctuations due to their fast response times.

Thus, a strategy is required to evaluate the feasibility of a hybrid locomotive capable of managing the power demands of a 1300 t freight train, incorporating an OESD design. This work does not try to optimize OESD efficiency or sizing but to demonstrate the advantages of integrating PBLM models in railway operations, specifically including storage models such as PBLM-EDLC (and other storage facilities).

3.5.1. Definition of a Power Profile for the Hybrid Locomotive in the Electrified Section

According to train simulations in our route (Section 3.2), the potential energy recovery can reach around 550 kWh by dynamic braking, basically in the negative grade track from stations RS and ER. Some of this potential cannot be recovered because pneumatic braking is also necessary. The hybridization also makes regenerative braking more complex because the "new S269" has half of its primary regenerative capacity (i.e., 1050 instead of 2100 kW). Our scenario presents 10 km of electrified track with a junction in ER station to Alicante with notable suburban, regional, and intercity passenger traffic to Madrid, Valencia, and Barcelone from Murcia and Cartagena (track section MC to ER, Figure 1). The railway track from Murcia to Alicante is very complex because 1435 mm gauge (a new route, basically for high-speed EMU) and 1668 mm gauge (suburban, regional, and freight) have different platforms and layouts.

The hypothesis assumes the worst possible scenario: the hybrid S269 should haul its 1300 t service without catenary and without the help of diesel 333. This scenario has a low probability of occurrence, but it is possible that OLE will be overcharged or train delays will require full power from the substation feeding the section ER-MC. On the other side, this scenario allows hybrid S269 to engage switcher tasks hauling EMUs on 1668 mm or dual 1668/1435 mm if the catenary fails (shunting and switcher units are almost unknown in this section because they have been decommissioned, perhaps due to low freight traffic).

Figure 18a illustrates the power requirements of the freight train, where the power profile is distributed between the EDLC and the battery. As previously mentioned, the EDLC handles the fluctuating component (analogous to an "AC" signal), while the battery manages the more stable component (similar to a "DC" or rectangular signal). From a mathematical or electrical engineering perspective, this demand profile can be interpreted as a combination of constant and alternating current components ("DC and AC"). Various techniques, such as the Fast Fourier Transform, Wavelet Transform, or Hilbert Transform [58], can be used to segment the profile, identify demand transitions [59], and decompose the power into its AC and DC components. Circuit simulators, functioning like oscilloscopes, can also perform this analysis. In our case, we implemented a straightforward algorithm that applies a boundary condition (initial SoC) and a constraint (final SoC of the EDLC), ensuring that the S269 unit retains enough charge for shunting tasks, such as repositioning locomotives from the rear to the front of the train.

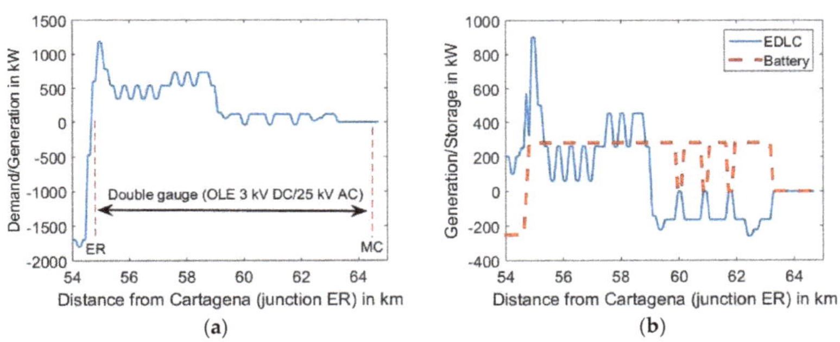

Figure 18. Test track for hybrid 269: (**a**) Power requirements of the 1300 t train from ER to MC; (**b**) Generation (positive) and Demand (negative) from EDLC and batteries.

Figure 18b shows results that enable a first approximation to EDLC size. For this example, the initial SoC is 100% (the train was using brakes more than 15 km before reaching ER), and the final SoC is around 30%. An EDLC capacity from 26 to 30 kW is estimated from Figure 18b and proposed boundary conditions and constraints. With this value, the user must consider other boundary conditions depicted in Table 9 to refine values in the next step. The most important remark is that physical-based models of EDLC are required for the right evaluation of voltage in DC buses, instantaneous power capacity, and energy requirements (including SoC). The next section deals with the validation of those constraints.

Table 9. Refine and testing of OESD capacity.

Constraint	Value/Range	Comments
OESD storage limits	500–550 kWh	Dynamic braking of the train
Demand peak in the route	3.0–3.5 MW	Max. Energy deployed 85 kWh
Peak demand covered by OESD	1000–1500 kW	Eq. to a single motor of S269.
Available volume for OESD	21–25 m^3	Rebuilt S269 locomotive [2]
Available load in 269 [1]	≈20 t	Without motor of bogie-2 [2]
Supply voltage to OESD	750 (500–900) V	Needs a PBLM of EDLC
OLE supply voltage	3 (2.2–3.6) kV DC	1668 mm and dual tracks
Efficiency in power conversion	0.95	Converters and inverters
Efficiency in OESD	0.9–0.99	Battery and EDLC
Battery SoC ranges	0.2–0.9	According to literature [52]
EDLC SoC ranges	(0.23–1.06)	DC bus voltage range

[1] Maximum load in Spanish tracks (1668/1435 mm): 22.5 t/axle. [2] See Figure 13.

3.5.2. Proposed EDLC (and Battery) Subsystems

The reader should consider that the objective of this work is validating the pre-feasibility of the proposed approach from a "common-sense" point of view and applying the PBLM-EDLC model proposed in Section 2 to validate the main hypotheses. At this stage, searching for an optimum for the OESD system is not considered. Otherwise, methodology and models can help in this task in future research works of authors.

With the hypotheses of Section 3.5.1, the OESD must include devices in series to define a row (for batteries S_{bat} batteries/rows, for EDLC S_{EDLC} units/row) according to DC bus rated voltage:

$$U_{BUS}^{min} = S_{EDLC} * U_{el-EDLC}^{min};\ U_{BUS}^{rated} = S_{BAT} * U_{el-bat}^{rated} \quad (24)$$

where the following is true:

U_{BUS}^{min}: is the minimum voltage at DC bus (e.g., 600 V).

U_{BUS}^{rated}: is the rated voltage at DC bus (e.g., 750 V).

$U_{el-EDLC}^{min}$: is the minimum voltage of elemental EDLC (500 F/16 V unit).

U_{el-bat}^{rated}: is the rated voltage of elemental battery (12.6 V/100 Ah).

Moreover, the EDLC set of OESD integrates P_{EDLC} parallel rows to reach EDLC capacity to deliver energy (in our case 28 kWh):

$$Cap_{EDLC} = S_{EDLC} * P_{EDLC} * Cap_{el-EDLC};\ Cap_{bat} = S_{bat} * P_{bat} * Cap_{el-bat} \quad (25)$$

where the following is true:

Cap_{EDLC}: is the proposed EDLC set rated capacity in kWh.

Cap_{bat}: is the proposed battery set rated capacity in kWh.

$Cap_{el-EDLC}$: is the EDLC set rated capacity (e.g., 0.018 kWh).

Cap_{el-bat}: is the elemental battery capacity (e.g., 1.28 kWh).

According to those equations, constraints, and the parameters of the elemental device parameters, the choice for the overall OESD is summarized in Table 10. Notice that Spain and other countries do not yet have a commercial market for EDLC devices suitable for the traction of the foreseen storage capacity. For this reason, it is difficult to obtain an accurate estimate of the costs of the proposed OESD. On the other hand, it can be stated that some basic cooling equipment facilities are still available in the S269 after hybridization (i.e., equipment for the motor and electronics being removed from the original unit).

Table 10. Nominal OESD properties (common-sense criteria). Notice that values for battery and EDLC modules were presented in Tables 2 and 3.

Property	Battery Pack	EDLC Pack
Module reference [1]	Victron 100 Ah	BMOD0500 P016
Volume (module)	0.0125 m^3	0.0092 m^3
Modules in series per row	62	56
Rows in parallel	10	40
Capacity (Ah)	500–1000 Ah	4000–6000 Ah
Nominal energy capacity (kWh)	740 kWh	40 kWh
Operational limits	SoC (0.2–0.9)	Voltage (8.5–16.5 V)
Available energy	520.8 kWh	27.4 (max 29.2) kWh
Weight	5.5 t	12.5 t
Volume	7.74 m^3	10 m^3
Capital cost	130 k€	1 M€ [3]
Additional power electronics [2]	500 kg	300 kg

[1] Example for simulation purposes.[2] According to reference [15]. [3] EDLC retail price.

3.6. Demand Management of EDLC and the Hybrid Locomotive Through PBLM

The instantaneous power p(t) and energy w(t) of EDLCs are strongly dependent on voltage and, consequently, the SoC of the overall unit proposed for the hybrid locomotive. It seems useful to remember the main equations for these magnitudes for an elemental EDLC device according to our model (see Figure 9 for the nomenclature of main magnitudes and parameters). The instantaneous power (generation criterium):

$$p(t) = U(t)I_{g0}(t) = u_c(t)I_{g0}(t) - R_S I_{go}^2(t) \tag{26}$$

The available stored energy, w(t), in the EDLC at time t of a simulation starting at t_0 (remember $K_P = 3K_V$) is

$$\begin{aligned} w(t) - w(t_0) &= \int_{t_0}^{t} u_c(\tau)\frac{dq_{EDLC}(\tau)}{d\tau}d\tau = \int_{u_c(t_0)}^{u_c(t)} u_c(\tau)\left[C_0 + K_P u_c^2(\tau)\right]du_c = \\ &= \left(\tfrac{1}{2}C_0 u_c^2 + \tfrac{1}{3}K_p u_c^3\right)\Big]_{u_c(t_0)}^{u_c(t)} = \left(\tfrac{1}{2}C_0 u_c^2 + K_v u_c^3\right)\Big]_{u_c(t_0)}^{u_c(t)} \end{aligned} \tag{27}$$

And SoC at the elemental (or overall) EDLC:

$$SoC(t) = \frac{w(t)}{w_{max}} = \frac{\frac{1}{2}C_0 u_c^2(t) + K_v u_c^3(t)}{\frac{1}{2}C_0 u_{c-max}^2 + K_v u_{c-max}^3} \tag{28}$$

Previous equations clearly show that an accurate voltage must be forecasted to achieve the best performance of both EDLC (OESD) and locomotives. The model can help in the achievement of some further targets, such as optimization of energy or EDLC size according to train requirements (both targets are not objectives in this paper).

The itinerary of freight trains is usually conditioned by passenger traffic and involves continuous stops. The energy management system of tandem locomotives should account for train energy requirements and the OESD state. Usually, several simulations must be applied according to changes in the itinerary. For this reason, commissioned and potential train stops define a partition of the route into several "route sections". An alternative is to consider signalization tracks. Moreover, the peak on the elevation profile also defines two sections between RS and ER stations. A simplified flow diagram is presented in Figure 19.

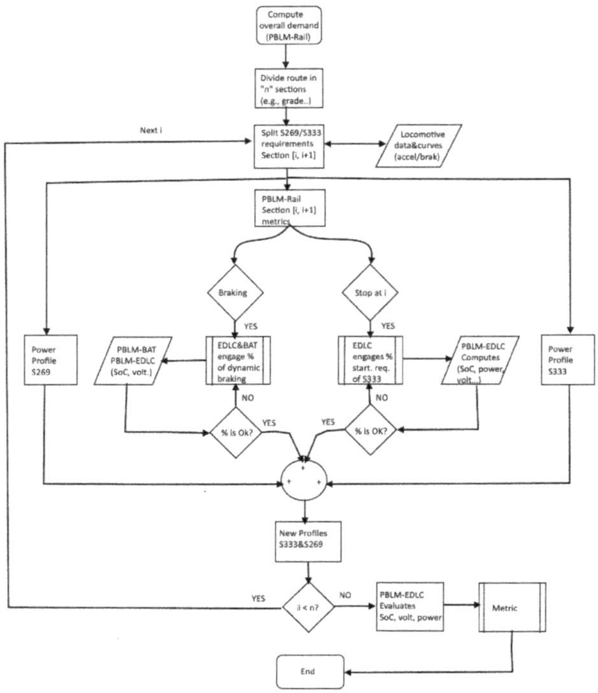

Figure 19. Flow diagram of the power and energy management for S333 and S269 through the aggregation of PBLM-Rail and PBM-EDLC.

A simplified premise for initiating energy management is illustrated in Figure 20a. Essentially, the S269 locomotive compensates for the power surplus above 2230 kW (the rated power of the diesel S333) and absorbs the surplus braking energy (1650 kW for the S333). Naturally, it means that the train's initial acceleration is limited without the support of the hybrid unit. During the initial acceleration phase, both units contribute their forces, with the EDLC operating at a state of charge (SoC) near 100% at the CT station stop, particularly at the bifurcation point between the freight and passenger stations.

Prior to departure, the proposed algorithm calculates the energy requirements for an upcoming segment (e.g., based on time from the starting point or target speed) and allocates a percentage of the total power demand to the EDLC (e.g., 50%) for the next section of track (e.g., Sec-1, from CT to an intermediate station located 12 km away, as shown in Figure 20b). The algorithm ensures that the EDLC can meet this power requirement (see Figure 20c, indicated by the green "checking arrow") and verifies that DC bus voltage remains within the acceptable range through the PBLM-EDLC. If these conditions are met, the EDLC replenishes its energy reserves before the end of the section.

This strategy has a practical rationale: the train may need to stop at an intermediate station to allow passenger units to cross. In such cases, the EDLC should return to a SoC near 100% to be ready for another start. In the second section (Sec-2, Figure 20b), the EDLC remains idle since the train is running at its programmed speed, and power demand (as depicted in Figure 15a) is managed entirely by the diesel unit. The process repeats during acceleration in Sec-3 (see Figure 20b), which corresponds to a scheduled speed increase near the TP station.

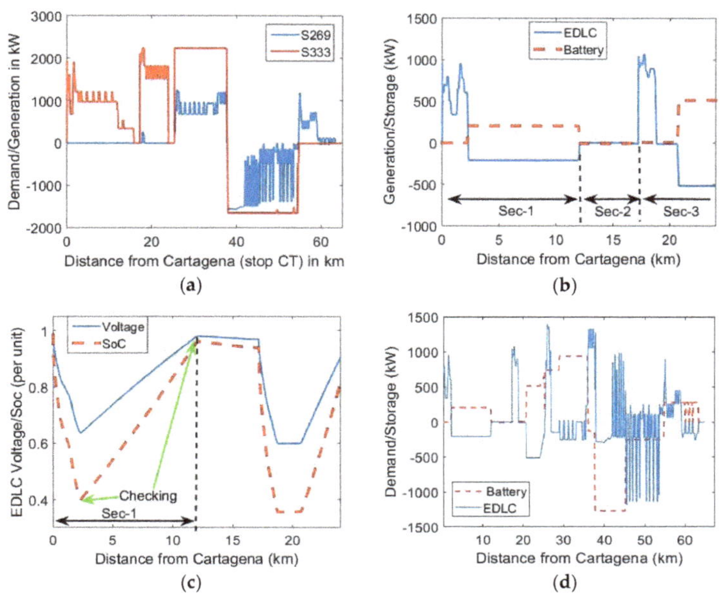

Figure 20. Main steps for the proposed management of EDLC: (**a**) Initial simplified profiles for diesel and hybrid locomotives: (**b**) Definition of EDLC duties in route sections 1 and 2; (**c**) testing of proposed duties through PBLM-EDLC; (**d**) EDLC and battery profiles after the evaluation of the overall route.

The same methodology is applied to the remaining eight sections of the track, with the EDLC handling power demands or storing energy during fluctuations as needed. The results of this process are shown in Figure 20d.

Finally, Figure 21 shows the main results for the overall route (from CT to MC) computed through PBLM-EDLC. Figure 21a shows EDLC voltage (notice that EDLC modules allow a certain increase with respect to rated voltage, e.g., 16.5 V over 16 V; see Table 2). Figure 21b shows SoC ranges (remember that they are design constraints; see Table 9), whereas Figure 21c shows currents from or to OESD (EDLC and batteries) to test overloading of the OESD. Figure 21 depicts a very important concern solved through the EDLC model: the accuracy of the power generated or consumed by the EDLC. This is different from the profile depicted in Figure 20d because the voltage of EDLC changes and voltage conditions the evaluation of the right instantaneous power.

As stated before, the objective of this work is not to refine or optimize the design of EDLC but to demonstrate through a plausible scenario the advantages and pre-feasibility of the integration and aggregation of PBLM models. Optimization and improvement of OESD tasks will be engaged in future works. To summarize the advantages of the hybrid locomotive, some representative power figures are presented in Table 9.

Regarding fuel consumption, we have used the throttle position given by the PBLM-Rail simulator (see Figure 16b) and early 80s data from historical EMD locomotives SD40 [59]. These data correspond to fuel consumption (gal/h) for each throttle position (1–8) and idle of the GP and SD series. Authors assume that the J26T model (the original EMD international platform for S333) corresponds to those series. With these hypotheses, our S333 demands around 2.7 L/km. Consequently, we reduce the consumption of one S333 and some additional energy demanded by the remaining unit that does not haul in the electrified section. Also, the reader should note that an electrical linkage between locomotives (e.g., close to the roofline used in high-speed trains, see, for example, [60]) enables the recovery of an additional 11.4% of braking energy from the diesel unit (Table 11).

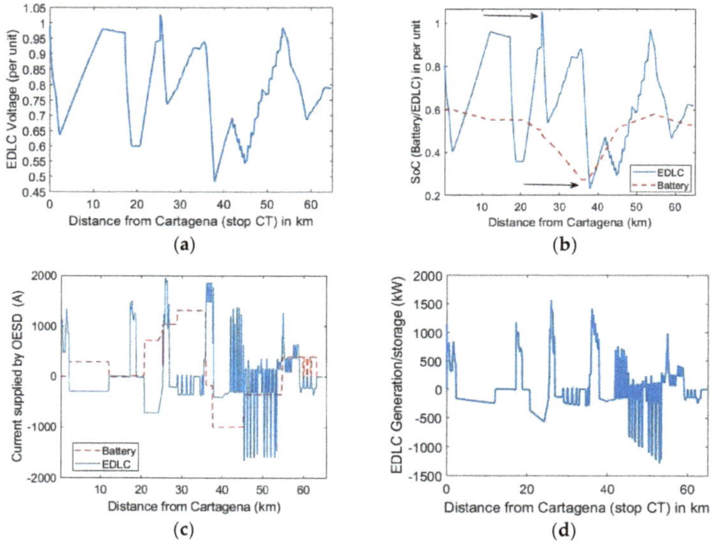

Figure 21. Results for the proposed management of OESD through PBLM: (**a**) EDLC voltage in p.u.; (**b**) EDLC SoC; (**c**) Currents in the EDLC; (**d**) Revisited power of EDLC.

Table 11. Main results from simulation (route Cartagena–Murcia).

Indicator	EDLC	Locomotive S269 (Hybridized)	Locomotive Diesel S333
Max Power	1570 kW	1450 kW	2200 kW
Δ Energy consumption (fuel to electricity)	N/A	150.9 kWh	−92.98 kWh [1]
Energy Recovered from resistive braking	5.45%	25.33%	11.36%
SoC range (max/min)	1.05/0.24	N/A	N/A
Δ Fuel demand/day	N/A	-	−349 L [2]
Δ Electricity demand/day [3]	140–160 kWh	220–250 kWh	-
Service in km/day [4] (km/year)	-	260 (78000)	260 (78000)
Δ Energy costs [4] (year)	-	6.48 k€	251.8 k€

[1] Unit coupled with S269. [2] Unit coupled and decommissioned. [3] Four services/day. [4] 300 days/year.

With respect to costs, the option seems feasible, with a service ranging 78,000 km/year; the hybrid unit can recover capital costs in 6–7 years without considering GHG savings and the benefits of having a 1.5 MW distributed generator in the RPS and a medium-power last-mile unit in this railway area to develop shunting and switching tasks.

4. Conclusions

This paper explores an alternative solution for the hybridization of freight locomotives operating on low-traffic routes, aiming to maintain their hauling capabilities while capitalizing on regenerative braking. The central hypothesis involves refurbishing mid-life conventional locomotives to extend their operational lifetime, an idea already proposed by some manufacturers for diesel-electric units. To achieve these objectives, an on-board energy storage device comprising electric double-layer capacitors and batteries is proposed. While EDLC technology has been applied to tramways, it remains largely unexplored for freight locomotives, with only a few battery-electric locomotives currently in operation.

To evaluate the feasibility of this approach, a simple, accurate, and physically based EDLC model was developed and tested in the laboratory. The results demonstrate that the model achieves an acceptable level of accuracy in capturing voltage and power fluctuations associated with EDLC charge and discharge cycles.

Using these models, the paper presents a practical example of OESD sizing for energy management in a hybrid locomotive (diesel-electric and hybrid-electric) hauling a 1300-ton freight train. The proposed methodology highlights the importance of integrating physical models of railway units with EDLC dynamics, offering a preliminary framework for validating hybrid locomotive performance in real-world scenarios. The sizing process incorporates the latest advancements in EDLC and battery technologies, ensuring that the proposed OESD specifications are both credible and achievable.

In conclusion, hybridizing locomotives with EDLC and batteries enables the recovery of approximately 40–50% of braking energy on routes with gradients of 1.0–1.2% while also reducing fuel consumption—and thus greenhouse gas emissions—by around 30%. This reduction could serve as a catalyst for shifting freight transport from road to rail. Additionally, on electrified tracks, hybrid units can act as mobile generation devices, supplying power to electric trains or restoring service during Overhead Line Equipment (OLE) failures.

Future work will focus on developing detailed physical models for batteries and refining the OESD design through additional simulations and modeling of storage devices.

Author Contributions: Conceptualization: A.G. (Antonio Gabaldón) and M.C.R.-A.; methodology: A.G. (Antonio Gabaldón), F.M. and M.C.R.-A.; software: A.G. (Antonio Gabaldón) and A.G. (Antonio Guillamón); validation: A.G. (Antonio Gabaldón) and A.G. (Antonio Guillamón); formal analysis: F.M. and M.C.R.-A.; investigation: all authors; resources: M.C.R.-A. and A.G. (Antonio Guillamón); writing—original draft preparation: all authors; writing—review and editing: A.G. (Antonio Gabaldón) and M.C.R.-A.; visualization: A.G. (Antonio Guillamón); supervision: A.G. (Antonio Gabaldón) and M.C.R.-A.; project administration: A.G. (Antonio Gabaldón) and M.C.R.-A.; funding acquisition: A.G. (Antonio Gabaldón) and M.C.R.-A. All authors have read and agreed to the published version of the manuscript.

Funding: Grant TED2021-129722B-C32 funded by MICIU/AEI/10.13039/501100011033 and by the European Union Next Generation EU/PRTR.

Institutional Review Board Statement: Not applicable.

Informed Consent Statement: Not applicable.

Data Availability Statement: The data presented in this study are available on request from the corresponding author. The data are not publicly available due to confidentiality concerns of train manufacturers and operators that have provided some data for simulation studies.

Acknowledgments: Authors are very grateful to the Agencia Estatal de Investigación (Spanish Government) for its financial support through the research project TED2021-129722B-C32 and to the European Union for providing additional funds and research support through the Next Generation/PRTR program.

Conflicts of Interest: The authors declare no conflicts of interest.

Abbreviations

BEL	Battery Electric Locomotive
BESS	Battery Energy Storage System
BMU	Battery Multiple Unit
DBEMU	Diesel Battery Electrical Multiple Unit
DER	Distributed Energy Resources

DMU	Diesel Multiple Unit
DG	Distributed Generation
DR	Demand Response
DTM	Differential Transformation Method
EDLC	Electric Double Layer Capacitor (or super/ultra capacitor)
EMD	Electro-Motive Division (1922–2004)
EMU	Electric Multiple Unit
ESS	Energy Storage Systems
DHEMU	Dual Hybrid Electric Multiple Unit
IVP	Initial Value Problem
Li-ion	Lithium-ion battery
LTO	Lithium Titanate Oxide
OESD	On-board Energy Storage Device
OLE	Overhead Line Equipment
PBLM	Physical-Based Load Modeling
PPS	Public Power System
RENFE	Spanish Railway Operator
RPS	Railway Power System
SoC	State of charge (battery/EDLC(OESD)

References

1. Statistics | Eurostat. Length of Railway Lines by Number of Tracks and Electrification of Line. Available online: https://ec.europa.eu/eurostat/databrowser/explore/all/transp?lang=en&subtheme=rail.rail_if&display=list&sort=category&extractionId=rail_if_line_tr (accessed on 16 September 2024).
2. Eurostat. Modal Split of Inland Freight Transport. Statistics | EUROSTAT. Modal Split of Inland Freight Transport. Available online: https://ec.europa.eu/eurostat/databrowser/view/tran_hv_frmod/default/table?lang=en (accessed on 13 February 2025).
3. Zero-Emission Locomotive Technologies: Pathways for U.S. Rail Decarbonization—International Council on Clean Transportation. Available online: https://theicct.org/publication/zero-emission-locomotive-technologies-pathways-for-u-s-rail-decarbonization-jun24/ (accessed on 23 January 2025).
4. European Union. Shift to Rail (S2R) European Rail Initiative. Available online: https://rail-research.europa.eu/about-shift2rail/ (accessed on 7 January 2025).
5. Eurostat. Transport Equipment Statistics—Statistics Explained. Available online: https://ec.europa.eu/eurostat/statistics-explained/index.php?title=Transport_equipment_statistics#Railway_transport_equipment:_more_electrical_energy.2C_fewer_seats (accessed on 1 October 2024).
6. National Commission on Markets and Competition (CNMC). Annual Report of Railways Transportation 2023 CNMC. (In Spanish). Available online: https://www.cnmc.es/expedientes/infdtsp03324 (accessed on 7 January 2025).
7. Saeed, M.; Briz, F.; Guerrero, J.M.; Larrazabal, I.; Ortega, D.; Lopez, V.; Valera, J.J. Onboard Energy Storage Systems for Railway: Present and Trends. *IEEE Open J. Ind. Appl.* **2023**, *4*, 238–259. [CrossRef]
8. Domínguez, M.; Fernández-Cardador, A.; Fernández-Rodríguez, A.; Cucala, A.; Pecharromán, R.; Sánchez, P.U.; Cortázar, I.V. Review on the use of energy storage systems in railway applications. *Renew. Sustain. Energy Rev.* **2025**, *207*, 114904. [CrossRef]
9. Kang, J.; Wen, J.; Jayaram, S.H.; Yu, A.; Wang, X. Development of an equivalent circuit model for electrochemical double layer capacitors (EDLCs) with distinct electrolytes. *Electrochim. Acta* **2014**, *115*, 587–598. [CrossRef]
10. Akkutriebwagen ETA 150, Battery Multiple Unit ETA 150. Available online: https://ralphhaussler.weebly.com/akkutriebwagen.html (accessed on 30 September 2024).
11. Caterpillar | Take a Ride Through the 100-Year History of EMD®. Available online: https://www.caterpillar.com/en/news/caterpillarNews/2022/emd-100-years.html (accessed on 22 January 2025).
12. Webinar on Alternatives to Fossil Diesel Use in Railways—Association of European Rail Rolling Stock Lessors. Available online: https://aerrl.eu/webinar-on-altnernatives-to-fossil-diesel-use-in-railways/ (accessed on 23 January 2025).
13. BNSF Railway Company. Zero-and Near Zero-Emission Freight Facilities Project (Zanzeff) Data Acquisition Support. Available online: https://ww2.arb.ca.gov/sites/default/files/2022-11/zanzeff-bnsf-belreport.pdf (accessed on 7 January 2025).
14. SRoyston, J.; Gladwin, D.T.; Stone, D.A.; Ollerenshaw, R.; Clark, P. Development and Validation of a Battery Model for Battery Electric Multiple Unit Trains. In Proceedings of the IECON 2019—45th Annual Conference of the IEEE Industrial Electronics Society, Lisbon, Portugal, 14–17 October 2019; pp. 4563–4568. [CrossRef]

15. Schenker, M.; Schirmer, T.; Dittus, H. Application and improvement of a direct method optimization approach for battery electric railway vehicle operation. *Proc. Inst. Mech. Eng. Part F J. Rail Rapid Transit* **2021**, *235*, 854–865. [CrossRef]
16. Magelli, M.; Boccardo, G.; Bosso, N.; Zampieri, N.; Farina, P.; Tosetto, A.; Mocera, F.; Somà, A. Feasibility study of a diesel-powered hybrid DMU. *Railw. Eng. Sci.* **2021**, *29*, 271–284. [CrossRef]
17. Abduraxman, B.M.; Harrison, T.; Ward, C.P.; Midgley, W.J. Optimal intermittent electrification and its effect on battery sizing and energy saving using a high-fidelity train model. *Proc. Inst. Mech. Eng. Part F J. Rail Rapid Transit* **2023**, *237*, 1205–1218. [CrossRef]
18. Spiryagin, M.; Wu, Q.; Wolfs, P.; Sun, Y.; Cole, C. Comparison of locomotive energy storage systems for heavy-haul operation. *Int. J. Rail Transp.* **2018**, *6*, 1–15. [CrossRef]
19. Gabaldón, A.; García-Garre, A.; Ruiz-Abellón, M.C.; Guillamón, A.; Molina, R.; Medina, J. Management of Railway Power System Peaks with Demand-Side Resources: An Application to Periodic Timetables. *Sustainability* **2023**, *15*, 2746. [CrossRef]
20. Allenbach, J.-M.; Chapas, P.; Comte, M.; Kaller, R. Traction électrique (EAN13: 9782880746742) | EPFL PRESS. Available online: https://www.epflpress.org/produit/364/9782880746742/traction-electrique (accessed on 7 January 2025).
21. Wu, Q.; Spiryagin, M.; Cole, C. Train energy simulation with locomotive adhesion model. *Railw. Eng. Sci.* **2020**, *28*, 75–84. [CrossRef]
22. Ogawa, T.; Manabe, S.; Yoshikawa, G.; Imamura, Y.; Kageyama, M. Method of Calculating Running Resistance by the Use of the Train Data Collection Device. *Q. Rep. RTRI* **2017**, *58*, 21–27. [CrossRef]
23. Rochard, B.P.; Schmid, F. A review of methods to measure and calculate train resistances. *Proc. Inst. Mech. Eng. Part F J. Rail Rapid Transit* **2000**, *214*, 185–199. [CrossRef]
24. Wu, Q.; Wang, B.; Spiryagin, M.; Cole, C. Curving resistance from wheel-rail interface. *Veh. Syst. Dyn.* **2022**, *60*, 1018–1036. [CrossRef]
25. Wu, Q.; Yang, X.; Jin, X. Hill-starting a heavy haul train with a 24-axle locomotive. *Proc. Inst. Mech. Eng. Part F J. Rail Rapid Transit* **2022**, *236*, 201–211. [CrossRef]
26. Lithium SuperPack 12,8V & 25,6V—Victron Energy. Available online: https://www.victronenergy.com/batteries/12,8v-lithium-superpack (accessed on 22 January 2025).
27. Ultracapacitor Modules—Maxwell Technologies. Available online: https://maxwell.com/products/ultracapacitors/modules/ (accessed on 22 January 2025).
28. Power Systems Group, UPCT. Demand Response Web Page. Available online: http://demandresponse.eu/ (accessed on 22 January 2025).
29. IP-Symcon: Automation Software. Available online: https://www.symcon.de/en/product/corporation/ (accessed on 22 January 2025).
30. Laadjal, K.; Cardoso, A.J.M. A review of supercapacitors modeling, SoH, and SoE estimation methods: Issues and challenges. *Int. J. Energy Res.* **2021**, *45*, 18424–18440. [CrossRef]
31. Zhang, L.; Hu, X.; Wang, Z.; Sun, F.; Dorrell, D.G. A review of supercapacitor modeling, estimation, and applications: A control/management perspective. *Renew. Sustain. Energy Rev.* **2018**, *81*, 1868–1878. [CrossRef]
32. Zubieta, L.; Bonert, R. Characterization of double-layer capacitors for power electronics applications. *IEEE Trans. Ind. Appl.* **2000**, *36*, 199–205. [CrossRef]
33. Nejad, S.; Gladwin, D.; Stone, D. A systematic review of lumped-parameter equivalent circuit models for real-time estimation of lithium-ion battery states. *J. Power Sources* **2016**, *316*, 183–196. [CrossRef]
34. Cui, X.; Zeng, J.; Zhang, H.; Yang, J.; Qiao, J.; Li, J.; Li, W. Optimization of the lumped parameter thermal model for hard-cased li-ion batteries. *J. Energy Storage* **2020**, *32*, 101758. [CrossRef]
35. Zhou, J.K. *Differential Transform and Its Applications for Electrical Circuits*; Huarjung University Press: Wuhan, China, 1986.
36. Chang, S.-H.; Chang, I.-L. A new algorithm for calculating one-dimensional differential transform of nonlinear functions. *Appl. Math. Comput.* **2008**, *195*, 799–808. [CrossRef]
37. Batiha, K.; Batiha, B. A New Algorithm for Solving Linear Ordinary Differential Equations. *World Appl. Sci. J.* **2011**, *15*, 1774–1779.
38. Ertürk, V.S. Differential Transformation Method for Solving Differential Equations of Lane-Emden Type. *Math. Comput. Appl.* **2007**, *12*, 135–139. [CrossRef]
39. Mukherjee, S.; Roy, B.; Chaterjee, P.K. Solution of Lane-Emden Equation by Differential Transform Method. 2011. Available online: https://www.semanticscholar.org/paper/Solution-of-Lane-Emden-Equation-by-Differential-Mukherjee-Roy/bf451be69930bfbd7d770f382ae0e4f641cf50ee (accessed on 23 January 2025).
40. Khan, Y.; Svoboda, Z.; Šmarda, Z. Solving certain classes of Lane-Emden type equations using the differential transformation method. *Adv. Differ. Equ.* **2012**, *2012*, 174. [CrossRef]
41. Baumgartner, J.P. *Prices and Costs in the Railway Sector*; École Polytechnique Fédérale de Lausanne: Lausanne, Switzerland, 2001.
42. Fighting Planned Obsolescence in Our TGVs | SNCF Group. Available online: https://www.groupe-sncf.com/en/commitments/sustainable-development/fighting-planned-obsolescence (accessed on 23 January 2025).

43. García-Garre, A.; Gabaldón, A. Analysis, Evaluation and Simulation of Railway Diesel-Electric and Hybrid Units as Distributed Energy Resources. *Appl. Sci.* **2019**, *9*, 3605. [CrossRef]
44. Ten Spanish Companies Join Forces to Apply Hydrogen Propulsion to a High-Speed Train for the First Time—Talgo. Available online: https://www.talgo.com/en/ten-spanish-companies-join-forces-to-apply-hydrogen-propulsion-to-a-high-speed-train-for-the-first-time (accessed on 30 September 2024).
45. Spanish Government. Transporte Ferroviario de Mercancías | OTLE. Available online: https://otle.transportes.gob.es/inform/es/2020/la-movilidad/-transporte-ferroviario/transporte-ferroviario-de-mercancias (accessed on 7 January 2025).
46. Mitsubishi Electric Railway Systems—Mitsubishi Electric Railway: Mitsubishi Electric Railway. Available online: https://www.mitsubishielectric-railway.eu/ (accessed on 22 January 2025).
47. RENFE Rolling Stock Hiring (Alquiler de Material Ferroviario). Available online: https://www.renfe.com/es/en/renfe-group/corporations/renfe-alquiler/rent-get-to-know-us (accessed on 22 January 2025).
48. BORDLINE CC for Diesel-Electric Multiple Units—BORDLINE Compact Converters | BORDLINE® Compact Converters | ABB. Available online: https://new.abb.com/electric-drivetrains/compact-converters/bordline-cc-for-diesel-electric-multiple-units (accessed on 23 January 2025).
49. Ding, D.; Wu, X.-Y. Hydrogen fuel cell electric trains: Technologies, current status, and future. *Appl. Energy Combust. Sci.* **2024**, *17*, 100255. [CrossRef]
50. Mitsubishi Electric Corporation. Locomotoras Mitsubishi BB Chopper de 3000V CC y 3100kW para RENFE. (In Spanish). Available online: https://www.aafmadrid.es/wp-content/uploads/2022/04/Folleto600lres.pdf (accessed on 7 January 2025).
51. Tank Wagon Zags 113 m^3 P27BH Wagons | Ermewa Wagon Hire. Available online: https://www.ermewa.com/en/wagons/zags-113m3-2/ (accessed on 30 December 2024).
52. Wang, J.; Purewal, J.; Liu, P.; Hicks-Garner, J.; Soukazian, S.; Sherman, E.; Sorenson, A.; Vu, L.; Tataria, H.; Verbrugge, M.W. Degradation of lithium ion batteries employing graphite negatives and nickel–cobalt–manganese oxide + spinel manganese oxide positives: Part 1, aging mechanisms and life estimation. *J. Power Sources* **2014**, *269*, 937–948. [CrossRef]
53. Fedele, E.; Iannuzzi, D.; Del Pizzo, A. Onboard energy storage in rail transport: Review of real applications and techno-economic assessments. *IET Electr. Syst. Transp.* **2021**, *11*, 279–309. [CrossRef]
54. Hybrid TER Now in Nouvelle-Aquitaine | SNCF Group. Available online: https://www.groupe-sncf.com/en/innovation/decarbonization-trains/hybrid-ter (accessed on 23 January 2025).
55. ÖBB. Cityjet Eco. The First Certified Battery-Powered Train in Europe. 2023. Available online: https://www.oebb.at/dam/jcr:b697320b-4493-48f2-a9f4-b8ff3af4d4ec/datenblatt-cityjet-eco.pdf (accessed on 18 February 2025).
56. ProgressRail | Rumo Invests in Hybrid Locomotives from Progress Rail, Reducing Environmental Impact of Freight Operations. Available online: https://www.progressrail.com/en/Company/News/PressReleases/RumoinvestsinhybridlocomotivesfromProgressRailreducingenvironmentalimpactoffreightoperations.html (accessed on 23 January 2025).
57. ProgressRail | EMD® Joule. Available online: https://www.progressrail.com/en/Segments/Locomotive/Locomotives/FreightLocomotives/EMDJoule.html (accessed on 23 January 2025).
58. Dual Voltage 750V/3000V Electric Traction Converter—Railway Systems CAF Power & Automation. Available online: https://www.cafpower.com/en/systems/electric-traction-systems/dc-converters/dual-750v-3000v-dc-converter (accessed on 23 January 2025).
59. Heritage Rail (USA). EMD Fuel Consumption Data for Various Models. Available online: https://heritagerail.org/wp-content/uploads/2022/09/EMD-TABLE-FUEL-CONS-NOTCH-HP-VARIOUS-MODELS.pdf (accessed on 6 January 2025).
60. Rail Roofline Systems, Products, and Service Solutions | TE Connectivity. Available online: https://www.te.com/en/industries/rail/applications/roofline-solutions.html (accessed on 23 January 2025).

applied sciences

MDPI

Article

PEMFC RUL Prediction for Non-Stationary Time Series Based on Crossformer Model

Ning Zhou [1], He Zeng [1], Zefei Zheng [1], Ke Wang [2] and Jianxin Zhou [1,*]

[1] School of Information Engineering, Wuhan University of Technology, Wuhan 430070, China; zhouning@whut.edu.cn (N.Z.); 334608@whut.edu.cn (H.Z.); 276240@whut.edu.cn (Z.Z.)

[2] School of Information Engineering, Wuhan Huaxia Institute of Technology, Wuhan 430223, China; 276028@whut.edu.cn

* Correspondence: zjx@whut.edu.cn

Abstract: Proton-Exchange Membrane Fuel Cells (PEMFCs), as efficient and environmentally friendly energy conversion devices, have wide application potential in areas such as transportation, mobile power, and distributed energy. However, the remaining useful life (RUL) issue of PEMFCs has been one of the main challenges limiting their commercialization. The RUL prediction problem of PEMFCs exhibits characteristics of time series forecasting, but its data possess multidimensional features and non-stationarity, which limits the applicability of classical time series forecasting models like the Transformer in solving the RUL prediction problem. In this paper, we propose a PEMFC RUL prediction model based on the Crossformer for non-stationary time series (De-stationary-Crossformer). Firstly, the overall architecture adopts the Crossformer model to extract dependencies between different features and temporal dependencies. Secondly, adaptive normalization is applied to the data to mitigate the non-stationarity in the original data, thereby increasing their predictability. Subsequently, a non-stationary attention mechanism is introduced in the model to simultaneously utilize the non-stationarity in the original data when extracting deep information. Additionally, manual features are introduced through mathematical statistics to enhance the predictive performance of the model. During the training process, the TILDE-Q loss function is used to focus on the similarity between the predicted sequence and the true sequence. The model proposed in this paper improves the MSE by 31% compared to the Transformer and 23% compared to the Crossformer in the experimental prediction of the RUL of PEMFCs in actual vehicles.

Keywords: PEMFC; RUL prediction; time series forecasting; multidimensional features; non-stationary

1. Introduction

With the continuous growth of global energy demand and increasing attention given to environmental sustainability, research on clean and efficient energy conversion technologies has become particularly important. PEMFCs have attracted attention due to their significant energy density, fast start-up and response, high efficiency, low pollution, and convenient fuel supply [1]. However, the RUL issue of PEMFCs limits their reliability and economic value in commercial applications. The RUL of PEMFCs is defined as the time interval from the current moment to the moment of fuel cell failure. Currently, most research selects the voltage of the fuel cell as the measure of fuel cell failure. For example, Wang et al. [2] use the voltage decay of PEMFCs to represent performance degradation in predicting the RUL of fuel cells. Due to the complexity of operating conditions during the operation of vehicles equipped with fuel cells, selecting the appropriate voltage decay to represent the RUL of fuel cells is also an urgent problem that needs to be addressed.

In recent years, many researchers have devoted themselves to the prediction of the steady-state voltage of PEMFCs. Research methods can be divided into two categories: model-driven and data-driven. Liu et al. [3] used a two-stage hybrid prediction method to predict the RUL of PEMFCs. In the first stage, the long-term degradation trend in PEMFCs was predicted using evolutionary algorithms and neuro-fuzzy inference methods. The neuro-fuzzy inference method employed was adaptive. In the second stage, based on the data from the first stage, a semi-empirical degradation model of PEMFCs and an adaptive Kalman filter algorithm were combined for RUL prediction. The feasibility of this method was demonstrated through experiments. Zhou et al. [4], while retaining the simplicity of the original gray model and the rapid prototyping design, proposed a new type of error-corrected gray prediction model using error correction factors to eliminate the inherent errors of the original gray model. This method demonstrated excellent effectiveness and robustness in aging experiments of lithium batteries and fuel cells and could achieve relatively accurate predictions even over longer time steps. Peng et al. [5] proposed predicting the RUL of PEMFCs using Convolutional Neural Network–Long Short-Term Memory (CNN-LSTM) and Convolutional Neural Network–Bidirectional Long Short-Term Memory (CNN-BiLSTM). They found that the addition of convolutional networks effectively improved the prediction accuracy of LSTM networks. Wang et al. [6], noting the suboptimal long-term predictive performance of LSTM neural networks, proposed Navigation Sequence-Driven LSTM (NSD-LSTM). This method demonstrated high accuracy in long-term prediction capabilities on aging experimental data from two different types of PEMFCs and under different operating conditions.

The prediction of the steady-state voltage of PEMFCs falls under the realm of time series forecasting. Recently, in the field of time series forecasting, Ragab et al. [7] utilized an attention-based sequence-to-sequence model (ATS2S) and proposed joint optimization with reconstruction loss to endow the model with higher predictive capabilities. Sano et al. [8] in 2021 applied LSTnet (long- and short-term time series network) to the domain of multidimensional time series forecasting. Hewage et al. [9] proposed the Time Convolutional Network (TCN) for rapidly processing long time series and extracting comprehensive information. Abbasimehr et al. [10] in 2022 introduced a time series forecasting model based on Long Short-Term Memory (LSTM) networks and the multi-head attention mechanism (LSTM-ATTN). The Transformer [11] neural network was initially designed to tackle Natural Language Processing (NLP) tasks. However, over time, Transformers have demonstrated outstanding performance in fields such as Computer Vision (CV), time series, and others. The Crossformer model proposed by Zhang et al. [12] effectively extracts cross-temporal dependencies and cross-dimensional dependencies through two-stage attention mechanisms.

In time series prediction, we generally consider stationary time series, but for PEMFC data, there exists strong non-stationarity. Passalis et al. [13] proposed DAIN, which adaptively stabilizes time series using nonlinear neural networks. Samal et al. [14] employed the Seasonal Autoregressive Integrated Moving Average (SARIMA) model to handle non-stationary time series with seasonality and trends. Kim et al. [15] utilized two-stage normalization to mitigate non-stationarity in the original data and enhance the predictability of the model. Liu et al. [16] introduced a non-stationarity attention mechanism to capture non-stationarity in data.

The main contributions of this paper lie in the following aspects:

(1) This paper proposes a remaining useful life (RUL) prediction model for Proton-Exchange Membrane Fuel Cells (PEMFCs) based on Crossformer. This model leverages a two-stage attention mechanism, deep adaptive normalization, and a non-

stationary attention mechanism to address the limitations of traditional Transformers in capturing dependencies between different features, the information loss caused by conventional normalization methods, and the traditional attention mechanism's neglect of non-stationarity information across different dimensions.

(2) Existing time series prediction models commonly use the mean squared error (MSE) as the loss function, which primarily focuses on the discrepancy between predicted and actual values at each time step, neglecting the overall trend and shape of the sequence. To address this limitation, this paper introduces TILDE-Q as the loss function, which accounts for both the gap in and shape of the predicted values, enabling the model to better capture the overall trend and structure of the sequence during training.

(3) Real datasets from logistics vehicles equipped with PEMFCs are utilized to compare the proposed model with five mainstream time series forecasting models.

2. Problem Formulation

This section will define concepts related to PEMFC lifetime prediction modeling and formalize the relevant research questions.

Definition 1. *The number of occurrences of fixed-range current values during PEMFC operation is statistically calculated. The range with the highest occurrence is defined as the steady-state current value, as shown in Equation (1).*

$$I_{steady} = MAX(f_I) \tag{1}$$

where I_{steady} represents the steady-state current and f_I denotes the frequency of occurrence of the current during operation. Figure 1 shows the frequency of different currents during the operation of the logistics vehicle's PEMFC, derived through mathematical statistical methods. The results indicate that currents in the range of 10–100 A occur less frequently, likely because the vehicle relies mainly on the battery for discharge during these conditions, and the fuel cell is either not operating or generating only minimal output. Additionally, while the current range of 100–110 A exhibits a relatively high occurrence frequency, it is more likely associated with low-speed or light-load operating conditions, where the fuel cell output is relatively unstable and less representative of overall system operation. Therefore, despite its high frequency, this range is not suitable for use as the steady-state current. In contrast, the current range of 198.4–199.4 A exhibits the highest occurrence frequency and better represents the stable output of the fuel cell under primary operating conditions. Thus, it is identified as the steady-state current of the PEMFC installed in the logistics vehicle.

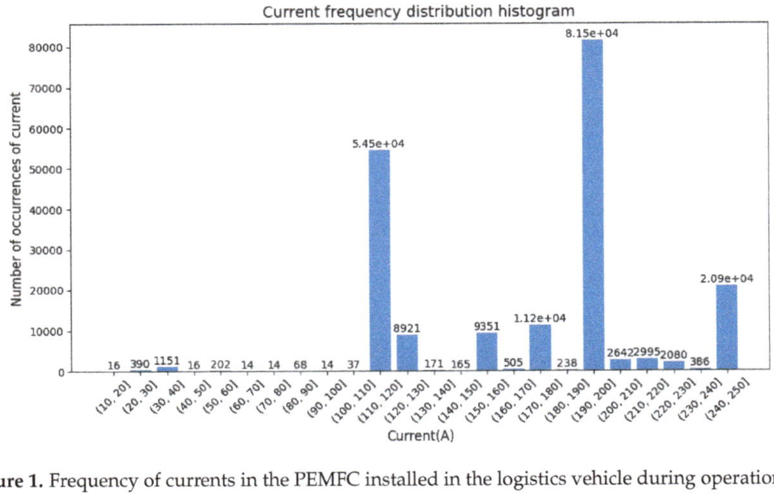

Figure 1. Frequency of currents in the PEMFC installed in the logistics vehicle during operation.

Definition 2. *The voltage corresponding to the steady-state current during operation is defined as the steady-state voltage, as shown in Equation* (2).

$$U_{steady} = U\left(I = I_{steady}\right) \tag{2}$$

where U_{steady} represents the steady-state voltage. Figure 2 illustrates the steady-state voltage values of the PEMFC installed in the logistics vehicle, with 198.4–199.4 A as the steady-state current, along with linear fitting. The primary purpose of this linear fitting is to verify the stability of the steady-state voltage. If the fitting results show significant fluctuations or an unclear trend, they may indicate instability or abnormalities in the operational state of the fuel cell. In such cases, it would be necessary to first address any anomalies or unstable data. However, as observed from the results in Figure 2, the steady-state voltage shows relatively stable variations, thus making it suitable for subsequent predictions of remaining useful life.

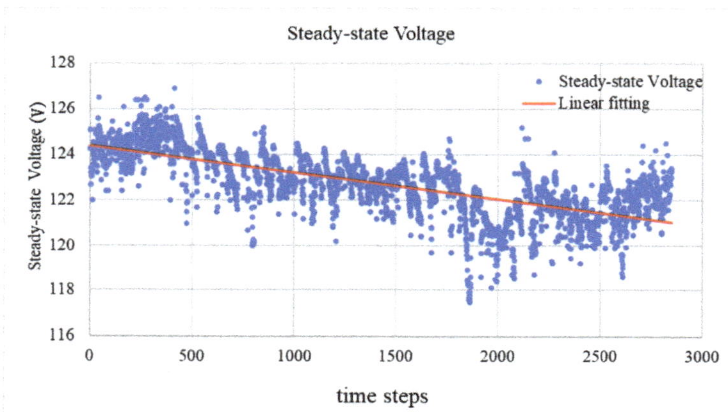

Figure 2. Steady-state voltage of the PEMFC installed in the logistics vehicle during operation and linear fitting.

Definition 3. *In this study, the steady-state voltage of the PEMFC during operation is adopted as the evaluation criterion for the degradation of remaining useful life, based on research findings of the demonstrated operation of various fuel cell vehicles in typical regions conducted by Tsinghua University. It is utilized as the label value for the dataset used in this study.*

When a PEMFC is initially activated for use, it undergoes an activation stage. After this stage, the steady-state voltage of the PEMFC will reach a peak, known as the initial voltage. This initial voltage represents 100% of the battery's life. Following the definition of failure in lithium-ion batteries [17], this paper considers a decrease in steady-state voltage to 80% of the initial voltage as a sign of PEMFC performance failure, indicating the depletion of its useful life and the need for timely shutdown and replacement.

Problem 1. *The essence of PEMFC RUL prediction is to predict the steady-state voltage values $y_{T+1:T+\tau} \in \mathbb{R}^{\tau \times 1}$ within a specified future time window, given a historical time series $x_{1:T} \in \mathbb{R}^{T \times D}$ [18]. Here, T represents the time steps in the historical sequence, and D denotes the dimensionality of the historical time series features, which includes data such as fuel cell voltage, current, hydrogen concentration, and other relevant parameters. $x_{1:T}$ represents the data of D feature dimensions at T time steps. τ represents the time steps of the prediction sequence. $y_{T+1:T+\tau}$ represents the output corresponding to a given historical time series.*

3. Methods

In this section, a prediction model based on Crossformer for non-stationary time series of RUL (De-stationary-Crossformer) is proposed, considering the multidimensional features and non-stationarity characteristics of the dataset in PEMFC RUL prediction, as illustrated in Figure 3.

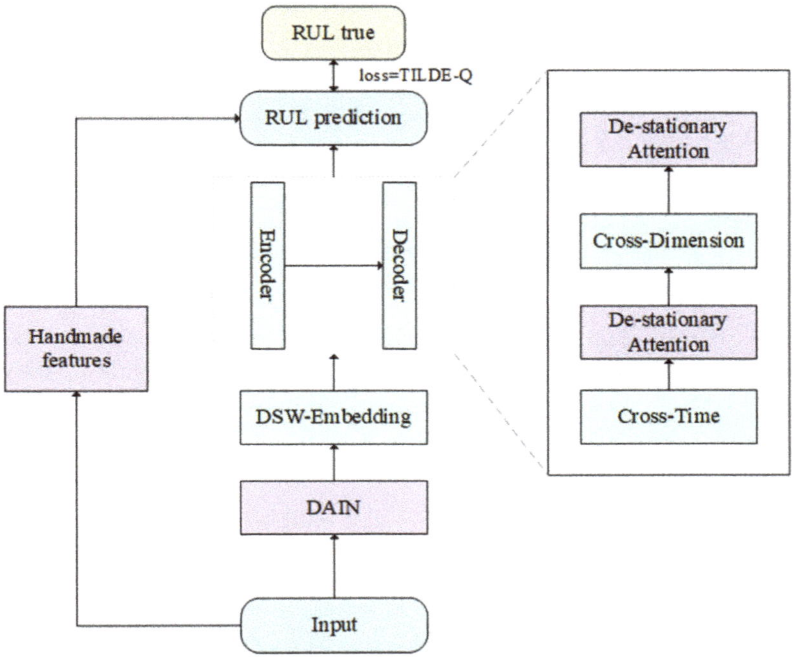

Figure 3. De-stationary-Crossformer model.

The data initially undergo deep adaptive instance normalization (DAIN) to mitigate non-stationarity and enhance predictability. Subsequently, the Crossformer model is employed as the main architecture to model the dependencies between feature dimensions and time dimensions. This architecture comprises Dimension-Segment-Wise Embedding (DSW Embedding) and an Encoder–Decoder with a Two-Stage Attention Layer. During attention computation, a De-stationary Attention (DSA) mechanism is used to extract non-stationarity information from the raw data. Additionally, handcrafted features are directly introduced from the original input data using mathematical statistics to enhance the predictive performance of the model. During model training, the TILDE-Q loss function is employed to encourage the model to consider the similarity between the predicted sequence and the true sequence, thereby enhancing prediction accuracy while preserving the non-stationarity characteristics of the original data.

3.1. DAIN

DAIN serves as a deep learning layer aimed at learning how to normalize based on the distribution of the data rather than relying on fixed normalization formulas. As depicted in Figure 4, this layer consists of three sub-layers. Its ultimate goal is to obtain the transformation factors α^i and β^i as described in Equation (3).

$$\widetilde{x}_{1:T} = \left(x_{1:T} - \alpha^i\right) \div \beta^i \quad (3)$$

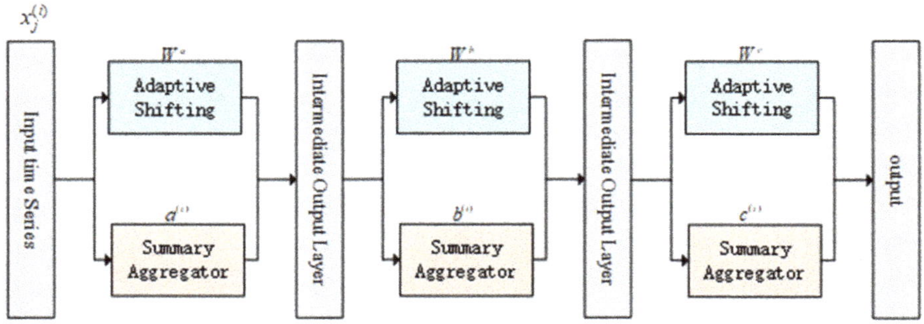

Figure 4. Structure of DAIN.

The first layer is primarily responsible for mapping the data to an appropriate region in the feature space. Through this step, it ensures that the input time series data have a suitable distribution in the feature space, aiding the model in better capturing the non-stationary characteristics of the data. For the input time series $x_{1:T} \in \mathbb{R}^{T \times D}$, the mean a^i within the T time range is initially calculated using Equation (4) to provide an initial estimate for the current time series, which is used to estimate the distribution of the sequence generated at the current time.

$$a^i = \frac{1}{T}\sum_{j=1}^{T} x_j \tag{4}$$

Then, the scaling factor α^i is calculated using Equation (5) and serves as the output of the first layer:

$$\alpha^i = W_a a^i \tag{5}$$

In this equation, W_a is a learnable parameter matrix. Subsequently, the output of the first layer is obtained through a linear transformation as given in Equation (6). This output is responsible for the moving measurements of the input time series in each dimension and is therefore also referred to as the output of the adaptive movement layer.

$$\hat{x}_{1:T} = x_{1:T} - W_a a^i \tag{6}$$

The task of the second layer is to linearly scale the data to increase or decrease their variance. This adjustment of the data's scale makes them more suitable for the learning process of the model. The process of linear scaling is dynamically adjusted based on the actual input time series, making it adaptive. Similarly, the standard deviation within the T time range is first calculated using Equation (7) as an initial estimate of the data offset.

$$b^i = \sqrt{\frac{1}{T}\sum_{j=1}^{T}\left(x - a^i\right)^2} \tag{7}$$

Then, the scaling factor β^i is calculated using Equation (8) to scale the moving measurements from the previous layer.

$$\beta^i = W_b b^i \tag{8}$$

In this equation, W_b is a learnable parameter matrix. The output of this layer can be calculated using Equation (9). This output is responsible for appropriately scaling the data based on the variance in the moving measurements.

$$\widetilde{x} = \left(x_{1:T} - W_a a^i\right) \div W_b b^i \tag{9}$$

Finally, the third layer performs gating operations, which nonlinearly suppress features unrelated to the target task. This helps the model focus more on task-relevant information, enhancing the model's ability to represent the input data. This layer uses the mean c^i of the output from the previous layer as an initial estimate, according to Equation (10):

$$c^i = \frac{1}{T}\sum_{j=1}^{T} \tilde{x}_j \tag{10}$$

Then, the sigmoid nonlinearity activation function is utilized to suppress features unrelated to the prediction. The specific calculation is performed using Equation (11) as follows:

$$\tilde{\tilde{x}} = \tilde{x} \odot sigmod\left(W_c c^i + d\right) \tag{11}$$

where W_c and d are learnable gating parameters. The entire process of deep adaptive instance normalization is adaptive, meaning that the normalization scheme depends on the actual time series input to the network, and this process is trainable. In this way, the model can dynamically adapt to the characteristics of different data distributions, enhancing its ability to model non-stationary data.

3.2. De-Stationary Attention

The loss of non-stationarity information in the data typically occurs within the depths of the model, especially during attention computation. To address this issue, this section introduces a non-stationarity attention mechanism [16]. The design objective of this mechanism is to capture the attention information of the data when they are non-stationary and to discover specific temporal dependencies in the original non-stationary data.

The calculation of the self-attention mechanism is performed as shown in Equation (12):

$$Attn(Q, K, V) = Softmax\left(\frac{QK^T}{\sqrt{d_k}}\right)V \tag{12}$$

where Q, K, V are obtained by the product of normalized data $\tilde{\tilde{x}}$ and different learnable weights, leading to a lack of non-stationary information in the process of attention information extraction. In the non-stationarity attention mechanism, they are computed as shown in Equation (13):

$$DeAttn\left(Q', K', V', \tau, \Delta\right) = Softmax\left(\frac{\tau Q' K'^T + \Delta^T}{\sqrt{d_k}}\right)V' \tag{13}$$

In this equation, Q', K', V' are also obtained by the product of normalized data and different learnable weights. To incorporate non-stationary information into attention computation, τ and Δ are introduced in this equation to compensate for the data's non-stationarity. These two non-stationarity factors are obtained from Equation (14):

$$\begin{cases} log\tau = MLP(\sigma_x, x) \\ \Delta = MLP(u_x, x) \end{cases} \tag{14}$$

In this equation, u_x and σ_x are statistical quantities extracted from the original unnormalized time series data. Specifically, u_x represents the mean of the input data x, while σ_x denotes its variance. These statistical measures are used to capture the underlying characteristics of the original data, which are essential for compensating for non-stationarity in the attention mechanism. The non-stationarity factors τ and Δ are shared by various neural network layers in the deep model that utilize the non-stationarity attention mechanism.

The non-stationarity attention mechanism is depicted in Figure 5. Through this mechanism, it is possible to learn temporal and cross-dimensional dependencies from the normalized stable data Q' and K', and to extract two non-stationarity factors τ and Δ from the original non-stationary data before normalization. These factors are then multiplied with the normalized values V', allowing them to benefit from the predictability of the stabilized sequence while retaining the inherent non-stationarity of the original time series.

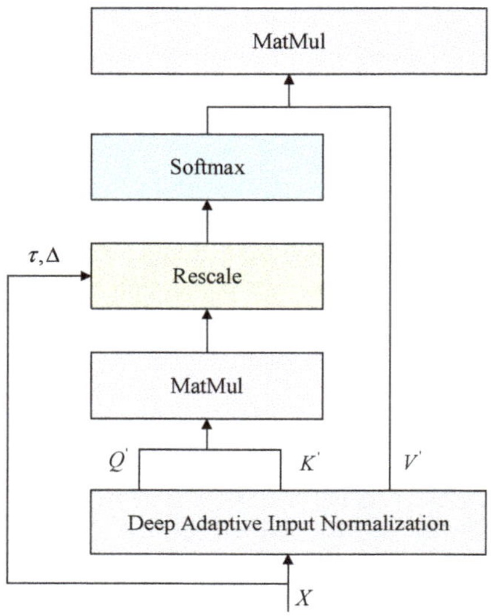

Figure 5. De-stationary Attention.

3.3. *Crossformer*

The Crossformer model is designed to address the modeling challenge of multidimensional features in the PEMFC dataset. It is composed of a multi-layer Encoder–Decoder architecture, which includes DSW Embedding and a Two-Stage Attention Layer.

In DSW Embedding, the input data $x_{1:T} \in \mathbb{R}^{T \times D}$ with multiple dimensions are segmented into time series segments of length L_{seq}. For computational convenience, L_{seq} is set to be a multiple of T, as shown in Equation (15):

$$x_{1:T} = \left\{ x_{i,d}^{(s)} | 1 \leq i \leq \frac{T}{L_{seq}}, 1 \leq d \leq D \right\} \tag{15}$$

The Two-Stage Attention Layer extracts dependencies between the temporal dimension and different feature dimensions, respectively.

In the cross-temporal dimension stage, attention information across the temporal dimension is extracted by applying Equation (16) to different times within each dimension, resulting in the attention information $\hat{Z}_{:,d}^{time}$:

$$\hat{Z}_{:,d}^{time} = LayerNorm\left(Z_{:,d} + DSA^{time}(Z_{:,d}, Z_{:,d}, Z_{:,d}) \right) \tag{16}$$

In this equation, $Z_{:,d}$ represents the vector of all time steps under dimension d. DSA denotes the computation using the non-stationary attention described in Section 3.2. LayerNorm indicates the application of standard normalization to the result. Subsequently, data information feature extraction is performed using Equation (17), where MLP denotes a multi-layer perceptron neural network.

$$Z^{time} = LayerNorm\left(\hat{Z}^{time} + MLP(\hat{Z}^{time})\right) \tag{17}$$

Utilizing the same approach, attention information across feature dimensions is extracted using Equation (18), resulting in $\hat{Z}_{i,:}^{dim}$:

$$\hat{Z}_{i,:}^{dim} = LayerNorm(Z_{i,:} + DSA(Z_{i,:}, Z_{i,:}, Z_{i,:})) \tag{18}$$

In this equation, $Z_{i,:}$ represents the vector of all feature dimensions at the *i-th* time step. Subsequently, information extraction is performed using a multi-layer perceptron neural network according to Equation (19):

$$Z^{dim} = LayerNorm\left(\hat{Z}^{dim} + MLP\left(\hat{Z}^{dim}\right)\right) \tag{19}$$

3.4. Handcrafted Features

This section extracts the trend coefficients of the mean and linear regression through mathematical statistics. The introduction of these handcrafted features aims to provide more intuitive and interpretable information through mathematical statistics, enhancing the model's capability in learning the original sequence shapes. In the experimental process, we applied the mean convolution kernel method to process the dataset containing multidimensional features such as fuel cell voltage, current, power, and total battery current and voltage. Specifically, we treated the multidimensional time series data as a combination of several univariate time series. To ensure the integrity of the boundary data, we performed padding on each time series by repeating the first and last values to extend the sequence length. Based on this, we applied a mean convolution kernel of size 3 to the time series to extract feature information reflecting the trend in the data. Figure 6 illustrates the mean of the original data. The mean serves as a measure of the overall level of the data, representing the average level of the data over a given time period. Through the introduction of this feature, the model is better able to grasp the overall performance of the system, capturing the overall trends and fluctuations in the data. This is crucial for predicting the RUL of a vehicle PEMFC, as the overall operation status of the system directly relates to changes in its RUL. During model training and inference, as shown in Figure 3, the decoder output is directly added to the handcrafted features before being fed into the RUL prediction module. This module consists of a single linear layer that ultimately generates the RUL prediction results.

Figure 6. Original data and mean.

Figure 7 illustrates the trend coefficients of linear regression for the data. These coefficients reflect the temporal change trend in the data, including the speed and direction of change. Through the introduction of this feature, the model can more accurately capture the dynamic change process of the RUL and identify the change trend in the system's state. This is crucial for improving prediction accuracy and understanding the system's state.

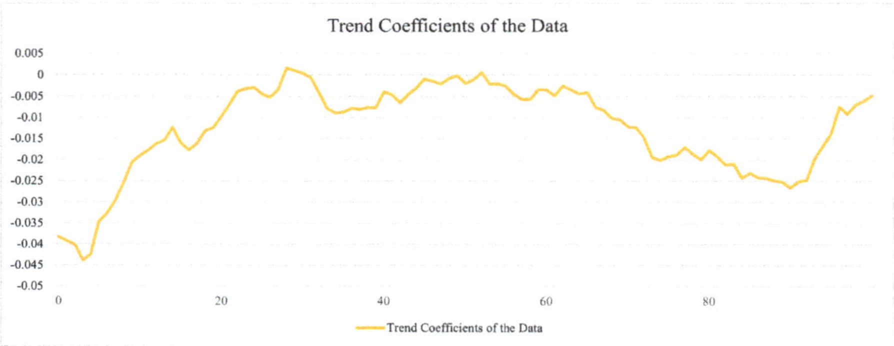

Figure 7. Trend coefficients of the data.

3.5. TILDE-Q

In order to enable the model to better focus on the shape similarity between the predicted sequence and the true sequence during the training process, this section employs the TILDE-Q loss function proposed by Lee et al. [19] as the training loss function for the model, as shown in Equation (20).

$$L_{TILDE-Q} = \alpha L_{amp}(Y, \hat{Y}) + (1-\alpha) L_{phase}(Y, \hat{Y}) + \gamma L_a(Y, \hat{Y}) \quad (20)$$

Here, $L_{amp}(Y, \hat{Y})$ represents the degree of amplitude deviation between the predicted sequence and the true sequence at each point, as shown in Equation (21).

$$L_{amp}(Y, \hat{Y}) = T \sum_{i=1}^{T} \left| \frac{1}{T} - Softmax(d(y_i, \hat{y}_i)) \right| \quad (21)$$

Here, *Softmax* denotes the proportion of the interval between the actual value y_i and the predicted value $\hat{y}_i$ at the i-th time point relative to the entire sequence length, as calculated in Equation (22).

$$Softmax(d(y_i, \hat{y}_i)) = \frac{e^{d(y_i, \hat{y}_i)}}{\sum_{i=1}^{T} e^{d(y_i, \hat{y}_i)}} \quad (22)$$

In Equation (20), L_{phase} represents the degree of phase shift between the predicted sequence and the true sequence, as shown in Equation (23).

$$L_{phase}(Y, \hat{Y}) = \begin{cases} \|\mathcal{F}(Y) - \mathcal{F}(\hat{Y})\|_p, & \text{dominant frequecy} \\ \|\mathcal{F}(\hat{Y})\|_p & \text{otherwise} \end{cases} \quad (23)$$

Here, $\mathcal{F}(\cdot)$ denotes the Fourier transform, and $||\cdot||_p$ is the p-norm, which is used to measure the magnitude between vectors.

In Equation (20), $L_a(Y, \hat{Y})$ employs autocorrelation and cross-correlation functions to assess the similarity in shape between the actual values and the predicted values, as shown in Equation (24).

$$L_a(Y, \hat{Y}) = \|\mathcal{R}(Y, Y) - \mathcal{R}(Y, \hat{Y})\|_p \tag{24}$$

In this equation, $R(Y, Y)$ represents the autocorrelation function of the actual values, measuring the similarity within the actual values themselves, while $R(Y, \hat{Y})$ represents the cross-correlation function between the actual values and the predicted values, measuring the similarity between the predicted and actual values.

4. Results and Discussion

4.1. Experiment

Datasets: The dataset used in this experiment was collected from logistics vehicles equipped with Proton-Exchange Membrane Fuel Cells (PEMFCs) in F City, G0 Province, China. The data collection period spanned from 16 October 2019 to 6 March 2021, with a sampling frequency of 30 s. The primary data collected include the PEMFC current, voltage, hydrogen concentration, and battery voltage. The experimental setup and data acquisition process are illustrated in Figure 8, which provides a schematic overview of the technical aspects of the experiment.

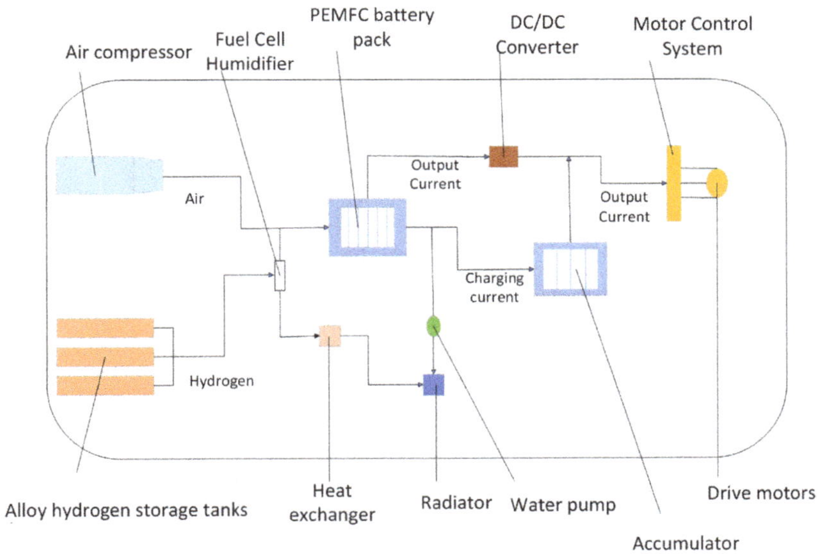

Figure 8. Data acquisition technology lab diagram.

Baselines: the experiment selected five models for comparative analysis—LSTnet [7], TCN [9], LSTM-ATTN [10], Transformer [11], and Crossformer [12].

Setup: The experiment adopts a ratio of 0.7:0.1:0.2 for dividing the dataset into a training set, validation set, and test set. The time step length for input data is set at 48, while the prediction time step lengths are variously set at 8, 16, 24, 32, 48, and 64. The experiment employs two evaluation metrics to assess model performance—mean squared error (MSE) and dynamic time warping (DTW).

4.2. Results

4.2.1. Network Model Comparative Experiment

Table 1 presents the comparative experimental results of the experimental vehicles at different time step lengths. In this study, the experimental subject was a fuel cell van with the VIN LDPGDBAD3KF*****. The bolded results indicate the best performance. The input time series length for the experiment was 48, and the Adam optimizer was employed to train and optimize the model, ensuring effective convergence and accurate trend prediction when processing time series data. Observations from the table reveal that the RUL prediction model designed in this study outperforms the other five models in terms of both evaluation metrics, MSE and DTW, in the task of PEMFC RUL prediction.

Table 1. Comparative experimental results of experimental vehicles at different time step lengths.

Models	LSTnet		TCN		LSTM-ATTN		Transformer		Crossformer		De-Stationary-Crossformer	
Steps	MSE	DTW	MSE	DTW	MSE	DTW	MSE	DTW	MSE	DTW	MSE	DTW
8	3.152	0.846	2.224	0.712	2.592	0.763	0.815	0.384	0.725	0.373	**0.556**	**0.324**
16	3.421	1.262	2.823	1.417	2.572	1.085	0.966	0.652	0.809	0.597	**0.717**	**0.551**
24	3.343	1.527	2.851	1.411	2.925	1.423	1.609	1.073	0.888	0.783	**0.778**	**0.743**
32	3.275	1.742	3.288	1.752	2.600	1.544	1.461	1.166	1.003	0.928	**0.791**	**0.862**
48	3.371	2.194	3.004	2.078	2.668	1.943	1.995	1.710	1.022	1.230	**0.886**	**1.136**
64	3.349	2.513	3.044	2.395	2.715	2.254	1.588	1.730	1.058	1.425	**0.795**	**1.237**

The proposed model, based on the Crossformer architecture, effectively addresses the issue of vanishing or exploding gradients that may occur in LSTM-based models. Additionally, it preserves the non-stationary information in the original time series while considering the dependencies between multiple features in multivariate time series. According to the analysis of the MSE metric, the model proposed in this paper significantly outperforms other mainstream models across prediction time steps ranging from 8 to 64 when the input sequence time step is set at 48. Specifically, compared to the Crossformer model, there is an average performance improvement of 17.73% across all prediction time steps; within the range of 8 to 24 time steps, the average improvement is 15.69%; and within the range of 32 to 64 time steps, the performance improves by an average of 19.77%. These results indicate that the model proposed in this study exhibits superior performance in the MSE, particularly over longer prediction time steps, in tasks related to PEMFC RUL prediction.

In this study, TILDE-Q was introduced as the loss function, which is more effective in handling unstable data, especially for the dataset used in this research, which was collected from real-world vehicles in operation. At the same time, dynamic time warping (DTW) was employed to measure the similarity between two distributions. According to the analysis of the DTW metric, as the time step length increases, the DTW values also show a significant upward trend in the PEMFC RUL prediction models used in the experiment. Notably, across different prediction time steps, the model proposed in this study continues to perform exceptionally well, outperforming other mainstream models. Compared to the Crossformer model, the performance of the proposed model improves by an average of 8.98% across prediction time steps ranging from 8 to 64. This indicates that the model proposed in this paper demonstrates superior performance in capturing the similarity between the predicted sequence and the actual sequence in PEMFC RUL prediction tasks.

In the combined MSE and DTW evaluation metrics, it is evident that the model proposed in this paper not only significantly enhances the accuracy of the predictions in PEMFC RUL prediction tasks but also improves the conformity of the predicted values to the actual shapes. This holds significant importance and value for both researchers and industry practitioners.

To provide a clearer understanding of the similarity between the predicted and actual sequences designed in this chapter, Figure 9 displays the performance of six comparative models on the test set for the experimental vehicles. Figure 9d shows a comparison between the predicted and actual values using the LSTNet model. It can be observed that while the model performs well in predicting smooth trends, it struggles to adequately fit the actual fluctuations in the dataset, which exhibits a certain degree of non-stationarity. Figures 9c and 9e show the results of the LSTM and TCN models, respectively, indicating that while these models have some success in predicting the overall trend, they fail to effectively capture short-term fluctuations. Figures 9a and 9f display the results of the Transformer and Crossformer models. The Transformer-based model demonstrates strong sensitivity to both long-term trends and short-term fluctuations, while the Crossformer model further improves the capture of short-term voltage fluctuations, though it may slightly deviate from the true values in some cases. Figure 9b shows the performance of the proposed model, which clearly outperforms the other five models in terms of trend prediction and accuracy. This model not only captures short-term details and long-term trends but also addresses non-stationarity in the data more effectively. For example, in the 0–200 time range in Figure 9b, the predicted values are clearly superior to those of the other models. Therefore, the proposed model effectively addresses the non-stationarity of PEMFC data while maintaining high prediction accuracy.

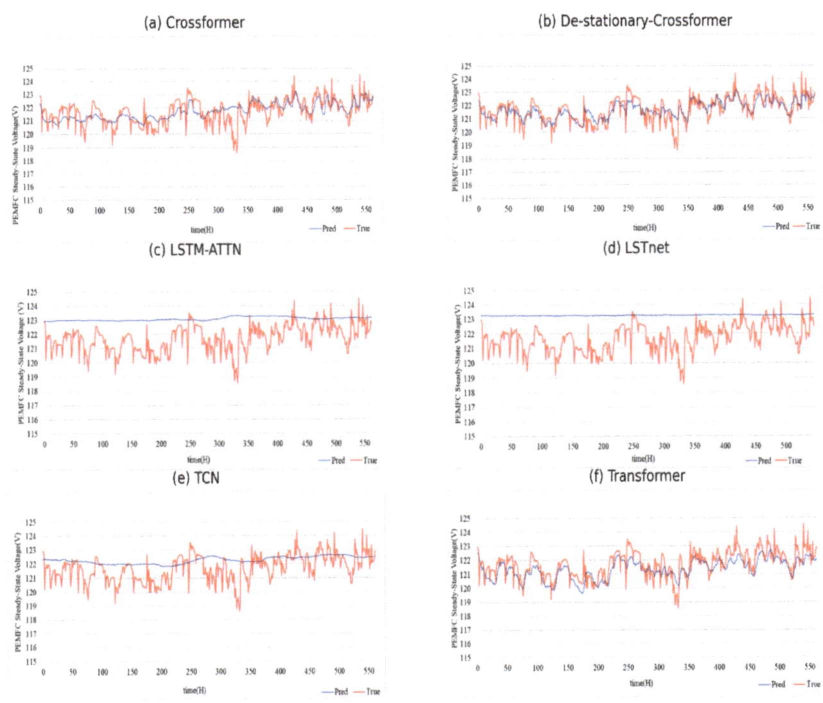

Figure 9. Comparison of predicted and actual values across models.

Figures 10 and 11 illustrate the error analysis of various models based on the mean squared error (MSE) and dynamic time warping (DTW) metrics, respectively. As is shown, the proposed model achieves the lowest MSE, with a highly concentrated distribution and almost no outliers, indicating high stability and low uncertainty in its error. This suggests that the proposed model's predictions are closely aligned with the actual values, demonstrating high accuracy. Similarly, the proposed model also excels in the DTW metric, displaying a narrow error distribution and few outliers, which highlights its consistency and accuracy in temporal alignment. In contrast, the MSE distribution for the Crossformer and Transformer models ranks second lowest, with a relatively concentrated error spread but a few outliers. This suggests that while these models perform well overall, they exhibit deviations in certain instances. Their performance on the DTW metric remains superior to that of other models, albeit with a slightly broader error range, indicating some variability, with the presence of outliers reflecting larger errors in specific samples. For the LSTM-ATTN, TCN, and LSTnet models, MSE values are comparatively higher, with a wider interquartile range, especially in the case of the TCN model, which exhibits greater uncertainty and variability. This reflects lower predictive accuracy for these models compared to the proposed model and other advanced models, particularly with less stable performance on certain samples. In summary, the proposed model demonstrates superior performance over other compared models in both the MSE and DTW metrics, exhibiting not only lower error values but also narrower error distributions, indicating greater stability and accuracy. This suggests that under experimental conditions, the proposed model is capable of providing more precise predictions with higher consistency and reliability.

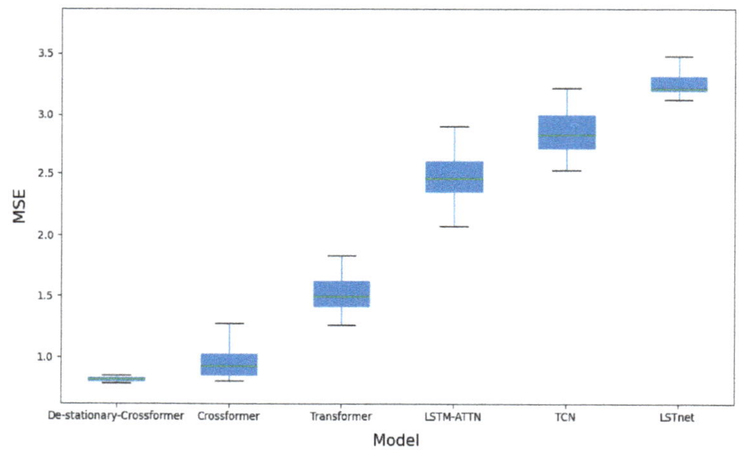

Figure 10. Model error analysis on MSE indicators.

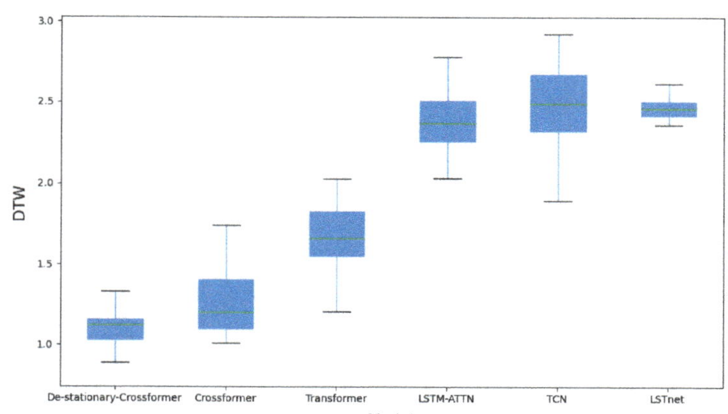

Figure 11. Model error analysis on DTW indicators.

Table 2 presents the experimental results of six models on four fuel cell vehicle datasets with an input length of 48 and a corresponding prediction length of 64. The bolded results indicate the best performance. The results demonstrate that the De-stationary-Crossformer achieves the best performance in terms of MSE and DTW, exhibiting superior prediction accuracy and sequence shape learning capability, with minimal variation, thus ensuring greater stability across datasets. The Crossformer, with a variation rate of approximately 30%, effectively captures complex temporal dependencies but shows slightly higher variability compared to the De-stationary-Crossformer. In contrast, the Transformer performs moderately, while LSTM-ATTN, TCN, and LSTnet exhibit higher error values and larger variability, highlighting their limitations in aligning time series data and learning sequence shapes. These findings underscore the inadequacy of traditional models in handling non-stationary time series data, emphasizing the need for more robust and effective modeling approaches.

Table 2. Prediction Results on fuel cell vehicle datasets.

Models	LSTnet	TCN	LSTM-ATTN	Transformer	Crossformer	De-Stationary-Crossformer
MSE	3.598 ± 1.621	2.401 ± 1.019	2.328 ± 1.470	2.425 ± 0.845	1.156 ± 0.357	**1.004 ± 0.209**
DTW	3.329 ± 2.233	2.950 ± 1.750	2.592 ± 1.337	2.138 ± 0.767	2.020 ± 0.600	**1.602 ± 0.367**

4.2.2. Network Model Ablation Studies

To validate the effectiveness of various components within the model architecture, an experiment was conducted that included four ablation models: Crossformer + features + MSE (Designed_1), Crossformer + features + DAIN + MSE (Designed_2), Crossformer + features + DAIN + De-stationary + MSE (Designed_3), and Crossformer + features + DAIN + De-stationary + TILDE-Q (Designed_4). The bolded results indicate the best performance. In the experiments, the input time series step length was fixed at 48, and ablation experiments were conducted with six different prediction step lengths in the range of 8 to 64. The Adam optimizer was used, with 100 training epochs, and early stopping was applied with a patience of 5. Table 3 presents the results of the ablation studies.

Table 3. Comparison of ablation study results at different time steps.

Models	Designed_1		Designed_2		Designed_3		Designed_4	
Steps	MSE	DTW	MSE	DTW	MSE	DTW	MSE	DTW
8	0.725	0.373	0.710	0.364	0.625	0.337	**0.556**	**0.324**
16	0.809	0.597	0.821	0.585	0.740	0.583	**0.717**	**0.551**
24	0.888	0.783	0.892	0.807	0.812	0.759	**0.778**	**0.743**
32	1.003	0.928	0.884	0.912	0.820	0.885	**0.791**	**0.862**
48	1.022	1.230	1.053	1.222	0.952	1.186	**0.886**	**1.136**
64	1.058	1.425	1.044	1.411	0.964	1.371	**0.795**	**1.237**

According to the analysis of the MSE metric, the Designed_2 model demonstrated a significant improvement over the Designed_1 model at longer time steps (32–64), enhancing performance by 3.31%. This improvement significantly surpasses the performance change observed at shorter time steps (8–24), which was only 0.04%. This indicates that DAIN has a more pronounced positive impact on prediction tasks at longer time steps. DAIN helps to mitigate internal covariate shift, enhancing the model's generalization capability, and thereby provides more effective performance gains for tasks involving long-term dependencies.

Following the introduction of the non-stationary attention mechanism and the TILDE-Q loss function, the MSE improvements for Designed_3 and Designed_4, compared to

Designed_2 and Designed_3, were 9.09% and 7.94%, respectively. These results suggest that incorporating these two structural components significantly enhances the model's ability to extract non-stationary information from data and to evaluate shape similarity, thereby reducing the error between predicted and actual values. This contributes to an improvement in the overall performance of the model, particularly when dealing with time series data that exhibit shape variability and non-stationarity, offering substantial practical value.

According to the analysis of the DTW metric, the introduction of the DAIN layer in Designed_2 resulted in a relatively minor improvement in overall performance, with only a 0.66% increase compared to Designed_1. However, the introduction of the non-stationary attention mechanism played a positive role in the prediction of PEMFC RUL. This mechanism enables the model to more effectively capture non-stationary features within the time series, making it more flexible and accurate in RUL prediction tasks. With the non-stationary attention mechanism included, Designed_3 exhibited a performance increase of 3.4% over Designed_2 across the prediction steps of 8–64. This indicates that the non-stationary attention mechanism helps the model better adapt to the complex characteristics of changes in PEMFC RUL, thereby enhancing the robustness of the predictions.

The introduction of the TILDE-Q loss function played a pivotal role in enhancing predictive performance for the RUL of PEMFC. This loss function, by incorporating a weighting term related to time series similarity, more comprehensively considers the correlations among data points within the series, thus aiding the model in more accurately capturing the trends in RUL changes. At prediction steps ranging from 8 to 64, the performance of Designed_4, which utilized the TILDE-Q loss function, showed a 5.23% improvement compared to Designed_3, which used MSE as its loss function. This indicates that the TILDE-Q loss function more effectively guided the model during the training process to learn information about the shape of the time series, increased sensitivity to changes in PEMFC RUL, and consequently significantly enhanced predictive performance.

5. Conclusions

In the context of the urgent global demand for sustainable development, hydrogen energy is receiving considerable attention due to its clean and efficient characteristics. Consequently, hydrogen-powered vehicles, which use PEMFCs as their power source, are rapidly gaining prominence. This paper focuses on PEMFCs as the subject of study, with the objective of predicting their RUL.

(1) This paper proposes a PEMFC RUL prediction model for non-stationary time series, called the De-stationary-Crossformer, which is an improved version of the Crossformer, incorporating non-stationary modeling techniques to enhance its capability in learning complex time series features. Compared to existing prediction methods based on LSTM, TCN, and the Transformer, the proposed model not only effectively captures short-term fluctuations but also leverages dependencies among multiple dimensions of the data, thereby improving prediction stability and accuracy. This study provides a new perspective on non-stationary time series modeling and fills the research gap in applying such methods to PEMFC lifetime prediction.

(2) Compared to the Crossformer model, the model proposed in this paper demonstrates superior performance in both MSE and DTW evaluation metrics. Specifically, our model effectively addresses the issues of gradient vanishing and explosion commonly encountered in traditional time series prediction models such as LSTM and TCN, while also improving the ability to capture short-term fluctuations. Moreover, it overcomes the limitations of Transformer-based models in utilizing the dependencies between different dimensions of multivariate time series and capturing the non-

stationarity present in real-world datasets. The experimental results show that the MSE and DTW of our model improved by 17.73% and 8.98%, respectively, compared to the Crossformer model.

(3) Through the plotting of the predicted and actual sequences for the experimental vehicles using five mainstream models and the De-stationary-Crossformer, it is observed that the model proposed in this paper significantly achieves the highest similarity between the predicted sequences and the actual sequences.

(4) Through network model ablation experiments, the contribution of each major component of the De-stationary-Crossformer to enhancing the overall performance of the model is validated.

Future work stemming from this paper could be approached from the following perspectives:

(1) To address existing limitations, the De-stationary-Crossformer improves the model's capability by adding or optimizing module structures, which leads to an increase in overall model parameters. However, this architectural design may still have limitations in handling extremely complex nonlinear features and long-range dependencies. Additionally, although the non-stationary attention mechanism enhances the model's ability to handle non-stationary time series data, it may become overly sensitive to noise and outliers in extreme cases, thus compromising the model's robustness. Future work will focus on optimizing and reducing model parameters to achieve a more lightweight model, improving computational efficiency. Furthermore, in terms of data processing, the more effective handling of outliers and noise is required to avoid the aforementioned extreme cases, thereby enhancing the model's practical applicability;

(2) When predicting data for vehicles with significant and frequent fluctuations, the accuracy of the results produced by the De-stationary-Crossformer still needs improvement. Future efforts should focus on enhancing the model's capability to handle strongly non-stationary data;

(3) This paper limits the scope of multi-step predictions in the PEMFC RUL forecasting task to 8–64 h. Future research could extend predictions to longer time steps.

Author Contributions: Conceptualization, N.Z.; Methodology, K.W. and J.Z.; Software, H.Z. and Z.Z.; Validation, H.Z., Z.Z. and J.Z.; Investigation, N.Z.; Resources, J.Z.; Data curation, H.Z., Z.Z. and K.W.; Writing—original draft, N.Z.; Writing—review & editing, N.Z. and J.Z.; Visualization, H.Z. and K.W.; Supervision, J.Z.; Project administration, K.W. All authors have read and agreed to the published version of the manuscript.

Funding: This research received no external funding.

Institutional Review Board Statement: Not applicable.

Informed Consent Statement: Not applicable.

Data Availability Statement: The original contributions presented in this study are included in the article. Further inquiries can be directed to the corresponding author.

Conflicts of Interest: The authors declare no conflict of interest.

References

1. Hosseini, S.E.; Wahid, M.A. Hydrogen from solar energy, a clean energy carrier from a sustainable source of energy. *Int. J. Energy Res.* **2020**, *44*, 4110–4131. [CrossRef]
2. Wang, F.K.; Mamo, T. A hybrid model based on support vector regression and differential evolution for remaining useful lifetime prediction of lithium-ion batteries. *J. Power Sources* **2018**, *401*, 49–54. [CrossRef]

3. Liu, H.; Chen, J.; Hissel, D.; Su, H. Remaining useful life estimation for proton exchange membrane fuel cells using a hybrid method. *Appl. Energy* **2019**, *237*, 910–919. [CrossRef]
4. Zhou, D.; Al-Durra, A.; Zhang, K.; Ravey, A.; Gao, F. A robust prognostic indicator for renewable energy technologies: A novel error correction grey prediction model. *IEEE Trans. Ind. Electron.* **2019**, *66*, 9312–9325. [CrossRef]
5. Peng, Y.; Chen, T.; Xiao, F.; Zhang, S. Remaining useful lifetime prediction methods of proton exchange membrane fuel cell based on convolutional neural network-long short-term memory and convolutional neural network-bidirectional long short-term memory. *Fuel Cells* **2023**, *23*, 75–87. [CrossRef]
6. Wang, C.; Li, Z.; Outbib, R.; Dou, M.; Zhao, D. A novel long short-term memory networks-based data-driven prognostic strategy for proton exchange membrane fuel cells. *Int. J. Hydrog. Energy* **2022**, *47*, 10395–10408. [CrossRef]
7. Ragab, M.; Chen, Z.; Wu, M.; Kwoh, C.-K.; Yan, R.; Li, X. Attention-based sequence to sequence model for machine remaining useful life prediction. *Neurocomputing* **2021**, *466*, 58–68. [CrossRef]
8. Sano, H.; Rokui, J. Multivariate time series forecasting accuracy improvement method based on LSTNet. *IEICE Tech. Rep.* **2021**, *121*, 71–76.
9. Hewage, P.; Behera, A.; Trovati, M.; Pereira, E.; Ghahremani, M.; Palmieri, F.; Liu, Y. Temporal convolutional neural (TCN) network for an effective weather forecasting using time-series data from the local weather station. *Soft Comput.* **2020**, *24*, 16453–16482. [CrossRef]
10. Abbasimehr, H.; Paki, R. Improving time series forecasting using LSTM and attention models. *J. Ambient Intell. Humaniz. Comput.* **2022**, *13*, 673–691. [CrossRef]
11. Yıldız, A.Y.; Koç, E.; Koç, A. Multivariate time series imputation with transformers. *IEEE Signal Process. Lett.* **2022**, *29*, 2517–2521. [CrossRef]
12. Zhang, Y.; Yan, J. Crossformer: Transformer utilizing cross-dimension dependency for multivariate time series forecasting. In Proceedings of the Eleventh International Conference on Learning Representations, Virtual, 25–29 April 2022.
13. Passalis, N.; Tefas, A.; Kanniainen, J.; Gabbouj, M.; Iosifidis, A. Deep adaptive input normalization for time series forecasting. *IEEE Trans. Neural Netw. Learn. Syst.* **2019**, *31*, 3760–3765. [CrossRef] [PubMed]
14. Samal KK, R.; Babu, K.S.; Das, S.K.; Acharaya, A. Time series based air pollution forecasting using SARIMA and prophet model. In Proceedings of the 2019 International Conference on Information Technology and Computer Communications, Singapore, 16–18 August 2019; pp. 80–85.
15. Kim, T.; Kim, J.; Tae, Y.; Park, C.; Choi, J.H.; Choo, J. Reversible instance normalization for accurate time-series forecasting against distribution shift. In Proceedings of the International Conference on Learning Representations, Vienna, Austria, 4 May 2021.
16. Liu, Y.; Wu, H.; Wang, J.; Long, M. Non-stationary transformers: Exploring the stationarity in time series forecasting. *Adv. Neural Inf. Process. Syst.* **2022**, *35*, 9881–9893.
17. Zhang, S.S. The effect of the charging protocol on the cycle life of a Li-ion battery. *J. Power Sources* **2006**, *161*, 1385–1391. [CrossRef]
18. Babu, G.S.; Zhao, P.; Li, X. Deep Convolutional Neural Network Based Regression Approach for Estimation of Remaining Useful Life. In Proceedings of the International Conference on Database Systems for Advanced Applications, Dallas, TX, USA, 16–19 April 2016.
19. Lee, H.; Lee, C.; Lim, H.; Ko, S. TILDE-Q: A Transformation Invariant Loss Function for Time-Series Forecasting. *arXiv* **2022**, arXiv:2210.15050.

applied sciences

MDPI

Article

A Case Study on the Validation of Renewable Energy Grid Code Compliance for a Large-Scale Wind Power Plant Grid-Connected Mode of Operation in Real-Time Simulation

Sinawo Nomandela *, Mkhululi E. S. Mnguni and Atanda K. Raji

Department of Electrical, Electronic and Computer Engineering, Cape Peninsula University of Technology, Bellville, P.O. Box 1906, Cape Town 7353, South Africa; mngunim@cput.ac.za (M.E.S.M.); rajia@cput.ac.za (A.K.R.)
* Correspondence: snomandela@wsu.ac.za

Abstract: In this study, the grid-connected mode of operation was evaluated, considering the renewable energy grid codes. First, the renewable energy grid code specifications have been revisited, focusing mainly on large-scale renewable power plants. An interconnected system has been developed by combining the traditional electrical power grid with a large-scale wind power plant. The modeling method used in this study and how the evaluation has been conducted can be used for other power system evaluation studies. The results obtained from the interconnected system show a significant improvement in substation (or busbar) voltages.

Keywords: large-scale wind power plant (LSWPP); grid-connected mode of operation; renewable energy grid code (REGC); wind turbine generator unit (WTGU); voltage stability; wind power plant (WPP) integration

1. Introduction

Wind power plants (WPPs) have continuously evolved globally and have become a significant participant in enhancing grid operation in most countries, showcasing the immense potential of renewable energy sources [1–8]. WPPs can be operated in one of two modes: standalone or grid-connected. In grid-connected mode, they provide extra power when the demand for loads in the power grid increases. In standalone mode, they act as an emergency supply for individual loads during power system outages [9,10].

WPPs are classified as small-scale or large-scale, depending on the size of the wind turbine generator units (WTGUs) they comprise. The size of these WTGUs is based on the rotor diameter (in meters), swept area (in square meters), and the standard power (in kilowatts) produced by the WTGU. Table 1 shows the lists of different sizes of WTGUs [11].

Large-scale wind power plants (LSWPPs) comprise large commercial wind turbine generator units (LCWTGUs).

On the other hand, WTGUs are classified into two by the axis at which wind turbine blades (WTs) rotate. They are horizontal-axis wind turbine generator units (HAWTGUs) and vertical-axis wind turbine generator units (VAWTGUs). HAWTGUs are the most popular class and are either upwind or downwind. Figures 1 and 2 show the upwind and downwind HAWTGUs [12]. Upwind HAWTGUs are the most popular class and are applied mostly in large-scale wind power plants (LSWPPs).

Table 1. Horizontal-axis wind turbine generator unit classification based on rotor diameter, swept area, and power rating.

		Rotor Diameter (m)	Swept Area (Square Meters)	Standard Power Rating (Kilowatts)
Small-scale	Micro	0.5–1.25	0.2–1.2	0.004–0.25
	Mini	1.25–3	1.2–7.1	0.25–1.4
	Household	3–10	7–79	1.4–16
Small commercial		10–20	79–314	25–100
Medium commercial		20–50	314–1963	100–1000
Large commercial		50–100	1963–7854	1000–3000

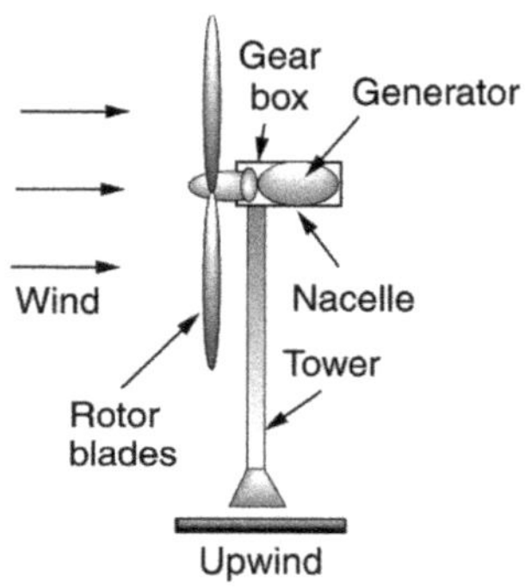

Figure 1. Upwind horizontal-axis wind turbine generator unit [12].

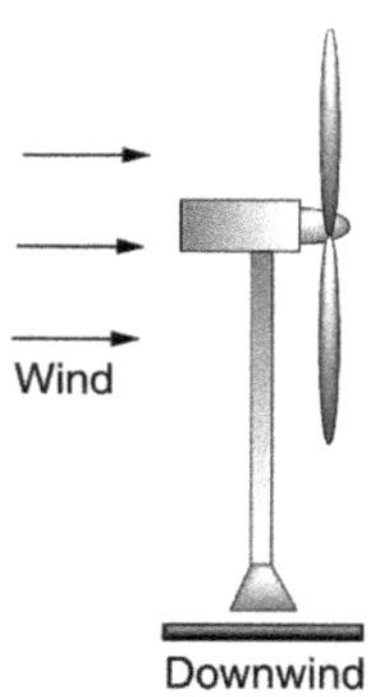

Figure 2. Downwind horizontal-axis wind turbine generator unit [12].

The operation of LSWPPs in a grid-connected mode has been leading over the past ten years [13–15]. Many studies have been conducted regarding LSWPPs in grid-connected modes of operation. However, few studies have examined the renewable energy grid code (REGC) specifications while integrating these LSWPPs into the power grid. Among the few, none have evaluated various operating conditions of the power grid into which the LSWPP is (or is to be) integrated.

For instance, a study in [16] conducted an assessment of the grid-connected mode of operation for an offshore LSWPP. The study compared the cost margin when the power transmission from an offshore LSWPP is alternating current (AC) or direct current (DC). The study aimed to prove little significance in choosing DC over AC when transmitting power from an offshore LSWPP to the power grid. However, it did not consider ensuring the LSWPP meets the grid code specifications before integration. This oversight underscores the crucial role of REGC in the operation of LSWPPs.

A study of integrating a LSWPP into a weak distribution network was conducted in [17]. A 6 MW single WTGU that was sufficiently sized to approximate the acceptable megawatt capacity of a WPP was used in the study. Additionally modeled in this study was a 9 MW single WTGU for grid integration using the static synchronous compensator (STATCOM) [18]. The impact integration of wind power plants (WPPs) and wind turbine generator units (WTGUs) has on power system stability has been studied in the references [19–21]. None of these models were initially performed independently for grid compliance tests as required by the REGCs.

Considering the models used in [17–21], it is understood that the computation of each component consumes time. However, these models may not provide adequate validation to align the REGCs. This violates the design and operational requirements of WPPs mentioned in [22–24].

An interconnected system was developed by integrating a LSWPP with the IEEE 9 Bus Power System in [25]. The study mainly focused on developing the point of common coupling (PCC) protection system that adapts to the conditions of the interconnected system and responds to the faulty conditions for both normal and increased loading conditions.

The work published in references [1,3,4,14,26] emphasized the significance of REGCs for LSWPPs in the grid-connected mode of operation.

The significant gap in all the studies reviewed above is that none have studied the LSWPP alone to ensure its operation meets the REGC specifications before integrating it into the considered power grid. This work stands out in the field due to its unique approach of combining REGC discussions with case studies. This approach provides a better understanding of the implications of integrating large-scale WPPs into the power grid. The significance of this work lies in its focus on the REGCs requirements, a crucial yet underexplored subject in the realm of renewable energy. Our revisit to these requirements, with the evaluation performed using a case study, promises to shed new light on their role in integrating LSWPPs, underscoring their importance in the field.

Following the introduction of this article, Section 2 revisits REGCs with an emphasis on WPPs. The overview of the proposed test system is in Section 3. This includes the modeling and simulation consisting of a few case studies of the power grid into which the LSWPP is integrated and the LSWPP itself. In contrast, the integration of the LSWPP into the power grid is dealt with in Section 4. Section 5 is dedicated to the crucial evaluation of the operation of the interconnected system, which also formulates an essential aspect of our research. Section 6 discusses the results. Finally, conclusions and future work are drawn and stated in Section 7.

2. Overview of Renewable Energy Grid Code Specifications

The REGCs are a set of technical standards and requirements to be followed when connecting renewable energy sources to the power grid [10,23,24].

Since the voltage level at which the RPPs are integrated into the power grid is significant, we first recap the standard voltage levels. We will then categorize the RPPs at a later stage. It should be noted that standard voltage levels differ depending on the country. For example, one of the Asian articles that categorized the voltage levels at which various WPP categories are integrated also provided some information on the Saudi Arabian standard voltages. This information is presented in Table 2 [1].

Table 2. Saudi Arabian standard grid voltages at which various types of wind power plants are integrated.

Category		Wind Power Plant Group	Voltage (kV)
Transmission	Normal transmission	Offshore: Group 1 or Group 2	400
			220
			132
	Sub-transmission	Onshore	66
Distribution		Small wind turbines	33
			11

However, this article defines voltage levels based on the South African power grid. The National Energy Regulator of South Africa (NERSA) [24] and ESKOM [27] categorizes the grid standard voltage as shown in Table 3.

Table 3. South African standard grid voltages.

Category		Voltage (kV)	Level
Transmission	Normal transmission	765	Extra-high
		400	
		275	
	Sub-transmission	132	High
Distribution		66	
		50	
		33	Medium
		25	
		22	
		11	

RPPs are categorized into three, based on their minimum and maximum apparent power production and the voltage level at which they can be integrated into the power grid. The first category is A (with subcategories A1, A2, and A3), followed by B and C. Table 4 lists the RPP categories and specifies the voltage levels at which the RPP category can be integrated [23].

Table 4. South African renewable power plant categories.

Category		Minimum Capacity (kVA)	Maximum Capacity (kVA)	Integration Level Voltage
A	A1	0	13.8000	LV
	A2	13.8000	100	LV
	A3	100	1000	LV
B		0	20,000	MV
C		>20,000	-	MV/HV

WPPs made of squirrel-cage induction generators (SCIGs) may have reactive power support in the power grid since their stator terminals are coupled with capacitor bank exciting circuits. However, this reactive power is minimal [1].

The work presented in this article mainly focuses on WPPs, whilst the discussions regarding renewable energy grid codes (REGCs) are based entirely on RPPS. Therefore, in this article, for discussions on REGCs, the term RPP instead of WPP is used.

2.1. Tolerance of Frequency and Voltage Deviations

2.1.1. Normal Operating Conditions

RPPs must be designed to operate continuously within the point of common coupling (PCC) voltage range specified in Table 5 [24,28], measured at the PCC.

Table 5. Minimum and maximum operating voltages at the point of common coupling.

Nominal Voltage (V_N) in Kilovolt	Minimum V_N (PU)	Maximum V_N (PU)
132.0000	0.9000	1.0985
88.0000	0.9000	1.0985
66.0000	0.9000	1.0985
44.0000	0.9000	1.0800
33.0000	0.9000	1.0800
22.0000	0.9000	1.0800
11.0000	0.9000	1.0800

RPPs must be designed to operate within the minimum range illustrated in Figure 3 during a system frequency disturbance. It must be disconnected from the power grid should it experience a frequency higher than 51.5000 Hz for longer than 4 s or a frequency lower than 47.0000 Hz for longer than 200 ms. Also, the RPPs must remain connected to the power grid when they experience a rate of frequency change with values of 1.5000 Hz per second as long its minimum frequency is still within the minimum and maximum frequencies of 49.0000 HZ and 51.0000 Hz [24,28].

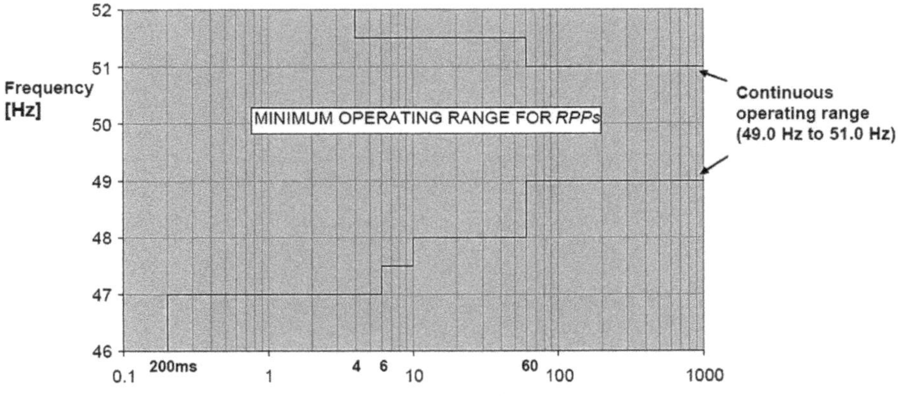

Figure 3. Minimum frequency operating range of renewable power plants during system frequency disturbances [24,28].

RPPs must only be connected to the power grid as early as 3 s after the following conditions have been met [24,28]:

- For transmission level-connected RPPs, the voltage at the PCC is within +/−5% of the nominal voltage.
- For distribution level-connected RPPs, the PCC voltage is within the minimum and maximum V_N specified in Table 5.
- Frequency in the RPP is within 49.0000 Hz and 50.2000 Hz or otherwise stated as agreed with the system operator (SO).

2.1.2. Abnormal Operating Conditions

RPPs must be built to tolerate abrupt phase changes of 20 degrees at the PCC without disconnecting or lowering its output. They must return to normal production within 5 s of the operating parameters of the PCC returning to normal following the settling time. They must also be designed to withstand and fulfill the PCC voltage conditions described in Table 5 and Figure 4. Area D in the figure is mainly for Category C RPPs. Furthermore, RPPs must withstand PCC voltage drops for a minimum duration of 0.150 s without disconnecting, as shown in Figure 4 [24,28].

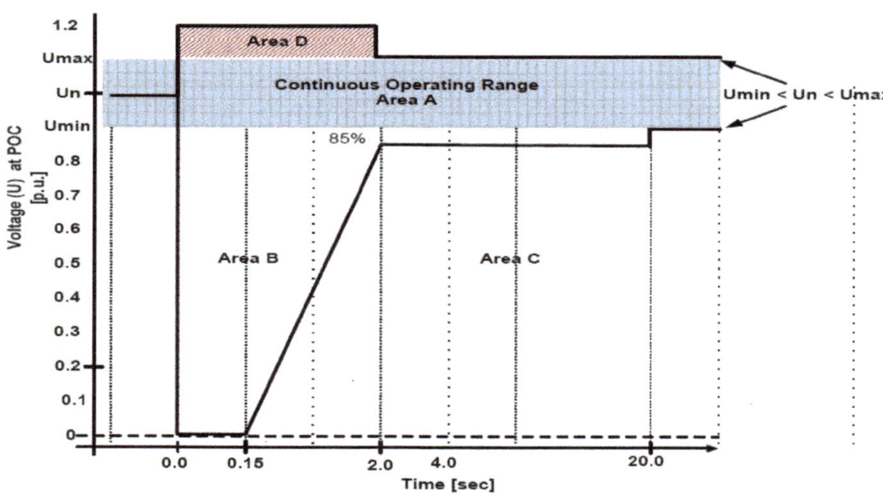

Figure 4. Voltage ride-through capability for renewable power plants of Category A3, B, and C [24,28].

RPPs must be designed to withstand voltage droops and peaks, as illustrated in Figure 5 [24,28]. As the figure illustrates, it must supply or absorb reactive current without disconnecting.

2.2. Frequency Response

RPPs must be built to offer a power–frequency response to stabilize the power grid frequency in the event of frequency deviations. RPPs must be able to meet the required active power reduction needs to stabilize the frequency during high-frequency operating situations (see Figure 6) [24,28].

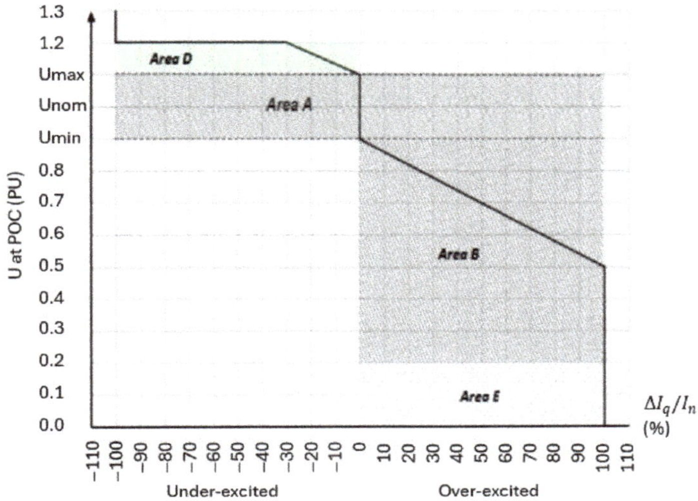

Figure 5. Requirements for reactive power support during the voltage drops and peaks at the point of common coupling [24,28].

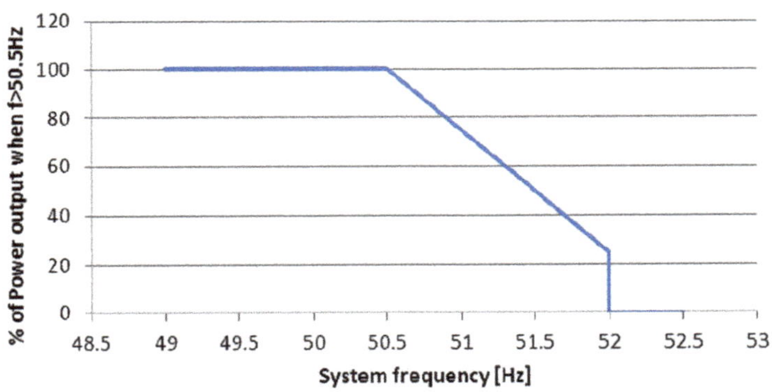

Figure 6. Power curtailment during over-frequency for renewable power plants [24,28].

The accuracy of grid frequency measuring must be at least +/−10 mHz. When the power grid frequency increases above 50.5000 Hz, RPPs must reduce the active power proportionate to the frequency change, as illustrated in Figure 6. RPPs must be disconnected if the frequency exceeds 51.5000 Hz for more than four seconds to protect the electrical grid [24,28].

2.3. *Reactive Power Capabilities for Category C*

RPPs must be designed to operate in one of three control modes: power factor, voltage, or reactive power (Q or Mvar). It is necessary to agree on the control operating mode (Q, power factor, or V control) and the network service provider (NSP). When operating between 5% and 100% of rated power Pn (MW), the RPPs must be able to adjust reactive power (Mvar) support at the PCC within the reactive power capacity ranges depicted in Figure 7, where Qmin and Qmax are voltage-dependent, as seen in Figure 8 [24,28].

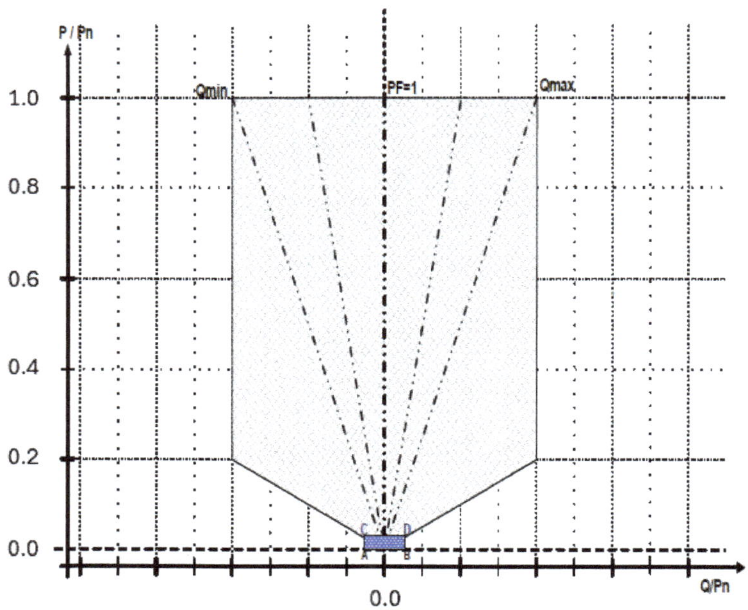

Figure 7. Reactive power requirements for renewable power plants of Category C at the point of common coupling (Qmin and Qmax are voltage-dependent, as defined in Figure 8) [24,28].

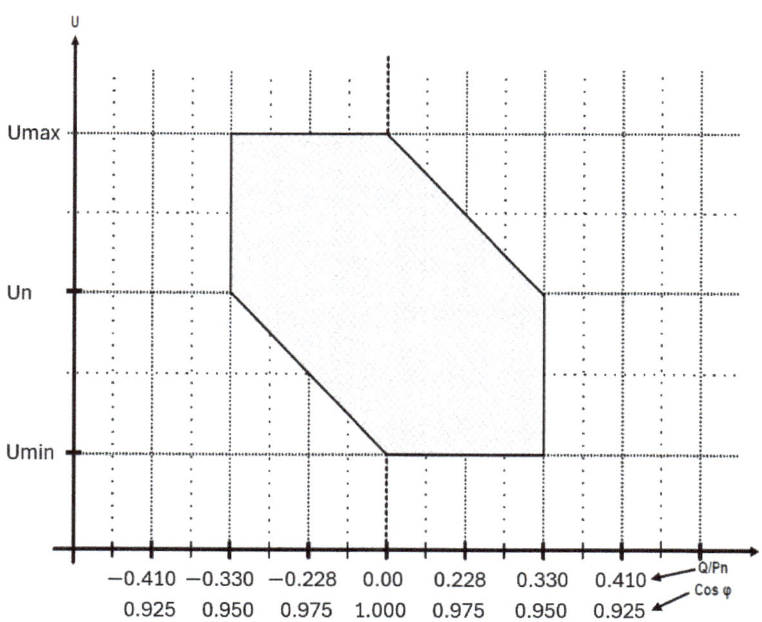

Figure 8. Requirements for reactive power and voltage control range for renewable power plants of Category C [24,28].

Figure 9 shows the necessary RPP reactive power capability (determined at the PCC) at nominal voltage. No reactive power capability is needed when operating at less than

5% of the rated power Pn (MW). As Figure 9 illustrates, the RPP can only function within the reactive power tolerance range not over +/−5% of rated power in Areas A, B, C, and D [24,28].

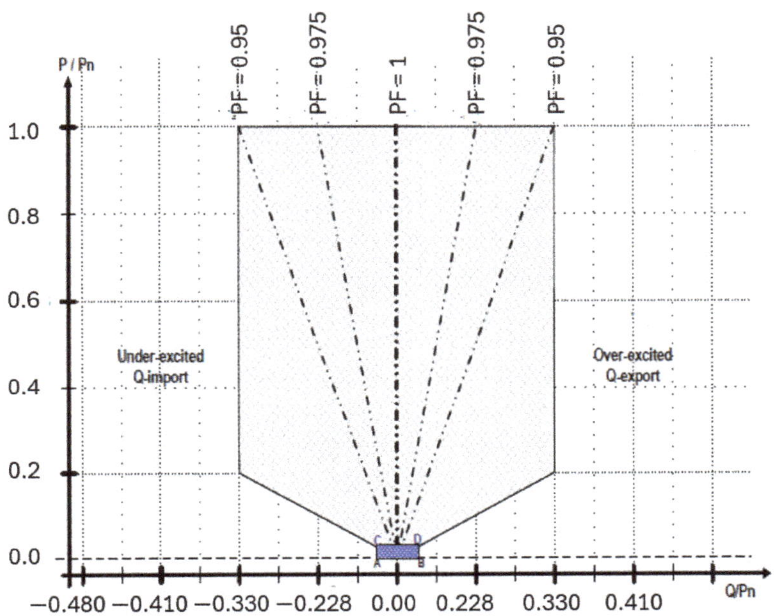

Figure 9. Reactive power requirements for renewable power plants of Category C (at the nominal voltage at the point of common coupling) [24,28].

2.4. *Voltage and Reactive Power Control Functions*

Reactive power control functions that can regulate the reactive power that the RPP supplies at the PCC and a voltage control function that can regulate the voltage at the PCC through commands utilizing setpoints and gradients are required for RPPs. Reactive power and voltage control are mutually incompatible, so the three functions, voltage control, power factor control, and reactive power control, can only be engaged simultaneously [24,28]. Moreover, the RPP generator will be employed to carry out the control function and apply parameter settings for reactive power and voltage control functions, which the NSP will decide in coordination with the SO. The operational agreement must include documentation of the agreed-upon control functions [24,28].

Reactive power control

Regardless of the active power and voltage, Q control regulates the reactive power supply and absorption at the PCC. Figure 10 shows this control function as a vertical line. The RPP generator must respond to any changes made to the Q control setpoint by the NSP, SO, or their agent within two seconds by updating its echo analog setpoint value. When an instruction to modify the setpoint is received, the RPP must react to the new setpoint within 30 s. The setpoint and the accuracy of the control must not differ by more than ±2% of the setpoint value or by ±0.5% of the maximum reactive power, whichever produces the highest tolerance. The RPP must receive a Q setpoint with a minimum precision of 1 kVAr [24,28].

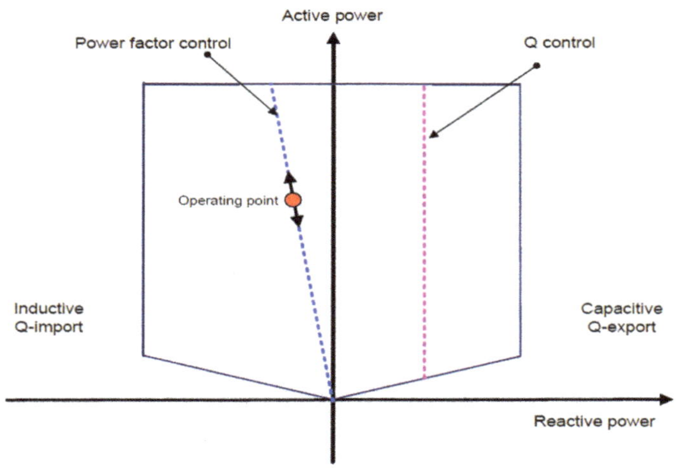

Figure 10. Reactive power control functions for a renewable power plant [24,28].

Power factor control

The power factor control function regulates reactive power at the PCC proportionately to active power. An example is a line with a constant gradient in Figure 10. If the NSP, SO, or their agent modifies the power factor setpoint, the RPP has to update its echo analog setpoint value within 2 s. When a command to change the setpoint is received, the RPP has 30 s to respond with the updated setpoint. The maximum difference between the control accuracy and the setpoint is ±0.02 [24,28].

Voltage control

One feature that regulates the voltage at the PCC is voltage control. If a modification is to be made to the voltage setpoint, it must start within 2 s and be finished within 30 s after receiving the command. Droop characteristics, as seen in Figure 11, require that the accuracy of the voltage setpoint be within ±0.5% of the nominal voltage and that the precision of the control be within ±2% of the required injection or absorption of reactive power [24,28].

With the droop setup as in Figure 11, each RPP can regulate within its dynamic range and voltage limit. Droop is defined here as the change in voltage (per unit) brought on by a change in reactive power (per unit). The control function will wait for potential overall control from the tap changer or other voltage control functions once the voltage control has reached the dynamic design limits of the RPP. Together with the SO, the NSP will be in charge of total voltage coordination [24,28].

2.5. Comparison with International Grid Codes

This study focused primarily on the renewable energy grid code (REGC) specifications of South Africa (SA), as regulated by the National Energy Regulator of South Africa (NERSA) and Electrical Supply Commission (ESKOM). Recognizing that REGCs for renewable power plant (RPP) integration vary across different electrical power systems worldwide is significant. Nonetheless, most international REGCs share fundamental operational requirements concerning continuous operating voltage and frequency ranges, active power control capabilities, reactive power support and control functions, fault ride-through (FRT) capabilities, frequency response, and stability requirements. For instance, the German BDEW Medium Voltage Directive, the UK National Grid Code, and the IEEE 1547 Standard for Interconnection of Distributed Energy Resources in the USA set comparable technical requirements for voltage and frequency tolerance, reactive power capability, and grid fault behavior [29–31].

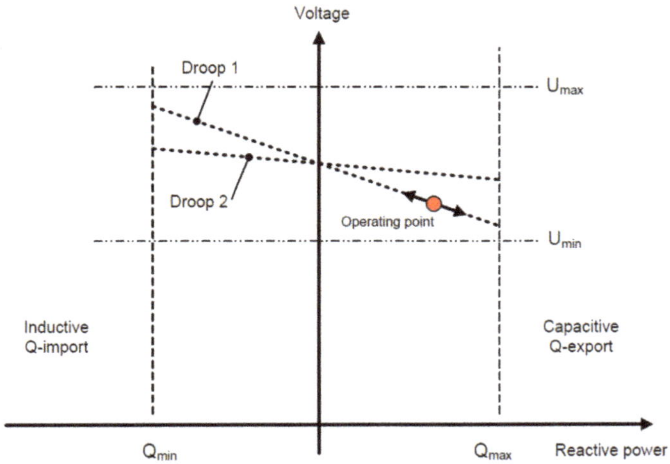

Figure 11. Voltage control of a renewable power plant [24,28].

However, differences typically arise in specific voltage ride-through curves, the permissible ranges of power factor adjustment, the exact frequency thresholds and response times, and the operational control modes, such as voltage versus power factor versus reactive power control priority.

The LSWPP model developed in this study was validated against REGCs of SA, which serve as a representative framework for Category C LSWPP. Given the similarity in core requirements across most international standards, the model can be adapted for other regions by adjusting the reactive power control ranges and setpoints, modifying the acceptable frequency and voltage operation limits, tuning the FRT capability curves and thresholds according to the REGCs of the target region.

This adaptability is possible because the model was built using flexible, parameterized control components within the Real-time Simulation Computer-Aided Design (RSCAD) simulation environment. Therefore, while the present study validated compliance based on SA standards, the methodology and simulation framework remain applicable for evaluating LSWPPs against other regional grid codes with minor parameter adjustments.

3. Overview and Development of the Proposed Test System Considered for the Study

This study evaluated the grid-connected mode of operation of the large-scale wind power plant (LSWPP). Operating the LSWPP in a grid-connected mode forms the modern power grid. On the other hand, a modern power grid combines any renewable power plant (RPP) and the traditional electrical power grid.

This section describes the traditional electrical power grid and the LSWPP considered in this study.

3.1. The Power Grid System Considered for the Study

The IEEE 9 Bus Power System is used in this study due to its simplicity, well-documented data availability, and established use in renewable integration studies. There are other publications in which the system has been used. For instance, in reference [19], it was used for another study dealing with renewable power plant (RPP) integration. The IEEE 9 Bus Power System allows the performance of power grid case studies that significantly complement the need for a large-scale wind power plant (LSWPP) integration.

3.1.1. Preparation of the IEEE 9 Bus Power System

The model of the IEEE 9 Bus Power System is available in the Real-time Simulator Computer-Aided Design (RSCAD) software, version FX 1.3 instruction manual from RTDS Technologies Inc., referenced in [32]. Below is a significant recap of how the components were modified from the original IEEE 9 Bus Power System in the RSCAD software instruction manual.

The components were rearranged and renamed, as shown in Table 6. The modified system is shown in Figure 12.

Table 6. Renaming of components for the modified IEEE 9 Bus Power System.

Original Name		Modified Names	
Busbar	**Component Connected to It**	**Busbar**	**Component Connected to It**
1	G1	G1Bus	Gen1
2	G2	G2Bus	Gen2
3	G3	G3Bus	Gen3
4	None	Bus1	None
5	Unnamed load 1	Bus2	DLoad1
6	Unnamed load 2	Bus3	DLoad2
7	None	Bus4	None
8	Unnamed load 3	Bus5	DLoad3
9	None	Bus6	None

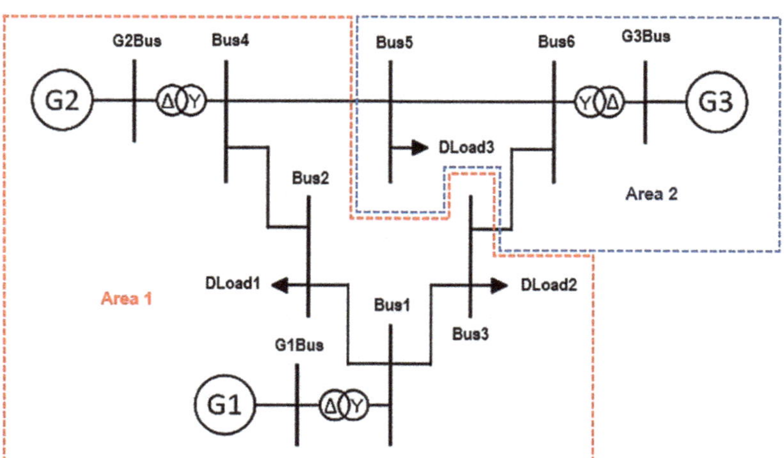

Figure 12. Modified IEEE 9 Bus Power System, adapted from the previous studies [32].

Area 1 and Area 2 in Figure 12 represent subsystems on RSCAD. These real-time simulations for an electrical system of two subsystems are performed in two RTDS hardware racks, Rack 1 and Rack 2. In the rest of this article, the two terms subsystems and racks are used interchangeably because they have the same meaning.

The transmission line parameters of the IEEE 9 Bus Power System have been listed in Table 7. All per unit values have been derived from a base apparent power of 100 MVA [32].

Table 7. IEEE 9 Bus Power System transmission line parameters.

Lines	Busbar to Which Line Is Connected	Length (km)	Resistance (PU)	Reactance (PU)	Susceptance (PU)
Line12	Bus1–Bus2	89.9300	0.0100	0.0850	0.1760
Line13	Bus1–Bus3	97.3360	0.0170	0.0920	0.1580
Line24	Bus2–Bus4	170.3380	0.0320	0.1610	0.3060
Line36	Bus3–Bus6	179.8600	0.0390	0.1700	0.3580
Line45	Bus4–Bus5	76.1760	0.0085	0.0720	0.1490
Line56	Bus5–Bus6	106.6460	0.0119	0.1008	0.2090

RSCAD consists of two modules, namely the draft and runtime modules. The draft module deals with the modeling of the system itself, while the runtime module deals with the real-time simulations of the modeled system. The modeling on the draft module deals with gathering components that make up a power system model. This includes control components, such as control switches, sliders, pushbuttons, and other operating systems.

After the completion of the model, the draft load flow simulation was executed to verify data accuracy and the initial conditions of the modeled system. Results were obtained and recorded in Table 8. These results are the same as those found in [32].

Table 8. Initial conditions of the IEEE 9 Bus Power System.

Bus	Bus Type	Voltage (PU)	P_G (MW)	Q_G (MVAr)	P_L (MW)	Q_L (MVAr)
G1Bus	Slack	$1.0400\angle 0.0000°$	71.7798	36.2675	-	-
G2Bus	PV	$1.0250\angle 8.3639°$	163.0000	11.2300	-	-
G3Bus	PV	$1.0250\angle 4.0221°$	85.0000	−3.7204	-	-
Bus1	PQ	$1.0207\angle 27.7678°$	-	-	-	-
Bus2	PQ	$0.9928\angle 26.3247°$	-	-	125.0000	50.0000
Bus3	PQ	$1.0064\angle 26.5655°$	-	-	90.0000	30.0000
Bus4	PQ	$1.0230\angle 32.7884°$	-	-	-	-
Bus5	PQ	$1.0128\angle 30.2929°$	-	-	100.0000	35.0000
Bus6	PQ	$1.0283\angle 31.3133°$	-	-	-	-

It has been stated in the references [33–35], that the reactive power transfer through substations (busbars) causes voltage stability challenges in the power grid. In this study, the assumption has been made that the increase in load demand means an increase in active and reactive power. We, therefore, configured the load increase simulation logic, as shown in Figure 13.

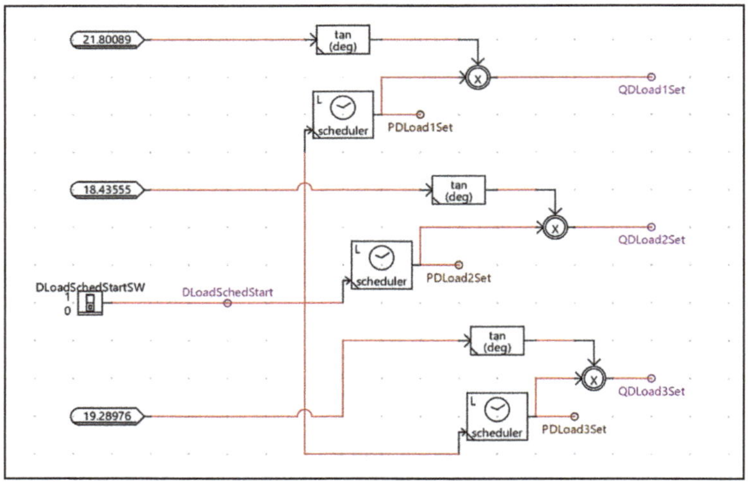

Figure 13. Logic for load schedular and direct proportionality of active and reactive power demand.

PDLoad1Set and *QDLoad1Set* in the figure are signal names to control the three loads, DLoad1, DLoad2, and DLoad3. *DLoadSchedStart* is a digital signal from a binary switch that controls the status (*ON/OFF*) of the scheduler component. The "scheduler" component multiplies the starting power levels by the specified multiplier to increase the load demand. The active power input signal of the dynamic loads is multiplied by the "tan (deg)" component. Reactive power to the load is shown by the output signal *QDload1Set*, which is based on the simultaneous rise in variables P and Q to maintain a constant load power factor. The power demand in the runtime module simulation platform will remain in its initial state until the switch DLoadSchedStartSW is turned on. The scheduler for each bulk load DLoad1, DLoad2, and DLoad3 is set to increase the load demand by 5.0000 MW every 0.1 s. An example is DLoad1, which has an initial active and reactive power of 125.0000 MW and 50.0000 MVAr. After 0.1 s, the load starts to draw 130.0000 MW and 51.9986 MVAr. And after 0.2 s, it starts to draw 135.0000 MW and 53.9986 MVAr.

To summarize the operation of the logic in Figure 13, the Expressions (1) to (3) are used.

Incremental increase in active power and reactive power:

$$P_{New} = P_{Old} + \Delta P \tag{1}$$

$\Delta P = 5$ MW every 0.1 s.

Reactive power is proportional to active power:

$$Q = P \times \tan(\varnothing) \tag{2}$$

The power factor $\cos(\varnothing)$ is calculated from the initial conditions. For example, for DLoad1 $P_0 = 125$ MW, $Q_0 = 50$ MVAr, then

$$\tan(\varnothing) = \frac{Q_0}{P_0} = 0.4$$

So,

$$Q_{New} = Q_{Old} + 0.4 \tag{3}$$

In real-world scenarios, this is when the loads connected to the electrical power grid increase. These simulations of load increase every 0.1 s. This simulation looks into what might happen regarding electrical power grid loading in the future.

3.1.2. IEEE 9 Bus Power System Simulations

Initial load demand simulations for the power grid

An initial load demand was simulated for the power grid. Tables 9 and 10 show the active and reactive power of the loads and busbars under initial load conditions, while Table 11 shows the voltages on the busbars Bus1, Bus2, Bus3, Bus4, Bus5, and Bus6.

Table 9. Active power on load busbars and their loads under initial load demand for the power grid.

	Load Active Power (MW)	Busbar Active Power (MW)
DLoad1 | Bus2	125.0000	125.0000
DLoad2 | Bus3	90.0000	90.0000
DLoad3 | Bua5	100.0000	100.0000

Table 10. Reactive power on load busbars and their loads under initial load demand for the power grid.

	Load Reactive Power (MVAr)	Busbar Reactive Power (MVAr)
DLoad1 | Bus2	50.0000	50.0000
DLoad2 | Bus3	30.0000	30.0000
DLoad3 | Bua5	35.0000	35.0000

Table 11. Busbar voltages under initial load demand for the power grid.

Busbars	Voltage (PU)
Bus1	1.0210
Bus2	0.9928
Bus3	1.0060
Bus4	1.0230
Bus5	1.0130
Bus6	1.0280

We used tables to present these results because the time parameter is insignificant since no disturbances were simulated.

Increased load demand simulations for the power grid

An increase in load demand was simulated to identify the voltage stability collapse point, which might be any voltage outside the range of 0.9500 and 1.0500 per unit based on power system grid codes. The voltages were observed in this simulation for busbars Bus1 to Bus6 in response to the load demand increase in different loads DLoad1, DLoad2, and DLoad3. It was identified that the voltage at Bus2, where DLoad1 is connected, could not recover and, therefore, is critical. This is because it settled and fell to 0.9486 per unit, out of the range of 0.9500 and 1.0500 per unit. This behavior is when the load demand of DLoad1 was raised by 5.0000 MW and 1.9999 MVAr every 0.1 s, which in total ended up in the value of 160.0000 MW and 65.0535 MVAr.

Figures 14–19 plot the active and reactive power demands of the loads DLoad1, DLoad2, and DLoad3 and the corresponding power at specific buses Bus2, Bus3, and Bus5. The voltage plots due to the active and reactive power demands of these loads are shown in Figures 20–25.

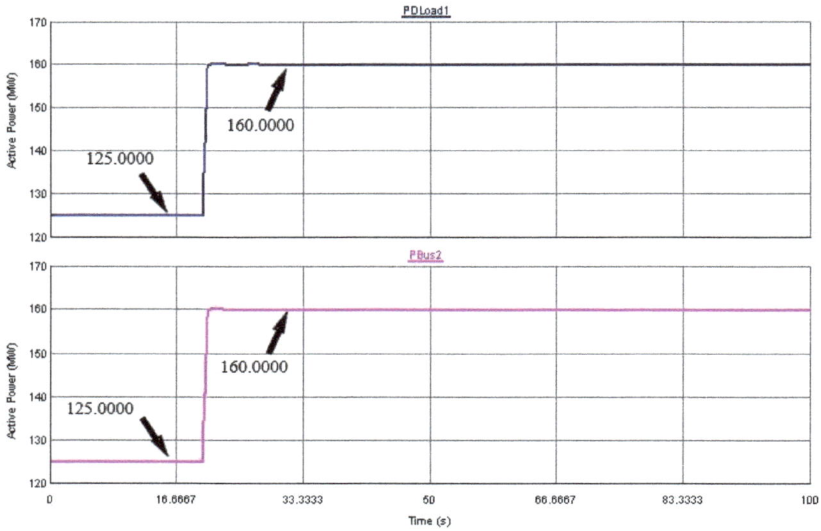

Figure 14. Increased load demand simulations for the power grid: Active power demand by DLoad1 and active power at Bus2.

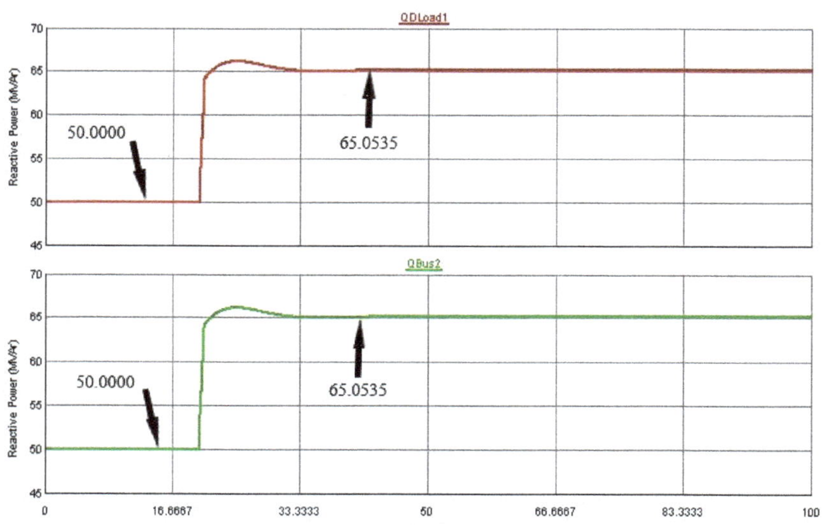

Figure 15. Increased load demand simulations for the power grid: Reactive power demand by DLoad1 and reactive power at Bus2.

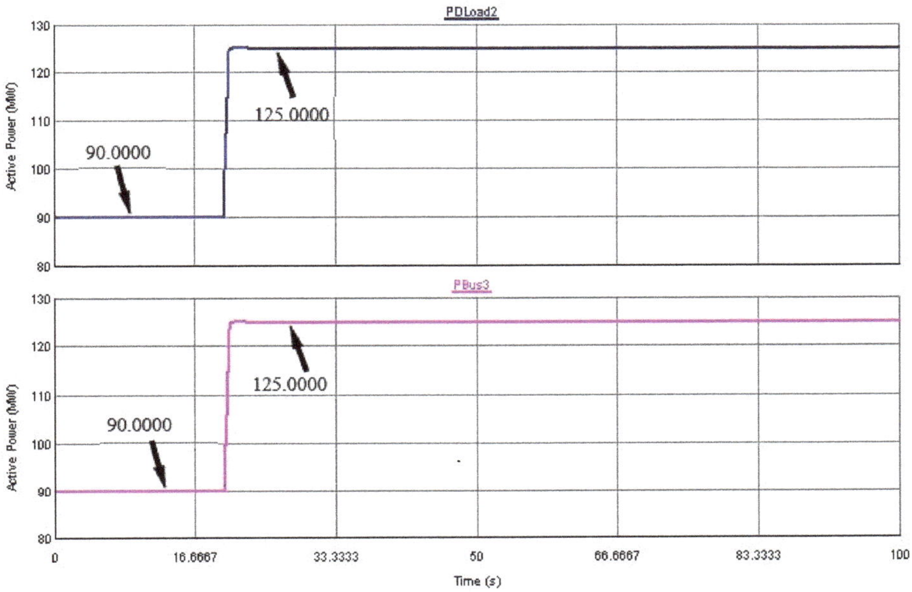

Figure 16. Increased load demand simulations for the power grid: Active power demand by DLoad2 and active power at Bus3.

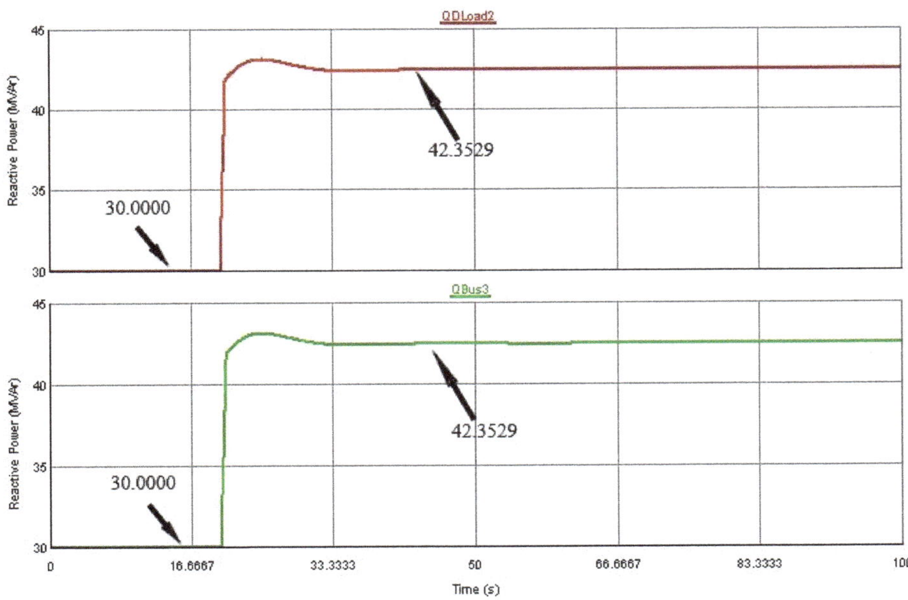

Figure 17. Increased load demand simulations for the power grid: Reactive power demand by DLoad2 and reactive power at Bus3.

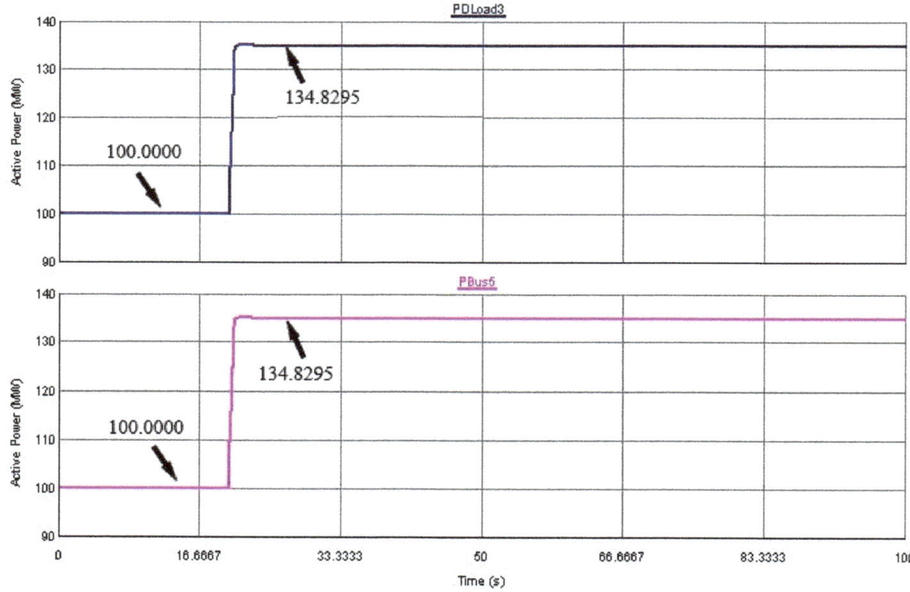

Figure 18. Increased load demand simulations for the power grid: Active power demand by DLoad3 and active power at Bus5.

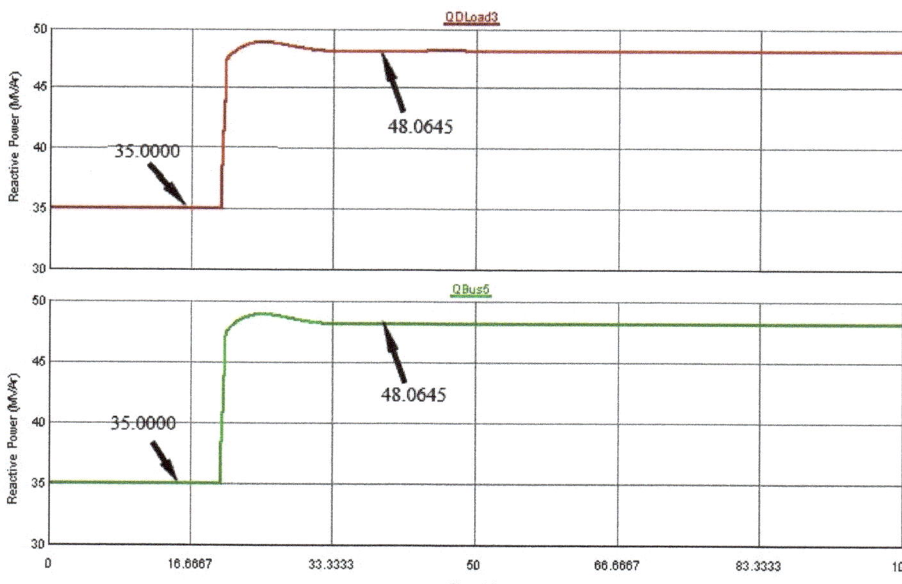

Figure 19. Increased load demand simulations for the power grid: Reactive power demand by DLoad3 and reactive power at Bus5.

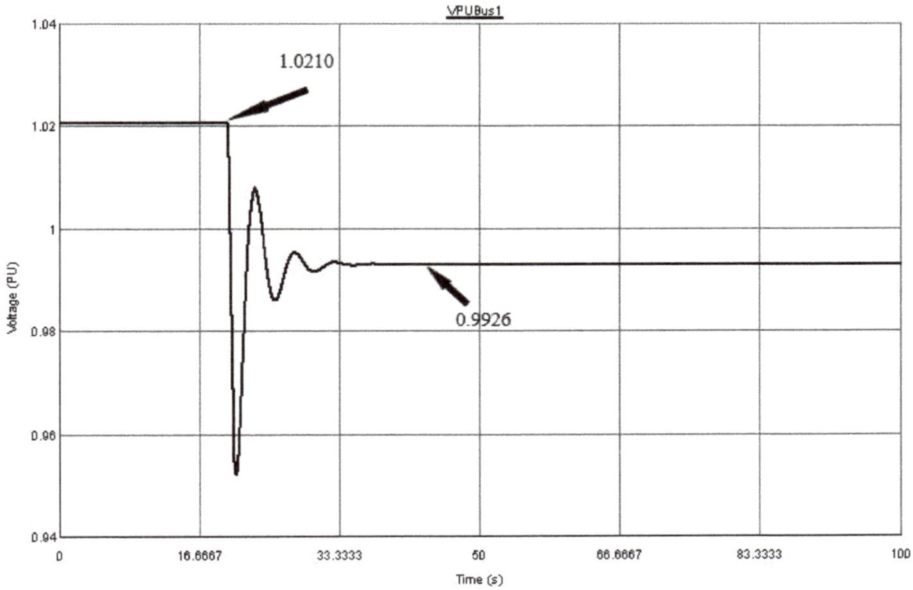

Figure 20. Increased load demand simulations for the power grid: Bus1 voltage.

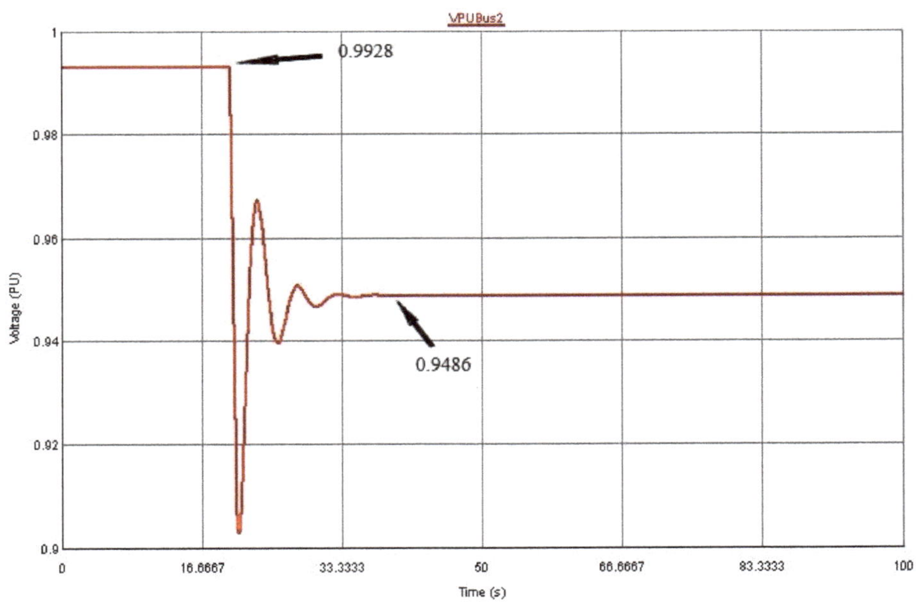

Figure 21. Increased load demand simulations for the power grid: Bus2 voltage.

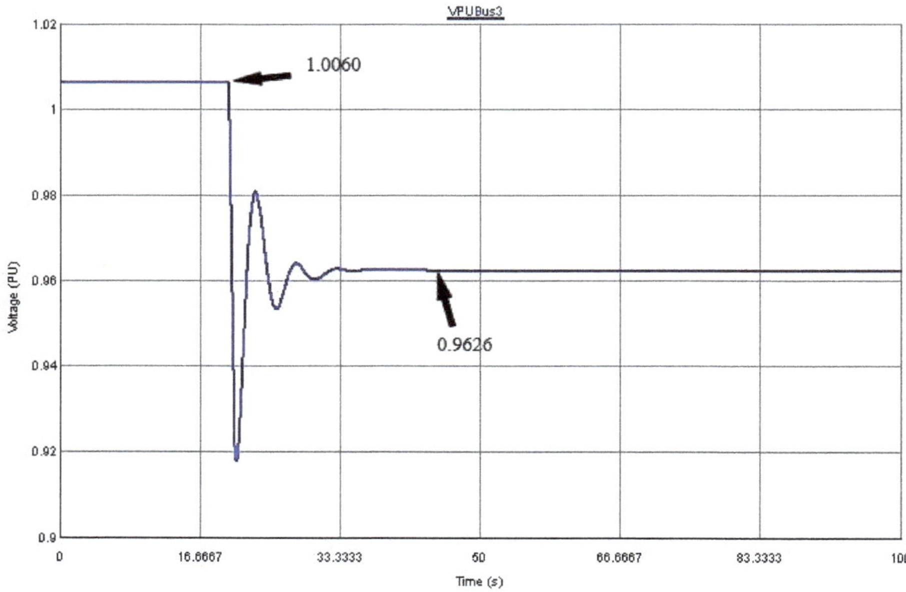

Figure 22. Increased load demand simulations for the power grid: Bus3 voltage.

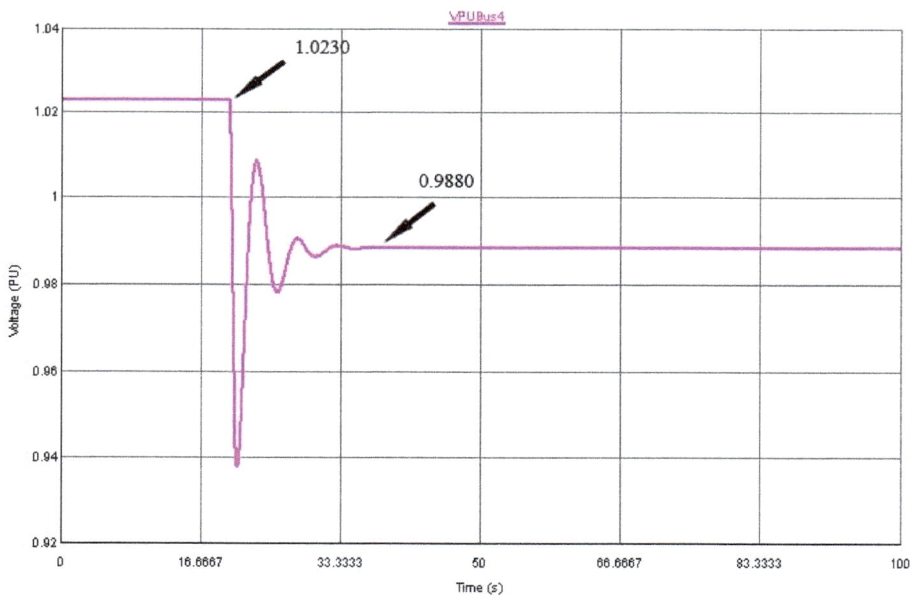

Figure 23. Increased load demand simulations for the power grid: Bus4 voltage.

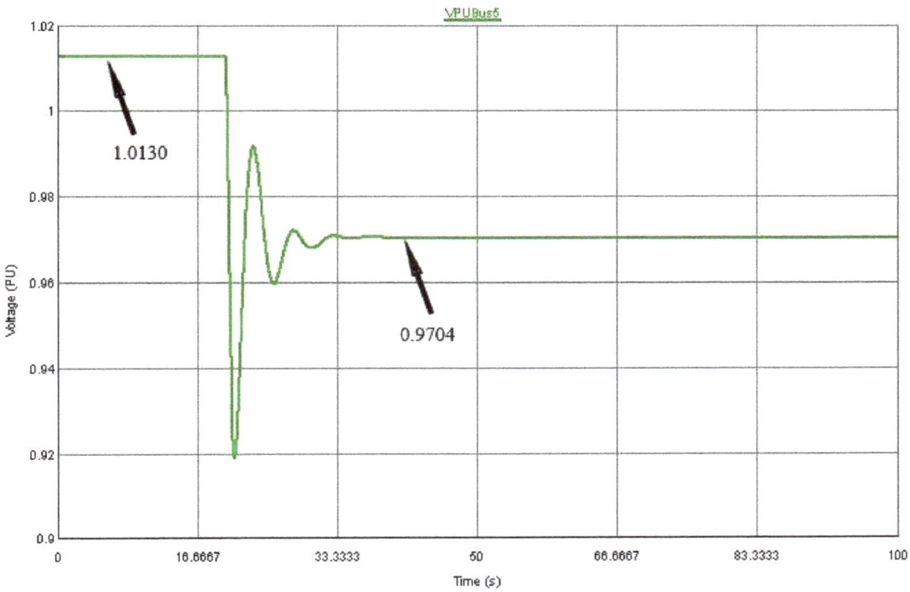

Figure 24. Increased load demand simulations for the power grid: Bus5 voltage.

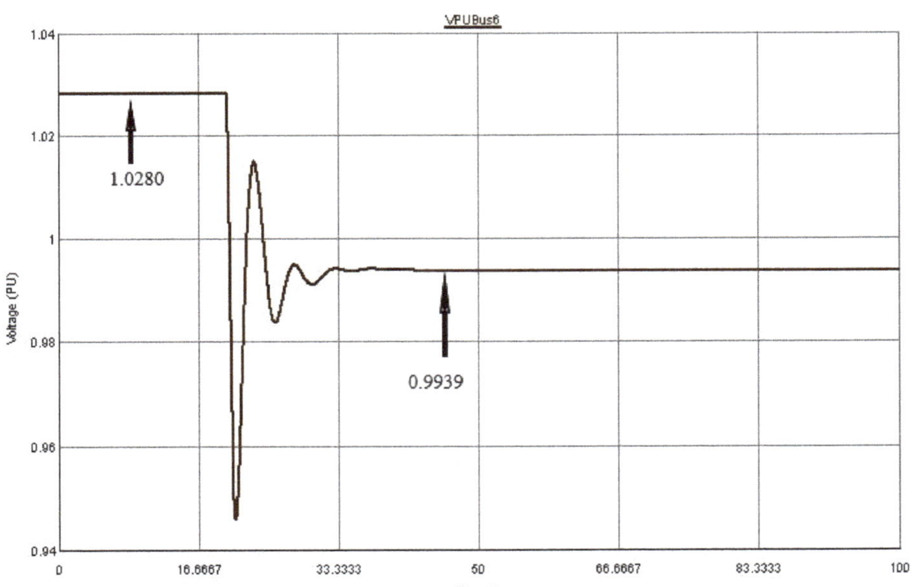

Figure 25. Increased load demand simulations for the power grid: Bus6 voltage.

In the simulation of the incremental load demand simulation performed above, we assessed the voltage stability thresholds and the compensation needed. This is how we studied the need to integrate the large-scale wind power plant (LSWPP) described below.

3.2. The Large-Scale Wind Power Plant Considered for the Study

The LSWPP used in this study is adapted from our previous study by the reference [36]. It consists of eighteen Type I upwind horizontal-axis wind turbine generator units (HAWT-GUs) modeled from a combination of V117-4.2 MW and 4.0000 MW squirrel-cage induction generators (SCIGs) from Vestas' brochure and Appendix B Generator Parameters, respectively [37,38]. In total, the LSWPP has a capacity of 72.0000 MW. This capacity complements Category C WPP, which this document already refers to as the LSWPP.

Significantly, only simulation-related control components and other operating systems are described in this section so that the simulation of the system itself can be easily understood.

3.2.1. Preparation of the Large-Scale Wind Power Plant

The simulation of the LSWPP requires fundamental tuning of wind speed (m/s) and pitch angle adjustment (degrees). See Figure 26 for logic adapted from the same reference [36]. The control components shown in the figure are logically connected to the wind turbine (WT) model and are accessible for control on the RSCAD runtime module. In the same figure, *WT1PitchAngle* is the signal name to control the change in the pitch angle of the wind turbine, while *WindSpd1(km/h)* is the input wind speed to the wind turbine. The pitch angle is adjusted using a dial selector switch WT1PitchAngleAdjust with figures or positions 1 to 5. The wind speed is adjusted with a slider labeled WindSpd1Adjust. This logic is the same for all other WTs in the LSWPP model.

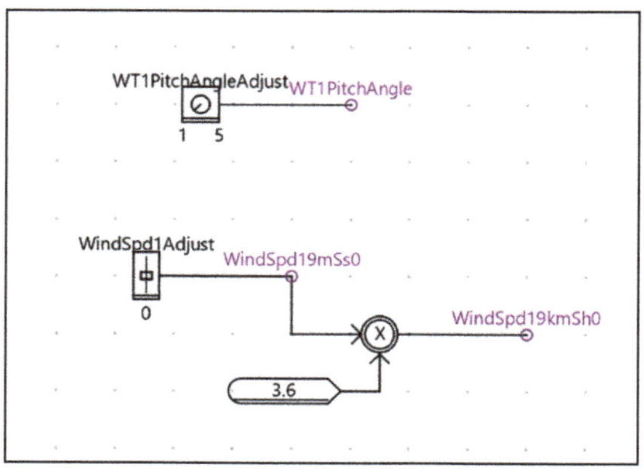

Figure 26. Logic for pitch angle and wind speed adjustments.

Other parameters like Default Air Temperature (DegC), Default Barometric Pressure (millibar), and Default Relative Humidity (%) do not need any additional control components to connect to them logically. Figure 27 shows the setting of these parameters on the WT model as viewed on the RSCAD draft module. By default, these parameters are accessible and can be tuned on the runtime module. Reference [36] can be visited for more information on how the LSWPP was modeled.

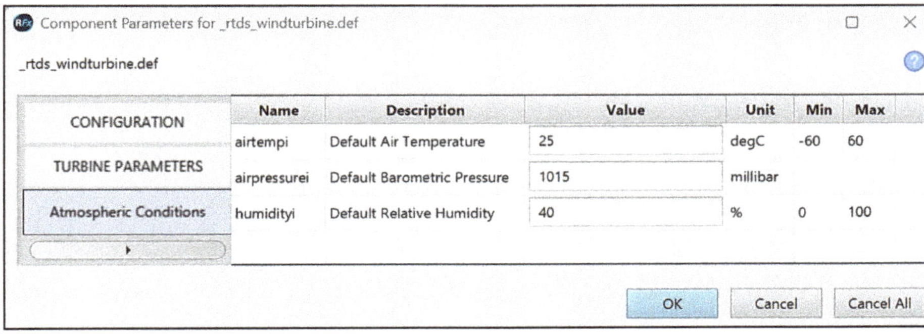

Figure 27. RSCAD draft wind turbine model parameters.

3.2.2. Large-Scale Wind Power Plant Simulations

The wind turbines (WTs) in the LSWPP model used in this study have operating temperatures between −20 °C and 45 °C. Nevertheless, WTs can be adjusted to generate the necessary output power almost equivalent to the rated output power as long as the temperature stays within the working range. Therefore, a temperature of 25 °C has been selected and used in this study.

The power coefficient, C_P, is also significantly influenced by air pressure and humidity. Humidity and air pressure were set to 30% and 1415 mbar, respectively, for the simulations. The sliders, dial selectors, and switches for these and other parameters in RSCAD runtime are shown in Figure 28.

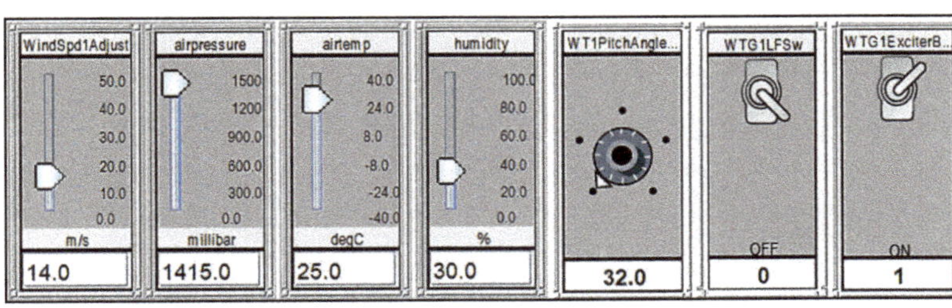

Figure 28. Sliders, dial selector switches, and control switches for a wind turbine. All other wind turbines in the model have the same setup of components for adjustments.

Simulations were performed for the LSWPP; the results are presented in Tables 12–14.

Table 12. Mechanical data for wind turbine generator units is monitored at steady-state power flow.

Quantities	Values
Wind speed	14.0000 m/s
Wind power	23.3100 MW
Wind turbine C_p	0.1720
Wind turbine power	4.0040 MW
Wind turbine rotor speed	9.9000 rpm
Wind turbine torque	3.8580 MN.m

Table 13. Wind turbine generator unit electrical data monitored at steady-state power flow.

Quantities	Values
WTGSU primary current	0.0160 kA
WTGSU secondary current	0.0010 kA
WTGU electrical torque	0.0100 PU
WTGU terminal voltage	1.0800 PU
WTGU exciter reactive power	2.1230 MVAr
WTGU active power (P) output	0.0240 MW
WTGU reactive power (Q) output	−2.0090 MVAr
WTGU reactive power circuit current	0.1600 kA

Table 14. Wind power plant sending-end data monitored at steady-state power flow.

Quantities	Values	
	Actual	Per Unit
Group 1 collector terminal voltage	26.0200 kV	1.0801 PU
Group 2 collector terminal voltage	26.0200 kV	1.0801 PU
Group 3 collector terminal voltage	26.0200 kV	1.0801 PU
WPP sending-end bus voltage	249.9000 kV	1.0801 PU
WPP receiving-end bus voltage	249.9000 kV	1.0801 PU
WPP sending-end frequency	50.0000 Hz	-
WPP receiving-end frequency	50.0000 Hz	-

3.2.3. Verifications of Renewable Energy Grid Code Specifications

The renewable energy grid codes (REGCs) specify the continuous operating per unit voltage level for Category C renewable power plants as 0.9000 (minimum) to 1.0985 (maximum). As shown in Table 14, under normal load flow conditions, the voltage measured on the high-voltage busbar of the modeled large-scale wind power plant (LSWPP) is at 1.0860 per unit, and is therefore acceptable. It is also stated that the RPP can run continuously at the frequency range between 49.0000 Hz and 51.0000 Hz. The frequency of the modeled LSWPP does not violate the continuous operating range specified by the grid codes, as it is at 50.0000 Hz under normal operating conditions. This means that the modeling and simulation (test) of the LSWPP model in this study were successful. This has also been attested in our previous study in reference [36].

4. Large-Scale Wind Power Plant Integration

Wind integration is the collection of all activities related to connecting the WPPs to the power grid. There are three high-level activities of grid integration, as listed in Table 15 [10].

This study focused on planning, as it uses the simulation-based technique in proposing and (or) setting up a significant method of evaluating the operation of an interconnected system consisting of a LSWPP. Among the activities listed under planning, short-circuit studies are out of the current scope of this study.

Table 15. High-level activities of grid integration.

Planning	Physical Connection	System Operations
Power flow, short-circuit, and system stability studies	Build transmission from WPP to substation	Unit commitment
System operations study	Connection at substation	Economic dispatch
Wind power interconnection code	-	Wind energy forecasting

The traditional electrical power grid and the LSWPP were studied individually in the previous sections. The LSWPP used has already been modeled with its transmission line for electrical power transmission to the busbar of selection (PCC) on the traditional power grid. Figure 29 illustrates the modern power grid formulated from the two systems studied above: the IEEE 9 Bus Power System and the LSWPP. The components WTGU1 to WTGU10 in the figure representing the first to the eighteenth wind turbine generator units. The LSWPP model and its components were modeled in subsystems (or racks) 3, 4, and 5. It was also shown in Section 3 that the IEEE 9 Bus Power System alone occupied subsystems 1 and 2.

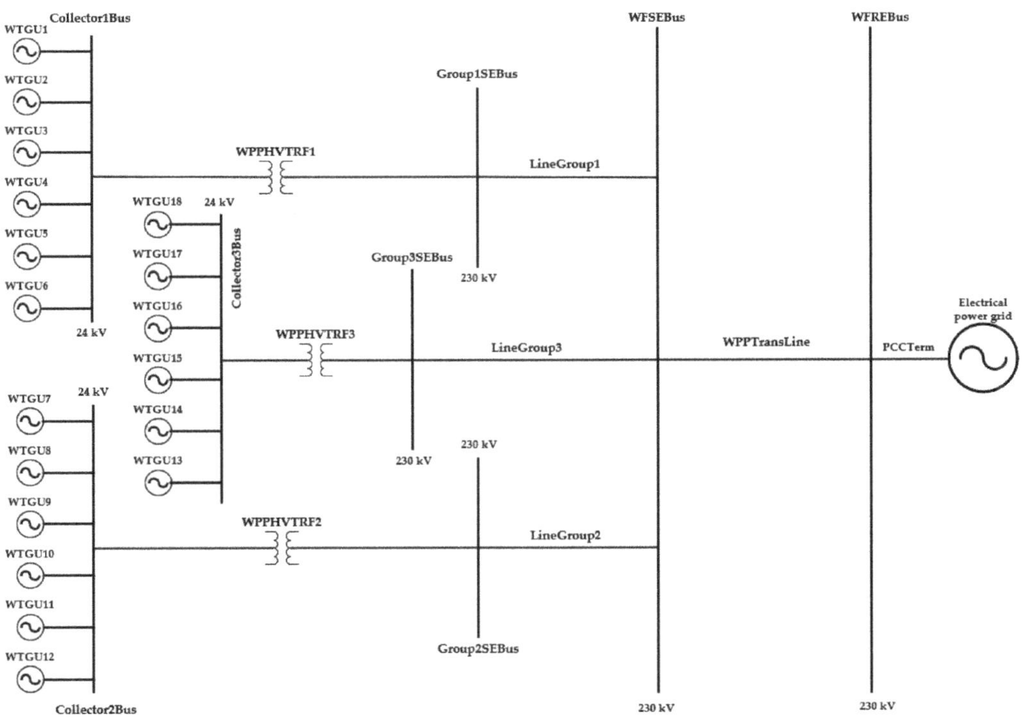

Figure 29. The resultant interconnected system (modern power grid) after coupling the LSWPP into Bus2 (PCC).

The WTGUs are a combination of wind turbines (WTs), squirrel-cage induction generators (SCIGs), and wind turbine generator substation unit transformers (WTGSUTs). It should be noted that the LSWPP used in this study consists of SCIGs whose output terminal voltage is 4 kV. These WTGSUTs step up to a medium voltage (MV) of 24 kV. This is to say that the voltage generated by each WTGU is considered MV and is distributed through the LSWPP MV circuit. This is the voltage where the high-voltage (HV) step-up power transformers (WPPHVTRFs) step up for transmission over the lines LineGroup1, LineGroup2, and LineGroup3, and then over the transmission line, WPPTransLine, to the grid point of connection (PCC). PCCTerm in the figure is the point of common coupling (PCC) terminal. It connects the LSWPP to the electrical power grid through Bus2, which was discussed in the early sections.

The parameters of the modern power grid (interconnected system) are listed in Appendix B of this article.

After integration, there is an additional transmission line circuit interconnecting the LSWPP, which, according to the renewable energy grid codes (REGCs), depending on the agreement between the power system operator (PSO) and the WPP operator (WPPO), this transmission line can either be operated by the PSO or the WPPO. At this point, attention was only given to monitoring from the sending-end busbar (WFSEBus) and receiving-end busbar (WFSEBus) up to the traditional electrical power grid side.

5. Evaluation of the Grid-Connected Mode of Operation of a Large-Scale Wind Power Plant

Case studies were executed to determine if the electrical quantities in the interconnected system were according to the renewable energy grid codes (REGCs). The case studies first looked at the initial load simulations to check if the busbar voltages were within the acceptable values. After that, the increased load demand was simulated to check if the voltages at the busbars had improved and were settling at acceptable values based on the power grid operating standards.

5.1. Initial Load Demand Simulations for the Modern Power Grid System

In this simulation, we put the load demand at its initial stage and monitored the load and busbar active and reactive power, as tabulated in Tables 16 and 17. We also recorded busbar voltage values, as shown in Table 18.

Table 16. Active power on load busbars and their loads under initial load demand for the modern power grid.

	Load Active Power (MW)	Busbar Active Power (MW)
DLoad1 ǀ Bus2	125.0000	125.0000
DLoad2 ǀ Bus3	90.0000	90.0000
DLoad3 ǀ Bua5	100.0000	100.0000

Table 17. Reactive power on load busbars and their loads under initial load demand for the modern power grid.

	Load Reactive Power (MVAr)	Busbar Reactive Power (MVAr)
DLoad1 ǀ Bus2	50.0000	50.0000
DLoad2 ǀ Bus3	30.0000	30.0000
DLoad3 ǀ Bua5	35.0000	35.0000

Table 18. Busbar voltage under initial load demand for the modern power grid.

Busbars	Voltage (PU)
Bus1	1.0210
Bus2	0.9928
Bus3	1.0060
Bus4	1.0230
Bus5	1.0130
Bus6	1.0280
WPPSEBus	0.9930
WPPREBus	0.9934

5.2. Increased Load Demand Simulations for the Modern Power Grid System

Our analysis in this section is based on Bus2 only since it has already been identified as the critical busbar in previous sections. While the attention was focused on Bus2, the load demand increase simulation was again performed, and the active power and reactive power for both DLoad1 and Bus2, along with voltage results pertinent to this busbar, have been recorded in Figures 30, 31, and 32, respectively.

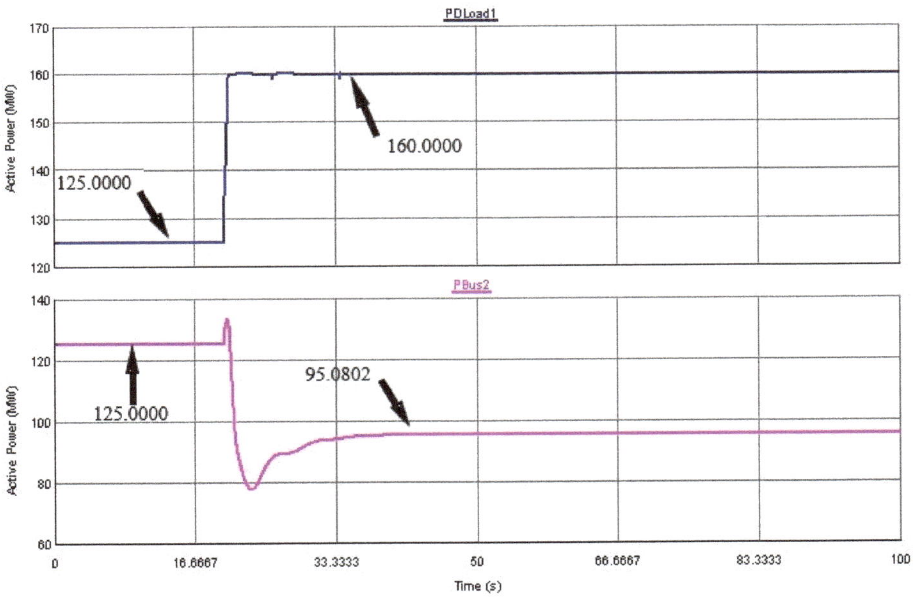

Figure 30. Increased load demand simulations for the modern power grid: Active power demand by DLoad1 and active power at Bus2.

The voltage Bus2 was again observed during this simulation in response to the load demand increase in DLoad1. With an increase in load demand, the voltage at Bus2 dropped less significantly compared to the standalone traditional power grid.

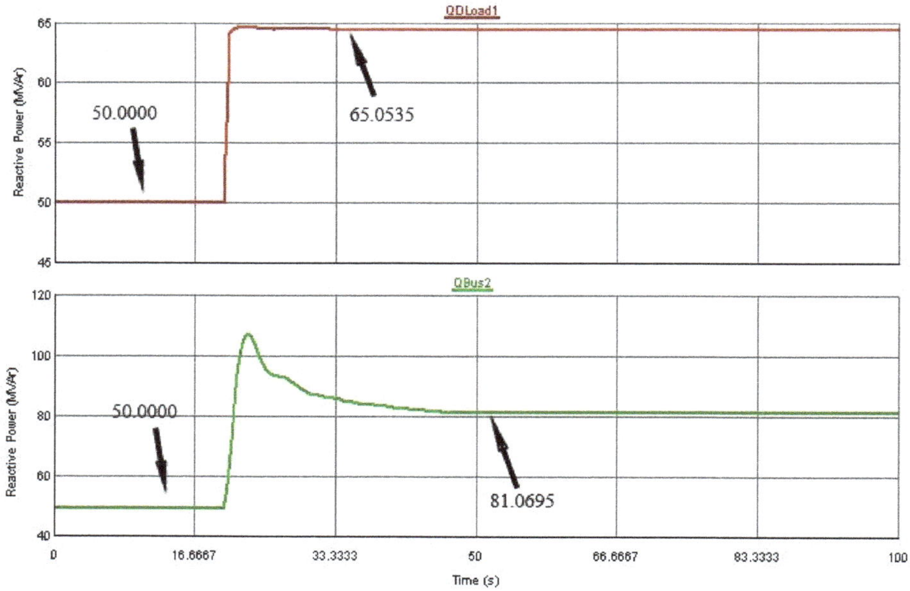

Figure 31. Increased load demand simulations for the modern power grid: Reactive power demand by DLoad1 and reactive power at Bus2.

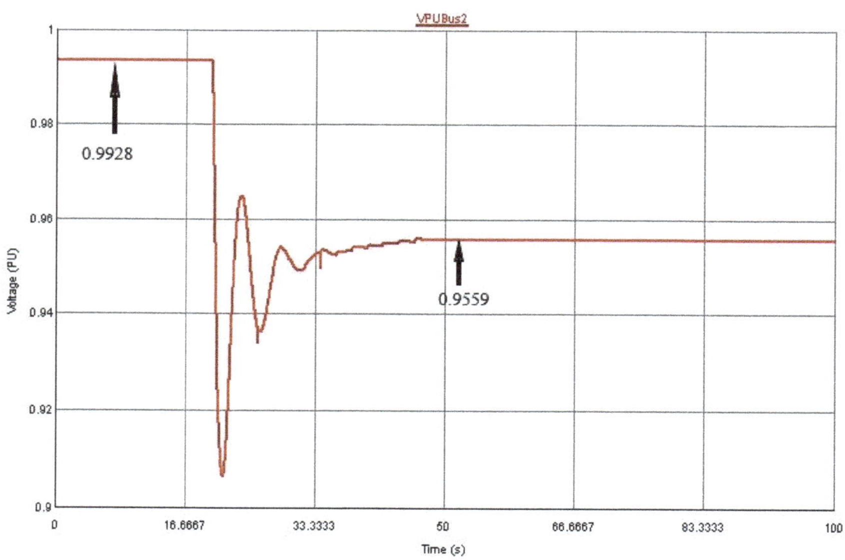

Figure 32. Increased load demand simulations for the modern power grid: Bus2 voltage.

6. Discussion of Results

When the load demand increase was implemented, a significant improvement in the voltage stability of the electrical power grid system was revealed after integrating the large-scale wind power plant (LSWPP) into this study. This finding is not new. However, the approach to reaching it is new in the following manner. The existing studies, like [19–21] rarely consider a step-by-step approach to obtaining to this finding. They did not consider evaluating the renewable energy grid code specifications before integrating the developed systems into the electrical power grid. Unlike the existing studies, this study aligned with the design and operational requirements of WPPs mentioned in [22–24]. This is due to its approach of first considering the simulations of the developed LSWPP alone to ensure its operation meets the REGC specifications before integrating it into the considered power grid. Furthermore, some studies on wind power plant (WPP) integration use a reasonable megawatt capacity of a single WTGU and treat it as a LSWPP, for instance, in reference [17]. Considering the models used in [17], and other models in [18–21], the computation time was saved. The challenge is that the models developed in these studies may not provide adequate validation to align the REGCs, whose significance has been mentioned in the work published in the references [1,3,4,14,26] for LSWPPs in the grid-connected mode of operation.

Lastly, using the Real-Time Digital Simulator (RTDS) devices with their software Real-time Simulation Computer-Aided Design (RSCAD) in this study enabled the simulation of load disturbance while monitoring them in real time. This has made the overall simulations in the study realistic compared to other studies.

7. Conclusions and Future Work

This work evaluated the large-scale wind power plant (LSWPP) grid-connected mode of operation by considering grid code specifications in real-time simulation. The studies associated with this topic have been explored. The interconnected system consisting of a LSWPP considered in this study has been carefully selected and, most importantly, described. Case studies have been conducted in consideration of the specifications of renewable energy grid codes (REGCs). The interconnected system developed and considered in this study effectively evaluates the grid-connected mode of operation of an LSWPP. It can be used for other large-scale renewable power plant (LSRPP) systems like photovoltaic (PV) systems.

It should be noted that in a WPP, not all wind turbine generator units (WTGUs) are operated simultaneously. This study considered the worst-case scenario where all WTGUs were required to operate simultaneously when the contribution of the entire WPP was urgently required. This is the part we will need to improve on when studying the need for WPP integration in the future, and we also look at developing a controller to ensure that the supply power required from the WPP side does not approach to the rated capacity.

Before this work, we had our previous work listed as follows:

(1) We first looked at the design of wind power plants (WPPs) and modeled a large-scale wind power (LSWPP) in [36].
(2) We performed a recap on the existing control methods for modern power grids consisting of WPPs.
(3) We evaluated the power grid voltage instability point using system overloading contingency in real-time simulation.
(4) Following the work in (3), we recently tested the busbar differential protection system, based on practical considerations, using a real-time digital simulator hardware-in-the-loop (HIL) testing technique.

Future work will consider a cross-comparison case study, applying the model developed in this study to a different international grid code to demonstrate its adaptability and robustness.

Author Contributions: Conceptualization, S.N., M.E.S.M. and A.K.R.; methodology, S.N., M.E.S.M. and A.K.R.; software, S.N., M.E.S.M. and A.K.R.; validation, S.N., M.E.S.M. and A.K.R.; formal analysis, S.N., M.E.S.M. and A.K.R.; investigation, S.N., M.E.S.M. and A.K.R.; resources, S.N., M.E.S.M. and A.K.R.; data curation, S.N., M.E.S.M. and A.K.R.; writing—original draft preparation, S.N., M.E.S.M. and A.K.R.; writing—review and editing, S.N., M.E.S.M. and A.K.R.; visualization, S.N., M.E.S.M. and A.K.R.; supervision, S.N., M.E.S.M. and A.K.R.; project administration, S.N., M.E.S.M. and A.K.R.; funding acquisition, S.N., M.E.S.M. and A.K.R. All authors have read and agreed to the published version of the manuscript.

Funding: This research received no external funding.

Institutional Review Board Statement: Not applicable.

Informed Consent Statement: Not applicable.

Data Availability Statement: The original contributions presented in this study are included in the article. Further inquiries can be directed to the corresponding author.

Acknowledgments: We thank the Center for Substation Automation and Energy Management Systems (CSAEMS). This research would not have been possible without their facilities.

Conflicts of Interest: The authors declare no conflict of interest.

Abbreviations

The following abbreviations are used in this manuscript.

WT	Wind turbine
WTGU	Wind turbine generator unit
WTGSUT	Wind turbine generator substation unit transformer
WFSEBus/WPPSEBus	Wind farm/wind power plant sending-end busbar
WPPHVTRF	Wind power plant high-voltage transformer
WFREBus/WPPREBus	Wind farm/wind power plant receiving-end busbar
G1Bus, G2Bus, and G3Bus	Generator 1 busbar, Generator 2 busbar, and Generator 3 busbar
G1, G2, and G3	Generator 1, Generator 2, and Generator 3
DLoad1, DLoad2, and DLoad3	Dynamic load 1, Dynamic load2, and Dynamic load 3
Bus1, Bus2, Bus3, Bus4, Bus5, and Bus6	Busbar 1, Busbar 2, Busbar 3, Busbar 4, Busbar 5, and Busbar 6
RSCAD	Real-time Simulation Computer-Aided Design
RTDS	Real-Time Digital Simulator
HIL	Hardware-In-the-Loop
WPP	Wind power plant
LSWPP	Large-scale wind power plant
REGC	Renewable energy grid code
RPP	Renewable power plant
PU	Per unit
WPPTransLine	Wind power plant transmission line
PCCTerm	Point of common coupling terminal
SEBus	Sending-end busbar
REBus	Receiving-end busbar

Appendix A

This section presents a flowchart of the overall study, as shown in Figure A1.

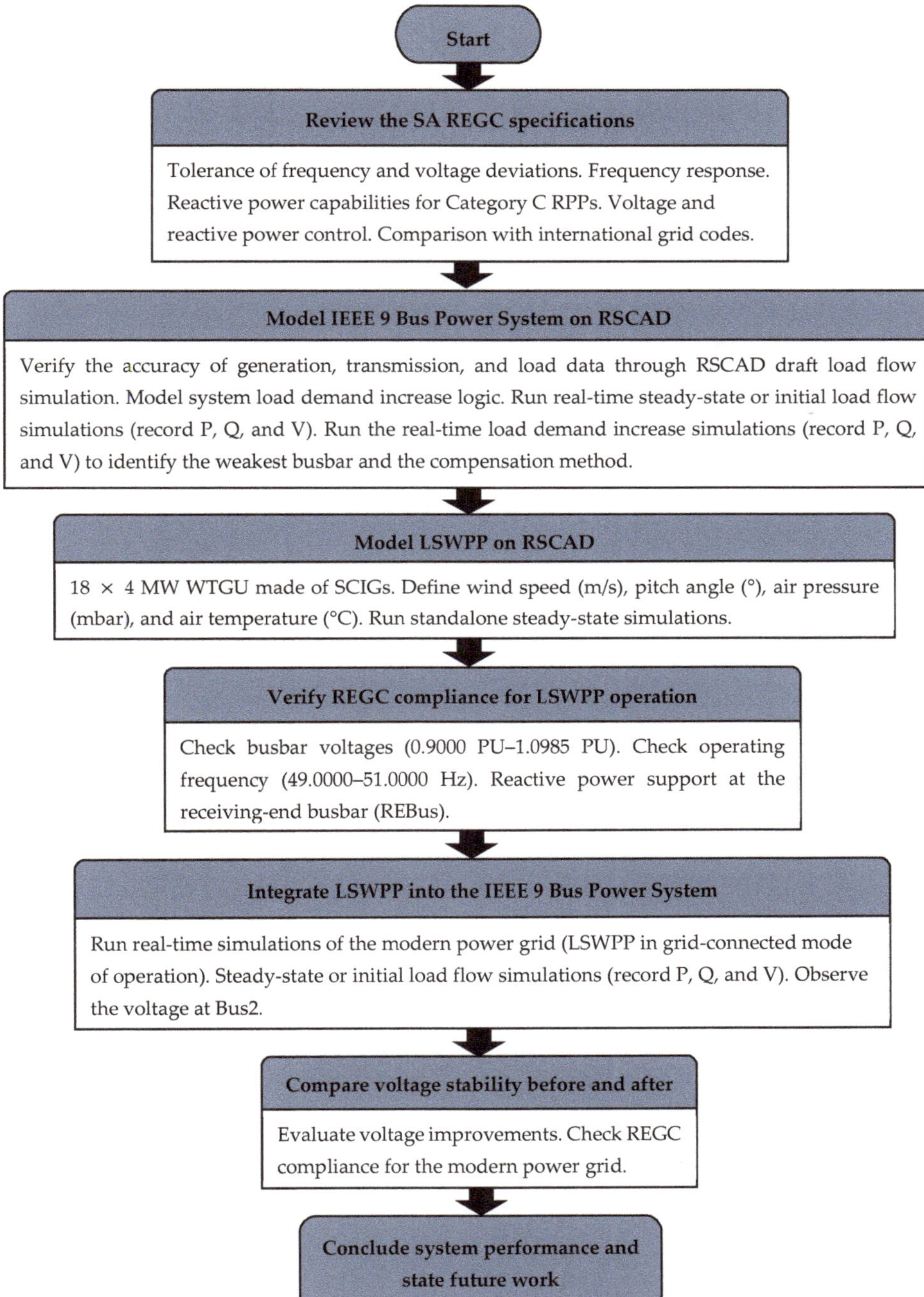

Figure A1. A flowchart showing all steps that were followed to conduct the study.

Appendix B

This section shows other parameters for the IEEE 9 Bus Power System combined with those of the large-scale wind power plant (LSWPP). All per unit values have been derived from a base apparent power of 100 MVA.

Table A1. Large-scale wind power plant transmission line parameters.

Lines	The Busbar to Which Line Is Connected	Length (km)	Resistance (PU)	Reactance (PU)	Susceptance (PU)
LineGroup1	Group1SEBus—WF1SEBus	15.0000	-	-	-
LineGroup2	Group2SEBus—WF1SEBus	15.0000	-	-	-
LineGroup3	Group3SEBus—WF1SEBus	15.0000	-	-	-
WPPTransLine					

All other wind turbine generator unit (WTGU) group lines, LineGroup2 and LineGroup3, are the same as the ones shown in Figure A2.

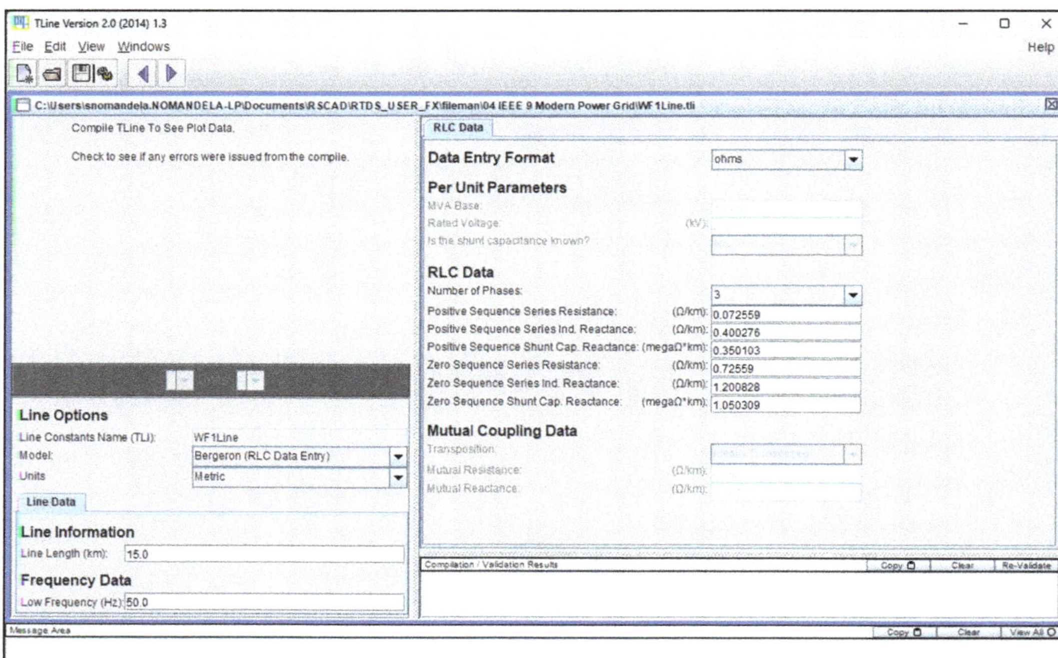

Figure A2. LSWPP LineGroup1 transmission line parameters.

Table A2. Power transformer parameters.

Transformer	The Busbar to Which the Power Transformer Is Connected	Voltage Rating (kV)	Resistance (PU)	Inductive Reactance (PU)	Capacity (MVA)
TRF1	G1Bus–Bus1	16.5/230	0.0	0.0576	100
TRF2	G2Bus–Bus4	18/230	0.0	0.0625	100
TRF3	G3Bus–Bus6	13.8/230	0.0	0.0586	100
WTGU1Trans	WTG1Terminal–Collector1Bus	4/24	-	-	5
Group1Trans	Collector1Bus–Group1Bus	24/230	-	-	30

The parameters for the transformer WTGU1Trans are the same as those of the WTGU2Trans and WTGU2Trans up to WTGU18Trans for other WTGUs. The other WTGU groups, Group2Trans and Group3Trans, are the same as Group1Trans. The IEEE 9 Bus Power System consists of three generators with their controllers, governors and exciter controllers. Tables A3 and A4 show the generator parameters.

Table A3. IEEE 9 Bus Power System generator parameters 1.

Generator	The Busbar to Which the Generator Is Connected	Xa (PU)	Xd (PU)	Xd′ (PU)	Xd″ (PU)	Xq (PU)	Xq′ (PU)	Xq″ (PU)
Gen1	G1Bus	0.0146	0.1460	0.0608	0.0600	0.1000	0.0969	0.0600
Gen2	G2Bus	0.0896	0.8958	0.1198	0.1100	0.8645	0.1969	0.1100
Gen3	G3Bus	0.1313	1.3125	0.1813	0.1800	1.2578	0.2500	0.1800

Table A4. IEEE 9 Bus Power System generator parameters 2.

Generator	The Busbar to Which the Generator Is Connected	Ra (PU)	Tdo′ (s)	Tdo″ (s)	Tqo′ (s)	Tqo″ (s)	H (s)	D (PU/PU)
Gen1	G1Bus	0.00013	8.9600	0.0100	0.3100	0.0100	23.6400	0.0
Gen2	G2Bus	0.00013	6.0000	0.0100	0.5350	0.0100	6.4000	0.0
Gen3	G3Bus	0.00013	5.8900	0.0100	0.6000	0.0100	3.0100	0.0

Table A5. Vestas V117-4.2 MW wind turbine and rotor data.

Parameter	Values
Rotor diameter (D_R)	117 m
Swept area (A_S)	10.7510 m^2
Minimum turbine angular speed, ω_{TRMin}	2.1000 rpm = 0.2200 rad/s
Nominal turbine angular speed, ω_{TRN}	9.9000 rpm = 1.0350 rad/s
Maximum turbine angular speed, ω_{TRMax}	17.7000 rpm = 1.8500 rad/s

Table A6. Vestas V117-4.2 MW wind turbine operating data [36].

Parameter	Values
Nominal power (P_N)	4.0000 MW/4.2000 MW
Cut-in-speed (v_{CI})	3.0000 m/s
Nominal speed (v_N)	14.0000 m/s
Cut-out-speed (v_{CO})	25.0000 m/s
Re-cut-in-speed (V_{R-CI})	23.0000 m/s
Wind class	EC IB-T/IEC IIA-T/IEC S-T
Standard operating temperatures	−20 °C to 45 °C with de-rating above 30 °C at 4 MW
Nominal power (P_N)	4.0000 MW/4.2000 MW
Cut-in-speed (v_{CI})	3.0000 m/s
Nominal speed (v_N)	14.0000 m/s

Table A7. Squirrel-cage induction generator parameters [36].

Parameter	Actual Values	Per Unit (PU) Values
Rated output power	4.0000 MW	-
Rated mechanical power	4.0606 MW	
Rated apparent power	4.8420 MVA	
Rated line-to-line voltage	40,000 V (rms)	
Rated phase voltage = Base voltage	2309.4000 V (rms)	
Rated stator current = Base current	698.8800 A (rms)	
Rated stator frequency	50.0000 Hz	
Rated power factor	0.8261	
Rated rotor speed	1510.5000 rpm	
Rated mechanical torque	25.6710 kN.m	
Rated stator flux linkage	7.3917 Wb (rms)	
Rated rotor flux linkage	6.7114 Wb (rms)	
Stator winding resistance	22.1040 mΩ	0.0067
Rotor winding resistance	23.1515 mΩ	0.0069
Stator leakage inductance	1.6980 mH	
Stator leakage reactance	j0.53344 Ω	j0.1615
Rotor leakage inductance	1.6980 mH	
Rotor leakage reactance	j0.53344 Ω	j0.1615
Magnetizing inductance	33.5970 mH	
Magnetizing reactance	j10.5548 Ω	j3.1946
Base flux linkage	7.3511 Wb (rms)	
Base impedance	3.3044 Ω	
Base inductance	10.5180 mH	
Base capacitance	963.2900 µF	

The letter j represents the reactance values of the machine.

References

1. Ahmed, S.D.; Al-Ismail, F.S.M.; Shafiullah, M.; Al-Sulaiman, F.A.; El-Amin, I.M. Grid Integration Challenges of Wind Energy: A Review. *IEEE Access* **2020**, *8*, 10857–10878. [CrossRef]
2. Osman, A.M.; Alsokhiry, F. Sliding Mode Control for Grid Integration of Wind Power System Based on Direct Drive PMSG. *IEEE Access* **2022**, *10*, 26567–26579. [CrossRef]
3. Ibrahim, H.; Ghandour, M.; Dimitrova, M.; Ilinca, A.; Perron, J. Integration of Wind Energy into Electricity Systems: Technical Challenges and Actual Solutions. *Energy Procedia* **2011**, *6*, 815–824. [CrossRef]
4. Yuan, X. Overview of Problems in Large-scale Wind Integrations. *J. Mod. Power Syst. Clean Energy* **2013**, *1*, 22–25. [CrossRef]
5. Abdelghany, M.B.; Mariani, V.; Liuzza, D.; Natale, O.R.; Glielmo, L. A Unified Control Platform and Architecture for the Integration of Wind-Hydrogen Systems into the Grid. *IEEE Trans. Autom. Sci. Eng.* **2023**, *21*, 4042–4057. [CrossRef]

6. Wu, D.; Seo, G.-S.; Xu, L.; Su, C.; Kocewiak, L.; Sun, Y.; Qin, Z. Grid Integration of Offshore Wind Power: Standards, Control, Power Quality and Transmission. *IEEE Open J. Power Electron.* **2024**, *5*, 583–604. [CrossRef]
7. Pandit, D.; Bera, A.; Saket, R.K. Probabilistic Sizing of Energy Storage Systems for Reliability and Frequency Security in Wind-Rich Power Grids. *IEEE Access* **2024**, *12*, 4608–4621. [CrossRef]
8. Li, L.; He, J.; Niu, J.; Liu, H.; Yu, X.; Huang, Y.; Zhang, X. A Phase-to-Phase Protection Algorithm With the Transient Information for the Wind Power Plant Transmission Line. *IEEE Access* **2024**, *12*, 100581–100590. [CrossRef]
9. Rekik, M.; Abdelkafi, A.; Krichen, L. Synchronization of wind farm power system to utility grid under voltage and frequency variations. *Int. J. Renew. Energy Res.* **2015**, *5*, 70–81.
10. Jain, P.; Wijayatunga, P. Grid Integration of Wind Power: Best Practices for Emerging Wind Markets. *Sustain. Dev. Work. Pap. Ser.* **2016**. Available online: https://www.adb.org/publications/grid-integration-wind-power-best-practices-wind-markets (accessed on 17 March 2025).
11. Tummala, A.; Velamati, R.K.; Sinha, D.K.; Indraja, V.; Krishna, V.H. A Review on Small Scale Wind Turbines. *Renew. Sustain. Energy Rev.* **2016**, *56*, 1351–1371. [CrossRef]
12. Masters, G.M. *Renewable and Efficient Electric Power Systems*; Wiley and Sons, Inc: Hoboken, NJ, USA, 2004. [CrossRef]
13. Stock, A.; Amos, L.; Alves, R.; Leithead, W. Design and Implementation of a Wind Farm Controller using Aerodynamics Estimated from LIDAR Scans of Wind Turbine Blades. In Proceedings of the 2021 American Control Conference (ACC), New Orleans, LA, USA, 25–28 May 2021; pp. 1790–1795. [CrossRef]
14. Blazquez, F.; Veganzones, C.; Martinez, S. Large Scale Integration of Wind Energy into Power Systems. *Electr. Power Qual. Utilis. Mag.* **2005**, *1*, 15–22.
15. Abbas, S.R.; Kazmi, S.A.A.; Naqvi, M.; Javed, A.; Naqvi, S.R.; Ullah, K.; Khan, T.-U.; Shin, D.R. Impact Analysis of Large-Scale Wind Farms Integration in Weak Transmission Grid from Technical Perspectives. *Energies* **2020**, *13*, 5513. [CrossRef]
16. Xu, D.; Mu, G.; Liu, H.; Yan, G. Assessment on Large-scale Offshore Wind Farm Integration. In Proceedings of the 2014 9th IEEE Conference on Industrial Electronics and Applications (ICIEA 2014), Hangzhou, China, 9–11 June 2014; IEEE: New York, NY, USA, 2014; pp. 556–559. [CrossRef]
17. Rana, V.; Gupta, R.; Kumar, N. Integration of Wind Farm into A Weak Distribution Network. *Change* **2014**, *4*, 404–409.
18. Hasnaoui, O.; Allagui, M. Dynamic performance improvement of wind farms equipped with three SCIG generators using STATCOM. *J. Energy South. Afr.* **2014**, *25*, 128–135. [CrossRef]
19. Benseddik, A.; Kouba, N.; Boudour, M.; Hasni, M.; Menaa, M. Wind Farm Integration Intermittency Impact on Power System Transient Stability. In Proceedings of the 2018 International Conference on Electrical Sciences and Technologies in Maghreb (CISTEM), Algiers, Algeria, 28–31 October 2018; pp. 1–6. [CrossRef]
20. Xia, S.; Zhang, Q.; Hussain, S.T.; Hong, B.; Zou, W. Impacts of Integration of Wind Farms on Power System Transient Stability. *Appl. Sci.* **2018**, *8*, 1289. [CrossRef]
21. Rachdi, T.; Saoudi, Y.; Mahmoud, I.; Florin, I.; Errachdi, A. The Impact of Integration Wind Turbine on the Stability of the Electricity Grid. In Proceedings of the 2022 IEEE Information Technologies & Smart Industrial Systems (ITSIS), Mankato, MN, USA, 19–21 May 2022; pp. 1–6. [CrossRef]
22. Niwas, S.; Singh, S.N.; Ostergaard, J.; Jain, N. Distributed Generation in Power Systems: An Overview and Key Issues. In Proceedings of the 24th Indian Engineering Congress (IEC), Kyoto, Japan, 10–13 December 2009. Available online: https://backend.orbit.dtu.dk/ws/portalfiles/portal/5202512/24IEC_paper.pdf (accessed on 17 March 2025).
23. Sewchurran, S.; Davidson, I.E. Introduction to the South African Renewable Energy Grid Code version 2.9 requirements (Part II—Grid code technical requirements). In Proceedings of the 2017 IEEE AFRICON, Cape Town, South Africa, 18–20 September 2017; pp. 1225–1230. [CrossRef]
24. National Energy Regulator of South Africa. Grid Connection Code for Renewable Power Plants Connected to the Electricity Transmission System or the Distribution System in South Africa. Vol. 1, No. January. National Energy Regulator of South Africa, Germiston, pp. 1–113. 2022. Available online: https://www.nersa.org.za/wp-content/uploads/bsk-pdf-manager/2022/01/SAGC-Requirements-for-Renewable-Power-Plants-Rev-3.1.pdf (accessed on 17 March 2025).
25. Nomandela, S.; Ratshitanga, M.; Mnguni, M.E.S. IEC 61850 Standard-Based Protection of the Coupling Point Between a Wind Farm and the Power Grid. Cape Peninsula University of Technology (CPUT). 2021. Available online: https://www.researchgate.net/publication/370501543_IEC_61850_Standard-based_Protection_of_the_Coupling_Point_Between_a_Wind_Farm_and_the_Power_Grid (accessed on 17 March 2025).
26. Meiqi, Y.; Liangzhong, Y. Integration of Large Scale Wind Farm into Electrical Grids. In Proceedings of the 2010 China International Conference on Electricity Distribution (CICED2010), Nanjing, China, 13–16 September 2010; IEEE: New York, NY, USA, 2010; pp. 1–5.
27. ESKOM. *Transmission and Distribution of Electricity, No. Revision 11*; ESKOM: Sandton, South Africa, 2021; pp. 1–3.

28. Davidson, I.E. Introduction to the South African Renewable Energy Grid Code Version 2.9 Requirements. In Proceedings of the 2017 IEEE AFRICON, Cape Town, South Africa, 18–20 September 2017; pp. 1263–1267. [CrossRef]
29. Plc, N.G.E.T. The Grid Code—UK, No. 5. 2017. Available online: https://www.neso.energy/document/85301/download (accessed on 17 March 2025).
30. Hooper, J.W.; Chester, R.O. *Technical Guidelines for Power Generating Units and Systems. Certification of the Electrical Characteristics of Power Generating Units and Systems in Medium-, High and Extra-High Voltage Grids*; Fördergesellschaft Windenergie und Andere Erneuerbare Energien: Berlin, Germany, 2016; pp. 51–136. [CrossRef]
31. *IEEE Std 1547-2018 (Revision of IEEE Std 1547-2003)*; IEEE Standard for Interconnection and Interoperability of Distributed Energy Resources with Associated Electric Power Systems Interfaces. IEEE: New York, NY, USA, 2016.
32. RTDS Technologies. *IEEE 9 Bus Power System*; RTDS Technologies: Winnipeg, MB, Canada, 2019; pp. 2–5.
33. Marison, G.; Gao, B.; Kundur, P. Voltage Stability Analysis Using Static and Dynamic Approaches. *IEEE Trans. Power Electron.* **1993**, *8*, 1159–1171. [CrossRef]
34. Kundur, P.; Paserba, J.; Ajjarapu, V.; Andersson, G.; Bose, A.; Canizares, C.; Hatziargyriou, N.; Hill, D.; Stankovic, A.; Taylor, C.; et al. Definition and Classification of Power System Stability. *IEEE Trans. Power Syst.* **2004**, *19*, 3271–3281. [CrossRef]
35. Kundur, P. *Power System Stability and Control*; McGraw: New York, NY, USA, 1993.
36. Nomandela, S.; Mnguni, M.E.S.; Raji, A.K. Modeling and Simulation of a Large-Scale Wind Power Plant Considering Grid Code Requirements. *Energies* **2023**, *16*, 2897. [CrossRef]
37. Vestas. 4 MW Wind Platform. Brocher. 2017. Available online: https://www.vestas.com/en/products/4-mw-platform (accessed on 17 March 2025).
38. Wu, B.; Lang, Y.; Zargari, N.; Kouro, S. Power Conversion and Control of Wind Energy Systems: Appendix B: Generator Parameters. In *Wind Energy*; John Wiley & Sons: Hoboken, NJ, USA, 2011; pp. 319–326.

applied sciences

MDPI

Article

Low-Carbon Economic Collaborative Scheduling Strategy for Aluminum Electrolysis Loads with a High Proportion of Renewable Energy Integration

Jingyu Li *, Yuanyu Chen *, Guangchen Liu and Ruyue Han

School of Electric Power, Inner Mongolia University of Technology, Hohhot 010080, China; liugc@imut.edu.com (G.L.); hanry1979@163.com (R.H.)
* Correspondence: ljy1986@imut.edu.cn (J.L.); yuanyuchen0919@163.com (Y.C.)

Abstract

In response to the challenges faced by high-energy-consuming enterprises in utilizing renewable energy and implementing low-carbon operations, this paper proposes a multi-objective optimization strategy based on source–storage–load collaborative scheduling. The strategy establishes a refined model of aluminum electrolysis load, thoroughly considering the coupling relationship between temperature, production output, and power consumption. Additionally, it develops a dynamic coupling model between multi-functional crane loads and aluminum electrolysis production to reveal the influence mechanism of auxiliary equipment on the main production process. Based on this foundation, this paper constructs a multi-objective optimization model that targets the minimization of operating costs, the minimization of carbon emissions, and the maximization of the renewable energy consumption rate. An improved heuristic intelligent optimization algorithm is employed to solve the model. The simulation results demonstrate that, under a renewable energy penetration of 67.8%, the proposed multi-objective optimization strategy achieves a maximum reduction in carbon emissions of 1677.35 t and an increase in renewable energy consumption rate of 12.11%, compared to the conventional single-objective economic optimization approach, while ensuring the stability of aluminum electrolysis production. Furthermore, when the renewable energy penetration is increased to 76.2%, the maximum reduction in carbon emissions reaches 8260.97 t, and the renewable energy consumption rate is improved by 18.86%.

Keywords: high-energy-consuming load; heuristic intelligent optimization algorithm; multi-objective optimization; renewable energy integration; source–storage–load coordination

1. Introduction

According to the International Energy Agency (IEA) report [1], renewable energy generation accounted for over 32% of global electricity production in 2024. In China, as the world's largest energy consumer, solar photovoltaic and wind power contributed nearly 20% of total power generation. However, wind and solar curtailment issues persist. Energy-intensive industries consumed more than 2 trillion kWh of electricity, accounting for approximately 20% of the country's total industrial electricity consumption [2]. These challenges have prompted the "14th Five-Year Plan for Renewable Energy Development" [3] to propose the construction of dedicated power supply lines and self-supplied power stations in industrial parks and other areas [4], promoting the replacement of coal-fired power plants with green electricity. However, the contradiction between rigid electricity demand and fluctuating supply has exacerbated the power supply–demand imbalance in remote areas, constraining both clean energy development and the efficiency of energy-intensive enterprises, necessitating urgent exploration of effective solutions for synergistic optimization between these two aspects [5].

In recent years, research on wind and solar energy integration in industrial parks has evolved from fundamental principles to system-level integration. The authors of Ref. [6] systematically analyzed the spatiotemporal characteristics of wind–solar complementarity, laying a theoretical foundation for subsequent modeling. The authors of Ref. [7] integrated advanced independent models from wind and photovoltaic fields, significantly enhancing the accuracy and applicability of wind–solar hybrid power station generation modeling. The authors of Ref. [8] introduced energy storage systems and, using LEAP and NEMO models, verified the comprehensive advantages of energy storage under new policy scenarios and high wind–solar capacity scenarios in terms of storage efficiency, carbon reduction, and cost savings. The authors of Ref. [9] developed a wind–solar–hydrogen energy system model, thereby optimizing system economy and renewable energy utilization rates. However, these strategies primarily focus on source-side analysis, with insufficient attention to load-side adjustability.

Domestically and internationally, studies on adjustable loads participating in wind–solar energy integration have been conducted [10]. As a typical high-energy-consumption industrial load, the aluminum electrolysis sector accounts for approximately 7% of China's total industrial electricity consumption [11]. Recent studies indicate that, under the premise of ensuring production stability, aluminum electrolysis loads exhibit certain frequency regulation potential [12]. Empirical research [13] demonstrates that electrolysis cells show significant load flexibility, adjustable upward by 5–10% and downward by 25–30% from the rated operating point. The authors of Ref. [14] construct scheduling models that incorporate aluminum electrolysis loads as key subjects, proposing day-ahead and intra-day joint economic scheduling methods to reduce power system operation costs. The authors of Ref. [13] establish a layered optimization model for aluminum electrolysis loads combined with peak regulation and deep thermal power regulation, aiming to minimize system wind curtailment and operating costs. The authors of Refs. [15–17] propose a source–load collaborative optimization configuration model considering aluminum electrolysis load demand response, with the goal of minimizing the construction costs of renewable energy stations and the operating costs of industrial park power systems.

However, existing research has significant limitations, failing to fully balance multi-dimensional constraints, especially in high-proportion renewable energy scenarios, which may result in insufficient energy absorption rates and carbon emission rebounds. At the same time, the advanced nature of three-objective algorithms has been widely recognized; for example, the authors of Ref. [18] introduce the NSGA-III algorithm in bridge network maintenance, compared with NSGA-II, to construct an optimization model that balances performance, cost, and low carbon; the authors of Ref. [19] proposed a three-level energy management system based on a multi-objective grey wolf optimization algorithm for optimizing AC/DC hybrid microgrids with renewable energy and hydrogen equipment, aiming to achieve optimal balance among cost, renewable energy utilization rate, and carbon emissions; the authors of Ref. [20] propose a three-objective smart grid optimization model integrating hybrid demand response, renewable energy, and energy storage, solved using algorithms such as NSGA-II, and demonstrating superior performance compared to other algorithms.

In the aluminum electrolysis industrial parks in western remote regions, these parks are adjacent to abundant hydropower, wind, and photovoltaic resources, presenting opportunities for renewable energy utilization while bearing the responsibility for promoting regional energy transition. In these areas, constructing an optimization framework that integrates coordinated scheduling of main and auxiliary equipment and balances the three-dimensional objectives of economy, environment, and renewable energy absorption rates not only pertains to enterprise sustainable development but also supports regional energy structure optimization and ecological civilization construction.

This paper aims to address the aforementioned limitations through three key innovations:

(1) Focusing on the power regulation capability of aluminum electrolysis, a refined coupling model of aluminum electrolysis load considering operational stability constraints was constructed, a dynamic constraint system for the aluminum electrolysis production process was established, and the stability boundaries under different regulation depths, rates, and durations were quantified.
(2) This research proposes the dynamic response relationship between electrolytic cell power regulation and multi-functional overhead crane operating characteristics, establishes a dynamic coupling model between electrolytic cells and multi-functional crane operating characteristics, and achieves coordinated scheduling optimization of primary and auxiliary equipment.
(3) Based on three key perspectives—the adjustable load characteristics of aluminum electrolysis, abundant renewable energy resources, and national energy transition strategy—this research develops a three-objective optimization strategy for source–storage–load coordinated scheduling in aluminum electrolysis parks under different renewable energy penetration rates.

In summary, the multi-objective collaborative optimization strategy proposed in this study not only helps improve renewable energy utilization in remote areas and reduce "wind and solar curtailment" ratios but also enhances the economic benefits and environmental performance of high-energy-consuming enterprises.

2. Adjustable Characteristics of Aluminum Electrolysis Loads

Aluminum electrolysis production [21] is a high-energy-consuming industry that employs the cryolite–alumina molten salt electrolysis method. This process primarily involves converting bauxite into alumina, which is then used to produce primary aluminum through high-temperature molten salt electrolysis. Throughout the production process, alumina serves as the electrolyte raw material, while cryolite functions as the solvent. Various raw and auxiliary materials, including alumina and fluoride salts, are fed into the electrolytic cell [22]. Alumina dissolves in the molten cryolite to form a homogeneous melt with excellent electrical conductivity. The production process utilizes carbon materials for both the cathode and anode. When direct current is applied, electrochemical reactions occur at both electrodes. At the anode, anode gases are produced, containing CO_2, CO, and small amounts of fluorides, which require treatment through a dry purification system (using alumina to adsorb fluorides) to prevent environmental pollution. At the cathode, liquid aluminum is deposited at the bottom of the electrolytic cell and is periodically extracted using vacuum ladles, then sent to the casting plant to produce aluminum ingots for remelting. The specific production process is illustrated in Figure 1, and the aluminum electrolytic cell model is shown in Figure 2.

During aluminum electrolysis production, auxiliary equipment load fluctuations exhibit significant correlation with electrolysis series loads, with multi-functional overhead cranes demonstrating particularly pronounced coupling characteristics. As critical material handling equipment in the electrolysis workshop, crane operational loads present a high positive correlation with electrolysis series loads. When the series current undergoes adjustment, the crane's operational parameters—including frequency, runtime, and transportation intensity—correspondingly transform, primarily manifesting in raw material transportation, anode replacement, and product logistics. Research substantiates that when electrolysis cell loads fluctuate within the 5–30% range [13], multi-functional crane loads dynamically adjust within a corresponding interval, with load variation curves demonstrating near-synchronous alignment with electrolysis series load variations. This auxiliary equipment load coupling characteristic unveils the systemic nature of aluminum electrolysis production systems, providing novel perspectives and methodological entry points for subsequent load regulation model research.

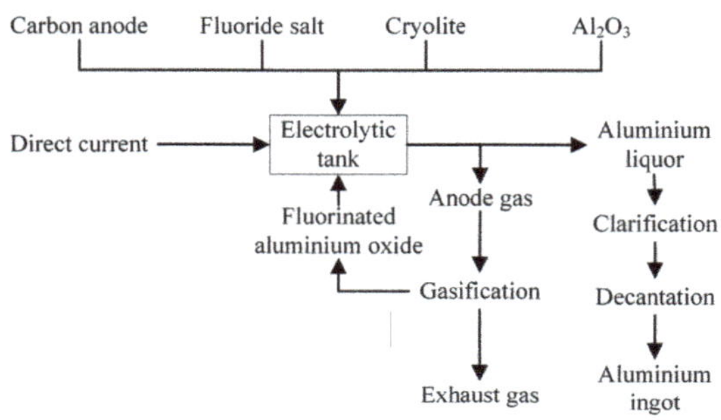

Figure 1. Process flow of aluminum electrolysis.

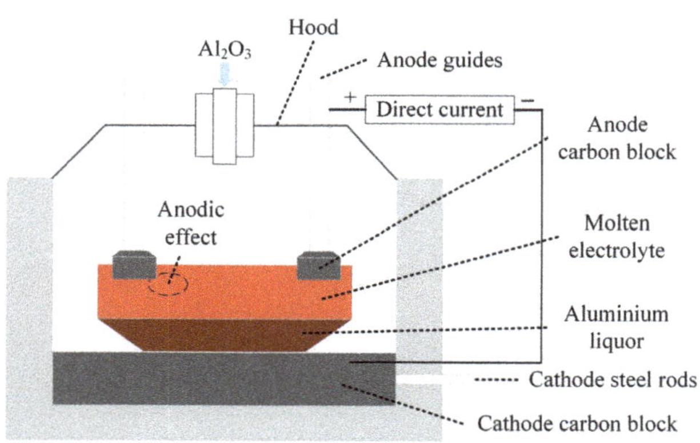

Figure 2. Model of an aluminum electrolysis cell.

(1) Power Constraints

To ensure the safety and stability of aluminum electrolysis production, power adjustments of aluminum electrolysis loads must remain within specified limits [23]:

$$P_{j,t,\min}^{\mathrm{al}} \leq P_{j,t}^{\mathrm{al}} \leq P_{j,t,\max}^{\mathrm{al}} \tag{1}$$

$$I^{\min} \leq I_{l,t} \leq I^{\max} \tag{2}$$

$$P_{j,t}^{\mathrm{al}} = U_l I_{l,t} \tag{3}$$

where $P_{j,t}^{\mathrm{al}}$ represents the power of aluminum electrolysis load j at time t; $P_{j,t,\max}^{\mathrm{al}}$ and $P_{j,t,\min}^{\mathrm{al}}$ represent the maximum and minimum values of aluminum electrolysis power, respectively; $I^{\max}$ and $I^{\min}$ are the upper and lower limits of current allowed to pass through the electrolytic cell; U_l is the series voltage of the electrolytic cell; and $I_{l,t}$ represents the current intensity passing through the electrolytic cell series l at time t.

(2) Temperature and Power Coupling Constraints

A certain magnitude of power variation will not cause serious damage to the thermal balance of aluminum electrolysis loads. When the power of an aluminum electrolysis load changes $\Delta P_{j,t}^{\mathrm{al}}$, the temperature also changes accordingly. The relationship between temperature change and power change can be expressed by the following formula [15]:

$$T_{j,t}^{\mathrm{al}} - T_{j,t-1}^{\mathrm{al}} = \frac{\left(P_{j,t}^{\mathrm{al}} - P_{j,t-1}^{\mathrm{al}}\right)\Delta t}{c_{\mathrm{al}} m_{\mathrm{al}}} \tag{4}$$

$$T_{\min}^{\mathrm{al}} \leq T_{j,t}^{\mathrm{al}} \leq T_{\max}^{\mathrm{al}} \tag{5}$$

where c_{al} and m_{al} represent the specific heat capacity coefficient and mass of the electrolyte, respectively; ΔT_{al} is the temperature change of the electrolytic cell; $T_{j,t}^{\mathrm{al}}$ represents the production temperature of the aluminum electrolysis load j; and $T_{\min}^{\mathrm{al}}$ and $T_{\max}^{\mathrm{al}}$ are the upper and lower temperature limits of the aluminum electrolysis load, generally set at 970 °C and 940 °C [24], respectively.

(3) Adjustable Time Constraints

In the production adjustment process of aluminum electrolysis loads, there are three states: power holding state, power up-regulation state, and power down-regulation state. It is assumed that within a day, the power of the electrolytic cell fluctuates around a reference power. The adjustable time for aluminum electrolysis loads can be expressed as follows [16]:

$$t_{\mathrm{adj,up}}^{i} \leq \frac{c_{\mathrm{al}} m_{\mathrm{al}}\left(T_{\max}^{al} - T_{j,\mathrm{rated}}\right)}{P_{j,t}^{\mathrm{al}} - P_{j,\mathrm{rated}}^{\mathrm{al}}} \tag{6}$$

$$t_{\mathrm{adj,down}}^{k} \leq \frac{c_{\mathrm{al}} m_{\mathrm{al}}\left(T_{j,\mathrm{rated}} - T_{\min}^{al}\right)}{P_{j,\mathrm{rated}}^{\mathrm{al}} - P_{j,t}^{\mathrm{al}}} \tag{7}$$

where $t_{\mathrm{adj,up}}^{i}$ is the time for the i-th upward adjustment of the power of the aluminum electrolysis load; $t_{\mathrm{adj,down}}^{k}$ is the time for the k-th downward adjustment of the power of the aluminum electrolysis load; $T_{j,\mathrm{rated}}$ is the rated temperature of the aluminum electrolysis load; and $P_{j,\mathrm{rated}}^{\mathrm{al}}$ is the rated power of the aluminum electrolysis load.

The constraints for the upward and downward adjustment times over the entire cycle can be expressed as follows [16]:

$$\sum_{t=1}^{T} t_{\mathrm{adj,up}}^{i} \leq T_{\mathrm{up}}^{\max} \tag{8}$$

$$\sum_{t=1}^{T} t_{\mathrm{adj,down}}^{k} \leq T_{\mathrm{down}}^{\max} \tag{9}$$

where $T_{\mathrm{up}}^{\max}$ is the maximum allowable time for upward power adjustments and $T_{\mathrm{down}}^{\max}$ is the maximum allowable time for downward power adjustments.

(4) Operational State Constraints

In the different operational phases of the aluminum electrolysis load, there must be a certain time interval between transitions [13]. For a series of electrolytic cells, the various adjustment phases, aside from normal operation, cannot switch continuously without a defined delay.

$$\gamma_{j,t-1}^{\mathrm{adj}} + \gamma_{j,t}^{\mathrm{adj}} = 1 \tag{10}$$

$$\gamma_{j,t}^{\mathrm{adj}} + \gamma_{j,t}^{\mathrm{keep}} = 1 \tag{11}$$

where $\gamma_{j,t}^{\mathrm{adj}}$ represents the adjustable state (0/1), and $\gamma_{j,t}^{\mathrm{keep}}$ represents the rated state (0/1), where a value of 1 indicates "yes" and a value of 0 indicates "no." Therefore, there is a unique state constraint at time t.

(5) Temperature and Efficiency Constraints

During normal operation, aluminum electrolysis loads have a rated current efficiency of approximately 94%. However, when the aluminum electrolysis load participates in system peak shaving optimization, power adjustments lead to temperature changes, which directly affect the current efficiency of the electrolytic cell. Relevant studies have confirmed that for every 10 °C decrease in the temperature of the electrolytic cell, the current efficiency increases by 1% to 1.5% [25]. Therefore, the production efficiency can be expressed as follows:

$$\eta_{j,t}^{\mathrm{al}} = \eta_{j,N}^{\mathrm{al}} - k_i \left(T_{j,t}^{\mathrm{al}} - T_{j,N}^{\mathrm{al}} \right) \tag{12}$$

where $\eta_{j,t}^{\mathrm{al}}$ represents the rated current efficiency of the aluminum electrolysis load j; $T_{j,N}^{\mathrm{al}}$ is the rated temperature for the production of aluminum electrolysis load j; and k_i is the ratio of current efficiency variation with temperature.

(6) Coupling Between Current Efficiency of Aluminum Electrolytic Cells and Aluminum Production

An aluminum electrolysis production series consists of dozens to hundreds of electrolytic cells connected in series, and the total production is the sum of the outputs from each cell. The production constraint is as follows [16]:

$$M_{j,t}^{\mathrm{al}} = n_{j,\mathrm{al}} K_{\mathrm{al}} I_{j,t}^{\mathrm{al}} \eta_{j,t}^{\mathrm{al}} \Delta t \tag{13}$$

$$M_{j,N}^{\mathrm{al}} = \sum_{t=1}^{T} M_{j,t}^{\mathrm{al}} \tag{14}$$

where $M_{j,N}^{\mathrm{al}}$ represents the total rated aluminum production of aluminum electrolysis load j within the scheduling period T; $n_{j,\mathrm{al}}$ is the number of electrolytic cells for aluminum electrolysis load j; and K_{al} is the electrochemical equivalent of aluminum, measured in g/(A·h), and is generally taken to be 0.3356.

(7) Coupling Constraints between Multi-functional Gantry Crane Load and Aluminum Production

The multifunctional crane is the core production service equipment that integrates six functions: "shell breaking, material feeding, anode changing, aluminum extraction, slag removal, and lifting." When considering only the electrolysis process, the multifunctional crane [26], as a representative adjustable auxiliary device, accounts for 5% of the total electricity load. By directly linking the power changes of the multifunctional crane to aluminum production, the model can more accurately reflect the dynamic changes in energy consumption during the actual production process, rather than simplifying the auxiliary equipment. Within the aluminum electrolysis plant, the formula for the material handling tasks of the multifunctional crane is as follows [27]:

$$M_{\mathrm{Al_2O_3}}(t) = M_{j,t}^{\mathrm{al}} \cdot k_{\mathrm{Al_2O_3}} \cdot \xi_{\mathrm{Al_2O_3}}(t) \tag{15}$$

$$M_{\mathrm{anode}}(t) = M_{j,t}^{\mathrm{al}} \cdot k_{\mathrm{anode}} \cdot \xi_{\mathrm{anode}}(t) \tag{16}$$

$$M_{\mathrm{metal}}(t) = M_{j,t}^{\mathrm{al}} \cdot \xi_{\mathrm{metal}}(t) \tag{17}$$

$$M_{\mathrm{spent}}(t) = M_{\mathrm{anode}}(t - \tau_{\mathrm{anode}}) \cdot \xi_{\mathrm{spent}}(t) \tag{18}$$

where $M_{\mathrm{Al_2O_3}}(t)$ represents the amount of alumina transported per unit time, $M_{\mathrm{anode}}(t)$ is the amount of anode replaced per unit time; $M_{\mathrm{metal}}(t)$ is the amount of aluminum liquid extracted from the cell per unit time; $M_{\mathrm{spent}}(t)$ is the amount of residual pole transport per unit time; k_i is the basic consumption coefficient for material i; $\xi_i(t)$ is the consumption correction coefficient for material i during power changes; and τ_{anode} is the anode usage cycle (typically 28–30 days).

Based on the previous analysis, the power demand of the multifunctional crane mainly comes from two components: the no-load power demand (the power required for the crane's movement) and the load power demand (the additional power required for lifting and transporting materials). The power calculation formula for the multifunctional crane is as follows [28]:

$$P_{\mathrm{crane}}(t) = P_{\mathrm{idle}} + \sum_{t=1}^{T} \sum_{i} \gamma_i \cdot M_i(t) \cdot f_i(t) \tag{19}$$

where P_{idle} represents the no-load power of the multifunctional crane, γ_i is the unit handling power coefficient for material i, $M_i(t)$ is the weight of material i transported per unit time, and $f_i(t)$ is the function representing the handling frequency of material i per unit time.

3. Coordinated Optimization Model for the Source–Load of Aluminum Electrolysis Load

3.1. System Architecture

This study focuses on industrial parks characterized by aluminum electrolysis enterprises, where electricity mainly comes from self-owned power plants, wind farms, and photovoltaic stations, as well as interconnection lines established between the park and the external power grid. The load side comprises aluminum electrolysis loads, multi-functional crane loads, and other auxiliary loads. To enhance renewable energy absorption and reduce operational costs, the industrial park has developed a multi-tiered electricity demand response regulation mechanism: aluminum electrolysis loads are dynamically adjusted according to wind and solar output fluctuation intervals, with captive power plants prioritizing output reduction or shutdown when renewable generation meets load demands under stable operating conditions. Energy storage systems perform flexible regulation, providing support when load demands cannot be immediately met by wind, solar, or thermal generation, and charging during excess renewable energy generation. The external grid serves as a backup power source during insufficient renewable and storage generation, with the potential for electricity sales when renewable generation is abundant and energy storage is fully charged. The source–storage–load coordinated dispatch architecture for aluminum electrolysis industrial parks is illustrated in Figure 3.

3.2. Objective Function

3.2.1. Objective Function 1

The total cost, which consists of the operating costs of the self-owned power plants, wind turbines, photovoltaics, and energy storage devices of the aluminum electrolysis enterprise, as well as the adjustment costs for scheduling loads and the interaction costs with the higher-level power grid, is minimized as Objective Function 1.

$$\min f_1 = C_{\text{op,re}} + C_{\text{op,the}} + C_{\text{op,al}} + C_{\text{op,crane}} + C_{\text{op,storage}} + C_{\text{op,ex}} \tag{20}$$

where $C_{\text{op,re}}$ represents the operating cost of renewable energy units, $C_{\text{op,the}}$ is the operating cost of thermal power units, $C_{\text{op,al}}$ is the adjustment cost for aluminum electrolysis loads, $C_{\text{op,crane}}$ is the adjustment cost for the multifunctional crane, $C_{\text{op,storage}}$ is the operating cost of energy storage, and $C_{\text{op,ex}}$ is the cost of power exchange with the grid.

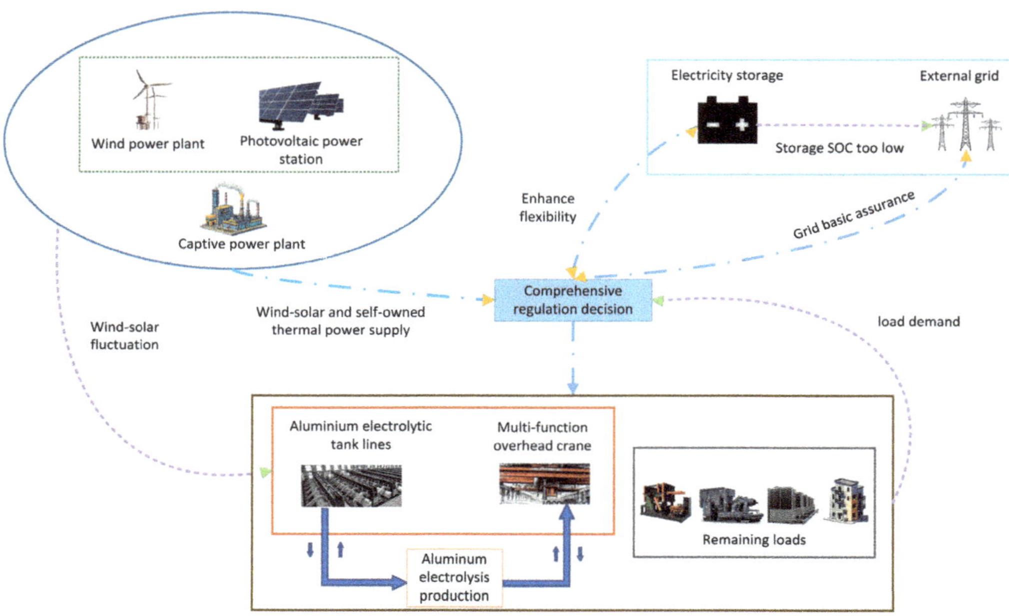

Figure 3. Coordinated Dispatch Architecture for Source–Storage–Load Integration in Aluminum Electrolysis Industrial Parks.

(1) Operating Cost of Renewable Energy Units

$$C_{\text{op,re}} = \sum_{t=1}^{T} \left(c_{\text{op}}^{\text{pv}} P_{\text{pv}}^{t} + c_{\text{op}}^{\text{wt}} P_{\text{wt}}^{t} \right) \tag{21}$$

where $c_{\text{op}}^{\text{pv}}$, $c_{\text{op}}^{\text{wt}}$ represent the operating costs of photovoltaic and wind power units, respectively, while P_{pv}^{t} and P_{wt}^{t} denote the grid-connected power of the photovoltaic station and wind farm during time period t.

(2) Operating Costs of Thermal Power Units

The operating cost of the self-owned power plant in the aluminum electrolysis park, which consists of thermal power units, is made up of the coal consumption cost and the startup and shutdown costs of the thermal power units.

$$C_{\text{op,the}} = \sum_{t=1}^{T} \sum_{k \in \Omega_{\text{the}}} [a_k (P_{\text{the},k}^t)^2 + b_k P_{\text{the},k}^t + c_k] + \sum_{t=1}^{T} \sum_{k \in \Omega_{\text{the}}} \sum_{q=1}^{n_{\text{the},k}} [\zeta_{\text{the},k,q}^t (1 - \zeta_{\text{the},k,q}^{t-1}) C_k^{\text{start}} + \zeta_{\text{the},k,q}^t (1 - \zeta_{\text{the},k,q}^{t-1}) C_k^{\text{stop}}] \tag{22}$$

where a_k, b_k, and c_k are the operating cost coefficients for the k-th type (referred to as type k) of thermal power units; C_k^{start} and C_k^{stop} represent the startup and shutdown costs for the k-th type of thermal power unit, respectively; $P_{\text{the},k}^t$ is the total output of the k-th type of thermal power unit during time period t; $\zeta_{\text{the},k,q}^t$ is a binary variable indicating the operational status of the thermal power unit, where $\zeta_{\text{the},k,q}^t = 1$ indicates that the unit of type k is in operation during time period t, and $\zeta_{\text{the},k,q}^t = 0$ indicates that it is shut down; $n_{\text{the},k}$ is the number of units of type k; and Ω_{the} is the set of thermal power unit types.

(3) Adjustment Cost of Aluminum Electrolysis Load

$$C_{\text{op,al}} = \sum_{t=1}^{T} c_{\text{al}} \Delta P_{j,t}^{\text{al}} \tag{23}$$

where c_{al} represents the adjustment cost of the aluminum electrolysis load, and $\Delta P_{j,t}^{\text{al}}$ is the adjustable capacity per unit time.

(4) Adjustment Cost of Multifunctional Overhead Crane Load

$$C_{\text{op,crane}} = \sum_{t=1}^{T} c_{\text{crane}} \Delta P_{j,t}^{\text{crane}} \tag{24}$$

where c_{crane} represents the adjustment cost of the aluminum electrolysis load, and $\Delta P_{j,t}^{\text{crane}}$ is the adjustable capacity per unit time.

(5) Operating Cost of Energy Storage

$$C_{\text{op,storage}} = \sum_{t=1}^{T} c_{\text{storage}} \left(P_{\text{charge}}^t + P_{\text{discharge}}^t \right) \tag{25}$$

where c_{storage} represents the operating cost coefficient of energy storage, P_{charge}^t is the amount of energy charged to the storage system per unit time, and $P_{\text{discharge}}^t$ is the amount of energy discharged from the storage system per unit time.

(6) Cost of Power Exchange with the Grid

$$C_{\text{op,ex}} = \sum_{t=1}^{T} \left(c_{\text{buy}} P_{\text{buy}}^t - c_{\text{sell}} P_{\text{sell}}^t \right) \tag{26}$$

where c_{buy} represents the electricity purchase price, P_{buy}^t is the amount of electricity purchased per unit time, c_{sell} is the electricity selling price, and P_{sell}^t is the amount of electricity sold per unit time.

3.2.2. Objective Function 2

In response to low-carbon emission reduction policies, objective function 2 aims to minimize the total carbon emissions, which consists of several components: carbon dioxide released during the consumption of prebaked carbon anodes as reducing agents in the aluminum electrolysis process; carbon emissions (perfluorocarbons) caused by anode effects during aluminum electrolysis [29]; carbon emissions from wind and solar power generation; and carbon emissions from coal consumption in both self-owned power plants and purchased electricity.

$$\min f_2 = T_{\text{ce,anode}} + T_{\text{ce,anode effect}} + T_{\text{ce,re}} + T_{\text{ce,the}} + T_{\text{ce,ex}} \tag{27}$$

where $T_{\text{ce,anode}}$ represents carbon emissions from the consumption of carbon anodes as raw materials; $T_{\text{ce,anode effect}}$ denotes carbon emissions (perfluorocarbons) caused by anode effects during production; $T_{\text{ce,re}}$ indicates carbon emissions from renewable energy units; $T_{\text{ce,the}}$ represents carbon emissions from thermal power units; and $T_{\text{ce,ex}}$ refers to carbon emissions from net purchased electricity from the power grid.

(1) Carbon Emissions from Carbon Anodes

CO_2 emissions primarily originate from fuel combustion, petroleum coke burning losses, and anode baking processes. The carbon emissions from carbon anode consumption as raw materials can be calculated using the following formula: Carbon Emissions from Carbon Anodes.

$$T_{\text{ce,anode}} = M^{\text{al}}_{j,t} \cdot B_{\text{anode}} \cdot (1 - C_{\text{anode}} - D_{\text{anode}}) \cdot \frac{E_{mol.wt}}{E_{amu}} \tag{28}$$

where B_{anode} represents the net consumption of carbon anodes per ton of aluminum; C_{anode} represents the average sulfur content of carbon anodes at 1%; D_{anode} represents the average ash content of carbon anodes at 1%; $E_{mol.wt}$ represents the molecular weight of CO_2; and E_{amu} represents the atomic weight of carbon.

(2) Carbon Emissions from Anode Effects

The carbon emissions (perfluorocarbons) caused by anode effects during the production process can be calculated using the following formula:

$$T_{\text{ce,anode effect}} = M^{\text{al}}_{j,t} \cdot \frac{B_{\text{effect}} \cdot D_{\text{effect}} - C_{\text{effect}} \cdot E_{\text{effect}}}{1000} \tag{29}$$

where B_{effect} represents the emission factor of CF_4 from anode effects; C_{effect} represents the emission factor of C_2F_6 from anode effects; D_{effect} represents the Global Warming Potential (GWP) of CF_4; and E_{effect} represents the Global Warming Potential (GWP) of C_2F_6.

(3) Carbon Emissions from Renewable Energy Units

$$T_{\text{ce,re}} = \sum_{t=1}^{T} \lambda_{\text{pv,factor}} P^t_{\text{pv}} + \sum_{t=1}^{T} \lambda_{\text{wt,factor}} P^t_{\text{wt}} \tag{30}$$

where $\lambda_{\text{pv,factor}}$ represents the carbon emission coefficient of photovoltaic power generation; P^t_{pv} represents the photovoltaic power generation per unit time; $\lambda_{\text{wt,factor}}$ represents the carbon emission coefficient of wind power generation; and P^t_{wt} represents the wind power generation per unit time.

(4) Carbon Emissions from Thermal Power Units

$$T_{\text{ce,the}} = \sum_{t=1}^{T} \sum_{k \in \Omega_{\text{the}}} \lambda_{\text{the,factor}} P_{\text{the},k}^{t} \tag{31}$$

where $\lambda_{\text{the,factor}}$ represents the carbon emission coefficient of thermal power units and $P_{\text{the},k}^{t}$ represents the power generation of a specific thermal power unit per unit time.

(5) Carbon Emissions from Grid Power Exchange

$$T_{\text{ce,ex}} = \lambda_{\text{ex,factor}} \sum_{t=1}^{T} P_{\text{buy}}^{t} \tag{32}$$

where $\lambda_{\text{ex,factor}}$ represents the carbon emission coefficient of grid electricity and P_{buy}^{t} represents the amount of electricity purchased from the grid per unit time.

3.2.3. Objective Function 3

To maximize the utilization of renewable energy sources such as wind and solar power, the objective function 3 aims to maximize the consumption of wind and photovoltaic power during aluminum electrolysis production, i.e., to maximize the renewable energy consumption rate:

$$\max f_3 = \frac{\sum_{t=1}^{T} P_{\text{res,used}}^{t}}{\sum_{t=1}^{T} P_{\text{res,predictable}}^{t}} \tag{33}$$

where $P_{\text{res,used}}^{t}$ represents the renewable energy output utilized per unit time and $P_{\text{res,predictable}}^{t}$ represents the predicted renewable energy output per unit time.

3.3. *Variable Constraints*

(1) Renewable Energy Constraints

$$0 \leq P_{\text{pv}}^{t} \leq P_{\text{pv}}^{t,\max} \tag{34}$$

$$0 \leq P_{\text{wt}}^{t} \leq P_{\text{wt}}^{t,\max} \tag{35}$$

where $P_{\text{pv}}^{t,\max}$ represents the maximum photovoltaic power output and $P_{\text{wt}}^{t,\max}$ represents the maximum wind power output.

(2) Thermal Power Unit Output Constraints

$$P_{\text{the,min}} \leq P_{\text{the}}^{t} \leq P_{\text{the,max}} \tag{36}$$

where $P_{\text{the,min}}$ represents the minimum output of thermal power units and $P_{\text{the,max}}$ represents the maximum output of thermal power units.

(3) Thermal Power Unit Ramp Rate Constraints

$$P_{\text{the},k}^{t} - P_{\text{the},k}^{t-1} \leq P_{\text{the},k,\text{up}} \tag{37}$$

$$P_{\text{the},k}^{t-1} - P_{\text{the},k}^{t} \leq P_{\text{the},k,\text{down}} \tag{38}$$

where $P_{\text{the},k}^{t-1}$ represents the active power output of conventional units during period $t-1$ and $P_{\text{the},k,\text{up}}$ and $P_{\text{the},k,\text{down}}$ represent the up and down ramp rates of thermal power units, respectively.

(4) Aluminum Electrolysis Load Regulation Constraints, refer to Equations (1)–(19).
(5) Grid Power Exchange Constraints

$$0 \leq P_{\text{buy}}^{t} \leq P_{\text{buy}}^{t,\max} \tag{39}$$

$$0 \leq P_{\text{sell}}^{t} \leq P_{\text{sell}}^{t,\max} \tag{40}$$

where $P_{\text{buy}}^{t,\max}$ represents the maximum allowable power purchase and $P_{\text{sell}}^{t,\max}$ represents the minimum allowable power purchase.

4. Model Solution

A multi-objective optimization model for collaborative dispatch in aluminum electrolysis industrial parks encompasses complex objectives. This optimization problem is further complicated by multidimensional nonlinear constraints, such as aluminum electrolysis regulation limitations, auxiliary equipment constraints that follow production variations, and operational constraints of wind–solar–thermal power units and energy storage charging/discharging processes. Traditional linear programming and deterministic optimization methods struggle to effectively solve such complex problems [30,31]. Consequently, intelligent algorithms—including improved particle swarm optimization and genetic algorithms—have emerged as viable alternatives. These approaches leverage their global search capabilities and advantages in handling high-dimensional nonlinear problems to simulate the complex dynamic evolution of the system. Among numerous multi-objective optimization algorithms, MOGWO combines the simplicity and efficiency of GWO with the powerful capabilities of multi-objective optimization. It can precisely simulate the load regulation characteristics of the aluminum electrolysis process and optimize the synergistic mechanism between renewable energy and the aluminum electrolysis system, making it an ideal choice for addressing complex engineering optimization problems (such as aluminum electrolysis load scheduling).

The source–storage–load coordinated scheduling optimization problem in aluminum electrolysis parks is a nonlinear multi-constraint problem. When solving with the original MOGWO algorithm, multiple constraints increase the computational time. To address this issue, a cosine convergence factor is introduced [32], significantly improving the algorithm's convergence speed. The multi-period cosine convergence factor enables more refined local search in later stages, enhancing global optimization capability and improving the quality of the Pareto front. The introduction of an elite selection strategy reduces the sparsity of solution sets [33], resulting in a more uniform distribution, particularly in displaying complete Pareto front profiles, and reducing optimization time.

Based on the above analysis, an improved multi-objective grey wolf optimization algorithm is adopted to solve the proposed multi-objective optimization dispatch model for source–load–storage coordination. The solution flowchart is shown in Figure 4.

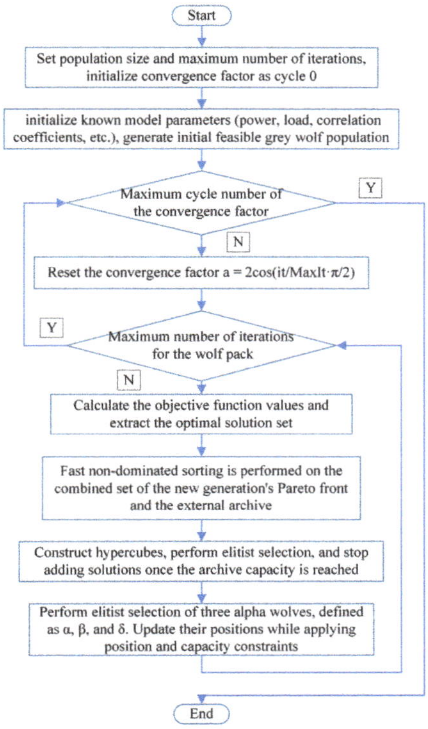

Figure 4. Solution process of the objective function.

The main steps of the IMOGWO algorithm are as follows:

(1) Step 1: Initialize grey wolf algorithm, set population size and maximum iterations, import known model parameters (power, load, correlation coefficients, etc.), initialize population matrix arrangement as position, fitness function value, Pareto rank, crowding distance, and initialize convergence factor as cycle 0.

(2) Step 2: Reset current iteration count for single convergence factor cycle wolf pack, and reset convergence factor as a = 2cos(it/MaxItpi/2), control search range, prepare for next iteration round.

(3) Step 3: Calculate objective function values: operating cost, carbon emissions, renewable energy consumption rate; perform non-dominated sorting on population, determine each wolf's superiority, establish Archive population, perform crowding distance sorting on Archive population, retain optimal and uniformly distributed solutions before entering the iteration process.

(4) Step 4: Construct a hypercube, place particles into the hypercube, then perform elite selection based on particle crowding degree, and stop adding when the accumulated storage capacity is reached.

(5) Step 5: Select three representative alpha wolves (Alpha, Beta, and Delta) from the repository, representing the current best, second-best, and third-best solutions. Update all grey wolves' positions based on alpha wolves' positions, process boundary violations of new positions to ensure variables remain within a reasonable range, proceed to the next round of objective function calculation and sorting.

(6) Step 6: Check if the current single cycle maximum iterations reached (inner loop), check if the maximum number of cycles reached (outer loop).
(7) Step 7: If the maximum number of cycles is reached, the algorithm terminates, output optimal solution set (Pareto front) for subsequent decision analysis.

5. Case Study

This section presents the simulation results, with the core algorithm (IMOGWO) implemented through custom functions. It should be noted that these simulations were conducted on a personal computer equipped with a 13th Gen Intel(R) Core(TM) i9-13900H (2.60 GHz) processor (Intel, Hohhot, Inner Mongolia Autonomous Region, China), 64 GB RAM, and 1 TB internal storage, using MATLAB 2022 software.

5.1. Case Introduction

To verify the effectiveness of the proposed model, based on actual operational data from an aluminum electrolysis park in Northeast China. The park contains 3 electrolytic cell series with rated powers of 210, 315, and 709.8 MW respectively, a multi-functional crane load of 77.175 MW, and other loads of 231.525 MW, totaling 1543.5 MW rated load. For renewable energy generation units, photovoltaic array parameters are detailed in reference [16], and wind turbines use MYSE7.5-193 units. The existing connected photovoltaic station capacity is 1120 MW, the wind farm capacity is 1200 MW, and their 24 h typical daily output data is shown in Figure A1. Fire power unit types and related equipment parameters are listed in Table A1. The tie line rated power is 800 MW, with an electricity purchase price of 0.485 yuan/kWh and a selling price of 0.28 yuan/kWh. Energy storage parameters are listed in Table A2, and carbon emission coefficients are listed in Table A3.

To verify the effectiveness and feasibility of the proposed model and algorithm, highlight the potential of aluminum electrolysis loads in grid regulation, and demonstrate the balance of multi-objective optimization in source–storage–load coordination and sustainability of green low-carbon transformation, comparative analyses are designed in the following aspects:

(1) Setting a basic scenario with economic benefits as the core objective, allowing daily production to fluctuate within 90–110% of rated production, which is significant for verifying the daily adjustability of aluminum electrolysis loads.
(2) In fixed-production scenario analysis, the study is divided into two directions: conventional economic optimization without considering adjustable characteristics, focusing on traditional cost-effectiveness; and three-objective collaborative optimization considering economics, carbon emissions, and renewable energy consumption rate.
(3) In fixed-production multi-objective scenario analysis, the three-objective optimization study focusing on economics, carbon emissions, and renewable energy consumption rate is further deepened by expanding the renewable energy integration scale.
(4) Based on the different emphasis of three-objective optimization analysis, pairwise analysis of economics, carbon emissions, and renewable energy consumption rate reveals the inherent correlations and trade-off characteristics between objectives.
(5) To verify the effectiveness and advancement of the proposed model and algorithm, comparisons with benchmark models highlight the superiority of the proposed model in application. Subsequently, IMOGWO is compared with mainstream algorithms such as NSGA-II and MOGWO.

5.2. Results Analysis

5.2.1. Analysis of Aluminum Electrolysis Load Regulation

Considering the adjustability of aluminum electrolysis loads and assuming that aluminum enterprises have product storage capability, daily aluminum production can be adjusted as needed, with daily production allowed to fluctuate between 90–110%. The load regulation and system output are shown in Figure 5.

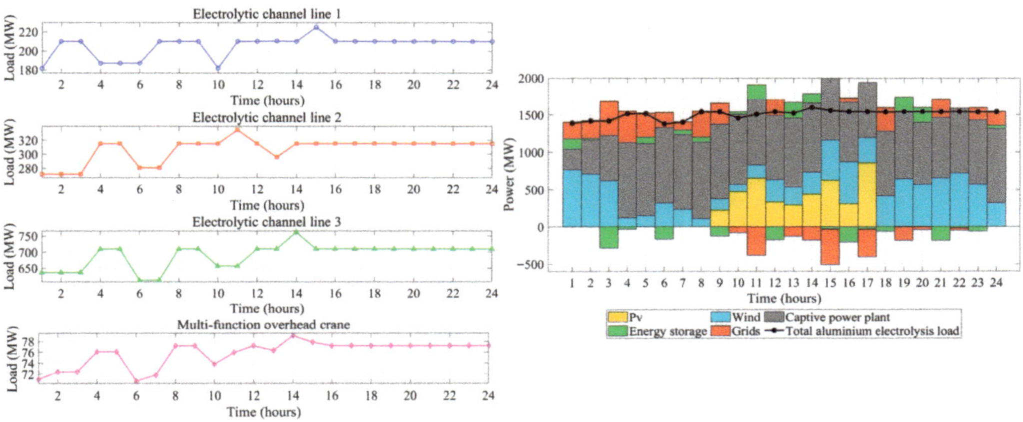

Figure 5. The 24 h variation curve of adjustable load and related power output in scenario 1.

According to Figure 5, the total adjustable capacity over 24 h is 995.024 MW·h, with the electrolytic cell load contributing 957.28 MW·h and the multi-functional crane load contributing 37.750 MW·h. This demonstrates the substantial regulation potential of aluminum electrolysis loads and auxiliary equipment, providing critical support for high-penetration renewable energy integration and grid peak shaving, while also laying the foundation for subsequent multi-objective optimization strategy analysis.

5.2.2. Multi-Objective Optimization Dispatch Study of Aluminum Electrolysis Load

Currently, the majority of aluminum electrolysis plants operate at constant power levels during their production processes. Consequently, in the fixed production scenario, the adjustability of aluminum electrolysis loads is not considered, with focus solely placed on traditional economic optimization as a single baseline scenario (Scenario 2). The load regulation and system output are shown in Figure 6.

When considering only economic optimization as a single objective, the cost is 9.5343 million yuan, carbon emissions are 24,314.04 t, and the renewable energy consumption rate is 65.71%.

Three-objective optimization provides a broader decision space compared to single economic optimization, better balancing enterprise development and social responsibility, offering more comprehensive solutions for the sustainable development of the aluminum electrolysis industry. Although the adjustable characteristics of aluminum electrolysis loads are restricted under fixed production conditions, multi-objective optimization methods can address environmental benefits and energy transition demands while ensuring economic benefits. In the algorithm settings, when the population size is set to 1000 with 100 iterations, the simulation output results achieve stability. The distribution of the Pareto solution set for three-objective optimization in the objective function space is shown in Figure 7.

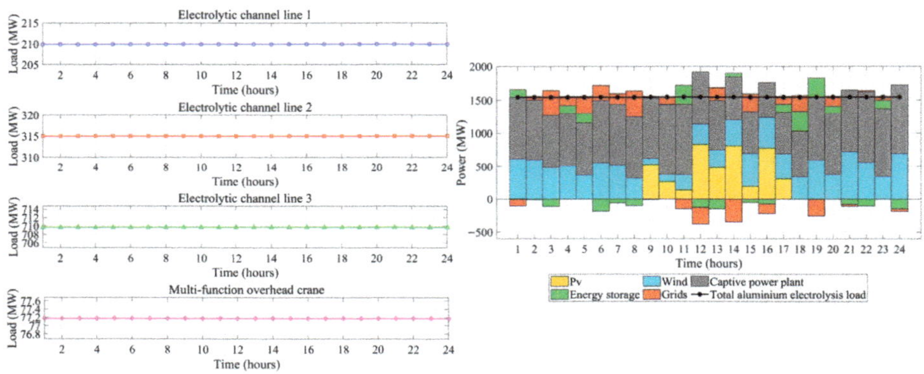

Figure 6. The 24 h curve of non-adjustable load and related power output in scenario 2.

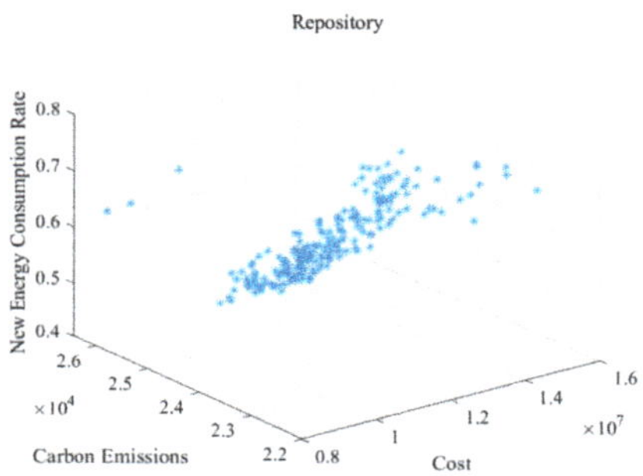

Figure 7. Distribution of Pareto optimal solution set.

To achieve a win-win-win situation for economic benefits, environmental protection, and energy transition, a series of optimization scheduling schemes with different emphases are provided. To prioritize economic factors, the aluminum electrolysis park sets cost minimization as the primary objective (40% weight), establishing Scenarios 3/6. This indicates that the park emphasizes short-term profitability or budget control in decision-making. For low-carbon considerations, the park prioritizes carbon emissions as the primary objective (40% weight), establishing Scenarios 4/7. This indicates that the park tends to prefer scheduling schemes with lower system carbon emissions in decision-making. When considering green corporate development, the park prioritizes renewable energy consumption as the primary objective (40% weight), establishing Scenarios 5/8. This indicates that the park tends to prefer scheduling schemes with higher renewable energy consumption rates in decision-making. The weight distributions are shown in Table 1.

Table 1. Weight distribution.

Algorithm	η_1	η_2	η_3
Scenario 3 vs. 6	0.4	0.3	0.3
Scenario 4 vs. 7	0.3	0.4	0.3
Scenario 5 vs. 8	0.3	0.3	0.4

From the load variation graph, the following can be concluded: In scenario 3, the adjustable load of three electrolytic cell lines is 134.239 MW·h, the multi-functional crane adjustable load is 3.336 MW·h, and the overall regulation capacity is 101.635 MW·h.

In scenario 4, the adjustable load of three electrolytic cell lines is 132.276 MW·h, the multi-functional crane adjustable load is 4.941 MWh, and the overall regulation capacity is 137.217 MW·h. In scenario 5, the adjustable load of three electrolytic cell lines is 98.299 MW·h, the multi-functional crane adjustable load is 3.336 MW·h, and the overall regulation capacity is 101.635 MW·h. By comparing the output conditions shown in Figures 6 and 8, it can be observed that the utilization rate of wind and solar power significantly increases in scenarios 3, 4, and 5.

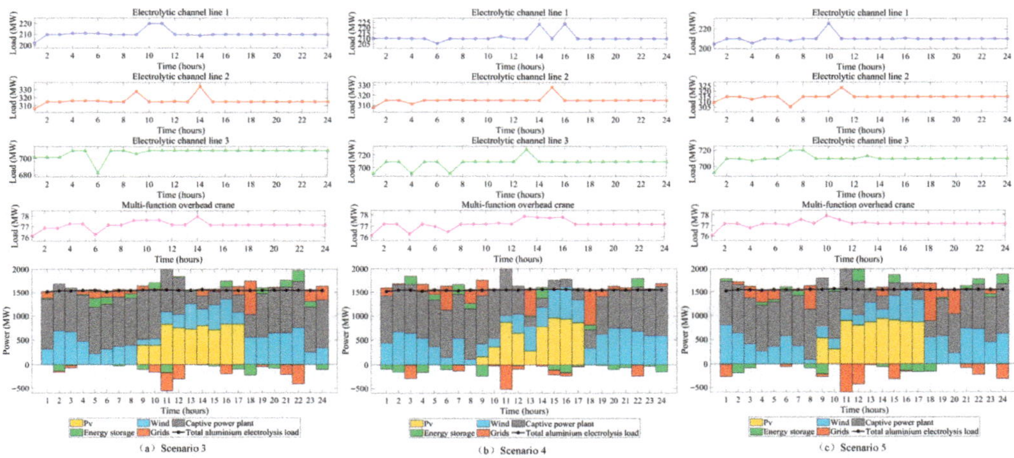

Figure 8. The 24 h variation curves of adjustable loads and related power outputs in scenarios 3, 4, and 5.

Analyzing the data from scenario 2 and scenarios 3, 4, and 5 in Table 2, from an economic perspective, scenario 2, which only considers economic optimization, performs best. However, compared to the other three scenarios, it lacks consideration of carbon emissions and the renewable energy consumption rate. In the three-objective optimization, scenario 4, which emphasizes carbon emissions, shows a reduction of 1677.35 t in carbon emissions compared to scenario 2. Scenario 5, which emphasizes the renewable energy consumption rate, demonstrates a 12.11% increase in the renewable energy consumption rate.

Table 2. Calculation results of scenarios 3, 4, and 5.

Renewable Energy Penetration Rate		Cost (10,000 CNY)	Carbon Emissions (t)	RE Consumption Rate
67.8% (1200 MW Wind + 1120 MW Solar)	Scenario 3	1199.63	23,283.58	73.36%
	Scenario 4	1292.60	22,636.69	76.74%
	Scenario 5	1260.26	22,873.14	77.82%

Through scenario comparison, the superiority of multi-objective optimization in source–load coordinated optimization mode is verified in terms of comprehensive benefits: improving system economics, reducing carbon emissions, and promoting renewable energy consumption rate.

5.2.3. Multi-Objective Optimization Dispatch Analysis with a Higher Proportion of Renewable Energy Integration

A scenario with a higher proportion of renewable energy integration is established by adding 1200 MW of wind power to the existing 1200 MW wind power and 1120 MW photovoltaic capacity. Through reasonable adjustment of aluminum electrolysis load to match renewable energy output, this further explores the application of high proportion wind and solar power in energy-intensive parks, facilitating enterprises' green and low-carbon transformation.

From the load variation graph, the following can be concluded: In scenario 6, the adjustable load of three electrolytic cell lines is 131.684 MW·h, the multi-functional crane adjustable load is 5.556 MW·h, and the overall regulation capacity is 137.24 MW·h. In scenario 7, the adjustable load of three electrolytic cell lines is 232.746 MW·h, the multi-functional crane adjustable load is 9.2 MWh, and the overall regulation capacity is 241.946 MW·h. In scenario 8, the adjustable load of three electrolytic cell lines is 81.53 MW·h, the multi-functional crane adjustable load is 3.202 MW·h, and the overall regulation capacity is 84.732 MW·h. By comparing the output conditions shown in Figures 8 and 9, it can be observed that wind and solar power output in scenarios 6, 7, and 8 have replaced most of the output from captive thermal power units.

Analyzing the results data from scenarios 6, 7, and 8 in Table 3 and scenarios 3, 4, and 5 in Table 2, significant improvements can be observed in all three aspects. From an economic perspective, scenario 6 saves 2.0807 million yuan compared to scenario 3; regarding carbon emissions, scenario 7 reduces emissions by 6583.62 t compared to scenario 4; in terms of renewable energy consumption rate, scenario 8 shows a 6.75% increase compared to scenario 5.

Through comparative analysis of optimization effects under different renewable energy integration scales, the significant advantages of multi-objective optimization under large-scale renewable energy integration conditions are highlighted. This optimization strategy not only achieves higher clean energy consumption rates but also reduces comprehensive energy costs through economies of scale while significantly decreasing carbon emissions. It provides a feasible pathway for the aluminum electrolysis industry's green and low-carbon transformation, and offers an important reference for production optimization in future scenarios with higher proportions of renewable energy integration.

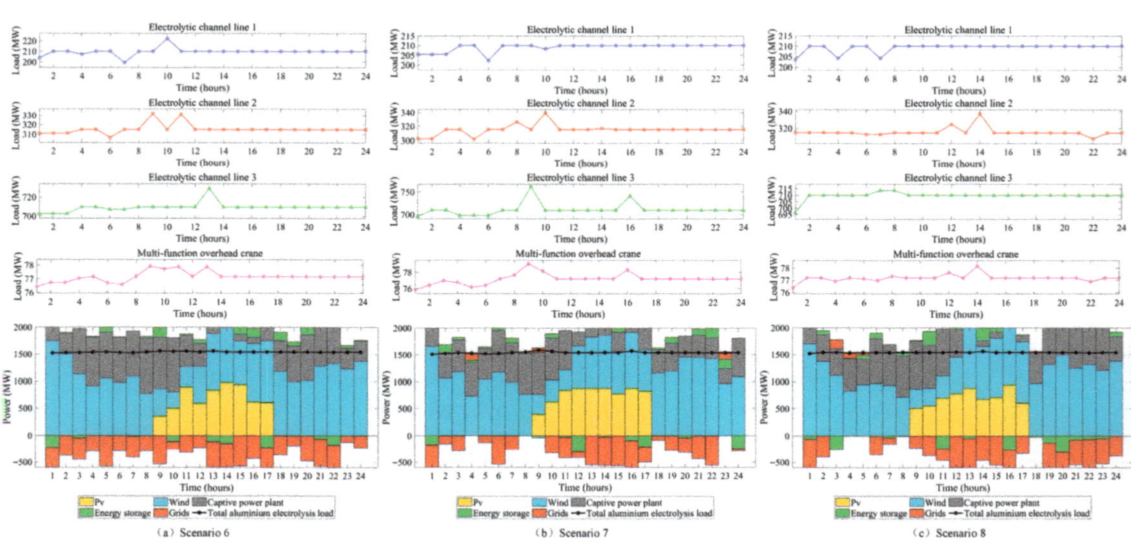

Figure 9. 24-h variation curves of adjustable loads and power outputs in scenarios 6, 7, and 8.

Table 3. Calculation results of scenarios 6, 7, and 8.

Renewable Energy Penetration Rate		Cost (10,000 CNY)	Carbon Emissions (t)	RE Consumption Rate
76.2% (2400 MW Wind + 1120 MW Solar)	Scenario 6	991.56	18,126.18	83.63%
	Scenario 7	1270.72	16,053.07	83.47%
	Scenario 8	1146.54	17,385.37	84.57%

5.2.4. Correlation Analysis Between Multiple Objectives

To identify the synergistic effects and conflicts among objectives, this analysis examines the mutual influences between different variables, thereby providing a reference for making appropriate adjustments to specific variables.

As shown in Figure 10a, there is a significant negative correlation between economic cost and carbon emissions objectives [34]. The fundamental reason is that low-carbon dispatching strategies, while reducing carbon emissions, incur two types of additional costs: first, increased start–stop and peak regulation costs of thermal power units to smooth renewable energy fluctuations; second, high-cost external power purchase expenses in specific periods due to reduced self-owned thermal power generation.

As shown in Figure 10b, economic cost and renewable energy consumption rate show a negative correlation. Pursuing extreme consumption rates significantly increases system costs, as it requires expensive flexibility resources (such as unit start–stop operations) to handle renewable energy fluctuations. Therefore, in certain periods, actively curtailing wind and solar power to avoid high regulation costs may exceed their power generation value, becoming a more economically optimal choice.

As shown in Figure 10c, carbon emissions and renewable energy consumption rate demonstrate a significant negative correlation, confirming that increasing renewable energy consumption is key to low-carbon transformation. This relationship essentially reflects the system's flexibility level: lower flexibility leads to greater dependence on thermal power for system balance, resulting in "high carbon emissions, low consumption"; conversely, higher flexibility enables a virtuous cycle of "low carbon emissions, high consumption".

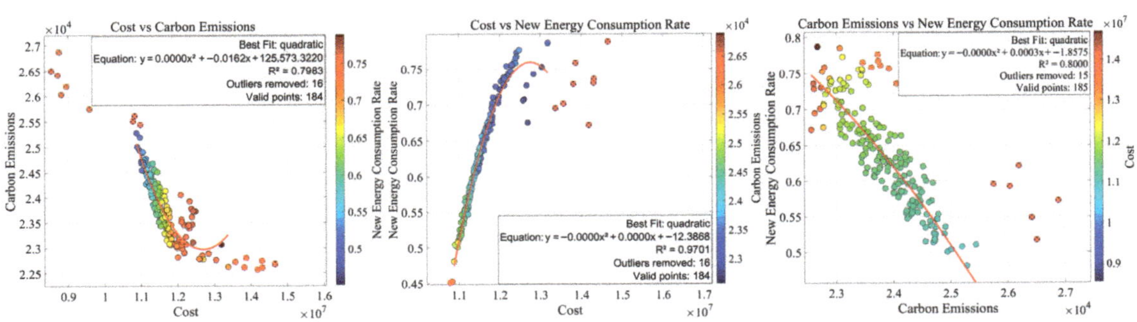

(a) The correlation between economy and carbon emissions has a fitting degree of 79.8%.

(b) The correlation between economy and renewable energy consumption rate has a fitting degree of 97.0%.

(c) The correlation between carbon emissions and renewable energy consumption rate has a fitting degree of 80%.

Figure 10. Correlation analysis between objectives.

5.2.5. Validation of Model and Algorithm Effectiveness

Model 1 represents the refined modeling considering multi-functional crane loads, while Model 2 represents modeling without considering multi-functional crane loads. To validate model effectiveness, both models were run in MATLAB 2022 software using an improved multi-objective grey wolf algorithm. The comparison of Pareto frontiers, along with algorithm solution time and efficiency comparisons, is shown in Figure 11.

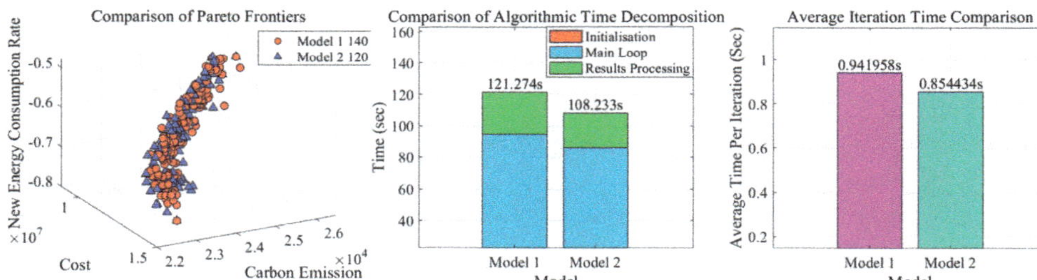

(a) Model 1 has more solutions with a broader range (b) Model 1 has a relatively longer solving time (c) Model 1 has longer average iterations

Figure 11. Model performance comparison.

According to Figure 11, when solving multi-objective problems, Model 1 produces 20 more solution sets than Model 2, but requires 13.041 s longer solving time, averaging 0.04 s slower per solution. While the solving efficiency between the two models is similar, Model 1's refined modeling, which considers the coupling relationship between multi-functional crane loads and production loads, discovers more feasible solution sets. This brings significant advantages in scheduling precision and resource allocation, providing strong support for the efficient operation of aluminum electrolysis enterprises.

To validate algorithm effectiveness, Model 1 was implemented in MATLAB 2022 software using three solving algorithms: multi-objective genetic algorithm, multi-objective grey wolf algorithm, and improved multi-objective grey wolf algorithm. The Pareto frontiers were obtained and compared in a three-dimensional graph showing the Pareto frontier sets of three objective functions. The parameter settings for the three algorithms are shown in Table 4, and the comparison of Pareto solution sets is shown in Figure 12.

Table 4. Algorithm parameter settings.

Algorithm	Population	Iterations	Specific Parameters
NSGA-II	500	100	pMutation = 0.4; pCrossover = 0.7
MOGWO	500	100	alpha = 0.1; nGrid = 10; beta = 4; gamma = 2;
IMOGWO	500	100	alpha = 0.1; nGrid = 10; beta = 4; gamma = 2;

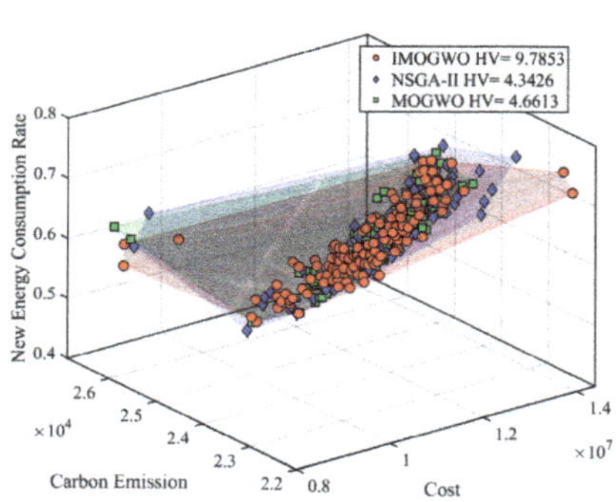

Figure 12. Comparison of Pareto frontiers.

According to Figure 12 and Table 5, when solving the multi-objective problem involving economics, carbon emissions, and renewable energy consumption rate for aluminum electrolysis enterprises, the improved multi-objective grey wolf algorithm demonstrates wider solution coverage across all three objective functions compared to the multi-objective genetic algorithm and the multi-objective grey wolf algorithm. Furthermore, the improved algorithm achieves the shortest solving time and generates the largest number of solution sets. Additionally, its highest HV value indicates better convergence and diversity of solutions, reflecting superior distribution range and uniformity in the objective space.

Table 5. Comparison of algorithm performance.

Algorithm	Number of Solution Sets	HV (Mean ± Std)	Runtime (s)
NSGA-II	97	4.3426	1653.38
MOGWO	98	4.6613	239.25
IMOGWO	135	9.783	96.7543

This indicates that the improved grey wolf algorithm demonstrates strong adaptability and solving efficiency for the source–storage–load coordinated scheduling multi-objective optimization model proposed in this paper.

6. Conclusions

This paper treats the aluminum electrolysis park as a load-side flexibility resource for system regulation, constructs a source–storage–load coordinated dispatch model. With objectives of minimizing economic cost and carbon emissions while maximizing renewable energy consumption rate, an improved multi-objective grey wolf optimization algorithm is applied to simulate the multi-objective optimization model based on source–storage–load coordination. The following conclusions can be drawn:

- Including operational characteristics of auxiliary equipment (multi-functional cranes) in aluminum electrolysis load modeling has significant practical implications. Considering the impact of multi-functional crane loads on overall system energy consumption reflects the dynamic load characteristics of the entire production system, enabling a more comprehensive evaluation of aluminum electrolysis enterprises' regulation capability.
- The multi-objective optimization strategy in the proposed source–storage–load coordinated optimization mode can systematically coordinate operating costs, overall carbon emissions, and renewable energy consumption of the aluminum electrolysis park, yielding decision solutions that balance overall economic and low-carbon objectives.
- The significant advantages of multi-objective optimization under large-scale renewable energy integration conditions are explored. This not only provides an effective pathway for the current green transformation of the aluminum electrolysis industry but also offers valuable reference for the sustainable development of other energy-intensive industries.

Author Contributions: Conceptualization, J.L. and Y.C.; methodology, J.L. and G.L.; software, Y.C.; validation, J.L. and Y.C.; formal analysis, Y.C. and R.H.; investigation, Y.C.; resources, Y.C.; data curation, J.L.; writing—original draft preparation, Y.C.; writing—review and editing, G.L., J.L. and Y.C.; visualization, J.L.; supervision, J.L., Y.C. and R.H.; project administration, J.L. and Y.C.; funding acquisition, J.L. All authors have read and agreed to the published version of the manuscript.

Funding: This research was funded by the Inner Mongolia Autonomous Region Science and Technology Breakthrough Project (2024KJTW0017), the National Key Research and Development Plan "Energy Storage and Smart Grid Technology" Project (2024YFB2408400), and the Inner Mongolia Natural Science Foundation under Grant (2025MS05012).

Institutional Review Board Statement: Not applicable.

Informed Consent Statement: Not applicable.

Data Availability Statement: All data are contained within the article. Additional data are available from the corresponding author upon reasonable request.

Conflicts of Interest: The authors declare no conflicts of interest.

Appendix A

Table A1. Parameters of thermal power units.

Unit Model Number	Number of Units	Maximum Output (MW)	Minimum Output (MW)	Start–Stop Cost (CNY)	Cost Parameters a/b/c (CNY/MW^2) (CNY/MW)(CNY)	Minimum Up/Down Time (h)	Maximum Ramp Rate (MW/h)	CO_2 Emission Intensity (N.m^3CO_2/kWh)
1	3	85	40	1820	0.00553/193.9/3360	3/3	80/80	0.46
2	3	130	65	3920	0.01477/115.5/4790	5/5	100/100	0.46
3	1	455	230	35000	0.00217/121.1/6790	5/5	200/200	0.46

Table A2. Energy storage system parameters.

	Values
Charging efficiency	0.92
Discharging efficiency	0.92
Maximum charge/discharge power (MW)	300
Maximum capacity (MWh)	1200
Initial capacity (MWh)	360
Minimum state of charge	0.1 × 1200
Maximum state of charge	0.9 × 1200
Operation and maintenance cost (CNY/kWh)	0.025

Table A3. Carbon emission coefficients of different equipment.

Energy Type	Carbon Emission Coefficient g/(kW·h)
Self-owned power plant	Calculated based on thermal power unit carbon emission intensity
Power grid	0.5703
Wind power generation	0.02362
Solar power generation	0.04191
Energy storage	0

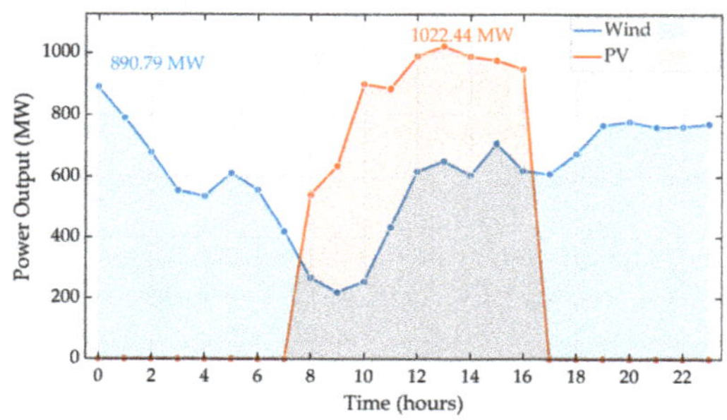

Figure A1. Predicted power output of wind and solar power in a typical day in the region.

References

1. Tregnago, G. Fifty years of change in the energy sector. *Nat. Energy* **2025**, *10*, 3–5. [CrossRef]
2. National Security Emergency Preparedness (NSEP) Network. Electricity Consumption Growth Ranking by Industry in China 2023. Available online: https://news.bjx.com.cn/html/20240223/1362664.shtml (accessed on 24 September 2025).
3. Hepburn, C.; Qi, Y.; Stern, N.; Ward, B.; Xie, C.; Zenghelis, D. Towards carbon neutrality and China's 14th Five-Year Plan: Clean energy transition, sustainable urban development, and investment priorities. *Environ. Sci. Ecotechnol.* **2021**, *8*, 100130. [CrossRef]
4. Chen, L.; Wang, C.; Wu, Z. Reinforcement Learning-Based Time of Use Pricing Design Toward Distributed Energy Integration in Low Carbon Power System. *IEEE Trans. Netw. Sci. Eng.* **2025**, *12*, 997–1010. [CrossRef]
5. Chen, Z.; Wu, Z.; Wei, L.; Yang, L.; Yuan, B.; Zhou, M. Understanding the synergy of energy storage and renewables in decarbonization via random forest-based explainable AI. *Appl. Energy* **2025**, *390*, 125891. [CrossRef]
6. Harrison-Atlas, D.; Murphy, C.; Schleifer, A.; Grue, N. Temporal complementarity and value of wind-PV hybrid systems across the United States. *Renew. Energy* **2022**, *201*, 111–123. [CrossRef]
7. Lindberg, O.; Arnqvist, J.; Munkhammar, J.; Lingfors, D. Review on power-production modeling of hybrid wind and PV power parks. *J. Renew. Sustain. Energy* **2021**, *13*, 042702. [CrossRef]
8. Xiao, Y.; Yang, C.; Chen, T.; Lei, M.; Wattana, S.; Wattana, B. Strategies of a Wind–Solar–Storage System in Jiangxi Province Using the LEAP–NEMO Framework for Achieving Carbon Peaking Goals. *Energies* **2025**, *18*, 1135. [CrossRef]
9. Xie, J.; Zhang, Y.; Zhang, Y.; Xiang, M. Optimized Configuration Method of Park-level Hydrogen Energy System for Coordinated Operation of Hybrid Electrolyzers. *Dianwang Jishu/Power Syst. Technol.* **2025**, *49*, 3256–3265. [CrossRef]
10. Li, C.; Li, R.; Wang, W.; Tang, Q.; Zhu, Y. Configuration optimization of distributed PV-storage system in industrial park considering load management. *J. Renew. Sustain. Energy* **2025**, *17*, 034102. [CrossRef]
11. Sun, H.; Chen, F. The impact of green finance on China's regional energy consumption structure based on system GMM. *Resour. Policy* **2022**, *76*, 102588. [CrossRef]
12. Li, S.W.; Hui, H.Y.; Bao, M.L. Quantified Adjustability of Aluminum Smelter Loads and Their Participation in Coordinated Scheduling of Power System. *Proc. CSU-EPSA* **2024**, *36*, 70–80. [CrossRef]
13. Yue, X.; Liao, S.; Fu, L.; Xu, J.; Ke, D.; Wang, H.; Li, L.; He, X. Hierarchical Optimization Strategy of High-penetration Wind Power System Considering Electrolytic Aluminum Energy-intensive Load Regulation and Deep Peak Shaving of Thermal Power. *Dianwang Jishu/Power Syst. Technol.* **2024**, *48*, 3186–3196. [CrossRef]
14. Zheng, W.; Yu, P.; Xu, Z.; Fan, T.; Liu, H.; Yu, M.; Zhang, J.; Li, J. Day-ahead Intra-day Economic Dispatch Methodology Accounting for the Participation of Electrolytic Aluminum Loads and Energy Storage in Power System Peaking. In Proceedings of the 2nd IEEE International Conference on Power Science and Technology, ICPST 2024, Dali, China, 9–11 May 2024; pp. 2201–2206. [CrossRef]
15. Ding, X.; Xu, J.; Sun, Y.; Liao, S. Source-load coordinated control strategy for smoothing wind power fluctuation in grid-connected high energy consuming electrolytic aluminum industrial power grid. *Dianli Zidonghua Shebei/Electr. Power Autom. Equip.* **2022**, *42*, 47–55. [CrossRef]
16. Wang, Y.; Xie, H.; Zhu, H. Source-load coordinated optimal planning method of electrolytic aluminum industrial park. *Dianli Zidonghua Shebei/Electr. Power Autom. Equip.* **2024**, *44*, 132–140. [CrossRef]
17. Yue, X.; Fu, L.; Liao, S.; Xu, J.; Ke, D.; Wang, H.; Feng, S.; Yang, J.; He, X. A source-load collaborative stochastic optimization method considering the electricity price uncertainty and industrial load peak regulation compensation benefit. *Int. J. Electr. Power Energy Syst.* **2025**, *167*, 110630. [CrossRef]
18. Dai, L.; Kang, Z.; Chen, R.; Wang, L. Study on multi-objective maintenance decision optimization for bridge networks based on NSGA-III. *Tumu Gongcheng Xuebao/China Civ. Eng. J.* **2024**, *57*, 41–52. [CrossRef]
19. El Qouarti, O.; Nasser, T.; Essadki, A.; Akarne, Y. AC/DC hybrid microgrid energy management optimization as a decisive factor towards De-Carbonization and rational integration of electrical self-generating units using three-objective grey wolf optimization algorithm-power to X and renewable energies solutions. *Int. J. Hydrogen Energy* **2025**, *138*, 1116–1130. [CrossRef]
20. Ullah, K.; Hafeez, G.; Khan, I.; Ullah, S.; Alghamdi, B.; Alsafran, A.S.; Kraiem, H. Energy optimization using hybrid demand response, renewable energy, and storage battery: A tri-objective optimization approach. *Sustain. Cities Soc.* **2025**, *122*, 106145. [CrossRef]
21. Haarberg, G.M. Electrowinning of Aluminum—Challenges and Possibilities for Reducing the Carbon Footprint. *Electrochemistry* **2024**, *92*, 043002. [CrossRef]
22. Zhang, X.; Hu, X.; Zhao, J.; Chen, C.; He, G.; Zhang, L. The study of thermal-electrical coupling numerical simulation of aluminum electrolytic cell anode assembly and parameter optimization of steel claws. *Results Eng.* **2025**, *26*, 104738. [CrossRef]

23. Xu, J.; Liao, S.; Sun, Y.; Ma, X.Y.; Gao, W.; Li, X.; Gu, J.; Dong, J.; Zhou, M. An Isolated Industrial Power System Driven by Wind-Coal Power for Aluminum Productions: A Case Study of Frequency Control. *IEEE Trans. Power Syst.* **2015**, *30*, 471–483. [CrossRef]
24. Gao, B.; Niu, H.; Guan, Y.; Wang, Z.; Liu, J.; Taylor, M.P.; Chen, J.J.J. Visualization of Anode Effect in Aluminum Electrolysis. *J. Electrochem. Soc.* **2022**, *169*, 013505. [CrossRef]
25. Morishige, T.; Haarberg, G.M.; Gudbrandsen, H.; Skybakmoen, E.; Solheim, A.; Takenaka, T. Effects of Composition and Temperature on Current Efficiency for Aluminium Electrolysis from Cryolite-Based Molten Alumina Electrolytes. *ECS Trans.* **2017**, *77*, 997. [CrossRef]
26. Tang, F.; Liu, B. The crane development of furnace tending assembly based on virtual design. *Light Met.* **2010**, *5*, 39–41. [CrossRef]
27. Lian, X.; Gao, H.; Shen, L.; Yu, Y.; Wang, Y.; Peng, Z. Life Cycle Assessment of Primary Aluminum Production. *Processes* **2025**, *13*, 419. [CrossRef]
28. Liu, Z.; Xie, J.; Liu, Y.; Gong, F. Demand-Side Flexibility Series: Overview of Flexibility Potential in Aluminum Electrolysis Industry. Available online: https://rmi.org.cn/insights/industrial-dsf/ (accessed on 15 September 2023).
29. Sun, K.; Li, J.; Zhang, H.; Li, T.; Li, J. First-Principles Study on the Mechanism of Greenhouse Gas Generation in Aluminum Electrolysis. *Metals* **2021**, *11*, 1118. [CrossRef]
30. Liu, T.; Wu, Y.; Xu, H. Multi-task assisted multi-objective optimization algorithm for autonomous underwater vehicle path planning. *Memetic Comput.* **2025**, *17*, 8. [CrossRef]
31. Chen, L.; Cao, L.; Wen, Y.; Chen, H.; Jiang, S.-L. A knowledge-based NSGA-II algorithm for multi-objective hot rolling production scheduling under flexible time-of-use electricity pricing. *J. Manuf. Syst.* **2023**, *69*, 255–270. [CrossRef]
32. Liu, J.; Liu, Z.; Wu, Y.; Li, K. MBB-MOGWO: Modified Boltzmann-Based Multi-Objective Grey Wolf Optimizer. *Sensors* **2024**, *24*, 1502. [CrossRef]
33. Shao, X.; Fan, Y. An Improved Beetle Antennae Search Algorithm Based on the Elite Selection Mechanism and the Neighbor Mobility Strategy for Global Optimization Problems. *IEEE Access* **2021**, *9*, 137524–137542. [CrossRef]
34. Maduka, A.C.; Ogwu, S.O.; Ekesiobi, C.S. Assessing the moderating effect of institutional quality on economic growth—Carbon emission nexus in Nigeria. *Environ. Sci. Pollut. Res.* **2022**, *29*, 64924–64938. [CrossRef] [PubMed]

applied sciences

Article

Performance Optimization of Vertical Axis Wind Turbines Through Passive Flow Control and Material Selection: A Dynamic Mesh Study

Ioana-Octavia Bucur [1,*], Daniel-Eugeniu Crunțeanu [1] and Mădălin-Constantin Dombrovschi [2]

[1] Doctoral School of Aerospace Engineering, National University of Science and Technology Polytechnic of Bucharest, 011061 Bucharest, Romania; daniel.crunteanu@upb.ro
[2] Romanian Research and Development Institute for Gas Turbines–COMOTI, 061126 Bucharest, Romania; madalin.dombrovschi@comoti.ro
* Correspondence: ioana96bucur@gmail.com

Featured Application

The findings of this study can be applied to the design of small- and medium-scale vertical axis wind turbines intended for urban or low-wind environments, where efficient start-up and improved energy capture are critical. The use of cavity-modified polycarbonate blades provides a practical pathway for lightweight, cost-effective, and easily manufacturable turbines, supporting the wider deployment of renewable energy solutions in distributed power generation and off-grid applications.

Abstract

Vertical axis wind turbines (VAWTs) have significant potential for renewable energy generation, yet their operational efficiency is often limited by reduced aerodynamic performance and difficulties during start-up. This study investigates the effect of passive flow control and material selection on the performance of H-Darrieus VAWT blades, with the aim of identifying design solutions that enhance start-up dynamics and overall efficiency. Two-dimensional numerical simulations were conducted using the Dynamic Mesh method with six degrees of freedom (6DOF) in ANSYS 19.2 Fluent, enabling a time-resolved assessment of rotor behavior under constant wind velocities. Two blade configurations were analyzed: a baseline NACA0012 geometry and a modified profile with inclined cavities on the extrados. In addition, the influence of blade material was examined by comparing 3D-printed resin blades with lighter 3D-printed polycarbonate blades. The results demonstrate that cavity-modified blades provide superior performance compared to the baseline, showing faster acceleration, higher tip speed ratios, and improved power coefficients, particularly at higher wind velocities. Furthermore, polycarbonate blades achieved more efficient energy conversion than resin blades, highlighting the importance of material properties in turbine optimization. These findings confirm that combining passive flow control strategies with advanced lightweight materials can significantly improve the aerodynamic and dynamic performance of VAWTs, offering valuable insights for future experimental validation and prototype development.

Keywords: renewable energy; computational fluid dynamics; vertical axis wind turbine; passive flow control

1. Introduction

The continuous increase in global energy demand, coupled with the urgent need to reduce greenhouse gas emissions, has intensified research on renewable energy technologies. Among these, wind energy represents one of the most promising alternatives due to its scalability and abundance. Unlike horizontal-axis wind turbines, VAWTs offer unique advantages for distributed generation, urban deployment, and turbulent wind environments due to their omni-directional operation and simpler structural design [1,2]. However, their widespread adoption remains limited by intrinsic aerodynamic challenges such as dynamic stall, low self-starting capability, and torque fluctuations [3,4]. These issues have been extensively discussed in critical reviews that emphasize the need for innovative aerodynamic control strategies and advanced numerical methodologies to unlock the full potential of VAWTs [1–3].

Recent comprehensive reviews have also highlighted the broader challenges and opportunities facing the wind-energy sector as a whole. Ligeza [5] outlined the current and forecasted technological trends in wind power generation, emphasizing the need for integrated design approaches, advanced control systems, and improved aerodynamic efficiency in both large-scale and small-scale turbines. Similarly, Roga et al. [6] reviewed the latest developments in turbine technology and energy-harvesting methods, underlining the importance of novel passive and hybrid flow-control concepts for enhancing performance at low and moderate wind velocities. Together, these works reinforce the necessity for continued innovation in aerodynamic optimization and flow-management strategies, providing the foundation for the approach proposed in the present study.

Several approaches have been proposed to enhance the aerodynamic performance and dynamic behavior of VAWTs. Passive flow control has attracted significant attention due to its simplicity and low operational cost. Strategies such as guided vane integration [7], leading-edge slot structures [8], and modifications including groove-flap and concave cavity designs [9] have demonstrated measurable improvements in lift, torque, and power coefficients. Other investigations have explored novel airfoil geometries, such as J-blades [10], confirming that careful aerodynamic tailoring can mitigate stall and boost efficiency. Comprehensive reviews [11–14] further highlight the effectiveness of passive devices while also stressing the risk of added drag and mechanical complexity, indicating that performance gains are strongly dependent on design optimization and operating conditions.

Parallel to geometric innovations, computational fluid dynamics (CFD) has emerged as a cornerstone for VAWT design, enabling detailed insight into complex unsteady flows. Recent advances include transient Dynamic Mesh simulations to capture lateral vortices and Strouhal number effects [15], systematic CFD optimization methods based on Taguchi design [16], and hybrid analytical–CFD frameworks for torque and power prediction [17]. Reviews of CFD methodologies stress the importance of turbulence modeling, with the k–ω SST model frequently employed for its balance between near-wall resolution and free-stream accuracy [18]. These approaches have been used to study dynamic stall, blade–vortex interaction, and flow-control effectiveness, offering increasingly accurate predictions of start-up and steady-state turbine behavior [18,19].

Despite these advances, the issue of self-starting remains one of the most critical barriers for small and medium-sized VAWTs. Strategies ranging from aerodynamic tailoring to passive and active flow control have been systematically reviewed [6], with many studies converging on the conclusion that improving self-starting requires balancing mass, inertia, and aerodynamic efficiency. Experimental studies [9] and CFD investigations [10,15] both confirm that material properties and blade mass distribution play a decisive role in determining the transient response during start-up. Additionally, experimental and numerical studies indicate that heavier blades, such as those made from resin composites, may hinder acceleration at low wind speeds, whereas lighter alternatives, including advanced polymers or 3D-printed materials, can improve dynamic responsiveness [15,19]. Nonetheless, systematic investigations that combine passive flow control with material optimization remain limited in the current literature.

The present study addresses this gap by performing a comparative numerical analysis of straight-bladed VAWTs using the Dynamic Mesh method in ANSYS Fluent. Two geometrical configurations were considered: a baseline NACA0012 profile and a modified profile incorporating inclined cavities on the extrados. The current investigation compares results for the mass of 3D printed resin blades and blades manufactured from modified polycarbonate filament produced also by 3D printing, enabling a direct assessment of how blade material influences start-up behavior, tip speed ratio, and power coefficient. The results highlight the combined effect of passive flow control and material selection, showing that cavity-modified polycarbonate blades exhibit superior start-up dynamics and enhanced aerodynamic performance compared to both resin blades and unmodified profiles.

With this approach, two complementary aspects—blade geometry and material selection—are examined simultaneously to capture their combined influence on start-up dynamics and energy conversion efficiency. Unlike most existing studies, which treat these parameters separately or under simplified steady-state assumptions, the current work integrates passive flow control through inclined cavities with mass–inertia effects derived from measured material properties within a unified 6DOF framework. This dual-parameter approach enables a physically realistic prediction of turbine behavior, from acceleration to steady operation, under varying wind conditions.

By integrating geometric and material considerations, this work contributes to the broader understanding of VAWT optimization. The findings provide actionable insights for the design of lightweight, high-performance turbines suited for urban and distributed renewable energy applications, while also informing future experimental validation and prototype development.

In recent years, the frontier of numerical and experimental research has increasingly shifted toward the integration of high-fidelity modeling, data-driven optimization, and multi-field coupling techniques. Examples include the use of virtual-sample calibration combined with autoencoder architectures for enhanced sensor accuracy in complex thermal systems [20], as well as multi-field particle–flow coupling and ultrasonic control methods applied in microreactor environments [21]. Although these studies address different engineering domains, they share a common objective: to achieve more adaptive, accurate, and physics-informed modeling frameworks. In alignment with these advances, the present work extends high-fidelity transient modeling to the renewable energy sector by coupling aerodynamic flow phenomena with real-time inertial dynamics in a six-degree-of-freedom framework, thereby contributing to the broader movement toward intelligent, fully coupled simulation methodologies.

2. Materials and Methods

This study evaluated the aerodynamic performance of a VAWT equipped with modified airfoil blades fabricated from polycarbonate filament in comparison with resin blades, using transient CFD simulations in ANSYS 19.2 Fluent.

2.1. Computational Domain and Grid

The turbine rotor was modeled as an H-Darrieus configuration with the following main dimensions, summarized in Table 1.

Table 1. The geometric characteristics of the studied H-Darrieus VAWT.

Parameter	Value	Unit
Airfoil	NACA0012	-
Blade chord	0.1	m
Turbine diameter	0.5	m
Blade height	0.5	m
Number of blades	3	-
Design velocity	10	m/s

Two blade geometries were studied: a baseline case–NACA0012 profile and a modified case–NACA0012 with tilted orifices introduced along the extrados. These are depicted in Figure 1.

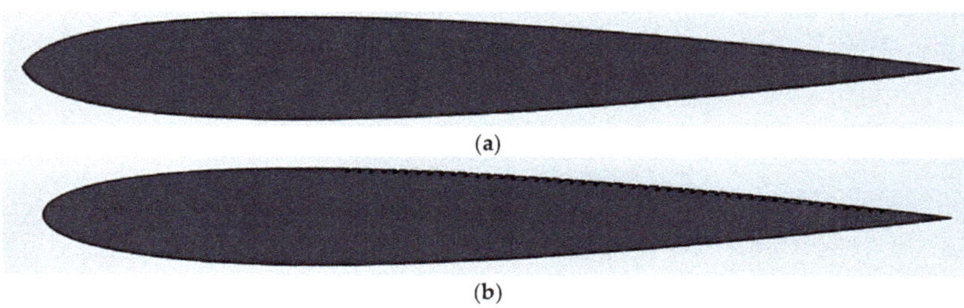

Figure 1. Studied blade geometries. (**a**) Baseline case—NACA0012; (**b**) Modified case—NACA0012 with tilted orifices on the last two-thirds of the chord (orifice opening = 0.003c; spacing between orifices = 0.01c; orifice depth = 0.005c).

The modified case was derived from a NACA0012 profile by introducing a series of 46 tilted orifices (45° inclination, promoting passive flow reattachment and delayed separation) distributed along the last two-thirds of the chord. Each orifice has an opening diameter of 0.0003 m, a spacing of 0.001 m, and a depth of 0.0005 m, positioned on the extrados surface. These details are illustrated in Figure 2.

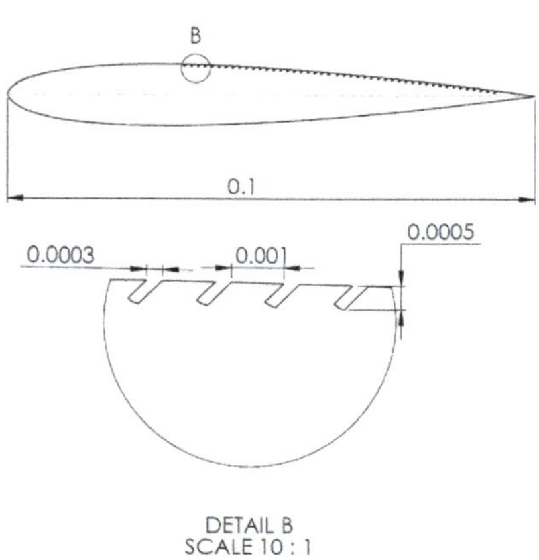

Figure 2. Modified NACA0012 profile.

The dimensions and layout of the tilted cavities were established based on the authors' previous parametric investigations of NACA0012 airfoil modified with inclined cavities [22,23]. Those studies evaluated several combinations of cavity size, spacing, and placement, concluding that the most effective configuration consisted of tilted cavities inclined at 45°, positioned on the last two-thirds of the upper surface. Each cavity featured an opening diameter of 0.003*c, a spacing of 0.01*c, and a depth of 0.005*c, where c is the blade chord. The results of these investigations, which included flow visualizations, demonstrated how the cavities promote flow reattachment and delay separation, leading to a significant lift increase without a noticeable drag penalty. Since the aerodynamic mechanism and local flow behavior around the cavities have already been illustrated and discussed extensively in those works, only turbine-scale aerodynamic results are presented in the current paper. Consequently, these validated geometric characteristics were adopted here as the reference passive-flow-control configuration.

The computational grid was generated in ANSYS Meshing and a hybrid grid was adopted: structured quadrilateral elements were applied around the airfoil blades to ensure boundary layer resolution, while a combination of quadrilateral and triangular unstructured elements was used in the rotor and stator zones. The computational domain was divided into two regions: rotor zone, including the blades, rotor fluid volume, and rotor/stator interface; and stator zone, representing undisturbed airflow, including inlet velocity, outlet pressure, symmetry planes, and the stator fluid volume. The computation domain is illustrated in Figure 3.

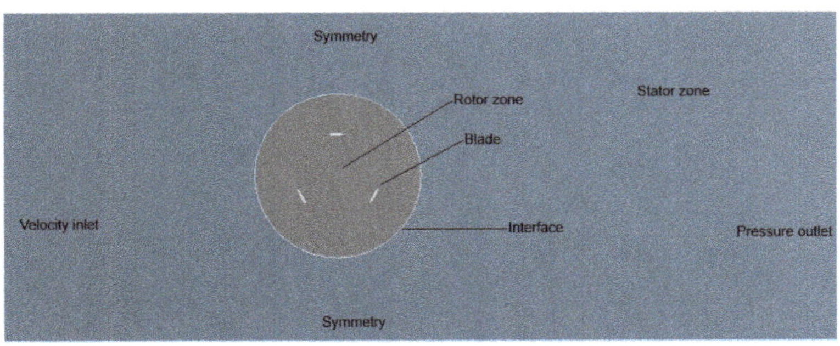

Figure 3. Computational domain.

The rotor/stator setup replicated wind tunnel conditions, with a unidirectional airflow from left to right. The "Quadrilateral Dominant" method was used for meshing, and inflation layers (10 boundary layers) were added around the blades to capture near-wall effects. The first cell height at the wall was determined using the flat-plate boundary layer theory [24], resulting in a calculated value of 6.822×10^{-5} m corresponding to y+ $\cong$ 1. For the final computational grids, a slightly refined first-layer spacing of $\Delta y = 1 \times 10^{-5}$ m was employed for both the airfoil and the cavity walls to ensure proper near-wall resolution. Furthermore, the mesh was locally refined along the airfoil and cavity walls, as well as around the orifice rims, using inflation layers and quadrilateral elements. The resulting mesh ensured high-quality boundary-layer capture and numerical stability throughout the unsteady simulations. To verify that the numerical solution is independent of the spatial discretization, a mesh-independence study was performed based on literature best practices for vertical axis wind turbines [25]. Three computational grids were generated for the baseline: a coarse grid with 90,488 elements and 94,216 nodes, a medium grid with 134,648 elements and 139,004 nodes, and a fine grid with 201,763 elements and

206,854 nodes. All grids maintained the same first-cell height near the wall while varying the number of elements along edges and surfaces. The baseline model was evaluated using, for the mesh independency study, Moving Mesh approach at 10 m/s wind speed for tip-speed-ratio values between 0.5 and 3.5. A comparison of the results showed negligible differences between the coarse, medium and fine grids throughout the entire TSR range, as depicted in Figure 4. The maximum variation in the moment coefficient was below 2%, confirming that the solution is mesh-independent. Consequently, the fine grid was adopted for all subsequent analyses.

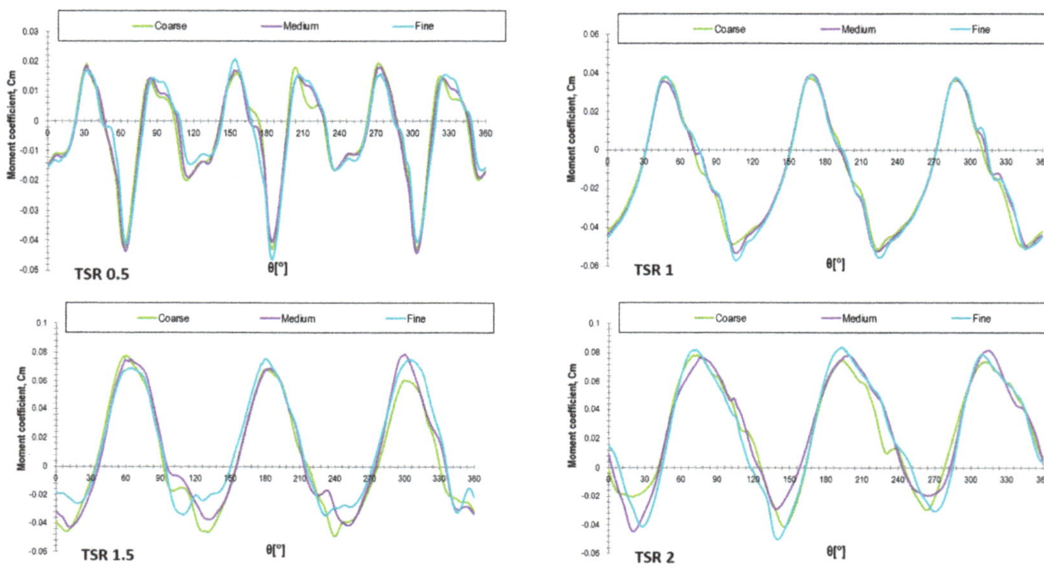

Figure 4. Moment coefficient variation for different grid sizes.

For the Dynamic Mesh study, the baseline model had 201,763 elements and 206,854 nodes, whereas the one for the blades with tilted cavities had 389,797 elements and 407,171 nodes. The generated computational grid is shown in Figures 5 and 6.

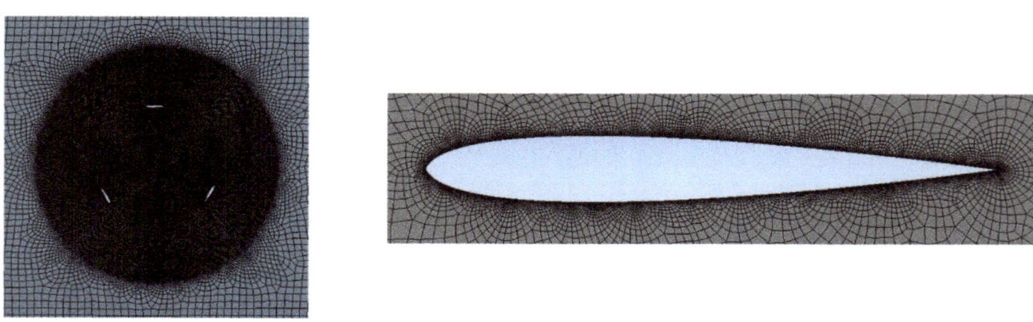

(**a**) (**b**)

Figure 5. Mesh for baseline wind turbine: (**a**) rotor and (**b**) near blade.

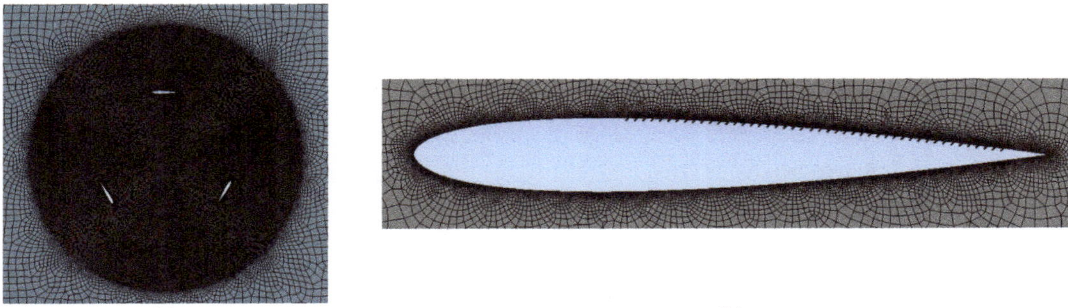

(a) (b)

Figure 6. Mesh for modified wind turbine: (**a**) rotor and (**b**) near blade.

2.2. Turbulence Model and Solver Settings

The k–ω Shear Stress Transport (SST) model was selected for the numerical analysis due to its proven capability to accurately predict separated and adverse-pressure-gradient flows in rotating aerodynamic systems. It combines the advantages of the standard k–ω model near walls with the k–ε formulation in fully turbulent regions, through a set of blending functions, ensuring robust performance across the entire computational domain. The k–ω SST approach has been extensively validated for wind-turbine applications, providing accurate predictions for VAWT aerodynamics [26,27]. The SIMPLE algorithm was used for pressure–velocity coupling, with second-order spatial discretization for momentum and turbulence quantities. Simulations employed a time step of 0.0005 s, with up to 50 iterations per step, corresponding to 120,000 time steps (≅60 s physical time) per case, sufficient for reaching steady periodic rotor motion.

2.3. Dynamic Mesh Setup

The turbine blades and rotor zone were defined as rigid bodies within the Dynamic Mesh framework with six degrees of freedom (6DOF). Blade motion was determined by their mass and inertia properties, derived from the fabricated 3D-printed polycarbonate blades and 3D-printed resin blades, respectively. This data is summarized in Table 2.

Table 2. Mass and inertia properties for the studied models.

Case	Blades Mass [kg]	Moment of Inertia [kg * m^2]
Polycarbonate Baseline	0.912	0.057
Polycarbonate Modified	0.876	0.054
Resin Baseline	0.969	0.061
Resin Modified	0.951	0.059

The rotor zone was coupled passively to blade motion, while the stator was fixed. The 6DOF solver tracked blade displacement and angular velocity in response to aerodynamic loads, allowing the turbine's self-starting behavior and dynamic equilibrium to be evaluated.

For each blade geometry and material configuration, five inlet wind speeds were investigated: 8, 10, 12, 14, and 16 m/s. Boundary conditions were applied as described in Table 3.

For each time step, blade angular displacement was recorded. The angular velocity was computed as the change in displacement over time, which was then used to determine the tip speed ratio (TSR), torque coefficient (cm), power coefficient (cp), and output power (P). The turbine was considered to have reached equilibrium once TSR values stabilized, corresponding to a constant angular velocity.

Table 3. Boundary conditions.

Boundary	Condition
Blades	Rigid body with dynamic motion
Rotor	Rigid body with passive motion (relative to the blades)
Stator	Stationary
Interface	Interface between rotor and stator zones

3. Results

By employing the Dynamic Mesh function, it was possible to record the blade position at each time step. Based on this data, the difference in position between consecutive steps was calculated, and once this difference became constant, the turbine was considered to have reached an equilibrium state, rotating with a constant angular velocity. The angular velocity (ω) was computed as the ratio between the change in blade position (in radians) and the time step (in seconds).

$$\omega = \frac{\Delta\theta}{0.0005}[\text{rad/s}], \tag{1}$$

Knowing the angular velocity, it was then possible to determine the velocity coefficient (λ) for each case using the following standard expression:

$$\lambda = \frac{\omega R}{V_\infty} \tag{2}$$

In parallel, the torque coefficient (c_M) was monitored throughout the simulation and numerically evaluated. Based on its values and the TSR, the power coefficient (c_P) was subsequently calculated, allowing a direct assessment of the aerodynamic efficiency of the turbine under different configurations and material properties.

$$c_P = c_M * \lambda \tag{3}$$

Having these coefficients available, the moment (torque—M) and the power of the turbine (P) could be directly determined, thus providing a comprehensive evaluation of its performance. The involved parameters are ρ [kg/m^3]—air density; A—rotor area [m^2]; V_∞ [m/s]—wind speed; R [m]—turbine radius.

$$M = \frac{\rho}{2} A V_\infty^2 R c_M \tag{4}$$

$$P = \frac{\rho}{2} A V_\infty^3 c_P \tag{5}$$

The results for each model under study are summarized in the following tables, which include the calculated values of angular velocity, tip speed ratio, torque coefficient, and power coefficient, as well as the corresponding torque and power. These datasets enable a direct comparison between the different geometrical configurations and blade materials, highlighting the influence of the introduced modifications on the turbine's overall performance.

To better visualize the results, Figure 7 illustrates the variation in the power coefficient as a function of wind speed for blades manufactured from resin and modified polycarbonate, both in the baseline configuration and in the configuration with orifices.

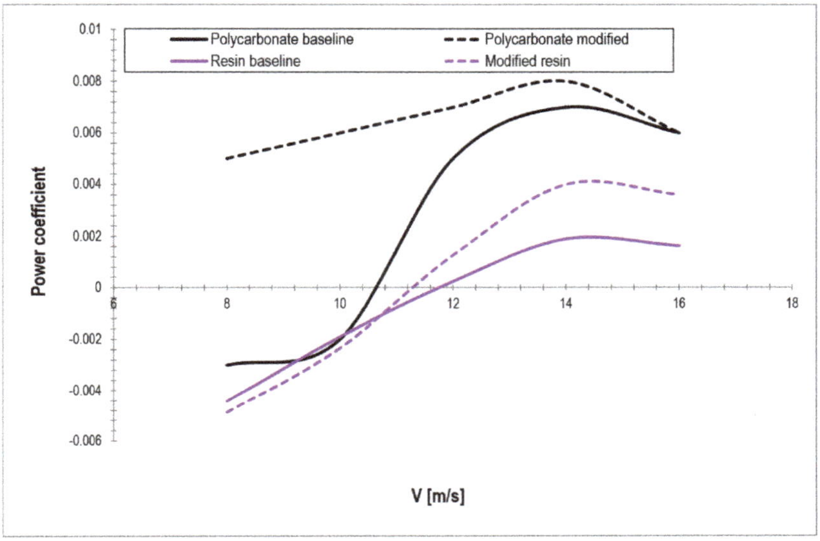

Figure 7. Power coefficient variation with wind speed.

Furthermore, Figure 8 presents the evolution of the tip speed ratio with wind speed for the same cases, offering a complementary perspective on the dynamic behavior of the turbine.

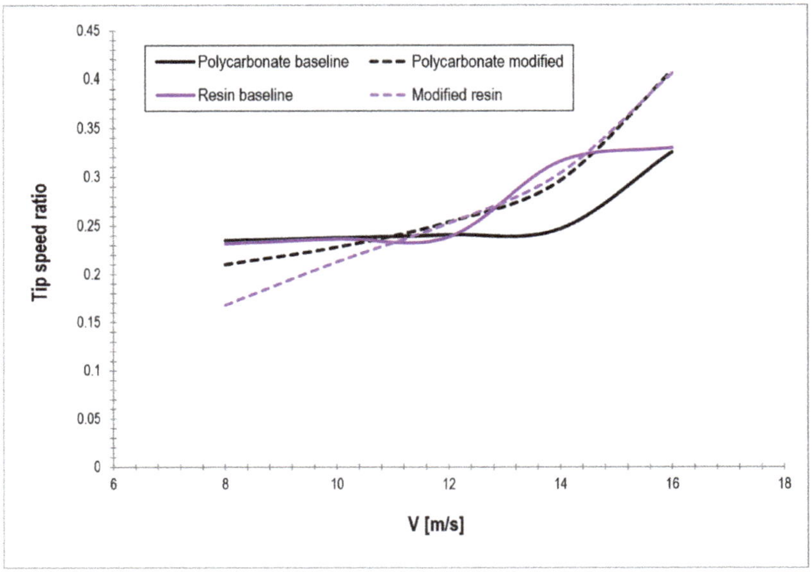

Figure 8. Tip speed ratio variation with wind speed.

To evaluate the dynamic behavior of the turbine, the time evolution of the tip speed ratio (TSR) was monitored at different wind speeds for both blade materials (resin and modified polycarbonate). The results, illustrated in Figure 9, allow for a direct comparison between the evolution and stabilization of TSR values for the two geometries, as influenced by the imposed flow conditions.

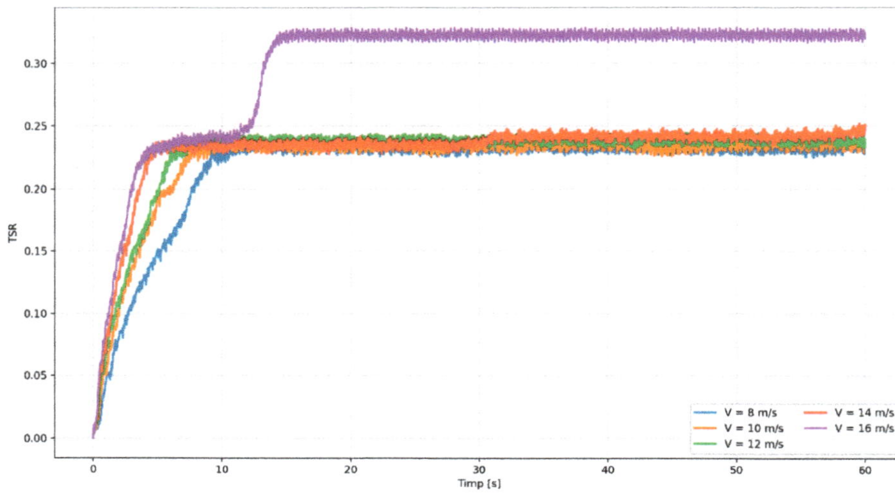

(**a**)

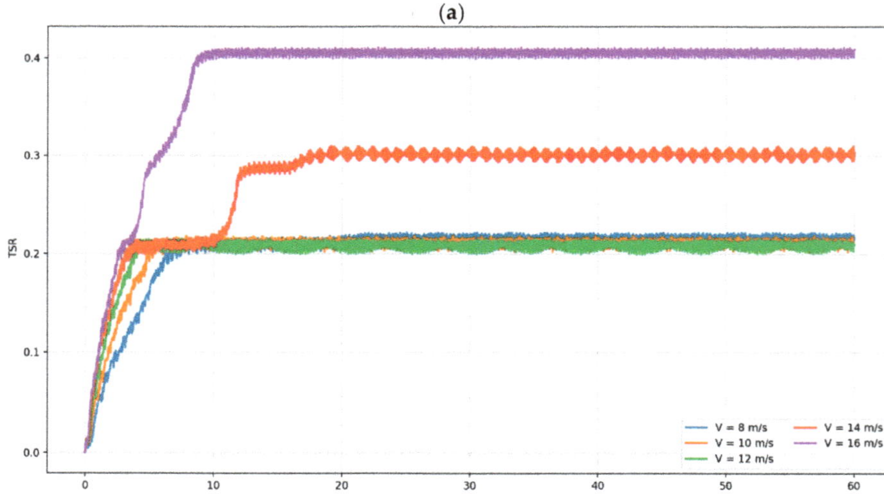

(**b**)

Figure 9. *Cont.*

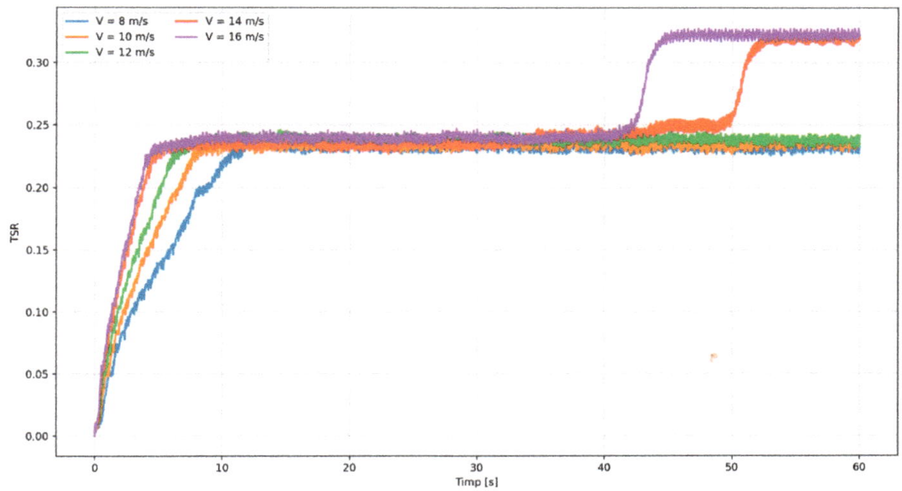

(c)

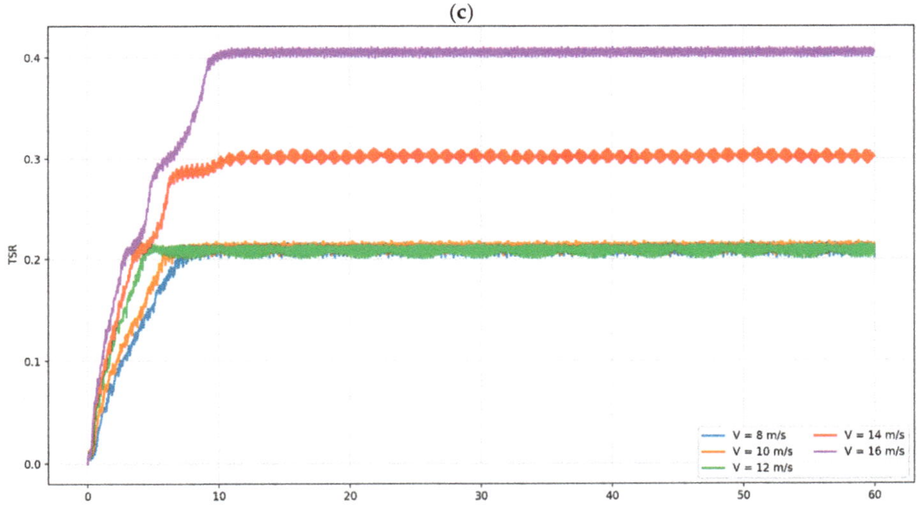

(d)

Figure 9. Tip speed ratio variation with time. (**a**) Polycarbonate blades—baseline case; (**b**) Polycarbonate blades—modified case; (**c**) Resin blades—baseline case; (**d**) Resin blades—modified case.

4. Discussion

The numerical results for the baseline and modified blades fabricated from polycarbonate and resin are summarized in Tables 4–7. For polycarbonate blades, the baseline configuration shows relatively low power coefficients at low wind speeds (negative values at 8–10 m/s), but performance improves significantly at higher speeds, reaching a maximum power coefficient of 0.007 at 14 m/s. These results are in accordance with those from similar studied for small VAWTs [7]. As mentioned, at lower wind velocities, the simulations produced negative torque and power coefficients. This behavior is characteristic of the self-starting phase of H-Darrieus VAWTs and indicates that the turbine operates temporarily in a drag-dominated regime. During this phase, the blades encounter highly unsteady flow conditions and dynamic stall. These transient effects lead to a negative torque until sufficient angular velocity is reached for the flow to stabilize around the airfoils, after which the turbine transitions into a lift-driven regime with positive torque and power generation. Such behavior has also been reported in previous experimental and numerical studies [6,10], confirming that negative torque and power coefficients at low wind speeds are a physically expected feature of small Darrieus-type turbines rather than a numerical error.

Table 4. Baseline blades—3D-printed polycarbonate blades.

V_∞ [m/s]	Δθ [Rad]	ω [Rad/s]	TSR-λ	c_m	c_p	P [W]	M [Nm]
8	0.004	7.522	0.235	−0.013	−0.003	−0.239	−0.0317
10	0.005	9.515	0.238	−0.01	−0.002	−0.389	−0.04
12	0.006	11.589	0.241	0.019	0.005	1.266	0.109
14	0.007	13.823	0.247	0.027	0.007	2.845	0.206
16	0.01	20.82	0.325	0.018	0.006	3.672	0.176

Table 5. Modified blades—3D-printed polycarbonate blades.

V_∞ [m/s]	Δθ [Rad]	ω [Rad/s]	TSR-λ	c_m	c_p	P [W]	M [Nm]
8	0.003	6.72	0.21	0.022	0.005	0.369	0.055
10	0.004	9.12	0.228	0.028	0.006	0.985	0.108
12	0.006	12.192	0.254	0.027	0.007	1.852	0.149
14	0.008	16.563	0.296	0.028	0.008	3.587	0.216
16	0.013	26.145	0.408	0.016	0.006	4.186	0.16

Table 6. Baseline blades—3D-printed resin blades.

V_∞ [m/s]	Δθ [Rad]	ω [Rad/s]	TSR-λ	c_m	c_p	P [W]	M [Nm]
8	0.004	7.424	0.232	−0.019	−0.004	−0.346	−0.046
10	0.005	9.48	0.237	−0.008	−0.0019	−0.290	−0.031
12	0.006	11.472	0.239	0.001	0.00024	0.063	0.005
14	0.009	17.698	0.316	0.006	0.0019	0.796	0.045
16	0.010	20.824	0.33	0.005	0.00163	1.02	0.049

Table 7. Modified blades—3D-printed resin blades.

V_∞ [m/s]	Δθ [Rad]	ω [Rad/s]	TSR-λ	c_m	c_p	P [W]	M [Nm]
8	0.003	5.376	0.168	−0.029	−0.00487	−0.382	−0.071
10	0.004	8.52	0.213	−0.011	−0.00234	−0.359	−0.042
12	0.006	12.144	0.253	0.005	0.001265	0.335	0.027
14	0.008	17.01	0.304	0.012	0.004	1.659	0.097
16	0.0208	41.6	0.406	0.009	0.0036	2.294	0.088

The modified polycarbonate blades with inclined orifices demonstrate higher performance across all velocities, with power coefficient peaking at 0.008. In comparison, the resin blades exhibit overall lower aerodynamic efficiency [6,10]. The baseline resin case shows negligible or even negative power at low wind speeds and modest improvement at higher velocities, with power coefficients not exceeding 0.0019. However, the modified resin blades show some performance recovery at higher velocities, reaching a power coefficient value of 0.004 at 14 m/s, but still remain below the levels of the polycarbonate models. These results confirm that both the material properties (lighter polycarbonate vs. heavier resin) and the passive flow control modification (orifices) strongly influence turbine self-starting capability and power generation.

To better visualize these results, Figure 5 illustrates the variation in the power coefficient with wind speed for both polycarbonate and resin blades, in baseline and modified configurations. It can be observed that polycarbonate blades consistently outperform resin blades, with modified designs achieving the highest values. The presence of orifices contributes to better startup characteristics and improved efficiency at moderate wind speeds, confirming the effectiveness of passive flow control.

Figure 6 shows the evolution of the tip speed ratio with wind speed. The modified polycarbonate blades achieve the highest TSR values, reaching 0.40 at 16 m/s, compared to 0.33 for the resin baseline. In general, TSR increases with wind speed for all cases, but polycarbonate blades demonstrate both higher peak values and more stable growth compared to resin blades.

To assess the dynamic response of the turbine, the time evolution of TSR was analyzed at different wind speeds (8, 10, 12, 14, and 16 m/s) for all configurations (Figure 7). For the polycarbonate baseline, TSR stabilizes quickly but at lower values, while the modified polycarbonate blades achieve faster acceleration and higher equilibrium TSR, especially at 14–16 m/s. Resin blades show slower stabilization and lower equilibrium values, with the modified design offering partial improvement but still underperforming compared to polycarbonate. The time-history evolution of the TSR reflects the interplay between aerodynamic torque generation and rotor inertia during the turbine's self-starting and steady-state phases. In the early seconds of each simulation, the rotor is nearly stationary and the blades encounter very high instantaneous angles of attack, resulting in drag-dominated torque and dynamic-stall cycles. This stage produces the steep initial rise in TSR visible for all wind speeds. As the rotational velocity increases, the relative inflow angle decreases and the blades progressively transition from separated to attached flow, marking the shift from a drag- to a lift-driven regime. The subsequent flattening of the TSR curves corresponds to the moment when aerodynamic torque and resistive torque (mainly viscous and inertial) reach equilibrium. The higher final TSR values observed for the modified blades arise from the tilted cavities, which delay separation and promote reattachment, increasing the net positive torque once the lift mechanism dominates. Differences between resin and polycarbonate blades are linked to inertia: the lighter polycarbonate blades accelerate faster and achieve higher steady TSR because of their lower rotational moment of inertia.

The results highlight two decisive factors in the aerodynamic performance of the studied VAWTs. First, the material influence: lighter polycarbonate blades exhibit faster acceleration and reach higher TSR values compared to resin blades, owing to their reduced inertia, which enhances the dynamic response of the turbine. Second, the geometric modification: the addition of inclined orifices on the extrados proves effective as a passive flow control strategy, improving startup behavior and sustaining higher aerodynamic efficiency across operating conditions.

5. Conclusions

This study presented a comparative numerical investigation of VAWTs equipped with baseline and cavity-modified blades fabricated from resin and polycarbonate. Using the Dynamic Mesh approach with six degrees of freedom in ANSYS Fluent, the analysis demonstrated that both blade geometry and material selection have a decisive influence on turbine performance.

The results showed that lighter polycarbonate blades achieved faster acceleration and higher tip speed ratios than their resin counterparts, highlighting the importance of reduced inertia for improving dynamic response and start-up capability. Furthermore, the introduction of inclined cavities along the extrados proved to be an effective passive flow control strategy, enhancing aerodynamic efficiency and increasing the power coefficient across a wide range of wind speeds. When combined, these two design considerations—lightweight material and cavity modification—produced the most favorable performance, indicating a promising pathway for optimizing small- and medium-scale VAWTs intended for urban and distributed energy applications.

While the present investigation provides valuable insight into the aerodynamic and dynamic behavior of modified VAWTs, it is limited to two-dimensional CFD simulations that neglect three-dimensional flow effects. Moreover, experimental validation has not yet been performed at this stage. Future work will focus on the experimental validation of the numerical findings through dedicated wind tunnel tests of the fabricated blade models. These tests will provide further insight into the real-world aerodynamic behavior, confirm the reliability of the numerical predictions, and support the development of scalable prototypes suitable for practical deployment.

Author Contributions: Conceptualization, I.-O.B. and D.-E.C.; methodology, I.-O.B. and M.-C.D.; software, I.-O.B.; validation, D.-E.C. and M.-C.D.; formal analysis, I.-O.B. and M.-C.D.; investigation, I.-O.B.; resources, D.-E.C.; data curation, I.-O.B.; writing—original draft preparation, I.-O.B.; writing—review and editing, D.-E.C. and M.-C.D.; supervision, D.-E.C. All authors have read and agreed to the published version of the manuscript.

Funding: This research was supported by the Doctoral School of Aerospace Engineering, Faculty of Aerospace Engineering, National University of Science and Technology Polytechnic of Bucharest. This article was funded by the PubArt program of the National University of Science and Technology Polytechnic of Bucharest.

Institutional Review Board Statement: Not applicable.

Data Availability Statement: The data supporting the reported results is available from the corresponding author on request.

Acknowledgments: This research was supported by the Doctoral School of Aerospace Engineering, Faculty of Aerospace Engineering, National University of Science and Technology Polytechnic of Bucharest.

Conflicts of Interest: The authors declare no conflicts of interest.

Abbreviations

The following abbreviations are used in this manuscript:

VAWT	Vertical Axis Wind Turbine
CFD	Computational Fluid Dynamics
TSR	Tip Speed Ratio
DOF	Degrees of Freedom
SST	Shear Stress Transport turbulence model

References

1. Kumar, R.; Raahemifar, K.; Fung, A.S. A critical review of vertical axis wind turbines for urban applications. *Renew. Sustain. Energy Rev.* **2018**, *89*, 281–291. [CrossRef]
2. Salvador-Gutierrez, B.; Sanchez-Cortez, L.; Hinojosa-Manrique, M.; Lozada-Pedraza, A.; Ninaquispe-Soto, M.; Montaño-Pisfil, J.; Gutiérrez-Tirado, R.; Chávez-Sánchez, W.; Romero-Goytendia, L.; Díaz-Aliaga, J.; et al. Vertical-Axis Wind Turbines in Emerging Energy Applications (1979–2025): Global Trends and Technological Gaps Revealed by a Bibliometric Analysis and Review. *Energies* **2025**, *18*, 3810. [CrossRef]
3. Abdolahifar, A.; Zanj, A. Addressing VAWT Aerodynamic Challenges as the Key to Unlocking Their Potential in the Wind Energy Sector. *Energies* **2024**, *17*, 5052. [CrossRef]

4. Gallegos-Molina, J.S.; Chavero-Navarrete, E. A Systematic Review of Technological Strategies to Improve Self-Starting in H-Type Darrieus VAWT. *Sustainability* **2025**, *17*, 7878. [CrossRef]
5. Ligeza, P. Basic, Advanced, and Sophisticated Approaches to the Current and Forecast Challenges of Wind Energy. *Energies* **2021**, *14*, 8147. [CrossRef]
6. Roga, S.; Bardhan, S.; Kumar, Y.; Dubey, S.K. Recent technology and challenges of wind energy generation: A review. *Sustain. Energy Technol. Assess.* **2022**, *52*, 102239. [CrossRef]
7. Aboelezz, A.; Ghali, H.; Elbayomi, G.; Madboli, M. A novel VAWT passive flow control numerical and experimental investigations: Guided Vane Airfoil Wind Turbine. *Ocean Eng.* **2022**, *257*, 111704. [CrossRef]
8. Zhang, R.; Zhan, L.; Kuang, L.; Yushan, R.; Tu, Y.; Zhu, H.; Su, J.; Han, Z.; Zhou, D. Passive flow control technique for enhancing power efficiency of vertical-axis wind turbines: Leading-edge slot structure. *J. Ocean Eng. Sci.* **2025**. [CrossRef]
9. Xue, J.; Chen, Y.; Song, L.; Xing, Y.; Wang, B.; Sun, Y. Experimental Study on the Influence of Groove-Flap and Concave Cavity on the Output Characteristics of Vertical Axis Wind Turbine. *Fluids* **2025**, *10*, 208. [CrossRef]
10. Abdallah, A.; William, M.A.; Moharram, N.A.; Zidane, I.F. Boosting H-Darrieus vertical axis wind turbine performance: A CFD investigation of J-Blade aerodynamics. *Results Eng.* **2025**, *27*, 106358. [CrossRef]
11. Syawitri, T.P.; Yao, Y.; Yao, J.; Chandra, B. A review on the use of passive flow control devices as performance enhancement of lift-type vertical axis wind turbines. *Wiley Interdiscip. Rev. Energy Environ.* **2022**, *11*, e435. [CrossRef]
12. Zhu, H.; Hao, W.; Li, C.; Ding, Q.; Wu, B. A critical study on passive flow control techniques for straight-bladed vertical axis wind turbine. *Energy* **2018**, *165*, 12–25. [CrossRef]
13. Nouh, F.; Ismaiel, A.; Eleashy, H. Enhancing Wind Turbine Performance Using Flow Control Techniques: A Mini Review. *Future Eng. J.* **2025**, *5*. Available online: https://digitalcommons.aaru.edu.jo/fej/vol5/iss1/3 (accessed on 31 January 2025).
14. Akhter, M.Z.; Omar, F.K. Review of Flow-Control Devices for Wind-Turbine Performance Enhancement. *Energies* **2021**, *14*, 1268. [CrossRef]
15. Priyadumkol, J.; Muangput, B.; Namchanthra, S.; Zin, T.; Phengpom, T.; Chookaew, W.; Suvanjumrat, C.; Promtong, M. CFD modelling of vertical-axis wind turbines using transient dynamic mesh towards lateral vortices capturing and Strouhal number. *Energy Convers. Manag. X* **2025**, *26*, 101022. [CrossRef]
16. Kalassov, N.; Baizhuma, Z.; Manatbayev, R.; Yershina, A.; Isataev, M.; Kalassova, A.; Seidulla, Z.; Bektibay, B.; Amir, B. Integrated Approach to Aerodynamic Optimization of Darrieus Wind Turbine Based on the Taguchi Method and Computational Fluid Dynamics (CFD). *Appl. Sci.* **2025**, *15*, 5739. [CrossRef]
17. Lisowski, F.; Augustyn, M. Analytical and Computational Fluid Dynamics Methods for Determining the Torque and Power of a Vertical-Axis Wind Turbine with a Carousel Rotor. *Appl. Sci.* **2025**, *15*, 208. [CrossRef]
18. Fazlizan, A.; Muzammil, W.K.; Al-Khawlani, N.A. A Review of Computational Fluid Dynamics Techniques and Methodologies in Vertical Axis Wind Turbine Development. *CMES—Comput. Model. Eng. Sci.* **2025**, *144*, 1371–1437. [CrossRef]
19. Battisti, L. Design Options to Improve the Dynamic Behavior and the Control of Small H-Darrieus VAWTs. *Appl. Sci.* **2021**, *11*, 9222. [CrossRef]
20. Sun, Z.; Yao, Q.; Jin, H.; Xu, Y.; Hang, W.; Chen, H.; Li, K.; Shi, L.; Gu, J.; Zhang, Q.; et al. A novel in-situ sensor calibration method for building thermal systems based on virtual samples and autoencoder. *Energy* **2024**, *297*, 131314. [CrossRef]
21. Li, L.; Xu, P.; Li, Q.; Yin, Z.; Zheng, R.; Wu, J.; Bao, J.; Bai, W.; Qi, H.; Tan, D. Multi-field coupling particle flow dynamic behaviors of the microreactor and ultrasonic control method. *Powder Technol.* **2025**, *454*, 120731. [CrossRef]
22. Bucur, I.O.; Crunteanu, D.E.; Dombrovschi, M.C. Numerical Analysis of Tilted Cavities Placement Effects on the Airfoils in Wind Turbine Systems. *IOP Conf. Ser. Earth Environ. Sci.* **2024**, *1375*, 012018. [CrossRef]
23. Bucur, I.O.; Crunteanu, D.E.; Dombrovschi, M. Numerical evaluation of airfoils with tilted cavities for vertical axis wind turbines applications. *AIP Conf. Proc.* **2025**, *3117*, 020012.
24. White, F.M. Flow Past Immersed Bodies. In *Fluid Mechanics*, 7th ed.; McGraw-Hill: New York, NY, USA, 2011.
25. Song, C.; Zheng, Y.; Zhao, Z.; Zhang, Y.; Li, C.; Jiang, H. Investigation of meshing strategies and turbulence models for computational fluid dynamics simulations of vertical axis wind turbines. *J. Renew. Sustain. Energy* **2015**, *7*, 033111. [CrossRef]
26. Guerri, O.; Sakout, A.; Bouhadef, K. Simulations of the Fluid Flow around a Rotating Vertical Axis Wind Turbine. *J. Wind Eng.* **2007**, *31*, 149–163. [CrossRef]
27. Rezaeiha, A.; Montazeri, H.; Blocken, B. On the accuracy of turbulence models for CFD simulations of vertical axis wind turbines. *Energy* **2019**, *180*, 838–857. [CrossRef]

MDPI AG
Grosspeteranlage 5
4052 Basel
Switzerland
Tel.: +41 61 683 77 34

Applied Sciences Editorial Office
E-mail: applsci@mdpi.com
www.mdpi.com/journal/applsci

www.ingramcontent.com/pod-product-compliance
Lightning Source LLC
Chambersburg PA
CBHW061507220926
51850CB00004B/192
* 9 7 8 3 7 2 5 8 6 2 4 4 3 *